AF591073

The Adenoviruses

THE VIRUSES

Series Editors
HEINZ FRAENKEL-CONRAT, *University of California*
Berkeley, California
ROBERT R. WAGNER, *University of Virginia School of Medicine*
Charlottesville, Virginia

THE HERPESVIRUSES, Volumes 1, 2, 3, and 4
Edited by Bernard Roizman

THE REOVIRIDAE
Edited by Wolfgang K. Joklik

THE PARVOVIRUSES
Edited by Kenneth I. Berns

THE ADENOVIRUSES
Edited by Harold S. Ginsberg

THE VIRUSES: Catalogue, Characterization, and Classification
Heinz Fraenkel-Conrat

The Adenoviruses

Edited by

HAROLD S. GINSBERG
College of Physicians and Surgeons of Columbia University
New York, New York

Springer Science+Business Media, LLC

Library of Congress Cataloging in Publication Data

Main entry under title:

The Adenoviruses.

(The Viruses)
Includes bibliographical references and index.
1. Adenoviruses. I. Ginsberg, Harold S., 1917- . II. Series.
QP396.A34 1984 576'.64 84-8264

DOI 10.1007/978-1-4684-7935-5

Originally published by Plenum Press, New York in 1984
MyCopy version of the original edition 1984

Contributors

Kenneth I. Berns, Department of Immunology and Medical Microbiology, University of Florida College of Medicine, Gainesville, Florida 32610

Harold S. Ginsberg, Department of Microbiology, College of Physicians and Surgeons of Columbia University, New York, New York 10032

F. L. Graham, Departments of Biology and Pathology, McMaster University, Hamilton, Ontario L8S 4K1, Canada

Masahide Ishibashi, Laboratory of Viral Oncology, Research Institute, Aichi Cancer Center, Chikusa-ku, Nagoya 464, Japan

Thomas J. Kelly, Jr., Department of Molecular Biology and Genetics, The Johns Hopkins University School of Medicine, Baltimore, Maryland 21205

Daniel F. Klessig, Department of Cellular, Viral and Molecular Biology, College of Medicine, University of Utah, Salt Lake City, Utah 84132

M. V. Nermut, National Institute for Medical Research, London NW7 1AA, England

Ulf Pettersson, Department of Medical Genetics, The Biomedical Centre, University of Uppsala, S-751 23 Uppsala, Sweden

Lennart Philipson, Department of Microbiology, The Biomedical Centre, University of Uppsala, S-751 23 Uppsala, Sweden; *Present address:* European Molecular Biology Laboratory, D-6900 Heidelberg, Federal Republic of Germany

Phillip A. Sharp, Center for Cancer Research and Department of Biology, Massachusetts Institute of Technology, Cambridge, Massachusetts 02139

T. Shenk, Department of Microbiology, State University of New York at Stony Brook, Stony Brook, New York 11794

Stephen E. Straus, Medical Virology Section, Laboratory of Clinical Investigation, National Institutes of Health, Bethesda, Maryland 20205

John S. Sussenbach, Laboratory for Physiological Chemistry, State University of Utrecht, 3521 GG Utrecht, The Netherlands

Hiroshi Yasue, Laboratory of Viral Oncology, Research Institute, Aichi Cancer Center, Chikusa-ku, Nagoya 464, Japan

C. S. H. Young, Department of Microbiology, College of Physicians and Surgeons of Columbia University, New York, New York 10032

Preface

The discovery of adenoviruses naturally induced a new interest in viruses of the human upper respiratory tract since previously unknown viruses infecting this portion of the human body had not been identified in 20 years, and their unique characteristics stimulated investigations into the biochemical events essential for replication of animal viruses. Indeed, the field of molecular virology has evolved during the period since their discovery, and adenoviruses have played a major role in this development. The exciting discoveries made with adenoviruses have had such a profound effect on knowledge in basic virology, molecular biology, viral genetics, human and animal infections, and cell transformation that this seemed a propitious time to have some of the major contributors review this field. This volume pays tribute to the late Wallace Rowe, Robert Huebner, and Maurice Hilleman whose initial discoveries of adenoviruses have tremendously enriched virology.

Harold S. Ginsberg

Contents

Chapter 3

The Structure of the Genome

John S. Sussenbach

Chapter 4

The Genetic System

C. S. H. Young, T. Shenk, and Harold S. Ginsberg

Chapter 5

Adenovirus Transcription

Phillip A. Sharp

Chapter 6

Structural and Nonstructural Adenovirus Proteins

Ulf Pettersson

Chapter 7

Adenovirus DNA Replication

Thomas J. Kelly, Jr.

Chapter 8

Adenovirus Assembly

Lennart Philipson

Chapter 9

Transformation by and Oncogenicity of Human Adenoviruses

F. L. Graham

Chapter 10

Adenovirus–Simian Virus 40 Interactions

Daniel F. Klessig

Chapter 11

Adenovirus Infections in Humans

Stephen E. Straus

Chapter 12

Adenoviruses of Animals

Masahide Ishibashi and Hiroshi Yasue

Chapter 13

Adeno-Associated Virus

Kenneth I. Berns

CHAPTER 1

An Overview

HAROLD S. GINSBERG

The discoveries of adenoviruses by Rowe *et al.* (1953) and by Hilleman and Werner (1954) aroused great interest and excitement among clinicians and virologists alike in that no new etiological agent of acute viral respiratory disease of humans had been isolated since the identification of influenza virus 20 years earlier (Smith *et al.*, 1933). During this interim period, extensive epidemiological studies had shown that acute viral respiratory infections are the most common cause of human disease and are responsible for a huge economic burden. Therefore, it is not surprising that the initial isolations of adenoviruses came from clinical stimuli. When Rowe *et al.* (1953) attempted to isolate "the common cold virus" using cultured, uninoculated control explants of human tonsils and adenoids as the host cells, they noted that the cells growing from the explants frequently underwent cytopathic changes. The rounding and grapelike clustering of the affected cells were shown to be caused by a new virus. Almost simultaneously, Hilleman and Werner, investigating an epidemic of influenzalike illness in army recruits, isolated a previously unknown virus in cultured human tracheal cells. The agent isolated by Hilleman and Werner was soon shown to be related to the virus isolated by Rowe and co-workers (Huebner *et al.*, 1954) and identical to the virus responsible for large epidemics of acute respiratory disease of recruits in the armed forces during World War II (Ginsberg *et al.*, 1954). This influenzalike illness was known by the abbreviation ARD, for "acute respiratory disease of recruits." Since that beginning, 41 distinct antigenic types of viruses have been found to infect humans, producing a variety of acute respiratory, ocular, gastrointestinal, and urinary diseases (see Chapter 11), and numerous other types infect a wide variety of lower

HAROLD S. GINSBERG • Department of Microbiology, College of Physicians and Surgeons of Columbia University, New York, New York 10032.

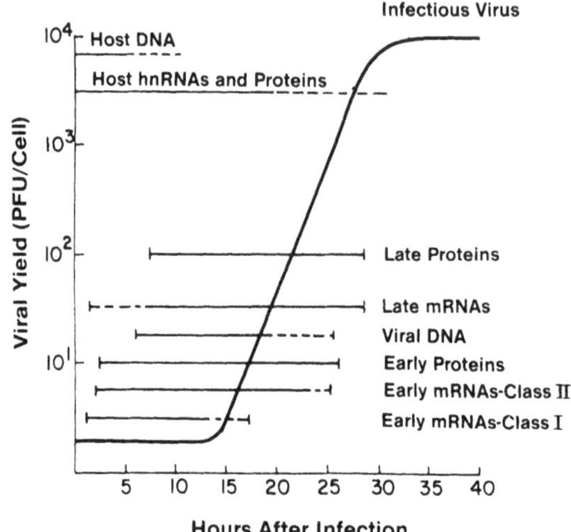

FIGURE 1. Initial cycle of replication of adenovirus type 5, and the sequential biosynthetic reactions that are essential for producing viral macromolecules. The effect of viral infection on synthesis of host proteins RNA and DNA are also diagramed. (hnRNA) Heterogeneous nuclear RNA; (mRNA) messenger RNA; (PFU) plaque-forming units.

animal species (see Chapter 12). These many viruses, which infect such a variety of animals, are related and classified into a single family on the basis of their many similar physical and chemical properties (Matthews, 1982). An immunological relationship further groups adenoviruses that infect mammals through a cross-reacting antigenic determinant on free hexons, the major capsid protein (Matthews, 1982; Ginsberg, 1979). This immunological assay remains as a rapid, convenient tool to identify a newly isolated virus as a member of the family Adenoviridae.

Although the impetus that led to the discovery of adenoviruses arose from clinical needs, it will become evident from the discussions presented in this volume that adenoviruses have had a great impact on uncovering previously inconceivable features of viral structure, replication, and cell transformation.

Infection encompases a well-ordered series of events, which is initiated by the attachment of an adenovirus particle to a susceptible cell via one of its fibers (see Chapters 2 and 6) and culminated by the assembly (Chapter 8) of approximately 10^4 infectious virions per cell (with adenovirus type 2 or 5). The chapters that follow reveal the complexities of this seemingly uncomplicated series of reactions, which are simply diagramed in Fig. 1. As a consequence of this infection, the affected cells may become damaged and die, producing disease (Chapter 11); they may become transformed, producing tumors (Chapter 9); or they may establish a latent infection of lymphoid cells such as uncovered by the initial discovery of adenoviruses by Rowe *et al.* (1953). The absence of a chapter that specifically discusses pathogenesis of adenovirus infections reflects our ignorance of the mechanisms that effect cell injury and disease. This scarcity of information not only denotes the lack of data concerning adenoviruses, but also reflects the general absence of detailed data either on mechanisms of blocking biosynthesis of host macromolecules in virus-infected cells (as sketched in Fig. 1) or on the molecular reactions that cause cell damage. Virology must in the near future turn its attention to this vital area of inquiry.

Of equal, or perhaps greater, significance to our understanding infection with a human pathogen, however, is that molecular studies of adenovirus replication have fulfilled the prophesies that detailed investigation of small DNA-containing viruses would probably furnish necessary clues toward understanding macromolecular synthesis in eukaryotic cells. It is further likely that an understanding of the reactions that regulate eukaryotic-cell macromolecular biosynthesis will also come from the wise construction of viral mutants and the subsequent genetic studies that are now being carried out with adenoviruses (see Chapter 4). Thus, studies of adenovirus biosynthesis revealed the surprising finding that large primary transcripts, such as the major late transcripts that encompass about 84% of the length of the genome, are processed by splicing, and that the regions of early transcription are noncontiguously situated in both strands of the genome, from one end to the other, but induction of their transcription is coordinately regulated (Chapter 5). Adenoviruses similarly uncovered unexpected features of viral DNA structure (Chapter 3) that must play roles not only in regulation of transcription but also in its unusual manner of semiconservative, asymmetrical DNA replication (Chapter 7).

Adenoviruses do not appear to be entirely self-centered, since they interact with other viruses, and these interrelationships may have epidemiological and ecological significance. Thus, the adeno-associated viruses (AAVs) are totally defective and depend on adenoviruses to provide a helper function for their relication (see Chapter 13). In turn, adenoviruses rely on simian virus 40 (SV_{40}), a papovavirus, to extend their host range and permit their unrestrained replication in monkey cells. Moreover, although adenoviruses and SV_{40} virus are without detectable genetic relationship, adenovirus–SV_{40} hybrids form as a consequence of their interactions during passage in monkey cells (see Chapter 10).

The chapters that follow provide the details of this broad view of adenoviruses, and they give appropriate recognition to the numerous talented virologists who have revealed the exciting plot of the adenovirus story. From these essays, the reader may also gain insights into the future experimental findings and concepts that will emerge from studies using adenoviruses. The knowledge that will accumulate will bear not only on the precise molecular reactions that regulate viral replication, transform cells, and effect disease, but also on many aspects of molecular and cell biology.

REFERENCES

Ginsberg, H.S., 1979, Adenovirus structural proteins, in: *Comprehensive Virology*, Vol. 13, (H. Fraenkel-Conrat and R.R. Wagner, eds.), pp. 409–457, Plenum Press, New York.

Ginsberg, H.S., Badger, G.F., Dingle, J.H., Jordan, W.S., Jr., and Katz, S., 1954, Etiologic relationship of the RI-67 agent to acute respiratory disease (ARD), *J. Clin. Invest.* **34:**1077–1086.

Hilleman, M.R., and Werner, J.R., 1954, Recovery of a new agent from patients with acute respiratory illness, *Proc. Soc. Exp. Biol. Med.* **85:**183–188.

Huebner, R.J., Rowe, W.P., Ward, T.G., Parrott, R.H., and Bell, J.A., 1954, Adenoidal–pharyngeal–conjunctival agents: A newly recognized group of common viruses of the respiratory system, *N. Engl. J. Med.* **251:**1077–1086.

Matthews, R.E.F., 1982, Classification and nomenclature of viruses, *Intervirology* **17:**59–61.

Rowe, W.P., Huebner, A.J., Gilmore, L.K., Parrott, R.N., and Ward, T.G., 1953, Isolation of a cytopathogenic agent from human adenoids undergoing spontaneous degeneration in tissue culture, *Proc. Soc. Exp. Biol. Med.* **84:**570–573.

Smith, W., Andrewes, C.H., and Laidlaw, P.P., 1933, A virus obtained from influenza patients, *Lancet* **2:**66–68.

CHAPTER 2

The Architecture of Adenoviruses

M. V. Nermut

I. INTRODUCTION

Viruses are "organized associations of macromolecules," and this definition predetermines our way of dealing with the problems of virion architecture. This review, therefore, is not a pure descriptive morphology, but aims at defining and describing the basic structural elements and complexes that constitute the adenovirion, their localization in the virion, and their mutual interactions. Since several reviews on the structural proteins of adenoviruses or virus morphology have been published over the past five years, the papers quoted herein have been selected on the basis of direct applicability to the aspect under discussion; in many cases, the reader is directed to the appropriate review for other references.

Structural research can be pursued at the level of molecular structure (primary through tertiary structure of macromolecules), of macromolecules (mainly the quarternary structures, i.e., morphology of isolated macromolecules—e.g., fibers, spikes), or of macromolecular assemblies (structure of macromolecular complexes and their mutual relationships within the virion). Electron microscopy (EM) has contributed to the studies of macromolecules (hexons, fibers) and their complexes (capsid, cores) since the very beginning of the "adenovirus era," and I will try to describe the present state of our knowledge of the architecture of this highly organized virus (Fig. 1).

EM studies of adenoviruses begun more than 20 years ago (for references, see Nermut, 1980a) have shown that the virion has an icosa-

M. V. NERMUT • National Institute for Medical Research, London NW7 1AA, England.

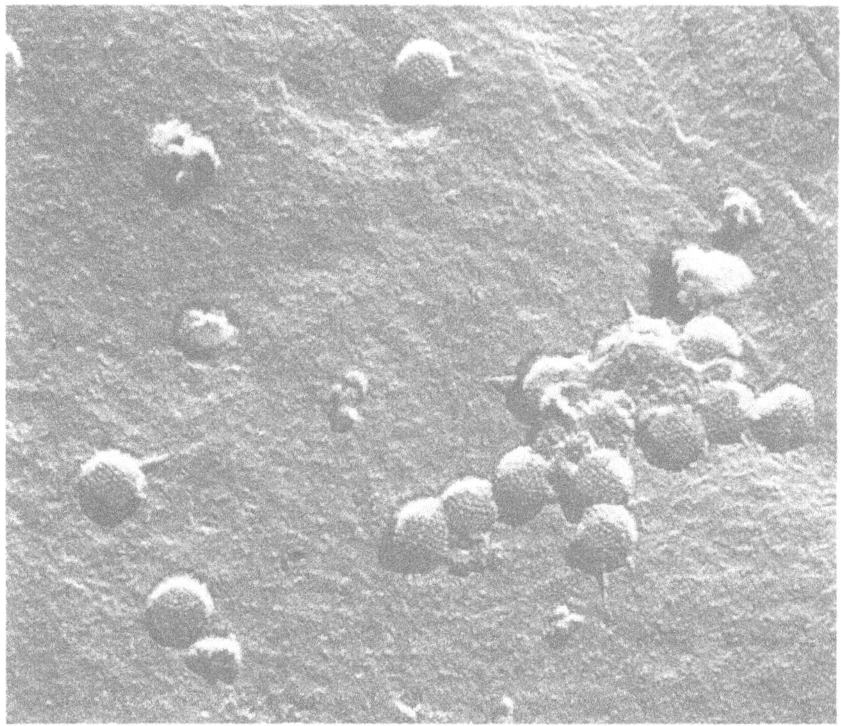

FIGURE 1. Human adenovirus invading a KB cell. Freeze–dry replica prepared 8 min post infection. ×100,000.

hedral shape and consists of two major structural complexes: the capsid—an icosahedral protein shell—and the core—an internal body that comprises the virus DNA and at least two major proteins. Before a detailed description of the individual structural elements (e.g., hexons or pentons) or complexes is presented, it is useful to recapitulate briefly what is known about the chemical composition of the adenovirus. It should be mentioned that most data used in this chapter stem from the studies of human adenovirus types 2 and 5.

II. CHEMICAL AND PHYSICAL PROPERTIES

Adenoviruses contain 11.6–13.5% of double-stranded DNA (ds DNA), the rest being protein. No lipid is present, and only about 1% is glycosylated (fiber). The molecular weight of mammalian adenovirus DNA is $20–25 \times 10^6$ [36,000 base pairs (bp)], whereas avian adenoviruses possess a DNA with a molecular weight of 30×10^6. The sedimentation constant of 31 S was reported for adenovirus type 5 (Ad 5) DNA (Harpst *et al.*, 1977) and 32 S for Ad2 DNA (Black and Center, 1979). The DNA molecule is linear—about 11–13 μm long (M. Green *et al.*, 1976a), but within the virion it may exist in a circular form mediated by the so-called DNA terminal protein (Robinson *et al.*, 1973; Rekosh *et al.*, 1977; Keegstra *et al.*, 1977). There are at least 11 species of polypeptides in the virion with molecular weights ranging from 3000 (3K) to 110K. Their properties and possible locations are summarized in Table I, which was compiled from various sources (see Philipson *et al.*, 1975; Ginsberg, 1979; Flint, 1980; for references see Akusjärvi and Persson, 1981). The topography of some virus polypeptides is still a matter of controversy and uncertainty, and these cases are discussed later. The molecular weight of the whole virion has been estimated using different methods, and values of 175–185 $\times$ 10^6 have been reported (M. Green *et al.*, 1967b; Devaux *et al.*, 1983). The bouyant density in CsCl is 1.33–1.35.

The size of the adenovirion is often reported as 600–700 Å in "diameter," but this is not really correct because the dimensions of an icosahedron are derived from its edge, i.e., the distance between two pentons, and given in terms of corner-to-corner distance (P = edge $\times$ 1.902) or edge-to-edge distance (E = edge $\times$ 1.618) or a "diameter" (D = edge 1.7) of the virus particle in a five-fold symmetry orientation (Mattern, 1969). For Ad5, the edge was estimated from electron micrographs to be 430 Å, and therefore $P \approx$ 820 Å, E = 696 Å, and D = 731 Å. A value of 736 Å for D was found after freeze–drying and shadowing (Nermut, 1975). Recently, Devaux *et al.* (1983) calculated the edge length from their neutron-diffraction data and obtained a value of 520 Å. This indicates that the D of a fully hydrated virus particle is about 880 Å.

III. VIRUS CAPSID: COMPOSITION AND ORGANIZATION

The icosahedral protein shell of the adenoviruses is made from 240 hexons and 12 pentons that each consist of a penton base and a fiber (Valentine and Pereira, 1965). One of the remarkable features of the adenovirus capsid is that it does not dissociate, after various treatments, into triangular facets of the icosahedron, as one would expect, but into "groups of nine" (GONs) hexons, leaving the remaining "peripentonal hexons" either attached to the pentons or free. However, it should be mentioned here that GONs are seen only rarely with some of the avian adenoviruses (Laver *et al.*, 1971).

The GONs are often seen in electron micrographs in two different orientations: either as left-handed (LH) or right-handed (RH) according to the definition by Pereira and Wrigley (1974). They differ in their appearance after negative staining, the LH GONs having a distinct central hole and a round profile and the RH GONs displaying three subunits separated by a tiny slot [$\approx$ 10 Å (Nermut, 1975)]. These features and the handedness are obvious in Fig. 2.

TABLE I. Adenovirus Proteins[a]

No.	Molecular weight	Number of copies/virion	Name	Location	Remarks
II	108,000	720	Hexon	Capsid	3 Polypeptides per hexon
III	85,000	36 (?)	Penton base	Capsid, vertices	Probably 3 polypeptides per penton base
IIIa	66,000	60 (?)	—	Peripentonal region	Phosphorylated
IV	62,000	24 (?)	Fiber	Capsid, vertices	Glycosylated
IVa_2	50,000	— (?)	—	Core-associated	DNA-binding
	55,000	2	DNA–terminal protein	5′ End of DNA	—
V	48,000	180	Core protein 1	Core shell	Moderately basic, weakly phosphorylated, DNA-binding
VI	23,400	420 (?)	—	Hexon-associated	Phosphorylated, DNA-binding
VII	18,000	1,070	Core protein 2	Nucleocapsid	Basic, DNA-binding
VIII	13,000	—	—	Hexon-associated	—
IX	11,500	300	—	Associated with GONs	—
X	7,000[b]	50	—	Internal	—
XI	4,500[b]	125	—	Internal	Possibly cleavage fragment from VII or identical with μ protein
XII	3,000[b]	—	—	Internal	Possibly cleavage fragment from VIII

[a] Compiled from various sources and based mainly on Ad2 and Ad5.
[b] Averaged figures.

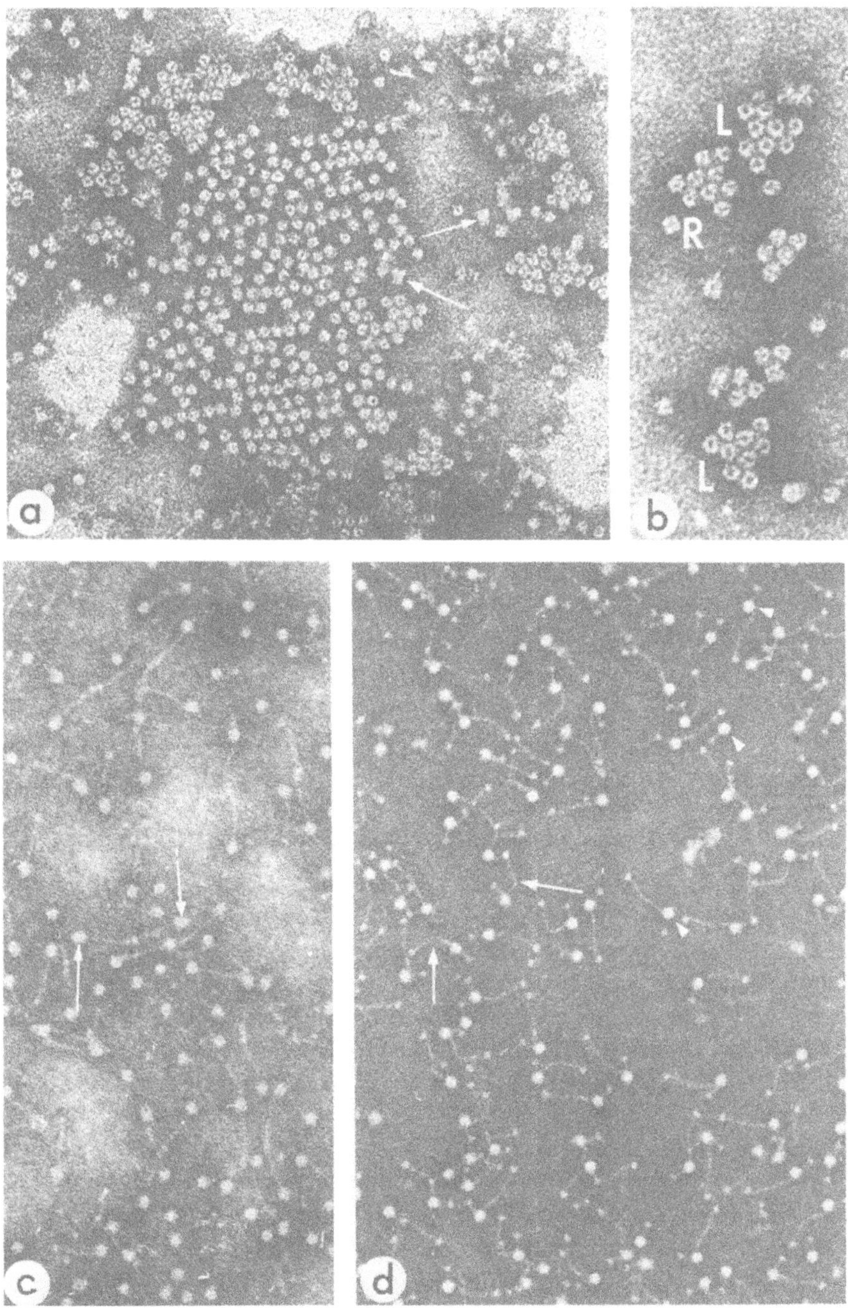

FIGURE 2. Adenovirus capsomers after negative staining (4% silicotungstate). (a) Hexons in a variety of orientations, either single or in "groups of nine" (GONs). (←) Longitudinally oriented hexons. (b) left-handed (L) and right-handed (R) GONs. Note the large central hole in GONs. Reprinted from Nermut (1980a) with permission. (c) Ad5 pentons. Note the conical shape of the penton base (↓). (d) Pentons with two fibers from an avian adenovirus (chick embryo lethal orphan virus). The longer fiber is often sharply bent (←). (◀) Pentagonal penton bases. Courtesy of Dr. N. G. Wrigley. (a, c, d) ×200,000; (b) ×300,000.

A. Hexon

There are 240 hexon macromolecules in the virus capsid and only 12 pentons. This underlines the main role of this capsomer, i.e., to form the surface protein shell. However, this structural function of the hexon could well be accompanied by more functional roles such as the control of the transport of solutes or larger molecules (enzymes) across the capsid and the behavior in, and interactions with, the environment and cellular membranes. How are these functions fulfilled and realized?

Obviously, they can be understood only if the three-dimensional (3D) structure of the hexon is known. During the past ten years, both the morphology and the molecular weight of the hexon polypeptide were a matter of controversial findings and conclusions. A molecular weight of 120K was eventually accepted by most workers in 1974 (Grütter and Franklin, 1974), but the most recent studies of the primary structure of the hexon polypeptide gave a molecular weight of 108,113 for Ad2 and about 103K for Ad5 (von Bahr-Lindström *et al.*, 1982).

Using physical and chemical methods, Grütter and Franklin (1974) determined the molecular weight of the hexon to be about 360K and concluded that there are three identical polypeptides per hexon (see their paper for references of previous molecular-weight estimates). Morphological evidence for this conclusion was provided by Crowther and Franklin (1972) using rotational filtering of GONs after negative staining. Three subunits have also been demonstrated by rotational averaging, but only in right-handed GONs (Nermut, 1975), indicating a morphological difference between the opposite ends of the hexon. Indeed, the left-handed GONs display a large "central hole" with no indication of three subunits. After the discovery that only a thin portion of the hexons (adjacent to the supporting film) is visualized by negative staining, a computer simulation technique was employed to build a 3D model of the hexon (Nermut and Perkins, 1979). This model (Fig. 3) has the following features: The low portion of the hexon, i.e., the innermost one, measures 75–80 Å in "diameter" and has an axial hole 25–30 Å across. The middle portion is "pseudohexagonal," with dimensions of 85 × 96 Å and the top (uppermost) portion is triangular (with a Y-shaped slot), measuring 85–90 Å between corner and edge. The corner of the upper triangle projects between the corners of the middle "hexagon"; i.e., the top is twisted by 30° in relation to the waist. The central "channel" is supposed to be very narrow inside, but is rather wide toward the base. The slots on the top, when visualized by negative staining, are 10–15 Å wide. The hexon is about 110 Å high.

At about the same time, an X-ray analysis of Ad2 hexon crystals was completed by Burnett *et al.* (1979) in Basel, and their model as described so far is similar in principle to the model described above, but shows finer structural details in the upper portion and has hexon dimensions as fol-

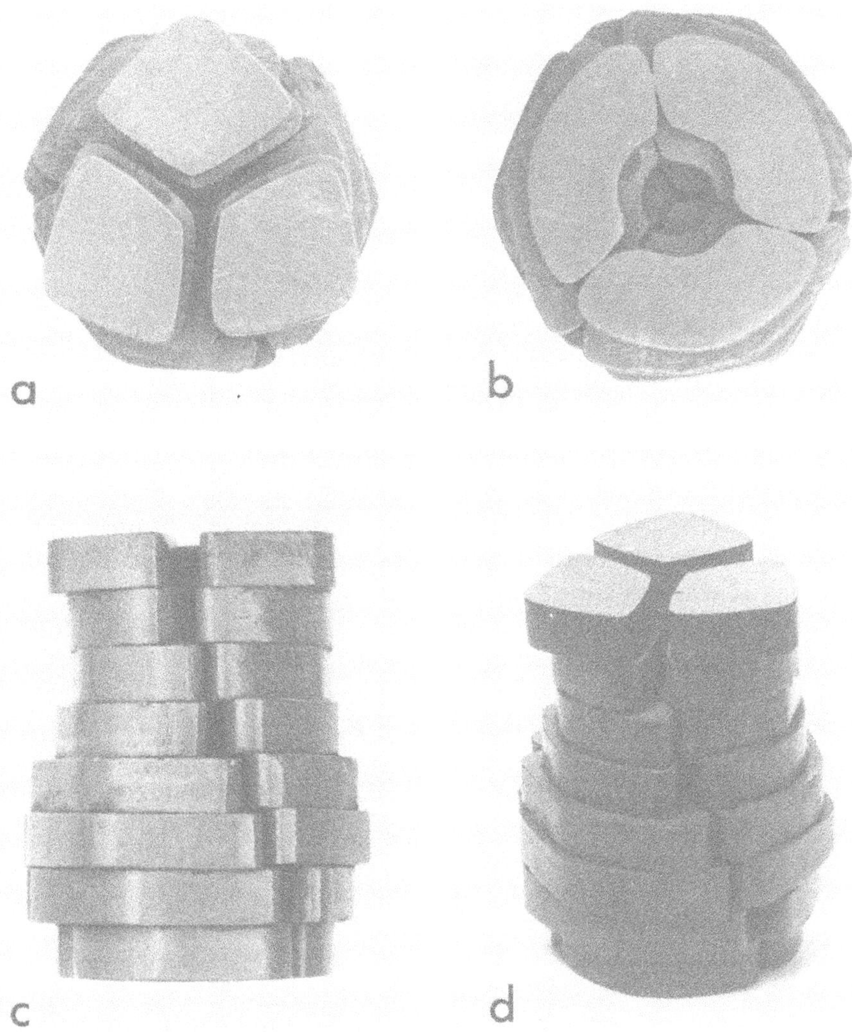

FIGURE 3. Hexon model as developed from EM and computer modeling (Nermut and Perkins, 1979). Views are of the triangular top (a), the round bottom portion with axial "hole" (b), the side (c), and a bird's-eye view (d). The pseudohexagonal shape is mainly at the third layer from the bottom.

lows: height = 105–110 Å, edge of the top equilateral triangle = 95 Å, and the hexagonal waist outer "diameter" = 105 Å. It also shows that the central channel is very narrow in the middle portion, limiting the passage of large molecules.

Thus, the hexon has unique structural properties; the three polypeptides form a tripartite body—round at the bottom, hexagonal in the middle portion, and triangular at the top. Moreover, there is also a phys-

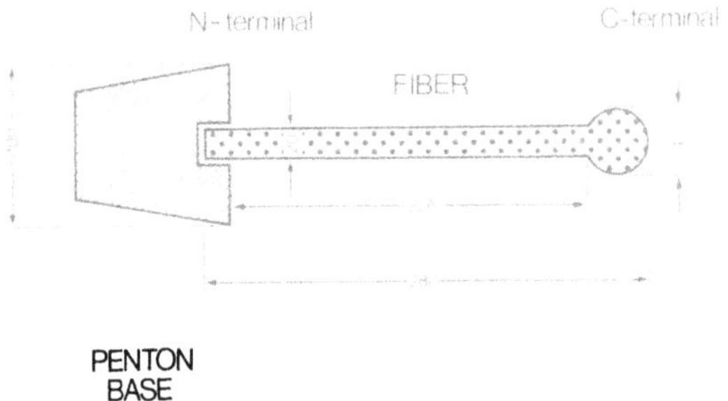

FIGURE 4. Diagram of Ad2 fiber. The dimensions are in angstrom units.

icochemical polarity in the hexon. The top of the hexon is hydrophilic and predominantly negatively charged, whereas the base shows hydrophobic properties (Nermut and Perkins, 1979). In electrophoresis on cellulose acetate, the hexon moves to the anode (always in front of the other virus components). The hydrophobic nature of the hexon base supports the concept (described in Section IV.A) that the capsid interacts with another protein shell—the core surface—most probably formed by polypeptide V. The top of the hexon bears the type-specific antigenic determinants, whereas the group specificity is determined by the bottom portion of the hexon (Norrby, 1969).

Those hexons that are in direct contact with the pentons are called *peripentonal*. Shortridge and Biddle (1970) suggested that their isoelectric point was lower (about 3.1) than that of the GONs (pI = 4.7). As yet, however, there is no evidence that the peripentonal hexons differ morphologically or structurally from the GONs. Recently, Devaux *et al.* (1982) calculated the mass of the hexons per virion to be 77.8×10^6, which is about 46% of the total mass.

B. Penton

The capsomers at the vertices of the icosahedron are called the pentons because they are surrounded by five neighbors, the peripentonal hexons (Ginsberg *et al.*, 1966). A penton is formed by a penton base and a thin antennalike projection called a *fiber* (see Figs. 2 and 4).

The molecular weight of the penton base polypeptide III is about 85K (Maizel *et al.*, 1968; Anderson *et al.*, 1973) and that of the fiber about 62K (Dorsett and Ginsberg, 1975; Hérissé *et al.*, 1981). The whole penton base was supposed to have a molecular weight of about 400–500K (Philipson *et al.*, 1975), but recent studies by Devaux *et al.* (1982) have shown that the whole penton has a molecular weight of 365K and the penton base of 245K, 120K being, then, the molecular weight of the fiber. This would mean that the penton base is a trimer, not a pentamer as usually thought. The fiber could then be a dimer, but since the molecular weight of the intact fiber determined by neutron-scattering and hydrodynamic measurements is about 160K (Devaux *et al.*, 1982), a trimeric organization is not ruled out. There is some EM evidence that the penton base is pentagonal (Pettersson and Höglund, 1969; Boudin *et al.*, 1979) and (Fig. 2d), but no structural studies have been done yet to show how such a pentagonal shape is formed by three subunits.

The most recent data obtained on the amino acid sequence as derived from the published DNA sequences helped to calculate the molecular weight of the fiber (= 62,294). Using these sequences, N.M. Green *et al.* (1983) have constructed a model of the fiber with overall dimensions as seen in Fig. 4. In this model, the shaft consists of two polypeptide chains in cross-β configuration. The hydrophobic bases of the sheets are opposed to each other [cf. T_4 tail fiber (Earnshaw *et al.*, 1979)].

Human adenoviruses have one fiber per penton with a length of 100–370 Å, whereas most avian strains possess two fibers of the same or very different lengths (see Fig. 2d) (Laver *et al.*, 1971). Serologically related fowl adenoviruses have two fibers, whereas the unrelated egg drop syndrome virus has only one (Gelderblom and Maichle-Läuppe, 1982). The fiber is terminated by a "knob" about 40 Å in diameter that is believed to carry the type-specific antigenic determinant (Norrby, 1969), whereas the subgroup-specific determinant is probably located on the fiber shaft. The Ad5 fiber has a sedimentation constant ranging from 5.8 to 6.2 S.

The penton can be rather easily dissociated into the base and the fiber using guanidine (Norrby, 1969), pyridine (Pettersson and Höglund, 1969), or deoxycholate (DOC) (Boudin *et al.*, 1979). This indicates that the interaction between them is hydrophobic in nature, certainly not covalent. However, fibers are often lost from virus particles after freezing and thawing and on treatment with antifiber antibody (Boudin and Boulanger, 1981). The fiber is assumed to be inserted into the penton base like a shaft. Some electron micrographs also show that the fiber is inserted into the wider end of the slightly conical base (Fig. 2c) (Laver *et al.*, 1969).

C. Other Virus Polypeptides Associated with the Capsid

Everitt *et al.* (1975), using cross-linking with tartryl-diazide, found that virus polypeptide (VP) VI is in close proximity to the hexon, but no further studies of this enigmatic protein had been reported until now. Recently, Russell and Precious (1982) found out that VP VI and IVa_2 bind to DNA as does VP V and VII. However, VP VI is usually lost from virus cores prepared with pyridine or DOC during gradient centrifugation (Everitt *et al.*, 1973; Nermut, 1979). The role of IVa_2 in the cores is not known. On the other hand, there is reasonable evidence that VP IX is associated with GONs, possibly in a ratio of 15 copies per GON (Boulanger *et al.*, 1979). However, the *in vitro* assembly experiments by Pereira and Wrigley (1974) were carried out with purified hexons, i.e., in the absence of VP IX. In addition, a deletion mutant of Ad5 has been described

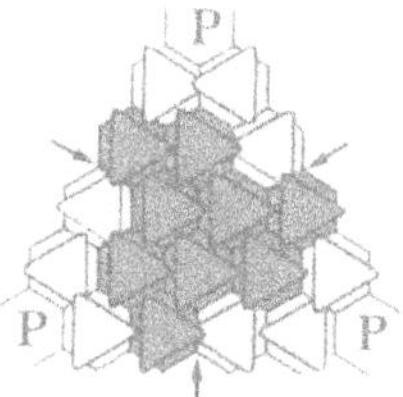

FIGURE 5. Diagram of one triangular facet of the capsid. (P) penton; (▲) tops of hexons; (○) lower waist of hexons. Note the corner-to-corner orientation between hexons belonging to different GONs (→), as opposed to the corner-to-corner edge orientation within one GON. Reprinted from Nermut and Perkins (1979) with the permission of *Micron*.

that was capable of assembling virus particles despite the absence of polypeptide IX (Colby and Shenk, 1981).

Another protein that might play an important structural role in the capsid is VP IIIa. Its association with the vertex region was postulated by Everitt *et al.* (1975) and recently by Devaux *et al.* (1982), who ascribed to this protein the function of a mediator between the penton base (supposed to be a trimer) and the five peripentonal hexons. Boudin *et al.* (1980) found that there are five copies of VP IIIa per vertex.

D. Organization of the Capsid

The design of the adenovirus capsid follows the requirements of the 5-, 3-, 2-fold symmetry and the postulate of the maximum possible bonding (= minimum free energy) between the capsomers. The adenovirus capsid belongs to the P-1 class, with a triangulation number $T = 25$.

As described in Section II.A, the hexon consists of three subunits, and the tight interaction with its six neighbors is accomplished through the pseudohexagonal shape of its "lower waist" portion. However, the orientation of the hexons in the capsid cannot be deduced from the model itself. Here, the observation that the triangular tops of hexons (within a GON) have a constant azimuth (60° toward the edge of the triangular facet) helped to understand why the capsid disintegrates preferentially into the GONs and not into large triangles or randomly. We have suggested (Nermut and Perkins, 1979) that the interactions within the GONs are stronger than those between two neighboring GONs, as can be seen from the diagram in Fig. 5. However, the presence of a linker protein in GONs is an alternative explanation for the existence of GONs (Boulanger *et al.*, 1979).

The orientation of the peripentonal hexons and their interaction with the pentons in Fig. 5 is tentative because solid information on this point is missing. Because the pentons can be removed by dialysis against low-ionic-strength buffers at pH 6.3–6.5, one could assume that the penton–hexon interaction is rather weak and possibly of electrostatic nature, although it is not ruled out that VP IIIa participates in this interaction (Devaux *et al.*, 1982).

The implications of the afore described model for the structure and function of the virus capsid can be summarized as follows: First, the capsid seems to be stable enough to exist without the internal contents, i.e., the core. This has been shown by the self-assembly experiments described by Pereira and Wrigley (1974) and is in line with the suggestion that the virus capsid is formed first during the assembly process (for references, see Philipson, 1979). The close contact between the hexons means that the capsid is completely sealed off for larger molecules (such as nucleases, molecular weight ≈ 16K), allowing the passage of solutes only into the virus interior. Thus, the capsid also fulfills the role of a "permeability barrier," most probably together with the underlying "core shell."

Since the hexon consists of three subunits only, the total number of structure units in the capsid is 720 + 60 (or 36), i.e., 780 or 756 (provided the penton has three subunits only). This figure is increased by 384 if the whole capsid complex including VP IIIa, IV, and IX is considered.

IV. VIRUS CORE

Virus particles devoid of the icosahedral capsid are called the *cores*. This name has now been in general use since 1968 (Laver *et al.*, 1968; Russell *et al.*, 1971), but definitions of and ideas about the core structure differ. The biochemical definition (e.g., Mirza and Weber, 1977) states that the cores consist of virus DNA plus two core proteins (VPs V and VII) and have a buoyant density in a CsC1 gradient of 1.58. Such a definition does not show any concern about the state of the complex—whether it is structurally native or not. The morphological definition describes the core as a dense, probably icosahedral body consisting of a surface protein shell and an internal DNA–protein complex—the nucleocapsid proper.*

There is a consensus here as far as the chemical composition of the cores is concerned (see Section 2.11), but very little agreement as to the organization and function of the two major proteins. For example, VP V has been proposed to be situated between VP VII and the hexon (Everitt *et al.*, 1973) or to be part of the nucleosomelike viral "chromatin" (Mirza and Weber, 1982) or to form a shell on the surface of the cores (Nermut, 1979). The function of polypeptide IVa_2, X, or μ is not yet fully understood.

The diversity in opinion discussed above might be due to the fact that the cores are rather fragile structures and are easily disintegrated by the preparation procedures. Although fresh DOC cores are usually well

* This follows the definition of the virus nucleocapsid as accepted by the International Committee for Nomenclature of Viruses (Wildy, 1971): "The nucleocapsid is nucleic acid plus the protein most closely in contact with it." However, nucleocapsid has been often used as a synonym for the virion (e.g., herpesviruses). To avoid misunderstanding, two terms are used throughout this chapter in describing the linear structures in the virus interior that contain DNA and one (or two) proteins: either the "DNA–protein complex" or the "nucleocapsid proper."

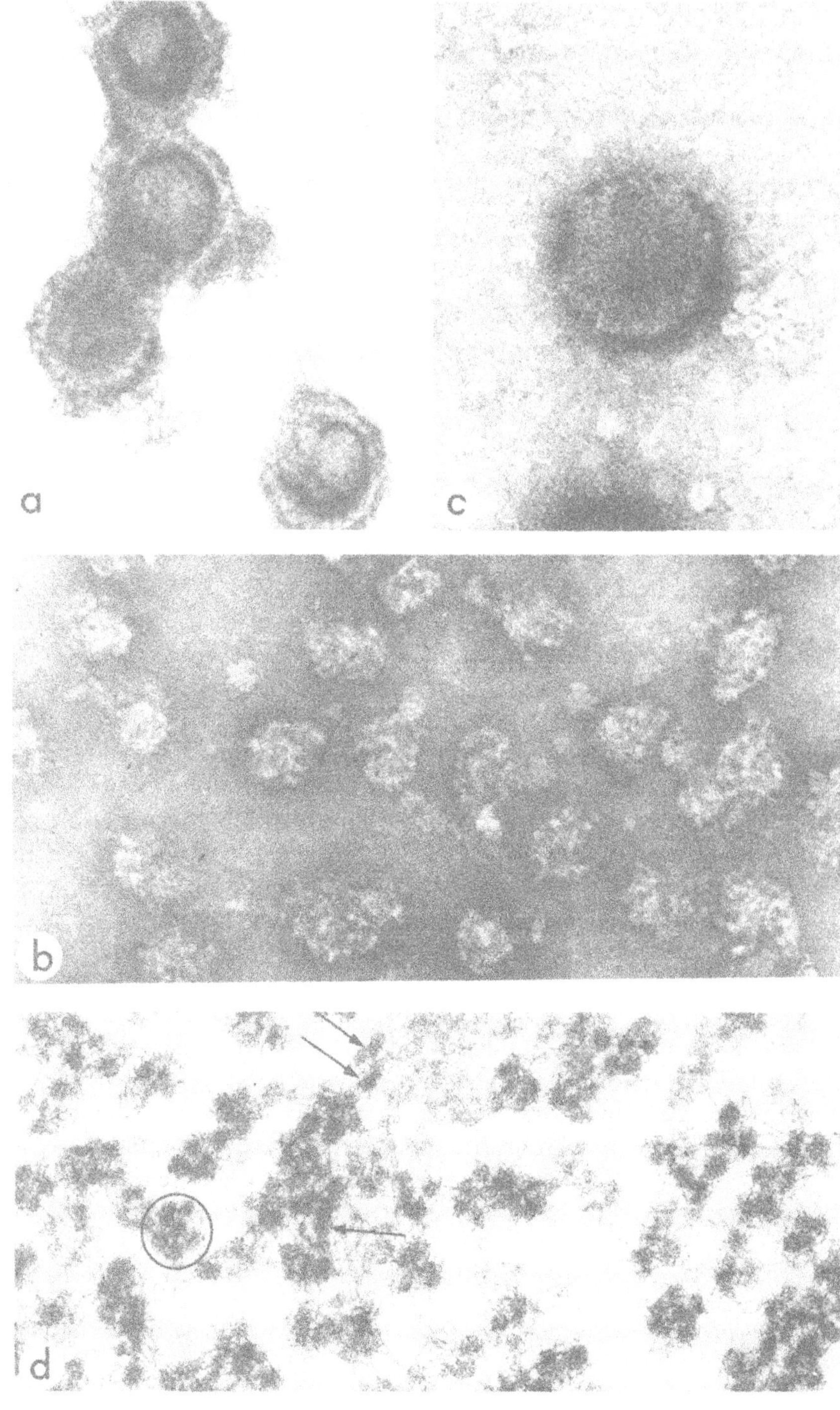
a
c
b
d

preserved, their sedimentation constant (S_{20w}) can range from 170 to 376 depending on the state of relaxation (Harpst *et al.*, 1977).

A. Evidence for the Core Shell

Virus cores prepared by mild treatments are solid 3D bodies with a smooth surface (D. T. Brown *et al.*, 1975; Nermut *et al.*, 1975). The way of preparation for, and also visualization by, EM is of paramount importance (Fig. 6). Freeze–fractured adenovirus particles sometimes reveal a smooth "icosahedral" body with an edge of 380 Å. This is the most valuable evidence for the existence of the core (D. T. Brown *et al.*, 1975; Nermut, 1978). However, fine details of the core surface can be seen better when freeze–fracture is followed by negative staining as shown in Fig. 6a. These micrographs reveal the surface shell beyond doubt, and since no chemical treatment was used in the case of freeze–fracture negative staining, we can assume that they show the native surface. Moreover, similar structural details have been observed after treatment of virus particles with DOC (Fig. 6c) or pyridine (D. T. Brown *et al.*, 1975). There is usually no sign of a surface dense track in ultrathin sections of virus particles, but this can be easily explained by the high density of the DNA-containing core. Thus, there seems to be a good deal of "morphological" evidence in favor of a surface shell, but which of the two core proteins would form such a shell? VP VII has been shown to be closely associated with DNA, so that the most probable candidate is VP V. This polypeptide has a molecular weight of 48K and is supposed to be present in 180 copies per virion (Everitt *et al.*, 1973), whereas VP VI, another possible candidate, is removed from the cores on a gradient (Everitt *et al.*, 1973; Nermut, 1979). High-resolution pictures of the cores revealed ringlike molecules 50–80 Å in diameter, the molecular weight of these being between 45 and 60K (Nermut, 1980a). Since the core is closely attached to the capsid, such a shell would be built as an icosahedron with a $T = 25$, but from protein molecules of smaller size than the hexon. Their number would be 240 (= without the vertex macromolecules) or 180 if peripentonal subunits were missing, although the electron micrographs obtained up to now do not support the latter figure of 180, since no large holes have been observed at the vertices.

←

FIGURE 6. Adenovirus cores prepared by various methods. (a) Freeze–fracture negative staining. Note that the cores are closely apposed to the capsid. ×300,000. (b) 0.5% DOC at 56°C, negatively stained with ammonium molybdate. ×130,000. (c) Same treatment as in (b), but negatively stained with uranyl acetate. The surface is covered with ringlike subunits. ×400,000. (d) Ultrathin section of DOC cores processed immediately after preparation. Round and elongated [rod-shaped (→)] structures are present. Some cores clearly show five peripheral "rods" around a central one (circled). ×100,000.

The vertices of this shell could very well be plugged by the penton bases, because if these bases are removed by dialysis, the virion becomes susceptible to nucleases. The vertex hole present after the removal of the penton is large enough to allow nucleases (the molecular weight of DNase is about 16,800) to penetrate into the virus interior.

The core shell could thus fulfill both a structural (increase in rigidity of the virion) and a physiological function, i.e., protecting the virus genome and together with the capsid regulating transport of water and salts in and out.

B. Organization of the DNA–Protein Complex (Nucleocapsid)

It has been shown by several groups of workers that the DNA–protein complexes prepared by a variety of procedures [e.g., DOC, pyridine, acetone, urea (for references, see Ginsberg, 1979)] consistently contain VP VII in close association with DNA. This protein contains about 22% of Arg (Laver, 1970), which could well neutralize the negative charges of DNA. Other proteins have been found in some preparations of cores. Those prepared by mild treatment, e.g., with 0.5% DOC (Russell *et al.*, 1971), contain VPs V, VII, and IVa_2. Vayda *et al.* (1983) found VPs V, VII, and X in their preparations of pyridine cores. A low-molecular-weight basic protein has been reported in Ad5 by Hosokawa and Sung (1976). However, VP V is easily lost from the core preparations (D. T. Brown *et al.*, 1975; Mirza and Weber, 1977; Vayda *et al.*, 1983) and does not seem to play a role equivalent to that of histone H_1 in eukaryotic chromatin. In addition, its molecular weight is 48K, which corresponds to a globular particle that would be about 40–60 Å in diameter and therefore easily visualized by EM if located between two "nucleosomes." No such evidence exists.

As preparation for a discussion of the organization of the DNA–protein complex, it is useful to sum up the present knowledge as obtained by EM and by biochemical studies of adenovirus cores.

1. Electron-Microscopic Observations on Virus Cores

As mentioned in Section IV.A, native cores are dense, smooth-looking "spherical" bodies. However, many core preparations, particularly after gradient centrifugation and depending on the preparation procedures used, contain cores in different degrees of relaxation or even disintegration (Fig. 7a). They often look like "spiders," showing a dense core with DNA loops all around (Brown *et al.*, 1975; Nermut *et al.*, 1975). Such cores are obviously little suited for decent structural or biochemical studies. It is therefore imperative to start any "relaxation" study with virus cores that are well preserved, possibly with no free DNA around.

Since most of the studies during the past few years have been carried out on virus cores prepared by treatment with Na-DOC (DOC cores) or pyridine, the results obtained are discussed separately.

DOC cores prepared under mild conditions (e.g., 0.4% DOC at 56°C for 40 sec) can be used as a starting material for controlled relaxation, which can be easily monitored by EM. It has been shown that treatment with rather high concentrations of ethyleneglycol bis(β-amino-ethyl ether)*N,N'*-tetraacetic acid (EGTA) at pH 7.5 relaxed the cores into "ribbons" that often had a beaded appearance (≈ 100 Å) (Nermut, 1979) (Fig. 7b).

A similar effect was observed after high-salt (0.5–2.0 M NaCl) or high-pH (e.g., 10) treatment, but several rodlike elements about 150 Å thick and up to 500 Å long appeared instead of ribbons (Fig. 7c). Because the final effect in all three cases may well be a withdrawal of calcium, it is conceivable to assume that the rodlike elements in native cores are closely packed by means of calcium ions that form bridges between DNA phosphates. When calcium is removed, the repulsive forces separate the "rods" from each other and make them visible. These findings are in good agreement with the observations of 5 + 1 rodlike elements in intact virus particles broken up by freeze–fracturing and also in ultrathin sections (see Fig. 6d) of DOC cores (Nermut, 1978). Occasionally, "beads on a string" have been observed in high-salt-treated cores (see Fig. 7d).

The pyridine cores are usually more disintegrated than the DOC cores, so that the starting material consists of thick filaments and lots of free DNA. In our hands, even 5% pyridine for half an hour was rather drastic, and round, smooth cores as reported by D. T. Brown *et al.* (1975) were exceptional. Beaded filaments have been described by Mirza and Weber (1982) and most recently by Vayda *et al.* (1983), who have been able to produce large amounts of "beaded filaments" by treating pyridine cores with 0.5–1.0 M NaCl. However, prolonged treatment with high salt resulted in protein-free DNA, presumably because of the breakdown of the electrostatic interaction between DNA and VP VII.

2. Biochemical Studies of "Relaxed" Cores

A series of papers appeared during the past six years making use of a specific property of an endonuclease from *Staphylococcus aureus* (micrococcal nuclease) to cleave only free dsDNA, leaving intact DNA fragments bound to protein molecules. This approach was first used by Corden *et al.* (1976) with adenovirus particles partially opened by dialysis against tris–maleate buffer at pH 6.3 (cornerless virus particles) or disrupted by means of heat or pyridine (virus cores). Treatment with micrococcal nuclease produced fragments of DNA about 200 bp long or multiples of 200 bp. Prolonged treatment led to the disappearance of large fragments, leaving behind only 200-bp fragments or smaller. Corden and co-workers suggested that the adenovirus "chromosome" is organized in

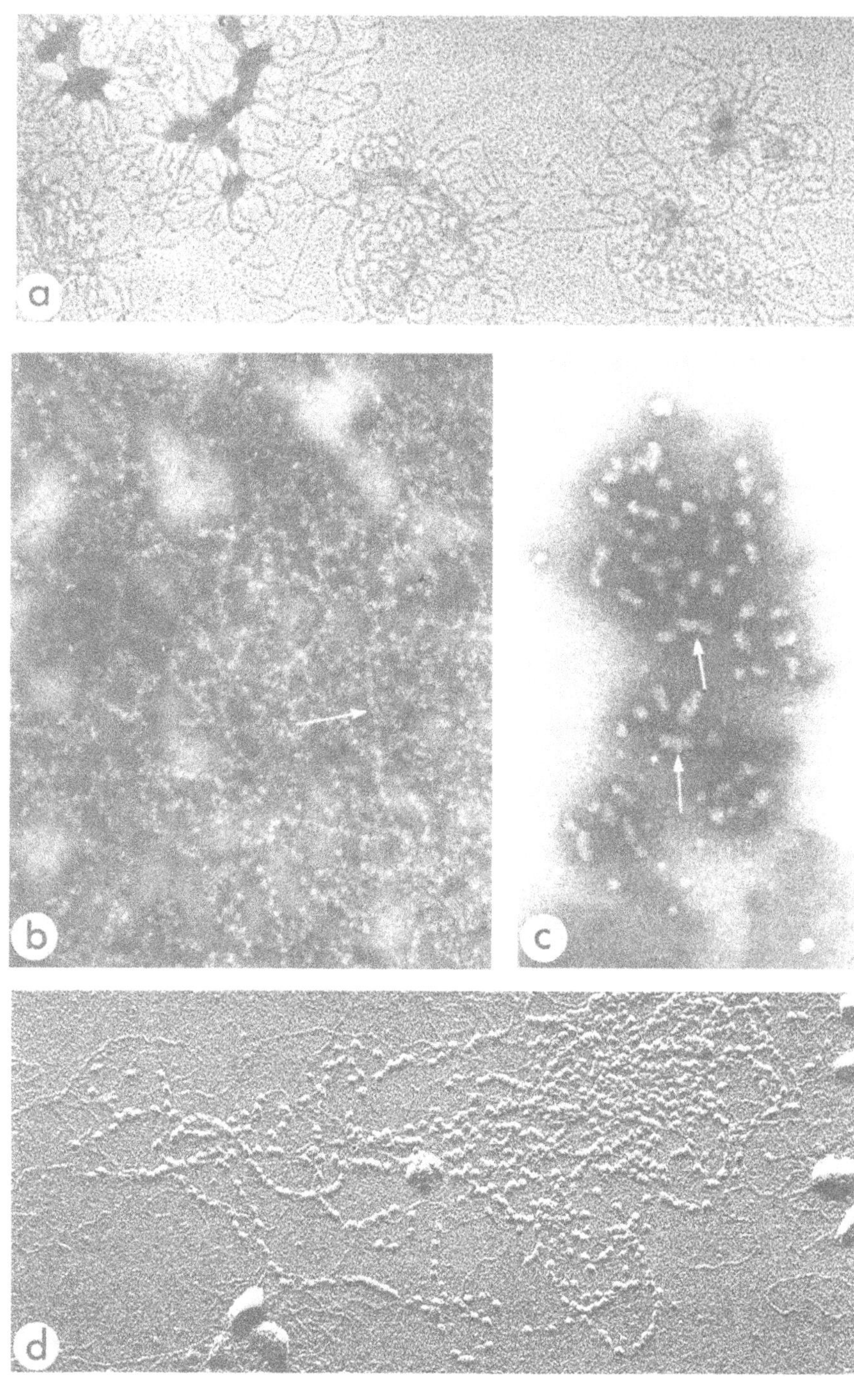
a
b
c
d

a similar way as cellular chromatin, i.e., in structures similar to nucleosomes with six copies of VP VII and one copy of VP V per nucleosome.

However, when a similar treatment was applied to DOC cores or pyridine cores in other laboratories, no distinct "protection pattern" was observed, usually only a long smear on the gels (M. Brown and Weber, 1980; Tate and Philipson, 1979). Only very recently, Mirza and Weber (1982) managed to isolate 11 S "monomers" from micrococcal-nuclease-digested cores that contained about 150 bp of DNA. However, "submonomeric" structures about 100 and 50 bp long have also been produced by prolonged digestion. These data are close to our findings that DOC cores relaxed with high salt or high pH can be digested with micrococcal nuclease down to 50 bp or less (Nermut, 1980b). Using [^{3}H]-Arg-labeled virus particles, Vayda *et al.* (1983) confirmed that pyridine cores contain VPs V, VII, and μ described by Hosokawa and Sung (1976). However, only VP VII was found in high-salt-treated cores that displayed the beaded-string appearance. Treatment of "high-salt cores" with micrococcal nuclease produced DNA some 130–400 bp long.

Interestingly, viral "chromatin" from virus-infected cells [3–6 hr postinfection (p.i.)] reveals a repeat size of 160–165 bp, as reported by several groups of workers (Sergeant *et al.*, 1979; Tate and Philipson, 1979; Corden and Pearson, 1979). This could be interpreted in the sense that VP VII is replaced by cellular histones early in the infection (Corden and Pearson, 1979) or that the cores are fully relaxed in the nucleus and therefore easily accessible, which is not the case with cornerless virus particles. However, viral "chromatin" from cell nuclei 18 hr p.i. reveals the same pattern as cornerless virions when treated with micrococcal nuclease (M. Brown and Weber, 1980).

Sato and Hosokawa (1981) used UV irradiation to cross-link viral DNA with the associated protein and found that only VP VII formed a complex with DNA and both monomers and dimers of VP VII were found on sodium dodecyl sulfate gels. Dimers of VP VII were also obtained after fixation of virus cores with glutaraldehyde (Mirza and Weber, 1982).

Finally, it seems appropriate to include some of the circular dichroism (CD) studies of adenovirus cores because they provide information on the conformation of DNA within the DNA–protein complex. These results can be briefly summarized as follows: (1) Adenovirus cores prepared by means of DOC show a CD spectrum similar to that of cellular

←

FIGURE 7. (a) DOC cores purified on a CsCl gradient, spread on distilled water, and rotary-shadowed. The large proportion of DNA is in the form of loops or supercoiled strands. ×75,000. (b) DOC cores treated with EGTA–DTT–NaCl as described by Nermut (1979) and spread on the surface of distilled water. Negative staining with uranyl acetate. Note the beaded appearance of some filaments (→). ×100,000. (c) DOC cores after treatment (on the grid) with 1 M NaCl. Negative staining with silicotungstate. Note the rod like elements (↑). ×100,000. (d) DOC cores after treatment with 1 M NaCl and freeze–dry shadowing. Beaded filaments as shown here are not frequent. ×100,000.

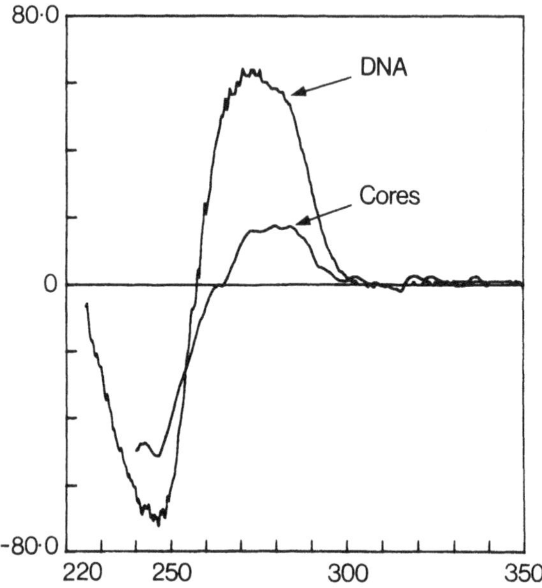

FIGURE 8. CD of adenovirus DNA and of DOC cores (0.4% DOC, 40 sec at 56°C). *Abscissa:* wavelength in nm; *ordinate:* differential absorption [$^{\Delta}A = (A_L - A_R) \times 10^6$]. (No quantitative relationship exists between the two CD spectra.)

chromatin, with a broad positive peak in the region between 280 and 285 nm (Nermut, 1979; Boulanger and Loucheux-Lefevbre, 1982) (Fig. 8). The presence of free DNA in pyridine cores (Tate, 1976) or in sarkosyl cores (Boulanger and Loucheux-Lefevbre, 1982) manifests itself as a second positive peak at about 275 nm in the same way as does protein-free DNA. (3) Withdrawal of calcium by means of EGTA does not alter the typical core spectrum, whereas high salt or dithiothreitol (DTT) abolishes the positive peak at 280 nm (Nermut, 1979).

These results allow the conclusion that the DNA conformation has been altered by virtue of its interaction with the core protein(s) and that the DNA–protein complex forms an ordered structure (Cowman and Fasman, 1978).

C. Tentative Model of the Adenovirus Nucleocapsid

The morphological and biochemical data discussed above provide sufficient background for a tentative model of the virus nucleocapsid. However, before an attempt is made to combine the results obtained in different laboratories into a "unifying" model, it is useful to start with theoretical considerations and to outline the basic principles of quarternary structure formation in this particular system, i.e., DNA–protein complex.

Undoubtedly, the major problem during virus assembly is how to pack a rather long DNA molecule into the available space. Any close packing of DNA molecules requires overcoming two specific properties of DNA: the rigidity of DNA, which in solution behaves like a stiff rod about 625 Å long (Hays *et al.*, 1969), and the dense negative charge of the phosphate groups on the surface of DNA. There is no doubt that an economical packing must be done in an organized fashion that will bring about the highest possible compaction ratio. This can be achieved only by a helical twist of the DNA around a protein core with a minimum diameter of about 80 Å. Energy calculations have shown that DNA can bend without kinking down to a radius of curvature of 40–50 Å (Finch *et al.*, 1977; Sussman and Trifonov, 1978).

Bending DNA into a helix and holding it in that configuration requires energy. It is presumed that the energy of the electrostatic bonds between a basic protein such as VP VII and the DNA phosphates would serve this purpose. If this is so, the DNA molecule would wind around a protein core that could be in the form of a helix or oligomeric units, e.g., hexamers or octamers. The remaining surface phosphates could be neutralized by any cations, or in the case of a close physical contact, they could be bridged over by divalent cations or polyamines.

The organization of the protein core depends on the properties of the DNA-associated protein. It was shown more than 30 years ago (Crane, 1950) that a linear structure made by identical asymmetrical units interacting with each other in an identical way has a helical organization. If two or more proteins are involved, more complex identical subunits can form that can be arranged as "beads on a string" or as a superhelical structure if further condensed.

Before applying these principles to the adenovirus nucleocapsid, let us sum up the available information: We deal with a 11 to 12-μm-long dsDNA (= 34,000–36,000 bp) and about 1100 copies of VP VII. The possibility of the participation of another protein is still open (some of the low-molecular-weight proteins, μ or X). Micrococcal-nuclease-digestion experiments suggest a repeating DNA fragment of about 150 bp. The cores contain linear structures approximately 150 Å or more thick and up to 0.3 μm long (Nermut, 1979). DNA filaments with beads have been observed particularly after pyridine treatment (Mirza and Weber, 1982; Vayda *et al.*, 1983). Six rodlike elements about 150 Å thick and 400–500 Å long fill the inner icosahedral space of the virion. The volume of this space is about 71,000 nm^3 (based on an inner edge of 320 Å). The volume of a spherical bead 90 Å in diameter is 380 nm^3.

In principal, there are two possible ways of organization of the DNA–protein complex in adenoviruses: as a "continuous" helical structure or as a "discontinuous" one. The decisive factor in this respect is the protein component. The presence of one DNA-associated protein would result in the "continuous" helix organization. If another protein is present that would interrupt the aforementioned "continuity" repeating subunits (e.g., nucleosomes) would form and the DNA superhelix would be "discontinuous." Until very recently, this issue was a matter of controversy, and both models are therefore discussed, though there is now more evidence in favor of the nucleosomelike model.

1. Continuous Helix Model

This model was based on the presence of one DNA-associated protein (VP VII) only. As already mentioned and explained *in extenso* earlier (Nermut, 1980a), this protein would form a linear helical filament that serve as a "core" for the DNA coiled around it. Such a filament would be about 80–90 Å thick and if 1 μm long would contain 1134 copies of VP VII and 11.5-μm-long DNA (= 36,000 bp) (Fig. 9a,b). Such filaments have in fact been observed (Nermut, 1980a). They could form superhelical structures—the "rods" (Fig. 9c,d). Six rods would measure about 0.3 μm. Some of the linear structures described earlier were up to 0.26 μm long (Nermut, 1980a). The close packing of the DNA–protein filaments would be mediated by calcium bridges between two opposite phosphate groups.

One turn of the DNA would encompass 82 bp (= 280 Å) , and the continuity of this helical arrangement would account for the absence of the typical nucleosome pattern in "encapsidated" nucleocapsids (Tate and Philipson, 1979; M. Brown and Weber, 1980; Mirza and Weber, 1982). One protein molecule would protect about 20 bp and two molecules aproximately 50 bp, the figure found with relaxed cores (Nermut, 1980b). On the other hand, the presence of 11 S "monomer" structures with about 150 bp would be difficult to account for on the basis of this model.

2. "Nucleosome" (Discontinuous Helix) Model

There are now two strong arguments in favor of this model: (1) the finding of a protection pattern after treatment with micrococcal nuclease as described first by Corden *et al.* (1976) and recently by Mirza and Weber (1982) and (2) the finding of "beads-on-a-string" in preparations of cores after various treatments (Nermut, 1979; Mirza and Weber, 1982; Vayda *et al.*, 1983). For the purpose of the model, we shall assume that the protected DNA fragment is about 150 bp long and that six copies of VP VII are contained in one bead (Corden *et al.*, 1976; Sato and Hosokawa, 1981; Mirza and Weber, 1982; Vayda *et al.*, 1983).

In this case, the nucleosomelike bead (Fig. 10). would be about 50 Å high and 80–90 Å in diameter. One turn of DNA would be approximately 280 Å, i.e., 82 bp, long and the pitch of the DNA helix about 25 Å. In a linear organization as shown in Fig. 10, two full turns of DNA (= 164 bp) would bind to one bead, and up to 100 Å (= 30 bp) linker DNA could connect two beads. In a superhelical (solenoidal) organization (Fig. 11), only about 1.8 turns of DNA (= 148 bp and 504 Å) would interact with the protein core, and the linker would be about 20 bp long. Several coils

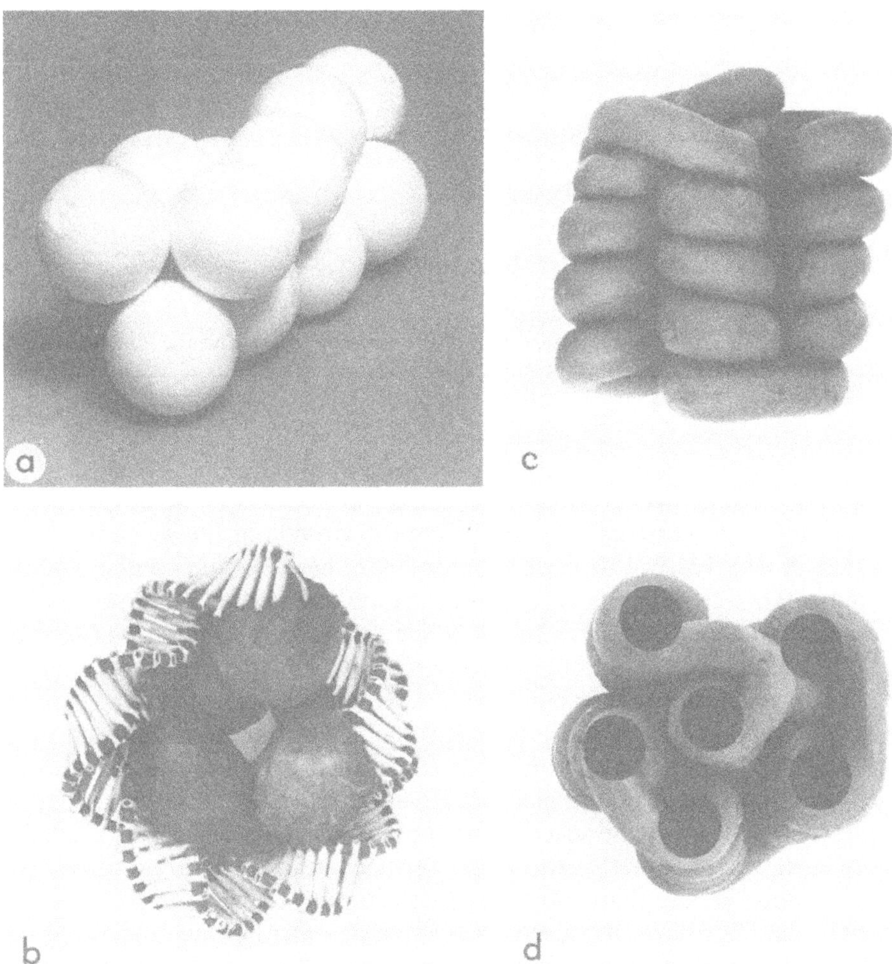

FIGURE 9. Model of the "continuous helix" organization of the DNA–protein complex. (a) Helical core from VP VII molecules. (b) End-on view of the protein core with DNA at the periphery. Three subunits interact with one turn of DNA. The outer diameter is 80–90 Å. (c) Superhelical organization of the DNA–protein filament into rod like elements. One "rod" is about 150 Å thick and 400 Å long. (d) Same as in (c) viewed end-on. Circles show density distribution as, for example, in sections (see Fig. 6d). Reprinted from Nermut (1980a) with permission.

of such a superhelix would in fact form a linear structure about 150 Å thick, perhaps an equivalent of the 5 + 1 rodlike elements mentioned in Section IV.C.

However, if we consider a pitch of 25 Å and five beads in one "superhelical turn" of the solenoidal structure (Fig. 11), there would be only two turns and ten beads per 500-Å-long "rod" (60 beads per core), and there would also be a lot of free space between the first and the fifth bead

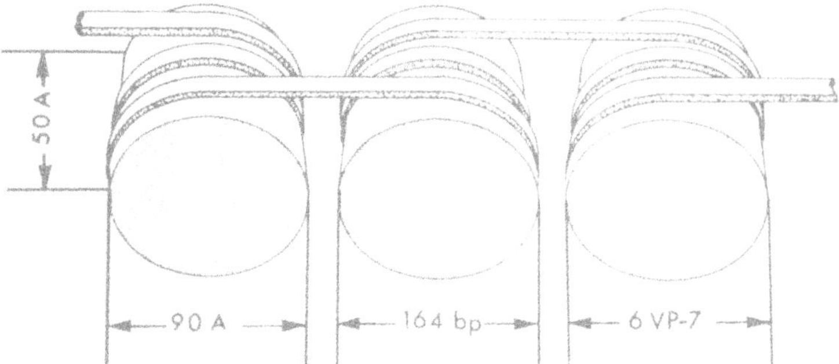

FIGURE 10. Diagrammatic representation of the pathway of DNA in the "nucleosome" model. In this linear arrangement, DNA makes two turns around the bead; the linker DNA would be 26–30 bp long.

(see Fig. 11a). However, this problem can be resolved if the beaded filament forms an antiparallel helix filling the grooves of the ascending superhelix (Fig. 12). In this case, the number of beads would double, and if connecters were taken into account, the number of beads per virion could amount to about 150. This figure is decisive for the calculations of the DNA length and the number of protein copies per virion. Several groups of authors determined the number of beads arithmetically from the length of DNA and the length of the "nucleosome" DNA fragment. For example, Mirza and Weber (1982) found that there must be 200 beads per virion to accommodate 36,000 bp of DNA. However, only 187 beads would completely fill the available space of the core interior provided the 90 Å beads are closely packed with very little free space. Obviously, the number of beads must be less than 200; one could estimate 150–180, depending on their size. The size of a bead used in these calculations, i.e., 90 Å in diameter and 50 Å high, is very close to that of the nucleosome of animal chromatin as determined by EM and X-ray crystallography (Richmond *et al.*, 1981). However, an octamer of four histone molecules forms the protein core, and the H_1 histone molecules plays the role of an "organizer" located between two nucleosomes. An equivalent of this histone has not yet been specified in adenoviruses. However, the recent work by Vayda *et al.* (1983) indicates that the μ protein could play this role. Calculations based on 150 and 180 beads per virion are presented in Table II. The compaction ratio of DNA for this model is about 6, also similar to the cellular chromatin.

The "nucleosome" model is supported by the EM evidence of beaded structures and the micrococcal-nuclease protection pattern. However, there are still many open questions. For example, why is it so difficult to relax the cores into the beaded filaments? Why is the micrococcal-nuclease digestion pattern atypical, i.e., without multiples of the "monomeric" DNA length? Is this method reliable? Recently, Pauli *et al.* (1982) reported that micrococcal nuclease produces a similar repeating pattern

TABLE II. DNA and Protein Contents per Adenovirion Calculated for 150 and 180 Nucleosomelike Beads

Beads per virion	Base pairs per bead + spacer	Total length of DNA		VP VII copies per virion	Length of beaded filament[a]	
		bp	μm		CTC = 90 Å	CTC = 100 Å
150	170	25,500	8.7	900	1.35 μ	1.5 μm
	180	27,000	9.2	900		
180	170	30,600	10.5	1,080	1.6 μm	1.8 μm
	180	32,400	12	1,080		

[a] (CTC) Center-to-center spacing of adjacent beads.

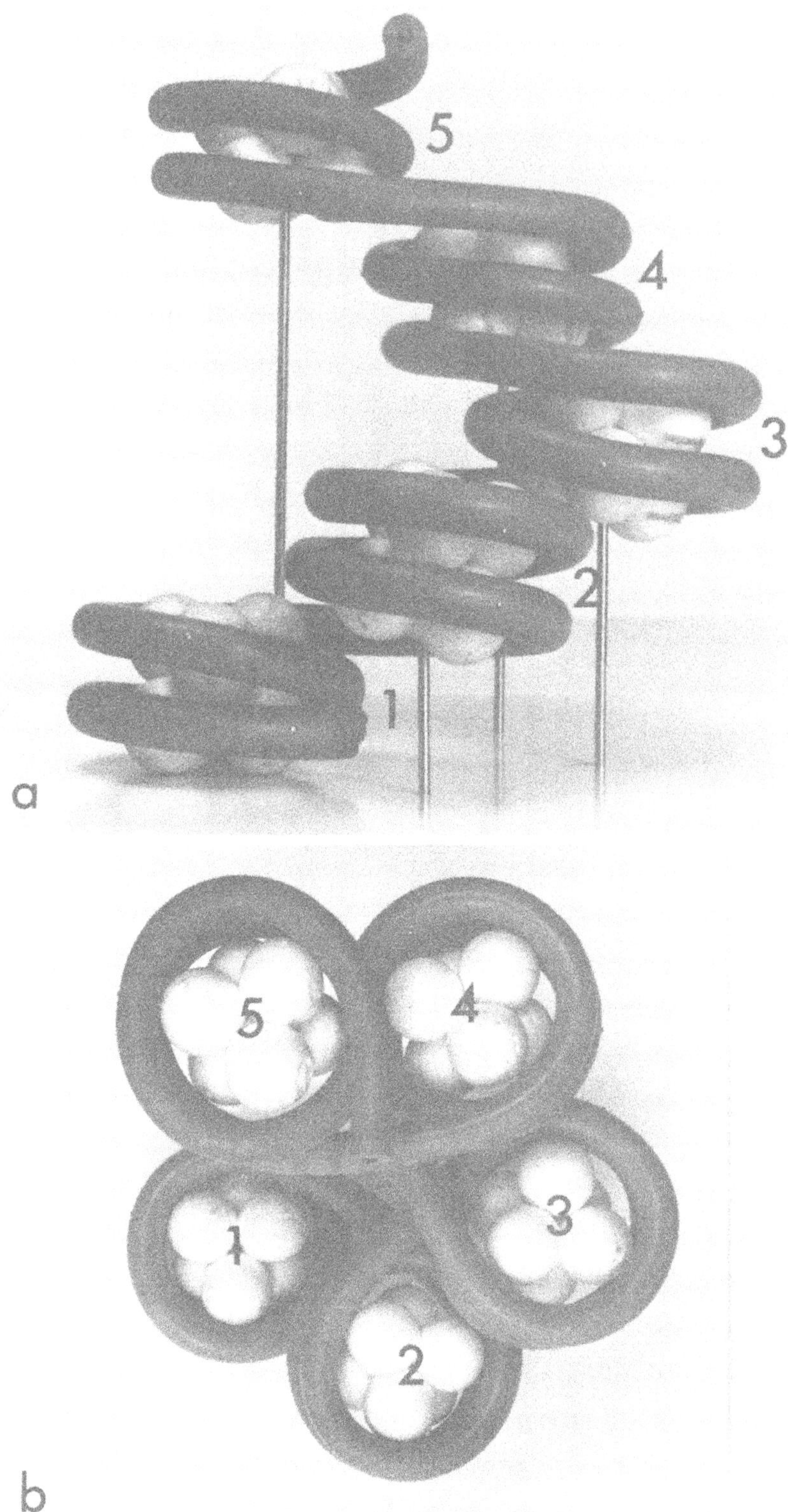
5
4
3
2
1
a
5
4
1
3
2
b

FIGURE 12. Model of one "rodlike" element formed by two antiparallel helices of nucleosomes. White balls represent an ascending helix (as in Fig. 11), black balls a descending one. The arrow indicates the place where they meet.

from naked DNA, probably because it can recognize some of the DNA sequences. In other words, the structure of the adenovirus nucleocapsid is far from being finally solved, and more biochemical and structural studies will be required. However, whichever of the two proposed models is correct, it remains without doubt that the adenovirus nucleocapsid—the DNA–protein(s) complex—is a helical structure. Recently, Devaux *et al.* (1983) reported a 29-Å spacing in X-ray diffractograms of adenovirus cores, which supports the aforestated conclusion.

V. MODEL OF THE ADENOVIRION

Figure 13 shows a physical model of the adenovirion viewed along the twofold symmetry axis (Fig. 13a) and with an insight into the interior (Fig. 13b). The "groups of nine" hexons of the left and right triangular facets meet on the edge so that the two hexons face each other via corner-to-corner, not corner-to-edge as within the GONs (also see Fig. 5).

In the open view, the internal nucleoprotein complex is presented as rodlike elements formed by a superhelically organized DNA–protein filament as shown in Fig. 9, without going into details of its substructure. Included in this model are only the well-defined structural complexes: the capsid, the core shell, and the nucleoprotein complex.

←

FIGURE 11. Superhelical organization of "nucleosomes." One helical turn is shown. In this case, only about 1.8 turns of DNA interact with six copies of VP VII. About 20 bp form the spacer. This is clearly shown in (b). The path of DNA and the height of one turn are better seen in (a). (1–5) Corresponding "nucleosomes" in (a) and (b).

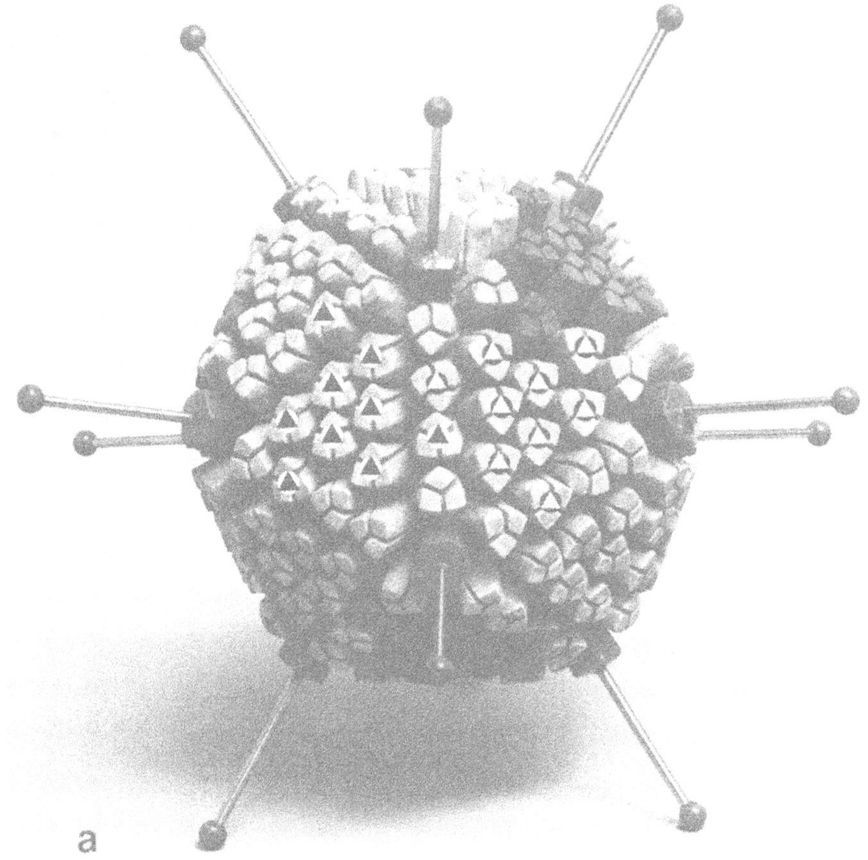

FIGURE 13. Model of adenovirion made in scale. (a) View along the twofold axis of symmetry showing how two "groups of nine" hexons meet at the edge. In the capsid, GONs are in a left-handed orientation. (b) Open view showing the "core shell" underneath the capsid and three of the superhelically organized rods of the nucleocapsid proper. (C) Capsid; (CS) core shell; (NP) DNA–protein complex.

Even so, this model shows the progress made over the past 20 years in our knowledge of the architecture of adenoviruses. Though the icosahedral shape and the existence of two different sorts of capsomers were recognized very early (Horne, 1962; Valentine and Pereira, 1965), the fine-structural details of the hexon and of the virus cores have been discovered only during the past five years thanks to a fruitful application of computer-assisted EM, X-ray crystallography, neutron diffraction, cryotechniques, and biochemistry.

If this interdisciplinary research continues, we shall soon know more about the molecular structure of the penton including the fiber and the function of the enigmatic VP VI. However, the most pressing problem is the organization of the DNA–protein complex. This knowledge should

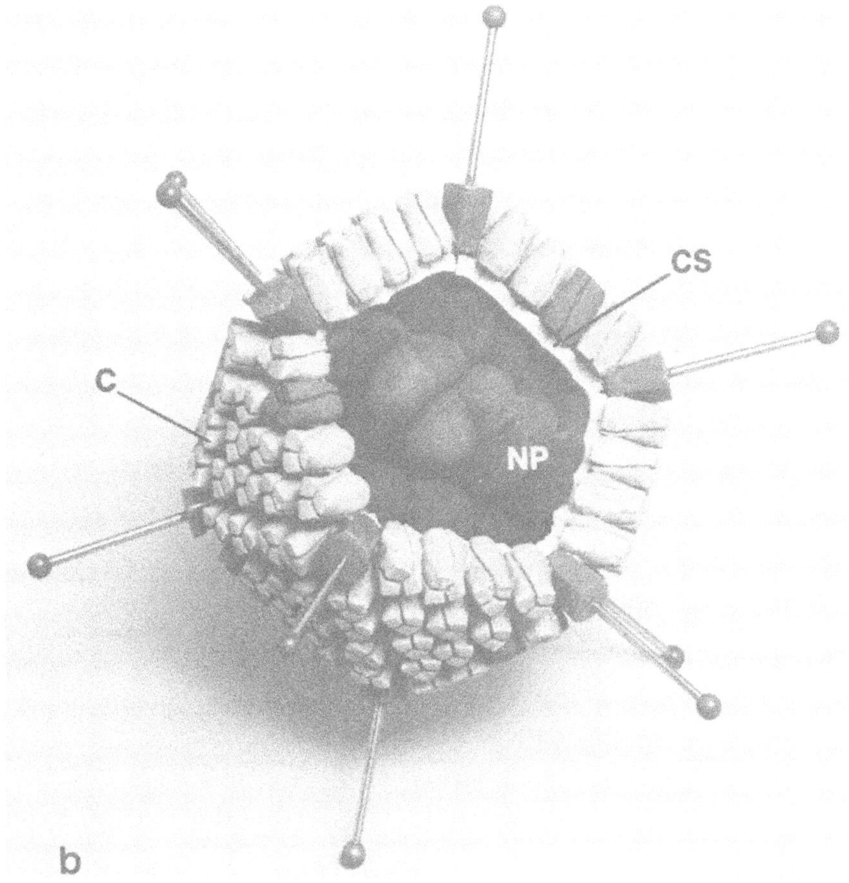

FIGURE 13 (*Continued*)

help us to understand why the core protein (VII) is possibly replaced by cellular histones before transcription and how the DNA–protein complex is assembled and packed into the icosahedral shell. Obviously, it is difficult to study assembly of a virus particle the architecture of which is not properly understood, and this is the major rationale behind the ultrastructural research.

ACKNOWLEDGMENTS The author wishes to thank Dr. W. C. Russell for valuable discussions and comments on the manuscript, Dr. Christiane Devaux and co-workers, Drs. S. J. Flint and M. E. Vayda, and Dr. N. M. Green for making their results available before publication. Thanks are also due to Miss L. D. Williams for technical assistance throughout my work on adenovirus structure and to Mrs. H. Burns for typing and editorial help.

REFERENCES

Akusjärvi, G., and Persson, H., 1981, Gene and mRNA for precursor polypeptide VI from adenovirus type 2, *J. Virol.* **38**:469.

Anderson, C.W., Baum, P.R., and Gesteland, R.F., 1973, Processing of adenovirus 2-induced proteins, *J. Virol.* **12**:241.

Black, B.C., and Center, M.S., 1979, DNA-binding properties of the major core protein of adenovirus 2, *Nucleic Acids Res.* **6**:2339.

Boudin, M.L., and Boulanger, P., 1981, Antibody-triggered dissociation of adenovirus penton capsomer, *Virology* **113**:781.

Boudin, M.L., Moncany, M., D'Halluin, J.-C., and Boulanger, P. A., 1979, Isolation and characterization of adenovirus type 2 vertex capsomer (penton base), *Virology* **92**:125.

Boudin, M.L., D'Halluin, J.C., Cousin, C., and Boulanger, P., 1980, Human adenovirus type 2 protein IIIa. II. Maturation and encapsidation, *Virology* **101**:144.

Boulanger, P., and Loucheux-Lefevbre, M.-H., 1982, Structure of adenovirus nucleoprotein core studied by circular dichroism and selective radiochemical labeling, *Biochem. Biophys. Res. Commun.* **107**:470.

Boulanger, P., Lemay, P., Blair, G. E., and Russell, W. C., 1979, Characterization of adenovirus protein IX, *J. Gen. Virol.* **44**:783.

Brown, D.T., Westphal, M., Burlingham, B. T., Winterhof, U., and Doefler, W., 1975, Structure and composition of adenovirus type 2 core, *J. Virol.* **16**:366.

Brown, M., and Weber, J., 1980, Virion core-like organization of intranuclear adenovirus chromatin late in infection, *Virology* **107**:306.

Burnett, R.M., Grütter, M.G., Markovic, Z., and White, J.L., 1979, The molecular envelope of adenovirus type 2 hexon and its interactions in the viral capsid, *J. Supramol. Struct. Suppl.* **3**:92.

Colby, W.W., and Shenk, T., 1981, Adenovirus type 5 virions can be assembled *in vivo* in the absence of detectable polypeptide IX, *J. Virol.* **39**:977.

Corden, J.L., and Pearson, G.D., 1979, Assembly of adenovirus chromatin, (Abstr.) ICRF Tumour Virus Meeting on SV-40, Polyoma and Adenoviruses, Cambridge, p. 70.

Corden, J., Engelking, M., and Pearson, G.D., 1976, Chromatin-like organization of the adenovirus chromosome, *Proc. Natl. Acad. Sci. U.S.A.* **73**:401.

Cowman, M.K., and Fasman, G.D., 1978, Circular dichroism analysis of mononucleosome DNA conformation, *Proc. Natl. Acad. Sci. U.S.A.* **75**:4759.

Crane, H.R., 1950, Principles and problems of biological growth, *Sci. Mon.* **70**:376.

Crowther, R.A., and Franklin, R.M., 1972, The structure of the groups of nine hexons from adenovirus, *J. Mol. Biol.* **68**:181.

Devaux, C., Zulauf, M., Boulanger, P., and Jacrot, B., 1982, Molecular weight of adenovirus serotype 2 capsomers: A new characterization, *J. Mol. Biol.* **156**:927.

Devaux, C., Timmins, P.A., and Berthet-Colominas, C., 1983, Structural studies of adenovirus type 2 by neutron and X-ray scattering, *J. Mol. Biol.* **167**:119.

Dorsett, P. H., and Ginsberg, H.S., 1975, Characterization of type 5 adenovirus fiber protein, *J. Virol.* **15**:208.

Earnshaw, W.C., Goldberg, E.B., and Crowther, R.A., 1979, The distal half of the tail fibre of bacteriophage T_4, *J. Mol. Biol.* **132**:101.

Everitt, E., Sundquist, B., Pettersson, N.U., and Philipson, L., 1973, Structural proteins of adenoviruses. X. Isolation and topography of low molecular weight antigens from the virion of adenovirus type 2, *Virology* **52**:130.

Everitt, E., Lutter, L., and Philipson, L., 1975, Structural proteins of adenovirus. XII. Location and neighbor relationship among proteins of adenovirion type 2 as revealed by enzymatic iodination, immunoprecipitation and chemical cross-linking, *Virology* **67**:197.

Finch, J.T., Lutter, L.C., Rhodes, D., Brown, R.S., Rushton, B., Levitt, M., and Klug, A., 1977, Structure of nucleosome core particles of chromatin, *Nature (London)* **269**:29.

Flint, S.J., 1980, Structure and genomic organization of adenoviruses, in: *The Molecular Biology of Tumor Viruses*, Part 2, *DNA Tumor Viruses*, 2nd rev. ed. (J. Tooze, ed.), pp. 383–442, Cold Spring Harbor Press, Cold Spring Harbor, New York.

Gelderblom, H., and Maichle-Läuppe, I., 1982, The fibers of fowl adenoviruses, *Arch. Virol.* **72**:289.

Ginsberg, H.S., 1979, Adenovirus structural proteins, in: *Comprehensive Virology*, Vol. 13, (H. Fraenkel-Conrat and R.R. Wagner, eds.), pp. 409–457, Plenum Press, New York.

Ginsberg, H.S., Pereira, H.G., Valentine, R.C., and Wilcox, W.C., 1966, A proposed terminology for the adenovirus antigens and virion morphological subunits, *Virology* **28**:782.

Green, M., Piña, M., Kemes, R.C., Wensick, P.C., MacHattie, L.A., and Thomas, C.A., Jr., 1967a, Adenovirus DNA. I. Molecular weight and conformation, *Proc. Natl. Acad. Sci. U.S.A.* **57**:1302.

Green, M., Piña, M., and Kimes, R.C., 1967b, Biochemical studies on adenovirus multiplication. XII. Plaquing efficiencies of purified human adenoviruses, *Virology* **31**:562.

Green, N.M., Wrigley, N.G., Russell, W.C., Martin, S.R., and McLachlan, A.D., 1983, Evidence for a repeating cross-beta sheet structure in the adenovirus fibre. *Eur. Mol. Biol. Org. J.* **2**:1357.

Grütter, M., and Franklin, R.M., 1974, Studies on the molecular weight of the adenovirus type 2 hexon and its subunit, *J. Mol. Biol.* **89**:163.

Harpst, J.A., Ennever, J.F., and Russell, W.C., 1977, Physical properties of nucleoprotein cores from adenovirus type 5, *Nucleic Acids Res.* **4**:477.

Hays, J.B., Magar, M.E., and Zimm, B.H., 1969, Persistence length of DNA, *Biopolymers* **8**:531.

Hérissé, J., Rigolet, M., Dupont de Dinechin, S., and Galibert, F., 1981, Nucleotide sequence of adenovirus 2 DNA fragment encoding for the carboxylic region of the fibre protein and the entire E_4 region, *Nucleic Acids Res.* **9**:4023.

Horne, R.W., 1962, The comparative structure of adenoviruses, *Ann. N.Y. Acad. Sci.* **101**:475.

Hosokawa, K., and Sung, M.T., 1976, Isolation and characterization of an extremely basic protein from adenovirus type 5, *J. Virol.* **17**:924.

Keegstra, W., Van Wielink, P.S., and Sussenbach, J.S., 1977, The visualization of a circular DNA–protein complex from adenovirions, *Virology* **76**:444.

Laver, W.G., 1970, Isolation of an arginine-rich protein from particles of adenovirus type 2, *Virology*, **41**:488.

Laver, W.G., Pereira, H.G., Russell, W.C., and Valentine, R.C., 1968, Isolation of an internal component from adenovirus type 5, *J. Mol. Biol.* **37**:379.

Laver, W.G., Wrigley, N.G., and Pereira, H.G., 1969, Removal of pentons from particles of adenovirus type 2, *Virology* **39**:599.

Laver, W.G., Younghusband, H.B., and Wrigley, N.G., 1971, Purification and properties of chick embryo lethal orphan virus (an avian adenovirus), *Virology* **45**:598.

Maizel, J.V., Jr., White, D.O., and Scharff, M.D., 1968, The polypeptides of adenovirus. II. Soluble proteins, cores, top components and the structure of the virion, *Virology* **36**:126.

Mattern, C.F.T., 1969, Virus architecture as determined by X-ray diffraction and electron microscopy, in: *Biochemistry of Viruses* (H.B. Levy, ed.), pp. 55–100, Marcel Dekker, New York and London.

Mirza, M.A., and Weber, J., 1977, Genetic analysis of adenovirus type 2. VII. Cleavage-modified affinity for DNA of internal virion proteins, *Virology* **80**:83.

Mirza, M.A., and Weber, J., 1982, Structure of adenovirus chromatin, *Biochim. Biophys. Acta* **696**:76.

Nermut, M.V., 1975, Fine structure of adenovirus type 5, *Virology* **65**:480.

Nermut, M.V., 1978, Structural elements in adenovirus cores: Studies by means of freeze–fracturing and ultrathin sectioning, *Arch. Virol.* **57**:323.

Nermut, M.V., 1979, Structural elements in adenovirus cores: Evidence for a "core shell" and linear structures in "relaxed" cores, *Arch. Virol.* **62**:101.

Nermut, M.V., 1980a, The architecture of adenoviruses: Recent views and problems, *Arch. Virol.* **64**:175.

Nermut, M.V., 1980b, Effects of micrococcal nuclease on "relaxed" cores of adenovirus type 5 on molecular biology of adenoviruses, (Abstr.) EMBO-Workshop, Peebles, p. 7.

Nermut, M.V., and Perkins, W.J., 1979, Consideration of the three dimensional structure of the adenovirus hexon from electron microscopy and computer modelling, *Micron* **10**:247.

Nermut, M.V., Harpst, J.A., and Russell, W.C., 1975, Electron microscopy of adenovirus cores, *J. Gen. Virol.* **28**:49.

Norrby, E., 1969, The structural and functional diversity of adenovirus capsid components, *J. Gen. Virol.* **5**:221.

Pauli, U.H., Seebeck, T., and Braun, R., 1982, Sequence specific cleavage of chromatin by staphylococcal nuclease can generate an atypical nucleosome pattern, *Nucleic Acids Res.* **10**:4121.

Pereira, H.G., and Wrigley, N.G., 1974, *In vitro* reconstitution, hexon bonding and handedness of incomplete adenovirus capsid, *J. Mol. Biol.* **85**:617.

Pettersson, U., and Höglund, S., 1969, Structural proteins of adenoviruses. III. Purification and characterization of the adenovirus type 2 penton antigen, *Virology* **39**:90.

Philipson, L., 1979, Adenovirus proteins and their messenger RNAs, *Adv. Virus Res.* **25**:357.

Philipson, L., Pettersson, U., and Lindberg, U., 1975, Molecular biology of adenoviruses, in: *Virology Monographs*, Vol. 14 (S. Gard and C. Hallauer, eds.),pp. 1–115, Springer-Verlag, Vienna and New York.

Rekosh, D.M.K., Russell, W.C., Bellét, A.J.D., and Robinson, A.J., 1977, Identification of a protein linked to the ends of adenovirus DNA, *Cell* **11**:283.

Richmond, T.J., Klug, A., Finch, J.T., and Lutter, L.C., 1981, The organization of DNA in the nucleosome core particle, in: *Biomolecular Stereodynamics*, Vol. II (R.H. Sarma, ed.), pp. 109–159, Adenine Press, New York.

Robinson, A.J., Younghusband, H.B., and Bellett, A.J.D., 1973, A circular DNA–protein complex from adenoviruses, *Virology* **56**:54.

Russell, W.C., and Precious, B., 1982, Nucleic acid-binding properties of adenovirus structural proteins, *J. Gen. Virol.* **63**:69.

Russell, W.C., McIntosh, K., and Skehel, J.J., 1971, The preparation and properties of adenovirus cores, *J. Gen. Virol.* **11**:35.

Sato, K., and Hosokawa, K., 1981, The structure of adenovirus chromatin revealed by ultraviolet light-induced cross-linking, *Biochem. Biophys. Res. Commun.* **101**:1318.

Seargeant, A., Tigges, M.A., and Raskas, H.J., 1979, Nucleosome-like structural subunits of intranuclear parental adenovirus type 2 DNA, *J. Virol.* **29**:888.

Shortridge, K.F., and Biddle, F., 1970, The proteins of adenovirus type 5, *Arch. Gesamte Virusforsch.* **29**:1.

Sussman, J.L., and Trifonov, E.N., 1978, Possibility of nonkinked packing of DNA in chromatin, *Proc. Natl. Acad. Sci. U.S.A.* **75**:103.

Tate, V.E., 1976, Structural and functional studies on DNA and nucleoprotein components of adenovirus type 5, Ph.D. thesis, Council for National Academic Rewards London.

Tate, V.E., and Philipson, L., 1979, Parental adenovirus DNA accumulates in nucleosome-like structures in infected cells, *Nucleic Acids Res.* **6**:2769.

Valentine, R.C., and Pereira, H.G., 1965, Antigens and structure of the adenovirus, *J. Mol. Biol.* **13**:13.

Vayda, M.E., Leong, K., and Flint, S.J., 1983, The structure of nucleoprotein cores released from adenoviruses, *Nucleic Acids Res.* **11**:441.

von Bahr-Lindström, Jörnvall, H., Althin, S., and Philipson, L., 1982, Structural differences between hexons from adenovirus types 2 and 5: Correlation with differences in size and immunological properties, *Virology* **118**:353.

Wildy, P., 1971, Classification and nomenclature of viruses, in: *Monographs in Virology*, Vol. 5 (J.L. Melnick, ed.), pp. 1–81, S. Karger, Basel.

CHAPTER 3

The Structure of the Genome

JOHN S. SUSSENBACH

I. INTRODUCTION

Adenovirus particles have a highly ordered structure and are composed of protein and DNA. Human adenoviruses contain about 87% protein and 13% DNA (Green and Piña, 1963), while the larger avian chick embryo lethal orphan (CELO) virus consists of 83% protein and 17% DNA (Laver *et al.*, 1971). In virions, the viral DNA is tightly associated with several virus-coded proteins. Disruption of virions with acetone, urea, or pyridine, or repeated freezing and thawing, releases the viral cores, which, in addition to the viral DNA, still contain about 18–20% of the total protein of the virions (Laver *et al.*, 1967, 1968; Maizel *et al.*, 1968; Prage *et al.*, 1968, 1970). The proteins found in viral cores are mainly two basic polypeptides. The major core protein is identical to polypeptide VII [molecular weight 18,000 (18K)], of which about 1000 copies are present in each viral particle. The minor core protein is polypeptide V (molecular weight 45.5K), of which each virion contains about 200 copies (Laver *et al.*, 1968; Prage *et al.*, 1968, 1970; Prage and Pettersson, 1971; Russell *et al.*, 1971; Everitt *et al.*, 1973; Laver, 1970). However, when cores are prepared by extraction of virions with sarkosyl, only polypeptide VII is found associated with the DNA (Brown *et al.*, 1975). The different protein compositions of pyridine and sarkosyl cores suggest that polypeptide VII is more intimately associated with the viral genome than is polypeptide V.

Corden *et al.* (1976) concluded that adenovirus DNA packed in virions has a chromatinlike structure. They found that digestion of disrupted virions with micrococcal nuclease cleaves the viral genome into fragments about 200 nucleotides long. However, these experiments could

JOHN S. SUSSENBACH • Laboratory for Physiological Chemistry, State University of Utrecht, 3521 GG Utrecht, The Netherlands.

not be repeated by Tate and Philipson (1979). Mirza and Weber (1982) proposed that although adenovirus DNA is indeed packed into subunits, its organization in the virion is not completely the same as that of eukaryotic chromatin. Partial deoxyribonuclease (DNase) digestion of eukaryotic chromatin leads to stretches of DNA with a length of 200 nucleotide pairs associated with histones. Mirza and Weber (1982) found that viral chromatin does indeed have a nucleosomelike structure, but that partial DNase digestion yields monomers of about 150 nucleotide pairs of DNA wrapped around three dimers of polypeptide VII. These monomers are linked by a variable length of DNA associated with one copy of polypeptide V.

Since adenovirus DNA is tightly associated with virion proteins, protein-free DNA can be obtained only by extensive digestion of virions or viral cores with proteolytic enzymes (papain, pronase, or proteinase K) followed by sodium dodecyl sulfate (SDS)–phenol extraction (van der Eb and van Kesteren, 1966; Green *et al.*, 1967; van der Eb *et al.*, 1969; Laver *et al.*, 1971). The DNA thus isolated has a linear structure and has been characterized in great detail.

An alternative isolation procedure for adenovirus DNA was first applied by Bellett and co-workers for CELO and adenovirus type 2 (Ad2) DNA (Robinson *et al.*, 1973; Robinson and Bellett, 1975a). These investigators isolated DNA in the absence of proteolytic enzymes, employing an extraction with 4 M guanidinium hydrochloride. The isolated DNA has in the electron microscope (EM) a circular structure, which can be converted into a linear configuration by digestion of the preparation with proteolytic enzymes (Robinson *et al.*, 1973). Similar studies have also been performed for Ad5 DNA (Keegstra *et al.*, 1977). The sensitivity of the circular structures for proteolytic enzymes suggests that the circular structures are maintained by a protein linker.

By *in vitro* labeling of the protein moiety with ^{125}I, it could be demonstrated that a polypeptide with a molecular weight of 55K is covalently attached to the 5′ end of each DNA strand (Rekosh *et al.*, 1977). This protein, designated terminal protein, has a hydrophobic character, which facilitates joining of the ends of the DNA–protein complexes, resulting in the formation of circular structures and concatemers. The properties of the linear deproteinized DNA as well as the characteristics of the circular DNA–protein complexes are discussed in more detail in the following sections.

II. GROUPING OF ADENOVIRUSES BASED ON DNA HOMOLOGY

The different human adenoviruses have been classified into subgroups on the basis of different criteria. Rosén (1960) originally proposed three subgroups based on differences in hemagglutinating capacity.

Hierholzer (1973) extended this classification system to ten subgroups. On the basis of the apparent molecular weights of virion polypeptides V, VI, and VII, Wadell (1978) arranged 20 human serotypes into five groups. A completely different type of classification is based on the oncogenicity of the human adenoviruses. The different serotypes have been subdivided into a highly oncogenic subgroup A (Ad12, Ad18, Ad31), a weakly oncogenic subgroup B (e.g., Ad3 and Ad7), and a nononcogenic subgroup C (e.g., Ad2 and Ad5) (Trentin *et al.*, 1962; Girardi *et al.*, 1964; Huebner *et al.*, 1962, 1965; Larson *et al.*, 1965; Pereira *et al.*, 1965; Green, 1970). It is interesting to note that there is a correlation between the guanine–cytosine (GC) content of the human adenovirus DNAs and the oncogenicity of the viruses. The GC content of the DNAs decreases with increasing oncogenicity (Piña and Green, 1965) (Table I). Probably this correlation has no physiological basis, since, in contrast to the human adenoviruses, the oncogenic simian adenoviruses tend to have slightly higher GC contents than the nononcogenic adenoviruses (Goodhearst, 1971). Further, the oncogenic simian serotypes have GC contents that are in general higher than those of the nononcogenic human serotypes.

The most meaningful and fundamental way to group adenoviruses is based on DNA sequence homology. Fortunately, the DNA homology grouping is in agreement with other groupings of human adenoviruses on the basis of oncogenicity, GC content, and molecular characteristics of viral proteins (Table I). Originally, Green *et al.* (1970) determined the homology among different DNAs employing filter hybridization. Recently, the classification was improved by employment of liquid-phase molecular hybridization with *in vitro*-labeled viral DNA. A total of 31 different human adenovirus serotypes were divided into five different subgroups, A–E (Green *et al.*, 1979b). In general, members of the same subgroup have genomes that are homologous for more than 90%. However, members of subgroup A share only 48–69% of their DNA sequences. The homology among members of different subgroups is less than 20% (Table I).

The major regions of least homology among DNAs of different human serotypes have been visualized by heteroduplex mapping (Garon *et al.*, 1973). Heteroduplexes of subgroups B and C DNAs contain two major regions of heterology located at positions 50–65 and 78–91 on the adenovirus genome map. Heteroduplexes of members of subgroup A show a more complex distribution of homologous and heterologous regions. However, in this case, too, heterology is found at the two positions mentioned above.

Using the single-strand specific endonuclease from *Neurospora crassa*, Bartok *et al.* (1974) were able to digest specifically the heterologous regions from heteroduplexes of Ad2 and Ad5 DNA and obtained three specific fragments, in agreement with the heteroduplex mapping. The heterologous regions contain the genetic information of the major coat proteins hexon and fiber, which play an important role in the se-

TABLE I. Properties of Human Adenovirus DNA Homology Groups A–E[a]

Groups	Types	DNA homology[b]	DNA MWs ($\times 10^{-6}$)[c]	DNA GC (%)[d]	ITR length[e]	Tumor induction[f]	Cell transformation[g]	HA group[h]
A	12, 18, 31	48–69% within group, 8–20% with other types	19.2–22.0	47–49	162/164 for Ad12, 165 for Ad18	High	+	3B
B	3, 7, 11, 14, 16, 21	89–94% within group, 9–20% with other types	22.7–23.0	49–52	136 for Ad3 and Ad7	Weak	+	1A, B
C	1, 2, 5, 6	99–100% within group, 10–16% with other types	23.0	57–59	102/103 for Ad2, 103 for Ad5	Nil	+	3A
D	8–10, 13, 15, 17, 19, 20, 22–30, 32, 33, 36, 37	94–99% within group, 4–17% with other types	?	57–59	?	Nil	?	2A–F
E	4	4–23% with other types	22.8	High	116	Nil	?	3A

[a] (MW) Molecular weight; (ITR) inverted terminal repetition; (HA) hemagglutination.
[b] Data from Green *et al.* (1979b).
[c] Data from Green and Piña (1964), van der Eb and van Kesteren (1966), and Tibbetts (1977).
[d] Data from Piña and Green (1965).
[e] Data from Steenbergh *et al.* (1977), Arrand and Roberts (1979), Shinagawa and Padmanabhan (1979), (1980), Tokunaga *et al.* (1982), Sugisaka *et al.* (1980), Garon *et al.* (1982), and Schwarz *et al.* (1982).
[f] Data from Girardi *et al.* (1964), Huebner *et al.* (1962, 1965), Larson *et al.* (1965), Pereira *et al.* (1965), and Trentine *et al.* (1962).
[g] Data from McBride and Weiner (1964), Sekikawa *et al.* (1978), Freeman *et al.* (1967), Gallimore (1974), van der Eb *et al.* (1977), and McAllister *et al.* (1969).
[h] Data from Hierholzer (1973).

rological classification of the different adenovirus serotypes. In addition, one of the heterologous regions codes for a group of nonvirion early proteins (see Section VII).

III. PHYSICOCHEMICAL PROPERTIES OF ADENOVIRUS DNA

DNA, extracted from adenovirus particles employing digestion with proteolytic enzymes, has a linear double-stranded structure (van der Eb and van Kesteren, 1966; Green *et al.*, 1967; van der Eb *et al.*, 1969; Younghusband and Bellett, 1971). The size of the viral genome varies from serotype to serotype. The molecular weights of the human adenovirus DNAs range from 19–22 $\times$ 10^6 for the highly oncogenic serotypes Ad12, Ad18, and Ad31 to 23–24 $\times$ 10^6 for the nononcogenic serotypes Ad1, Ad2, and Ad5 (Green *et al.*, 1967) (Table I). On the basis of nucleotide sequence data and the sum of restriction fragments, it has been inferred that the genome of Ad2 and Ad5 is about 36,000 nucleotide pairs and that Ad12 DNA is 34,300 nucleotide pairs long. The sizes of the genomes of nonhuman serotypes are comparable to those of their human counterparts [that of mouse serotype FL DNA being 20.7 $\times$ 10^6 (Temple *et al.*, 1981) and of simian adenovirus SA7 DNA being 22 $\times$ 10^6 (Burnett and Harrington, 1968)]. On the other hand, the genome of the avian chick embryo lethal orphan (CELO) virus is much larger, measuring 30 $\times$ 10^6 (Younghusband and Bellett, 1971; Laver *et al.*, 1971).

When native adenovirus DNA is digested with *Escherichia coli* exonuclease III and is subsequently examined under the EM, no circularization of the linear genome is observed, indicating that adenovirus DNA is not terminally redundant as T7 DNA (Green *et al.*, 1967; Younghusband and Bellett, 1971). On the other hand, when double-stranded DNA (dsDNA) is denatured and reannealed at low DNA concentrations, both strands of human as well as of avian adenovirus DNA are able to form single-stranded circles (Garon *et al.*, 1972; Wolfson and Dressler, 1972; Robinson and Bellett, 1975b). The formation of single-stranded circles indicates that adenovirus DNA contains an inverted terminal repetition. This inverted terminal repetition is discussed in more detail in Section V.

The distribution of adenine–thymine (AT) and GC base pairs in adenovirus DNA has been investigated by partial thermal denaturation mapping. The unique thermal denaturation patterns of DNAs from Ad2, Ad5, and Ad12, the avian CELO virus, and the mouse strain FL indicate that adenovirus DNA is not circularly permuted as T7 DNA, but that all DNA molecules from the same serotype have an identical nucleotide sequence (Doerfler and Kleinschmidt, 1970; Younghusband and Bellett, 1971; Doerfler *et al.*, 1972; Ellens *et al.*, 1974; Temple *et al.*, 1981). In most denaturation patterns, the distribution of AT and GC base pairs along the DNA molecule is asymmetrical. By convention, the AT-rich half of an adenovirus DNA molecule has been designated the right-hand half of the molecule (Doerfler and Kleinschmidt, 1970). In some cases (Ad2 and Ad5), the AT- and GC-rich halves of the DNA molecules can be separated by CsCl or $HgCl_2$–Cs_2SO_4 gradient centrifugation of sheared DNA (Kimes and Green, 1970; Doerfler and Kleinschmidt, 1970; Horwitz, 1974; Graham *et al.*, 1974b). However, due to the more even distribution of AT and GC base pairs in Ad12 DNA, separation of the left and right halves of Ad12 DNA by this procedure is not possible (Doerfler *et al.*, 1972).

Separation of the complementary strands of adenovirus DNA can be performed by complexing of the single strands of denatured native DNA with poly(I:G) or poly(U:G). Intact complementary strands have been obtained for Ad2, Ad5, Ad7, and Ad12 DNA (Kubinski and Rose, 1967; Landgraf-Leurs and Green, 1971; Patch *et al.*, 1972; Tibbetts *et al.*, 1974; Vlak *et al.*, 1975). Since the two complementary strands bind unequal amounts of the copolymers, the two strands can be separated by equilibrium density-gradient centrifugation or by gel electrophoresis (Goldbach *et al.*, 1978). Complementary strands of Ad2 and Ad5 DNA have also been separated by alkaline CsCl equilibrium density-gradient centrifugation (Sussenbach *et al.*, 1973; Sharp *et al.*, 1975). The buoyant densities of the two strands in alkaline CsCl differ by 2–4 mg/ml, which is sufficient for separation. The heavy strands of Ad2 and Ad5 DNA obtained by poly(U:G)–CsCl gradient centrifugation have the lower density in alkaline CsCl (Tibbetts *et al.*, 1974; Vlak *et al.*, 1975).

Tibbetts *et al.*, (1973) showed that Ad2 single-stranded DNA (ssDNA) is retained by hydroxyapatite columns under conditions generally used for selective retention of dsDNA, probably due to partialy complementary regions in the single strands. Other indications for regions of complementarity in adenovirus ssDNA were obtained by EM. Under suitable conditions, an extended region of secondary structure is observed at position 73 on the conventional adenovirus map (Wu *et al.*, 1977). Regions that contain complementary sequences were also detected at the molecular termini (Padmanabhan and Green, 1976; Wu *et al.*, 1977). Digestion of native Ad2 DNA with exonuclease III followed by repair synthesis of the exposed single-stranded ends with DNA polymerase I revealed the presence of self-complementary sequences about 50 nucleotides long, located at a distance of about 180 nucleotides from each molecular end (Padmanabhan and Green, 1976). Nucleotide sequence analysis of the termini confirmed the existence of self-complementary sequences in these regions.

IV. COORDINATE SYSTEM

To come to an unambiguous nomenclature for the two complementary strands of adenovirus DNA, it has been proposed to adopt a nomenclature that is based on the direction of transcription, rather than on physical properties, e.g., densities. By convention, the AT-rich half of the DNA molecule is oriented to the right and the strand transcribed to the right is called the r-strand, while the leftward-transcribed strand is designated the l-strand.* The r-strand appears to be identical to the strand with the higher density in alkaline CsCl and to the strand with lower density in poly(U:G)–CsCl (see the proposal in *J. Virol.* **22**:830, 1977). Further, it is agreed to divide the adenovirus DNA into 100 map units (m.u.) from left to right on the viral genome.

The agreement on a unique orientation of adenovirus DNA molecules formed the basis for an unambiguous mapping of significant landmarks on the adenovirus genome. With the discovery and the purification of restriction endonucleases, powerful tools became available to dissect the adenovirus genome in distinct specific fragments (for a review of available enzymes, see Roberts, 1981). These fragments have been used to unravel the organization of the adenovirus genome in detail. For many adenovirus serotypes, accurate restriction endonuclease cleavage maps of the viral genome are available, and with the increasing knowledge of the nucleotide sequences of several adenovirus DNAs, this number is still growing. A summary of restriction endonuclease cleavage maps is presented in Appendix A.

Many restriction fragments have been inserted into prokaryotic plasmids employing recombinant DNA techniques (Stenlund *et al.*, 1980). These adenovirus DNA-containing plasmids are very useful for obtaining large amounts of specific fragments, especially of poorly growing serotypes. They have frequently been used for nucleotide sequence analysis and site-directed mutagenesis. The two complementary strands of restriction fragments have been separated by annealing denatured fragments in the presence of an excess of one of the intact complementary strands followed by separation of the partial duplex and the remaining single strand. Strand separation has also been obtained by gel electrophoresis of denatured restriction fragments (Tibbetts and Pettersson, 1974; Sharp *et al.*, 1975; Sussenbach *et al.*, 1973; Goldbach *et al.*, 1978). These single strands have frequently been used to isolate specific messenger RNA (mRNA) species.

The most detailed information on the structure of the adenovirus genome and the positions of important landmarks became available by nucleotide sequence analysis of DNAs from different adenovirus serotypes (see Appendix B). The most extended sequences have been established for Ad2 DNA, of which about 70% has been sequenced (Arrand and Roberts, 1979; Zain and Roberts, 1979; Zain *et al.*, 1979a,b; Shinagawa and Padmanabhan, 1979; Galibert *et al.*, 1979; Akusjärvi and Pettersson, 1978a,b, 1979a,b; Hérissé *et al.*, 1980, 1981; Akusjärvi *et al.*,

It should be noted that r-strand transcripts are equivalent to l-strand DNA sequences and that l-strand transcripts are homologous to r-strand sequences.

1980, 1981; Shinagawa *et al.*, 1980; Hérissé and Galibert, 1981; Aleström *et al.*, 1980, 1982; Akusjärvi and Persson, 1981a; Kruijer *et al.*, 1982; Gingeras *et al.*, 1982). This allows the positioning of many landmarks on the Ad2 genome at the nucleotide level. Comparison of the Ad2 nucleotide sequence and the restriction maps revealed that the nucleotide equivalent of 1% of the genome depends on the particular location on the Ad2 genome (Gingeras *et al.*, 1982). It was derived that a value of 365 nucleotides for 1% gives the best fit for the left end, while a value of 357 nucleotides for 1% is the best fit for the right end. The differences in nucleotide equivalent for 1% are probably caused by the differences in nucleotide composition between the right and left halves of the Ad2 genome.

V. INVERTED TERMINAL REPETITION

The existence of an inverted terminal repetition (ITR) in adenovirus DNA was discovered when denatured DNA was reannealed at low concentrations and examined under the EM. A high percentage of the single strands were present in a circular form, indicating that adenoviral DNA contains an ITR (Garon *et al.*, 1972; Wolfson and Dressler, 1972). So far, ITRs have been detected in every serotype investigated, although the length of the repetitions may vary (Table I). The general occurrence of an ITR in adenovirus DNA suggests very strongly that this feature plays an important role in viral propagation.

The single-stranded circular structures have a rather high thermal stability, which is consistent with a highly ordered base-pairing between the terminal sequences (Garon *et al.*, 1972; Wolfson and Dressler, 1972). It also suggests that the ITRs must be of considerable length. Circularization of adenovirus ssDNA can be abolished by digestion with exonuclease III, and this treatment has been used to estimate the size of the terminal repetitions. Garon *et al.* (1972) concluded that the length of the terminal repetition ranged from 350 base pairs (bp) for Ad2 to 1400 bp for Ad31. However, since inverted repeats of these sizes can be visualized under the EM and no double-stranded regions were detected in the single-stranded circles, it was concluded that the exonuclease III experiments obviously lead to an overestimation of the lengths of the ITRs. An exceptionally long ITR was detected in Ad18 DNA (Garon *et al.*, 1975). In single-stranded circles of this serotype, a double-stranded panhandle with a mean length of 0.31 μm was seen, equivalent to 3% of the genome length.

A more accurate estimate of the size of the ITR of Ad2 DNA was obtained by restriction enzyme analysis of end-labeled DNA. When a restriction enzyme cleaves within the repeated sequence, both molecular ends will yield a fragment of the same size, while cleavage outside the repeated sequence will yield fragments of different size. Employing this approach, Roberts *et al.* (1974) estimated that the terminal repetition of Ad2 DNA is between 100 and 140 nucleotides long (also see Arrand *et al.*, 1975).

Recently, nucleotide sequence analysis has been used to determine exactly the size and composition of several adenovirus serotypes (Appendix B). Some general features of the adenovirus ITRs can be demonstrated in the ITR of Ad5 DNA, the first sequenced repetition. The ITR of Ad5 is 103 bp long (Steenbergh *et al.*, 1977). Its sequence is unique and does not contain extended self-complementary regions. A striking property of the Ad5 terminal repetition is the asymmetrical distribution of GC and AT base pairs. The first 50 bp contain 72% AT, while the next 50 bp have only 27% AT. Although the lengths of inverted repeats of other serotypes may differ considerably, they all show the same asymmetrical distribution of base pairs. As for a function of this property, it is not unlikely that the high AT content of the first half of terminal repetitions is of relevance for a rapid unwinding of the molecular ends during initiation of DNA replication.

Comparison of the inverted repetitions of serotypes from the same subgroup shows a high degree of homology (see Appendix B). The repetitions of Ad2 and Ad5 both have a length of 103 bp and are completely identical (Steenbergh *et al.*, 1977; Shinagawa and Padmanabhan, 1979), although the repetition of a particular Ad2 strain has been described that is 102 bp long (Arrand and Roberts, 1979). The terminal repetitions of Ad3 and Ad7 strain Greider both have a length of 136 bp and differ at 7 positions (Tolun *et al.*, 1979; Shinagawa and Padmanabhan, 1980). Comparison of two Ad7 strains (Greider and Gomen) reveals that both repeats are 136 bp long but differ at 5 positions (Dijkema and Dekker, 1979; Shinagawa and Padmanabhan, 1980). Similar strain differences have also been found for Ad12. The length of the Ad12 ITR varies between 162 (Shinagawa and Padmanabhan, 1980) and 164 bp (Sugisaki *et al.*, 1980; Schwarz *et al.*, 1982). In all ITRs determined except one, a dCMP residue has been found at the 5′ ends of adenovirus DNA. The exception is chick enbryo lethal orphan (CELO) DNA, which has at its 5′ end a dGMP residue (Aleström *et al.*, 1982a). In the ITRs of all human adenovirus DNAs, the sequence ATAATATACCTTAT (nucleotides 9–22) is present (Tolun *et al.*, 1979); the regions of the inverted repetitions beyond nucleotide 50 show a low degree of homology, although in all serotypes an asymmetrical distribution of base pairs is found. Comparison of the DNAs of the human serotypes with mouse strain FL DNA (Temple *et al.*, 1981) reveals that they have the sequence ATAATATAC (nucleotides 9–17) in common, while the homologous region between human adenovirus DNAs and CELO DNA is located between positions 9 and 15 (ATAATAT) (Aleström, *et al.*, 1982a). It is very likely that the conserved sequences 9–15 and 9–17 play a crucial role in the initiation of DNA replication and are probably involved in recognition of the site of initiation by the precursor of the terminal protein. In this respect, it is interesting to note that mouse adenovirus strain FL DNA can be replicated in an *in vitro* DNA replication system of Ad2 DNA (Temple *et al.*, 1981). Shinagawa and Padmanabhan (1980) have pointed out that in Ad2, Ad3, Ad5, Ad7, and Ad12 DNA, an additional region of interesting homology is present. In these serotypes, the hexanucleotide TGACGT is found at or near the site where the sequences beyond the ITR begin to diverge. The function of this homology is unknown.

VI. TERMINAL PROTEIN

The presence of protein at the termini of adenovirus DNA was originally detected by Bellett and co-workers, employing DNA isolation procedures that avoid proteolytic digestion (Robinson *et al.*, 1973; Robinson and Bellett, 1975a). These investigators observed that the DNA–protein complex obtained is resistant to boiling and treatment with SDS, indicating that the protein is probably covalently linked to the DNA (Robinson *et al.*, 1973; Sharp *et al.*, 1976; Carusi, 1977; Padmanabhan and Padmanabhan, 1977).

When the buoyant densities of Ad2 and Ad5 DNA–protein complexes are compared with the densities of the corresponding DNAs isolated by digestion with pronase, a small difference of 2–10 mg/ml is found. This corresponds to an amount of protein present in the DNA–protein complex of a maximal 0.3% of the total virion protein (Robinson and Bellett, 1975a; Keegstra *et al.*, 1977). By gel electrophoresis of labeled DNA-free terminal protein (TP), it could be established that TP has an apparent molecular weight of 55K (Rekosh *et al.*, 1977).

Due to the hydrophobic character of TP, DNA–protein complexes aggregate very easily. As a result of this aggregation, DNA–protein complexes accumulate on tops of agarose and polyacrylamide gels during electrophoresis. It has been observed that when DNA–protein complexes are digested with restriction endonucleases and the digestion products are separated by gel electrophoresis, the terminal fragments carrying TP preferentially stay on top of the gel, while internal fragments conventionally run into the gel (Brown *et al.*, 1975; Sharp *et al.*, 1976). Another way to separate the DNA–protein complexes from protein-free DNA is based on differential binding of these compounds to glass-fiber filters (Coombs and Pearson, 1978; Coombs *et al.*, 1978).

To establish the nature of the DNA–protein linkage, deproteinized DNA and DNA–protein complexes have been subjected to enzymatic and nonenzymatic treatments. Both types of DNA are inaccessible to phosphatase, DNA polynucleotide kinase, and λ-exonuclease VII (Carusi, 1977; Sharp *et al.*, 1976), indicating that the 5′ ends of adenovirus DNA are blocked. On the other hand, the 3′ ends can freely be labeled with terminal transferase and are accessible to exonuclease III. These results are most easily explained assuming that in the DNA–protein complex,

TP is covalently attached to the 5′ ends of the two complementary strands. The inaccessibility of deproteinized DNA is probably due to the fact that the 5′ ends are still linked to short peptides. Treatment of DNA–protein complexes or deproteinized DNA with alkali or piperidine removes these peptides and makes the DNA freely accessible for enzymes (Robinson *et al.*, 1973; Carusi, 1977; Tolun *et al.*, 1979; Rekosh, 1981). TP can also be separated from adenovirus DNA by digestion with nuclease S1 (Ariga *et al.*, 1979; Roninson and Padmanabhan, 1980; Rijnders *et al.*, 1983). The DNA–protein complex is cleaved in close proximity to the protein–DNA linkage and yields a protein with a molecular weight of 55K (Rijnders *et al.*, 1983). Recently, Rekosh (1981) showed that treatment of the Ad2 DNA–protein complex with piperidine releases a protein with a molecular weight of 52K. This observation suggests that after DNase I or S1 digestion, the TP isolated still contains a few nucleotide residues.

The nature of the linkage between TP and the DNA molecule has been elucidated by Desiderio and Kelly (1981). Their experiments clearly indicate that Ad2 TP is bound to DNA by a phosphodiester bond between the hydroxyl group of a Ser residue of TP and the 5′-phosphate group of the terminal deoxycytidine residue of the two complementary strands of adenovirus DNA. The particular Ser residue in the TP amino acid sequence involved in the linkage of TP to DNA has recently been identified (Smart and Stillman, 1982).

The origin of TP has been uncertain for many years. Green *et al.* (1979c) showed by tryptic fingerprinting of TPs of five different human serotypes that these proteins were very similar in structure. On the other hand, Rekosh (1981) found different sizes for the TPs of different human serotypes, suggesting that TP is not of cellular origin. He concluded that TP is a highly conserved virus-coded protein. The viral origin of TP was unambiguously proved by Stillman *et al.* (1981), who showed that cell-free translation of mRNAs selected from a region between coordinates 11 and 31.5 on the viral l-strand (see Section IV) leads to synthesis of proteins with apparent molecular weights of 105, 87, and 75K. The 87K protein appeared to be identical to an 80K protein (Challberg *et al.*, 1980) that is covalently attached to the 5′ ends of growing Ad2 DNA strands synthesized in an *in vitro* DNA replication system (Challberg and Kelly, 1979a,b). The 80K protein is structurally related to TP, suggesting that TP is synthesized as an 80K precursor TP (pTP) and that pTP is the active form of TP in adenovirus DNA replication. The different molecular weights found for pTP (80 and 87K) are due to the use of different molecular-weight markers. The 80/87K protein appears to be identical to the protein that is covalently attached to the DNA from temperature-sensitive (*ts*) mutant Ad2*ts*1 virions grown at the nonpermissive temperature (Stillman *et al.*, 1981; Challberg and Kelly, 1981). Ad2*ts*1 is a mutant that cannot cleave virus-coded precursor proteins to their mature counterparts during virion maturation (Bégin and Weber, 1975; Weber *et al.*, 1975).

The mapping of pTP on the virus genome led to the definition of a new early transcription unit, designated E2b. The structure of this region is discussed in detail in Section VII.B.3.

Evidence has been presented that TP plays an essential role in the initiation of adenovirus DNA replication. Analysis of the *in vitro* DNA replication system developed by Challberg and Kelly (1979a,b), in which the DNA–TP complex is used as a template, showed that the first step in the replication of adenovirus DNA is the linkage of dCMP to pTP. The protein probably recognizes a specific sequence within the inverted terminal repetition, which might be involved in binding of pTP to the DNA (Tamanoi and Stillman, 1982). It is likely that the conserved sequence 9–22 in different adenovirus serotypes functions as such a recognition sequence. The presence of TP in the DNA–TP complex might stabilize the initiation complex. Recently, it was shown that the protein is dispensable (Tamanoi and Stillman, 1982), since adenovirus DNA devoid of TP or remaining amino acids can also be used as template in an *in vitro* DNA replication system. It has been proposed that the presence of TP in the DNA–TP complex protects the viral DNA against nucleolytic degradation.

A protecting function of TP has also been proposed to explain the high infectivity of DNA–protein complexes. Deproteinized DNA is infectious when assayed by the calcium coprecipitation procedure (Nicolson and McAllister, 1972; Graham and van der Eb, 1973). However, the infectivity of DNA–TP complexes is 50–100 times higher (Sharp *et al.*, 1976; Chinnadurai *et al.*, 1978; van Wielink, 1978). Although the difference in infectivity might be due to a protective function of TP, it cannot be excluded that the presence of TP on the template is essential for accurate positioning of the pTP on the DNA during the first stage of initiation of adenovirus DNA replication. The role of TP in DNA replication is discussed extensively in Chapter 7.

VII. ORGANIZATION OF THE ADENOVIRUS GENOME

For the unraveling of the organization of the adenovirus genome, a great variety of techniques have been employed, i.e., DNA–RNA hybridization, R-loop mapping, genetic mapping of mutants, translation of preselected mRNA species, and nucleotide sequence analysis (for details, see Mautner *et al.*, 1975; Sambrook *et al.*, 1975; Grodzicker *et al.*, 1975, 1977; Chow *et al.*, 1977b, 1979a,b; Berk and Sharp, 1977a, 1978; Westphal *et al.*, 1976; Westphal and Lai, 1977; Kitchingman *et al.*, 1977; Kitchingman and Westphal, 1980; Miller *et al.*, 1980) (for sequences, see Appendix B). Despite a substantial nucleotide sequence divergence, all adenovirus serotypes studied so far show the general genetic organization (see Appendix B). Since the genomes of the highly homologous types Ad2 and Ad5 have been investigated most extensively, the organization of the adenovirus genome is discussed employing for the most part data obtained with these particular serotypes. The precise location of major landmarks at the nucleotide level is indicated in the Ad2 sequence (Appendix B), unless otherwise stated. During the productive infection cycle of adenoviruses, the different viral genes are expressed in a rather complex pattern (Tooze, 1981; Persson and Philipson, 1982).

Traditionally, the adenovirus genes are subdivided into early genes, which are expressed before the onset of viral DNA replication, and late genes, which are transcribed after replication of adenovirus DNA has started. However, a group of intermediate genes has also been distinguished. These genes are expressed at intermediate times in infection in the absence of DNA synthesis and are also easily detected at late times. The complex transcription pattern of adenovirus DNA is discussed extensively in Chapter 5. A summary of the major RNA transcripts and the corresponding proteins is presented in Figs. 1 and 2. These diagrams demonstrate that the adenovirus genetic information is scattered over the

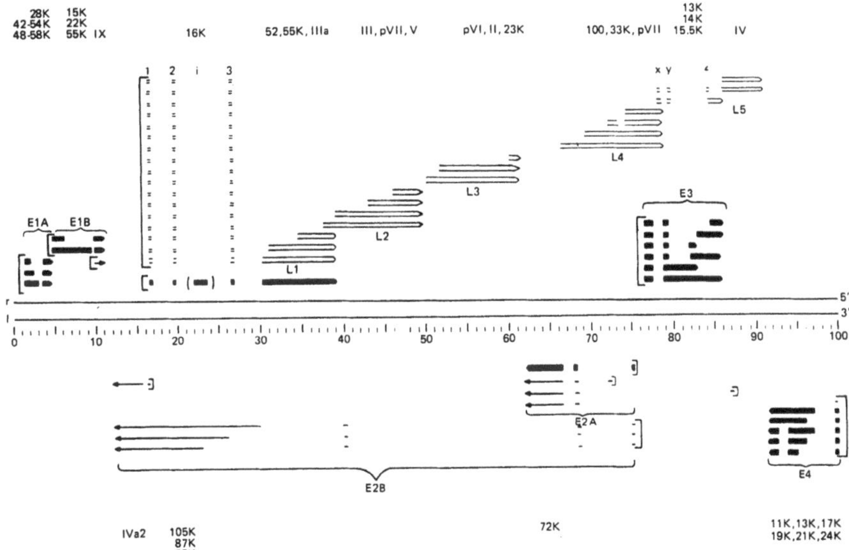

FIGURE 1. Transcriptional organization of the Ad2 genome. The genome is divided into 100 map units. The r-strand is rightward-transcribed into RNA and the l-strand leftward. The direction of transcription is indicated by arrows. The capped 5′ ends of the cytoplasmic RNA indicate the positions of transcriptional promoters, while the arrowheads represent the 3′ polyadenylation sites. Gaps in arrows indicate intervening sequences, which have been removed from the cytoplasmic RNA by splicing. The RNA shown in bold lines can be detected early in infection before the onset of DNA replication (regions E1a, E1b, E2a, E3, E4; also the late promoter at 16.5 units is active early in infection, leading to transcription to 39 units). The light lines represent intermediate RNAs synthesized at early as well as at late times in the infection cycle (E2a, E2b, polypeptide IX). The double-lined arrows indicate late RNA species. Correlations of mRNAs with encoded proteins are based on cell-free translation of selected RNA species and RNA mapping data. Proteins are designated by their molecular weights in kilodaltons (K) or by roman numerals (virion components).

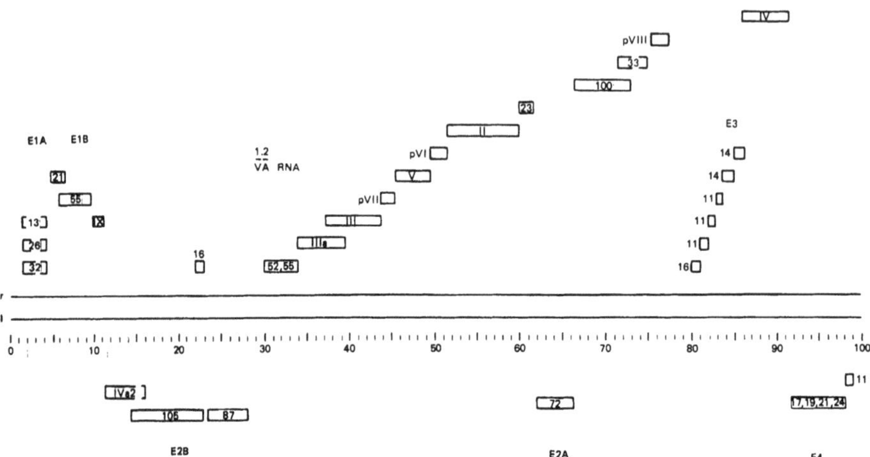

FIGURE 2. Protein-coding regions of the Ad2 genome. The regions on the adenovirus genome that code for protein have been determined by hybrid-arrest translation, by *in vitro* translation of preselected mRNAs, by RNA mapping, and by direct DNA and RNA sequence analysis. The identified proteins are designated by their apparent or theoretical molecular weights in kilodaltons or by roman numerals (virion components). Regions pVI, pVII, and pVIII indicate the positions of the precursors of polypeptides VI, VII, and VIII. Interrupted coding regions indicate discontinuous genes.

two complementary strands. About 69% of all genetic information is located on the rightward-transcribed strand (r-strand), while only 31% of the coding sequences are present on the leftward-transcribed strand (l-strand).

The positions of promoters and starts of transcription have been mapped via a variety of methods (Berk and Sharp, 1977b; Pettersson and Mathews, 1977; Spector *et al.*, 1978; Seghal *et al.*, 1979; Wilson *et al.*, 1979; Chow *et al.*, 1979a,b; Shaw and Ziff, 1980; Akusjärvi and Persson, 1981a; Stillman *et al.*, 1981). Many of the positions of promoters have been correlated with sequences generally indicated as TATA or Goldberg–Hogness boxes. These AT-rich sequences are considered to represent a constitutive part of promoter signals (see Chapter 5). The genes expressed early in infection are transcribed from six different promoters (r-strand: positions 1.3, 4.6, 16.5, and 76.6; l-strand: 75.1 and 99.1). The intermediate genes are transcribed from promoters located at positions 9.7 on the r-strand and 16.1 and 75.1 on the viral l-strand. The long late transcription unit uses the major late promoter at map position 16.5 on the viral r-strand. All primary transcription products of adenovirus DNA are processed in the nucleus before entering the cytoplasm. They are capped with $^{7me}G5'pppN$ at the 5′ end, and they are polyadenylated at the 3′ end. With one exception (polypeptide IX mRNA), all primary transcription products are processed into families of related mRNAs that share common 5′ and 3′ ends, but differ by alternative splicing (early regions E1a, E1b, E2a, E3, and E4, intermediate regions E2b and IVa_2, and late regions L1, L2, L3, L4, and L5). It should be noted that in fact, analysis of the late transcription unit of adenovirus led to the original discovery of the phenomenon of RNA splicing. A detailed analysis of the transcription of the adenovirus genome is presented in Chapter 5. The organization of the transcriptional units of the adenovirus genome will now be described systematically from left to right. Since the organization of the Ad2 and Ad5 genomes has been investigated most extensively, these genomes are used for illustration.

The positions of major landmarks of the transcription units are indicated in Figs. 3–6 and Appendix B in the r- and l-strand sequences. It should be borne in mind that sequences of the r-strand of DNA are equivalent to RNA transcribed from the l-strand and that sequences of the l-strand of the genome are equivalent to mRNA transcribed from the r-strand. Unfortunately, the entire nucleotide sequences of Ad2 and Ad5 are not yet available, only a number of noncontiguous regions having been sequenced. Therefore, the numbering of the base pairs in Fig. 3–6 and Appendix B has not been added, but the sequence of each specific region starts from the left with base pair number 1.

A. Early Region E1 (1.3–11.2)

Early region E1 is transcribed from the leftmost part of the viral r-strand. It contains genes involved in cell transformation (Graham *et al.*, 1974a,b; van der Eb *et al.*, 1979) and regulation of transcription (Berk *et al.*, 1979; Jones and Shenk, 1979a; Nevins, 1981). The complete nucleotide sequence of this region has been established for human serotypes Ad2, Ad5, Ad7, and Ad12 (van Ormondt *et al.*, 1978, 1980a,b; Sugisaka *et al.*, 1980; Dijkema *et al.*, 1980a,b, 1981; Bos *et al.*, 1981; Kimura *et al.*, 1981; Gingeras *et al.*, 1982). The overall organization of this region appears to be very similar for the different serotypes (van Ormondt *et al.*, 1980b; Dijkema *et al.*, 1982). The region between 1.3 and 11.2 m.u. can be subdivided into three transcription units designated E1a, E1b, and region IX (Kitchingman *et al.*, 1977; Berk and Sharp, 1977a, 1978; Chow *et al.*, 1979a,b). The mRNAs derived from region E1 have been characterized by EM mapping, *in vitro* translation, and sequence analysis. It appears that all mRNAs except protein IX mRNA have a spliced structure and code for a variety of proteins, some of which are structurally related.

1. Early Region E1a (1.3–4.6)

Early region E1a is transcribed from the r-strand between 1.3 and 4.6 m.u. and codes for proteins that are involved in initiation of transformation (van der Eb *et al.*, 1979) and regulation of early gene expression

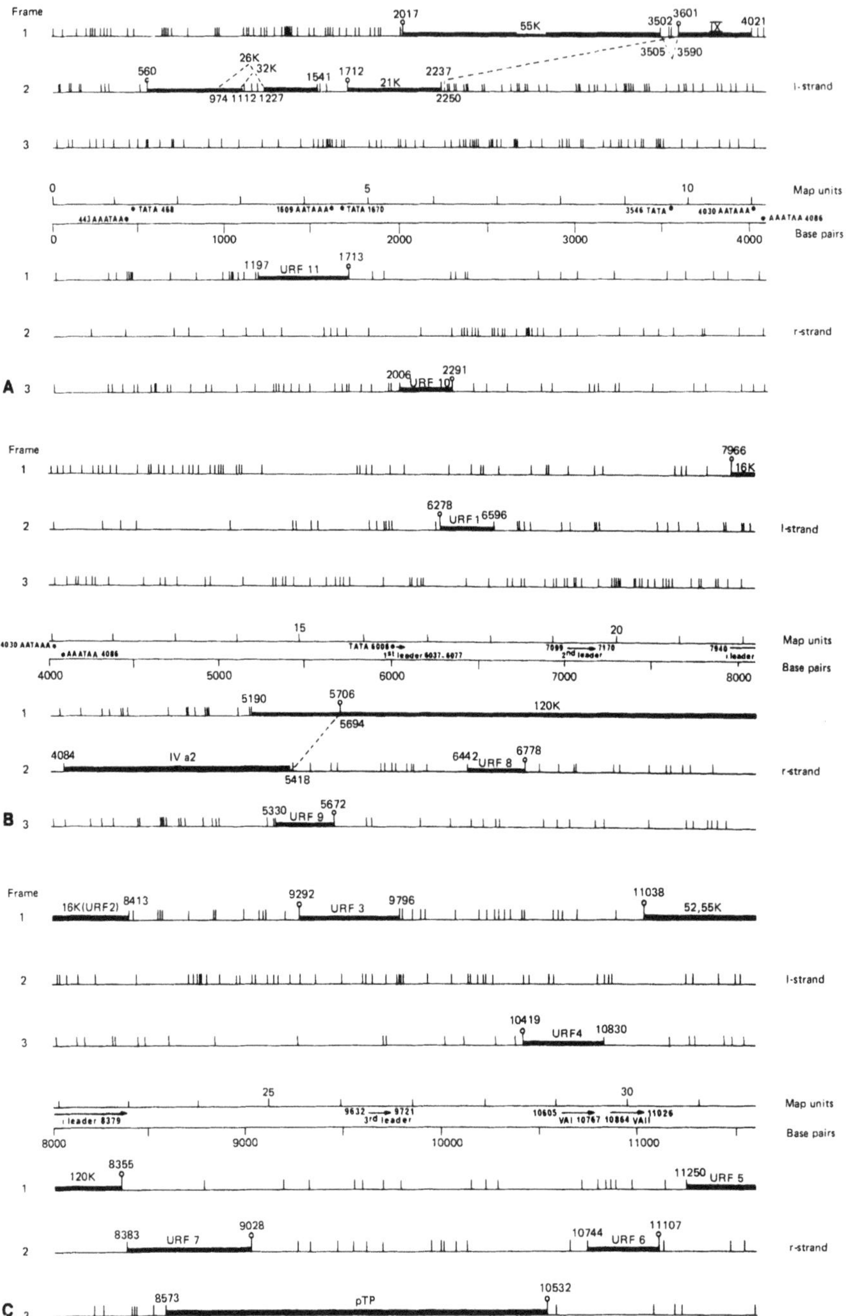

FIGURE 3A–C. Structural organization of the region between coordinates 0.0 and 31.7 on the Ad2 genome. The analysis of the structural organization is based on the nucleotide sequence shown in Fig. 18 (Appendix B), and indicated positions refer to this sequence. The l-strand of the DNA is homologous to r-strand transcripts, while the r-strand is homologous to l-strand transcripts. Here and in Figs. 4–6 and Appendix B: Termination codons (TAA, TGA, and TAG) are indicated in the three frames of the l- and r-strands by short vertical

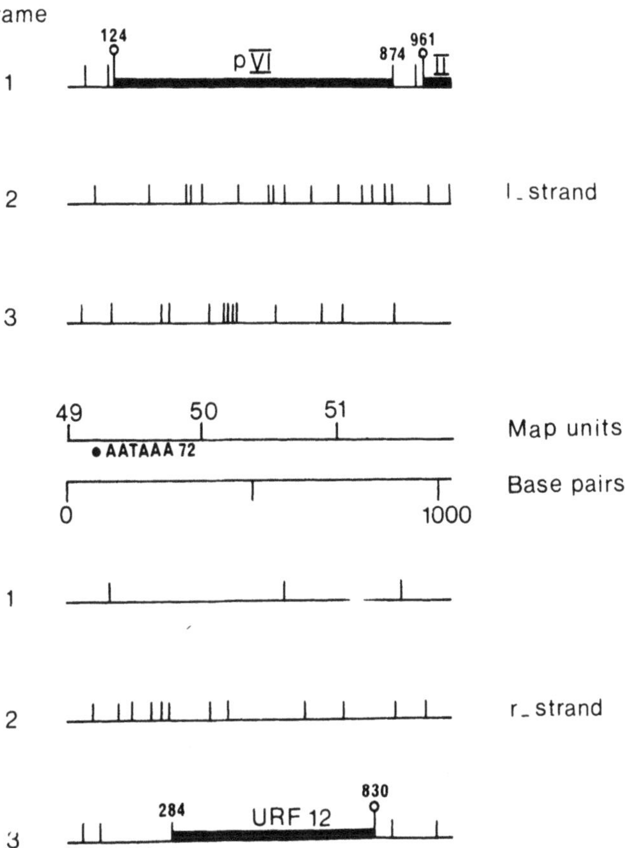

FIGURE 4. Structural organization of the region between coordinates 49.0 and 51.8 on the Ad2 genome. This analysis is based on the nucleotide sequence shown in Fig. 19 (Appendix B). This region mainly codes for the precursor of polypeptide VI. For explanation of the symbols, see the Fig. 3 caption.

(Jones and Shenk, 1979a; Berk *et al.*, 1979) (see Fig. 3). The promoter of this region has been mapped at position 1.3 (Wilson *et al.*, 1979). Analysis of the Ad2 sequence reveals that at position 468 [see Fig. 18 (Appendix B)], the TATA box TATTTATA is present. Baker and Ziff (1980, 1981) have characterized the position where transcription of the E1a RNA is initiated. They found that all mRNAs start with a capped dAMP residue

---

lines, while the initiation codon ATG is indicated by the symbol ♀. The coding regions that have been correlated with known proteins are shown by bold lines and are designated by molecular weights of the corresponding proteins or by roman numerals. Unidentified reading frames [(URF) initiating with ATG and terminating with one of the termination codons] or open reading frames [(ORF) regions between two termination codons] longer than 300 nucleotides are also indicated. Between the scales for Map units and Base pairs, the positions of TATA boxes, polyadenylation signals, and leader sequences are indicated. At some positions along the genome, splicing may occur. These positions are indicated by interrupted lines.

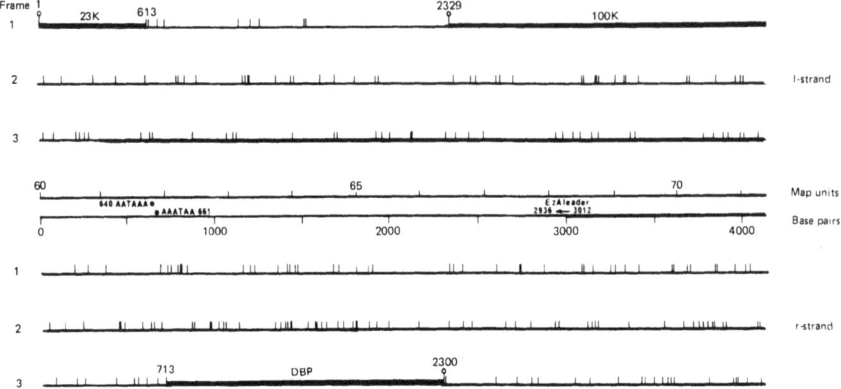

FIGURE 5. Structural organization of the region between coordinates 59.9 and 71.4 on the Ad5 genome. This analysis is based on the nucleotide sequence shown in Fig. 24 (Appendix B). This region codes for a 23K protein, DNA-binding protein (DBP), and a part of the 100K protein. For explanation of the symbols, see the Fig. 3 caption.

derived from position 499. Three mRNA species have been identified from region E1a with sedimentation coefficients of 13, 12, and 9 S. These mRNAs share the same 5′ and 3′ termini and differ only in the size of the RNA fragment removed by splicing during the processing of nuclear RNA (Kitchingman *et al.*, 1977; Berk and Sharp, 1977a, 1978; Chow *et al.*, 1979a,b; Perricaudet *et al.*, 1979). The splice points of the 13 S RNA have been mapped at nucleotide positions 1112 and 1227 and of the 12 S mRNA at positions 974 and 1227 (Perricaudet *et al.*, 1979). The donor splice site of the 9 S mRNA species has not been determined yet. The 3′ ends of the mRNAs are located at nucleotide position 1630, while the polyadenylation signal AATAAA is found at position 1609 (Perricaudet *et al.*, 1979; Fraser *et al.*, 1982).

Since the reading frames in the E1a mRNAs are the same, the proteins derived from these mRNAs share their N-terminal and C-terminal segments and differ only in the number of intervening amino acids. From the DNA sequence, the complete amino acid sequences of the proteins specified by the 13 and 12 S mRNA species can be predicted. Both proteins must be rich in Pro and Glu residues and have theoretical molecular weights of 32 and 26K, respectively. The protein derived from the 9 S mRNA has an estimated molecular weight of 13K. These proteins have been correlated with proteins produced during cell-free translation of isolated mRNAs (Lewis *et al.*, 1976; Pettersson and Mathews, 1977; Harter and Lewis, 1978; Green *et al.*, 1979a; Esche *et al.*, 1980; Spector *et al.*, 1980a,b; van der Eb *et al.*, 1979; Lupker *et al.*, 1980). These translation products with apparent molecular weights of 48–58, 42–54, and 28K are structurally related, which is in agreement with the nucleotide sequence of this region. The discrepancy between the theoretical and apparent molecular weights probably reflects the extremely high Pro contents of these proteins, which lead to aberrant migration in gels.

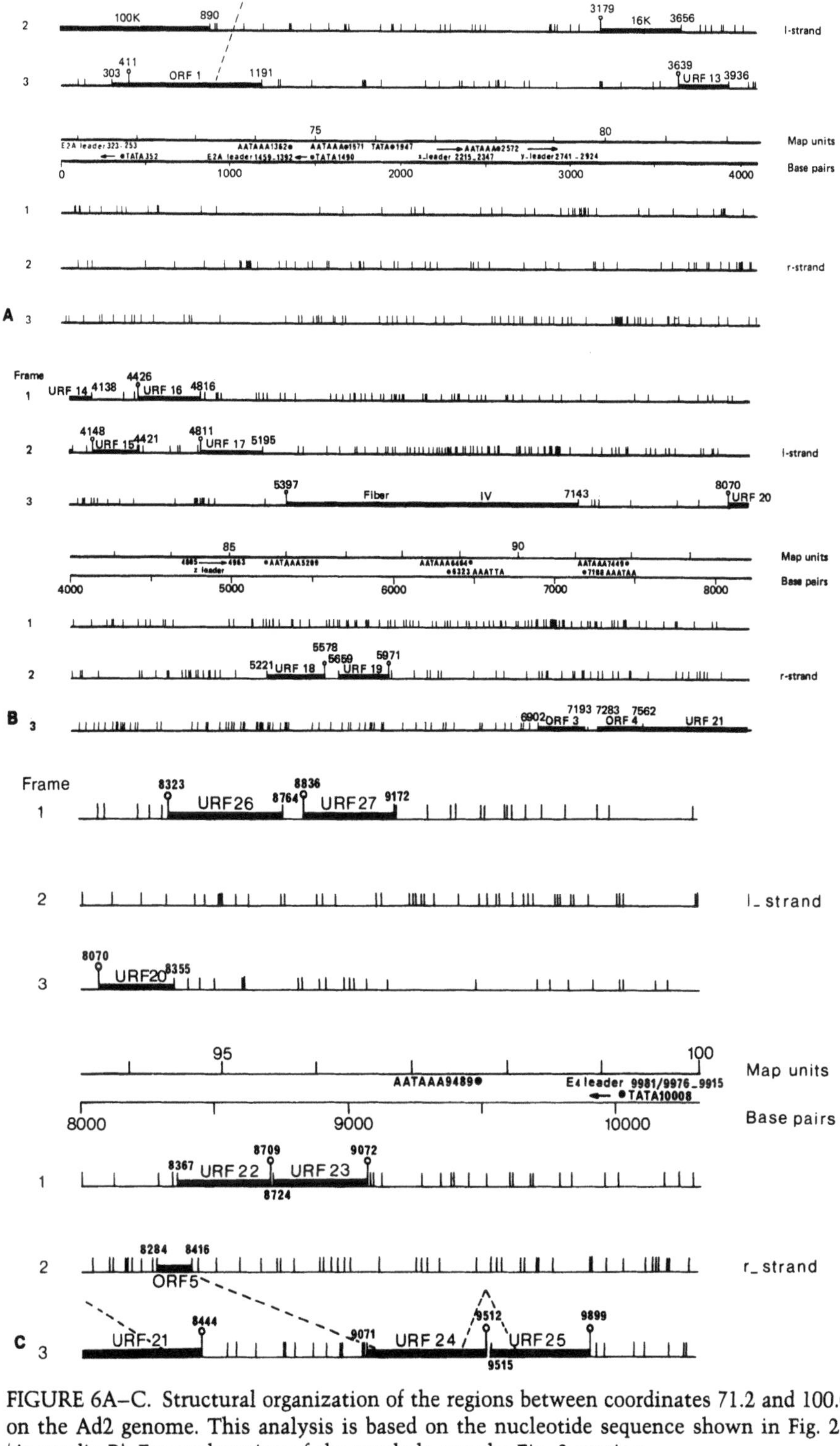

FIGURE 6A–C. Structural organization of the regions between coordinates 71.2 and 100.0 on the Ad2 genome. This analysis is based on the nucleotide sequence shown in Fig. 21 (Appendix B). For explanation of the symbols, see the Fig. 3 caption.

As mentioned before, the E1a regions of Ad2, Ad5, Ad7, and Ad12 show very similar organization. In all serotypes, three spliced mRNA species are synthesized. Recently, it was shown that the protein encoded by the 13 S mRNA governs early gene expression (Montell *et al.*, 1982).

2. Early Region E1b (4.6–11.2)

Early region E1b is transcribed from the viral r-strand between map coordinates 4.6 and 11.2 [see Figs. 3 and 18 (Appendix B)]. The proteins encoded by this region are involved in transformation and play an important role in oncogenesis; during lytic infection, these proteins are involved in DNA replication (Harrison *et al.*, 1977; Frost and Williams, 1978; Jones and Shenk, 1979a,b; van der Eb *et al.*, 1979; Bernards *et al.*, 1982; van den Elsen *et al.*, 1982). Little is known about the precise role of these proteins. Studies of cells transformed by DNA fragments of different length have suggested that region E1a is able to immortalize cells, while region E1b is required for full expression of the typical phenotype of adenovirus-transformed cells (van der Eb *et al.*, 1979; Houweling *et al.*, 1980).

The promoter of early region E1b is located at map position 4.6, where, at nucleotide 1670, a Goldberg–Hogness box TATATAA is found (Fig. 18). Transcription may start at position 1700 or 1702 (Baker and Ziff, 1981) and proceeds until nucleotide 4061 (Perricaudet *et al.*, 1980; Fraser *et al.*, 1982). The polyadenylation signal of region E1b is located at nucleotide 4030. The primary transcription product of region E1b is processed by splicing into a 22 and a 13 S mRNA species. Both species share a 3′-terminal segment from nucleotide 3590 to a polyadenylation site at nucleotide 4061. Both species also contain a 5′-terminal sequence from 1700 or 1702 to a donor splice site at nucleotide 2250. In the 13 S mRNA, nucleotide 2250 is joined to an acceptor splice site at 3590, whereas the 22 S mRNA includes nucleotide 2250 to a second donor splice site at nucleotide 3505. Nucleotide 3505 of the 22 S mRNA is ligated to the common acceptor splice site at nucleotide 3590. From these points, the mRNA sequence continues to the polyadenylation site near nucleotide 4061 (Perricaudet *et al.*, 1980; Aleström *et al.*, 1980). *In vitro* translation experiments have shown that two major proteins with molecular weights of 55–65 and 15–19K can be assigned to this transcription unit (Lewis *et al.*, 1976; Harter and Lewis, 1978; van der Eb *et al.*, 1979; Brackmann *et al.*, 1980). This observation is in agreement with the fact that the two mRNA species contain information for two major tumor (T) antigens with theoretical molecular weights of 21 and 55K, which are encoded by two overlapping reading frames. The 22 S mRNA codes for both proteins depending on which particular ATG triplet serves as the start codon. The 21K protein initiates at the 5′-proximal ATG (position 1712), while the 55K protein initiates at the second ATG (nucleotide 2017) in another reading frame (Anderson and Lewis, 1980; Bos *et al.*, 1981). In addition, the 21K protein can also be synthesized from the 13 S mRNA. Peptide mapping has shown that the small-t and the large-T antigens do not share tryptic peptides, in accordance with the nucleic acid sequence data (Bos *et al.*, 1981).

Similar organization of region E1b has been found for Ad2, Ad7, and Ad12 (Bos *et al.*, 1981; Kimura *et al.*, 1981; Dijkema *et al.*, 1982; Gingeras *et al.*, 1982). This does not exclude small differences between mRNAs from different serotypes. Comparison of the E1b mRNAs of Ad5 and Ad12 has revealed that the Ad12 mRNA contains additional splices in the 3′ noncoding part of the mRNA (Virtanen *et al.*, 1982a). The precise functions of the 21K and 55K proteins are still unknown.

The 22 and 13 S mRNAs both contain information for protein IX, a protein that has been mapped between 9.7 and 11.2 map units (Chow *et al.*, 1977b; Pettersson and Mathews, 1977; Esche *et al.*, 1980). However, this information is not translated from these messengers. Instead, a unique short mRNA is synthesized from an independent transcription unit between coordinates 9.7 and 11.2 (Wilson *et al.*, 1979; Chow *et al.*, 1977a,b; Pettersson and Mathews, 1977). The sequences of the genes that encode the Ad2 and Ad5 polypeptides IX have been established, which allowed the identification of transcription and translation signals (Maat *et al.*, 1980; Aleström *et al.*, 1980). The polypeptide IX TATA box is located at position 3546, and transcription starts at nucleotide position 3575 or 3577 (map position 9.7) in the Ad2 sequence [Fig. 18 (Appendix B)]. Its 3′ end has been located at nucleotide position 4061 (map position 11.2) (Aleström *et al.*, 1980; Fraser *et al.*, 1982), while the polyadenylation signal AATAAA is located at position 4030. The same polyadenylation signal is also used for processing of the large and the small E1b T antigen mRNAs. The RNA synthesized is not processed and represents the only known unspliced adenovirus mRNA. The mRNA contains a continuous open reading frame that codes for a protein of 14K. Protein IX (apparent molecular weight 12.5K) is found in virions and was therefore originally classified as a late protein (Pettersson and Mathews, 1977). Later experiments showed that protein IX is also synthesized in the absence of viral DNA replication, indicating that it is an intermediate protein (Persson *et al.*, 1978). The complete nucleotide sequence of the polypeptide IX gene has been determined for human serotypes Ad2, Ad3, Ad5, Ad7, and Ad12 (Maat *et al.*, 1980; Aleström *et al.*, 1980; Dijkema *et al.*, 1981; Kimura *et al.*, 1981; Engler, 1981). Within the same group, the protein IX genes exhibit a striking similarity, but the genes of serotypes from different groups are much less homologous.

3. Unidentified Reading Frames

In the l-strand transcripts, a number of unidentified reading frames (URFs) have been detected. The URFs larger than 300 nucleotides are indicated in Figs. 3 and 18 (Appendix B). However, recently it could be shown that in transformed cells and infected cells, an l-strand transcript is synthesized that spans the E1a–E1b junction and codes for a protein with a molecular weight of 11K (Katze *et al.*, personal communication). This transcript might very well be derived from URF 11 located between nucleotides 1713 and 1197 on the viral l-strand. At position 443, the sequence AATAAA is found, which might function as a polyadenylation signal. This indicates that it is certainly not impossible that later some of these will appear to be expressed during the infection cycle, albeit at a very low frequency.

B. Late and Intermediate Genes in the Region between Coordinates 11.2 and 31

1. Major Late Promoter and Tripartite Leader

The region between 11.2 and 31 contains a mosaic of different strategic regions in both complementary strands [see (Figs. 3 and 18 (Appendix B)]. The major late promoter has been mapped on the r-strand at position 16.5 (Evans *et al.*, 1977; Ziff and Evans, 1978). This promoter is also active early in infection (Shaw and Ziff, 1980; Akusjärvi and Persson, 1981b). In the nucleotide sequence at this position, there is a TATA box TATAAA at nucleotide position 6006, and transcription starts from position 6037 (Baker and Ziff, 1981). During early times in infection, transcription proceeds no further than map position 39, while at late times, transcription proceeds to map position 99.0 (Fraser *et al.*, 1979). Messenger RNAs derived from r-strand transcripts starting at position 16.5 contain a common tripartite leader (Berget *et al.*, 1977, 1978; Chow *et al.*, 1977a,b; Akusjärvi and Pettersson, 1979a,b; Zain *et al.*, 1979a,b; Ziff and Evans, 1978). The sequence of the tripartite leader of late Ad2 RNA has been determined by sequencing complementary DNA (cDNA) transcribed from hexon mRNA and a cDNA clone of fiber mRNA (Zain *et al.*, 1979a; Akusjärvi and Pettersson, 1979b). The tripartite leader sequences have been established for a number of serotypes [Ad2 (Ziff and Evans, 1978; Akusjärvi and Pettersson, 1979a; Zain *et al.*, 1979a), Ad5 (van Beveren *et al.*, 1981), Ad3 and Ad7 (Engler *et al.*, 1981)].

The overall length of the Ad2 tripartite leader is 203 nucleotides, comprising 41 nucleotides from the promoter region at map position 16.5, 72 nucleotides from position 19.6, and 90 nucleotides from position 26.5 on the genome. Examination of the sequence reveals that the tripartite leader does not contain an AUG triplet, suggesting that translation of late adenoviral mRNA does not initiate within the tripartite leader. In some intermediate and late transcripts, an additional leader fragment (i-leader) has been detected by R-loop mapping, which maps at coordinates 21.5–23.0 (Chow *et al.*, 1979a). Sequence analysis has shown that in contrast to the tripartite leader, the i-leader (nucleotides 7940–8379) contains an open reading frame for a hypothetical protein of 15.9 kilodaltons (kd).

In vitro translation of mRNA selected on DNA fragments that contain i-leader sequences does indeed lead to synthesis of a hitherto unknown protein (URF2) with an apparent molecular weight of 13.6–16K (Lewis *et al.*, 1979; Lewis and Mathews, 1980; Virtanen *et al.*, 1982b). The termination codon for the 15.9-kd protein is not present in the i-leader, but is probably located within the third leader. The function of the 15.9-kd protein is still unknown.

2. Virus-Associated RNAs

At positions 28.8 and 29.5 on the genome, the genetic information for two low-molecular-weight RNAs is located, these RNAs being designated virus-associated (VA) RNAs VA-RNAI and VA-RNAII (Söderland *et al.*, 1976; Mathews and Pettersson, 1978) (Fig. 3). In contrast to all other genes, the VA genes are transcribed by RNA polymerase III instead of RNA polymerase II (Price and Penman, 1972; Weinman *et al.*, 1974, 1976; Söderland *et al.*, 1976). The VA-RNAs are probably synthesized from two separate promoter sites in the r-strand and do not undergo post-transcriptional processing. The genes and the RNA products have been subjected to nucleotide sequence analysis (Ohe and Weissman, 1970, 1971; Ohe, 1972; Pan *et al.*, 1977; Celma *et al.*, 1977a,b; Akusjärvi *et al.*, 1980). The nucleotide sequence of VA-RNAI was determined by Ohe and Weissman (1971) to be 157–160 nucleotides long (nucleotides 10,608–10,764/10,767). Vennström *et al.* (1978a,b) demonstrated that the 5′ end of VA-RNAI is heterogeneous and may start at nucleotide 10,605 or 10,608 [Fig. 18 (Appendix B)]. The length of VA-RNAII is 158–163 nucleotides (nucleotides 10,864–11,021/11,026), and the two VA-RNAs are separated by a spacer about 98 nucleotides long. The function of these RNAs is still unknown; so far, no proteins derived from them have been found. It has been suggested that these RNAs play a role in splicing or stabilization of late mRNA (Murray and Holliday, 1979; Mathews, 1980). It is interesting to note that the VA-RNAs can form almost identical secondary structures with high stability. The structures show similarities to transfer RNA (Zain *et al.*, 1979b; Akusjärvi *et al.*, 1980).

3. Early Region E2b and Protein IVa_2 (11.2–30.2)

For a long time, it has been thought that the l-strand transcripts between map units 11 and 30 coded only for the intermediate protein IVa_2 (molecular weight 50K), a protein that is involved in the morphogenesis of virions (Persson *et al.*, 1979a). The gene of this protein has been mapped between coordinates 11.3 and 16.1 (Lewis *et al.*, 1975, 1977) [see Figs. 3 and 18 (Appendix B)]. Transcription of the IVa_2 gene starts from a promoter located at map position 16.1. Nucleotide sequences of this region reveal that although no regular TATA box is located in this region, the sequence TCCTT, which may resemble a TATA box, is present at nucleotide 5859. RNA synthesis starts at position 5826 or 5824 and proceeds to nucleotide 4051 (Aleström *et al.*, 1980; Baker and Ziff, 1981; Fraser *et al.*, 1982) (Fig. 18). The messengers from this region contain an intron located between nucleotides 5419 and 5693 (Chow *et al.*, 1977a,b; Broker *et al.*, 1977; Kilpatrick *et al.*, 1979; van Beveren *et al.*, 1981). The mRNA contains a long open reading frame (ORF) corresponding to 445 amino acids of which the first 4 N-terminal amino acids are coded by RNA upstream from the donor splice site and the remaining amino acid residues by RNA downstream from the acceptor splice site. It is noteworthy that the reading frame in which these 4 N-terminal amino acids lie is part of a much longer reading frame that codes for a protein of 120 kd (see below). Another interesting feature of the IVa_2 gene is that the 3′ end of the message overlaps the end of the E1b and polypeptide IX mRNAs with 9 nucleotides. Also, the IVa_2 termination codon TAA (nucleotide 4084) forms a part of the IVa_2 polyadenylation signal AATAAA (nucleotide 4086). The IVa_2 genes of serotypes Ad2, Ad5, and Ad7 have all been sequenced and show the same structural organization (van Beveren *et al.*, 1981; Engler and van Bree, 1982; Gingeras *et al.*, 1982; Aleström *et al.*, 1982b). The IVa_2 nucleotide sequences of Ad7 and Ad5 are 78% homologous.

A new class of mRNAs from the region between 11 and 30 m.u. was identified by Stillman *et al.* (1981). The promoter of these transcripts has been mapped at position 75.1 and is probably identical to the promoter of early region E2a. Transcripts of this region, which is designated E2b, contain, in addition to the 75.1-m.u. leader, additional leaders from 68.5 and 39 m.u. Region E2b has been classified as an intermediate transcription unit (Fig. 3). The main bodies of messages derived from this transcription unit may start at positions 30, 26, and 23, respectively, and continue to position 11.2. *In vitro* translation of preselected mRNAs derived from the region between 11.2 and 31.5 led to synthesis of proteins with molecular weights of 105, 87, and 75K (Stillman *et al.*, 1981; Binger *et al.*, 1982). The 87K protein is identical to the precursor terminal protein (pTP) with a molecular weight of 80K described by Challberg *et al.* (1980) (see Section VI). Nucleotide sequence analysis of this region has indicated the presence of two long ORFs located between 28.9 and 23.5 m.u. and 24.1 and 14.2 m.u. [Fig. 18 (Appendix B)]. The region between 28.9 and 23.5 m.u. beginning at nucleotide 10,577 has the first ATG at nucleotide 10,532 and continues to a terminator at nucleotide 8573. This frame codes for a protein with a minimum molecular weight of 74.5K. The second large ORF begins at nucleotide 8793, has the first ATG at 8355, and continues to a terminator TAG at nucleotide 5190. The total coding capacity of this reading frame is 132.1kd, while the capacity from the first ATG to the terminator is 120.4kd (Gingeras *et al.*, 1982; Aleström *et al.*, 1982; Engler *et al.*, 1983). Since the precise structure of the spliced E2b mRNAs is still unknown, it cannot be excluded that a part of the leader from map position 39 is part of the coding sequences of E2b mRNAs. EM mapping of E2b mRNAs has indicated that the 3′ ends of the messengers map at position 11.2, the same position where the 3′ end of IVa_2 mRNA is located. It is therefore likely that the mRNAs of pTP and the 120kd polypeptide have the same 3′ end and polyadenylation site as the IVa_2 mRNA (Aleström *et al.*, 1980; Stillman *et al.*, 1981). Smart and Stillman (1982) showed by analysis of tryptic peptides from the terminal protein and its precursor that the ORF between 28.9 and 23.5 codes for pTP. Very recently, the ORF from 24.1 to 14.2 was assigned to an adenovirus-specific DNA polymerase (Kelly, Stillman, and Hurwitz, personal communications). This polymerase has an apparent molecular weight of 140K, copurifies with pTP, and is able to complement a defective *in vitro* DNA replication system of the DNA-synthesis-negative temperature-sensitive (*ts*) mutant Ad5*ts*36 (Enomoto *et al.*, 1981; Lichy *et al.*, 1982; Kelly and Stillman, personal communications). The mutant Ad5*ts*36 has been mapped between 18.5 and 22.0 m.u. (Galos *et al.*, 1979). In addition to these two proteins, all E2b messengers contain genetic information for the IVa_2 protein, but this information is probably not translated from the E2b messengers.

4. Unidentified Reading Frames

Several unidentified shorter reading frames are present in this region of the viral genome (Fig. 3). However, no correlation with known proteins or gene functions has been discovered yet. In this respect, it should be noted that translation *in vitro* of early mRNA selected by hybridization to fragments of DNA derived from this region has identified mRNA species that encode additional proteins (Lewis and Mathews, 1980). A DNA fragment from 17.0 to 21.5 m.u. selects an mRNA that is complementary to the r-strand and codes for a 13.5-kd protein (Lewis *et al.*, 1979; Lewis and Mathews, 1980). Further, two polypeptides of 16.5 and 17.0kd have been described, translated from mRNAs that are selected by DNA fragments lying between 11.6 and 17.0 m.u. (Lewis *et al.*, 1979).

C. Late Regions L1, L2, and L3 (31.0–61.7)

A major event in the infection cycle of adenoviruses is the activation of the entire late transcription unit. As mentioned in Section VII.B.1, the promoter of the late transcription unit is located at map position 16.5, and this promoter is already active early in infection. However, during the early phase, transcription does not proceed further than map position 39 (Shaw and Ziff, 1980; Akusjärvi and Persson, 1981b). In the late phase, transcription continues to map position 99.0 (Fraser *et al.*, 1979, 1982). The transcription product ranging from map positions 16.5 to 99.0 is considerably processed, leading to the production of five families of late mRNAs (L1–L5) (Chow *et al.*, 1977b; McGrogan and Raskas, 1978; Chow and Broker, 1978; Nevins and Darnell, 1978). Each of the five classes expresses more than one protein and contains mRNAs with a common 3′ end (Ziff and Fraser, 1978; Nevins and Darnell, 1978; Fraser and Ziff, 1978). At the 5′ end, all these mRNAs contain the tripartite leader.

The region on the Ad2 genome between 30.2 and 61.7 m.u. contains the genes for the families L1–L3. As mentioned above, the L1 family of RNAs is already expressed early in infection. This family consists of three mRNAs that have a common 3′ end mapping at 39 m.u. At the same position, the polyadenylation site of the L1 family has been mapped (Fraser *et al.*, 1979, 1982). The L1 mRNAs code for two structurally related proteins of 52 and 55K (Lewis and Mathews, 1980; Miller *et al.*, 1980) and polypeptide IIIa (molecular weight 66K). Since nucleotide sequences from the left-hand end of Ad2 DNA have not been established further than position 31.5, only the initiation codon of the 52,55K protein has been identified unambiguously (Akusjärvi *et al.*, 1980). The function of the 52,55K protein is still unknown. The L1 family further contains genetic information for protein IIIa, which has been mapped by hybrid-arrest translation between 34.3 and 39.3 m.u. This protein has a molecular weight of 66K and is present in virions associated with the hexon polypeptides.

Located from positions 39 to 50 is the L2 family, consisting of three mRNA species that code for polypeptide III (molecular weight 85K), the precursor of polypeptide VII (20K), and polypeptide V (48.5K). These proteins are all constituents of adenovirus particles. One of these, the precursor of polypeptide VII, is processed during maturation of virions to mature polypeptide VII (molecular weight 18.5K). This protein is identical to the major core protein. The genes for protein III, the precursor of protein VII, and protein V have been mapped by R-loop mapping and hybrid-arrest translation at 37.4–43.9, 43.9–45.4, and 45.3–49.6, respectively (Miller *et al.*, 1980).

Fraser *et al.* (1982) have mapped the polyadenylation site of the L2 family at position 50. This fits well with the fact that in the nucleotide sequence from the region between coordinates 49.0 and 51.8 [Fig. 19 (Appendix B)], the polyadenylation site of the L2 family has been identified at nucleotide 92, while an AATAAA signal is present at nucleotide 72 (Akusjärvi and Persson, 1981a).

The nucleotide sequence data from region 49.0–51.8 make it possible to pinpoint exactly some landmarks of the L3 family of late mRNAs (see Figs. 4 and 19). Three species of mRNAs have been identified that can be translated into the precursor of polypeptide VI (pVI), hexon (polypeptide II), and a 23K protein. The gene for polypeptide pVI is located from 49.1 to 51.2 and has been sequenced completely (Miller *et al.*, 1980; Akusjärvi and Persson, 1981a). Also, the acceptor splice site at which the 5′ leader sequences are joined to the pVI message has been determined (nucleotide 123) (Fig. 19). This splice site is situated very close to the start codon (nucleotide 124). The gene for polypeptide pVI codes for a protein with a theoretical molecular weight of 27K. This protein is cleaved during maturation of young virions, resulting in the formation of polypeptide VI (molecular weight 24K), which is part of the adenovirion. With the help of nucleotide sequence analysis, the N-terminal end of the hexon polypeptide has been mapped at coordinate 51.6, while the C terminus is located at 59.7 (Akusjärvi and Pettersson, 1978a,b). The hexon polypeptide is translated from start codon 961 of an mRNA that contains, in addition to the tripartite leader, a main body starting at nucleotide 925 in the sequence of Fig. 19 (Appendix B) to nucleotide 836 in the sequence of Fig. 20.1. The common polyadenylation site of the L3 RNAs has been mapped at the same position. In accord with other polyadenylation sites, the sequence AATAAA is located close to this addition site (nucleotide 812) (Fig. 20.1). The total nucleotide sequence of the hexon gene has not been established yet; only stretches of nucleotides have been determined (Jörnvall *et al.*, 1981b). However, by combination of nucleotide sequence and amino acid sequence data, the complete amino acid sequence of the Ad2 hexon polypeptide has been established (Jörnvall *et al.*, 1981a). It appears that the hexon polypeptide of Ad2 consists of 966 amino acid residues. It is the largest viral protein and has a calculated molecular weight of 108K and an apparent molecular weight of 120K.

From positions 59.9 to 61.7, r-strand transcripts code for a protein of molecular weight 23K (Kruijer *et al.*, 1980; Akusjärvi *et al.*, 1981) [see Figs. 5 and 20.2 (Appendix B)]. A minor RNA species consisting of the tripartite leader and a main body corresponding to this region has been identified and translated. A protein with a molecular weight of 23K is synthesized from this messenger. Since the Ad2 mutant *ts*1 has been mapped in the L3 region and is hampered in proteolytic cleavage of precursors of polypeptides VI, VII, and VIII, it has been suggested that the 23K protein is identical to a virus-coded protease (Bhatti and Weber, 1979).

D. Early Region E2a (61.5–75.1)

Early region E2a codes for the single-strand-specific, DNA-binding protein (DBP) (Figs. 5 and 6). This protein, discovered by van der Vliet and Levine (1973), is phosphorylated, has an apparent molecular weight of 72K, and is involved in DNA replication, in regulation of early and late gene expression, and in cell transformation (Ginsberg *et al.*, 1974; van der Vliet *et al.*, 1975, 1977; van der Vliet and Sussenbach, 1975; Carter and Ginsberg, 1976; Horwitz, 1978; Mayer and Ginsberg, 1977; Carter and Blanton, 1978; Nevins and Jensen-Winkler, 1980; Klessig and Grodzicker, 1979). The DBP genes of Ad2 and Ad5 have been analyzed in most detail. Therefore, the positions of strategic signals in the DBP gene are described in these sequences [Figs. 21 and 24 (Appendix B)]. It should be pointed out that the main bodies of the Ad2 and Ad5 genes are highly homologous. The promoter for region E2a is located at 75.1 m.u. on the viral l-strand and is used early in infection (Baker *et al.*, 1979). At this position, the sequence TCCTTAA (nucleotide 1490) (Fig. 21) is found, which is an aberrant type of TATA box. This promoter is probably also used for transcription of the E2b transcription unit. At later times in infection, transcription of the E2a region starts from a promoter at map position 72.0, where the TATA box TACAAATTT is found (nucleotide 352) (Fig. 21). A minor start of transcription at intermediate and late times is found at 87 m.u. Recently, an additional promoter sequence was identified about 26 nucleotides upstream from the major early promoter (Mathis *et al.*, 1981). The function of the minor promoter sequence is still unknown.

Depending on the time in infection, mRNA species from the E2a region contain two different short leaders. Depending on the time postinfection, one is derived from position 75.1 [nucleotides 1392–1458/1459 (67/68 nucleotides long)] or 72.0 (nucleotides 253–321/323 (69/71 nucleotides long)] [Fig. 21 (Appendix B)]. The other is derived from position 68.8 [nucleotides 2936–3012 (77 nucleotides long)] (Baker *et al.*, 1979; Kruijer *et al.*, 1981, 1983) [Fig. 24 (Appendix B)]. The main body of the E2a mRNAs is located between map positions 66.5 and 61.5 [Fig. 24 (nucleotides 2309–642)] (Kruijer *et al.*, 1981; Akusjärvi *et al.*, 1981). The site of polyadenylation has been localized at nucleotide 642, while the sequence AATAAA is found at position 661 (Akusjärvi *et al.*, 1981; Fraser *et al.*, 1982). From the nucleotide sequence of the E2a region and the structure of DBP mRNAs, it can be derived that all coding sequences of these RNAs are located within the main body (Kruijer *et al.*, 1981, 1982). Translation starts at ATG 2300 and runs to stop codon 713. The Ad2 and Ad5 mRNAs code for a protein of 529 amino acids (molecular weight 59K), while Ad12 DBP is 484 amino acid residues long (molecular weight 54K). Comparison of the Ad2 and Ad5 DBP nucleotide sequences reveals a high degree of homology, with only 9 amino acid differences in the corresponding amino acid sequences. However, Ad5 and Ad12 DBPs differ considerably in nucleotide and amino acid sequences. These differences are mainly located in the N-terminal part of the DBP molecule. In contrast, the C-terminal regions of the DBP molecules show a high degree of homology (80%) (Kruijer *et al.*, 1983). It is especially this part of the molecule that is involved in DNA replication (Ariga *et al.*, 1980; Kruijer *et al.*, 1981). The terminal part of DBP is involved in regulation of late expression (Klessig and Grodzicker, 1979; Kruijer *et al.*, 1981).

E. Late Region L4 (66.5–77.3)

This region includes a set of r-strand transcripts that code for a 100-kd protein (66.5–73.1), a 33-kd protein (71.5–74.0), and the precursor of polypeptide VIII (molecular weight 26K) (75.5–77.3) (Figs. 5 and 6). The indicated map positions have been determined by hybrid-arrest translation (Miller *et al.*, 1980). Polypeptide VIII (molecular weight 13K) is produced by proteolytic cleavage of its precursor during maturation of virions and is in virions associated with the hexon capsomers. The 100-kd protein is involved with folding of the hexon polypeptide chains into trimers (Ginsberg, personal communication), while the function of the 33-kd protein is still unknown. The four mRNAs that code for these proteins form the L4 family of late mRNAs and share the 3′-terminal sequences. The common polyadenylation site has been mapped at 78 map units.

Nucleotide sequences of this region have been determined in Ad2 and Ad5 DNA (Galibert *et al.*, 1979; Hérissé *et al.*, 1980; Kruijer *et al.*, 1981, 1982). Therefore, the strategic landmarks of the L4 proteins can be indicated at the nucleotide level. The acceptor splice point of the Ad5 100-kd polypeptide has been determined by reverse transcription of 100-kd mRNA and is located at nucleotide 2316 [Fig. 24 (Appendix B)] (Kruijer *et al.*, 1983). The polyadenylation site of the L4 mRNAs is mapped close to the sequence AATAAA at nucleotide 2572 [Fig. 21 (Appendix B)] (Fraser *et al.*, 1982). Comparison of the Ad5 sequence, which extends to coordinate 71.4, with the sequence of Ad2 indicates that nucleotides 3855–4107 of the Ad5 sequence (Fig. 24) are colinear with nucleotides 1–253 of the Ad2 sequence (Fig. 21). The frames in the overlapping sequences are identical and code, with a single exception, for identical amino acids. Using the combined sequences, it is possible to construct a hybrid 100-kd protein consisting of an amino-terminal part from Ad5 and a carboxy-terminal part of Ad2. The hypothetical hybrid protein consists of 805 amino acids and has an actual molecular weight of 89K.

The coding sequences of the 100 and 33-kd proteins partially overlap. However, since these proteins do not share tryptic peptides (Gambke and Deppert, 1981), it is most likely that they are encoded by r-strand transcripts in different ORFs. While the information for the 100-kd protein terminates at nucleotide 890, two ORFs (ORFs 1 and 2) can be distinguished in the other two reading frames, viz., ORF 1 from nucleotides 306 to 1191 (between stop codons 303 and 1191) and ORF 2 from nucleotides 1006 to 1492 (between stop codons 1003 and 1492 (Fig. 21). An ATG is present at nucleotide 411. Since one of the L4 mRNAs contains an internal splice that maps reasonably well in the region where these two ORFs overlap, it is likely that these regions code for the 33-kd protein. However, this has still to be proved by experimental data. One of the three short additional leaders for the fiber mRNA (x-leader) is also transcribed in this region from the r-strand (77.2–77.6). The x-leader has not been sequenced yet, but employing EM mapping data and typical RNA splice-site sequences, it has been inferred that this leader is transcribed from the r-strand from nucleotides 2215 to 2347. The l-strand between 66.5 and 77.3 units codes for the DBP mRNA leaders from positions 75.1,

72.0, and 68.8, respectively. The structure of the corresponding TATA boxes and individual leaders was described in Section VII.D.

F. Early Region E3 (76.6–86.0)

This region, located between coordinates 76.6 and 86.0, codes for a large number of r-strand transcripts and polypeptides (Fig. 6). At least six major species of mRNAs have been identified, coding for proteins of 13, 14, 15.5–16, and 19–21 kd, respectively (Lewis *et al.*, 1976; Harter *et al.*, 1976; Green *et al.*, 1979d; Ross *et al.*, 1980). The polypeptides of 19–21 kd are glycoproteins, which are associated with the membrane fraction (Persson *et al.*, 1979b, 1980a). Tryptic peptide analysis has shown that the 16-kd polypeptide is the unglycosylated precursor of the 19-kd protein (Persson *et al.*, 1980b).

The mRNAs from this region share sequences at their 5′ ends from coordinates 76.6 to 77.6, which are ligated to sequences starting at 78.6 m.u. The 3′ ends of the transcripts may vary.

Nucleotide sequence analysis of this region has revealed that a TATA box of the structure TATAA is located at nucleotide 1947 (76.7 m.u.), while transcription starts at nucleotide 1976/1978 (Baker and Ziff, 1981) [Fig. 21 (Appendix B)]. In region E3, two polyadenylation sites are present, one of which has been mapped at the nucleotide level (nucleotide 4148). Examination of the sequence of this region reveals that the sequence ATTAAA is found at position 4136. This sequence differs from the common hexanucleotide AATAAA that is found in all other Ad2 mRNAs associated with the polyadenylation site. In the sequence of region E3, the sequence AATAAA is located at nucleotide 5209, which fits very well with EM mapping data of some E3 mRNA species. However, for these messengers, the polyadenylation site has not yet been determined in detail.

The first ATG in the E3 region is found at position 2266, which suggests that E3 mRNAs have a 290-nucleotide-long untranslatable region at their 5′ ends. About 80 nucleotides downstream from this ATG lies a potential splice site, and this site fits very well with the position where the common leader sequence of E3 mRNAs has been mapped (positions 76.6–77.6). This leader sequence may code for 27 amino acid residues, which would be common to all E3 proteins. However, determination of the amino-terminal sequence of the unglycosylated 16-kd protein has shown that translation of the coding sequence of this protein starts at nucleotide 3179 and continues to nucleotide 3656. This codes for a protein of 159 amino acids with a molecular weight of 18.4K. Obviously, the ATG at position 2266 present in all E3 mRNAs is not recognized during translation. If the 3′ splice point of the first E3 intervening sequence is located around position 2840 (Hérissé *et al.*, 1980), this implies that the mRNA for the 16-kd protein has an untranslated region more than 700 nucleotides long. Region E3 contains a number of short URFs. A hypothetical organization of translation is indicated in Fig. 6. Unfortunately, no data are available to assign the URFs unambiguously to individual proteins. As described above, the only exception is the 16-kd protein. The function of the E3 proteins is completely obscure. In some adenovirus–simian virus 40 hybrids, this region is absent without affecting the viability of the virus. Apparently this region is nonessential for viral multiplication (for a review, see Tooze, 1981). In addition to the E3 proteins, this region codes for two additional leaders of the fiber mRNAs, viz., the y-leader (78.6–79.2) and the z-leader (84.7–85.1) (Chow and Broker, 1978). Only the y-leader has been sequenced and appears to be located at nucleotides 2741–2924 (Zain *et al.*, 1979a). Employing EM mapping data and the common sequences of RNA splice sites, it has been inferred that the z-leader is located at nucleotides 4805–4963 (Hérissé *et al.*, 1980).

G. Late Region L5 (86.0–91.3)

The L5 family of late transcripts consists of two major mRNA species that code for a single virion protein, the fiber (polypeptide IV). The main bodies of these RNAs map between coordinates 86.0 and 91.3 (Miller *et al.*, 1980) (Fig. 6). RNA from this region differs from all other late messengers in that it may contain, in addition to the common tripartite leader, additional leader sequences (x, y, and z) from map positions 77.2, 78.6, and 84.7 (Chow and Broker, 1978; Zain *et al.*, 1979a). The y-leader is the most abundant additional leader of fiber mRNA; however, even this leader is not present in all RNA species. It has been shown that the presence or absence of the y-leader does not influence the translation of fiber mRNA. Even in the absence of the y-leader, the mRNA can be translated normally to fiber protein in an *in vitro* translation system (Dunn *et al.*, 1978). The nucleotide sequence of this leader has been established to be 184 nucleotides long, and although an ATG is present in this sequence, it is obviously not employed and not required for appropriate translation of fiber mRNA.

The complete nucleotide sequence of region L5 has been established (Zain *et al.*, 1979a; Zain and Roberts, 1979; Hérissé and Galibert, 1981; Hérissé *et al.*, 1981; Gingeras *et al.*, 1982) [Fig. 21 (Appendix B)]. The 5′ end of the main body of the fiber mRNA is located at nucleotide 5395, adjacent to the codon of fiber mRNA at position 5397 (Zain and Roberts, 1979; Zain *et al.*, 1979a). The termination codon of the fiber gene is located at nucleotide 7143 and is part of the polyadenylation signal AATAAA at position 7141. The mRNA codes for 582 amino acid residues that contitute a protein with a theoretical molecular weight of 61.9K, which agrees very well with the apparent molecular weight of the fiber protein of 62K.

H. Early Region E4 (91.3–99.2)

Early region E4 messengers are transcribed from the viral l-strand between coordinates 91.3 and 99.0 and code for a large set of polypeptides (Fig. 6). The promoter of this region has been mapped at 99.2 m.u., while the 3′ ends of E4 RNAs have been localized at 91.3 m.u. (Berk and Sharp, 1978; Chow *et al.*, 1979a,b; Baker and Ziff, 1981; Hashimoto *et al.*, 1981).

All E4 mRNAs share their 5′- and 3′-terminal nucleotide sequences, but vary in the location of splice points (Berk and sharp, 1978; Chow *et al.*, 1979a; Kitchingman and Westphal, 1980). These messengers code for a number of polypeptides with molecular weights of 11, 13, 17, 19, 21, and 24K (Lewis *et al.*, 1976; Green *et al.*, 1979d; Ross *et al.*, 1980). As yet, these proteins have not been assigned unambiguously to individual mRNA species. Only the position of the acidic 11K polypeptide has been correlated to a specific region in the nucleotide sequence of this region (Hérissé *et al.*, 1981).

Besides the fact that the synthesis of the E4 proteins starts about 2 hr after infection, reaches a maximum around 3 hr, and then declines, these proteins seem to be nonessential for DNA replication, and their role is at present unknown.

Recently, the complete Ad2 nucleotide sequence of this region has been established (Shinagawa *et al.*, 1980; Hérissé *et al.*, 1981; Gingeras *et al.*, 1982) [Fig. 21 (Appendix B)], while for Ad5, the region between 97 and 100 m.u. has been determined (Steenbergh and Sussenbach, 1979) [Fig. 25.1 (Appendix B)]. At nucleotide 10,008 in the Ad2 sequence, a TATA box with the structure TATATATA can be recognized as part of a promoter sequence. Transcription begins with the sequence TTTTTA at nucleotides 9981–9976, leading to a heterogeneous array of starts (Baker and Ziff, 1981) (Fig. 21). All major species of mRNAs contain a leader sequence starting at the cap sites and probably terminating at nucleotide 9915, where a potential 5′ splice site is located. This leader sequence is devoid of ATG able to play a role in initiation of translation. Therefore, such a signal should be located in the body of the various mRNA species spliced to this leader sequence. At the other end of the sequence, transcription terminates close to an AATAAA sequence, which is located at position 7188. This is consistent with EM mapping data of E4 RNAs. It should be pointed out that transcription sometimes proceeds beyond this point to coordinate 61.5, leading to the production of a minor species of E2a mRNA (see Fig. 1).

The nucleotide sequence of the E4 region reveals that a large number of short URFs are present in all three reading frames.

Comparison of the nucleotide sequence and the mRNA mapping data indicates that there is a reasonably good correlation between the mapping data and potential donor and acceptor splice sites in the sequence. From the predicted structure of the various spliced mRNA species, a hypothetical translation pattern has been proposed (Hérissé *et al.*, 1981; Gingeras *et al.*, 1982). However, only in the case of the acidic 11K protein could its coding region be deduced with reasonable certainty from the nucleotide sequence to be located in URF 23. Further nucleotide sequence analysis of mRNAs and translation of individual mRNA species is required to determine unambiguously the relationship between individual RNAs and the corresponding proteins.

I. Unidentified Reading Frames

In addition to the URFs of early region E4, an additional ORF with a coding capacity of 12kd (ORF 3) is found in the viral l-strand transcripts (Fig. 6). This region is located between stop codons at positions 7193 and 6902 and starts with AAA (7190) (Fig. 21). At nucleotide 7166, the first ATG codon is found, while at nucleotide 6323, even the sequence ATTAAA is present, which resembles an aberrant type of polyadenylation signal also present in early region E3. It should be noted that although the major E4 transcription termination site has been mapped at 91.3 m.u., Nevins *et al.* (1980) have calculated that transcription termination takes place at 88.4 m.u., which corresponds very well with the sequence ATTAAA at nucleotide 6323 (Hérissé *et al.*, 1981). However, no mRNA species derived from this region are currently known. The same holds for two URFs in r-strand transcripts that code for proteins with theoretical molecular weights of 10.6 and 12K (URFs 26 and 27).

VIII. COMPARISON OF GENOMES AND CONCLUDING REMARKS

The organization of the adenovirus genome as described in Section VII has mainly been restricted to Ad2 because the most detailed information is available for this serotype. However, it should be emphasized that for all serotypes the structure of which has been investigated, the same overall organization has been observed. For a number of serotypes, nucleotide sequence data are available. These data are compiled in Appendix B, including the analysis of these sequences. For a number of genes, the nucleotide sequences have been compared, as well as the amino acid sequences of the corresponding proteins. Van Ormondt *et al.* (1980b) have analyzed the homology among the E1a regions of Ad5, Ad7, and Ad12, while Bos *et al.* (1981) and Kimura *et al.* (1981) have studied the homology of the E1b regions of Ad5 and Ad12. The IVa_2 and polypeptide IX genes of Ad2, Ad3, Ad5, and Ad7 have been compared (Dijkema *et al.*, 1981; Engler, 1981; Engler and van Bree, 1982), as well as the late leaders of Ad2, Ad3, and Ad7 (Engler *et al.*, 1981) and the E2b regions of Ad2 and Ad7 (Engler *et al.*, 1983). The redundancies of different serotypes were analyzed by Tolun *et al.* (1979) and Shinagawa and Padmanabhan (1980), while the DNA-binding protein genes of Ad2, Ad5, and Ad12 were compared by Kruijer *et al.* (1981, 1982, 1983).

Detailed analysis of the organization of the adenovirus genome reveals that the available coding information of this virus is used in a very economical fashion. Unraveling of the information at the nucleotide level reveals all kinds of peculiar properties in its organization. There are spliced and unspliced mRNA species (e.g., hexon and polypeptide IX RNA), overlapping termination codons and AATAAA signals (e.g., fiber and IVa_2 RNA), overlapping genes (e.g., the 33- and 100-kd proteins), and symmetrical transcription (120-kd protein and the 16-kd i-leader product). There are classic TATA boxes (e.g., E1a proteins) and polyadenylation signals (AATAAA) (hexon RNA) and aberrant sequences with the same function [TATA box TCCTT (E2a early promoter) and polyadenylation signal ATTAAA (region E3)].

In conclusion, the adenovirus genome is a microuniverse in itself, and the study of its organization and regulation of expression is a great joy and satisfaction for every scientist who dedicates herself or himself to the unraveling of its secrets.

ACKNOWLEDGMENTS. The author gratefully acknowledges the very valuable assistance of Mr. O. van Hien for providing computer facilities and Dr. T. Broker for maps and other information. Without their help, this chapter would never have been completed. He also thanks M. M. Kwant, M. G. ter Braak-Kuijk, W. van Driel, F. M. A. van Schaik, E. Simon, W. Kruijer, A. W. M. Rijnders, J. van der Rijst, and H. Laanen for technical assistance and Dr. P. C. van der Vliet for critical reading of the manuscript. He gratefully acknowledges the fact that his colleagues Drs. J. Engler, R. J. Roberts, K. Fujinaga, M. Horwitz, U. Pettersson, H. van Ormondt, R. Padmanabhan, B. Stillman, E. Ziff, and F. Galibert have made available new data prior to publication.

APPENDIX A: RESTRICTION ENDONUCLEASE CLEAVAGE MAPS

This appendix contains a compilation of restriction maps of the genomes of different adenovirus serotypes (Figs. 7–17). These maps have partially been published and partially been presented as personal communications. Most of these maps have been compiled before by Tooze (1981) and are redrawn with permission from the Cold Spring Harbor Laboratory Publication Department. The coordinates of the Ad1, Ad2, and Ad5 maps have been recalculated (Gingeras *et al.*, 1982). Details on the origin of the maps are indicated in Tooze (1981), unless otherwise stated.

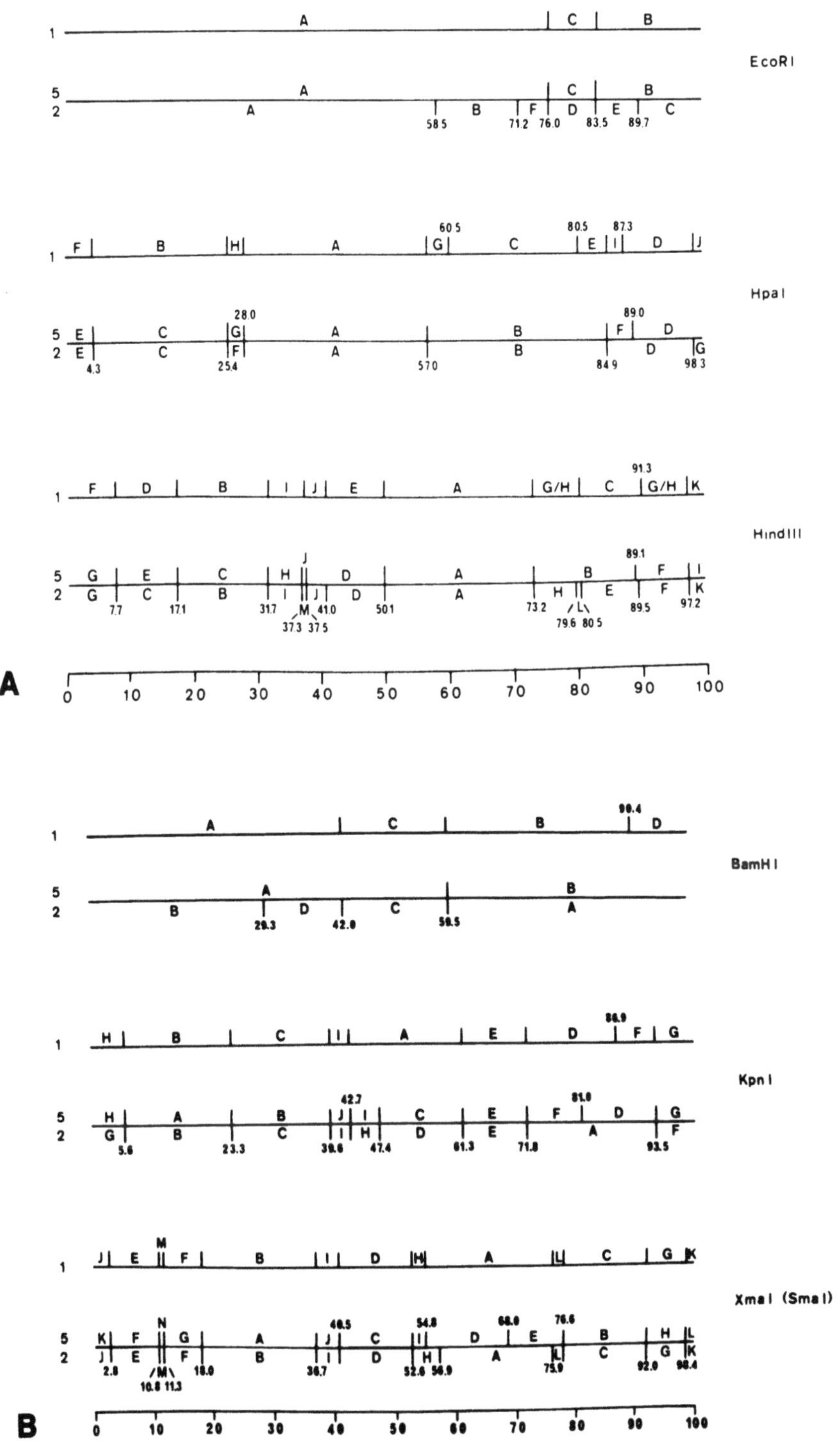

FIGURE 7A–D. Restriction endonuclease cleavage maps of Group C Ad1, Ad2, and Ad5.

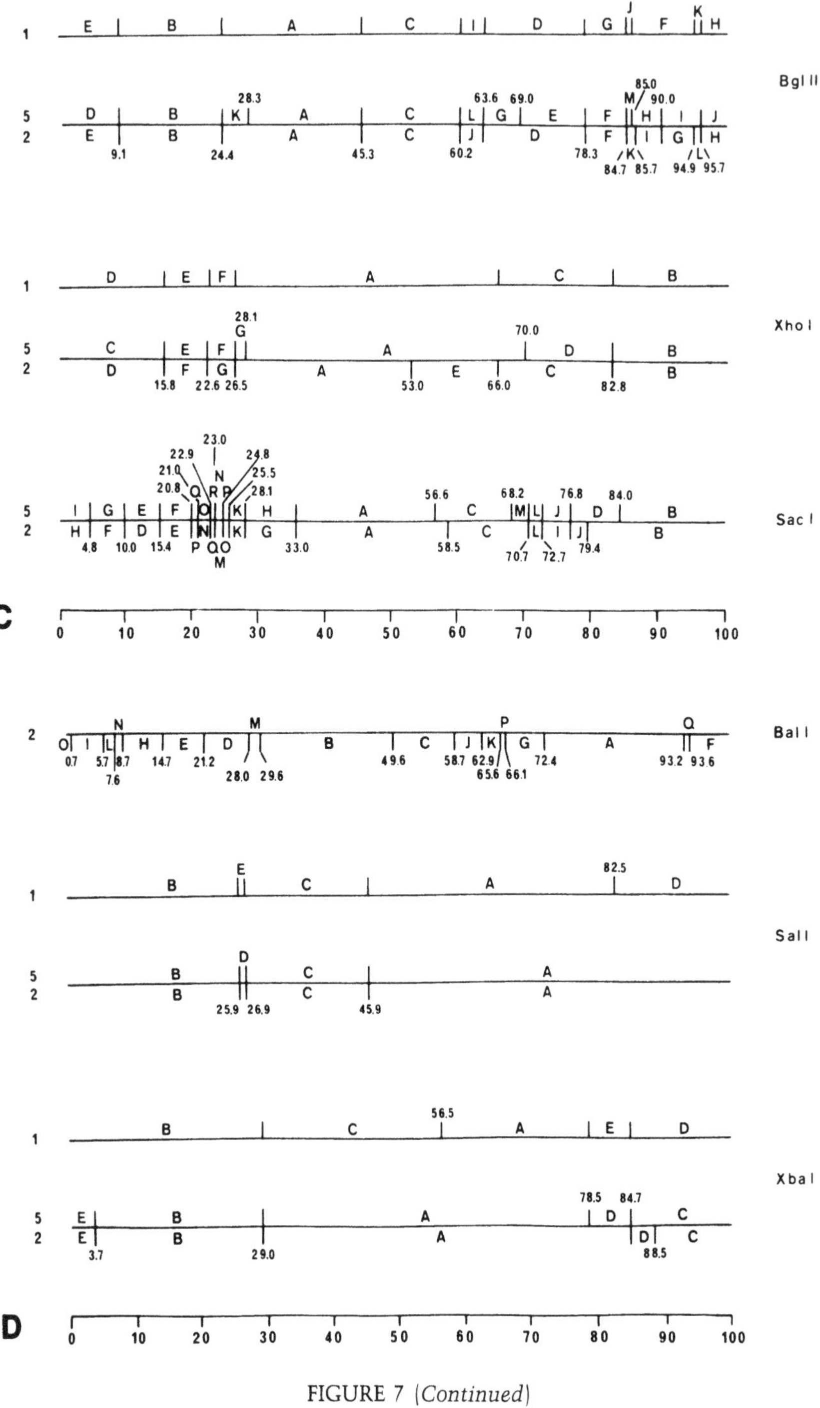

FIGURE 7 *(Continued)*

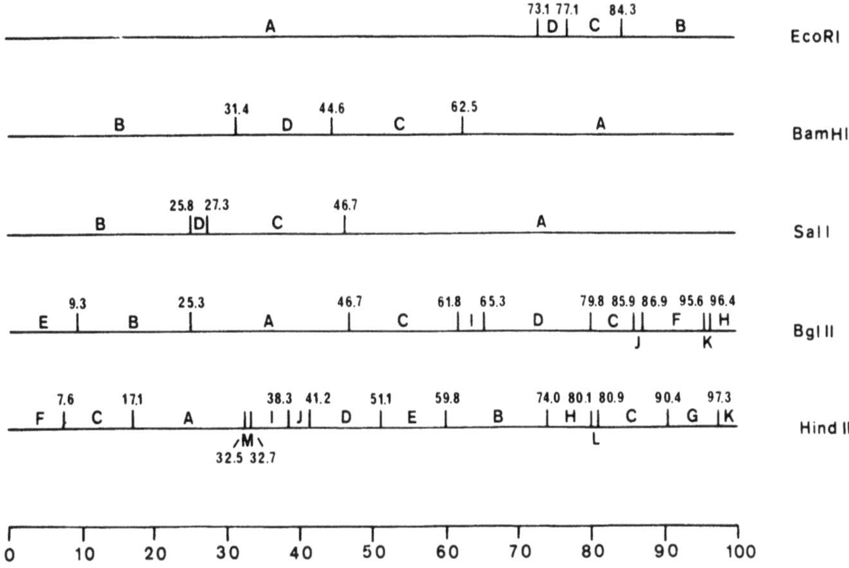

FIGURE 8. Restriction endonuclease cleavage maps of Group C Ad6. The maps were determined by Naroditsky *et al.* (1980) and oriented such that the transforming region is located at the left. The *Eco*RI map was determined by Forsblom *et al.* (1976).

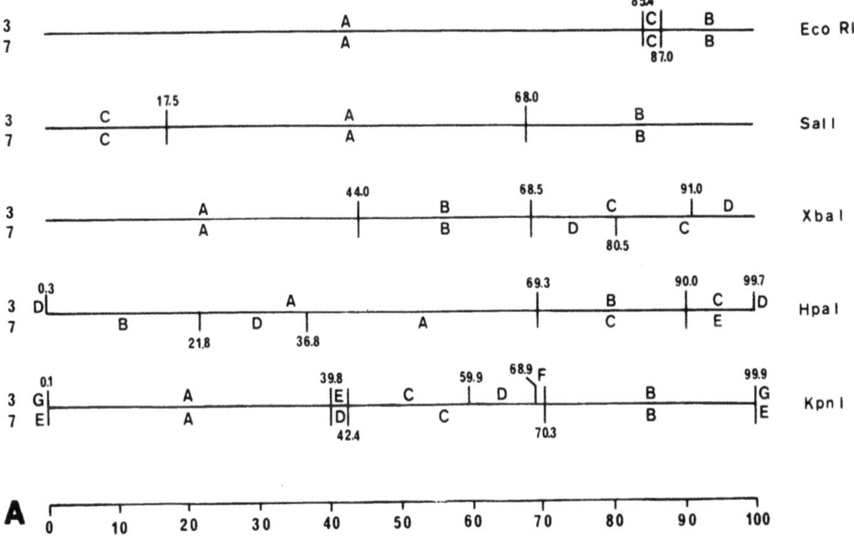

FIGURE 9A, B. Restriction endonuclease cleavage maps of Group B Ad3 and Ad7. The *Bst*EII and *Bcl*I maps were determined by R. Padmanabhan (personal communication).

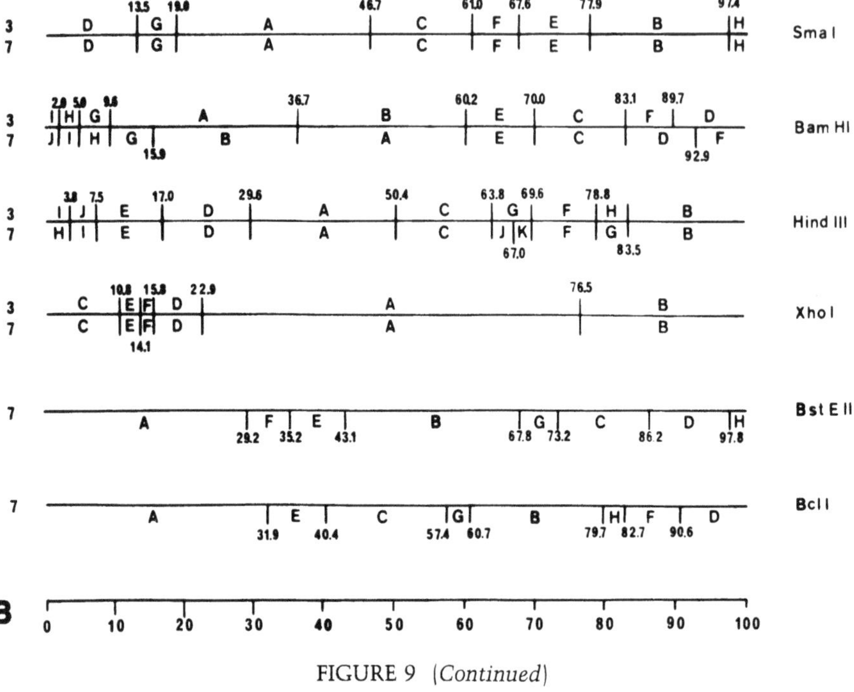

FIGURE 9 (*Continued*)

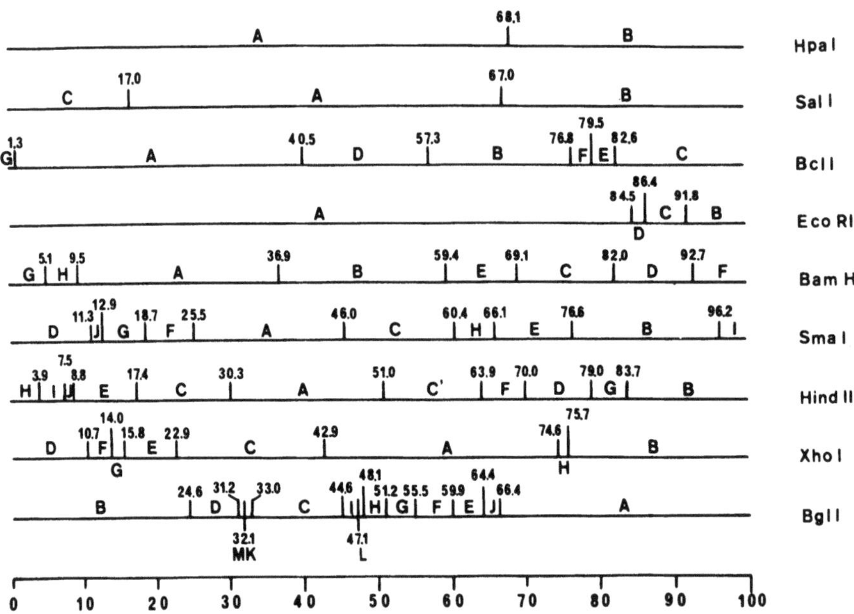

FIGURE 10. Restriction endonuclease cleavage maps of Group B Ad16 (Chang 79). These maps were determined by Varsanyi *et al.* (1977), Winberg and Hammarskjöld (1980), and Hammarskjöld and Winberg (personal communication).

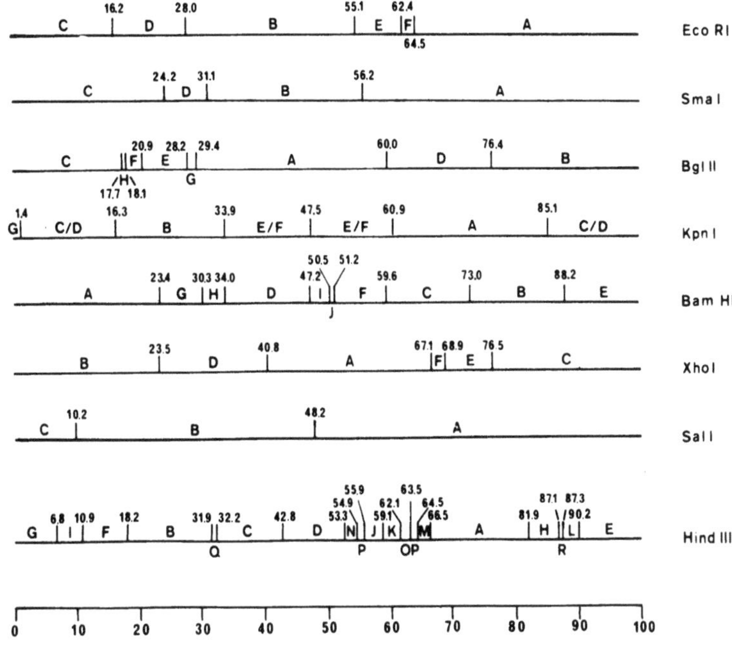

FIGURE 11. Restriction endonuclease cleavage maps of Group A Ad12 (Huie).

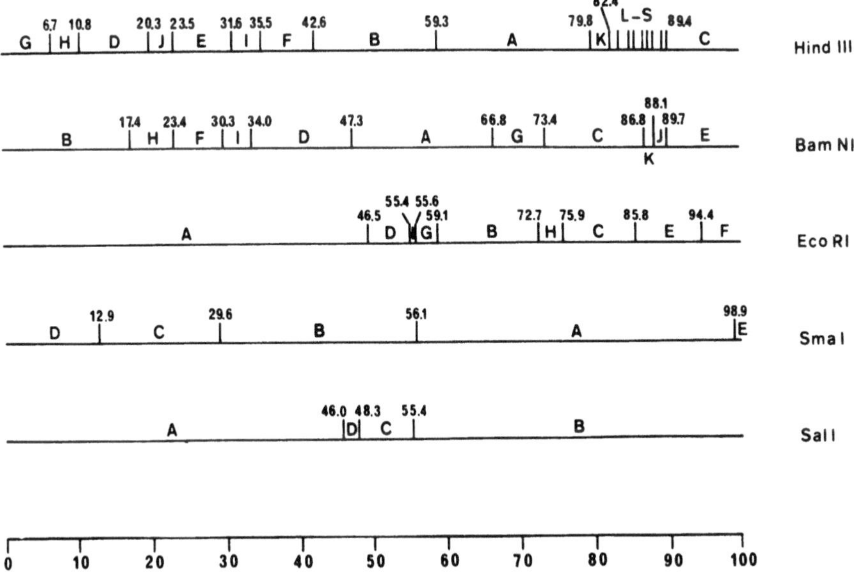

FIGURE 12. Restriction endonuclease cleavage maps of Group A Ad31 (strain 1315). The maps were determined by Y. Sawada, Y. Yamashita, F. Kamda, K. Sekikawa, and K. Fujinaga (personal communication).

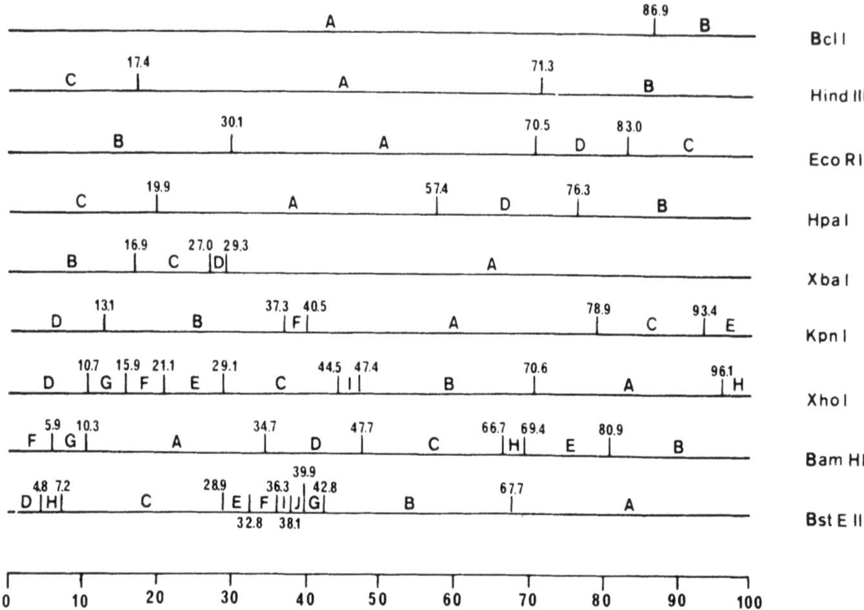

FIGURE 13. Restriction endonuclease cleavage maps of Group E Ad4. These maps were determined by Tokunaga *et al.* (1982).

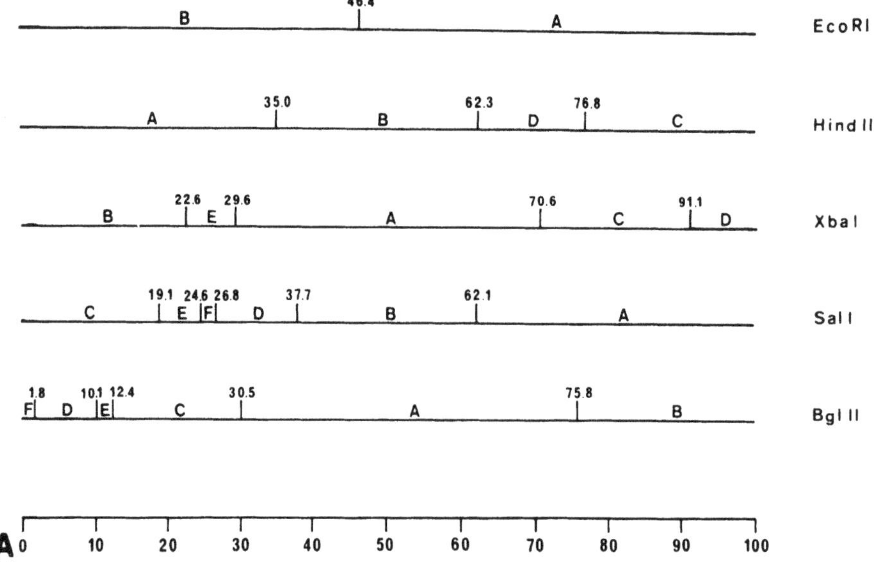

FIGURE 14A, B. Restriction endonuclease cleavage maps of simian adenovirus type 7. The *Eco*RI, *Sal*I, and *Bgl*II maps of simian adenovirus (strain C8) were determined by Naroditsky *et al.* (1980) and oriented with respect to the conventional genetic map by Ponomareva *et al.* (1979), who located the transforming region to the left. The other maps were determined by T. I. Tikchonenko and colleagues (personal communication).

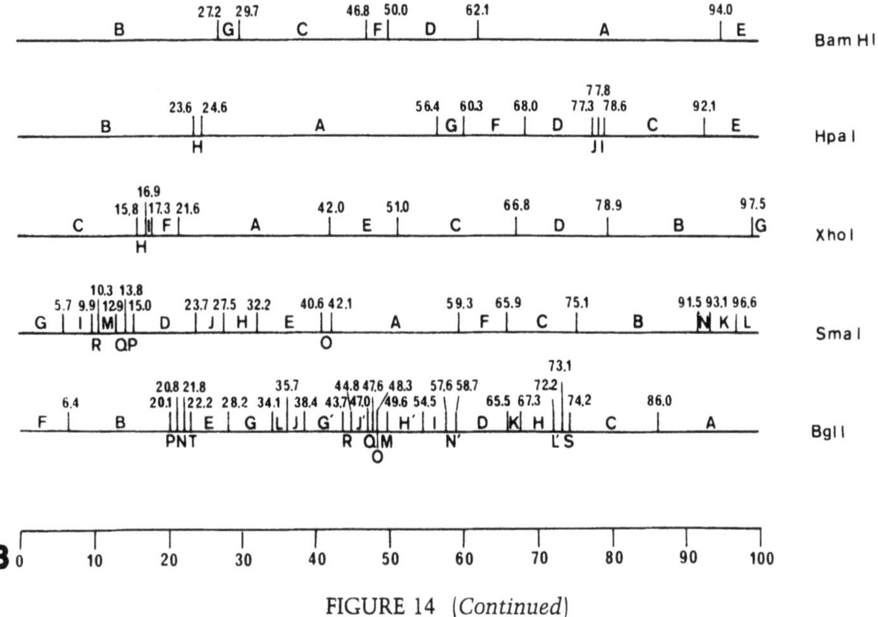

FIGURE 14 (*Continued*)

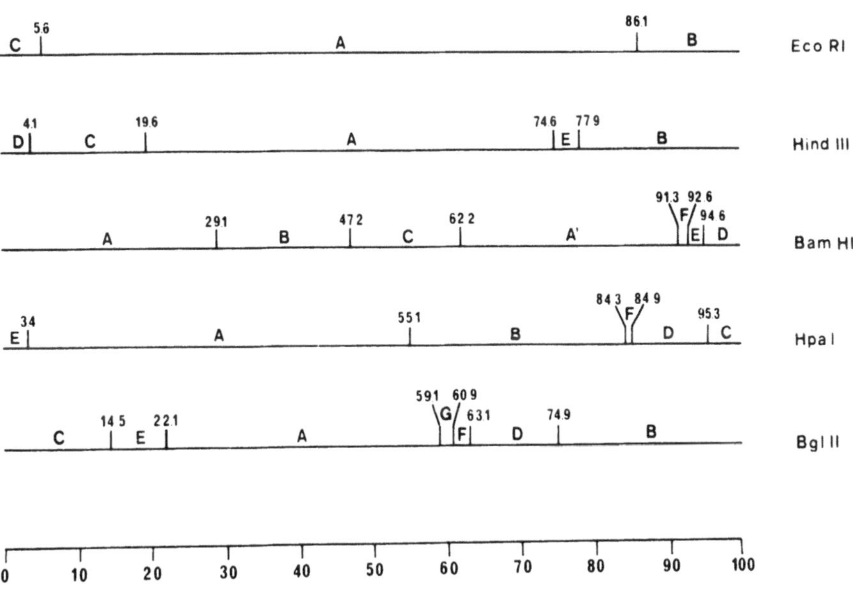

FIGURE 15. Restriction endonuclease cleavage maps of simian adenovirus type 20. These maps were determined by T. I. Tikchonenko and colleagues (personal communication).

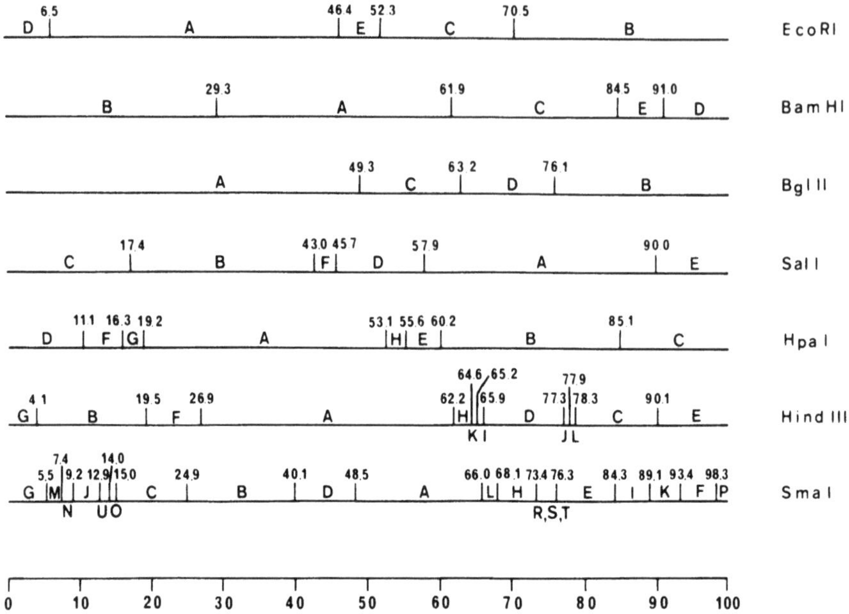

FIGURE 16. Restriction endonuclease cleavage maps of simian adenovirus type 30. The *Eco*RI and *Bgl*II maps were determined by Dimitrov *et al.* (1979). They were originally reported to be those of simian adenovirus type 38, and identification subsequently revised by Tikchonenko and colleagues (personal communication), who also determined the other maps.

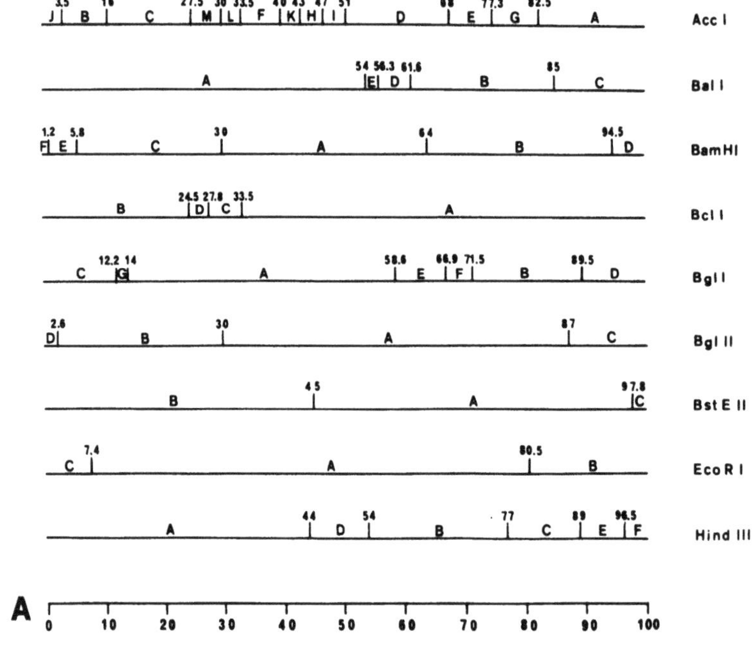

FIGURE 17A, B. Restriction endonuclease cleavage maps of mouse adenovirus type FL. These maps were determined by Larsen *et al.* (1979). For the orientation, see Larsen *et al.* (1979) and Temple *et al.* (1981).

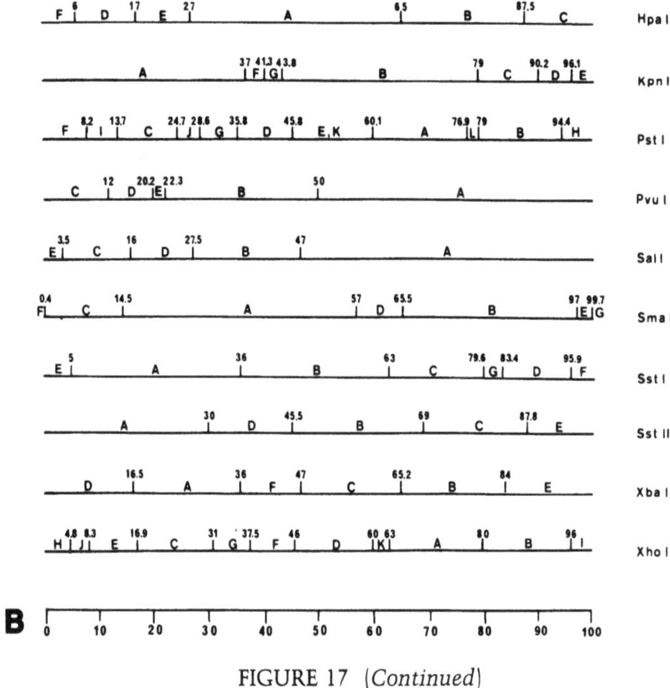

FIGURE 17 (*Continued*)

APPENDIX B: NUCLEOTIDE SEQUENCES

This appendix contains a compilation of nucleotide sequences partially published and partially presented as personal communications (Figs. 18–29). Since r-strand transcripts are homologous to the l-strand, the positions of important landmarks for r-strand transcripts are indicated in the l-strand sequence. Likewise, strategic sequences for l-strand transcripts are indicated in the r-strand sequence. The sequences of Ad2 and Ad5 are very homologous. Therefore, it has been supposed that specific signals identified in the sequence of one serotype also indicate the positions of these signals in the sequence of the other serotype. The positions of the inverted terminal repetition boundaries and start and termination codons of known coding regions are indicated, as well as the positions of 5′ and 3′ ends of mRNAs, splice points, and TATA boxes. The latter signals are supposed to be a constitutive part of transcriptional promoters. The sequences AATAAA and ATTAAA, which are found within about 30 nucleotides from the 3′ end of the mRNAs, are underlined. These sequences have been associated with polyadenylation. Open reading frames (ORFs), defined as regions between two termination codons in the same frame, have been indicated when the size exceeds 300 nucleotides. The same holds for unidentified reading frames (URFs) (regions that start with an ATG codon and terminate with one of the termination codons).

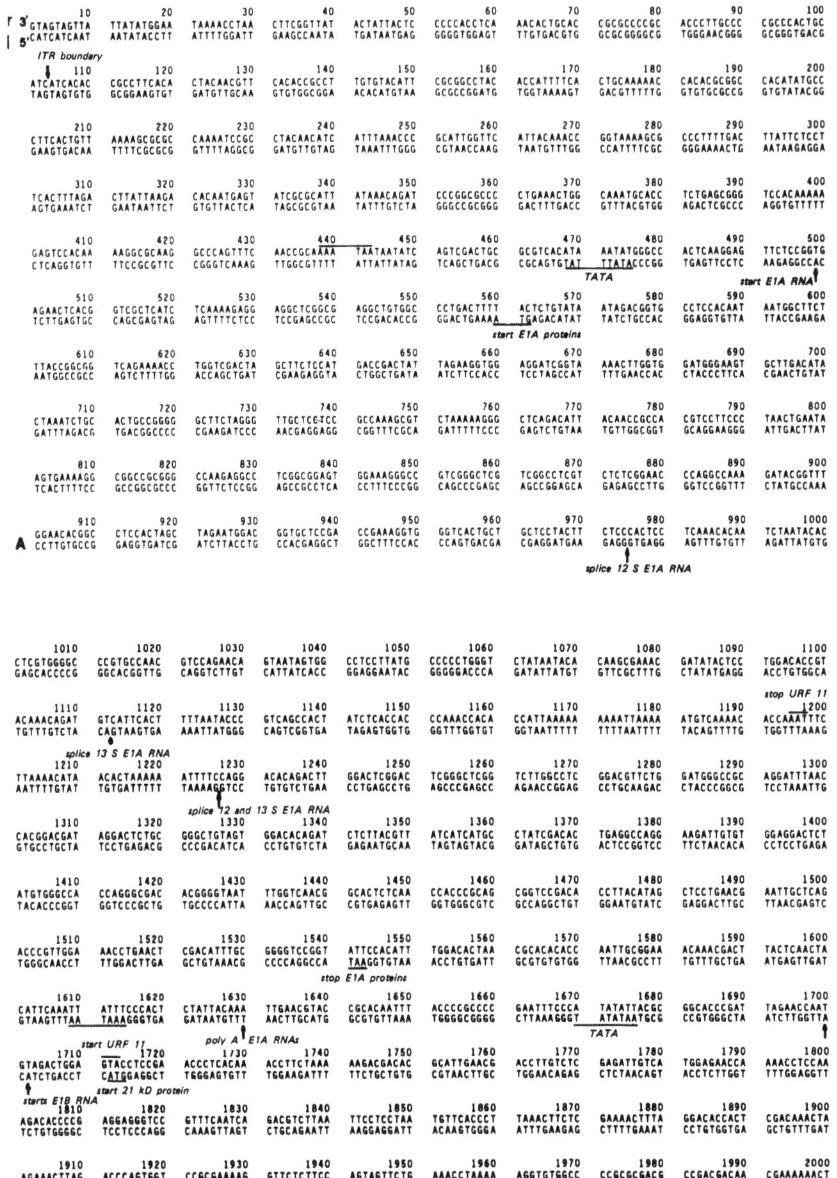

```
               10          20          30          40          50          60          70          80          90         100
3' GTAGTAGTTA  TTATATGGAA  TAAAACCTAA  CTTCGGTTAT  ACTATTACTC  CCCCACCTCA  AACACTGCAC  CGCGCCCCGC  ACCCTTGCCC  CGCCCACTGC
5' CATCATCAAT  AATATACCTT  ATTTTGGATT  GAAGCCAATA  TGATAATGAG  GGGGTGGAGT  TTGTGACGTG  GCGCGGGGCG  TGGGAACGGG  GCGGGTGACG
ITR boundary
          110         120         130         140         150         160         170         180         190         200
ATCATCACAC  CGCCTTCACA  CTACAACGTT  CACACCGCCT  TGTGTACATT  CGCGGCCTAC  ACCATTTTCA  CTGCAAAAAC  CACACGCGGC  CACATATGCC
TAGTAGTGTG  GCGGAAGTGT  GATGTTGCAA  GTGTGGCGGA  ACACATGTAA  GCGCCGGATG  TGGTAAAAGT  GACGTTTTTG  GTGTGCGCCG  GTGTATACGG

          210         220         230         240         250         260         270         280         290         300
CTTCACTGTT  AAAAGCGCGC  CAAAATCCGC  CTACAACATC  ATTTAAACCC  GCATTGGTTC  ATTACAAACC  GGTAAAAGCG  CCCTTTTGAC  TTATTCTCCT
GAAGTGACAA  TTTTCGCGCG  GTTTTAGGCG  GATGTTGTAG  TAAATTTGGG  CGTAACCAAG  TAATGTTTGG  CCATTTTCGC  GGGAAAACTG  AATAAGAGGA

          310         320         330         340         350         360         370         380         390         400
TCACTTTAGA  CTTATTAAGA  CACAATGAGT  ATCGCGCATT  ATAAACAGAT  CCCGGCGCCC  CTGAAACTGG  CAAATGCACC  TCTGAGCGGG  TCCACAAAAA
AGTGAAATCT  GAATAATTCT  GTGTTACTCA  TAGCGCGTAA  TATTTGTCTA  GGGCCGCGGG  GACTTTGACC  GTTTACGTGG  AGACTCGCCC  AGGTGTTTTT

          410         420         430         440         450         460         470         480         490         500
GAGTCCACAA  AAGGCGCAAG  GCCCAGTTTC  AACCGCAAAA  TAATAATATC  AGTCGACTGC  GCGTCACATA  AATATGGGCC  ACTCAAGGAG  TTCTCCGGTG
CTCAGGTGTT  TTCCGCGTTC  CGGGTCAAAG  TTGGCGTTTT  ATTATTATAG  TCAGCTGACG  CGCAGTGTAT  TTATACCCGG  TGAGTTCCTC  AAGAGGCCAC
                                                                               TATA                  start E1A RNA
          510         520         530         540         550         560         570         580         590         600
AGAACTCACG  GTCGCTCATC  TCAAAAGAGG  AGGCTCGGCG  AGGCTGTGGC  CCTGACTTTT  ACTCTGTATA  ATAGACGGTG  CCTCCACAAT  AATGGCTTCT
TCTTGAGTGC  CAGCGAGTAG  AGTTTTCTCC  TCCGAGCCGC  TCCGACACCG  GGACTGAAAA  TGAGACATAT  TATCTGCCAC  GGAGGTGTTA  TTACCGAAGA
                                                                 start E1A proteins
          610         620         630         640         650         660         670         680         690         700
TTACCGGCGG  TCAGAAAACC  TGGTCGACTA  GCTTCTCCAT  GACCGACTAT  TAGAAGGTGG  AGGATCGGTA  AAACTTGGTG  GATGGGAAGT  GCTTGACATA
AATGGCCGCC  AGTCTTTTGG  ACCAGCTGAT  CGAAGAGGTA  CTGGCTGATA  ATCTTCCACC  TCCTAGCCAT  TTTGAACCAC  CTACCCTTCA  CGAACTGTAT

          710         720         730         740         750         760         770         780         790         800
CTAAATCTGC  ACTGCCGGGG  GCTTCTAGGG  TTGCTCCTCC  GCCAAAGCGT  CTAAAAAGGG  CTCAGACATT  ACAACCGCCA  CGTCCTTCCC  TAACTGAATA
GATTTAGACG  TGACGGCCCC  CGAAGATCCC  AACGAGGAGG  CGGTTTCGCA  GATTTTTCCC  GAGTCTGTAA  TGTTGGCGGT  GCAGGAAGGG  ATTGACTTAT

          810         820         830         840         850         860         870         880         890         900
AGTGAAAAGG  CGGCCGCGGG  CCAAGAGGCC  TCGGCGGAGT  GGAAAGGGCC  GTCGGGCTCG  TCGGCCTCGT  CTCTCGGAAC  CCAGGCCAAA  GATACGGTTT
TCACTTTTCC  GCCGGCGCCC  GGTTCTCCGG  AGCCGCCTCA  CCTTTCCCGG  CAGCCCGAGC  AGCCGGAGCA  GAGAGCCTTG  GGTCCGGTTT  CTATGCCAAA

          910         920         930         940         950         960         970         980         990        1000
  GGAACACGGC  CTCCACTAGC  TAGAATGGAC  GGTGCTCCGA  CCGAAAGGTG  GGTCACTGCT  GCTCCTACTT  CTCCCACTCC  TCAAACACAA  TCTAATACAC
A CCTTGTGCCG  GAGGTGATCG  ATCTTACCTG  CCACGAGGCT  GGCTTTCCAC  CCAGTGACGA  CGAGGATGAA  GAGGGTGAGG  AGTTTGTGTT  AGATTATGTG
                                                                                     splice 12 S E1A RNA

         1010        1020        1030        1040        1050        1060        1070        1080        1090        1100
CTCGTGGGGC  CCGTGCCAAC  GTCCAGAACA  GTAATAGTGG  CCTCCTTATG  CCCCCTGGGT  CTATAATACA  CAAGCGAAAC  GATATACTCC  TGGACACCGT
GAGCACCCCG  GGCACGGTTG  CAGGTCTTGT  CATTATCACC  GGAGGAATAC  GGGGGACCCA  GATATTATGT  GTTCGCTTTG  CTATATGAGG  ACCTGTGGCA
                                                                                                          stop URF 11
         1110        1120        1130        1140        1150        1160        1170        1180        1190        1200
ACAAACAGAT  GTCATTCACT  TTTAATACCC  GTCAGCCACT  ATCTCACCAC  CCAAACCACA  CCATTAAAAA  AAAATTAAAA  ATGTCAAAAC  ACCAAATTTC
TGTTTGTCTA  CAGTAAGTGA  AAATTATGGG  CAGTCGGTGA  TAGAGTGGTG  GGTTTGGTGT  GGTAATTTTT  TTTTAATTTT  TACAGTTTTG  TGGTTTAAAG
     splice 13 S E1A RNA
         1210        1220        1230        1240        1250        1260        1270        1280        1290        1300
TTAAAACATA  ACACTAAAAA  ATTTTCCAGG  ACACAGACTT  GGACTCGGAC  TCGGGCTCGG  TCTTGGCCTC  GGACGTTCTG  GATGGGCCGC  AGGATTTAAC
AATTTTGTAT  TGTGATTTTT  TAAAAGGTCC  TGTGTCTGAA  CCTGAGCCTG  AGCCCGAGCC  AGAACCGGAG  CCTGCAAGAC  CTACCCGGCG  TCCTAAATTG
                   splice 12 and 13 S E1A RNA
         1310        1320        1330        1340        1350        1360        1370        1380        1390        1400
CACGGACGAT  AGGACTCTGC  GGGCTGTAGT  GGACACAGAT  CTCTTACGTT  ATCATCATGC  CTATCGACAC  TGAGGCCAGG  AAGATTGTGT  GGAGGACTCT
GTGCCTGCTA  TCCTGAGACG  CCCGACATCA  CCTGTGTCTA  GAGAATGCAA  TAGTAGTACG  GATAGCTGTG  ACTCCGGTCC  TTCTAACACA  CCTCCTGAGA

         1410        1420        1430        1440        1450        1460        1470        1480        1490        1500
ATGTGGGCCA  CCAGGGCGAC  ACGGGGTAAT  TTGGTCAACG  GCACTCTCAA  CCACCCGCAG  CGGTCCGACA  CCTTACATAG  CTCCTGAACG  AATTGCTCAG
TACACCCGGT  GGTCCCGCTG  TGCCCCATTA  AACCAGTTGC  CGTGAGAGTT  GGTGGGCGTC  GCCAGGCTGT  GGAATGTATC  GAGGACTTGC  TTAACGAGTC

         1510        1520        1530        1540        1550        1560        1570        1580        1590        1600
ACCCGTTGGA  AACCTGAACT  CGACATTTGC  GGGGTCCGGT  ATTCCACATT  TGGACACTAA  CGCACACACC  AATTGCGGAA  ACAAACGACT  TACTCAACTA
TGGGCAACCT  TTGGACTTGA  GCTGTAAACG  CCCCAGGCCA  TAAGGTGTAA  ACCTGTGATT  GCGTGTGTGG  TTAACGCCTT  TGTTTGCTGA  ATGAGTTGAT
                                                stop E1A proteins
         1610        1620        1630        1640        1650        1660        1670        1680        1690        1700
CATTCAAATT  ATTTCCCACT  CTATTACAAA  TTGAACGTAC  CGCACAATTT  ACCCCGCCCC  GAATTTCCCA  TATATTACGC  GGCACCCGAT  TAGAACCAAT
GTAAGTTTAA  TAAAGGGTGA  GATAATGTTT  AACTTGCATG  GCGTGTTAAA  TGGGGCGGGG  CTTAAAGGGT  ATATAATGCG  CCGTGGGCTA  ATCTTGGTTA
                                                                                  TATA
       start URF 11            poly A E1A RNAs
         1710        1720        1730        1740        1750        1760        1770        1780        1790        1800
GTAGACTGGA  GTACCTCCGA  ACCCTCACAA  ACCTTCTAAA  AAGACGACAC  GCATTGAACG  ACCTTGTCTC  GAGATTGTCA  TGGAGAACCA  AAACCTCCAA
CATCTGACCT  CATGGAGGCT  TGGGAGTGTT  TGGAAGATTT  TTCTGCTGTG  CGTAACTTGC  TGGAACAGAG  CTCTAACAGT  ACCTCTTGGT  TTTGGAGGTT
            start 21 kD protein
starts E1B RNA
         1810        1820        1830        1840        1850        1860        1870        1880        1890        1900
AGACACCCCG  AGGAGGGTCC  GTTTCAATCA  GACGTCTTAA  TTCCTCCTAA  TGTTCACCCT  TAAACTTCTC  GAAAACTTTA  GGACACCACT  CGACAAACTA
TCTGTGGGGC  TCCTCCCAGG  CAAAGTTAGT  CTGCAGAATT  AAGGAGGATT  ACAAGTGGGA  ATTTGAAGAG  CTTTTGAAAT  CCTGTGGTGA  GCTGTTTGAT

         1910        1920        1930        1940        1950        1960        1970        1980        1990        2000
  AGAAACTTAG  ACCCAGTGGT  CCGCGAAAAG  GTTCTCTTCC  AGTAGTTCTG  AAACCTAAAA  AGGTGTGGCC  CCGCGCGACG  CCGACGACAA  CGAAAAAACT
B TCTTTGAATC  TGGGTCACCA  GGCGCTTTTC  CAAGAGAAGG  TCATCAAGAC  TTTGGATTTT  TCCACACCGG  GGCGCGCTGC  GGCTGCTGTT  GCTTTTTTGA
```

FIGURE 18A–L. Nucleotide sequence of a region between coordinates 0.0 and 31.7 on Ad2 DNA. This sequence was determined by Gingeras *et al.* (1982b) and Aleström *et al.* (1982) (nucleotides 5776–11,558). Although base pair 9 (AT) is absent in the sequence determined by Gingeras and co-workers, most other investigators do find this base pair in their Ad2 strains. In the latter case, the terminal sequences of Ad2 and Ad5 are identical. To allow comparison of Ad2 and Ad5 sequences, this base pair was included in the sequence presented, which leads to a numbering one nucleotide higher than that of Gingeras and co-workers. Short stretches of this sequence were previously determined by Arrand and Roberts (1979), Shinagawa and Padmanabhan (1979), Perricaudet *et al.* (1979), Aleström *et al.* (1980), and Baker and Ziff (1980). The identification of strategic sequences is indicated in Sections VII.A and VII.B.

```
stop URF 10
      2010        2020        2030        2040        2050        2060        2070        2080        2090        2100
CAAAATATTT  CCTATTTACC  TCGCTTCTTT  GGGTAGACTC  GCCCCCCATG  GACGACCTAA  AAGACCGGTA  CGTAGACACC  TCTCGCCACC  ACTCTGTGTT
GTTTTATAAA  GGATAAATGG  AGCGAAGAAA  CCCATCTGAG  CGGGGGGTAC  CTGCTGGATT  TTCTGGCCAT  GCATCTGTGG  AGAGCGGTGG  TGAGACACAA
            start 55 kD protein

      2110        2120        2130        2140        2150        2160        2170        2180        2190        2200
CTTAGCGGAC  GATGACAACA  GAAGGCAGGC  GGGCCGTTAT  TATGGCTGCC  TCCTCGTTGT  CGTCCTCCTT  CGGTCCGCCG  CCGCCGCCGT  CCTCGTCTCG
GAATCGCCTG  CTACTGTTGT  CTTCCGTCCG  CCCGGCAATA  ATACCGACGG  AGGAGCAACA  GCAGGAGGAA  GCCAGGCGGC  GGCGGCGGCA  GGAGCAGAGC

                                                                                                start URF 10
      2210        2220        2230        2240        2250        2260        2270        2280        2290        2300
GGTACCTTGG  GCTCTCGGCC  GGACCTGGGA  GCCCTTACTT  ACAACATGTC  CACCGACTTG  ACAAAGGTCT  TGACTCTGCG  TAAAATTGGT  AATTGCTCCT
CCATGGAACC  CGAGAGCCGG  CCTGGACCCT  CGGGAATGAA  TGTTGTACAG  GTGGCTGAAC  TGTTTCCAGA  ACTGAGACGC  ATTTTAACCA  TTAACGAGGA
                                    stop 21 kD protein    splice 13 S E1B RNA

      2310        2320        2330        2340        2350        2360        2370        2380        2390        2400
ACCCGTCCCC  GATTTCCCCC  ATTTCTCCCT  CGCCCCCCGA  AGACTCCGAT  GTCTCCTCCG  ATCCTTAGAT  TGAAAATCGA  ATTACTGGTC  TGTGGCAGGA
TGGGCAGGGG  CTAAAGGGGG  TAAAGAGGGA  GCGGGGGGCT  TCTGAGGCTA  CAGAGGAGGC  TAGGAATCTA  ACTTTTAGCT  TAATGACCAG  ACACCGTCCT

      2410        2420        2430        2440        2450        2460        2470        2480        2490        2500
CTCACACAAT  GAAAAGTCGT  CTAATTCCTA  TTAACGCGAT  TACTCGAACT  AGACGACCGC  GTCTTCATAA  GGTATCTCGT  CGACTGGTGA  ATGACCGACG
GAGTGTGTTA  CTTTTCAGCA  GATTAAGGAT  AATTGCGCTA  ATGAGCTTGA  TCTGCTGGCG  CAGAAGTATT  CCATAGAGCA  GCTGACCACT  TACTGGCTGC

      2510        2520        2530        2540        2550        2560        2570        2580        2590        2600
TCGGTCCCCT  ACTAAAACTC  CTCCGATAAT  CCCATATACG  TTTCCACCGT  GAATCCGGTC  TAACGTTCAT  GTTCTAATCG  TTTGAACATT  TATAGTCCTT
AGCCAGGGGA  TGATTTTGAG  GAGGCTATTA  GGGTATATGC  AAAGGTGGCA  CTTAGGCCAG  ATTGCAAGTA  CAAGATTAGC  AAACTTGTAA  ATATCAGGAA

      2610        2620        2630        2640        2650        2660        2670        2680        2690        2700
AACAACGATG  TAAAGACCCT  TGCCCCGGCT  CCACCTCTAT  CTATGCCTCC  TATCCCACCG  GAAATCTACA  TCGTACTATT  TATACACCGG  CCCCCACGAA
TTGTTGCTAC  ATTTCTGGGA  ACGGGGCCGA  GGTGGAGATA  GATACGGAGG  ATAGGGTGGC  CTTTAGATGT  AGCATGATAA  ATATGTGGCC  GGGGGTGCTT

      2710        2720        2730        2740        2750        2760        2770        2780        2790        2800
CCGTACCTGC  CCCACCAATA  ATACTTACAC  TCCAAATGAC  CAGGGTTAAA  ATCGCCATGC  CAAAAGGACC  GGTTATGGTT  AGAATAGGAT  GTGCCACATT
GGCATGGACG  GGGTGGTTAT  TATGAATGTG  AGGTTTACTG  GTCCCAATTT  TAGCGGTACG  GTTTTCCTGG  CCAATACCAA  TCTTATCCTA  CACGGTGTAA

      2810        2820        2830        2840        2850        2860        2870        2880        2890        2900
CGAAGATACC  CAAATTGTTA  TGGACACACC  TTCGGACCTG  GCTACATTCC  CAAGCCCCGA  CACGGAAAAT  GACGACGACC  TTCCCCCACC  ACACAGCGGG
GCTTCTATGG  GTTTAACAAT  ACCTGTGTGG  AAGCCTGGAC  CGATGTAAGG  GTTCGGGGCT  GTGCCTTTTA  CTGCTGCTGG  AAGGGGGTGG  TGTGTCGCCC

      2910        2920        2930        2940        2950        2960        2970        2980        2990        3000
GTTTTCGTCC  CGAAGTTAAT  TCTTTACGGA  CAAACTTTCC  ACATGGAACC  CATAGGACAG  ACTCCCATTG  AGGTCCCACG  CGGTGTTACA  CCGGAGGCTG
C CAAAAGCAGG  GCTTCAATTA  AGAAATGCCT  GTTTGAAAGG  TGTACCTTGG  GTATCCTGTC  TGAGGGTAAC  TCCAGGGTGC  GCCACAATGT  GGCCTCCGAC


      3010        3020        3030        3040        3050        3060        3070        3080        3090        3100
ACACCAACGA  AGTACGATCA  CTTTTCGCAC  CGACACTAAT  TCGTATTGTA  CCACACACCG  TTGACGCTCC  TGTCCCGGAG  AGTCTACGAC  TGGACGAGCC
TGTGGTTGCT  TCATGCTAGT  GAAAAGCGTG  GCTGTGATTA  AGCATAACAT  GGTGTGTGGC  AACTGCGAGG  ACAGGGCCTC  TCAGATGCTG  ACCTGCTCGG

      3110        3120        3130        3140        3150        3160        3170        3180        3190        3200
TGCCGTTGAC  AGTGAACGAC  TTCTGGTAAG  TGCATCGGTC  GGTGAGAGCG  TTCCGGACCG  GTCACAAACT  CGTGTTGTAT  GACTGGGCGA  CAAGGAACGT
ACGGCAACTG  TCACTTGCTG  AAGACCATTC  ACGTAGCCAG  CCACTCTCGC  AAGGCCTGGC  CAGTGTTTGA  GCACAACATA  CTGACCCGCT  GTTCCTTGCA

      3210        3220        3230        3240        3250        3260        3270        3280        3290        3300
AAACCCATTG  TCCTCCCCCC  ACAAGGATGG  AATGGTTACG  TTAAACTCAG  TGTGATTCTA  TAACGAACTC  GGGCTCTCGT  ACAGGTTCCA  CTTGGACTTG
TTTGGGTAAC  AGGAGGGGGG  TGTTCCTACC  TTACCAATGC  AATTTGAGTC  ACACTAAGAT  ATTGCTTGAG  CCCGAGAGCA  TGTCCAAGGT  GAACCTGAAC

      3310        3320        3330        3340        3350        3360        3370        3380        3390        3400
CCCCACAAAC  TGTACTGGTA  CTTCTAGACC  TTCCACGACT  CCATGCTACT  CTGGGCGTGG  TCCACGTCTG  GGACGCTCAC  ACCGCCATTT  GTATAATCCT
GGGGTGTTTG  ACATGACCAT  GAAGATCTGG  AAGGTGCTGA  GGTACGATGA  GACCCGCACC  AGGTGCAGAC  CCTGCGAGTG  TGGCGGTAAA  CATATTAGGA

      3410        3420        3430        3440        3450        3460        3470        3480        3490        3500
TGGTCGGACA  CTACGACCTA  CACTGGCTCC  TCGACTCCGG  GCTAGTGAAC  CACGACCGGA  CGTGGGCGCG  ACTCAAACCG  AGATCGCTAC  TTCTATGTCT
ACCAGCCTGT  GATGCTGGAT  GTGACCGAGG  AGCTGAGGCC  CGATCACTTG  GTGCTGGCCT  GCACCCGCGC  TGAGTTTGGC  TCTAGCGATG  AAGATACAGA

      3510        3520        3530        3540        3550        3560        3570        3580        3590        3600
AACTCCATGA  CTTTACACAC  CCGCACCGAA  TTCCCACCCT  TTCTTATATA  TTCCACCCCC  AGAGTACATC  AAAACATAGA  CAAAACGTCG  TCGGCGGCGG
TTGAGGTACT  GAAATGTGTG  GGCGTGGCTT  AAGGGTGGGA  AAGAATATAT  AAGGTGGGGG  TCTCATGTAG  TTTTGTATCT  GTTTTGCAGC  AGCCGCCGCC
stop   splice 22 S E1B RNA                                TATA                 starts protein IX  RNA   splice 13 and 22 E1B RNA
55 kD
protein 3610      3620        3630        3640        3650        3660        3670        3680        3690        3700
TACTCGCGGT  TGAGCAAACT  ACCTTCGTAA  CACTCGAGTA  TAAACTGTTG  CGCGTACGGG  GGTACCCGGC  CCCACGCAGT  CTTACACTAC  CCGAGGTCGT
ATGAGCGCCA  ACTCGTTTGA  TGGAAGCATT  GTGAGCTCAT  ATTTGACAAC  GCGCATGCCC  CCATGGGCCG  GGGTGCGTCA  GAATGTGATG  GGCTCCAGCA
start protein IX

      3710        3720        3730        3740        3750        3760        3770        3780        3790        3800
AACTACCAGC  GGGGCAGGAC  GGGCGTTTGA  GATGATGGAA  CTGGATGCTC  TGGCACAGAC  CTTGCGGCAA  CCTCTGACGT  CGGAGGCGGC  GGCGAAGTCG
TTGATGGTCG  CCCCGTCCTG  CCCGCAAACT  CTACTACCTT  GACCTACGAG  ACCGTGTCTG  GAACGCCGTT  GGAGACTGCA  GCCTCCGCCG  CCGCTTCAGC

      3810        3820        3830        3840        3850        3860        3870        3880        3890        3900
GCGACGTCGG  TGGCGGGCGC  CCTAACACTG  ACTGAAACGA  AAGGACTCGG  GCGAACGTTC  GTCACGTCGA  AGGGCAAGTA  GGCGGGCGCT  ACTGTTCAAC
CGCTGCAGCC  ACCGCCCGCG  GGATTGTGAC  TGACTTTGCT  TTCCTGAGCC  CGCTTGCAAG  CAGTGCAGCT  TCCCGTTCAT  CCGCCCGCGA  TGACAAGTTG

      3910        3920        3930        3940        3950        3960        3970        3980        3990        4000
TGCCGAGAAA  ACCGTGTTAA  CCTAAGAAAC  TGGGCCCTTG  AATTACAGCA  AAGAGTCGTC  GACAACCTAG  ACGCGGTCGT  CCAAAGACGG  GACTTCCGAA
D ACGGCTCTTT  TGGCACAATT  GGATTCTTTG  ACCCGGGAAC  TTAATGTCGT  TTCTCAGCAG  CTGTTGGATC  TGCGCCAGCA  GGTTTCTGCC  CTGAAGGCTT
```

FIGURE 18 (Continued)

```
                                                            poly A protein IVa2 RNA                              stop protein IVa2
      4010       4020       4030       4040       4050       4060       4070       4080       4090       4100
GGAGGGGAGG GTTACGCCAA ATTTTGTATT TATTTTTGGT CTGAGACAAA CCTAAAACTA GTTCGTTCAC AGAACGACAG AAATAAATCC CCAAAACGCG
CCTCCCCTCC CAATGCGGTT TAAAACATAA ATAAAAACCA GACTCTGTTT GGATTTTGAT CAAGCAAGTG TCTTGCTGTC TTTATTTAGG GGTTTTGCGC
                      stop protein IX                              poly A E1B RNAs

      4110       4120       4130       4140       4150       4160       4170       4180       4190       4200
CGCGCCATCC GGGCCCTGGT CGCCAGAGCC AGCAACTCCC AGGACACATA AAAAAGGTCC TGCACCATTT CCACTGAGAC CTACAAGTCT ATGTACCCGT
GCGCGGTAGG CCCGGGACCA GCGGTCTCGG TCGTTGAGGG TCCTGTGTAT TTTTTCCAGG ACGTGGTAAA GGTGACTCTG GATGTTCAGA TACATGGGCA

      4210       4220       4230       4240       4250       4260       4270       4280       4290       4300
ATTCGGGCAG AGACCCCACC TCCATCGTGG TGACGTCTCG AAGTACGACG CCCCACCACA ACATCTACTA GGTCAGCATC GTCCTCGCGA CCCGCACCAC
TAAGCCCGTC TCTGGGGTGG AGGTAGCACC ACTGCAGAGC TTCATGCTGC GGGGTGGTGT TGTAGATGAT CCAGTCGTAG CAGGAGCGCT GGGCGTGGTG

      4310       4320       4330       4340       4350       4360       4370       4380       4390       4400
GGATTTTTAC AGAAAGTCAT CGTTCGACTA ACGGTCCCCG TCCGGGAACC ACATTCACAA ATGTTTCGCC AATTCGACCC TACCCACGTA TGCACCCCTA
CCTAAAAATG TCTTTCAGTA GCAAGCTGAT TGCCAGGGGC AGGCCCTTGG TGTAAGTGTT TACAAAGCGG TTAAGCTGGG ATGGGTGCAT ACGTGGGGAT

      4410       4420       4430       4440       4450       4460       4470       4480       4490       4500
TACTCTACGT AGAACCTGAC ATAAAAATCC AACCGATACA AGGGTCGGTA TAGGGAGGCC CCTAAGTACA ACACGTCTTG GTGGTCGTGT CACATAGGCC
ATGAGATGCA TCTTGGACTG TATTTTTAGG TTGGCTATGT TCCCAGCCAT ATCCCTCCGG GGATTCATGT TGTGCAGAAC CACCAGCACA GTGTATCCGG

      4510       4520       4530       4540       4550       4560       4570       4580       4590       4600
ACGTGAACCC TTTAAACAGT ACATCGAATC TTCCTTTACG CACCTTCTTG AACCTCTGCG GGAACACTGG AGGCTCTAAA AGGTACGTAA GCAGGTATTA
TGCACTTGGG AAATTTGTCA TGTAGCTTAG AAGGAAATGC GTGGAAGAAC TTGGAGACGC CCTTGTGACC TCCGAGATTT TCCATGCATT CGTCCATAAT

      4610       4620       4630       4640       4650       4660       4670       4680       4690       4700
CTACCGTTAC CCGGGTGCCC GCCGCCGGAC CCGCTTCTAT AAAGACCCTA GTGATTGCAG TATCAACACA AGGTCCTACT CTAGCAGTAT CCGGTAAAAA
GATGGCAATG GGCCCACGGG CGGCGGCCTG GGCGAAGATA TTTCTGGGAT CACTAACGTC ATAGTTGTGT TCCAGGATGA GATCGTCATA GGCCATTTTT

      4710       4720       4730       4740       4750       4760       4770       4780       4790       4800
TGTTTCGCGC CCGCCTCCCA CGGTCTGACG CCATATTACC AAGGTAGGCC GGGTCCCCGC ATCAATGGGA GTGTCTAAAC GTAAAGGGTG CGAAACTCAA
ACAAAGCGCG GGCGGAGGGT GCCAGACTGC GGTATAATGG TTCCATCCGG CCCAGGGGCG TAGTTACCCT CACAGATTTG CATTTCCCAC GCTTTGAGTT

      4810       4820       4830       4840       4850       4860       4870       4880       4890       4900
GTCTACCCCC CTAGTACAGA TGGACGCCCC GCTACTTCTT TTGGCAAAGG CCCCATCCCC TCTAGTCGAC CCTTCTTTCG TCCAAGGACT CGTCGACGCT
CAGATGGGGG GATCATGTCT ACCTGCGGGG CGATGAAGAA AACCGTTTCC GGGGTAGGGG AGATCAGCTG GGAAGAAAGC AGGTTCCTGA GCAGCTGCGA

      4910       4920       4930       4940       4950       4960       4970       4980       4990       5000
GAATGGCGTC GGCCACCCGG GCATTTAGTG TGGATAATGG CCGACGTTGA CCATCAATTC TCTCGACGTC GACGGCAGTA GGGACTCGTC CCCCCGGTGA
E CTTACCGCAG CCGGTGGGCC CGTAAATCAC ACCTATTACC GGCTGCAACT GGTAGTTAAG AGAGCTGCAG CTGCCGTCAT CCCTGAGCAG GGGGGCCACT


      5010       5020       5030       5040       5050       5060       5070       5080       5090       5100
AGCAATTCGT ACAGGGACTG AACGTACAAA AGGGACTGGT TTACGCGGTC TTCCGCGAGC GGCGGGTCGC TATCGTCAAG AACGTTCCTT CGTTTCAAAA
TCGTTAAGCA TGTCCCTGAC TTGCATGTTT TCCCTGACCA AATGCGCCAG AAGGCGCTCG CCGCCCAGCG ATAGCAGTTC TTGCAAGGAA GCAAAGTTTT
                                                                                            stop 120 kD protein
      5110       5120       5130       5140       5150       5160       5170       5180       5190       5200
AGTTGCCAAA CTCCGGCAGG CGGCATCCGT ACGAAAACTC GCAAACTGGT TCGTCAAGGT CCGCCAGGGT GTCGAGCCAG TGCACGAGAT GCCGTAGAGC
TCAACGGTTT GAGGCCGTCC GCCGTAGGCA TGCTTTTGAG CGTTTGACCA AGCAGTTCCA GGCGGTCCCA CAGCTCGGTC ACGTGCTCTA CGGCATCTCG

      5210       5220       5230       5240       5250       5260       5270       5280       5290       5300
TAGGTCGTAT AGAGGAGCAA AGCGCCCAAC CCCGCCGAAA GCGACATGCC GTCATCAGCC ACGAGCAGGT CTGCCCGGTC CCAGTACAGA AAGGTGCCCG
ATCCAGCATA TCTCCTCGTT TCGCGGGTTG GGGCGGCTTT CGCTGTACGG CAGTAGTCGG TGCTCGTCCA GACGGGCCAG GGTCATGTCT TTCCACGGGC
                      stop URF 9
      5310       5320       5330       5340       5350       5360       5370       5380       5390       5400
CGTCCCAGGA GCAGTCGCAT CAGACCCAGT GCCACTTCCC CACGCGAGGC CCGACGCGCG ACCGGTCCCA CGCGAACTCC GACCAGGACG ACCACGACTT
GCAGGGTCCT CGTCAGCGTA GTCTGGGTCA CGGTGAAGGG GTGCGCTCCG GGCTGCGCGC TGGCCAGGGT GCGCTTGAGG CTGGTCCTGC TGGTGCTGAA
           splice IVa2 RNA
      5410       5420       5430       5440       5450       5460       5470       5480       5490       5500
CGCGACGGCC AGAAGCGGGA CGCGCAGCCG GTCCATCGTA AACTGGTACC ACAGTATCAG GTGGGGAGGC GCGCACCGGG AACCGGCGTC GAACGGGAAC
GCGCTGCCGG TCTTCGCCCT GCGCGTCGGC CAGGTAGCAT TTGACCATGG TGTCATAGTC CACCCCTCCG CGCGTGGCCC TTGGCCGCAG CTTGCCCTTG

      5510       5520       5530       5540       5550       5560       5570       5580       5590       5600
CTCCTCCGCG GCGTGCTCCC CGTCACGTCT GAAAATTCCC GCATCTCGAA CCCGCGCTCT TTATGGCTAA GGCCCCTCAT CCGTAGGCGC GGCGTCCGGG
GAGGAGGCGC CGCACGAGGG GCAGTGCAGA CTTTTAAGGG CGTAGAGCTT GGGCGCGAGA AATACCGATT CCGGGGAGTA GGCATCCGCG CCGCAGGCCC
                                                                        start URF 9            splice IVa2 RNA
      5610       5620       5630       5640       5650       5660       5670       5680       5690       5700
GCGTCTGCCA GAGCGTAAGG TGCTCGGTCC ACTCGAGACC GGCAAGCCCC AGTTTTTGGT CCAAAGGGGG TACGAAAAAC TACGCAAAGA ATGGAGACCA
CGCAGACGGT CTCGCATTCC ACGAGCCAGG TGAGCTCTGG CCGTTCGGGG TCAAAAACCA GGTTTCCCCC ATGCTTTTTG ATGCGTTTCT TACCTCTGGT
start IVa2 protein
      5710       5720       5730       5740       5750       5760       5770       5780       5790       5800
AAGGTACTCG * GCCACAGGTG CGAGCCACTG CTTTTCCGAC AGGCACAGGG GCATATGTCT GAACTCTCCG GACAGGAGCT CGCCACAAGG CGCCAGGAGG
TTCCATGAGC CGGTGTCCAC GCTCGGTGAC GAAAAGGCTG TCCGTGTCCC CGTATACAGA CTTGAGAGGC CTGTCCTCGA GCGGTGTTCC GCGGTCCTCC
                      starts IVa2 RNA
      5810       5820       5830       5840       5850       5860       5870       5880       5890       5900
AGCATATCTT TGAGCCTGGT GAGACTCTGC TTCCGAGCGC AGGTCCGGTC GTGCTTCCTC CGATTCACCC TCCCCATCGC CAGCAACAGG TGATCCCCCA
TCGTATAGAA ACTCGGACCA CTCTGAGACG AAGGCTCGCG TCCAGGCCAG CACGAAGGAG GCTAAGTGGG AGGGGTAGCG GTCGTTGTCC ACTAGGGGGT

      5910       5920       5930       5940       5950       5960       5970       5980       5990       6000
GGTGAGCGAG GTCCCACACT TCTGTGTACA GCGGGAGAAG CCGTAGTTCC TTCCACTAAC CAAATATCCA CATCCGGTGC ACTGGCCCAC AAGGACTTCC
F CCACTCGCTC CAGGGTGTGA AGACACATGT CGCCCTCTTC GGCATCAAGG AAGGTGATTG GTTTATAGGT GTAGGCCACG TGACCGGGTG TTCCTGAAGG
```

FIGURE 18 (*Continued*)

```
       6010        6020        6030        6040        6050        6060        6070        6080        6090        6100
 CCCCGATATT  TTCCCCCACC  CCCGCGCAAG  CAGGAGTGAG  AGAAGGCGTA  GCGACAGACG  CTCCCGGTCG  ACAACCCCAC  TCATGAGGGA  GAGTTTTCGC
 GGGGCTATAA  AAGGGGGTGG  GGGCGCGTTC  GTCCTCACTC  TCTTCCGCAT  CGCTGTCTGC  GAGGGCCAGC  TGTTGGGGTG  AGTACTCCCT  CTCAAAAGCG
       TATA                      start 1st late leader                          splice 1st leader
       6110        6120        6130        6140        6150        6160        6170        6180        6190        6200
 CCGTACTGAA  GACGCGATTC  TAACAGTCAA  AGGTTTTTGC  TCCTCCTAAA  CTATAAGTGG  ACCGGGCGCC  ACTACGGAAA  CTCCCACCGG  CGCAGGTAGA
 GGCATGACTT  CTGCGCTAAG  ATTGTCAGTT  TCCAAAAACG  AGGAGGATTT  GATATTCACC  TGGCCCGCGG  TGATGCCTTT  GAGGGTGGCC  GCGTCCATCT

       6210        6220        6230        6240        6250        6260        6270        6280        6290        6300
 CCAGTCTTTT  CTGTTAGAAA  AACAACAGTT  CGAACCACCG  TTTGCTGGGC  ATCTCCCGCA  ACCTGTCGTT  GAACCGCTAC  CTCGCGTCCC  AAACCAAAAA
 GGTCAGAAAA  GACAATCTTT  TTGTTGTCAA  GCTTGGTGGC  AAACGACCCG  TAGAGGGCGT  TGGACAGCAA  CTTGGCGATG  GAGCGCAGGG  TTTGGTTTTT
                                                                                      start URF 1
       6310        6320        6330        6340        6350        6360        6370        6380        6390        6400
 CAGCGCTAGC  CGCGCGAGGA  ACCGGCGCTA  CAAATCGACG  TGCATAAGCG  CGCGTTGCGT  GGCGGTAAGC  CCTTTCTGCC  ACCACGCGAG  CAGCCCGTGG
 GTCGCGATCG  GCGCGCTCCT  TGGCCGCGAT  GTTTAGCTGC  ACGTATTCGC  GCGCAACGCA  CCGCCATTCG  GGAAAGACGG  TGGTGCGCTC  GTCGGGCACC
                                           stop URF 8
       6410        6420        6430        6440        6450        6460        6470        6480        6490        6500
 TCCACGTGCG  CGGTTGGCGC  CAACACGTCC  CACTGTTCCA  GTTGCGACCA  CCGATGGAGA  GGCGCATCCG  CGAGCAACCA  GGTCGTCTCC  GCCGGCGGGA
 AGGTGCACGC  GCCAACCGCG  GTTGTGCAGG  GTGACAAGGT  CAACGCTGGT  GGCTACCTCT  CCGCGTAGGC  GCTCGTTGGT  CCAGCAGAGG  CGGCCGCCCT

       6510        6520        6530        6540        6550        6560        6570        6580        6590        6600
 ACGCGCTTGT  CTTACCGCCA  TCACCCAGAT  CGACGCAGAG  CAGGCCCCCC  AGACGCAGGT  GCCATTTCTG  GGGCCCGTCG  TCCGCGCGCA  GCTTCATCAG
 TGCGCGAACA  GAATGGCGGT  AGTGGGTCTA  GCTGCGTCTC  GTCCGGGGGG  TCTGCGTCCA  CGGTAAAGAC  CCCGGGCAGC  AGGCGCGCGT  CGAAGTAGTC
                                                                                                             stop URF 1
       6610        6620        6630        6640        6650        6660        6670        6680        6690        6700
 ATAGAACGTA  GGAACGTTCA  GATCGCGGAC  GACGGTACGC  GCCCGCCGTT  CGCGCGCGAG  CATACCCAAC  TCACCCCCTG  GGGTACCGTA  CCCCACCCAC
 TATCTTGCAT  CCTTGCAAGT  CTAGCGCCTG  CTGCCATGCG  CGGGCGGCAA  GCGCGCGCTC  GTATGGGTTG  AGTGGGGGAC  CCCATGGCAT  GGGGTGGGTG
                                                                                  start URF 8
       6710        6720        6730        6740        6750        6760        6770        6780        6790        6800
 TCGCGCCTCC  GCATGTACGG  CGTTTACAGC  ATTTGCATCT  CCCCGAGAGA  CTCATAAGGT  TCTATACATC  CCATCGTAGA  AGGTGGCGCC  TACGACCGCG
 AGCGCGGAGG  CGTACATGCC  GCAAATGTCG  TAAACGTAGA  GGGGCTCTCT  GAGTATTCCA  AGATATGTAG  GGTAGCATCT  TCCACCGCGG  ATGCTGGCGC

       6810        6820        6830        6840        6850        6860        6870        6880        6890        6900
 CGTGCATTAG  CATATCAAGC  ACGCTCCCTC  GCTCCTCCAG  CCCTGGCTCC  AACGATGCCC  GCCCGACGAG  ACGAGCCTTC  TGATAGACGG  ACTTCTACCG
 GCACGTAATC  GTATAGTTCG  TGCGAGGGAG  CGAGGAGGTC  GGGACCGAGG  TTGCTACGGG  CGGGCTGCTC  TGCTCGGAAG  ACTATCTGCC  TGAAGATGGC

       6910        6920        6930        6940        6950        6960        6970        6980        6990        7000
 TACACTCAAC  CTACTATACC  AACCTGCGAC  CTTCTGCAAC  TTCGACCGCA  GACACTCTGG  ATGGCGCAGT  GCGTGCTTCC  TCCGCATCCT  CAGCGCGTCG
G ATGTGAGTTG  GATGATATGG  TTGGACGCTG  GAAGACGTTG  AAGCTGGCGT  CTGTGAGACC  TACCGCGTCA  CGCACGAAGG  AGGCGTAGGA  GTCGCGCAGC


       7010        7020        7030        7040        7050        7060        7070        7080        7090        7100
 AACAACTGGT  CGAGCCGCCA  CTGGACGTGC  AGATCCCGCG  TCATCAGGTC  CCAAAGGAAC  TACTACAGTA  TGAATAGGAC  AGGGAAAAAA  AAGGTGTCGA
 TTGTTGACCA  GCTCGGCGGT  GACCTGCACG  TCTAGGGCGC  AGTAGTCCAG  GGTTTCCTTG  ATGATGTCAT  ACTTATCCTG  TCCCTTTTTT  TTCCACAGCT
                                                                                               splice 2nd leader
       7110        7120        7130        7140        7150        7160        7170        7180        7190        7200
 GCGCCAACTC  CTGTTTGAGA  AGCGCCAGAA  AGGTCATGAG  AACCTAGCCT  TTGGGCAGCC  GGAGGCTTGC  CATTCTCGGA  TCGTACATCT  TGACCAACTG
 CGCGGTTGAG  GACAAACTCT  TCGCGGTCTT  TCCAGTACTC  TTGGATCGGA  AACCCGTCGG  CCTCCGAACG  GTAAGAGCCT  AGCATGTAGA  ACTGGTTGAC
                                                                              splice 2nd leader
       7210        7220        7230        7240        7250        7260        7270        7280        7290        7300
 CCGGACCATC  CGCGTCGTAG  GGAAAAGATG  CCCATCGCGC  ATACGGACGC  GCCGGAAGGC  CTCGCTCCAC  ACCCACTCGC  GTTTCCACAG  GGATTGGTAC
 GGCCTGGTAG  GCGCAGCATC  CCTTTTCTAC  GGGTAGCGCG  TATGCCTGCG  CGGCCTTCCG  GAGCGAGGTG  TGGGTGAGCG  CAAAGGTGTC  CCTAACCATG

       7310        7320        7330        7340        7350        7360        7370        7380        7390        7400
 TGAAACTCCA  TGACCATAAA  CTTCAGTCAC  AGCAGCGTAG  GCGGGACGAG  GGTCTCGTTT  TTCAGGCACG  CGAAAAACCT  TGCGCCCAAA  CCGTCCCGCT
 ACTTTGAGGT  ACTGGTATTT  GAAGTCAGTG  TCGTCGCATC  CGCCCTGCTC  CCAGAGCAAA  AAGTCCGTGC  GCTTTTTGGA  ACGCGGGTTT  GGCAGGGCGA

       7410        7420        7430        7440        7450        7460        7470        7480        7490        7500
 TCCACTGTAG  CAACTTTTCA  TAGAAAGGGC  GCGCTCCGTA  TTTCAACGCA  CACTACGCCT  TCCCAGGGCC  GTGGAGCCTT  GCCAACAATT  AATGGACCCG
 AGGTGACATC  GTTGAAAAGT  ATCTTTCCCG  CGCGAGGCAT  AAAGTTGCGT  GTGATGCGGA  AGGGTCCCGG  CACCTCGGAA  CGGTTGTTAA  TTACCTGGGC

       7510        7520        7530        7540        7550        7560        7570        7580        7590        7600
 GCGCTCGTGC  TAGAGCAGCT  TCGGCAACTA  CAACACCGGG  TGCTACATTT  CAAGGTTCTT  CGCGCCCCAC  GGGAACTACC  TCCCGTTAAA  AAATTCAAGG
 CGCGAGCACG  ATCTCGTCGA  AGCCGTTGAT  GTTGTGGCCC  ACGATGTAAA  GTTCCAAGAA  GCGCGGGGTG  CCCTTGATGG  AGGGCAATTT  TTTAAGTTCC

       7610        7620        7630        7640        7650        7660        7670        7680        7690        7700
 AGCATCCACT  CGAGGAGTCC  CCTCGACTCG  GGCACAAGAC  TGTCCCGGGT  CAGACGTTCT  ACTCCCAACC  TTCGCTGCTT  ACTCGAGGTG  TCCAGTGCCC
 TCGTAGGTGA  GCTCCTCAGG  GGAGCTGAGC  CCGTGTTCTG  ACAGGGCCCA  GTCTGCAAGA  TGAGGGTTGG  AAGCGACGAA  TGAGCTCCAC  AGGTCACGGG

       7710        7720        7730        7740        7750        7760        7770        7780        7790        7800
 GGTAATCGTA  AACGTCCACC  AGCGCTTTCC  AGGATTTGAC  CGCTGGATAC  CGGTAAAAAA  GACCCCACTA  CGTCATCTTC  CATTCGCCCA  GAACAAGGGT
 CCATTAGCAT  TTGCAGGTGG  TCGCGAAAGG  TCCTAAACTG  GCGACCTATG  GCCATTTTTT  CTGGGGTGAT  GCAGTAGAAG  GTAAGCGGGT  CTTGTTCCCA

       7810        7820        7830        7840        7850        7860        7870        7880        7890        7900
 CGCCAGGGTA  GGTTCCAGGT  GCCGATCCAG  AGCGCGCCGC  CAGTGGTCTC  CGAGTAGAGG  CGGCTTGAAG  TATTGGTCGT  ACTTCCCGTG  CTCGACGAAG
 GCGGTCCCAT  CCAAGGTCCA  CGGCTAGGTC  TCGCGCGGCG  GTCACCAGAG  GCTCATCTCC  GCCGAACTTC  ATAACCAGCA  TGAAGGGCAC  GAGCTGCTTC

       7910        7920        7930        7940        7950        7960        7970        7980        7990        8000
 GGTTTCCGGG  GGTAGGTTCA  TATCCAGAGA  TGTAGCATCC  ACTGTTTCTC  TGCGAGCCAC  GCTCCTACGC  TCGGCTAGCC  CTTCTTGACC  TAGAGGGCGG
H CCAAAGGCCC  CCATCCAAGT  ATAGGTCTCT  ACATCGTAGG  TGACAAAGAG  ACGCTCGGTG  CGAGGATGCG  AGCCGATCGG  GAAGAACTGG  ATCTCCCGCC
                                                                      start URF 2
                                   splice i leader
```

FIGURE 18 (*Continued*)

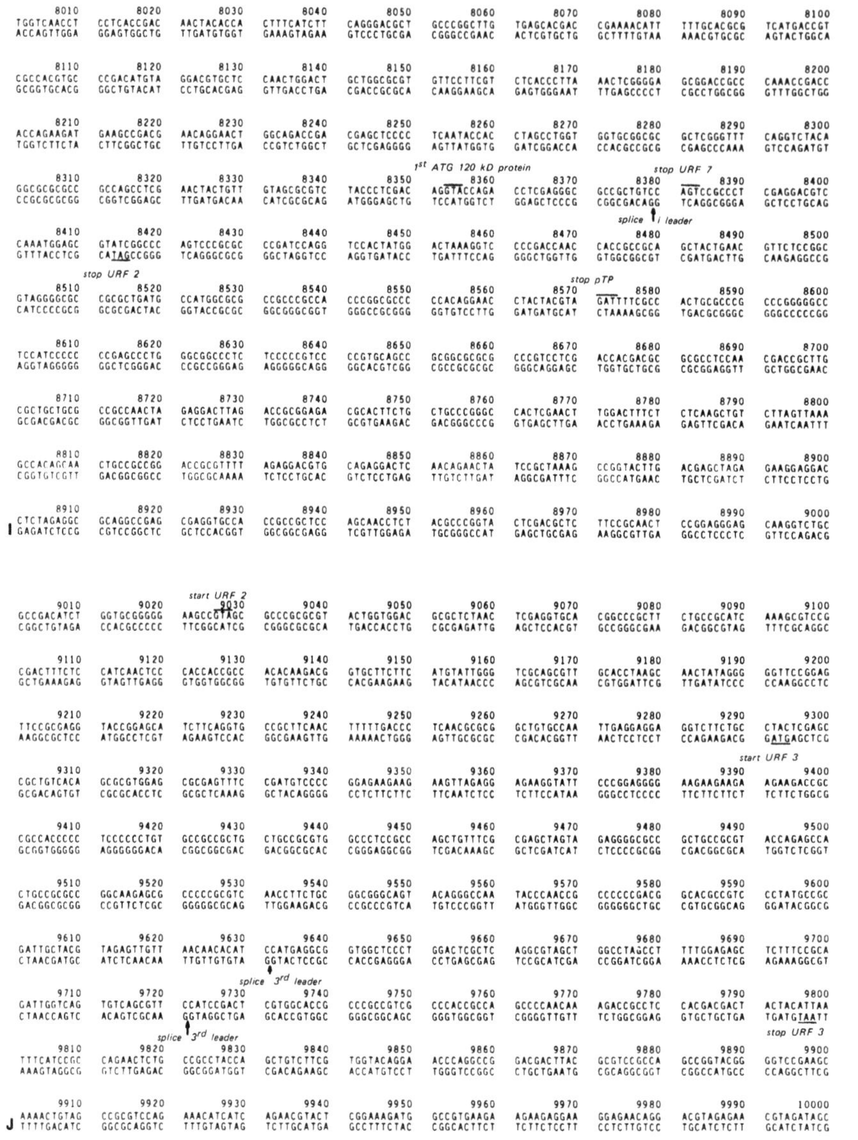

```
      8010       8020       8030       8040       8050       8060       8070       8080       8090       8100
TGGTCAACCT CCTCACCGAC AACTACACCA CTTTCATCTT CAGGGACGCT GCCCGGCTTG TGAGCACGAC CGAAAACATT TTTGCACGCG TCATGACCGT
ACCAGTTGGA GGAGTGGCTG TTGATGTGGT GAAAGTAGAA GTCCCTGCGA CGGGCCGAAC ACTCGTGCTG GCTTTTGTAA AAACGTGCGC AGTACTGGCA

      8110       8120       8130       8140       8150       8160       8170       8180       8190       8200
CGCCACGTGC CCGACATGTA GGACGTGCTC CAACTGGACT GCTGGCGCGT GTTCCTTCGT CTCACCCTTA AACTCGGGGA GCGGACCGCC CAAACCGACC
GCGGTGCACG GGCTGTACAT CCTGCACGAG GTTGACCTGA CGACCGCGCA CAAGGAAGCA GAGTGGGAAT TTGAGCCCCT CGCCTGGCGG GTTTGGCTGG

      8210       8220       8230       8240       8250       8260       8270       8280       8290       8300
ACCAGAAGAT GAAGCCGACG AACAGGAACT GGCAGACCGA CGAGCTCCCC TCAATACCAC CTAGCCTGGT GGTGCGGCGC GCTCGGGTTT CAGGTCTACA
TGGTCTTCTA CTTCGGCTGC TTGTCCTTGA CCGTCTGGCT GCTCGAGGGG AGTTATGGTG GATCGGACCA CCACGCCGCG CGAGCCCAAA GTCCAGATGT

                                                  1st ATG 120 kD protein                stop URF 7
      8310       8320       8330       8340       8350       8360       8370       8380       8390       8400
GGCGCGCGCC GCCAGCCTCG AACTACTGTT GTAGCGCGTC TACCCTCGAC AGGTACCAGA CCTCGAGGGC GCCGCTGTCC AGTCCGCCCT CGAGGACGTC
CCGCGCGCGG CGGTCGGAGC TTGATGACAA CATCGCGCAG ATGGGAGCTG TCCATGGTCT GGAGCTCCCG CGGCGACAGG TCAGGCGGGA GCTCCTGCAG
                                                                                 splice ↑ i leader
      8410       8420       8430       8440       8450       8460       8470       8480       8490       8500
CAAATGGAGC GTATCGGCCC AGTCCCGCGC CCGATCCAGG TCCACTATGG ACTAAAGGTC CCCGACCAAC CACCGCCGCA GCTACTGAAC GTTCTCCGGC
GTTTACCTCG CATAGCCGGG TCAGGGCGCG GGCTAGGTCC AGGTGATACC TGATTTCCAG GGGCTGGTTG GTGGCGGCGT CGATGACTTG CAAGAGGCCG
     stop URF 2                                                            stop pTP
      8510       8520       8530       8540       8550       8560       8570       8580       8590       8600
GTAGGGGCGC CGCGCTGATG CCATGGCGCG CCGCCCGCCA CCCGGCGCCC CCACAGGAAC CTACTACGTA GATTTTCGCC ACTGCGCCCG CCCGGGGGCC
CATCCCCGCG GCGCGACTAC GGTACCGCGC GGCGGGCGGT GGGCCGCGGG GGTGTCCTTG GATGATGCAT CTAAAAGCGG TGACGCGGGC GGGCCCCCGG

      8610       8620       8630       8640       8650       8660       8670       8680       8690       8700
TCCATCCCCC CCGAGCCCTG GGCGGCCCTC TCCCCCGTCC CCGTGCAGCC GCGGCGCGCG CCCGTCCTCG ACCACGACGC GCGCCTCCAA CGACCGCTTG
AGGTAGGGGG GGCTCGGGAC CCGCCGGGAG AGGGGGCAGG GGCACGTCGG CGCCGCGCGC GGGCAGGAGC TGGTGCTGCG CGCGGAGGTT GCTGGCGAAC

      8710       8720       8730       8740       8750       8760       8770       8780       8790       8800
CGCTGCTGCG CCGCCAACTA GAGGACTTAG ACCGCGGAGA CGCACTTCTG CTGCCCGGGC CACTCGAACT TGGACTTTCT CTCAAGCTGT CTTAGTTAAA
GCGACGACGC GGCGGTTGAT CTCCTGAATC TGGCGCCTCT GCGTGAAGAC GACGGGCCCG GTGAGCTTGA ACCTGAAAGA GAGTTCGACA GAATCAATTT

      8810       8820       8830       8840       8850       8860       8870       8880       8890       8900
GCCACAGCAA CTGCCGCCGG ACCGCGTTTT AGAGGACGTG CAGAGGACTC AACAGAACTA TCCGCTAAAG CCGGTACTTG ACGAGCTAGA GAAGGAGGAC
CGGTGTCGTT GACGGCGGCC TGGCGCAAAA TCTCCTGCAC GTCTCCTGAG TTGTCTTGAT AGGCGATTTC GGCCATGAAC TGCTCGATCT CTTCCTCCTG

        8910       8920       8930       8940       8950       8960       8970       8980       8990       9000
  CTCTAGAGGC GCAGGCCGAG CGAGGTGCCA CCGCCGCTCC AGCAACCTCT ACGCCCGGTA CTCGACGCTC TTCCGCAACT CCGGAGGGAG CAAGGTCTGC
I GAGATCTCCG CGTCCGGCTC GCTCCACGGT GGCGGCGAGG TCGTTGGAGA TGCGGGCCAT GAGCTGCGAG AAGGCGTTGA GGCCTCCCTC GTTCCAGACG

                      start URF 2
      9010       9020       9030       9040       9050       9060       9070       9080       9090       9100
GCCGACATCT GGTGCGGGGG AAGCCGTAGC GCCCGCGCGT ACTGGTGGAC GCGCTCTAAC TCGAGGTGCA CGGCCCGCTT CTGCCGCATC AAAGCGTCCG
CGGCTGTAGA CCACGCCCCC TTCGGCATCG CGGGCGCGCA TGACCACCTG CGCGAGATTG AGCTCCACGT GCCGGGCGAA GACGGCGTAG TTTCGCAGGC

      9110       9120       9130       9140       9150       9160       9170       9180       9190       9200
CGACTTTCTC CATCAACTCC CACCACCGCC ACACAAGACG GTGCTTCTTC ATGTATTGGG TCGCAGCGTT GCACCTAAGC AACTATAGGG GGTTCCGGAG
GCTGAAAGAG GTAGTTGAGG GTGGTGGCGG TGTGTTCTGC CACGAAGAAG TACATAACCC AGCGTCGCAA CGTGGATTCG TTGATATCCC CCAAGGCCTC

      9210       9220       9230       9240       9250       9260       9270       9280       9290       9300
TTCCGCGAGG TACCGGAGCA TCTTCAGGTG CCGCTTCAAC TTTTTGACCC TCAACGCGCG GCTGTGCCAA TTGAGGAGGA GGTCTTCTGC CTACTCGAGC
AAGGCGCTCC ATGGCCTCGT AGAAGTCCAC GGCGAAGTTG AAAAACTGGG AGTTGCGCGC CGACACGGTT AACTCCTCCT CCAGAAGACG GATGAGCTCG
                                                                                              start URF 3
      9310       9320       9330       9340       9350       9360       9370       9380       9390       9400
CGCTGTCACA GCGCGTGGAG CGCGAGTTTC CGATGTCCCC GGAGAAGAAG AAGTTAGAGG AGAAGGTATT CCCGGAGGGG AAGAAGAAGA AGAAGACCGC
GCGACAGTGT CGCGCACCTC GCGCTCAAAG GCTACAGGGG CCTCTTCTTC TTCAATCTCC TCTTCCATAA GGGCCTCCCC TTCTTCTTCT TCTTCTGGCG

      9410       9420       9430       9440       9450       9460       9470       9480       9490       9500
CGCCACCCCC TCCCCCCTGT GCCGCCGCTG CTGCCGCGTG GCCCTCCGCC AGCTGTTTCG CGAGCTAGTA GAGGGGCGCC GCTGCCGCGT ACCAGAGCCA
GCGGTGGGGG AGGGGGGACA CGGCGGCGAC GACGGCGCAC CGGGAGGCGG TCGACAAAGC GCTCGATCAT CTCCCCGCGG CGACGGCGCA TGGTCTCGGT

      9510       9520       9530       9540       9550       9560       9570       9580       9590       9600
CTGCCGCGCC GGCAAGAGCG CCCCCGCGTC AACCTTCTGC GGCGGGCAGT ACAGGGCCAA TACCCAACCG CCCCCCGACG GCACGCCGTC CCTATGCCGC
GACGGCGCGG CCGTTCTCGC GGGGGCGCAG TTGGAAGACG CCGCCCGTCA TGTCCCGGTT ATGGGTTGGC GGGGGGCTGC CGTGCGGCAG GGATACGGCG

      9610       9620       9630       9640       9650       9660       9670       9680       9690       9700
GATTGCTACG TAGAGTTGTT AACAACACAT CCATGAGGCG GTGGCTCCCT GGACTCGCTC AGGCGTAGCT GGCCTAGCCT TTTGGAGAGC TCTTTCCGCA
CTAACGATGC ATCTCAACAA TTGTTGTGTA GGTACTCCGC CACCGAGGGA CCTGAGCGAG TCCGCATCGA CCGGATCGGA AAACCTCTCG AGAAAGGCGT
                                 splice ↓ 3rd leader
      9710       9720       9730       9740       9750       9760       9770       9780       9790       9800
GATTGGTCAG TGTCAGCGTT CCATCCGACT CGTGGCACCG CCCGCCGTCG CCCACCGCCA GCCCCAACAA AGACCGCCTC CACGACGACT ACTACATTAA
CTAACCAGTC ACAGTCGCAA GGTAGGCTGA GCACCGTGGC GGGCGGCAGC GGGTGGCGGT CGGGGTTGTT TCTGGCGGAG GTGCTGCTGA TGATGTAATT
                      splice ↑ 3rd leader                                                     stop URF 3
      9810       9820       9830       9840       9850       9860       9870       9880       9890       9900
TTTCATCCGC CAGAACTCTG CCGCCTACCA GCTGTCTTCG TGGTACAGGA ACCCAGGCCG GACGACTTAC GCGTCCGCCA GCCGGTACGG GGTCCGAAGC
AAAGTAGGCG GTCTTGAGAC GGCGGATGGT CGACAGAAGC ACCATGTCCT TGGGTCCGGC CTGCTGAATG CGCAGGCGGT CGGCCATGCC CCAGGCTTCG

        9910       9920       9930       9940       9950       9960       9970       9980       9990      10000
  AAAACTGTAG CCGCGTCCAG AAACATCATC AGAACGTACT CGGAAAGATG GCCGTGAAGA AGAAGAGGAA GGAGAACAGG ACGTAGAGAA CGTAGATAGC
J TTTTGACATC GGCGCAGGTC TTTGTAGTAG TCTTGCATGA GCCTTTCTAC CGGCACTTCT TCTTCTCCTT CCTCTTGTCC TGCATCTCTT GCATCTATCG
```

FIGURE 18 (*Continued*)

→

FIGURE 19. Nucleotide sequence of a region between coordinates 49.0 and 51.8 on the Ad2 genome. This sequence was determined by Akusjärvi and Persson (1981a). The positions of strategic signals were determined by Akusjärvi and Pettersson (1979a) and Akusjärvi and Persson (1981a).

```
     10010      10020      10030      10040      10050      10060      10070      10080      10090      10100
GATGCCGCCG CCGCCTCAAA CCGGCATCCA CCGCGGGAGA AGGAGGGTAC GCACACTGGG GCTTCGGGGA GTAGCCGACT TCGTCCCGGT CCAGCCGCTG
CTACGGCGGC GGCGGAGTTT GGCCGTAGGT GGCGCCCTCT TCCTCCCATG CGTGTGACCC CGAAGCCCCT CATCGGCTGA AGCAGGGCCA GGTCGGCGAC

     10110      10120      10130      10140      10150      10160      10170      10180      10190      10200
TTGCGCGAGC CGATTATACC GGACGACGTG GACGCACTCC CATCTGACCT TCAGTAGGTA CAGGTGTTTC GCCACCATAC GCGGGCACAA CTACCACATT
AACGCGCTCG GCTAATATGG CCTGCTGCAC CTGCGTGAGG GTAGACTGGA AGTCATCCAT GTCCACAAAG CGGTGGTATG CGCCCGTGTT GATGGTGTAA

     10210      10220      10230      10240      10250      10260      10270      10280      10290      10300
CACGTCAACC GGTATTGCCT GGTCAATTGC CAGACCACTG GGCCGACGCT CTCGAGCCAC ATGGACTCTG CGCTCATTCG GGAACTCAGT TTCTGCATCA
GTGCAGTTGG CCATAACGGA CCAGTTAACG GTCTGGTGAC CCGGCTGCGA GAGCTCGGTG TACCTGAGAC GCGAGTAAGC CCTTGAGTCA AAGACGTAGT

     10310      10320      10330      10340      10350      10360      10370      10380      10390      10400
GCAACGTTCA GGCGTGGTCC ATGACTATAG GGTGGTTTTT CACGCCGCCG CCGACCGCCA TCTCCCCGGT CGCATCCCAC CGGCCCCGAG GCCCCCGCTC
CGTTGCAAGT CCGCACCAGG TACTGATATC CCACCAAAAA GTGCGGCGGC GGCTGGCGGT AGAGGGGCCA GCGTAGGGTG GCCGGGGCTC CGGGGGCGAG

     10410      10420      10430      10440      10450      10460      10470      10480      10490      10500
CAGAAGGTTG TATTCCGCTA CTATAGGCAT CTACATGGAC CTGTAGGTCC ACTACGGCCG CCGCCACCAC CTCCGCGCGC CTTTCAGCGC CTGCGCCAAG
GTCTTCCAAC ATAAGGCGAT GATATCCGTA GATGTACCTG GACATCCAGG TGATGCCGGC GGCGGTGGTG GAGGCGCGCG GAAAGTCGCG GACGCGGTTC
              start URF 4      1st ATG pTP
     10510      10520      10530      10540      10550      10560      10570      10580      10590      10600
GTCTACAACG CGTCGCCGTT TTTCACGAGG TACCAGCCCT GCGAGACCGG CCACTCCGCA CGCGTCAGCA ACTGCGAGAT CTGGCACGTT TTCCTCTCGG
CAGATGTTGC GCAGCGGCAA AAAGTGCTCC ATGGTCGGGA CGCTCTGGCC GGTGAGGCGT GCGCAGTCGT TGACGCTCTA GACCGTGCAA AAGGAGAGCC

     10610      10620      10630      10640      10650      10660      10670      10680      10690      10700
ACATTCGCCC GTGAGAAGGC ACCAGACCAC CTATTTAAGC GTTCCCATAG TACCGCCTGC TGGCCCCAAG CTTGGGGCCT AGGCCGGCAG GCGGCACTAG
TGTAAGCGGG CACTCTTCCG TGGTCTGGTG GATAAATTCG CAAGGGTATC ATGGCGGACG ACCGGGGTTC GAACCCCGGA TCCGGCCGTC CGCCGTGATC
  ↑  ↑ starts VAI RNA                       stop URF 6
     10710      10720      10730      10740      10750      10760      10770      10780      10790      10800
GTACGCCAAT GGCGGGCGCA CAGCTTGGGT CCACACGCTG CAGTCTGTTG CCCCCTCGCG AGGAAAACCG AAGGAAGGTC CGCGCCGCCG ACGACGCGAT
CATGCGGTTA CCGCCCGCGT GTCGAACCCA GGTGTGCGAC GTCAGACAAC GGGGGAGCGC TCCTTTTGGC TTCCTTCCAG GCGCGGCGGC TGCTGCGCTA
                                                                stops ↑↑↑↑ VAI RNA
    .10810      10820      10830      10840      10850      10860      10870      10880      10890      10900
CGAAAAAACC GGTGACCGGC GCGCGCCGCA TTCGCCAATC CGACCTTTCG CTTTCGTAAT TCACCGAGCG AGGGACATCG GCCTCCCAAT AAAAGGTTCC
GCTTTTTTGG CCACTGGCCG CGCGCGGCGT AAGCGGTTAG GCTGGAAAGC GAAAGCATTA AGTGGCTCGC TCCCTGTAGC CGGAGGGTTA TTTTCCAAGG
                             stop URF 4                          start ↑ VAII RNA
     10910      10920      10930      10940      10950      10960      10970      10980      10990      11000
 CAACTCAGCG TCCTGGGGGC CAAGCTCAGA GCCCGGCCGG CCTGACGCCG CTTGCCCCCA AACGGAGGGG CAGTACGTTC TGGGGCGAAC GTTTAAGGAG
K GTTGAGTCGC AGGACCCCCG GTTCGAGTCT CGGGCCGGCC GGACTGCGGC GAACGGGGGT TTGCCTCCCC GTCATGCAAG ACCCCGCTTG CAAATTCCTC


     11010      11020      11030      11040      11050      11060      11070      11080      11090      11100
GCCTTTGTCC CTGCTCGGGG AAAAAACGAA AAGGGTCTAC GTAGGCCACG ACGCCGTCTA CGCGGGGGGA GGAGTCGTCG CCGTTCTCGT TCTCGTCGCC
CGGAAACAGG GACGAGCCCC TTTTTTGCTT TTCCCAGATG CATCCGGTGC TGCGGCAGAT GCGCCCCCCT CCTCAGCAGC GGCAAGAGCA AGAGCAGCGG
                  stops ↑↑↑↑↑↑ VAII RNA  ↑ start 52, 55 kD proteins
start URF 6                          splice
     11110      11120      11130      11140      11150      11160      11170      11180      11190      11200
GTCTGTACGT CCCGTGGGAG GGGAAGAGGA TGGCGCAGTC CTCCCCGTTG TAGGCGCCGA CTGCGCCGCC GTCTACCACT AATGCTTGGG GGCGCCGCGG
CAGACATGCA GGGCACCCTC CCCTTCTCCT ACCGCGTCAG GAGGGGCAAC ATCCGCGGCT GACGCGGCGG CAGATGGTGA TTACGAACCC CCGCGGCGCC
                                              stop URF 5
     11210      11220      11230      11240      11250      11260      11270      11280      11290      11300
CCCGGGCCGT GATGGACCTG AACCTCCTCC CGCTCCCGGA CCGCGCCGAT CCTCGCGGGA GAGGACTCGC TGTGGGTTCC CACGTCGACT TCGCACTGTG
GGGCCCGGCA CTACCTGGAC TTGGAGGAGG GCGAGGGCCT GGCGCGGCTA GGAGCGCCCT CTCCTGAGCG ACACCCAAGG GTGCAGCTGA AGCGTGACAC

     11310      11320      11330      11340      11350      11360      11370      11380      11390      11400
CGCGCTCCGC ATGCACGGCG CCGTCTTGGA CAAAGCGCTG GCGCTCCCTC TCCTCGGGCT CCTCTACGCC CTAGCTTTCA AGGTGCGTCC CGCGCTCAAC
GCGCGAGGCG TACGTGCCGC GGCAGAACCT GTTTCGCGAC CGCGAGGGAG AGGAGCCCGA GGAGATGCGG GATCGAAAGT TCCACGCAGG GCGCGAGTTG

     11410      11420      11430      11440      11450      11460      11470      11480      11490      11500
GCCGTACCGG ACTTGGCGCT CGCCAACGAC GCGCTCCTCC TGAAACTCGG GCTGCGCGCC TGGCCCTAAT CAGGGCGCGC GCGTGTGCAC CGCCGGCGGC
CGGCATGGCC TGAACCGCGA GCGGTTGCTG CGCGAGGAGG ACTTTGAGCC CGACGCGCGG ACCGGGATTA GTCCCGCGCG CGCACACGTG GCGGCCGCCG

     11510      11520      11530      11540      11550      11560
 TGGACCATTG GCGCATGCTC GTCTGCCACT TGGTCCTCTA ATTGAAAGTT TTTTCGAA
L ACCTGGTAAC CGCGTACGAG CAGACGGTGA ACCAGGAGAT TAACTTTCAA AAAAGCTT
```

FIGURE 18 (*Continued*)

```
        10         20         30         40         50         60         70         80         90        100
r3'GGCCTTAACG TAGGCACCGG AACGTCCGCG TCTCTGTGAC TAATTTTTGT TCAACGTACA CCTTTTTAGT TTTATTTTTC AGACCTCAGA GTGCGAGCGA
|5'CCGGAATTGC ATCCGTGGCC TTGCAGGCGC AGAGACACTG ATTAAAAACA AGTTGCATGT GGAAAAATCA AAATAAAAAG TCTGGAGTCT CACGCTCGCT
                                                                                                polyA ↑ L2 RNA
       110        120        130        140        150        160        170        180        190        200
ACCAGGACAT TGATAAAACA TCTTACCTTC TGTAGTTGAA ACGCAGAGAC CGGGGCGCTG TGCCGAGCGC GGGCAAGTAC CCTTTGACCG TTCTATAGCC
TGGTCCTGTA ACTATTTTGT AGAATGGAAG ACATCAACTT TGCGTCTCTG GCCCCGCGAC ACGGCTCGCG CCCGTTCATG GGAAACTGGC AAGATATCGG
        splice pVI RNA ↑ start pVI
       210        220        230        240        250        260        270        280 stop URF 12 290      300
GTGGTCGTTA TACTCGCCAC CGCGGAAGTC GACCCCGAGC GACACCTCGC CGTAATTTTT AAAGCCAAGG TGGTAATTCT TGATACCGTC GTTCCGGACC
CACCAGCAAT ATGAGCGGTG GCGCCTTCAG CTGGGGCTCG CTGTGGAGCG GCATTAAAAA TTTCGGTTCC ACCATTAAGA ACTATGGCAG CAAGGCCTGG

       310        320        330        340        350        360        370        380        390        400
TTGTCGTCGT GTCCGGTCTA CGACTCCCTG TTCAACTTTC TCGTTTTAAA GGTTGTTTTC CACCATCTAC CGGACCGGAG ACCGTAATCG CCCCACCACC
AACAGCAGCA CAGGCCAGAT GCTGAGGGAC AAGTTGAAAG AGCAAAATTT CCAACAAAAG GTGGTAGATG GCCTGGCCTC TGGCATTAGC GGGGTGGTGG

       410        420        430        440        450        460        470        480        490        500
TGGACCGGTT GGTCCGTCAC GTTTTATTCT AATTGTCATT CGAACTAGGG GCGGGAGGGC ATCTCCTCGG AGGTGGCCGG CACCTCTGTC ACAGAGGTCT
ACCTGGCCAA CCAGGCAGTG CAAAATAAGA TTAACAGTAA GCTTGATCCC CGCCCTCCCG TAGAGGAGCC TCCACCGGCC GTGGAGACAG TGTCTCCAGA

       510        520        530        540        550        560        570        580        590        600
CCCCGCACCG CTTTTCGCAG GCGCCGGGCT GTCCCTTCTT TGAGACCACT GCGTTTATCT ACTCGGAGGG AGCATGCTCC TCCGTGATTT CGTTCCGGAC
GGGGCGTGGC GAAAAGCGTC CGCGGCCCGA CAGGGAAGAA ACTCTGGTGA CGCAAATAGA TGAGCCTCCC TCGTACGAGG AGGCACTAAA GCAAGGCCTG

       610        620        630        640        650        660        670        680        690        700
GGGTGGTGGG CAGGGTAGCG CGGGTACCGA TGGCCTCACG ACCCGGTCGT GTGTGGACAT TGCGACCTGG ACGGAGGGGG GCGACTGTGG GTCGTCTTTG
CCCACCACCC GTCCCATCGC GCCCATGGCT ACCGGAGTGC TGGGCCAGCA CACACCTGTA ACGCTGGACC TGCCTCCCCC CGCTGACACC CAGCAGAAAC

       710        720        730        740        750        760        770        780        790        800
GACACGACGG TCCCGGCAGG CGGCAACAAC ATTGGGCGGG ATCGGCGCGC AGGGACGCGG CACGGCGGTC GCCAGGCGCT AGCTACGCCG GGCATCGGTC
CTGTGCTGCC AGGGCCGTCC GCCGTTGTTG TAACCCGCCC TAGCCGCGCG TCCCTGCGCC GTGCCGCCAG CGGTCCGCGA TCGATGCGGC CCGTAGCCAG
                          start URF 12
       810        820        830        840        850        860        870        880        890        900
ACCGTTGACC GTTTCGTGTG ACTTGTCGTA GCACCCAGAC CCCCACGTTA GGGACTTCGC GGCTGCTACG AAGATTTATC GATTGCACAG CATACACAGT
TGGCAACTGG CAAAGCACAC TGAACAGCAT CGTGGGTCTG GGGGTGCAAT CCCTGAAGCG CCGACGATGC TTCTAAATAG CTAACGTGTC GTATGTGTCA
                                                                             stop pVI
       910        920        930        940        950        960        970        980        990       1000
ACATACGCAG GTACAGCGGC GGTCTCCTCG ACGACTCGGC GGCACGCGGG CGAAAGGTTC TACCGATGGG GAAGCTACTA CGGCGTCACC AGAATGTACG
TGTATGCGTC CATGTCGCCG CCAGAGGAGC TGCTGAGCCG CCGTGCGCCC GCTTTCCAAG ATGGCTACCC CTTCGATGAT GCCGCAGTGG TCTTACATGC
                 splice ↑ hexon RNA                          start hexon
      1010       1020       1030       1040
TGTAGAGCCC GGTCCTGCGG AGCCTCATGG ACTCGGGCCC
ACATCTCGGG CCAGGACGCC TCGGAGTACC TGAGCCCGGG
```

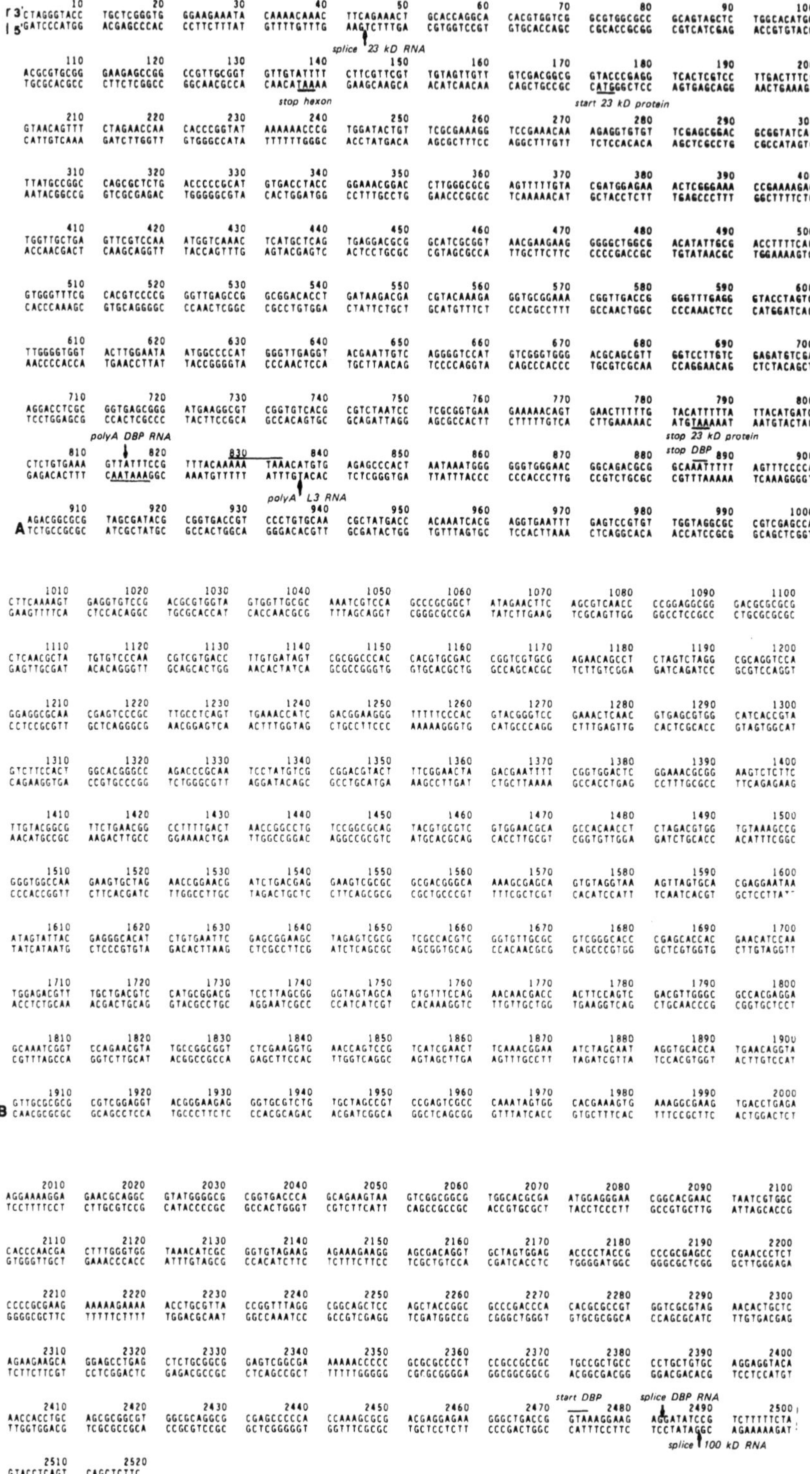

FIGURE 20.1A–C. Nucleotide sequence of a region between coordinates 59.5 and 66.4 on the Ad2 genome. This sequence and the positions of strategic sequences were determined by Akusjärvi *et al.* (1981) (nucleotides 1–1164) and Kruijer *et al.* (1982) (nucleotides 858–2514).

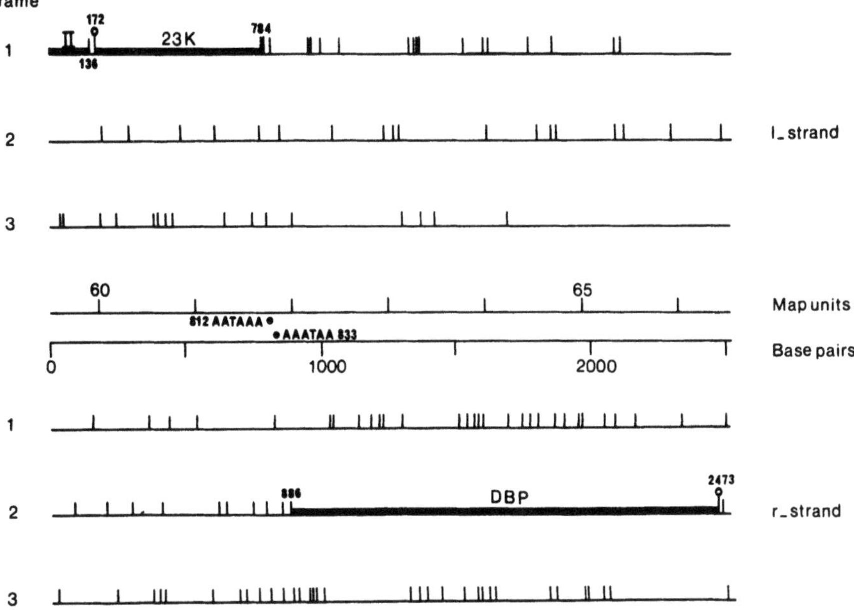

FIGURE 20.2. Structural organization of a region between coordinates 59.5 and 66.4 on the Ad2 genome. This map is derived from the nucleotide sequence in Fig. 20.1. For explanation of the symbols, see the Fig. 3 caption (Section VII).

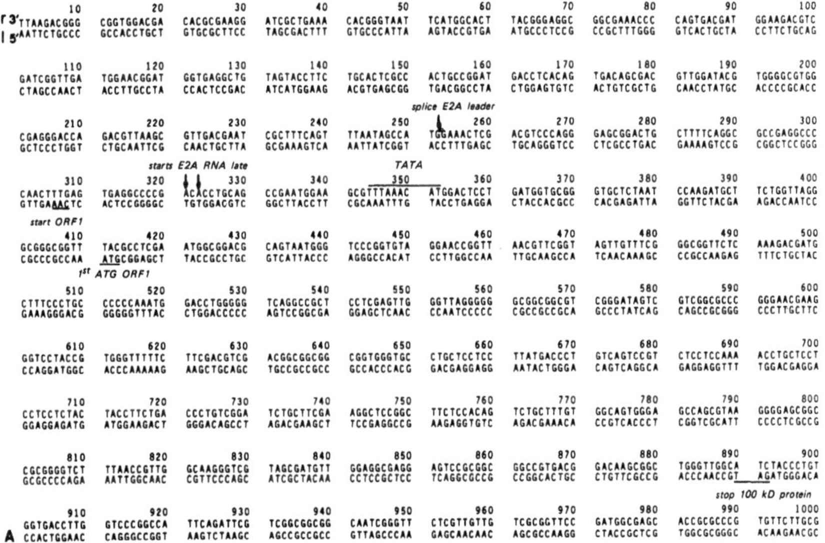

FIGURE 21A–K. Nucleotide sequence of a region between coordinates 70.7 and 100.0 on the Ad2 genome. This sequence was established by Galibert *et al.* (1979), Hérissé *et al.* (1980), and Hérissé and Galibert (1981). Short sequences were also determined by Zain *et al.* (1979a,b), Zain and Roberts (1979), Baker and Ziff (1980, 1981), Arrand and Roberts (1979), and Shinagawa *et al.* (1980). The region between 89.5 and 100 was also determined by Gingeras *et al.* (1982).

1010	1020	1030	1040	1050	1060	1070	1080	1090	1100
GTATCAACGA	ACGAACGTTC	TGACACCCCC	GTTGTAGAGG	AAGCGGGCGG	CGAAAGAAGA	GATGGTAGTG	CCGCACCGGA	AGGGGGCATT	GTAGGACGTA
CATAGTTGCT	TGCTTGCAAG	ACTGTGGGGG	CAACATCTCC	TTCGCCCGCC	GCTTTCTTCT	CTACCATCAC	GGCGTGGCCT	TCCCCCGTAA	CATCCTGCAT
start ORF 2									
1110	**1120**	**1130**	**1140**	**1150**	**1160**	**1170**	**1180**	**1190**	**1200**
ATGATGGCAG	TAGAGATGTC	GGGGATGACG	TGGCCGCCGT	CGCCGTCGCC	GTCGTTGTCG	TCGCCAGTGT	GTCTTCGTTT	CCGCTGGCCT	ATCGTTCTGA
TACTACCGTC	ATCTCTACAG	CCCCTACTGC	ACCGGCGGCA	GCGGCAGCGG	CAGCAACAGC	AGCGGTCACA	CAGAAGCAAA	GGCGACCGGA	TAGCAAGACT
								stop ORF 1	
1210	**1220**	**1230**	**1240**	**1250**	**1260**	**1270**	**1280**	**1290**	**1300**
GACTGTTTCG	GGTTCTTTAG	GTGTCGCCGC	CGTCGTCGTC	CTCCTCCTCG	CGACGCAGAC	CGCGGGTTGC	TTGGGCATAG	CTGGGCGCTC	GAATCTTTAT
CTGACAAAGC	CCAAGAAATC	CACAGCGGCG	GCAGCAGCAG	GAGGAGGAGC	GCTGCGTCTG	GCGCCCAACG	AACCCGTATC	GACCCGCGAG	CTTAGAAATA
								splice E2A leader	
1310	**1320**	**1330**	**1340**	**1350**	**1360**	**1370**	**1380**	**1390**	**1400**
CCTAAAAAGG	GTGAGACATA	CGATATAAAG	TTGTTTCGTC	CCCGGTTCTT	GTTCTCGACT	TTTATTTTTT	GTCCAGAGAC	GCGAGGGAGT	GGGCGTCGAC
GGATTTTTCC	CACTCTGTAT	GCTATATTTC	AACAAAGCAG	GGGCCAAGAA	CAAGAGCTGA	AAATAAAAAA	CAGGTCTCTG	CGCTCCCTCA	CCCGCAGCTG
					start E2A RNA early			*TATA*	
1410	**1420**	**1430**	**1440**	**1450**	**1460**	**1470**	**1480**	**1490**	**1500**
GGACATAGTG	TTTTCGCTTC	TAGTCGAAGC	CGCGTGCGAC	CTTCTGCGCC	TCCGAGAGAA	GTCGTTTATG	ACGCGCGACT	GAGAATTCCT	GATCAAAGCG
CCTGTATCAC	AAAAGCGAAG	ATCAGCTTCG	GCGCACGCTG	GAAGACGCGG	AGGCTCTCTT	CAGCAAATAC	TGCGCGCTGA	CTCTTAAGGA	CTAGTTTCGC
								stop ORF 2	
1510	**1520**	**1530**	**1540**	**1550**	**1560**	**1570**	**1580**	**1590**	**1600**
CGGGAAAGAG	TTTAAATTCG	CGCTTTTGAT	GCAGTAGAGG	TCGCCGGTGT	GGGCCGCGGT	CGTGGACAGC	AGTCGCGGTA	ATACTCGTTC	CTTTAAGGGT
GCCCTTTCTC	AAATTTAAGC	GCGAAAACTA	CGTCATCTCC	AGCGGCCACA	CCCGGCGCCA	GCACCTGTCG	TCAGCGCCAT	TATGAGCAAG	GAAATTCCCA
								start protein pVIII	
1610	**1620**	**1630**	**1640**	**1650**	**1660**	**1670**	**1680**	**1690**	**1700**
GCGGGATGTA	CACCTCAATG	GTCGGTGTTT	ACCCTGAACG	CCGACCTCGA	CGGGTTCTGA	TGAGTTGGGC	TTATTTGATG	TACTCGCGCC	CTGGGGTGTA
CGCCCTACAT	GTGGAGTTAC	CAGCCACAAA	TGGGACTTGC	GGCTGGAGCT	GCCCAAGACT	ACTCAACCCG	AATAAACTAC	ATGAGCGCGG	GACCCCACAT
1710	**1720**	**1730**	**1740**	**1750**	**1760**	**1770**	**1780**	**1790**	**1800**
CTATAGGGCC	CAGTTGCCTT	AGGCGCGGGT	GGCTTTGGCT	TAAGAGGAGC	TTGTCCGCCG	ATAATGGTGG	TGTGGAGCAT	TATTGGAATT	AGGGGCATCA
GATATCCCGG	GTCAACGGAA	TCCGCGCCCA	CCGAAACCGA	ATTCTCCTCG	AACAGGCGGC	TATTACCACC	ACACCTCGTA	ATAACCTTAA	TCCCCGTAGT
1810	**1820**	**1830**	**1840**	**1850**	**1860**	**1870**	**1880**	**1890**	**1900**
ACCGGGCGAC	GGGACCACAT	GGTCCTTTCA	GGGCGAGGGT	GGTGACACCA	TGAAGGGTCT	CTGCGGGTCC	GGCTTCAAGT	CTACTGATTG	AGTCCCCGCG
TGGCCCGCTG	CCCTGGTGTA	CCAGGAAAGT	CCCGCTCCCA	CCACTGTGGT	ACTTCCCAGA	GACGCCCAGG	CCGAAGTTCA	GATGACTAAC	TCAGGGGCGC
1910	**1920**	**1930**	**1940**	**1950**	**1960**	**1970**	**1980**	**1990**	**2000**
TCGAACGCCC	GCCGAAAGCA	GTGTCCCACG	CCAGCGGGCC	CGTCCCATAT	TGAGTGGACT	TTTAGTCTCC	CGCTCCATAA	GTCGAGTTGC	TGCTCAGCCA
B AGCTTGCGGG	CGGCTTTCGT	CACAGGGTGC	GGTCGCCCGG	GCAGGGTATA	ACTCACCTGA	AAATCAGAGG	GCGAGGTATT	CAGCTCAACG	ACGAGTCGGT
				TATA			*start E3 RNA*		
2010	**2020**	**2030**	**2040**	**2050**	**2060**	**2070**	**2080**	**2090**	**2100**
CTCGAGGAGA	GAACCAGAGG	CAGGCCTGCC	CTGTAAAGTC	TAGCCGCCGC	GACCGGCGAG	AAGTAAATGC	GGGGCAGTCC	GCTAGGATTG	AGACGTCTGG
GAGCTCCTCT	CTTGGTCTCC	GTCCGGACGG	GACATTTCAG	ATCGGCGGCG	CTGGCCGCTC	TTCATTTACG	CCCCGTCAGG	CGATCCTAAC	TCTGCAGACC
2110	**2120**	**2130**	**2140**	**2150**	**2160**	**2170**	**2180**	**2190**	**2200**
AGCAGGAGCC	TCGGCGCGAG	GCCTCCGTAA	CCTTGAGATG	TTAAATAACT	CCTCAAGCAC	GGAAGCCAAA	TGAAGTTGGG	GAAAAGACCT	GGAGGGCCGG
TCGTCCTCGG	AGCCGCGCTC	CGGAGGCATT	GGAACTCTAC	AATTTATTGA	GGAGTTCGTG	CCTTCGGTTT	ACTTCAACCC	CTTTTCTGGA	CCTCCCGGCC
2210	**2220**	**2230**	**2240**	**2250**	**2260**	**2270**	**2280**	**2290**	**2300**
TGATGGGCCT	GGTCAAATAA	GGGTTGAAAC	TGCGCCACTT	TCTGAGCCGC	CTGCCGATGC	TGACTTACTG	GTCACCTCTC	CGTCTCGCTG	ACGCGGACTG
ACTACCCGGA	CCAGTTTATT	CCCAACTTTG	ACGCGGTGAA	AGACTCGGCG	GACGGCTACG	ACTGAATGAC	CAGTGGAGAG	GCAGAGCGAC	TGCGCCTGAC
	splice x leader					*stop protein pVIII*			
2310	**2320**	**2330**	**2340**	**2350**	**2360**	**2370**	**2380**	**2390**	**2400**
TGTGGAGCTG	GTGACGGCGG	CGGTGTTCAC	GAAACGGGCG	CCGAGGCCAC	TCAAAACAAT	GAAACTTAAC	GGGCTTCTCG	TATAGCTCCC	GGGCCGCGTG
ACACCTCGAC	CACTGCCGCC	GCCACAAGTG	CTTTGCCCGC	GGCTCCGGTG	AGTTTTGTTA	CTTTGAATTG	CCCGAAGAGC	ATATCGAGGG	CCCGGCGCAC
				splice x leader					
2410	**2420**	**2430**	**2440**	**2450**	**2460**	**2470**	**2480**	**2490**	**2500**
CCGCAGGCCG	AGTGGTGGGT	CCATCTCGAA	TGTGCATCGG	ACTAAGCCCT	CAAATGGTTC	GCGGGGGACG	ATCACCTCGC	CCTCGCCCCA	GGGACACAAG
GGCGTCCGGC	TCACCACCCA	GGTAGAGCTT	ACACGTAGCC	TGATTCGGGA	GTTTACCAAG	CGCCCCCTGC	TAGTGGAGCG	GGAGCGGGGT	CCCTGTGTTC
2510	**2520**	**2530**	**2540**	**2550**	**2560**	**2570**	**2580**	**2590**	**2600**
ACTGGCACCA	AACGTTGACA	GGATTGGGAC	CTAATGTAGT	TCTAGAAACA	ACAGTAGAGA	CACGACTCAT	ATTATTTATG	TCTTTAATCT	TAGATGACCC
TGACCGTGGT	TTGCAACTGT	CCTAACCCTG	GATTACATCA	AGATCTTTGT	TGTCATCTCT	GTGCTGAGTA	TAATAAATAC	AGAAATTAGA	ATCTACTGGG
2610	**2620**	**2630**	**2640**	**2650**	**2660**	**2670**	**2680**	**2690**	**2700**
CGAGGACAGC	GGTAGGACAC	TTGCGGTGGC	AAAAATGGGT	GGGTTTCGTC	TGGTTTCGTT	TGGAGTGGAG	GCCAAACGTG	TTCGCCCGGT	TATTCATGGA
GCTCCTGTCG	CCATCCTGTG	AACGCCACCG	TTTTTACCCA	CCCAAAGCAG	ACCAAAGCAA	ACCTCACCTC	CGGTTTGCAC	AAGCGGGCCA	ATAAGTACCT
2710	**2720**	**2730**	**2740**	**2750**	**2760**	**2770**	**2780**	**2790**	**2800**
ATGGACCATG	AAATTGCCGA	GAAGTAAACA	TTAAATGTTG	TCAAAGGTCG	CTCTGCTTCA	TTCAAACGGT	GTGTTGGAAG	AGCCGAAGTT	GATGTGGCAG
TACCTGGTAC	TTTAACGGCT	CTTCATTTGT	AATTTACAAC	AGTTTCCAGC	GAGACGAAGT	AAGTTTGCCA	CACAACCTTC	TCGGCTTCAA	CTACACCGTC
				splice y leader					
2810	**2820**	**2830**	**2840**	**2850**	**2860**	**2870**	**2880**	**2890**	**2900**
TTCTTTTTGT	GGTGGTGGTG	GTGGGAGGAG	TGGACGGCCC	TTGCATGCTC	ACGCAGTGGC	CAACGACGCG	GGTGTGGATG	TCGGACTCGC	ATTGGTCTGT
AAGAAAAACA	CCACCACCAC	CACCCTCCTC	ACCTGCCGGG	AACGTACGAG	TGCGTCACCG	GTTGCTGCGC	CCACACCTAC	AGCCTGAGCG	TAACCAGACA
2910	**2920**	**2930**	**2940**	**2950**	**2960**	**2970**	**2980**	**2990**	**3000**
AATGAGGGTA	AAAAGGTTTT	GTCCTCCACT	CGAGTTGAGG	GCCTTGAGTC	CAGTTTTTTC	GTAAAACGCC	CCACGACCCT	AAAAAATTAA	TTCATATACT
C TTACTCCCAT	TTTTCCAAAA	CAGGAGGTGA	GCTCAACTCC	CGGAACTCAG	GTCAAAAAAG	CATTTTGCGG	GGTGCTGGGA	TTTTTTAATT	AAGTATATGA
		splice y leader							

FIGURE 21 (*Continued*)

```
      3010        3020        3030        3040        3050        3060        3070        3080        3090        3100
CGTTAAGTTC  ATTGAGATGT  TCGAACAGAT  TAAAAAGACC  TTAACCCCAG  CCCCAATAGG  AATGAGAACA  TTAAGACAAA  TAAGAATATG  ATCGTGAAGA
GCAATTCAAG  TAACTCTACA  AGCTTGTCTA  ATTTTTCTGG  AATTGGGGTC  GGGGTTATCC  TTACTCTTGT  AATTCTGTTT  ATTCTTATAC  TAGCACTTCT

      3110        3120        3130        3140        3150        3160        3170        3180        3190        3200
CACGGAATCC  CAACGGCGGA  CGACGTGCGT  GCAAACATGG  ATAACAGTCG  AAAAATTTGC  GACCCCCGTT  GTAGGTTCTA  CTCCATGTAC  TAAAATCCGA
GTGCCTTAGG  GTTGCCGCCT  GCTGCACGCA  CGTTTGTACC  TATTGTCAGC  TTTTTAAACG  CTGGGGGCAA  CATCCAAGAT  GAGGTACATG  ATTTTAGGCT
                                                                                      start E3 16 kD protein

      3210        3220        3230        3240        3250        3260        3270        3280        3290        3300
ACGAGCGGGA  ACGCCGTCAG  ACGTCGCGAC  GGTTTTTCCA  ACTCAAATTC  CTTGGTCGAA  CGTTACAATG  TAAATTTAGT  CTTCGATTAC  TTACGTGATG
TGCTCGCCCT  TGCGGCAGTC  TGCAGCGCTG  CCAAAAAGGT  TGAGTTTAAG  GAACCAGCTT  GCAATGTTAC  ATTTAAATCA  GAAGCTAATG  AATGCACTAC

      3310        3320        3330        3340        3350        3360        3370        3380        3390        3400
AGAATATTTT  ACGTGGTGTC  TTGTACTTTT  CGAATAATAA  GCGGTGTTTC  TGTTTTAACC  GTTCATACGA  CATATACGAT  AAACCGTCGG  TCCACTGTGA
TCTTATAAAA  TGCACCACAG  AACATGAAAA  GCTTATTATT  CGCCACAAAG  ACAAAATTGG  CAAGTATGCT  GTATATGCTA  TTTGGCAGCC  AGGTGACACT

      3410        3420        3430        3440        3450        3460        3470        3480        3490        3500
TTGCTGATAT  TACAGTGTCA  GAAGGTTCCA  CTTTTAGCAT  TTTGAAAATA  CATATTTAAA  GGTAAAATAC  TTTACACGCT  ATAATGGTAC  ATGTACTCGT
AACGACTATA  ATGTCACAGT  CTTCCAAGGT  GAAAATCGTA  AAACTTTTAT  GTATAAATTT  CCATTTTATG  AAATGTGCGA  TATTACCATG  TACATGAGCA

      3510        3520        3530        3540        3550        3560        3570        3580        3590        3600
TTGTCATGTT  CAACACCGGG  GGTGTTTTCA  CAAATCTCTT  GTGACCGTGG  AAAACAAGGT  GGCGAGACGA  ATAATGTCCC  GAACGAAACC  ATACATGGAA
AACAGTACAA  GTTGTGGCCC  CCACAAAAGT  GTTTAGAGAA  CACTGGCACC  TTTTGTTCCA  CCGCTCTGCT  TATTACAGCG  CTTGCTTTGG  TATGTACCTT

      3610        3620        3630        3640        3650        3660        3670        3680        3690        3700
TGAAATAGAG  TTTATGTTTT  CGTCTGCGTC  AAAATAACTA  CTTTTCTTTT  ACGGAACTAA  AAGGCGAACG  AACATAAGGG  GACCTGTTAA  ATGAGATACA
ACTTTATCTC  AAATACAAAA  GCAGACGCAG  TTTTATTGAT  GAAAAGAAAA  TGCCTTGATT  TTCCGCTTGC  TTGTATTCCC  CTGGACAATT  TACTCTATGT
                                    start URF 13            stop E3 16 kD protein

      3710        3720        3730        3740        3750        3760        3770        3780        3790        3800
CCCTATACGA  GGTCCGCCCG  TTCTAATATG  GGTGTTGGAA  GTTTAGTTTG  AAAGGACCTG  CAATCGCGGA  CTAAAGACGG  TCGCGGACGT  GACGTTTAAA
GGGATATGCT  CCAGGCGGGC  AAGATTATAC  CCACAACCTT  CAAATCAAAC  TTTCCTGGAC  GTTAGCGCCT  GATTTCTGCC  AGCGCCTGCA  CTGCAAATTT

      3810        3820        3830        3840        3850        3860        3870        3880        3890        3900
CTAGTTTGGG  TCGAAGTCGA  ACGGACGAGG  TCTCTACTGG  CCGAGTTGGT  AGCGCGGGTG  TTGCCTGATA  GCGTTGTGGT  GACGATGGCC  TGATTGTAGA
GATCAAACCC  AGCTTCAGCT  TGCCTGCTCC  AGAGATGACC  GGCTCAACCA  TCGCGCCCAC  AACGGACTAT  CGCAACACCA  CTGCTACCGG  ACTAACATCT
                                    start URF 14

      3910        3920        3930        3940        3950        3960        3970        3980        3990        4000
 CGGGATTTAA  ATGGGGTTCA  AGTACGGAAA  CAGTTACTGA  CCCGCTCGAA  CCTGTACACC  ACCAAAAGGT  -ATCGCGAATA  CAAACAAACG  GAATAATAAT
D GCCCTAAATT  TACCCCAAGT  'CATGCCTTT  GTCAATGACT  GGGCGAGCTT  GGACATGTGG  TGGTTTTCCA  TAGCGCTTAT  GTTTGTTTGC  CTTATTATTA
                                    stop URF 13

      4010        4020        4030        4040        4050        4060        4070        4080        4090        4100
ACACCGAATA  AACAACGGAT  TTCGCGTCTG  CGCGGTCTGG  GGGGTAGATA  TCCGGATAGT  AACACGAGTT  GGGTGTGTTA  CTTTTTTAAG  TATCTAACCT
TGTGGCTTAT  TTGTTGCCTA  AAGCGCAGAC  GCGCCAGACC  CCCCATCTAT  AGGCCTATCA  TTGTGCTCAA  CCCACACAAT  GAAAAAATTC  ATAGATTGGA

      4110        4120        4130        4140        4150        4160        4170        4180        4190        4200
GCCAGACTTT  GGTACAAGAG  AAGAAAATGT  CATACTAATT  TACTCTGTAC  TAAGGAGCTC  AAGAATATAA  TAACTGGGAA  CAACGCGAAA  AGACACGCAC
CGGTCTGAAA  CCATGTTCTC  TTCTTTTACA  GTATGATTAA  ATGAGACATG  ATTCCTCGAG  TTCTTATATT  ATTGACCCTT  GTTGCGCTTT  TCTGTGCGTG
                                    stop URF 14   start URF 15

      4210        4220        4230        4240        4250        4260        4270        4280        4290        4300
GAGATGTAAC  CGGCGCCAGC  GAGTGTAGCT  TCATCTAACG  TAGGGTGGAA  AGTGTCAAAT  GGACGAAATG  CCTAAACAGT  GGGAATAGGA  GTAGACGTCG
CTCTACATTG  GCCGCGGTCG  CTCACATCGA  AGTAGATTGC  ATCCCACCTT  TCACAGTTTA  CCTGCTTTAC  GGATTTGTCA  CCCTTATCCT  CATCTGCAGC

      4310        4320        4330        4340        4350        4360        4370        4380        4390        4400
GAGCAGTGAC  ATCAGTAGCG  GAAGTAAGTC  AAGTAACTGA  CCCAAACACA  CGCGTAACGC  ATGGAGTCCG  TGGTAGGCGT  TATGTCTCTG  TCCTGATATC
CTCGTCACTG  TAGTCATCGC  CTTCATTCAG  TTCATTGACT  GGGTTTGTGT  GCGCATTGCG  TACCTCAGGC  ACCATCCGCA  ATACAGAGAC  AGGACTATAG

      4410        4420        4430        4440        4450        4460        4470        4480        4490        4500
GACTAGAAGA  GTCTTAAGAA  ATTAATACTT  TGCCTCACAG  TAAAAACAAA  ACGACTAAAA  AACGCGGGAT  GGACACGAAA  CGAGGGTTTG  GAGTCGCGGA
CTGATCTTCT  CAGAATTCTT  TAATTATGAA  ACGGAGTGTC  ATTTTTGTTT  TGCTGATTTT  TTGCGCCCTA  CCTGTGCTTT  GCTCCCAAAC  CTCAGCGCCT
                  stop URF 15   start URF 16

      4510        4520        4530        4540        4550        4560        4570        4580        4590        4600
GGGTTTTCTG  TATAAAGGAC  GTCTAAGTGA  GTTTATACCT  TGTAAGGGTC  GACGATGTTG  TTTGTCTCGC  TAAACAGTCT  TCGGACCAAT  ATGCGGTAGT
CCCAAAAGAC  ATATTTCCTG  CAGATTCACT  CAAATATGGA  ACATTCCCAG  CTGCTACAAC  AAACAGAGCG  ATTTGTCAGA  AGCCTGGTTA  TACGCCATCA

      4610        4620        4630        4640        4650        4660        4670        4680        4690        4700
AGAGACAGTA  CCAAAAAACG  TCATGGTAAA  AACGGGATCG  GTATATAGGT  ATGGAACTGT  AACCGACCTT  ACGGTATCTA  CGGTACTTGG  TGGGATGAAA
TCTCTGTCAT  GGTTTTTTGC  AGTACCATTT  TTGCCCTAGC  CATATATCCA  TACCTTGACA  TTGGCTGGAA  TGCCATAGAT  GCCATGAACC  ACCCTACTTT

      4710        4720        4730        4740        4750        4760        4770        4780        4790        4800
GGGTCACGGG  CGACAGTATG  GTGACGTTGT  CCAATAACGG  GGTTAGTTAG  TCGGAGCGGG  GGGAAGAGGG  TGGGGGTGAC  TCTAATCGAT  GAAATTAAAC
CCCAGTGCCC  GCTGTCATAC  CACTGCAACA  GGTTATTGCC  CCAATCAATC  AGCCTCGCCC  CCCTTCTCCC  ACCCCCACTG  AGATTAGCTA  CTTTAATTTG

      4810        4820        4830        4840        4850        4860        4870        4880        4890        4900
TGTCCACCTC  TACTGACTTA  GAGATCTAGA  TCTTAACCTA  CCTTAATTGT  GGCTTGTCGC  GGATGATCTT  TCCGCGTTCC  GCCGCAGGCT  CGCTCTTGCG
ACAGGTGGAG  ATGACTGAAT  CTCTAGATCT  AGAATTGGAT  GGAATTAACA  CCGAACAGCG  CCTACTAGAA  AGGCGCAAGG  CGGCGTCCGA  GCGAGAACGC
splice    start URF 17   stop URF 16
z leader
      4910        4920        4930        4940        4950        4960        4970        4980        4990        5000
 GATTTTGTTC  TTCAACTTCT  GTACCAATTG  GATGTGGTCA  CATTTTCTCC  ATAGAAAACA  CACCAGTTCG  TCCGGTTTGA  ATGGATGCTT  TTTTGGTGAT
E CTAAAACAAG  AAGTTGAAGA  CATGGTTAAC  CTACACCAGT  GTAAAAGAGG  TATCTTTTGT  GTGGTCAAGC  AGGCCAAACT  TACCTACGAA  AAAACCACTA
                                                                      splice z leader
```

FIGURE 21 (*Continued*)

5010	5020	5030	5040	5050	5060	5070	5080	5090	5100
GGCCGTTGGC	GGAGTCGATG	TTCGATGGGT	GGGTCGCGGT	TTTTGACCAC	GAATACCACC	CTCTTTTTGG	ATAGTGGCAG	TGGGTCGTGA	GCCGTCTTTG
CCGGCAACCG	CCTCAGCTAC	AAGCTACCCA	CCCAGCGCCA	AAAACTGGTG	CTTATGGTGG	GAGAAAAACC	TATCACCGTC	ACCCAGCACT	CGGCAGAAAC
5110	5120	5130	5140	5150	5160	5170	5180	5190	5200
TCTCCCGACG	GACGTGAAGG	GGATAGTCCC	AGGTCTCCTG	GAGACGTGAG	AATAATTTTG	GTACACACCA	TAATCTCTAG	AATAAGGTAA	GTTGATTGTA
AGAGGGCTGC	CTGCACTTCC	CCTATCAGGG	TCCAGAGGAC	CTCTGCACTC	TTATTAAAAC	CATGTGTGGT	ATTAGAGATC	TTATTCCATT	CAACTAACAT
	stop URF 18								*stop URF 17*
5210	5220	5230	5240	5250	5260	5270	5280	5290	5300
TTTGTGTGTT	ATTTAATGAA	TGAATTTTAG	TCAGTCGTTT	AGAAACAGGT	CGAATAAGTC	GTAGTGGAGG	AAAGGAAGGA	GGGTTGAGAC	CATAGAGTCG
AAACACACAA	TAAATTACTT	ACTTAAAATC	AGTCAGCAAA	TCTTTGTCCA	GCTTATTCAG	CATCACCTCC	TTTCCTTCCT	CCCAACTCTG	GTATCTCAGC
5310	5320	5330	5340	5350	5360	5370	5380	5390	5400
GCGGAAAATC	GACGTTTGAA	AGAGGTTTCA	AATTTACCCT	ACAGTTTAAG	GAGTACAAGA	ACAGGGAGGC	GTGGGTGATA	GAAGTATAAC	AACGTCTACT
CGCCTTTTAG	CTGCAAACTT	TCTCCAAAGT	TTAAATGGGA	TGTCAAATTC	CTCATGTTCT	TGTCCCTCCG	CACCCACTAT	CTTCATATTG	TTGCAGATGA
								splice L5 RNA	*start fiber*
5410	5420	5430	5440	5450	5460	5470	5480	5490	5500
TTGCGCGGTC	TGGCAGACTT	CTGTGGAAGT	TGGGGCACAT	AGGTATACTG	TGTCTTTGGC	CCGGAGGTTG	ACACGGGAAA	GAATGGGGAG	GTAAACAAAG
AACGCGCCAG	ACCGTCTGAA	GACACCTTCA	ACCCCGTGTA	TCCATATGAC	ACAGAAACCG	GGCCTCCAAC	TGTGCCCTTT	CTTACCCCTC	CATTTGTTTC
							start URF 18		
5510	5520	5530	5540	5550	5560	5570	5580	5590	5600
TGGGTTACCA	AAGGTTCTTT	CAGGGGGACC	TCAAGAGAGA	GATGCGCAGA	GGCTTGGAAA	CCTGTGGAGG	GTGCCGTACG	AACGCGAATT	TTACCCGTCG
ACCCAATGGT	TTCCAAGAAA	GTCCCCCTGG	AGTTCTCTCT	CTACGCGTCT	CCGAACCTTT	GGACACCTCC	CACGGCATGC	TTGCGCTTAA	AATGGGCAGC
					stop URF 19				
5610	5620	5630	5640	5650	5660	5670	5680	5690	5700
CCAGAATGGG	ATCTGTTCCG	GCCTTTGGAG	TGGAGGGTTT	TACATTGGTG	ACAATGAGTC	GGTGAATTTT	TTTGTTTCAG	TTTGTATTCA	AACCTGTGGA
GGTCTTACCC	TAGACAAGGC	CGGAAACCTC	ACCTCCCAAA	ATGTAACCAC	TGTTACTCAG	CCACTTAAAA	AAACAAAGTC	AAACATAAGT	TTGGACACCT
5710	5720	5730	5740	5750	5760	5770	5780	5790	5800
GGCGTGGTGA	ATGTTAATGG	AGTCCGCGGG	ATTGTCACCG	TTGGTGGCGA	GGAGACTATC	AATGATCGCC	GCGAGAATCG	CATGTCAGTG	TTCGGGGTGA
CCGCACCACT	TACAATTACC	TCAGGCGCCC	TAACAGTGGC	AACCACCGCT	CCTCTGATAG	TTACTAGCGG	CGCTCTTAGC	GTACAGTCAC	AAGCCCCACT
5810	5820	5830	5840	5850	5860	5870	5880	5890	5900
CTGGCACGTT	CTGAGGTTTG	ATTCGTAACG	ATGATTTCCC	GGGTAATGTC	ACAGTCTACC	TTTCGATCGG	GACGTTTGTA	GTCGGGGGGA	GAGACCGTCA
GACCGTGCAA	GACTCCAAAC	TAAGCATTGC	TACTAAAGGG	CCCATTACAG	TGTCAGATGG	AAAGCTAGCC	CTGCAAACAT	CAGCCCCCCT	CTCTGGCAGT
						start URF 19			
5910	5920	5930	5940	5950	5960	5970	5980	5990	6000
CTGTCGCTGT	GGGAATGACA	TTGACGTAGT	GGGGGCGATT	GATGACGGTG	CCCATCGAAC	CCGTAATTGT	ACCTTCTAGG	ATAAATACAT	TTATTACCTT
F GACAGCGACA	CCCTTACTGT	AACTGCATCA	CCCCCGCTAA	CTACTGCCAC	GGGTAGCTTG	GGCATTAACA	TGGAAGATCC	TATTTATGTA	AATAATGGAA

6010	6020	6030	6040	6050	6060	6070	6080	6090	6100
TTTATCCTTA	ATTTTATTCG	CCAGGAAACG	TTCATCGTGT	TTTGAGGCTA	TGTGATTGTC	ATCAATGACC	TGGTCCACAG	TGGCAACTTG	TTTTGAGGGA
AAATAGGAAT	TAAAATAAGC	GGTCCTTTGC	AAGTAGCACA	AAACTCCGAT	ACACTAACAG	TAGTTACTGG	ACCAGGTGTC	ACCGTTGAAC	AAAACTCCCT
6110	6120	6130	6140	6150	6160	6170	6180	6190	6200
ATCTTGGTTT	CAACGTCCTC	GATAACCAAT	ACTAAGTAGT	TTGTTGTACC	TTTAATTTTG	CCCGCCACCG	TACGCATATT	TATTGTTGAA	CAATTAAGAT
TAGAACCAAA	GTTGCAGGAG	CTATTGGTTA	TGATTCATCA	AACAACATGG	AAATTAAAAC	GGGCGGTGGC	ATGCGTATAA	ATAACAACTT	GTTAATTCTA
6210	6220	6230	6240	6250	6260	6270	6280	6290	6300
CTACACCTAA	TGGGTAAACT	ACGAGTTTGT	TTTGATGCAG	AATTTGACCC	CGTCCCTGGG	GACATATAAT	TACGTAGAGT	ATTGAACCTG	TATTTGATAT
GATGTGGATT	ACCCATTTGA	TGCTCAAACA	AAACTACGTC	TTAAACTGGG	GCAGGGACCC	CTGTATATTA	ATGCATCTCA	TAACTTGGAC	ATAAACTATA
6310	6320	6330	6340	6350	6360	6370	6380	6390	6400
TGTCTCCGGA	TATGGAAAAA	TTACGTAGTT	TGTTATGATT	TTTTGACCTT	CAATCGTATT	TTTTTAGGTC	ACCTGATTTG	AAACTATTAT	GACGGTATCG
ACAGAGGCCT	ATACCTTTTT	AATGCATCAA	ACAATACTAA	AAAACTGGAA	GTTAGCATAA	AAAAATCCAG	TGGACTAAAC	TTTGATAATA	CTGCCATAGC
6410	6420	6430	6440	6450	6460	6470	6480	6490	6500
ATATTTACGT	CCTTTCCCAG	ACCTCAAACT	ATGTTTGTGT	AGACTCAGAG	GTCTATAGTT	GGGTTATTTT	TGATTTTAAC	CGAGACCGTA	ACTAATGTTA
TATAAATGCA	GGAAAGGGTC	TGGAGTTTGA	TACAAACACA	TCTGAGTCTC	CAGATATCAA	CCCAATAAAA	ACTAAAATTG	GCTCTGGCAT	TGATTACAAT
6510	6520	6530	6540	6550	6560	6570	6580	6590	6600
CTTTTGCCAC	GGTACTAATG	ATTTGAACCT	CGCCCAAATT	CGAAACTGTT	GAGTCCCCGG	TAATGTTATC	CTTTGTTTTT	ACTACTGTTT	GAATGGGACA
GAAAACGGTG	CCATGATTAC	TAAACTTGGA	GCGGGTTTAA	GCTTTGACAA	CTCAGGGGCC	ATTACAATAG	GAAACAAAAA	TGATGACAAA	CTTACCCTGT
6610	6620	6630	6640	6650	6660	6670	6680	6690	6700
CCTGTTGGGG	TCTGGGTAGA	GGATTGACGT	CTTAAGTAAG	TCTATTACTG	ACGTTTAAAT	GAAACCAAGA	ATGTTTTACA	CCCTCAGTTC	ATGATCGATG
GGACAACCCC	AGACCCATCT	CCTAACTGCA	GAATTCATTC	AGATAATGAC	TGCAAATTTA	CTTTGGTTCT	TACAAAATGT	GGGAGTCAAG	TACTAGCTAC
6710	6720	6730	6740	6750	6760	6770	6780	6790	6800
ACATCGACGA	AACCGACATA	GACCTCTAGA	AAGTAGGTAC	TGTCCGTGGC	AACGTTCACA	ATCATATAAG	GAATCTAAAC	TGGTTTTGCC	ACAAGATTAC
TGTAGCTGCT	TTGGCTGTAT	CTGGAGATCT	TTCATCCATG	ACAGGCACCG	TTGCAAGTGT	TAGTATATTC	CTTAGATTTG	ACCAAAACGG	TGTTCTAATG
6810	6820	6830	6840	6850	6860	6870	6880	6890	6900
CTCTTGAGGA	GTGAATTTTT	TGTAATGACC	TTGAAATCTT	TACCCTTGAG	TTGATTACGT	TTAGGTATGT	GTTTACGTCA	ACCTAAATAC	GGATTGGAAG
GAGAACTCCT	CACTTAAAAA	ACATTACTGG	AACTTTAGAA	ATGGGAACTC	AACTAATGCA	AATCCATACA	CAAATGCAGT	TGGATTTATG	CCTAACCTTC
stop ORF 3									
6910	6920	6930	6940	6950	6960	6970	6980	6990	7000
ATCGGATAGG	TTTTTGGGTT	TCAGTTTGAC	GATTTTTATT	GTAACAGTCA	GTTCAAATGA	ACGTACCACT	ATTTTGATTT	GGATACTATG	AATGGTAATG
G TAGCCTATCC	AAAAACCCAA	AGTCAAACTG	CTAAAAATAA	CATTGTCAGT	CAAGTTTACT	TGCATGGTGA	TAAAACTAAA	CCTATGATAC	TTACCATTAC

FIGURE 21 (*Continued*)

7010	7020	7030	7040	7050	7060	7070	7080	7090	7100
TGAATTACCG	TGATCACTTA	GGTGTCTTTG	ATCGCTCCAT	TCGTGAATGA	GATACAGAAA	ATGTACCAGG	ACCCTTTCAC	CTTTTATGTG	GTGACTTTGA
ACTTAATGGC	ACTAGTGAAT	CCACAGAAAC	TAGCGAGGTA	AGCACTTACT	CTATGTCTTT	TACATGGTCC	TGGGAAAGTG	GAAAATACAC	CACTGAAACT
								start ORF 3	
7110	7120	7130	7140	7150	7160	7170	7180	7190	7200
AAACGATGGT	TGAGAATGTG	GAAGAGGATG	TAACGGGTCC	TTATTTCTTA	GCACTTGGAC	AACGTACAAT	ACAAAGTTGC	ACAAATAAAA	AGTTAACGTC
TTTGCTACCA	ACTCTTACAC	CTTCTCCTAC	ATTGCCCAGG	AATAAAGAAT	CGTGAACCTG	TTGCATGTTA	TGTTTCAACG	TGTTTATTTT	TCAATTGCAG
				stop fiber				*stop ORF 4*	
7210	7220	7230	7240	7250	7260	7270	7280	7290	7300
TTTTAAAGTT	CAGTAAAAAG	TAAGTCATCA	TATCGGGGTG	GTGGTGTATC	GAATATAACT	AGTGGCATGG	AATTAGTTTG	AGTGTCTTGG	GATCATAAGT
AAAATTTCAA	GTCATTTTTC	ATTCAGTAGT	ATAGCCCCAC	CACCACATAG	CTTATATTGA	TCACCGTACC	TTAATCAAAC	TCACAGAACC	CTAGTATTCA
7310	7320	7330	7340	7350	7360	7370	7380	7390	7400
TGGACGGTGG	AGGGAGGGTT	GTGTGTCTCA	TGTGTCAGGA	AAGAGGGGCC	GACCGGAATT	TTTCGTAGTA	TAGTACCCAT	TGTCTGTATA	AGAATCCACA
ACCTGCCACC	TCCCTCCCAA	CACACAGAGT	ACACAGTCCT	TTCTCCCCGG	CTGGCCTTAA	AAAGCATCAT	ATCATGGGTA	ACAGACATAT	TCTTAGGTGT
7410	7420	7430	7440	7450	7460	7470	7480	7490	7500
ATATAAGGTG	TGCCAAAGGA	CAGCTCGGTT	TGCGAGTAGT	CACTATAATT	ATTTGAGGGG	CCCGTCGAGC	GAATTCAAGT	ACAGCGACAG	GTCGACGACT
TATATTCCAC	ACGGTTTCCT	GTCGAGCCAA	ACGCTCATCA	GTGATATTAA	TAAACTCCCC	GGGCAGCTCG	CTTAAGTTCA	TGTCGCTGTC	CAGCTGCTGA
					start ORF 4	*stop URF 21*			
7510	7520	7530	7540	7550	7560	7570	7580	7590	7600
CGGTGTCCGA	CGACAGGTTG	AACGCCAACG	AGTTGCCCGC	CGCTTCCCCT	TCAGGTGCGG	ATGTACCCCC	ATCTCAGTAT	TAGCACGTAG	TCCTATCCCG
GCCACAGGCT	GCTGTCCAAC	TTGCGGTTGC	TCAACGGGCG	GCGAAGGGGA	AGTCCACGCC	TACATGGGGG	TAGAGTCATA	ATCGTGCATC	AGGATAGGGC
7610	7620	7630	7640	7650	7660	7670	7680	7690	7700
CCACCACGAC	GTCGTCGCGC	GCTTATTTGA	CGACGGCGGC	GGCGAGGCAG	GACGTCCTTA	TGTTGTACCG	TCACCAGAGG	AGTCGCTACT	AAGCGTGGCG
GGTGGTGCTG	CAGCAGCGCG	CGAATAAACT	GCTGCCGCCG	CCGCTCCGTC	CTGCAGGAAT	ACAACATGGC	AGTGGTCTCC	TCAGCGATGA	TTCGCACCGC
7710	7720	7730	7740	7750	7760	7770	7780	7790	7800
GGCGTCGTAC	TCTGCGGAAC	AGGAGGCCCG	TGTCGTCGCG	TGGGACTAGA	GTGAATTTAG	TCGTGTCATT	GACGTCGTGT	CGTGGTGTTA	TAACAAGTTT
CCGCAGCATG	AGACGCCTTG	TCCTCCGGGC	ACAGCAGCGC	ACCCTGATCT	CACTTAAATC	AGCACAGTAA	CTGCAGCACA	GCACCACAAT	ATTGTTCAAA
7810	7820	7830	7840	7850	7860	7870	7880	7890	7900
TAGGGTGTCA	CGTTCCGCGA	CATAGGTTTC	GAGTACCGCC	CCTGGTGTCT	TGGGTGCACC	GGTAGTATGG	TGTTCGCGTC	CATCTAATTC	ACCGCTGGGG
ATCCCACAGT	GCAAGGCGCT	GTATCCAAAG	CTCATGGCGG	GGACCACAGA	ACCCACGTGG	CCATCATACC	ACAAGCGCAG	GTAGATTAAG	TGGCGACCCC
7910	7920	7930	7940	7950	7960	7970	7980	7990	8000
AGTATTTGTG	CGACCTGTAT	TTGTAATGGA	GAAAACCGTA	CAACATTAAG	TGGTGGAGGG	CCATGGTATA	TTTGGAGACT	AATTTGTACC	GCGGTAGGTG
TCATAAACAC	GCTGGACATA	AACATTACCT	CTTTTGGCAT	GTTGTAATTC	ACCACCTCCC	GGTACCATAT	AAACCTCTGA	TTAAACATGG	CGCCATCCAC

H

8010	8020	8030	8040	8050	8060	8070	8080	8090	8100
GTGGTAGGAT	TTGGTCGACC	GGTTTTGGAC	GGGCGGCCGA	TACGTGACGT	CCCTTGGCCC	TGACCTTGTT	ACTGTCACCT	CTCGGGTCCT	GAGCATTGGT
CACCATCCTA	AACCAGCTGG	CCAAAACCTG	CCCGCCGGCT	ATGCACTGCA	GGGAACCGGG	ACTGGAACAA	TGACAGTGGA	GAGCCCAGGA	CTCGTAACCA
						start URF 20			
8110	8120	8130	8140	8150	8160	8170	8180	8190	8200
ACCTAGTAGT	ACGAGCAGTA	CTATAGTTAC	AACCGTGTTG	TGTCCGTGTG	CACGTATGTG	AAGGAGTCCT	AATGTTCGAG	GAGGGCGCAG	TCTTGGTATA
TGGATCATCA	TGCTCGTCAT	GATATCAATG	TTGGCACAAC	ACAGGCACAC	GTGCATACAC	TTCCTCAGGA	TTACAAGCTC	CTCCCGCGTC	AGAACCATAT
								stop ORF5	
8210	8220	8230	8240	8250	8260	8270	8280	8290	8300
GGGTCCCTTG	TTGGGTAAGG	ACTTAGTCGC	ATTTAGGGTG	TGACGTCCCT	TCTGGAGCGT	GCATTGAGTG	CAACACGTAA	CAGTTTCACA	ATGTAAGCCC
CCCAGGGAAC	AACCCATTCC	TGAATCAGCG	TAAATCCCAC	ACTGCAGGGA	AGACCTCGCA	CGTAACTCAC	GTTGTGCATT	GTCAAAGTGT	TACATTCGGG
						stop URF 22			
8310	8320	8330	8340	8350	8360	8370	8380	8390	8400
GTCGTCGCCT	ACTAGGAGGT	CATACCATCG	CGCCCAGAGA	CAGAGTTTTC	CTCCATCCGC	TAGGGATGAC	ATGCCTCACG	CGGCTCTGTT	GGCTCTAGCA
CAGCAGCGGA	TGATCCTCCA	GTATGGTAGC	GCGGGTCTCT	GTCTCAAAAG	GAGGTAGGCG	ATCCCTACTG	TACGGAGTGC	GCCGAGACAA	CCGAGATCGT
start ORF 5		*start URF 26*		*start URF 21*	*stop URF 20*				
8410	8420	8430	8440	8450	8460	8470	8480	8490	8500
CAACCAGCAT	CACAGTACGG	TTTACCTTGC	GGCCTGCATC	AGTATAAAGG	ACTTCGTTTT	GGTCCACGCC	CGCACTGTTT	GTCTAGACGC	AGAGGCCAGA
GTTGGTCGTA	GTGTCATGCC	AAATGGAACG	CCGGACGTAG	TCATATTTCC	TGAAGCAAAA	CCAGGTGCGG	GCGTGACAAA	CAGATCTGCG	TCTCCGGTCT
8510	8520	8530	8540	8550	8560	8570	8580	8590	8600
GCAGCGAATC	GAGCGAGACA	CATCATCAAC	ATCATATAGG	TGAGAGAGTT	TCGTAGGTCC	GCGGGGGACC	GAAGCCCAAG	ATACATTTGA	GGAAGTACGC
CGTCGCTTAG	CTCGCTCTGT	GTAGTAGTTG	TAGTATATCC	ACTCTCTCAA	AGCATCCAGG	CGCCCCCTGG	CTTCGGGTTC	TATGTAAACT	CCTTCATGCG
8610	8620	8630	8640	8650	8660	8670	8680	8690	8700
GGCGACGGGA	CTATTGTAGG	TGGTGGCGTC	TTATTCGGTG	TGGGTCGGTT	GGATGTGTAA	GCAAGACGCT	CAGTGTGTGC	CCTCCTCGCC	CTTCTCGACC
CCGCTGCCCT	GATAACATCC	ACCACCGCAG	AATAAGCCAC	ACCCAGCCAA	CCTACACATT	CGTTCTGCGA	GTCACACACG	GGAGGAGCGG	GAAGAGCTGG
start URF 22		*stop URF 23*							
8710	8720	8730	8740	8750	8760	8770	8780	8790	8800
TTCTTGGTAC	AAAAAAAAAA	AAATAAGGTT	TTCTAATAGG	TTTTGGAGTT	TTACTTCTAG	ATAATTCACT	TGCGCGAGGG	GAGGCCACCG	CACCAGTTTG
AAGAACCATG	TTTTTTTTTT	TTTATTCCAA	AAGATTATCC	AAAACCTCAA	AATGAAGATC	TATTAAGTGA	ACGCGCTCCC	CTCCGGTGGC	GTGGTCAAAC
						stop URF 26			
8810	8820	8830	8840	8850	8860	8870	8880	8890	8900
AGATGTCGGT	TTCTTGTCTA	TTACCGTAAA	CATTCTACAA	CGTGTTACCG	AAGGTTTTCC	GTTTGACGGG	AGTGCAGGTT	CACCTGCATT	TCCGATTTGG
TCTACAGCCA	AAGAACAGAT	AATGGCATTT	GTAAGATGTT	GCACAATGGC	TTCCAAAAGG	CAAACTGCCC	TCACGTCCAA	GTGGACGTAA	AGGCTAAACC
			start URF 27						
8910	8920	8930	8940	8950	8960	8970	8980	8990	9000
GAAGTCCCAC	TTAGAGGAGA	TATTTGTAAG	GTCGTGGAAG	TTGGTACGGG	TTTATTAAAA	GTAGAGCGGT	GGAATAGTTA	TACAGAGATT	CGTTTAGGGC
CTTCAGGGTG	AATCTCCTCT	ATAAACATTC	CAGCACCTTC	AACCATGCCC	AAATAATTTT	CATCTCGCCA	CCTTATCAAT	ATGTCTCTAA	GCAAATCCCG

FIGURE 21 (*Continued*)

```
                                                                                       stop URF 24 start URF 23
      9010        9020        9030        9040        9050        9060        9070        9080        9090        9100
TTATAATTCA  GGCCGGTAAC  ATTTTTAGAC  GAGGTCTCGC  GGGAGGTGGA  AGTCGGAGTT  CGTCGCTTAG  TACTAACGTT  TTTAAGTCCA  AGGAGTGTCT
AATATTAAGT  CCGGCCATTG  TAAAAATCTG  CTCCAGAGCG  CCCTCCACCT  TCAGCCTCAA  GCAGCGAATC  ATGATTGCAA  AAATTCAGGT  TCCTCACAGA

      9110        9120        9130        9140        9150        9160        9170        9180        9190        9200
GGACATATTC  TAAGTTTTCG  CCTTGTAATT  GTTTTTATGG  CGCTAGGGCA  TCCAGGGAAG  CGTCCCGGTC  GACTTGTATT  AGCACGTCCA  GACGTGCCTG
CCTGTATAAG  ATTCAAAAGC  GGAACATTAA  CAAAAATACC  GCGATCCCGT  AGGTCCCTTC  GCAGGGCCAG  CTGAACATAA  TCGTGCAGGT  CTGCACGGAC
                                                                                      stop URF 27
      9210        9220        9230        9240        9250        9260        9270        9280        9290        9300
GTCGCGCCGG  TGAAGGGGCG  GTCCTTGGTA  CTGTTTTCTT  GGGTGTGACT  AATACTGTGC  GTATGAGCCT  CGATACGATT  GGTCGCATCG  GGGATACATT
CAGCGCGGCC  ACTTCCCCGC  CAGGAACCAT  GACAAAAGAA  CCCACACTGA  TTATGACACG  CATACTCGGA  GCTATGCTAA  CCAGCGTAGC  CCCTATGTAA

      9310        9320        9330        9340        9350        9360        9370        9380        9390        9400
CGAACAACGT  ACCCGCCGCT  ATATTTTACG  TTCCACGACG  AGTTTTTTAG  TCCGTTTCGG  AGCGCGTTTT  TTCGTTCGTG  TAGCATCAGT  ACGAGTACGT
GCTTGTTGCA  TGGGCGGCGA  TATAAAATGC  AAGGTGCTGC  TCAAAAAATC  AGGCAAAGCC  TCGCGCAAAA  AAGCAAGCAC  ATCGTAGTCA  TGCTCATGCA

      9410        9420        9430        9440        9450        9460        9470        9480        9490        9500
CTATTTCCGT  CCATTCAAGG  CCTTGGTGGT  GTCTTTTTCT  GTGGTAAAAA  GAGAGTTTGT  ACAGACGCCC  AAGGACGTAA  TTTGTGTTTT  ATTTTATTGT
GATAAAGGCA  GGTAAGTTCC  GGAACCACCA  CAGAAAAAGA  CACCATTTTT  CTCTCAAACA  TGTCTGCGGG  TTCCTGCATT  AAACACAAAA  TAAAATAACA
start URF 24 stop URF 25
      9510        9520        9530        9540        9550        9560        9570        9580        9590        9600
TTTTTTTTTG  TAAATTTGTA  ATCTTCGGAC  AGAATGTTGT  CCTTTTTGTT  GGGAATATTC  GTATTCTGCC  TGATGCCGGT  ACGGCCGCAC  TGGCATTTTT
AAAAAAAAAC  ATTTAAACAT  TAGAAGCCTG  TCTTACAACA  GGAAAAACAA  CCCTTATAAG  CATAAGACGG  ACTACGGCCA  TGCCGGCGTG  ACCGTAAAAA

      9610        9620        9630        9640        9650        9660        9670        9680        9690        9700
TTGACCAGTG  GCACTAATTT  TTCGTGGTGG  CTGTCAAGGA  GCCAGTACAG  GCCTCAGTAT  TACATTCTGA  GCCATTTGTG  TAGTCCAACC  AATTGTAGCC
AACTGGTCAC  CGTGATTAAA  AAGCACCACC  GACAGTTCCT  CGGTCATGTC  CGGAGTCATA  ATGTAAGACT  CGGTAAACAC  ATCAGGTTGG  TTAACATCGG

      9710        9720        9730        9740        9750        9760        9770        9780        9790        9800
AGTCACGATT  TTTCGCTGGC  TTTATCGGGC  CCCCTTATGT  ATGGGCGTCC  GCATCTCTGT  TGTAATGTCG  GGGGTATCCT  CCATATTGTT  TTAATTATCC
TCAGTGCTAA  AAAGCGACCG  AAATAGCCCG  GGGGAATACA  TACCCGCAGG  CGTAGAGACA  ACATTACAGC  CCCCATAGGA  GGTATAACAA  AATTAATAGG
                                                                                                        start URF 25
      9810        9820        9830        9840        9850        9860        9870        9880        9890        9900
TCTCTTTTTG  TGTATTTGTG  GACTTTTTGG  GAGGACGGAT  CCGTTTTATC  GTGGGAGGGC  GAGGTCTTGT  TGTATGTCGC  GAAGGTGTCG  CCGTCGGTAT
AGAGAAAAAC  ACATAAACAC  CTGAAAAACC  CTCCTGCCTA  GGCAAAATAG  CACCCTCCCG  CTCCAGAACA  ACATACAGCG  CTTCCACAGC  GGCAGCCATA
            splice E4 leader                                                        starts E4 RNAs
      9910        9920        9930        9940        9950        9960        9970                    9990       10000
TGTCAGTCGG  AATGGTCATT  TTTTTGGATA  ATTTTTTGTG  GTGAGCTGTG  CCGTGGTCGA  GTTAGTCAGT  GTCACATTTT  TCCCGGTTCA  TGTCTCGCTC
ACAGTCA[illegible]  TTACCAGTAA  AAAAACCTAT  TAAAAAACAC  CACTCGACAC  GGCACCAGCT  CAATCAGTCA  CAGTGTAAAA  AGGGCCAAGT  ACAGAGCGAG


 TATA
     10010       10020       10030       10040       10050       10060       10070       10080       10090       10100
ATATATATCC  TGATTTTTTA  CTGCATTGCC  AATTTCAGGT  GTTTTTTGTG  GGTCTTTTGG  CGTGCGCTTG  GATGCGGGTC  TTTGCTTTCG  GTTTTTTGGG
TATATATAGG  ACTAAAAAAT  GACGTAACGG  TTAAAGTCCA  CAAAAAACAC  CCAGAAAACC  GCACGCGAAC  CTACGCCCAG  AAACGAAAGC  CAAAAAACCC

     10110       10120       10130       10140       10150       10160       10170       10180       10190       10200
TGTTGAAGGA  GTTTAGAAGT  GAAGGCAAAA  GGGTGCTATG  CAGTGAAGGG  TAAAATTTTT  TTGATGTTAA  GGGTTATGTA  CGTTCAATGA  GGCGGGATTT
ACAACTTCCT  CAAATCTTCA  CTTCCGTTTT  CCCACGATAC  GTCACTTCCC  ATTTTAAAAA  AACTACAATT  CCCAATACAT  GCAAGTTACT  CCGCCCTAAA
ITR boundary
     10210       10220       10230       10240       10250       10260       10270       10280       10290       10300
TGGATGCAGT  GGGCGGGGCA  AGGGTGCGGG  GCGCGGTGCA  GTGTTTGAGG  TGGGGGAGTA  ATAGTATAAC  CGAAGTTAGG  TTTTATTCCA  TATAATAACT
ACCTACGTCA  CCCGCCCCGT  TCCCACGCCC  CGCGCCACGT  CACAAACTCC  ACCCCCTCAT  TATCATATTG  GCTTCAATCC  AAAATAAGGT  ATATTATTGA

     10310
  ACTAC
L TGATG
```

FIGURE 21 (*Continued*)

```
          10          20          30          40          50          60          70          80          90         100
r 3' CCCCTGAAAG  TCCAACCATT  CCACCCGTCT  AACCCATTTA  AAACAATTAA  AGACAGAACG  TCGACGGTAC  TCACCTTCGC  GAAGAAAACT  CCCCCCTCAT
l 5' GGGGACTTTC  AGGTTGGTAA  GGTGGGCAGA  TTGGGTAAAT  TTTGTTAATT  TCTGTCTTGC  AGCTGCCATG  AGTGGAAGCG  CTTCTTTTGA  GGGGGGAGTA
                                                                        start protein IX
         110         120         130         140         150         160         170         180         190         200
AAATCGGGAA  TAGACTGCCC  GTCCGAGGGT  GGTACCCGTC  CTCAAGCAGT  CTTACAGTAC  CCTAGGTGAC  ACCTACCCTC  TGGGCAGGTC  GGGCGGTTAA
TTTAGCCCTT  ATCTGACGGG  CAGGCTCCCA  CCATGGGCAG  GAGTTCGTCA  GAATGTCATG  GGATCCACTG  TGGATGGGAG  ACCCGTCCAG  CCCGCCAATT

         210         220         230         240         250         260         270         280         290         300
GGAGTTGCGA  CTGGATACGG  TGAAACTCAA  GCAGTGGTAA  CCTACGTCGA  CGTCGGCGGC  GGCGATGACG  ACGGCGGTTG  TGGTAGGAAC  CTTACCCGAT
CCTCAACGCT  GACCTATGCC  ACTTTGAGTT  CGTCACCATT  GGATGCAGCT  GCAGCCGCCG  CCGCTACTGC  TGCCGCCAAC  ACCATCCTTG  GAATGGGCTA

         310         320         330         340         350         360         370         380         390         400
AATACCTTCG  TAGCAACGGT  TAAGGTCAAG  GAGATTATTG  GGAAGTTGGG  ACCGACTCCT  GTTCGATGAA  CAAGAGAACC  GCGTCGAGCT  CCGGAATTGG
TTATGGAAGC  ATCGTTGCCA  ATTCCAGTTC  CTCTAATAAC  CCTTCAACCC  TGGCTGAGGA  CAAGCTACTT  GTTCTCTTGG  CGCAGCTCGA  GGCCTTAACC

         410         420         430         440         450         460         470         480         490         500
GTTGCGAATC  CGCTTGACAG  ATTCGTCCAC  GGGGTCAACG  CACTCGTTTG  ACTCAGACGA  CAACGGTGTC  GTTTCAGATT  TATTTCTAGA  GTTTAGTTAT
CAACGCTTAG  GCGAACTGTC  TAAGCAGGTG  GCCCAGTTGC  GTGAGCAAAC  TGAGTCTGCT  GTTGCCACAG  CAAAGTCTAA  ATAAAGATCT  CAAATCAATA
                                                                                                 stop protein IX
                                     stop IVa2 protein
         510         520         530         540         550
TTATTTCTTT  ATGAACTATA  TTTTGTTTAC  TTACAAATAA  ACTAAAAAGC  GCG
AATAAAGAAA  TACTTGATAT  AAAACAAATG  AATGTTTATT  TGATTTTTCG  CGC
```

FIGURE 22. Nucleotide sequence of a region between coordinates 9.6 and 11.2 on the Ad3 genome. This sequence was established by Engler (1981). The region codes for polypeptide IX. For the positioning of strategic signals, see Fig. 3 (Section VII).

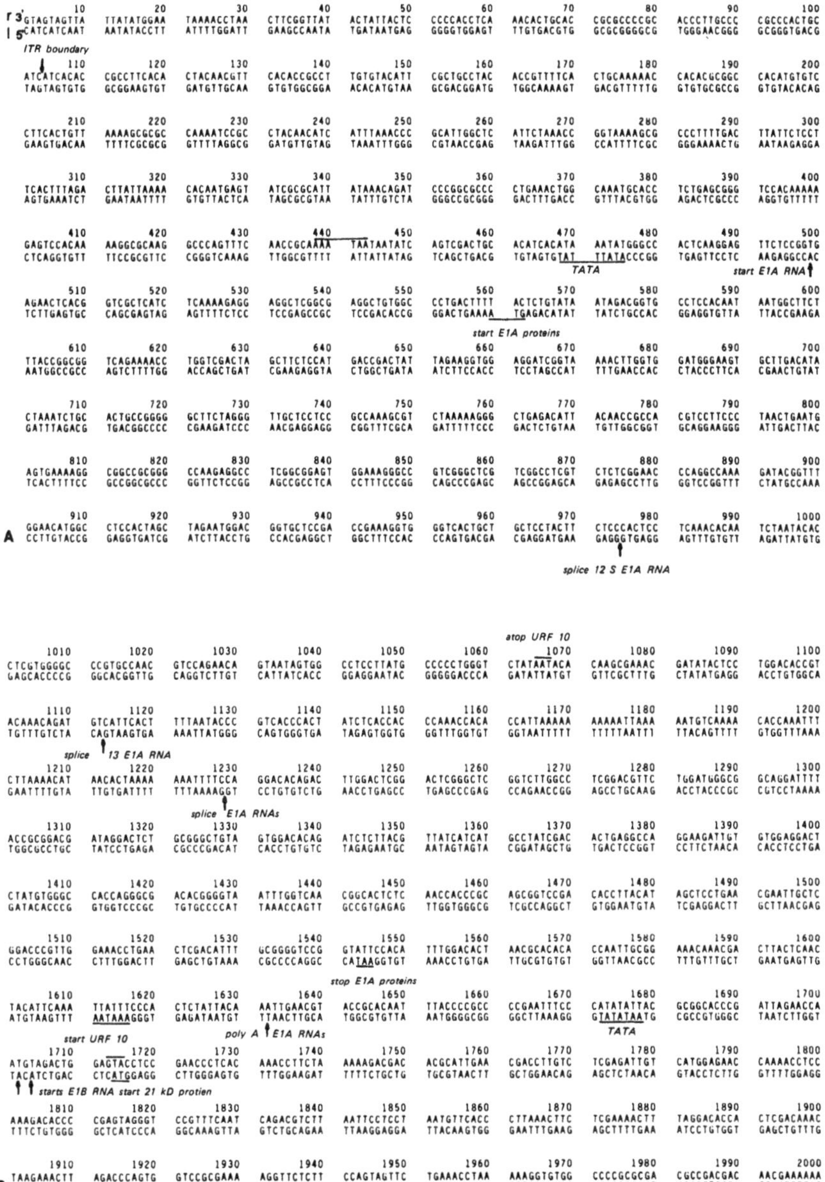

FIGURE 23.1A–L. Nucleotide sequence of a region between coordinates 0.0 and 31.7 on the Ad5 genome. This sequence was established by Steenbergh *et al.* (1977), van Ormondt *et al.* (1978), Maat and van Ormondt (1979), Maat *et al.* (1980), van Beveren *et al.* (1981), Bos *et al.* (1981), and H. van Ormondt and B. M. M. Dekker (personal communication). For interpretations, see van der Eb *et al.* (1979) and van Ormondt *et al.* (1980a,b).

```
      2010        2020        2030        2040        2050        2060        2070        2080        2090        2100
CTCAAAATAT  TTCCTATTTA  CCTCGCTTCT  TTGGGTAGAC  TCGCCCCCCA  TGGACGACCT  AAAAGACCGG  TACGTAGACA  CCTCTCGCCA  ACACTCTGTG
GAGTTTTATA  AAGGATAAAT  GGAGCGAAGA  AACCCATCTG  AGCGGGGGGT  ACCTGCTGGA  TTTTCTGGCC  ATGCATCTGT  GGAGAGCGGT  TGTGAGACAC
                 start 55 kD protein

      2110        2120        2130        2140        2150        2160        2170        2180        2190        2200
TTCTTAGCGG  ACGATGACAA  CAGAAGGCAG  GCGGGCCGCT  ATTATGGCTG  CCTCCTCGTC  GTCGTCGTCG  TCCTCCTTCG  GTCCGCCGCC  GCCGTCCTCG
AAGAATCGCC  TGCTACTGTT  GTCTTCCGTC  CGCCCGGCGA  TAATACCGAC  GGAGGAGCAG  CAGCAGCAGC  AGGAGGAAGC  CAGGCGGCGG  CGGCAGGAGC

      2210        2220        2230        2240        2250        2260        2270        2280        2290        2300
TCTCGGGTAC  CTTGGGCTCT  CGGCCGGACC  TGGGAGCCCT  TACTTACAAC  ATGTCCACCG  ACTTGACATA  GGTCTTGACT  CTGCGTAAAA  CTGTTAATGT
AGAGCCCATG  GAACCCGAGA  GCCGGCCTGG  ACCCTCGGGA  ATGAATGTTG  TACAGGTGGC  TGAACTGTAT  CCAGAACTGA  GACGCATTTT  GACAATTACA
                                        stop 21 kD protein  splice 13 S E1B RNA

      2310        2320        2330        2340        2350        2360        2370        2380        2390        2400
CTCCTACCCG  TCCCCGATTT  CCCCCATTTC  TCCCTCGCCC  CCCGAACACT  CCGATGTCTC  CTCCGATCCT  TAGATCGAAA  ATCGAATTAC  TGGTCTGTGG
GAGGATGGGC  AGGGGCTAAA  GGGGGTAAAG  AGGGAGCGGG  GGGCTTGTGA  GGCTACAGAG  GAGGCTAGGA  ATCTAGCTTT  TAGCTTAATG  ACCAGACACC

      2410        2420        2430        2440        2450        2460        2470        2480        2490        2500
CAGGACTCAC  ATAATGAAAA  GTTGTCTAGT  TCCTATTAAC  GCGATTACTC  GAACTAGACG  ACCGCGTCTT  CATAAGGTAT  CTCGTCGACT  GGTGAATGAC
GTCCTGAGTG  TATTACTTTT  CAACAGATCA  AGGATAATTG  CGCTAATGAG  CTTGATCTGC  TGGCGCAGAA  GTATTCCATA  GAGCAGCTGA  CCACTTACTG

      2510        2520        2530        2540        2550        2560        2570        2580        2590        2600
CGACGTCGGT  CCCCTACTAA  AACTCCTCCG  ATAATCCCAT  ATACGTTTCC  ACCGTGAATC  CGGTCTAACG  TTCATGTTCT  AGTCGTTTGA  ACATTTATAG
GCTGCAGCCA  GGGGATGATT  TTGAGGAGGC  TATTAGGGTA  TATGCAAAGG  TGGCACTTAG  GCCAGATTGC  AAGTACAAGA  TCAGCAAACT  TGTAAATATC

      2610        2620        2630        2640        2650        2660        2670        2680        2690        2700
TCCTTAACAA  CGATGTAAAG  ACCCTTGCCC  CGGCTCCACC  TCTATCTATG  CCTCCTATCC  CACCGGAAAT  CTACATCGTA  CTATTTATAC  ACCGGCCCCC
AGGAATTGTT  GCTACATTTC  TGGGAACGGG  GCCGAGGTGG  AGATAGATAC  GGAGGATAGG  GTGGCCTTTA  GATGTAGCAT  GATAAATATG  TGGCCGGGGG

      2710        2720        2730        2740        2750        2760        2770        2780        2790        2800
ACGAACCGTA  CCTGCCCCAC  CAATAATACT  TACATTCCAA  ATGACCGGGG  TTAAAATCGC  CATGCCAAAA  GGACCGGTTA  TGGTTGGAAT  AGGATGTGCC
TGCTTGGCAT  GGACGGGGTG  GTTATTATGA  ATGTAAGGTT  TACTGGCCCC  AATTTTAGCG  GTACGGTTTT  CCTGGCCAAT  ACCAACCTTA  TCCTACACGG

      2810        2820        2830        2840        2850        2860        2870        2880        2890        2900
ACATTCGAAG  ATACCCAAAT  TGTTATGGAC  ACACCTTCGG  ACCTGGCTAC  ATTCCCAAGC  CCCGACACGG  AAAATGACGA  CGACCTTCCC  CCACCACACA
TGTAAGCTTC  TATGGGTTTA  ACAATACCTG  TGTGGAAGCC  TGGACCGATG  TAAGGGTTCG  GGGCTGTGCC  TTTTACTGCT  GCTGGAAGGG  GGTGGTGTGT

       2910        2920        2930        2940        2950        2960        2970        2980        2990        3000
 GCGGGGTTTT  CGTCCCGAAG  TTAATTCTTT  ACGGAGAAAC  TTTCCACATG  GAACCCATAG  GACAGACTCC  CATTGAGGTC  CCACGCGGTG  TTACACCGGA
C CGCCCCAAAA  GCAGGGCTTC  AATTAAGAAA  TGCCTCTTTG  AAAGGTGTAC  CTTGGGTATC  CTGTCTGAGG  GTAACTCCAG  GGTGCGCCAC  AATGTGGCCT


      3010        3020        3030        3040        3050        3060        3070        3080        3090        3100
GGCTGACACC  AACGAAGTAC  GATCACTTTT  CGCACCGACA  CTAATTCGTA  TTGTACCATA  CACCGTTGAC  GCTCCTGTCC  CGGAGAGTCT  ACGACTGGAC
CCGACTGTGG  TTGCTTCATG  CTAGTGAAAA  GCGTGGCTGT  GATTAAGCAT  AACATGGTAT  GTGGCAACTG  CGAGGACAGG  GCCTCTCAGA  TGCTGACCTG

      3110        3120        3130        3140        3150        3160        3170        3180        3190        3200
GAGCCTGCCG  TTGACAGTGG  ACGACTTCTG  GTAAGTGCAT  CGGTCGGTGA  GAGCGTTCCG  GACCGGTCAC  AAACTCGTAT  TGTATGACTG  GGCGACAAGG
CTCGGACGGC  AACTGTCACC  TGCTGAAGAC  CATTCACGTA  GCCAGCCACT  CTCGCAAGGC  CTGGCCAGTG  TTTGAGCATA  ACATACTGAC  CCGCTGTTCC

      3210        3220        3230        3240        3250        3260        3270        3280        3290        3300
AACGTAAACC  CATTGTCCTC  CCCCCACAAG  GATGGAATGG  TTACGTTAAA  CTCAGTGTGA  TTCTATAACG  AACTCGGGCT  CTCGTACAGG  TTCCACTTGG
TTGCATTTGG  GTAACAGGAG  GGGGGTGTTC  CTACCTTACC  AATGCAATTT  GAGTCACACT  AAGATATTGC  TTGAGCCCGA  GAGCATGTCC  AAGGTGAACC

      3310        3320        3330        3340        3350        3360        3370        3380        3390        3400
ACTTGCCCCA  CAAACTGTAC  TGGTACTTCT  AGACCTTCCA  CGACTCCATG  CTACTCTGGG  CGTGGTCCAC  GTCTGGGACG  CTCACACCGC  CATTTGTATA
TGAACGGGGT  GTTTGACATG  ACCATGAAGA  TCTGGAAGGT  GCTGAGGTAC  GATGAGACCC  GCACCAGGTG  CAGACCCTGC  GAGTGTGGCG  GTAAACATAT

      3410        3420        3430        3440        3450        3460        3470        3480        3490        3500
ATCCTTGGTC  GGACACTACG  ACCTACACTG  GCTCCTCGAC  TCCGGGCTAG  TGAACCACGA  CCGGACGTGG  GCGCGACTCA  AACCGAGATC  GCTACTTCTA
TAGGAACCAG  CCTGTGATGC  TGGATGTGAC  CGAGGAGCTG  AGGCCCGATC  ACTTGGTGCT  GGCCTGCACC  CGCGCTGAGT  TTGGCTCTAG  CGATGAAGAT

      3510        3520        3530        3540        3550        3560        3570        3580        3590        3600
TGTCTAACTC  CATGACTTTA  CACACCCGCA  CCGAATTCCC  ACCCTTTCTT  ATATATTCCA  CCCCCAGAAT  ACATCAAAAC  ATAGACAAAA  CGTCGTCGGC
ACAGATTGAG  GTACTGAAAT  GTGTGGGCGT  GGCTTAAGGG  TGGGAAAGAA  TATATAAGGT  GGGGGTCTTA  TGTAGTTTTG  TATCTGTTTT  GCAGCAGCCG
stop 55 kD   splice 22 S E1B RNA                              TATA                    starts protein IX RNA   splice E1B RNAs
protein
      3610        3620        3630        3640        3650        3660        3670        3680        3690        3700
GGCGGCGGTA  CTCGTGGTTG  AGCAAACTAC  CTTCGTAACA  CTCGAGTATA  AACTGTTGCG  CGTACGGGGG  TACCCGGCCC  CACGCAGTCT  TACACTACCC
CCGCCGCCAT  GAGCACCAAC  TCGTTTGATG  GAAGCATTGT  GAGCTCATAT  TTGACAACGC  GCATGCCCCC  ATGGGCCGGG  GTGCGTCAGA  ATGTGATGGG
start protein IX  stop 8.3 kD protein

      3710        3720        3730        3740        3750        3760        3770        3780        3790        3800
GAGGTCGTAA  CTACCAGCGG  GGCAGGACGG  GCGTTTGAGA  TGATGGAACT  GGATGCTCTG  GCACAGACCT  TGCGGCAACC  TCTGACGTCG  GAGGCGGCGG
CTCCAGCATT  GATGGTCGCC  CCGTCCTGCC  CGCAAACTCT  ACTACCTTGA  CCTACGAGAC  CGTGTCTGGA  ACGCCGTTGG  AGACTGCAGC  CTCCGCCGCC

      3810        3820        3830        3840        3850        3860        3870        3880        3890        3900
CGAAGTCGGC  GACGTCGGTG  GCGGGCGCCC  TAACACTGAC  TGAAACGAAA  GGACTCGGGC  GAACGTTCGT  CACGTCGAAG  GGCAAGTAGG  CGGGCGCTAC
GCTTCAGCCG  CTGCAGCCAC  CGCCCGCGGG  ATTGTGACTG  ACTTTGCTTT  CCTGAGCCCG  CTTGCAAGCA  GTGCAGCTTC  CCGTTCATCC  GCCCGCGATG

       3910        3920        3930        3940        3950        3960        3970        3980        3990        4000
 TGTTCAACTG  CCGAGAAAAC  CGTGTTAACC  TAAGAAACTG  GGCCCTTGAA  TTACAGCAAA  GAGTCGTCGA  CAACCTAGAC  GCGGTCGTCC  AAAGACGGGA
D ACAAGTTGAC  GGCTCTTTTG  GCACAATTGG  ATTCTTTGAC  CCGGGAACTT  AATGTCGTTT  CTCAGCAGCT  GTTGGATCTG  CGCCAGCAGG  TTTCTGCCCT
```

FIGURE 23.1 (*Continued*)

```
                                                                        polyA protein IVa2 RNA                    stop protein IVa2
      4010        4020        4030        4040        4050        4060        4070        4080        4090        4100
CTTCCGAAGG  AGGGGAGGGT  TACGCCAAAT  TTTGTATTTA  TTTTTTGGTC  TGAGACAAAC  CTAAACCTAG  TTCGTTCACA  GAACGACAGA  AATAAATCCC
GAAGGCTTCC  TCCCCTCCCA  ATGCGGTTTA  AAACATAAAT  AAAAAACCAG  ACTCTGTTTG  GATTTGGATC  AAGCAAGTGT  CTTGCTGTCT  TTATTTAGGG
                    stop protein IX                                         polyA E1B RNAs

      4110        4120        4130        4140        4150        4160        4170        4180        4190        4200
CAAAACGCGC  GCGCCATCCG  GGCCCTGGTC  GCCAGAGCCA  GCAACTCCCA  GGACACATAA  AAAAGGTCCT  GCACCATTTC  CACTGAGACC  TACAAGTCTA
GTTTTGCGCG  CGCGGTAGGC  CCGGGACCAG  CGGTCTCGGT  CGTTGAGGGT  CCTGTGTATT  TTTTCCAGGA  CGTGGTAAAG  GTGACTCTGG  ATGTTCAGAT

      4210        4220        4230        4240        4250        4260        4270        4280        4290        4300
TGTACCCGTA  TTCGGGCAGA  GACCCCACCT  CCATCGTGGT  GACGTCTCGA  AGTACGACGC  CCCACCACAA  CATCTACTAG  GTCAGCATCG  TCCTCGCGAC
ACATGGGCAT  AAGCCCGTCT  CTGGGGTGGA  GGTAGCACCA  CTGCAGAGCT  TCATGCTGCG  GGGTGGTGTT  GTAGATGATC  CAGTCGTAGC  AGGAGCGCTG

      4310        4320        4330        4340        4350        4360        4370        4380        4390        4400
CCGCACCACG  GATTTTTACA  GAAAGTCATC  GTTCGACTAA  CGGTCCCCGT  CCGGGAACCA  CATTCACAAA  TGTTTCGCCA  ATTCGACCCT  ACCCACGTAT
GGCGTGGTGC  CTAAAAATGT  CTTTCAGTAG  CAAGCTGATT  GCCAGGGGCA  GGCCCTTGGT  GTAAGTGTTT  ACAAAGCGGT  TAAGCTGGGA  TGGGTGCATA

      4410        4420        4430        4440        4450        4460        4470        4480        4490        4500
GCACCCCTAT  ACTCTACGTA  GAACCTGACA  TAAAAATCCA  ACCGATACAA  GGGTCGGTAT  AGGGAGGCCC  CTAAGTACAA  CACGTCTTGG  TGGTCGTGTC
CGTGGGGATA  TGAGATGCAT  CTTGGACTGT  ATTTTTAGGT  TGGCTATGTT  CCCAGCCATA  TCCCTCCGGG  GATTCATGTT  GTGCAGAACC  ACCAGCACAG

      4510        4520        4530        4540        4550        4560        4570        4580        4590        4600
ACATAGGCCA  CGTGAACCCT  TTAAACAGTA  CATCGAATCT  TCCTTTACGC  ACCTTCTTGA  ACCTCTGCGG  GAACACTGGA  GGTTCTAAAA  GGTACGTAAG
TGTATCCGGT  GCACTTGGGA  AATTTGTCAT  GTAGCTTAGA  AGGAAATGCG  TGGAAGAACT  TGGAGACGCC  CTTGTGACCT  CCAAGATTTT  CCATGCATTC

      4610        4620        4630        4640        4650        4660        4670        4680        4690        4700
CAGGTATTAC  TACCGTTACC  CGGGTGCCCG  CCGCCGGACC  CGCTTCTATA  AAGACCCTAG  TGATTGCAGT  ATCAACACAA  GGTCCTACTC  TAGCAGTATC
GTCCATAATG  ATGGCAATGG  GCCCACGGGC  GGCGGCCTGG  GCGAAGATAT  TTCTGGGATC  ACTAACGTCA  TAGTTGTGTT  CCAGGATGAG  ATCGTCATAG

      4710        4720        4730        4740        4750        4760        4770        4780        4790        4800
CGGTAAAAAT  GTTTCGCGCC  CGCCTCCCAC  GGTCTGACGC  CATATTACCA  AGGTAGGCCG  GGTCCCCGCA  TCAATGGGAG  TGTCTAAACG  TAAAGGGTGC
GCCATTTTTA  CAAAGCGCGG  GCGGAGGGTG  CCAGACTGCG  GTATAATGGT  TCCATCCGGC  CCAGGGGCGT  AGTTACCCTC  ACAGATTTGC  ATTTCCCACG

      4810        4820        4830        4840        4850        4860        4870        4880        4890        4900
GAAACTCAAG  TCTACCCCCC  TAGTACAGAT  GGACGCCCCG  CTACTTCTTT  TGCCAAAGGC  CCCATCCCCT  CTAGTCGACC  CTTCTTTCGT  CCAAGGACTC
CTTTGAGTTC  AGATGGGGGG  ATCATGTCTA  CCTGCGGGGC  GATGAAGAAA  ACGGTTTCCG  GGGTAGGGGA  GATCAGCTGG  GAAGAAAGCA  GGTTCCTGAG

        4910        4920        4930        4940        4950        4960        4970        4980        4990        5000
  GTCGACGCTG  AATGGCGTCG  GCCACCCGGG  CATTTAGTGT  GGATAATGGC  CCACGTTGAC  CATCAATTCT  CTCGACGTCG  ACGGCAGTAG  GGACTCGTCC
E CAGCTGCGAC  TTACCGCAGC  CGGTGGGCCC  GTAAATCACA  CCTATTACCG  GGTGCAACTG  GTAGTTAAGA  GAGCTGCAGC  TGCCGTCATC  CCTGAGCAGG


      5010        5020        5030        5040        5050        5060        5070        5080        5090        5100
CCCCGGTGAA  GCAATTCGTA  CAGGGACTGA  GCGTACAAAA  GGGACTGGTT  TAGGCGGTCT  TCCGCGAGCG  GCGGGTCGCT  ATCGTCAAGA  ACGTTCCTTC
GGGGCCACTT  CGTTAAGCAT  GTCCCTGACT  CGCATGTTTT  CCCTGACCAA  ATCCGCCAGA  AGGCGCTCGC  CGCCCAGCGA  TAGCAGTTCT  TGCAAGGAAG

                                                                                                        stop 120 kD protein
      5110        5120        5130        5140        5150        5160        5170        5180        5190        5200
GTTTCAAAAA  GTTGCCAAAC  TCTGGCAGGC  GGCATCCGTA  CGAAAACTCG  CAAACTGGTT  CGTCAAGGTC  CGCCAGGGTG  TCGAGCCAGT  GGACGAGATG
CAAAGTTTTT  CAACGGTTTG  AGACCGTCCG  CCGTAGGCAT  GCTTTTGAGC  GTTTGACCAA  GCAGTTCCAG  GCGGTCCCAC  AGCTCGGTCA  CCTGCTCTAC

      5210        5220        5230        5240        5250        5260        5270        5280        5290        5300
CCGTAGAGCT  AGGTCGTATA  GAGGAGCAAA  GCGCCCAACC  CCGCCGAAAG  CGACATGCCG  TCATCAGCCA  CGAGCAGGTC  TGCCCGGTCC  CAGTACAGAA
GGCATCTCGA  TCCAGCATAT  CTCCTCGTTT  CGCGGGTTGG  GGCGGCTTTC  GCTGTACGGC  AGTAGTCGGT  GCTCGTCCAG  ACGGGCCAGG  GTCATGTCTT

                                    stop URF 9
      5310        5320        5330        5340        5350        5360        5370        5380        5390        5400
AGGTGCCCGC  GTCCCAGGAG  CAGTCGCATC  AGACCCAGTG  CCACTTCCCC  ACGCGAGGCC  CGACGCGCGA  CCGGTCCCAC  GCGAACTCCG  ACCAGGACGA
TCCACGGGCG  CAGGGTCCTC  GTCAGCGTAG  TCTGGGTCAC  GGTGAAGGGG  TGCGCTCCGG  GCTGCGCGCT  GGCCAGGGTG  CGCTTGAGGC  TGGTCCTGCT

                        splice IVa2 RNA
      5410        5420        5430        5440        5450        5460        5470        5480        5490        5500
CCACGACTTC  GCGACGGCCA  GAAGCGGGAC  GCGCAGCCGG  TCCATCGTAA  ACTGGTACCA  CAGTATCAGG  TCGGGGAGGC  GCCGCACCGG  GAACCGCGCG
GGTGCTGAAG  CGCTGCCGGT  CTTCGCCCTG  CGCGTCGGCC  AGGTAGCATT  TGACCATGGT  GTCATAGTCC  AGCCCCTCCG  CGGCGTGGCC  CTTGGCGCGC

      5510        5520        5530        5540        5550        5560        5570        5580        5590        5600
TCGAACGGGA  ACCTCCTCCG  CGGCGTGCTC  CCCGTCACGT  CTGAAAACTC  CCGCATCTCG  AACCCGCGCT  CTTTATGGCT  AAGGCCCCTC  ATCCGTAGGC
AGCTTGCCCT  TGGAGGAGGC  GCCGCACGAG  GGGCAGTGCA  GACTTTTGAG  GGCGTAGAGC  TTGGGCGCGA  GAAATACCGA  TTCCGGGGAG  TAGGCATCCG

                                                                                    start URF 9
      5610        5620        5630        5640        5650        5660        5670        5680        5690        5700
GCGGCGTCCG  GGGCGTCTGC  CAGAGCGTAA  GGTGCTCGGT  CCACTCGAGA  CCGGCAAGCC  CCAGTTTTTG  GTCCAAAGGG  GGTACGAAAA  ACTACGCAAA
CGCCGCAGGC  CCCGCAGACG  GTCTCGCATT  CCACGAGCCA  GGTGAGCTCT  GGCCGTTCGG  GGTCAAAAAC  CAGGTTTCCC  CCATGCTTTT  TGATGCGTTT

splice IVa2 RNA  start IVa2 protein
      5710        5720        5730        5740        5750        5760        5770        5780        5790        5800
GAATGGAGAC  CAAAGGTACT  CGGCCACAGG  TGCGAGCCAC  TGCTTTTCCG  ACAGGCACAG  GGGCATATGT  CTGAACTCTC  CGGACAGGAG  CTCGCCACAA
CTTACCTCTG  GTTTCCATGA  GCCGGTGTCC  ACGCTCGGTG  ACGAAAAGGC  TGTCCGTGTC  CCCGTATACA  GACTTGAGAG  GCCTGTCCTC  GAGCGGTGTT

                                    starts IVa2 RNA
      5810        5820        5830        5840        5850        5860        5870        5880        5890        5900
GGCGCCAGGA  GGAGCATATC  TTTGAGCCTG  GTGAGACTCT  GTTTCCGAGC  GCAGGTCCGG  TCGTGCTTCC  TCCGATTCAC  CCTCCCCATC  GCCAGCAACA
CCGCGGTCCT  CCTCGTATAG  AAACTCGGAC  CACTCTGAGA  CAAAGGCTCG  CGTCCAGGCC  AGCACGAAGG  AGGCTAAGTG  GGAGGGGTAG  CGGTCGTTGT

        5910        5920        5930        5940        5950        5960        5970        5980        5990        6000
  GGTGATCCCC  CAGGTGAGCG  AGGTCCCACA  CTTCTGTGTA  CAGCGGGAGA  AGCCGTAGTT  CCTTCCACTA  ACCAAACATC  CACATCCGGT  GCACTGGCCC
F CCACTAGGGG  GTCCACTCGC  TCCAGGGTGT  GAAGACACAT  GTCGCCCTCT  TCGGCATCAA  GGAAGGTGAT  TGGTTTGTAG  GTGTAGGCCA  CGTGACCGGG
```

FIGURE 23.1 (*Continued*)

```
      6010        6020        6030        6040        6050        6060        6070        6080        6090        6100
ACAAGGACTT  CCCCCCGATA  TTTTCCCCCA  CCCCCGCGCA  AGCAGGAGTG  AGAGAAGGCG  TAGCGACAGA  CGCTCCCGGT  CGACAACCCC  ACTCATGAGG
TGTTCCTGAA  GGGGGGCTAT  AAAAGGGGGT  GGGGGCGCGT  TCGTCCTCAC  TCTCTTCCGC  ATCGCTGTCT  GCGAGGGCCA  GCTGTTGGGG  TGAGTACTCC
                   TATA                        start 1st leader                                 splice 1st leader

      6110        6120        6130        6140        6150        6160        6170        6180        6190        6200
GAGACTTTTC  GCCCGTACTG  AAGACGCGAT  TCTAACAGTC  AAAGGTTTTT  GCTCCTCCTA  AACTATAAGT  GGACCGGGCG  CCACTACGGA  AACTCCCACC
CTCTGAAAAG  CGGGCATGAC  TTCTGCGCTA  AGATTGTCAG  TTTCCAAAAA  CGAGGAGGAT  TTGATATTCA  CCTGGCCCGC  GGTGATGCCT  TTGAGGGTGG

      6210        6220        6230        6240        6250        6260        6270        6280        6290        6300
GGCGTAGGTA  GACCAGTCTT  TTCTGTTAGA  AAAACAACAG  TTCGAACCAC  CGTTTGCTGG  GCATCTCCCG  CAACCTGTCG  TTGAACCGCT  ACCTCGCGTC
CCGCATCCAT  CTGGTCAGAA  AAGACAATCT  TTTTGTTGTC  AAGCTTGGTG  GCAAACGACC  CGTAGAGGGC  GTTGGACAGC  AACTTGGCGA  TGGAGCGCAG
                                                                                                   start URF 1

      6310        6320        6330        6340        6350        6360        6370        6380        6390        6400
CCAAACCAAA  AACAGCGCTA  GCCGCGCGAG  GAACCGGCGC  TACAAATCGA  CGTGCATAAG  CGCGCGTTGC  GTGGCGGTAA  GCCCTTTCTG  CCACCACGCG
GGTTTGGTTT  TTGTCGCGAT  CGGCGCGCTC  CTTGGCCGCG  ATGTTTAGCT  GCACGTATTC  GCGCGCAACG  CACCGCCATT  CGGGAAAGAC  GGTGGTGCGC

                                                          stop URF 8
      6410        6420        6430        6440        6450        6460        6470        6480        6490        6500
AGCAGCCCGT  GGTCCACGTG  CGCGGTTGGC  GCCAACACGT  CCCACTGTTC  CAGTTGCGAC  CACCGATGGA  GAGGCGCATC  CGCGAGCAAC  CAGGTCGTCT
TCGTCGGGCA  CCAGGTGCAC  GCGCCAACCG  CGGTTGTGCA  GGGTGACAAG  GTCAACGCTG  GTGGCTACCT  CTCCGCGTAG  GCGCTCGTTG  GTCCAGCAGA

      6510        6520        6530        6540        6550        6560        6570        6580        6590        6600
CCGCCGGCGG  GAACGCGCTC  GTCTTACCGC  CATCCCCCAG  ATCGACGCAG  AGCAGGCCCC  CCAGACGCAG  GTGCCATTTC  TGGGGCCCGT  CGTCCGCGCG
GGCGGCCGCC  CTTGCGCGAG  CAGAATGGCG  GTAGGGGGTC  TAGCTGCGTC  TCGTCCGGGG  GGTCTGCGTC  CACGGTAAAG  ACCCCGGGCA  GCAGGCGCGC

                                                                                                             start URF 8
      6610        6620        6630        6640        6650        6660        6670        6680        6690        6700
CAGCTTCATC  AGATAGAACG  TAGGAACGTT  CAGATCGCGG  ACGACGGTAC  GCGCCCGCCG  TTCGCGCGCG  AGCATACCCA  ACTCACCCCC  TGGGGTACCG
GTCGAAGTAG  TCTATCTTGC  ATCCTTGCAA  GTCTAGCGCC  TGCTGCCATG  CGCGGGCGGC  AAGCGCGCGC  TCGTATGGGT  TGAGTGGGGG  ACCCCATGGC
   stop URF 1

      6710        6720        6730        6740        6750        6760        6770        6780        6790        6800
TACCCCACCC  ACTCGCGCCT  CCGCATGTAC  GGCGTTTACA  GCATTTGCAT  CTCCCCGAGA  GACTCATAAG  GTTCTATACA  TCCCATCGTA  GAAGGTGGCG
ATGGGGTGGG  TGAGCGCGGA  GGCGTACATG  CCGCAAATGT  CGTAAACGTA  GAGGGGCTCT  CTGAGTATTC  CAAGATATGT  AGGGTAGCAT  CTTCCACCGC

      6810        6820        6830        6840        6850        6860        6870        6880        6890        6900
CCTACGACCG  CGCGTGCATT  AGCATATCAA  GCACGCTCCC  TCGCTCCTCC  AGCCCTGGCT  CCAACGATGC  CCGCCCGACG  AGACGAGCCT  TCTGATAGAC
GGATGCTGGC  GCGCACGTAA  TCGTATAGTT  CGTGCGAGGG  AGCGAGGAGG  TCGGGACCGA  GGTTGCTACG  GGCGGGCTGC  TCTGCTCGGA  AGACTATCTG

      6910        6920        6930        6940        6950        6960        6970        6980        6990        7000
 GGACTTCTAC  CGTACACTCA  ACCTACTATA  CCAACCTGCG  ACCTTCTGCA  ACTTCGACCG  CAGACACTCT  GGATGGCGCA  GTGCGTGCTT  CCTCCGCATC
GCCTGAAGATG  GCATGTGAGT  TGGATGATAT  GGTTGGACGC  TGGAAGACGT  TGAAGCTGGC  GTCTGTGAGA  CCTACCGCGT  CACGCACGAA  GGAGGCGTAG

      7010        7020        7030        7040        7050        7060        7070        7080        7090        7100
CTCAGCGCGT  CGAACAACTG  GTCGAGCCGC  CACTGGACGT  GCAGATCCCG  CGTCATCAGG  TCCCAAAGGA  ACTACTACAG  TATGAATAGG  ACAGGGAAAA
GAGTCGCGCA  GCTTGTTGAC  CAGCTCGGCG  GTGACCTGCA  CGTCTAGGGC  GCAGTAGTCC  AGGGTTTCCT  TGATGATGTC  ATACTTATCC  TGTCCCTTTT

      7110        7120        7130        7140        7150        7160        7170        7180        7190        7200
AAAAGGTGTC  GAGCGCCAAC  TCCTGTTTGA  GAAGCGCCAG  AAAGGTCATG  AGAACCTAGC  CTTTGGGCAG  CCGGAGGCTT  GCCATTCTCG  GATCGTACAT
TTTTCCACAG  CTCGCGGTTG  AGGACAAACT  CTTCGCGGTC  TTTCCAGTAC  TCTTGGATCG  GAAACCCGTC  GGCCTCCGAA  CGGTAAGAGC  CTAGCATGTA
    splice 2nd leader                                                                          splice 2nd leader

      7210        7220        7230        7240        7250        7260        7270        7280        7290        7300
CTTGACCAAC  TGCCGGACCA  TCCGCGTCGT  AGGGAAAAGA  TGCCCATCGC  GCATACGGAC  GCGCCGGAAG  GCCTCGCTCC  ACACCCACTC  GCGTTTCCAC
GAACTGGTTG  ACGGCCTGGT  AGGCGCAGCA  TCCCTTTTCT  ACGGGTAGCG  CGTATGCCTG  CGCGGCCTTC  CGGAGCGAGG  TGTGGGTGAG  CGCAAAGGTG

      7310        7320        7330        7340        7350        7360        7370        7380        7390        7400
AGGGACTGGT  ACTGAAACTC  CATGACCATA  AACTTCAGTC  ACAGCAGCGT  AGGCGGGACG  AGGGTCTCGT  TTTTCAGGCA  CGCGAAAAAC  CTTGCGCCTA
TCCCTGACCA  TGACTTTGAG  GTACTGGTAT  TTGAAGTCAG  TGTCGTCGCA  TCCGCCCTGC  TCCCAGAGCA  AAAAGTCCGT  GCGCTTTTTG  GAACGCGGAT

      7410        7420        7430        7440        7450        7460        7470        7480        7490        7500
AACCGTCCCG  CTTCCACTGT  AGCAACTTCT  CATAGAAAGG  GCGCGCTCCG  TATTTCAACG  CACACTACGC  CTTCCCAGGG  CCGTGGAGCC  TTGCCAACAA
TTGGCAGGGC  GAAGGTGACA  TCGTTGAAGA  GTATCTTTCC  CGCGCGAGGC  ATAAAGTTGC  GTGTGATGCG  GAAGGGTCCC  GGCACCTCGG  AACGGTTGTT

      7510        7520        7530        7540        7550        7560        7570        7580        7590        7600
TTAATGGACC  CGCCGCTCGT  GCTAGAGCAG  TTTCGGCAAC  TACAACACCG  GGTGTTACAT  TTCAAGGTTC  TTCGCGCCCT  ACGGGAACTA  CCTTCCGTTA
AATTACCTGG  GCGGCGAGCA  CGATCTCGTC  AAAGCCGTTG  ATGTTGTGGC  CCACAATGTA  AAGTTCCAAG  AAGCGCGGGA  TGCCCTTGAT  GGAAGGCAAT

      7610        7620        7630        7640        7650        7660        7670        7680        7690        7700
AAAAATTCAA  GGAGCATCCA  CTCGAGAAGT  CCCCTCGACT  CGGGCACGAG  ACTTTCCCGG  GTCAGACGTT  CTACTCCCAA  CCTTCGCTGC  TTACTCGAGG
TTTTTAAGTT  CCTCGTAGGT  GAGCTCTTCA  GGGGAGCTGA  GCCCGTGCTC  TGAAAGGGCC  CAGTCTGCAA  GATGAGGGTT  GGAAGCGACG  AATGAGCTCC

      7710        7720        7730        7740        7750        7760        7770        7780        7790        7800
TGTCCAGTGC  CCGGTAATCG  TAAACGTCCA  CCAGCGCTTT  CCAGGATTTG  ACCGCTGGAT  ACCGGTAAAA  AAGACCCCAC  TACGTCATCT  TCCATTCGCC
ACAGGTCACG  GGCCATTAGC  ATTTGCAGGT  GGTCGCGAAA  GGTCCTAAAC  TGGCGACCTA  TGGCCATTTT  TTCTGGGGTG  ATGCAGTAGA  AGGTAAGCGG

      7810        7820        7830        7840        7850        7860        7870        7880        7890        7900
CAGAACAAGG  GTCGCCAGGG  TAGGTTCCAA  GCGCCGATCC  AGAGCGCGCC  GTCAGTGATC  TCCGAGTAGA  GGCGGCTTGA  AGTACTGGTC  GTACTTCCCG
GTCTTGTTCC  CAGCGGTCCC  ATCCAAGGTT  CGCGGCTAGG  TCTCGCGCGG  CAGTCACTAG  AGGCTCATCT  CCGCCGAACT  TCATGACCAG  CATGAAGGGC

      7910        7920        7930        7940        7950        7960        7970        7980        7990        8000
 TGCTCGACGA  AGGGTTTCCG  GGGGTAGGTT  CATATCCAGA  GATGTAGCAT  CCACTGTTTC  TCTGCGAGCC  ACGCTCCTAC  GCTCGGCTAG  CCCTTCTTGA
HACGAGCTGCT  TCCCAAAGGC  CCCCATCCAA  GTATAGGTCT  CTACATCGTA  GGTGACAAAG  AGACGCTCGG  TGCGAGGATG  CGAGCCGATC  GGGAAGAACT
                                                      splice i leader          start URF 2
```

FIGURE 23.1 (*Continued*)

8010 8020 8030 8040 8050 8060 8070 8080 8090 8100
CCTAGAGGGC GGTGGTTAAC CTCCTCACCG ATAACTACAC CACTTTCATC TTCAGGGACG CTGCCCGGCT TGTGAGCACG ACCGAAAACA TTTTTGCACG
GGATCTCCCG CCACCAATTG GAGGAGTGGC TATTGATGTG GTGAAAGTAG AAGTCCCTGC GACGGGCCGA ACACTCGTGC TGGCTTTTGT AAAAACGTGC

8110 8120 8130 8140 8150 8160 8170 8180 8190 8200
CGTCATGACC GTCGCCACGT GCCCGACATG TAGGACGTGC TCCAACTGGA CTGCTGGCGC GTGTTCCTTC GTCTCACCCT TAAACTCGGG GAGCGGACCG
GCAGTACTGG CAGCGGTGCA CGGGCTGTAC ATCCTGCACG AGGTTGACCT GACGACCGCG CACAAGGAAG CAGAGTGGGA ATTTGAGCCC CTCGCCTGGC

8210 8220 8230 8240 8250 8260 8270 8280 8290 8300
CCCAAACCGA CCACCAGAAG ATGAAGCCGA CGAACAGGAA CTGGCAGACC GACGAGCTCC CCTCAATGCC ACCTAGCCTG GTGGTGCGGC GCGCTCGGGT
GGGTTTGGCT GGTGGTCTTC TACTTCGGCT GCTTGTCCTT GACCGTCTGG CTGCTCGAGG GGAGTTACGG TGGATCGGAC CACCACGCCG CGCGAGCCCA

1st ATG 120 kD proteins

8310 8320 8330 8340 8350 8360 8370 8380 8390 8400
TTCAGGTCTA CAGGCGCGCG CCGCCAGCCT CGAACTACTG TTGTAGCGCG TCTACCCTCG ACAGGTACCA GACCTCGAGG GCGCCGCAGT CCAGTCCGCC
AAGTCCAGAT GTCCGCGCGC GGCGGTCGGA GCTTGATGAC AACATCGCGC AGATGGGAGC TGTCCATGGT CTGGAGCTCC CGCGGCGTCA GGTCAGGCGG

stop URF 7 *splice i leader*

8410 8420 8430 8440 8450 8460 8470 8480 8490 8500
CTCGAGGACG TCCAAATGGA GCGTATCTGC CCAGTCCCGC GCCCGATCTA GGTCCACTAT GGATTAAAGG TCCCCGACCA ACCACCGCCG CAGCTACCGA
GAGCTCCTGC AGGTTTACCT CGCATAGACG GGTCAGGGCG CGGGCTAGAT CCAGGTGATA CCTAATTTCC AGGGGCTGGT TGGTGGCGGC GTCGATGGCT

stop URF 2 *stop pTP*

8510 8520 8530 8540 8550 8560 8570 8580 8590 8600
ACGTTCTCCG GCGTAGGGGC GCCGCGCTGA TGCCATGGCG CGCCGCCCGC CACCCGGCGC CCCCACAGGA ACCTACTACG TAGATTTTCG CCACTGCGCC
TGCAAGAGGC CGCATCCCCG CGGCGCGACT ACGGTACCGC GCGGCGGGCG GTGGGCCGCG GGGGTGTCCT TGGATGATGC ATCTAAAAGC GGTGACGCGG

8610 8620 8630 8640 8650 8660 8670 8680 8690 8700
CGCTCGGGGG CCTCCATCCC CCCCGAGGCC TGGGCGGCCC TCTCCCCCGT CCCCGTGCAG CCGCGGCGCG CGCCCGTCCT CGACCACGAC GCGCGCATCC
GCGAGCCCCC GGAGGTAGGG GGGGCTCCGG ACCCGCCGGG AGAGGGGGCA GGGGCACGTC GGCGCCGCGC GCGGGCAGGA GCTGGTGCTG CGCGCGTAGG

8710 8720 8730 8740 8750 8760 8770 8780 8790 8800
AACGACCGCT TGCGCTGCTG CGCCGCCAAC TAGAGGACTT AGACCGCGGA GACGCACTTC TGCTGCCCGG GCCACTCGAA CTCGGACTTT CTCTCAAGCT
TTGCTGGCGA ACGCGACGAC GCGGCGGTTG ATCTCCTGAA TCTGGCGCCT CTGCGTGAAG ACGACGGGCC CGGTGAGCTT GAGCCTGAAA GAGAGTTCGA

8810 8820 8830 8840 8850 8860 8870 8880 8890 8900
GTCTTAGTTA AAGCCACAGC AACTGCCGCC GGACCGCGTT TTAGAGGACG TGCAGAGGAC TCAACAGAAC TATCCGCTAG AGCCGGTACT TGACGAGCTA
CAGAATCAAT TTCGGTGTCG TTGACGGCGG CCTGGCGCAA AATCTCCTGC ACGTCTCCTG AGTTGTCTTG ATAGGCGATC TCGGCCATGA ACTGCTCGAT

8910 8920 8930 8940 8950 8960 8970 8980 8990 9000
GAGAAGGAGG ACCTCTAGAG GCGCAGGCCG AGCGAGGTGC CACCGCCGCT CCAGCAACCT TTACGCCCGG TACTCGACGC TCTTCCGCAA CTCCGGAGGG
I CTCTTCCTCC TGGAGATCTC CGCGTCCGGC TCGCTCCACG GTGGCGGCGA GGTCGTTGGA AATGCGGGCC ATGAGCTGCG AGAAGGCGTT GAGGCCTCCC

start URF 7

9010 9020 9030 9040 9050 9060 9070 9080 9090 9100
AGCAAGGTCT GCGCCGACAT CTGGTGCGGG GGAAGCCGTA GCGCCCGCGC GTACTGGTGG ACGCGCTCTA ACTCGAGGTG CACGGCCCGC TTCTGCCGCA
TCGTTCCAGA CGCGGCTGTA GACCACGCCC CCTTCGGCAT CGCGGGCGCG CATGACCACC TGCGCGAGAT TGAGCTCCAC GTGCCGGGCG AAGACGGCGT

9110 9120 9130 9140 9150 9160 9170 9180 9190 9200
TCAAAGCGTC CGCGACTTTC TCCATCAACT CCCACCACCG CCACACAAGA CGGTGCTTCT TCATGTATTG GGTCGCAGCG TTGCACCTAA GCAACTATAG
AGTTTCGCAG GCGCTGAAAG AGGTAGTTGA GGGTGGTGGC GGTGTGTTCT GCCACGAAGA AGTACATAAC CCAGCGTCGC AACGTGGATT CGTTGATATC

9210 9220 9230 9240 9250 9260 9270 9280 9290 9300
GGGGTTCCGG AGTTCCGCGA GGTACCGGAG CATCTTCAGG TGCCGCTTCA ACTTTTTGAC CCTCAACGCG CGGCTGTGCC AATTGAGGAG GAGGTCTTCT
CCCCAAGGCC TCAAGGCGCT CCATGGCCTC GTAGAAGTCC ACGGCGAAGT TGAAAAACTG GGAGTTGCGC GCCGACACGG TTAACTCCTC CTCCAGAAGA

9310 9320 9330 9340 9350 9360 9370 9380 9390 9400
GCCTACTCGA GCCGCTGTCA CAGCGCGTGG AGCGCGAGTT TCCGATGTCC CCGGAGAAGA AGAAGAAGTT AGAGGAGAAG GTATTCCCGG AGGGGAAGAA
CGGATGAGCT CGGCGACAGT GTCGCGCACC TCGCGCTCAA AGGCTACAGG GGCCTCTTCT TCTTCTTCAA TCTCCTCTTC CATAAGGGCC TCCCCTTCTT

start URF 3

9410 9420 9430 9440 9450 9460 9470 9480 9490 9500
GAAGAAGACC GCCGCCACCC CCTCCCCCCT GTGCCGCCGC TGCTGCCGCG TGGCCCTCCG CCAGCTGTTT CGCGAGCTAG TAGAGGGGCG CCGCTGCCGC
CTTCTTCTGG CGGCGGTGGG GGAGGGGGGA CACGGCGGCG ACGACGGCGC ACCGGGAGGC GGTCGACAAA GCGCTCGATC ATCTCCCCGC GGCGACGGCG

9510 9520 9530 9540 9550 9560 9570 9580 9590 9600
GTACCAGAGC CACTGCCGCG CCGGCAAGAG CGCCCCCGCG TCAACCTTCT GCGGCGGGCA GTACAGGGCC AATACCCAAC CGCCCCCCGA CGGTACGCCG
CATGGTCTCG GTGACGGCGC GGCCGTTCTC GCGGGGGCGC AGTTGGAAGA CGCCGCCCGT CATGTCCCGG TTATGGGTTG GCGGGGGGCT GCCATGCGGC

9610 9620 9630 9640 9650 9660 9670 9680 9690 9700
TCCCTATGCC GCGATTGCTA CGTAGAGTTG TTAACAACAC ATCCATGAGG CGGCGGCTCC CTGGACTCGC TCAGGCGTAG CTGGCCTAGC CTTTTGGAGA
AGGGATACGG CGCTAACGAT GCATCTCAAC AATTGTTGTG TAGGTACTCC GCCGCCGAGG GACCTGAGCG AGTCCGCATC GACCGGATCG GAAAACCTCT

splice 3rd leader

9710 9720 9730 9740 9750 9760 9770 9780 9790 9800
GCTCTTTCCG CAGATTGGTC AGTGTCAGCG TTCCATCCGA CTCGTGGCAC CGCCCGCCGT CGCCCGCCGC CAGCCCCAAC AAAGACCGCC TCCACGACGA
CGAGAAAGGC GTCTAACCAG TCACAGTCGC AAGGTAGGCT GAGCACCGTG GCGGGCGGCA GCGGGCGGCG GTCGGGGTTG TTTCTGGCGG AGGTGCTGCT

splice 3rd leader

9810 9820 9830 9840 9850 9860 9870 9880 9890 9900
CTACTACATT AATTTCATCC GCCAGAACTC TGCCGCCTAC CAGCTGTCTT CGTGGTACAG GAACCCAGGC CGGACGACTT ACGCGTCCGC CAGCCGGTAC
GATGATGTAA TTAAAGTAGG CGGTCTTGAG ACGGCGGATG GTCGACAGAA GCACCATGTC CTTGGGTCCG GCCTGCTGAA TGCGCAGGCG GTCGGCCATG

stop URF 3

9910 9920 9930 9940 9950 9960 9970 9980 9990 10000
GGGGTCCGAA GCAAAACTGT AGCCGCGTCC AGAAACATCA TCAGAACGTA CTCGGAAAGA TGGCCGTGAA GAAGAAGAGG AAGGAGAACA GGACGTAGAG
J CCCCAGGCTT CGTTTTGACA TCGGCGCAGG TCTTTGTAGT AGTCTTGCAT GAGCCTTTCT ACCGGCACTT CTTCTTCTCC TTCCTCTTGT CCTGCATCTC

FIGURE 23.1 (*Continued*)

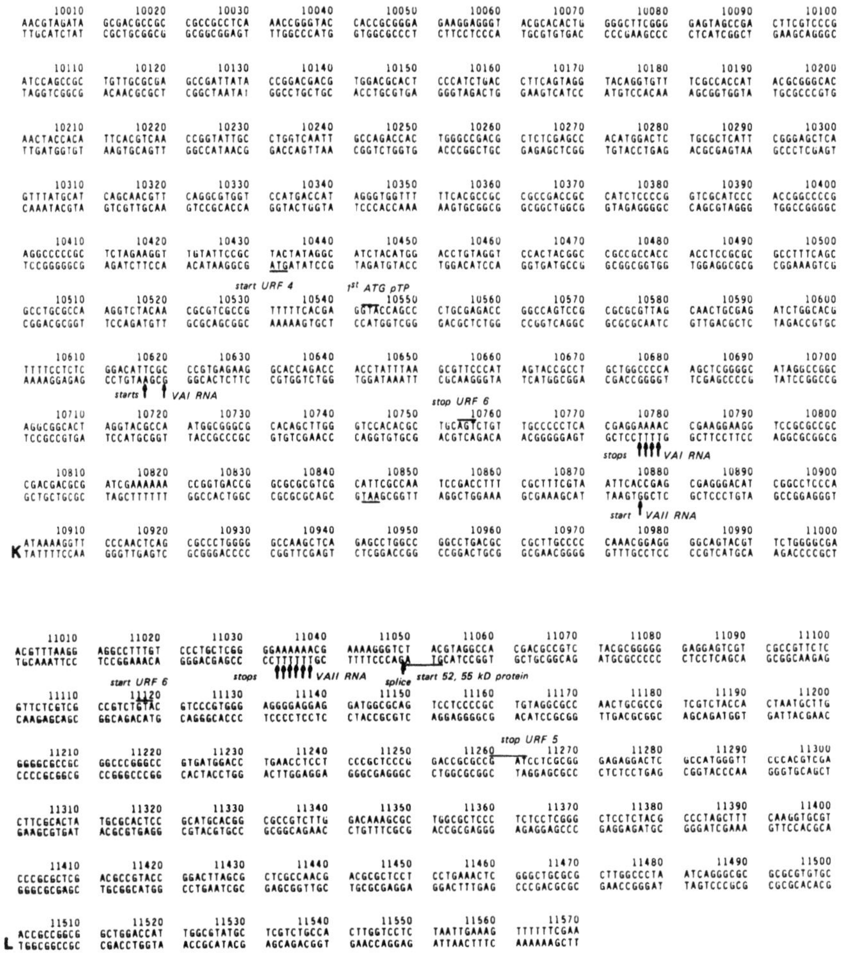

```
      10010       10020       10030       10040       10050       10060       10070       10080       10090       10100
 AACGTAGATA  GCGACGCCGC  CGCCGCCTCA  AACCGGGTAC  CACCGCGGGA  GAAGGAGGGT  ACGCACACTG  GGGCTTCGGG  GAGTAGCCGA  CTTCGTCCCG
 TTGCATCTAT  CGCTGCGGCG  GCGGCGGAGT  TTGGCCCATG  GTGGCGCCCT  CTTCCTCCCA  TGCGTGTGAC  CCCGAAGCCC  CTCATCGGCT  GAAGCAGGGC

      10110       10120       10130       10140       10150       10160       10170       10180       10190       10200
 ATCCAGCCGC  TGTTGCGCGA  GCCGATTATA  CCGGACGACG  TGGACGCACT  CCCATCTGAC  CTTCAGTAGG  TACAGGTGTT  TCGCCACCAT  ACGCGGGCAC
 TAGGTCGGCG  ACAACGCGCT  CGGCTAATAT  GGCCTGCTGC  ACCTGCGTGA  GGGTAGACTG  GAAGTCATCC  ATGTCCACAA  AGCGGTGGTA  TGCGCCCGTG

      10210       10220       10230       10240       10250       10260       10270       10280       10290       10300
 AACTACCACA  TTCACGTCAA  CCGGTATTGC  CTGGTCAATT  GCCAGACCAC  TGGGCCGACG  CTCTCGAGCC  ACATGGACTC  TGCGCTCATT  CGGGAGCTCA
 TTGATGGTGT  AAGTGCAGTT  GGCCATAACG  GACCAGTTAA  CGGTCTGGTG  ACCCGGCTGC  GAGAGCTCGG  TGTACCTGAG  ACGCGAGTAA  GCCCTCGAGT

      10310       10320       10330       10340       10350       10360       10370       10380       10390       10400
 GTTTATGCAT  CAGCAACGTT  CAGGCGTGGT  CCATGACCAT  AGGGTGGTTT  TTCACGCCGC  CGCCGACCGC  CATCTCCCCG  GTCGCATCCC  ACCGGCCCCG
 CAAATACGTA  GTCGTTGCAA  GTCCGCACCA  GGTACTGGTA  TCCCACCAAA  AAGTGCGGCG  GCGGCTGGCG  GTAGAGGGGC  CAGCGTAGGG  TGGCCGGGGC

      10410       10420       10430       10440       10450       10460       10470       10480       10490       10500
 AGGCCCCCGC  TCTAGAAGGT  TGTATTCCGC  TACTATAGGC  ATCTACATGG  ACCTGTAGGT  CCACTACGGC  CGCCGCCACC  ACCTCCGCGC  GCCTTTCAGC
 TCCGGGGGCG  AGATCTTCCA  ACATAAGGCG  ATGATATCCG  TAGATGTACC  TGGACATCCA  GGTGATGCCG  GCGGCGGTGG  TGGAGGCGCG  CGGAAAGTCG
                              start URF 4      1st ATG pTP
      10510       10520       10530       10540       10550       10560       10570       10580       10590       10600
 GCCTGCGCCA  AGGTCTACAA  CGCGTCGCCG  TTTTTCACGA  GGTACCAGCC  CTGCGAGACC  GGCCAGTCCG  CGCGCGTTAG  CAACTGCGAG  ATCTGGCACG
 CGGACGCGGT  TCCAGATGTT  GCGCAGCGGC  AAAAAGTGCT  CCATGGTCGG  GACGCTCTGG  CCGGTCAGGC  GCGCGCAATC  GTTGACGCTC  TAGACCGTGC

      10610       10620       10630       10640       10650       10660       10670       10680       10690       10700
 TTTTCCTCTC  GGACATTCGC  CCGTGAGAAG  GCACCAGACC  ACCTATTTAA  GCGTTCCCAT  AGTACCGCCT  GCTGGCCCCA  AGCTCGGGGC  ATAGGCCGGC
 AAAAGGAGAG  CCTGTAAGCG  GGCACTCTTC  CGTGGTCTGG  TGGATAAATT  CGCAAGGGTA  TCATGGCGGA  CGACCGGGGT  TCGAGCCCCG  TATCCGGCCG
             starts ↑ ↑ VAI RNA                             stop URF 6
      10710       10720       10730       10740       10750       10760       10770       10780       10790       10800
 AGGCGGCACT  AGGTACGCCA  ATGGCGGGCG  CACAGCTTGG  GTCCACACGC  TGCAGTCTGT  TGCCCCCTCA  CGAGGAAAAC  CGAAGGAAGG  TCCGCGCCGC
 TCCGCCGTGA  TCCATGCGGT  TACCGCCCGC  GTGTCGAACC  CAGGTGTGCG  ACGTCAGACA  ACGGGGGAGT  GCTCCTTTTG  GCTTCCTTCC  AGGCGCGGCG
                                                                                stops ↑↑↑↑ VAI RNA
      10810       10820       10830       10840       10850       10860       10870       10880       10890       10900
 CGACGACGCG  ATCGAAAAAA  CCGGTGACCG  GCGCGCGTCG  CATTCGCCAA  TCCGACCTTT  CGCTTTCGTA  ATTCACCGAG  CGAGGGACAT  CGGCCTCCCA
 GCTGCTGCGC  TAGCTTTTTT  GGCCACTGGC  CGCGCGCAGC  GTAAGCGGTT  AGGCTGGAAA  GCGAAAGCAT  TAAGTGGCTC  GCTCCCTGTA  GCCGGAGGGT
                                                                                 start ↑ VAII RNA
      10910       10920       10930       10940       10950       10960       10970       10980       10990       11000
 ATAAAAGGTT  CCCAACTCAG  CGCCCTGGGG  GCCAAGCTCA  GAGCCTGGCC  GGCCTGACGC  CGCTTGCCCC  CAAACGGAGG  GGCAGTACGT  TCTGGGGCGA
KTATTTTCCAA  GGGTTGAGTC  GCGGGACCCC  CGGTTCGAGT  CTCGGACCGG  CCGGACTGCG  GCGAACGGGG  GTTTGCCTCC  CCGTCATGCA  AGACCCCGCT


      11010       11020       11030       11040       11050       11060       11070       11080       11090       11100
 ACGTTTAAGG  AGGCCTTTGT  CCCTGCTCGG  GGAAAAAACG  AAAAGGGTCT  ACGTAGGCCA  CGACGCCGTC  TACGCGGGGG  GAGGAGTCGT  CGCCGTTCTC
 TGCAAATTCC  TCCGGAAACA  GGGACGAGCC  CCTTTTTTGC  TTTTCCCAGA  TGCATCCGGT  GCTGCGGCAG  ATGCGCCCCC  CTCCTCAGCA  GCGGCAAGAG
                                 stops ↑↑↑↑↑↑ VAII RNA  splice ↑ start 52, 55 kD protein
              start URF 6
      11110       11120       11130       11140       11150       11160       11170       11180       11190       11200
 GTTCTCGTCG  CCGTCTGTAC  GTCCCGTGGG  AGGGGAGGAG  GATGGCGCAG  TCCTCCCCGC  TGTAGGCGCC  AACTGCGCCG  TCGTCTACCA  CTAATGCTTG
 CAAGAGCAGC  GGCAGACATG  CAGGGCACCC  TCCCCTCCTC  CTACCGCGTC  AGGAGGGGCG  ACATCCGCGG  TTGACGCGGC  AGCAGATGGT  GATTACGAAC
                                                                    stop URF 5
      11210       11220       11230       11240       11250       11260       11270       11280       11290       11300
 GGGGCGCCGC  GGCCCGGGCC  GTGATGGACC  TGAACCTCCT  CCCGCTCCCG  GACCGCGCCG  ATCCTCGCGG  GAGAGGACTC  GCCATGGGTT  CCCACGTCGA
 CCCCGCGGCG  CCGGGCCCGG  CACTACCTGG  ACTTGGAGGA  GGGCGAGGGC  CTGGCGCGGC  TAGGAGCGCC  CTCTCCTGAG  CGGTACCCAA  GGGTGCAGCT

      11310       11320       11330       11340       11350       11360       11370       11380       11390       11400
 CTTCGCACTA  TGCGCACTCC  GCATGCACGG  CGCCGTCTTG  GACAAAGCGC  TGGCGCTCCC  TCTCCTCGGG  CTCCTCTACG  CCCTAGCTTT  CAAGGTGCGT
 GAAGCGTGAT  ACGCGTGAGG  CGTACGTGCC  GCGGCAGAAC  CTGTTTCGCG  ACCGCGAGGG  AGAGGAGCCC  GAGGAGATGC  GGGATCGAAA  GTTCCACGCA

      11410       11420       11430       11440       11450       11460       11470       11480       11490       11500
 CCCGCGCTCG  ACGCCGTACC  GGACTTAGCG  CTCGCCAACG  ACGCGCTCCT  CCTGAAACTC  GGGCTGCGCG  CTTGGCCCTA  ATCAGGGCGC  GCGCGTGTGC
 GGGCGCGAGC  TGCGGCATGG  CCTGAATCGC  GAGCGGTTGC  TGCGCGAGGA  GGACTTTGAG  CCCGACGCGC  GAACCGGGAT  TAGTCCCGCG  CGCGCACACG

      11510       11520       11530       11540       11550       11560       11570
 ACCGCCGGCG  GCTGGACCAT  TGGCGTATGC  TCGTCTGCCA  CTTGGTCCTC  TAATTGAAAG  TTTTTTCGAA
LTGGCGGCCGC  CGACCTGGTA  ACCGCATACG  AGCAGACGGT  GAACCAGGAG  ATTAACTTTC  AAAAAAGCTT
```

FIGURE 23.1 (*Continued*)

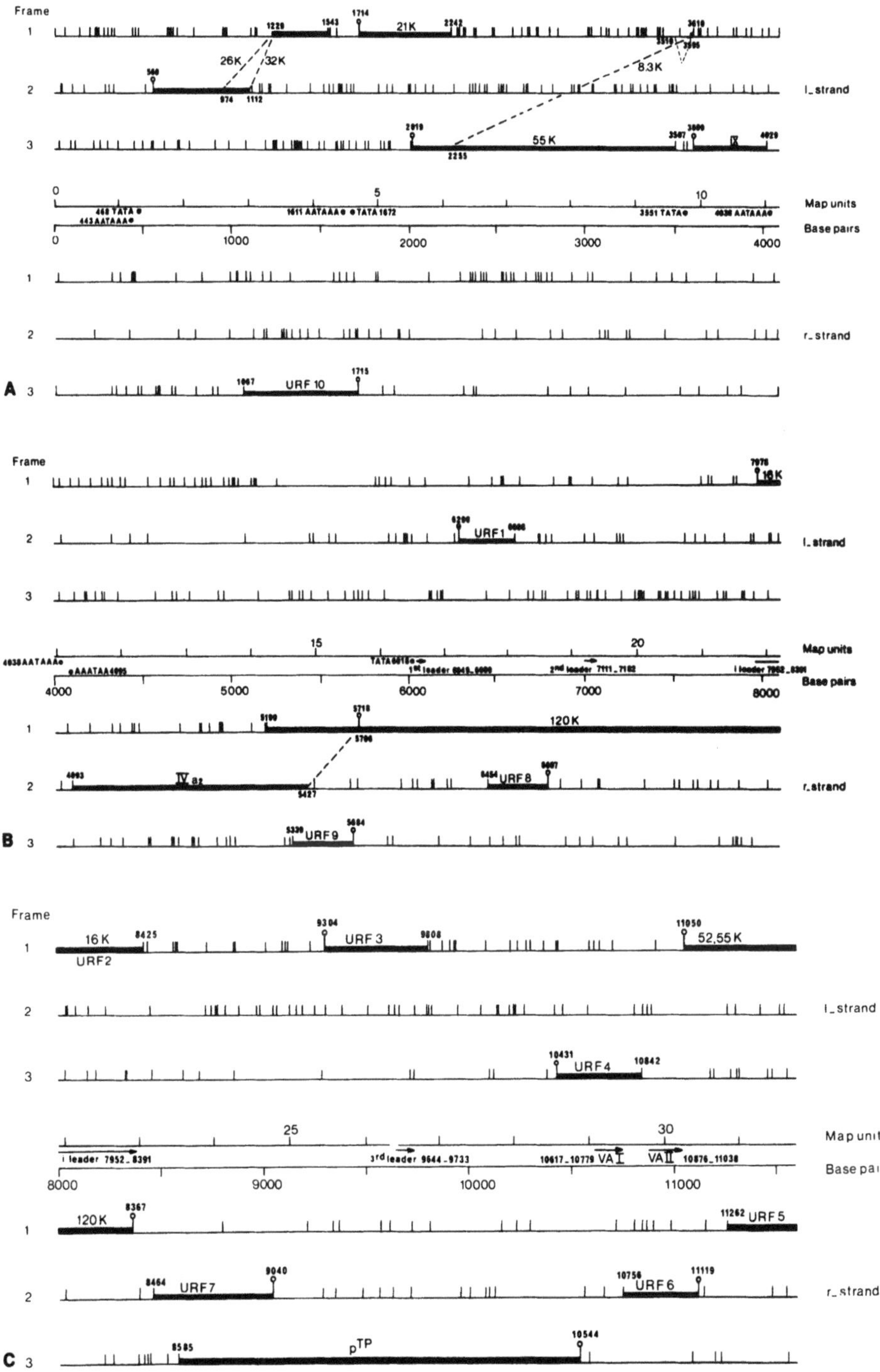

FIGURE 23.2A–C. Structural organization of a region between coordinates 0.0 and 31.7 on the Ad5 genome. This map is derived from the nucleotide sequence in Fig. 23.1. For the positioning of strategic signals, see Fig. 3 (Section VII).

```
          10          20          30          40          50          60          70          80          90         100
3' TACCCGAGGT  CACTCGTCCT  TGACTTTCGG  TAACAGTTTC  TAGAACCAAC  ACCCGGTATA  AAAAACCCGT  GGATACTGTT  CGCGAAAGGT  CCGAAACAAA
5' ATGGGCTCCA  GTGAGCAGGA  ACTGAAAGCC  ATTGTCAAAG  ATCTTGGTTG  TGGGCCATAT  TTTTTGGGCA  CCTATGACAA  GCGCTTTCCA  GGCTTTGTTT
   start 23 kD protein

         110         120         130         140         150         160         170         180         190         200
   GAGGTGTGTT  CGAGCGGACG  CGGTATCAGT  TATGCCGGCC  AGCGCTCTGA  CCCCCGCATG  TGACCTACCG  GAAACCGACC  TTGGGCGTGA  GTTTTTGTAC
   CTCCACACAA  GCTCGCCTGC  GCCATAGTCA  ATACGGCCGG  TCGCGAGACT  GGGGGCGTAC  ACTGGATGGC  CTTTGGCTGG  AACCCGCACT  CAAAAACATG

         210         220         230         240         250         260         270         280         290         300
   GATGGAGAAA  CTCGGGAAAC  CGAAAAGACT  GGTCGCTGAG  TTCGTCCAAA  TGGTCAAACT  CATGCTCAGT  GAGGACGCGG  CATCGCGGTA  ACGAAGAAGG
   CTACCTCTTT  GAGCCCTTTG  GCTTTTCTGA  CCAGCGACTC  AAGCAGGTTT  ACCAGTTTGA  GTACGAGTCA  CTCCTGCGCC  GTAGCGCCAT  TGCTTCTTCC

         310         320         330         340         350         360         370         380         390         400
   GGGCTGGCGA  CATATTGCGA  CCTTTTCAGG  TGGGTTTCGC  ATGTCCCCGG  GTTGAGCCGG  CGGACACCTG  ATAAGACGAC  GTACAAAGAG  GTGCGGAAAC
   CCCGACCGCT  GTATAACGCT  GGAAAAGTCC  ACCCAAAGCG  TACAGGGGCC  CAACTCGGCC  GCCTGTGGAC  TATTCTGCTG  CATGTTTCTC  CACGCCTTTG

         410         420         430         440         450         460         470         480         490         500
   GGTTGACCGG  GGTTTGAGGG  TACCTAGTGT  TGGGGTGGTA  CTTGGAATAA  TGGCCCCATG  GGTTGAGGTA  CGAGTTGTCA  GGGGTCCATG  TCGGGTGGGA
   CCAACTGGCC  CCAAACTCCC  ATGGATCACA  ACCCCACCAT  GAACCTTATT  ACCGGGGTAC  CCAACTCCAT  GCTCAACAGT  CCCCAGGTAC  AGCCCACCCT

         510         520         530         540         550         560         570         580         590         600
   CGCAGCGTTG  GTCCTTGTCG  AGATGTCGAA  GGACCTCGCG  GTGAGCGGGA  TGAAGGCGTC  GGTGTCACGC  GTCTAATCCT  CGCGGTGAAG  AAAAACAGTG
   GCGTCGCAAC  CAGGAACAGC  TCTACAGCTT  CCTGGAGCGC  CACTCGCCCT  ACTTCCGCAG  CCACAGTGCG  CAGATTAGGA  GCGCCACTTC  TTTTTGTCAC

                                              polyA DBP RNA
         610         620         630         640          650         660         670         680         690         700
   AACTTTTTGT  ACATTTTTAT  TACATGATCT  CTGTGAAAGT  TATTTCCGTT  TACGAAAATA  AACATGTGAG  AGCCCACTAA  TAAATGGGGG  TGGGAACGGC
   TTGAAAAACA  TGTAAAAATA  ATGTACTAGA  GACACTTTCA  ATAAAGGCAA  ATGCTTTTAT  TTGTACACTC  TCGGGTGATT  ATTTACCCCC  ACCCTTGCCG
               stop 23 kD protein                                        polyA L3 RNA
         710  stop DBP 720         730         740         750         760         770         780         790         800
   AGACGCGGCA  AATTTTTAGT  TTCCCCAAGA  CGGCGCGTAG  CGATACGCGG  TGACCGTCCC  TGTGCAACGC  TATGACCACA  AATCACGAGG  TGAATTTGAG
   TCTGCGCCGT  TTAAAAATCA  AAGGGGTTCT  GCCGCGCATC  GCTATGCGCC  ACTGGCAGGG  ACACGTTGCG  ATACTGGTGT  TTAGTGCTCC  ACTTAAACTC

         810         820         830         840         850         860         870         880         890         900
   TCCGTGTTGG  TAGGCGCCGT  CGAGGCACTT  CAAAAGTGAG  GTGTCCGACG  CGTGGTAGTG  GTTGCGCAAA  TCGTCCAGCC  CGCGGCTATA  GAACTTCAGC
   AGGCACAACC  ATCCGCGGCA  GCTCCGTGAA  GTTTTCACTC  CACAGGCTGC  GCACCATCAC  CAACGCGTTT  AGCAGGTCGG  GCGCCGATAT  CTTGAAGTCG

         910         920         930         940         950         960         970         980         990        1000
   GTCAACCCCG  GAGGCGGGAC  GCGCGCGCTC  AACGCTATGT  GTCCCAACGT  CGTGACCTTG  TGATAGTCGC  GGCCCACCAC  GTGCGACCGG  TCGTGCGAGA
A  CAGTTGGGGC  CTCCGCCCTG  CGCGCGCGAG  TTGCGATACA  CAGGGTTGCA  GCACTGGAAC  ACTATCAGCG  CCGGGTGGTG  CACGCTGGCC  AGCACGCTCT


        1010        1020        1030        1040        1050        1060        1070        1080        1090        1100
   ACAGCCTCTA  GTCTAGGCGC  AGGTCCAGGA  GGCGCAACGA  GTCCCGCTTG  CCTCAGTTGA  AACCATCGAC  GGAAGGGTTT  TTCCCGCGCA  CGGGTCCGAA
   TGTCGGAGAT  CAGATCCGCG  TCCAGGTCCT  CCGCGTTGCT  CAGGGCGAAC  GGAGTCAACT  TTGGTAGCTG  CCTTCCCAAA  AAGGGCGCGT  GCCCAGGCTT

        1110        1120        1130        1140        1150        1160        1170        1180        1190        1200
   ACTCAACGTG  AGCGTGGCAT  CACCGTAGTT  TTCCACTGGC  ACGGGCCAGA  CCCGCAATCC  TATGTCGCGG  ACGTATTTTC  GGAACTAGAC  GAATTTTCGG
   TGAGTTGCAC  TCGCACCGTA  GTGGCATCAA  AAGGTGACCG  TGCCCGGTCT  GGGCGTTAGG  ATACAGCGCC  TGCATAAAAG  CCTTGATCTG  CTTAAAAGCC

        1210        1220        1230        1240        1250        1260        1270        1280        1290        1300
   TGGACTCGGA  AACGCGGAAG  TCTCTTCTTG  TACGGCGTTC  TGAACGGCCT  TTTGACTAAC  CGGCCTGTCC  GGCGCAGCAC  GTGCGTCGTG  GAACGCAGCC
   ACCTGAGCCT  TTGCGCCTTC  AGAGAAGAAC  ATGCCGCAAG  ACTTGCCGGA  AAACTGATTG  GCCGGACAGG  CCGCGTCGTG  CACGCAGCAC  CTTGCGTCGG

        1310        1320        1330        1340        1350        1360        1370        1380        1390        1400
   ACAACCTCTA  GACGTGGTGT  AAAGCCGGGG  TGGCCAAGAA  GTGCTAGAAC  CGGAACGATC  TGACGAGGAA  GTCGCGCGCG  ACGGGCAAAA  GCGAGCAGTG
   TGTTGGAGAT  CTGCACCACA  TTTCGGCCCC  ACCGGTTCTT  CACGATCTTG  GCCTTGCTAG  ACTGCTCCTT  CAGCGCGCGC  TGCCCGTTTT  CGCTCGTCAC

        1410        1420        1430        1440        1450        1460        1470        1480        1490        1500
   TAGGTAAAGT  TAGTGCACGA  GGAATAAATA  GTATTACGAA  GGCACATCTG  TGAATTCGAG  CGGAAGCTAG  AGTCGCGTCG  CCACGTCGGT  GTTGCGCGTC
   ATCCATTTCA  ATCACGTGCT  CCTTATTTAT  CATAATGCTT  CCGTGTAGAC  ACTTAAGCTC  GCCTTCGATC  TCAGCGCAGC  GGTGCAGCCA  CAACGCGCAG

        1510        1520        1530        1540        1550        1560        1570        1580        1590        1600
   GGGCACCCGA  GCACTACGAA  CATCCAGTGG  AGACGTTTGC  TGACGTCCAT  GCGGACGTCC  TTAGCGGGGT  AGTAGCAGTG  TTTCCAGAAC  AACGACCACT
   CCCGTGGGCT  CGTGATGCTT  GTAGGTCACC  TCTGCAAACG  ACTGCAGGTA  CGCCTGCAGG  AATCGCCCCA  TCATCGTCAC  AAAGGTCTTG  TTGCTGGTGA

        1610        1620        1630        1640        1650        1660        1670        1680        1690        1700
   TCCAGTCGAC  GTTGGGCGCC  ACGAGGAGCA  AGTCGGTCCA  GAACGTATGC  CGGCGGTCTC  GAAGGTGAAC  CAGTCCGTCA  TCAAACTTCA  AGCGGAAATC
   AGGTCAGCTG  CAACCCGCGG  TGCTCCTCGT  TCAGCCAGGT  CTTGCATACG  GCCGCCAGAG  CTTCCACTTG  GTCAGGCAGT  AGTTTGAAGT  TCGCCTTTAG

        1710        1720        1730        1740        1750        1760        1770        1780        1790        1800
   TAGCAATAGG  TGCACCATGA  ACAGGTAGTC  GCGCGCGCGT  CGGAGGTACG  GGAAGAGGGT  GCGTCTGTGC  TAGCCGTGTG  AGTCGCCCAA  GTAGTGGCAT
   ATCGTTATCC  ACGTGGTACT  TGTCCATCAG  CGCGCGCGCA  GCCTCCATGC  CCTTCTCCCA  CGCAGACACG  ATCGGCACAC  TCAGCGGGTT  CATCACCGTA

        1810        1820        1830        1840        1850        1860        1870        1880        1890        1900
   TAAAGTGAAA  GGCGAAGCGA  CCCGAGAAGG  AGAAGGAGAA  CGCAGGCGTA  TGGTGCGCGG  TGACCCAGCA  GAAGTAAGTC  GGCGGCGTGA  CACGCGAATG
   ATTTCACTTT  CCGCTTCGCT  GGGCTCTTCC  TCTTCCTCTT  GCGTCCGCAT  ACCACGCGCC  ACTGGGTCGT  CTTCATTCAG  CCGCCGCACT  GTGCGCTTAC

        1910        1920        1930        1940        1950        1960        1970        1980        1990        2000
   GAGGAAACGG  TACGAACTAA  TCGTGGCCAC  CCAACGACTT  TGGGTGGTAA  ACATCGCGGT  GTAGAAGAGA  AAGAAGGAGC  GACAGGTGCT  AATGGAGACC
   CTCCTTTGCC  ATGCTTGATT  AGCACCGGTG  GGTTGCTGAA  ACCCACCATT  TGTAGCGCCA  CATCTTCTCT  TTCTTCCTCG  CTGTCCACGA  TTACCTCTGG
```

FIGURE 24A–D. Nucleotide sequence of a region between coordinates 59.9 and 71.4 on the Ad5 genome. This sequence and the positions of splice points and leaders were determined by Kruijer *et al.* (1980, 1981, 1983). A schematic presentation of this sequence is shown in Fig. 5 (Section VII).

```
      2010       2020       2030       2040       2050       2060       2070       2080       2090       2100
ACTACCGCCC GCGAGCCCGA ACCCTCTTCC CGCGAAGAAA AAGAAGAACC CGCGTTACCG GTTTAGGCGG CGGCTCCAGC TACCGGCGCC CGACCCACAC
TGATGGCGGG CGCTCGGGCT TGGGAGAAGG GCGCTTCTTT TTCTTCTTGG GCGCAATGGC CAAATCCGCC GCCGAGGTCG ATGGCCGCGG GCTGGGTGTG

      2110       2120       2130       2140       2150       2160       2170       2180       2190       2200
GCGCCGTGGT CGCGCAGAAC ACTACTCAGA AGGAGCAGGA GCCTGAGCTA TGCGGCGGAG TAGGCGAAAA AACCCCCGCG GGCCCCTCCG CCGCCGCTGC
CGCGGCACCA GCGCGTCTTG TGATGAGTCT TCCTCGTCCT CGGACTCGAT ACGCCGCCTC ATCCGCTTTT TTGGGGGCGC CCGGGGAGGC GGCGGCGACG

                                                                                                     start
      2210       2220       2230       2240       2250       2260       2270       2280       2290       2300
CCCTGCCCCT GCTGTGCAGG AGGTACCAAC CCCCTGCAGC GCGGCGTGGC GCAGGCGCGA GCCCCCACCA AAGCGCGACG AGGAGAAGGG CTGACCGGTA
GGGACGGGGA CGACACGTCC TCCATGGTTG GGGGACGTCG CGCCGCACCG CGTCCGCGCT CGGGGGTGGT TTCGCGCTGC TCCTCTTCCC GACTGGCCAT

splice DBP RNA
      2310       2320       2330       2340       2350       2360       2370       2380       2390       2400
AAGGAAGAGG ATATCCGTCT TTTTCTAGTA CCTCAGTCAG CTCTTCTTCC TGTCGGATTG GCGGGGGAGA CTCAAGCGGT GGTGGCGGAG GTGGCTACGG
TTCCTTCTCC TATAGGCAGA AAAAGATCAT GGAGTCAGTC GAGAAGAAGG ACAGCCTAAC CGCCCCCTCT GAGTTCGCCA CCACCGCCTC CACCGATGCC
           splice 100 kD RNA     start 100 kD protein

      2410       2420       2430       2440       2450       2460       2470       2480       2490       2500
CGGTTGCGCG GATGGTGGAA GGGGCAGCTC CGTGGGGGCG AACTCCTCCT CCTTCACTAA TAGCTCGTCC TGGGTCCAAA ACATTCGCTT CTGCTGCTCC
GCCAACGCGC CTACCACCTT CCCCGTCGAG GCACCCCCGC TTGAGGAGGA GGAAGTGATT ATCGAGCAGG ACCCAGGTTT TGTAAGCGAA GACGACGAGG

      2510       2520       2530       2540       2550       2560       2570       2580       2590       2600
TGGCGAGTCA TGGTTGTCTC CTATTTTTCG TTCTGGTCCT GTTGCGTCTC CGTTTGCTCC TTGTTCAGCC CGCCCCCCTG CTTTCCGTAC CGCTGATGGA
ACCGCTCAGT ACCAACAGAG GATAAAAAGC AAGACCAGGA CAACGCAGAG GCAAACGAGG AACAAGTCGG GCGGGGGGAC GAAAGGCATG GCGACTACCT

      2610       2620       2630       2640       2650       2660       2670       2680       2690       2700
TCTACACCCT CTGCTGCACG ACAACTTCGT AGACGTCGCG GTCACGCGGT AATAGACGCT GCGCAACGTT CTCGCGTCGC TACACGGGGA GCGGTATCGC
AGATGTGGGA GACGACGTGC TGTTGAAGCA TCTGCAGCGC CAGTGCGCCA TTATCTGCGA CGCGTTGCAA GAGCGCAGCG ATGTGCCCCT CGCCATAGCG

      2710       2720       2730       2740       2750       2760       2770       2780       2790       2800
CTACAGTCGG AACGGATGCT TGCGGTGGAT AAGAGTGGCG CGCATGGGGG GTTTGCGGTT CTTTTGCCGT GTACGCTCGG GTTGGGCGCG GAGTTGAAGA
GATGTCAGCC TTGCCTACGA ACGCCACCTA TTCTCACCGC GCGTACCCCC CAAACGCCAA GAAAACGGCA CATGCGAGCC CAACCCGCGC CTCAACTTCT

      2810       2820       2830       2840       2850       2860       2870       2880       2890       2900
TGGGGCATAA ACGGCACGGT CTCCACGAAC GGTGGATAGT GTAGAAAAAG GTTTTGACGT TCTATGGGGA TAGGACGGCA CGGTTGGCGT CGGCTCGCCT
ACCCCGTATT TGCCGTGCCA GAGGTGCTTG CCACCTATCA CATCTTTTTC CAAAACTGCA AGATACCCCT ATCCTGCCGT GCCAACCGCA GCCGAGCGGA

                                 splice E2A internal leader
      2910       2920       2930       2940       2950       2960       2970       2980       2990       3000
GTTCGTCGAC CGGAACGCCG TCCCGCGACA GTATGGACTA TAGCGGAGCG AGTTGCTTCA CGGTTTTTAG AAACTCCCAG AACCTGCGCT GCTCTTCGCG
CAAGCAGCTG GCCTTGCGGC AGGGCGCTGT CATACCTGAT ATCGCCTCGC TCAACGAAGT GCCAAAAATC TTTGAGGGTC TTGGACGCGA CGAGAAGCGC
C
```

```
splice E2A internal leader
      3010       3020       3030       3040       3050       3060       3070       3080       3090       3100
CGCCGTTTGC GAGACGTTGT CCTTTTGTCG CTTTTACTTT CAGTGAGACC TCACAACCAC CTTGAGCTCC CACTGTTGCG CGCGGATCGG CATGATTTTG
GCGGCAAACG CTCTGCAACA GGAAAACAGC GAAAATGAAA GTCACTCTGG AGTGTTGGTG GAACTCGAGG GTGACAACGC GCGCCTAGCC GTACTAAAAC

      3110       3120       3130       3140       3150       3160       3170       3180       3190       3200
CGTCGTAGCT CCAGTGGGTG AAACGGATGG GCCGTGAATT GGATGGGGGG TTCCAGTACT CGTGTCAGTA CTCACTCGAC TAGCACGCGG CACGCGTCGG
GCAGCATCGA GGTCACCCAC TTTGCCTACC CGGCACTTAA CCTACCCCCC AAGGTCATGA GCACAGTCAT GAGTGAGCTG ATCGTGCGCC GTGCGCAGCC

      3210       3220       3230       3240       3250       3260       3270       3280       3290       3300
GGACCTCTCC CTACGTTTAA ACGTTCTTGT TTGTCTCCTC CCGGATGGGC GTCAACCGCT GCTCGTCGAT CGCGCGACCG AAGTTTGCGC GCTCGGACGG
CCTGGAGAGG GATGCAAATT TGCAAGAACA AACAGAGGAG GGCCTACCCG CAGTTGGCGA CGAGCAGCTA GCGCGCTGGC TTCAAACGCG CGAGCCTGCC

      3310       3320       3330       3340       3350       3360       3370       3380       3390       3400
CTGAACCTCC TCGCTGCGTT TGATTACTAC CGGCGTCACG AGCAATGGCA CCTCGAACTC ACGTACGTCG CCAAGAAACG ACTGGGCCTC TACGTCGCGT
GACTTGGAGG AGCGACGCAA ACTAATGATG GCCGCAGTGC TCGTTACCGT GGAGCTTGAG TGCATGCAGC GGTTCTTTGC TGACCCGGAG ATGCAGCGCA

      3410       3420       3430       3440       3450       3460       3470       3480       3490       3500
TCGATCTCCT TTGTAACGTG ATGTGGAAAG CTGTCCCGAT GCATGCGGTC CGGACGTTCT AGAGGTTGCA CCTCGAGACG TTGGACCAGA GGATGGAACC
AGCTAGAGGA AACATTGCAC TACACCTTTC GACAGGGCTA CGTACGCCAG GCCTGCAAGA TCTCCAACGT GGAGCTCTGC AACCTGGTCT CCTACCTTGG

      3510       3520       3530       3540       3550       3560       3570       3580       3590       3600
TTAAAACGTG CTTTTGGCGG AACCCGTTTT GCACGAAGTA AGGTGCGAGT TCCCGCTCCG CGCGGCGCTG ATGCAGGCGC TGACGCAAAT GAATAAAGAT
AATTTTGCAC GAAAACCGCC TTGGGCAAAA CGTGCTTCAT TCCACGCTCA AGGGCGAGGC GCGCCGCGAC TACGTCCGCG ACTGCGTTTA CTTATTTCTA

      3610       3620       3630       3640       3650       3660       3670       3680       3690       3700
ACGATGTGGA CCGTCTGCCG GTACCCGCAA ACCGTCGTCA CGAACCTCCT CACGTTGGAG TTCCTCGACG TCTTTGACGA TTTCGTTTTG AACTTCCTGG
TGCTACACCT GGCAGACGGC CATGGGCGTT TGGCAGCAGT GCTTGGAGGA GTGCAACCTC AAGGAGCTGC AGAAACTGCT AAAGCAAAAC TTGAAGGACC

      3710       3720       3730       3740       3750       3760       3770       3780       3790       3800
ATACCTGCCG GAAGTTGCTC GCGAGGCACC GGCGCGTGGA CCGCCTGTAG TAAAAGGGGC TTGCGGACGA ATTTTGGGAC GTTGTCCCAG ACGGTCTGAA
TATGGACGGC CTTCAACGAG CGCTCCGTGG CCGCGCACCT GGCGGACATC ATTTTCCCCG AACGCCTGCT TAAAACCCTG CAACAGGGTC TGCCAGACTT

      3810       3820       3830       3840       3850       3860       3870       3880       3890       3900
GTGGTCAGTT TCGTACAACG TCTTGAAATC CTTGAAATAG GATCTCGCGA GTCCTTAGAA CGGGCGGTGG ACGACACGTG AAGGATCGCT GAAACACGGG
CACCAGTCAA AGCATGTTGC AGAACTTTAG GAACTTTATC CTAGAGCGCT CAGGAATCTT GCCCGCCACC TGCTGTGCAC TTCCTAGCGA CTTTGTGCCC

      3910       3920       3930       3940       3950       3960       3970       3980       3990       4000
TAATTCATGG CGCTTACGGG AGGCGGCGAA ACCCCGGTGA CGATGGAAGA CGTCGATCGG TTGATGGAAC GGATGGTGAG ACTGTATTAC CTTCTGCACT
ATTAAGTACC GCGAATGCCC TCCGCCGCTT TGGGGCCACT GCTACCTTCT GCAGCTAGCC AACTACCTTG CCTACCACTC TGACATAATG GAAGACGTGA

      4010       4020       4030       4040       4050       4060       4070       4080       4090       4100
CGCCACTGCC AGATGACCTC ACAGTGACAG CGACGTTGGA TACGTGGGGC GTGGCGAGGG ACCAAACGTT AAGCGTCGAC GAATTGCTTT CAGTTTAATA
GCGGTGACGG TCTACTGGAG TGTCACTGTC GCTGCAACCT ATGCACCCCG CACCGCTCCC TGGTTTGCAA TTCGCAGCTG CTTAACGAAA GTCAAATTAT

      4110
GCCATGG
CGGTACC
D
```

FIGURE 24 (Continued)

FIGURE 25.1. Nucleotide sequence of a region between coordinates 97.0 and 100.0 on the Ad5 genome. This sequence was determined by Steenbergh *et al.* (1977) and Steenbergh and Sussenbach (1979). The strategic sequences were determined by Baker and Ziff (1980, 1981) and further derived from the Ad2 sequence of this region (Fig. 21).

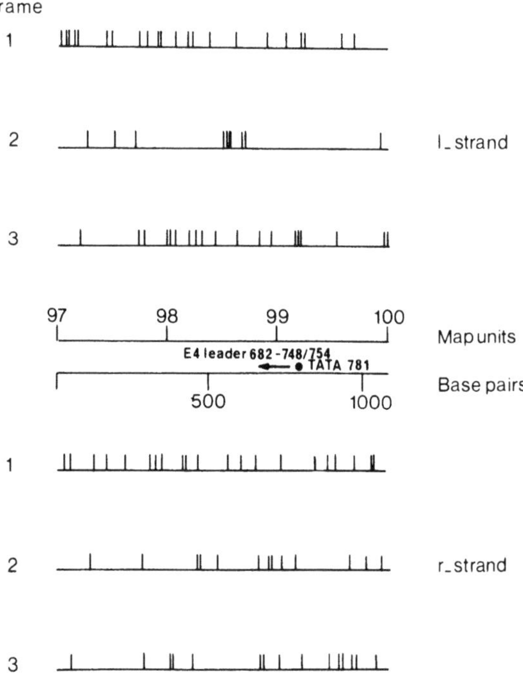

FIGURE 25.2. Structural organization of a region between coordinates 97.0 and 100.0 on the Ad5 genome. This map is derived from the nucleotide sequence in Fig. 25.1.

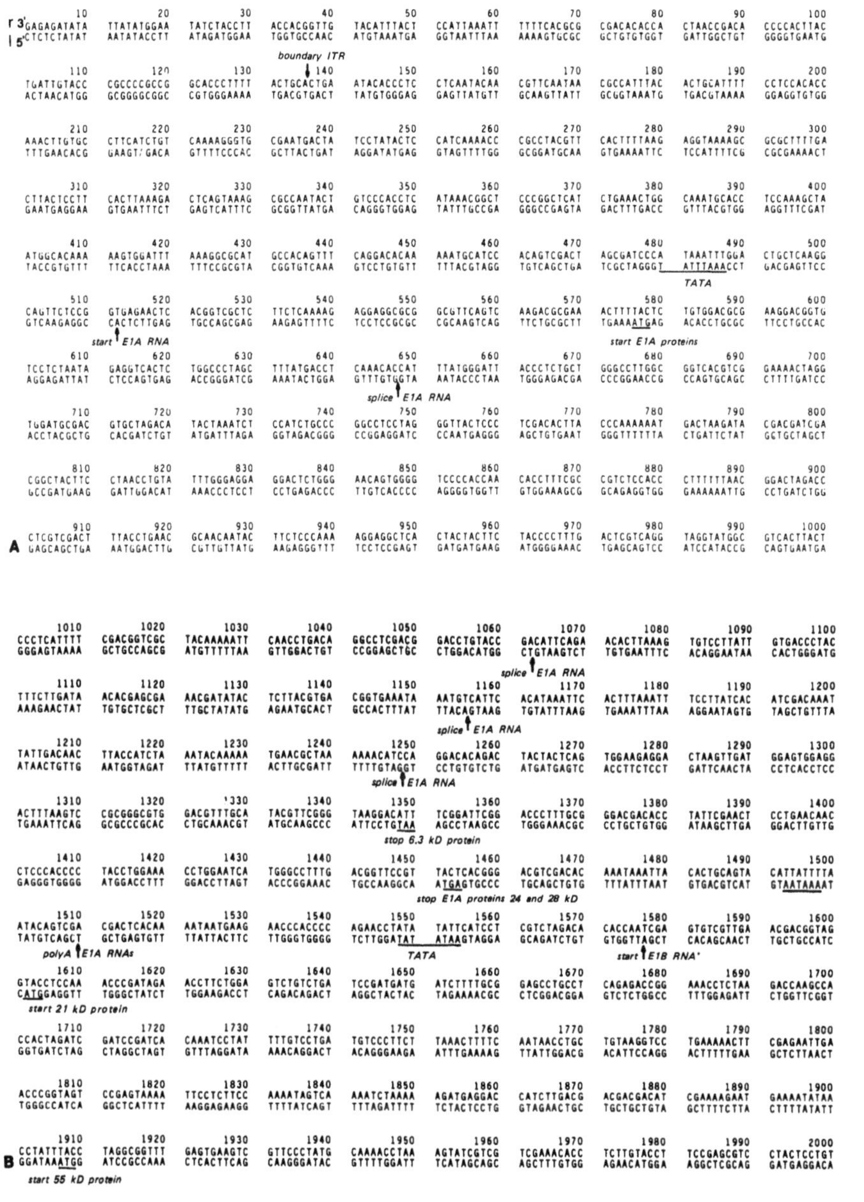

```
      10         20         30         40         50         60         70         80         90        100
r 3' GAGAGATATA TTATATGGAA TATCTACCTT ACCACGGTTG TACATTTACT CCATTAAATT TTTTCACGCG CGACACACCA CTAACCGACA CCCCACTTAC
l 5' CTCTCTATAT AATATACCTT ATAGATGGAA TGGTGCCAAC ATGTAAATGA GGTAATTTAA AAAAGTGCGC GCTGTGTGGT GATTGGCTGT GGGGTGAATG
                                  boundary ITR
     110        120        130      ↓ 140        150        160        170        180        190        200
TGATTGTACC CGCCCCGCCG GCACCCTTTT ACTGCACTGA ATACACCCTC CTCAATACAA CGTTCAATAA CGCCATTTAC ACTGCATTTT CCTCCACACC
ACTAACATGG GCGGGGCGGC CGTGGGAAAA TGACGTGACT TATGTGGGAG GAGTTATGTT GCAAGTTATT GCGGTAAATG TGACGTAAAA GGAGGTGTGG

     210        220        230        240        250        260        270        280        290        300
AAACTTGTGC CTTCATCTGT CAAAAGGGTG CGAATGACTA TCCTATACTC CATCAAAACC CGCCTACGTT CACTTTTAAG AGGTAAAAGC GCGCTTTTGA
TTTGAACACG GAAGT/GACA GTTTTCCCAC GCTTACTGAT AGGATATGAG GTAGTTTTGG GCGGATGCAA GTGAAAATTC TCCATTTTCG CGCGAAAACT

     310        320        330        340        350        360        370        380        390        400
CTTACTCCTT CACTTAAAGA CTCAGTAAAG CGCCAATACT GTCCCACCTC ATAAACGGCT CCCGGCTCAT CTGAAACTGG CAAATGCACC TCCAAAGCTA
GAATGAGGAA GTGAATTTCT GAGTCATTTC GCGGTTATGA CAGGGTGGAG TATTTGCCGA GGGCCGAGTA GACTTTGACC GTTTACGTGG AGGTTTCGAT

     410        420        430        440        450        460        470        480        490        500
ATGGCACAAA AAGTGGATTT AAAGGCGCAT GCCACAGTTT CAGGACACAA AAATGCATCC ACAGTCGACT AGCGATCCCA TAAATTTGGA CTGCTCAAGG
TACCGTGTTT TTCACCTAAA TTTCCGCGTA CGGTGTCAAA GTCCTGTGTT TTTACGTAGG TGTCAGCTGA TCGCTAGGGT ATTTAAACCT GACGAGTTCC
                                                                                      TATA
     510        520        530        540        550        560        570        580        590        600
CAGTTCTCCG GTGAGAACTC ACGGTCGCTC TTCTCAAAAG AGGAGGCGCG GCGTTCAGTC AAGACGCGAA ACTTTTACTC TGTGGACGCG AAGGACGGTG
GTCAAGAGGC CACTCTTGAG TGCCAGCGAG AAGAGTTTTC TCCTCCGCGC CGCAAGTCAG TTCTGCGCTT TGAAAATGAG ACACCTGCGC TTCCTGCCAC
     start ↑ E1A RNA                                                        start E1A proteins
     610        620        630        640        650        660        670        680        690        700
TCCTCTAATA GAGGTCACTC TGGCCCTAGC TTTATGACCT CAAACACCAT TTATGGGATT ACCCTCTGCT GGGCCTTGGC GGTCACGTCG GAAAACTAGG
AGGAGATTAT CTCCAGTGAG ACCGGGATCG AAATACTGGA GTTTGTGGTA AATACCCTAA TGGGAGACGA CCCGGAACCG CCAGTGCAGC CTTTTGATCC
                                             splice ↑ E1A RNA
     710        720        730        740        750        760        770        780        790        800
TGGATGCGAC GTGCTAGACA TACTAAATCT CCATCTGCCC GGCCTCCTAG GGTTACTCCC TCGACACTTA CCCAAAAAAT GACTAAGATA CGACGATCGA
ACCTACGCTG CACGATCTGT ATGATTTAGA GGTAGACGGG CCGGAGGATC CCAATGAGGG AGCTGTGAAT GGGTTTTTTA CTGATTCTAT GCTGCTAGCT

     810        820        830        840        850        860        870        880        890        900
CGGCTACTTC CTAACCTGTA TTTGGGAGGA GGACTCTGGG AACAGTGGGG TCCCCACCAA CACCTTTCGC CGTCTCCACC CTTTTTAAC  GGACTAGACC
GCCGATGAAG GATTGGACAT AAACCCTCCT CCTGAGACCC TTGTCACCCC AGGGGTGGTT GTGGAAAGCG GCAGAGGTGG GAAAAAATTG CCTGATCTGG

     910        920        930        940        950        960        970        980        990       1000
CTCGTCGACT TTACCTGAAC GCAACAATAC TTCTCCCAAA AGGAGGCTCA CTACTACTTC TACCCCTTTG ACTCGTCAGG TAGGTATGGC GTCACTTACT
A GAGCAGCTGA AATGGACTTG CGTTGTTATG AAGAGGGTTT TCCTCCGAGT GATGATGAAG ATGGGGAAAC TGAGCAGTCC ATCCATACCG CAGTGAATGA
```

```
    1010       1020       1030       1040       1050       1060       1070       1080       1090       1100
CCCTCATTTT CGACGGTCGC TACAAAAATT CAACCTGACA GGCCTCGACG GACCTGTACC GACATTCAGA ACACTTAAAG TGTCCTTATT GTGACCCTAC
GGGAGTAAAA GCTGCCAGCG ATGTTTTTAA GTTGGACTGT CCGGAGCTGC CTGGACATGG CTGTAAGTCT TGTGAATTTC ACAGGAATAA CACTGGGATG
                                                           splice ↑ E1A RNA
    1110       1120       1130       1140       1150       1160       1170       1180       1190       1200
TTTCTTGATA ACACGAGCGA AACGATATAC TCTTACGTGA CGGTGAAATA AATGTCATTC ACATAAATTC ACTTTAAATT TCCTTATCAC ATCGACAAAT
AAAGAACTAT TGTGCTCGCT TTGCTATATG AGAATGCACT GCCACTTTAT TTACAGTAAG TGTATTTAAG TGAAATTTAA AGGAATAGTG TAGCTGTTTA
                                                splice ↑ E1A RNA
    1210       1220       1230       1240       1250       1260       1270       1280       1290       1300
TATTGACAAC TTACCATCTA AATACAAAAA TGAACGCTAA AAAACATCCA GGACACAGAC TACTACTCAG TGGAAGAGGA CTAAGTTGAT GGAGTGGAGG
ATAACTGTTG AATGGTAGAT TTATGTTTTT ACTTGCGATT TTTTGTAGGT CCTGTGTCTG ATGATGAGTC ACCTTCTCCT GATTCAACTA CCTCACCTCC
                                         splice ↑ E1A RNA
    1310       1320       1330       1340       1350       1360       1370       1380       1390       1400
ACTTTAAGTC CGCGGGCGTG GACGTTTGCA TACGTTCGGG TAAGGACATT TCGGATTCGG ACCCTTTGCG GGACGACACC TATTCGAACT CCTGAACAAC
TGAAATTCAG GCGCCCGCAC CTGCAAACGT ATGCAAGCCC ATTCCTGTAA AGCCTAAGCC TGGGAAACGC CCTGCTGTGG ATAAGCTTGA GGACTTGTTG
                                           stop 6.3 kD protein
    1410       1420       1430       1440       1450       1460       1470       1480       1490       1500
CTCCCACCCC TACCTGGAAA CCTGGAATCA TGGGCCTTTG ACGGTTCCGT TACTCACGGG ACGTCGACAC AAATAAATTA CACTGCAGTA CATTATTTTA
GAGGGTGGGG ATGGACCTTT GGACCTTAGT ACCCGGAAAC TGCCAAGGCA ATGAGTGCCC TGCAGCTGTG TTTATTTAAT GTGACGTCAT GTAATAAAAT
                                              stop E1A proteins 24 and 28 kD
    1510       1520       1530       1540       1550       1560       1570       1580       1590       1600
ATACAGTCGA CGACTCACAA AATAATGAAG AACCCACCCC AGAACCTATA TATTCATCCT CGTCTAGACA CACCAATCGA GTGTCGTTGA ACGACGGTAG
TATGTCAGCT GCTGAGTGTT TTATTACTTC TTGGGTGGGG TCTTGGATAT ATAAGTAGGA GCAGATCTGT GTGGTTAGCT CACAGCAACT TGCTGCCATC
   polyA ↑ E1A RNAs                                 TATA                      start ↑ E1B RNA'
    1610       1620       1630       1640       1650       1660       1670       1680       1690       1700
GTACCTCCAA ACCCGATAGA ACCTTCTGGA GTCTGTCTGA TCCGATGATG ATCTTTTGCG GAGCCTGCCT CAGAGACCGG AAACCTCTAA GACCAAGCCA
CATGGAGGTT TGGGCTATCT TGGAAGACCT CAGACAGACT AGGCTACTAC TAGAAAACGC CTCGGACGGA GTCTCTGGCC TTTGGAGATT CTGGTTCGGT
 start 21 kD protein
    1710       1720       1730       1740       1750       1760       1770       1780       1790       1800
CCACTAGATC GATCCGATCA CAAATCCTAT TTTGTCCTGA TGTCCCTTCT TAAACTTTTC AATAACCTGC TGTAAGGTCC TGAAAAACTT CGAGAATTGA
GGTGATCTAG CTAGGCTAGT GTTTAGGATA AAACAGGACT ACAGGGAAGA ATTTGAAAAG TTATTGGACG ACATTCCAGG ACTTTTTGAA GCTCTTAACT

    1810       1820       1830       1840       1850       1860       1870       1880       1890       1900
ACCCGGTAGT CCGAGTAAAA TTCCTCTTCC AAAATAGTCA AAATCTAAAA AGATGAGGAC CATCTTGACG ACGACGACAT CGAAAAGAAT GAAAATATAA
TGGGCCATCA GGCTCATTTT AAGGAGAAGG TTTTATCAGT TTTAGATTTT TCTACTCCTG GTAGAACTGC TGCTGCTGTA GCTTTTCTTA CTTTTATATT

    1910       1920       1930       1940       1950       1960       1970       1980       1990       2000
CCTATTTACC TAGGCGGTTT GAGTGAAGTC GTTCCCTATG CAAAACCTAA AGTATCGTCG TCGAAACACC TCTTGTACCT TCCGAGCGTC CTACTCCTGT
B GGATAAATGG ATCCGCCAAA CTCACTTCAG CAAGGGATAC GTTTTGGATT TCATAGCAGC AGCTTTGTGG AGAACATGGA AGGCTCGCAG GATGAGGACA
 start 55 kD protein
```

FIGURE 26.1A–K. Nucleotide sequence of a region between coordinates 0 and 31.7 on the Ad7 genome. This sequence and the positions of strategic sequences were established by Dijkema and Dekker (1979), Dijkema *et al.* (1980a,b, 1981, 1982), van Beveren *et al.* (1981), Engler (1981), Engler *et al.* (1981, 1983), and Engler and van Bree (1982).

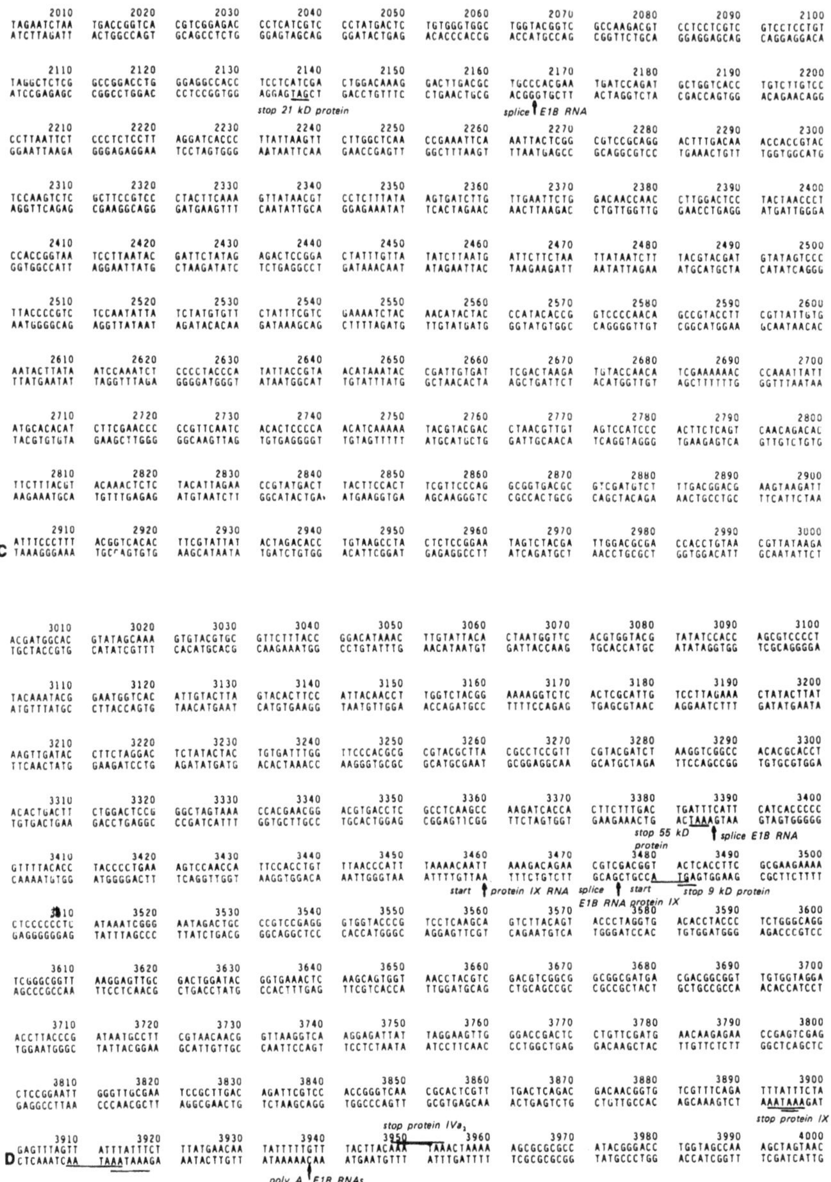

FIGURE 26.1 (*Continued*)

```
      4010       4020       4030       4040       4050       4060       4070       4080       4090       4100
TCTTGAGCCA CCTAGAAAAG GTCATGGGAC ATTTCCACCC TAACTTACAA ATCTATGTAC CCGTAATCAG GCAGAGCCCC CACCTCTACT GAGGTAACTT
AGAACTCGGT GGATCTTTTC CAGTACCCTG TAAAGGTGGG ATTGAATGTT TAGATACATG GGCATTAGTC CGTCTCGGGG GTGGAGATGA CTCCATTGAA

      4110       4120       4130       4140       4150       4160       4170       4180       4190       4200
CTCGGAGAAC GAGGCCCCAT CACAATATTT AGTGGGTCAG TATCGTTCCA GCCTCACGTA CCACAACGTG TTATAGAAAA TCCTCGTCTG ATTAACGTTG
GAGCCTCTTG CTCCGGGGTA GTGTTATAAA TCACCCAGTC ATAGCAAGGT CGGAGTGCAT GGTGTTGCAC AATATCTTTT AGGAGCAGAC TAATTGCAAC

      4210       4220       4230       4240       4250       4260       4270       4280       4290       4300
CCCCTCCGGG AATCACATCC ACAAATGTTT AGACAACTCG ACCCTGCCCA CGTAGGCCCC ACTTTAATAT ACGTAAAACC TGACCTAGAA CTCCAACCGT
GGGGAGGCCC TTAGTGTAGG TGTTTACAAA TCTGTTGAGC TGGGACGGGT GCATCCGGGG TGAAATTATA TGCATTTTGG ACTGGATCTT GAGGTTGGCA

      4310       4320       4330       4340       4350       4360       4370       4380       4390       4400
TACAACGGCG GATCTAGGGC AGAGCCCAAG TATAACACGT CCTGGTGGTT CTGTCACATA GGCCACGTGA ACCCTTTAGA TAGTACGTCG AATCTCCCTT
ATGTTGCCGC CTAGATCCCG TCTCGGGTTC ATATTGTGCA GGACCACCAA GACAGTGTAT CCGGTGCACT TGGGAAATCT ATCATGCAGC TTAGAGGGAA
                                                                                         start URF 1

      4410       4420       4430       4440       4450       4460       4470       4480       4490       4500
TTCGTACTTT TTTAAACCTC TGCGGAAACA CTGGGGGGTC TAAGAGGTAC GTGAGTAGGT ATTACTATCG CTACCCCGGC ACCCGTCGCC GTGCCCGCTT
AAGCATGAAA AAATTTGGAG ACGCCTTTGT GACCCCCCAG ATTCTCCATG CACTCATCCA TAATGATAGC GATGGGGCCG TGGGCAGCGG CACGGGCGAA

      4510       4520       4530       4540       4550       4560       4570       4580       4590       4600
GTGCAAGGCC CCTAGTGATT GCAGTATCAA CACAAGGTCC TACTCTAGCA GTATCCGGTA AAAATGTTTG AAACCCGCCT CCCACGGTCT AACCCCCTAC
CACGTTCCGG GGATCACTAA CGTCATAGTT GTGTTCCAGG ATGAGATCGT CATAGGCCAT TTTTACAAAC TTTGGGCGGA GGGTGCCAGA TTGGGGGATG

      4610       4620       4630       4640       4650       4660       4670       4680       4690       4700
TTTCAAGGGA CACCGGGCCC TCGTATCAAA GGGAGTGTAT AAACGTAAAG GGTCCGAAAG TCAAGTCTCC CCCCCTAGTA CAGGTGGACG CCCCGATATT
AAAGTTCCCT GTGGCCCGGG AGCATAGTTT CCCTCACATA TTTGCATTTC CCAGGCTTTC AGTTCAGAGG GGGGGATCAT GTCCACCTGC GGGGCTATAA

      4710       4720       4730       4740       4750       4760       4770       4780       4790       4800
TTTTATGGCA AAGACCTCGG CCCCACTAAT TGACCCTACT CTCGTTTAAG GATTCGTCGA CTCTGAACGG CGTGGGCCAC CCTGGCATTT ACTGGGGTTA
AAAATACCGT TTCTGGAGCC GGGGTGATTA ACTGGGATGA GAGCAAATTC CTAAGCAGCT GAGACTTGCC GCACCCGGTG GGACCGTAAA TGACCCCAAT
                         stop URF 1

      4810       4820       4830       4840       4850       4860       4870       4880       4890       4900
ATGCCCAACG TCTACCATCA AATCCCTCGC TGTCGACGGC AGGAGGGCCT CGTCCCCCCG GTGAAGCAAG TAGTAAAGGG AATGTACCTA TAAAAGGGCG
TACGGGTTGC AGATGGTAGT TTAGGGAGCG ACAGCTGCCG TCCTCCCGGA GCAGGGGGGC CACTTCGTTC ATCATTTCCC TTACATGGAT ATTTTCCCGC

        4910       4920       4930       4940       4950       4960       4970       4980       4990       5000
  TGGTTCAGGC AATCCTCCGC GAGAGGGGGT TCACTATCTT CGAGGACCTC GCTCCTCTTC AAAAAGTCGC CGAAGTCGGG CAGTCGGTAC CCGTAAAACC
E ACCAAGTCCG TTAGGAGGCG CTCTCCCCCA AGTGATAGAA GCTCCTGGAG CGAGGAGAAG TTTTTCAGCG GCTTCAGCCC GTCAGCCATG GGCATTTTGG


                                                      stop URF 9
      5010       5020       5030       5040       5050       5060       5070       5080       5090       5100
TTTCTCAGAC AACGTTCTCG AGCTCGGCCA GGGTCTCGAG CCACTACACG AGATACCGTA GAGCTAGGTC GTCTGGAGGA GCAAAGCGCC CAACCCTGCC
AAAGAGTCTG TTGCAAGAGC TCGAGCCGGT CCCAGAGCTC GGTGATGTGC TCTATGGCAT CTCGATCCAG CAGACCTCCT CGTTTCGCGG GTTGGGACGG

      5110       5120       5130       5140       5150       5160       5170       5180       5190       5200
GAGGACCTCA TCCCTTAGTC TGCTACCCGC AGGTCGCGAC GGTCCCAGGC TAGGAAGGTA CCAGCGTCGC AGGCTCAGTC CCAACAAAGG CAGTGCCACT
CTCCTGGAGT AGGGAATCAG ACGATGGGCG TCCAGCGCTG CCAGGGTCCG ATCCTTCCAT GGTCGCAGCG TCCGAGTCAG GGTTGTTTCC GTCACGGTGA
                 start URF 2
                                                                                      splice IVa2 RNA
      5210       5220       5230       5240       5250       5260       5270       5280    ↓  5290       5300
TCCCCACGCG CGGACCAACC CGCGAACGCT CCCACGCGAA GTCTGAGTAG GACGACCAGC TCTTGGCGAC GGCTAGCCGC GGGACGTACA GCCGGTCCAT
AGGGGTGCGC GCCTGGTTGG GCGCTTGCGA GGGTGCGCTT CAGACTCATC CTGCTGGTCG AGAACCGCTG CCGATCGGCG CCCTGCATGT CGGCCAGGTA

      5310       5320       5330       5340       5350       5360       5370       5380       5390       5400
CGTCAAATGG TATTCAAGCA TCAACTCGCG GAGCCGGCGC ACCGGAAACC GTGCCTCGAA TGGAAACCTT CAAAATACCG TCCGTCCCGT CATCTATGTA
GCAGTTTACC ATAAGTTCGT AGTTGAGCGC CTCGGCCGCG TGGCCTTTGG CACGGAGCTT ACCTTTGGAA GTTTTATGGC AGGCAGGGCA GTAGATACAT

      5410       5420       5430       5440       5450       5460       5470       5480       5490       5500
AACTCCCGTA TGTCGAACCC GCGCTCCTTT TACCTAAGCC CCCTCATACG TAGGCGTGGC GTCCTCTGCG TCTGCCAAAG CGTGAGGTGT TCGGTCCAGT
TTGAGGGCAT ACAGCTTGGG CGCGAGGAAA ATGGATTCGG GGGAGTATGC ATCCGCACCG CAGGAGACGC AGACGGTTTC GCACTCCACA AGCCAGGTCA
                                                         splice IVa2 RNA     start IVa2 protein
      5510       5520       5530       5540       5550       5560 ↓     5570       5580       5590       5600
CTAGGCCGAG TAGTCCCAGT TTTTGTTCAA AAGGCGGTAC AAAAAACTAC GCAAAGAATG GAAACCAAAG GTACTCAAGC ACAGGTGCGA CCCACTGTTT
GATCCGGCTC ATCAGGGTCA AAAACAAGTT TTCCGCCATG TTTTTTGATG CGTTTCTTAC CTTTGGTTTC CATGAGTTCG TGTCCACGCT GGGTGACAAA
                                                                                                starts IVa2 RNA
      5610       5620       5630       5640       5650       5660       5670       5680       5690  ↓ ↓  5700
CTCCGACAGG CACAGGGGCA TCTGGCTGAA ATACCCGGAC AGGAGCTCGC CTCACGGAGC CAGGAGAAGC ATCTCCTTAG GTCGGGTGAG ACTATGTTTT
GAGGCTGTCC GTGTCCCCGT AGACCGACTT TATGGGCCTG TCCTCGAGCG GAGTGCCTCG GTCCTCTTCG TAGAGGAATC CAGCCCACTC TGATACAAAA
                                                                                               stop URF 2

      5710       5720       5730       5740       5750       5760       5770       5780       5790       5800
CGCGCACAGG TCCGGTCGTG TTTCCTCCGG TGCACCCTCC CCATCGCCAG CAACAGTTGG TCCCCTAGGT GGAAGAGATG CCATACATTT GTGTACAGGG
GCGCGTGTCC AGGCCAGCAC AAAGGAGGCC ACGTGGGAGG GGTAGCGGTC GTTGTCAACC AGGGGATCCA CCTTCTCTAC GGTATGTAAA CACATGTCCC

      5810       5820       5830       5840       5850       5860       5870       5880       5890       5900
GGAGGAGGTG TAGGTTCTTA CACTAACCGA ACATTCACAT CCGGTGCACT GGTCCCCAGG GGCGGCCCCC CCATATTTTC CCCCGCCTGG AGACAAGCAG
CCTCCTCCAC ATCCAAGAAT GTGATTGGCT TGTAAGTGTA GGCCACGTGA CCAGGGGTCC CCGCCGGGGG GGTATAAAAG GGGGCGGACC TCTGTTCGTC
                                                                                TATA

        5910       5920       5930       5940       5950       5960       5970       5980       5990       6000
  GAGTGACAGA AGGCCTAGCG ACAGGTCCTC GCGGTCGACA ACCCCATCCA TAAGGGAGAG CTTACGCCCG TACTGGAGAC GTGAGTCCAA CAGTCAAAGA
F CTCACTGTCT TCCGGATCGC TGTCCAGGAG CGCCAGCTGT TGGGGTAGGT ATTCCCTCTC GAATGCGGGC ATGACCTCTG CACTCAGGTT GTCAGTTTCT
  ↑ start 1st leader                                 splice ↑ 1st leader
```

FIGURE 26.1 (*Continued*)

```
      6010       6020       6030       6040       6050       6060       6070       6080       6090       6100
TCCTTGCTCC TCCTAAACTA TAACTGTCAT GGTCGTCTCT ACGGAAAATA TTCTGAGAGC AGGTAGACCA GTCTTTTGTG TTAGAAGAAC AACAGGTCGA
AGGAACGAGG AGGATTTGAT ATTGACAGTA CCAGCAGAGA TGCCTTTTAT AAGACTCTCG TCCATCTGGT CAGAAAACAC AATCTTCTTG TTGTCCAGCT

      6110       6120       6130       6140       6150       6160       6170       6180       6190       6200
ACCACCGTTT ACTAGGTATC TCCCGCAACC TATCTTCGAA CCGCTACCTC GCGTACCAAA CCAAGAAAAG GGACAGGCGC GCGAGGAACC GCCGCTACAA
TGGTGGCAAA TGATCCATAG AGGGCGTTGG ATAGAAGCTT GGCGATGGAG CGCATGGTTT GGTTCTTTTC CCTGTCCGCG CGCTCCTTGG CGGCGATGTT
                                            start URF 3
      6210       6220       6230       6240       6250       6260       6270       6280       6290       6300
TTCGACCTGC ATGAGCGCGC GGTGTGTAAA GGTAAGTCCT TTCTACCAAC AGTCAAGTAG GCCTTGATAA GACTGAGCGG TAGGGGATAA CACGTCCCAA
AAGCTGGACG TACTCGCGCG CCACACATTT CCATTCAGGA AAGATGGTTG TCAGTTCATC CGGAACTATT CTGACTCGCC ATCCCCTATT GTGCAGGGTT

      6310       6320       6330       6340       6350       6360       6370       6380       6390       6400
TAGTCTAGGT GTGACCACCG GTGGAGCGGA GCCTCCCCGA GTAACCAGGT CGTCTCAGCT GGAGGAAAAG AACTTGTCTT TCCCCCCTCC CCCAGATCGT
ATCAGATCCA CACTGGTGGC CACCTCGCCT CGGAGGGGCT CATTGGTCCA GCAGAGTCGA CCTCCTTTTC TTGAACAGAA AGGGGGGAGG GGGTCTAGCA
                                                                                        stop URF 7
      6410       6420       6430       6440       6450       6460       6470       6480       6490       6500
ACTTGAGTAG TCCCCCCAGG CGTAGATACC ATTTATAAGG GCCATCGTTT AGAAACAGTT TTATCGACTA CCACCGTCCT AGTAGGTTCC AGTAGACGGT
TGAACTCATC AGGGGGGTCC GCATCTATGG TAAATATTCC CGGTAGCAAA TCTTTGTCAA AATAGCTGAT GGTGGCAGGA TCATCCAAGG TCATCTGCCA
                                                             stop URF 3
      6510       6520       6530       6540       6550       6560       6570       6580       6590       6600
AAGAGCTTGA CGGTCGCGCG CGAGTATCCC CAATTCTCCC CACGGGGTCC CGTACCCCAC CCACTCGCGC CTCCGTATGT ACGGTGTCTA TAGCATCTGT
TTCTCGAACT GCCAGCGCGC GCTCATAGGG GTTAAGAGGG GTGCCCCAGG GCATGGGGTG GGTGAGCGCG GAGGCATACA TGCCACAGAT ATCGTAGACA

      6610       6620       6630       6640       6650       6660       6670       6680       6690       6700
ATCTCCCCGA GAAGCTCTTA CGGCTACATT CACCCTATTG TCGCGGGGGG AGACTACGAA CGAGCGTGTA TCAGTATCTC AAGTACACTC CCCCGCTCCT
TAGAGGGGCT CTTCGAGAAT GCCGATGTAA GTGGGATAAC AGCGCCCCCC TCTGATGCTT GCTCGCACAT AGTCATAGAG TTCATGTGAG GGGGCGAGGA

      6710       6720       6730       6740       6750       6760       6770       6780       6790       6800
CTGGGCCCGG GTCTAACCAC GCCAACCCAA AAAGGCGGGA CATTTGCTAA ACCGCTTTCT ACCGTACCCT TAACCTTCTT TATCATCCAG AGACCTTATA
GACCCGGGCC CAGATTGGTG CGGTTGGGTT TTTCCGCCCT GTAAACGATT TGGCGAAAGA TGGCATGGGA ATTGGAAGAA ATAGTAGGTC TCTGGAATAT
    start URF 7                                                  stop URF 8
      6810       6820       6830       6840       6850       6860       6870       6880       6890       6900
CAATTTTACT TGTACTCCAT CCGGATGTCT CAGAGAATAC TTCACCCGTA TACTGAGAAC GTCGAACCGA TGGTCGAGAC GCCACTGCTC ATGTAGGTCC
GTTAAAATGA ACATGAGGTA GGCCTACAGA GTCTCTTATG AAGTGGGCAT ATGACTCTTG CAGCTTGGCT ACCAGCTCTG CGGTGACGAG TACATCCAGG

      6910       6920       6930       6940       6950       6960       6970       6980       6990       7000
 CGTGTCATCA ACTCTCAAAG GACCTACTAC AGTATTGCGC CAACCGAAAA GAAAAGGGTG TCGAGCGCCA ACTCTTCCAT AAGAAGCACT AGGAAGGTCA
G GCACAGTAGT TGAGAGTTTC CTGGATGATG TCATAACGCG GTTGGCTTTT CTTTTCCCAC AGCTCGCGGT TGAGAAGGTA TTCTTCGTGA TCCTTCCAGT
                                                               splice ↑2nd leader

      7010       7020       7030       7040       7050       7060       7070       7080       7090       7100
TGAGAAGCTC CCCTTTGGGC AGAAAAAGAC GTGCCATTCT CGGGTTGTAC ATCTTGACTA ACTGACGGAA CATCCCTGTC GTAGGGAAGA GGTGACCCTC
ACTCTTCGAG GGGAAACCCG TCTTTTTCTG CACGGTAAGA GCCCAACATG TAGAACTGAT TGACTGCCTT GTAGGGACAG CATCCCTTCT CCACTGGGAG
                                 splice ↑2nd leader
      7110       7120       7130       7140       7150       7160       7170       7180       7190       7200
TCTCATACGA ACCCGACGTA ACGCGTCGCT CCATACTCAC TCCCGTTTTC ACAGGGACTG GTACTGAAAC TCCTTAACTA TGAACTTCAG CTACAGTAGT
AGAGTATGCT TGGGCTGCAT TGCGCAGCGA GGTATGAGTG AGGGCAAAAG TGTCCCTGAC CATGACTTTG AGGAATTGAT ACTTGAAGTC GATGTCATCA

      7210       7220       7230       7240       7250       7260       7270       7280       7290       7300
GTCCGGGGGA CAAGGGTCTC AACCTTCAGG CGGGCGAAGA ACATCCGCCC TAACCCGTTT CGCTTTCATT GTAGTAACTT CTCCTAGAGT GGCCGGGACC
CAGGCCCCCT GTTCCCAGAG TTGGAAGTCC GCCCGCTTCT TGTAGGCGGG ATTGGGCAAA GCGAAAGTAA CATCATTGAA GAGGATCTCA CCGGCCCTGG
                                                                                     start URF 8
      7310       7320       7330       7340       7350       7360       7370       7380       7390       7400
CGTACTTTAA AGCCCACTAA AATTTTCCGA CTCCCTGGAG ACGAGCCAAT AACTATTGGA CTCGCCGGTT CTGCTAGAGT AGTTTCGGTA ACTACAACAC
GCATGAAATT TCGGGTGATT TTAAAAGGCT GAGGGACCTC TGCTCGGTTA TTGATAACCT GAGCGGCCAA GACGATCTCA TCAAAGCCAT TGATGTTGTG

      7410       7420       7430       7440       7450       7460       7470       7480       7490       7500
GGGGTGATAC ATGTCAAGAT TCTTAGCTCC CCACGGGGAC TGTACTCCGT CGAAGAACTC AAGAAGTTTT CACTCTAGAC ATCCCAGTCA CTCTCGTATC
CCCCACTATG TACAGTTCTA AGAATCGAGG GGTGCCCCTG ACATGAGGCA GCTTCTTGAG TTCTTCAAAA GTGAGATCTG TAGGGTCAGT GAGAGCATAG

      7510       7520       7530       7540       7550       7560       7570       7580       7590       7600
ACAAGCTCCC GGGTAAGCAC GTACACTCCC AAGCGAAACT CCTTCCTCCT GGTCTCCAGG TGACGGTCAC GACAAACATT GACCAGGGCC ATGACTGCTT
TGTTCGAGGG CCCATTCGTG CATGTGAGGG TTCGCTTTGA GGAAGGAGGA CCAGAGGTCC ACTGCCAGTG CTGTTTGTAA CTGGTCCCGG TACTGACGAA

      7610       7620       7630       7640       7650       7660       7670       7680       7690       7700
TTACGACAGG CTGACGGTAG AAAAGACCCC ACTACGTTAT CTTCCAAACC CCCAGGACGG CGGTCGCTAG GGTGAACTCA AAGTACCGCT ACAGTATCCG
AATGCTGTCC GACTGCCATC TTTTCTGGGG TGATGCAATA GAAGGTTTGG GGGTCCTGCC GCCAGCGATC CCACTTGAGT TTCATGGCGA TGTCATAGGC

      7710       7720       7730       7740       7750       7760       7770       7780       7790       7800
CTACAATTGC TCGGCGACCA GAGGTCTCTC AAAGTACTGG TCGTACTTCC CCTAATCGAC GAACGGTTTC CGGGGGTAGG TCCATATCCA GAGATGTAGC
GATGTTAACG AGCCGCTGGT CTCCAGAGAG TTTCATGACC AGCATGAAGG GGATTAGCTG CTTGCCAAAG GCCCCCATCC AGGTATAGGT CTCTACATCG

      7810       7820       7830       7840       7850       7860       7870       7880       7890       7900
ATCCACTCCT TCTCGGAAAG ACACGCTCCT ACTCTCGGCT AGCCCTTCTT GACCTAAAGG ACGGTGGTCA ACCTCCTTAC CGACAACTAC ACTACCTTCA
TAGGTGAGGA AGAGCCTTTC TGTGCGAGGA TGAGAGCCGA TCGGGAAGAA CTGGATTTCC TGCCACCAGT TGGAGGAATG GCTGTTGATG TGATGGAAGT
↑ start i leader           start URF 4
      7910       7920       7930       7940       7950       7960       7970       7980       7990       8000
 TCTTGAGGGA CGCTGCGCGG CTCGTAAGTA CGAACACGAA CATGTCTGCC GGCGTCATGA GCGTCGCTAA GTGCCCTACG TGGAGTACTT ACTCAACATG
H AGAACTCCCT GCGACGCGCC GAGCATTCAT GCTTGTGCTT GTACAGACGG CCGCAGTACT CGCAGCGATT CACGGGATGC ACCTCATGAA TGAGTTGTAC
```

FIGURE 26.1 (*Continued*)

```
       8010       8020       8030       8040       8050       8060       8070       8080       8090       8100
  GACTGAAGGA AACTGCTCTT TAAAGTCACC TTTTAACTCC GAACCGCGAA CATGGAGCGC GAGATGATAC AACAGACGTA GCCGTACTGG TAGAAGACAG
  CTGACTTCCT TTGACGAGAA ATTTCAGTGG AAAATTGAGG CTTGGCGCTT GTACCTCGCG CTCTACTATG TTGTCTGCAT CGGCATGACC ATCTTCTGTC

       8110       8120       8130       8140       8150       8160       8170       8180       8190       8200
  AGCTACCACC AGTACGACTG CTCGGGAGCG CCCTCCGTTC AGGTCTGGAG CCGCGCCGTC CCCGCCTCGA GCTCCTGCTC TCGCGCGTCC GGCCTCGACA
  TCGATGGTGG TCATGCTGAC GAGCCCTCGC GGGAGGCAAG TCCAGACCTC GGCGCGGCAG GGGCGGAGCT CGAGGACGAG AGCGCGCAGG CCGGAGCTGT

       8210       8220       8230       8240       8250       8260       8270       8280       8290       8300
  GGTCCCAGGA CTCTGCGACG CCTCAGTCCA ATCATCCGTC ACAGTCCTCT AACTGAACGT ACTAGAAAAG CTCCCGCACT CCCTCCAAGT CTACCATGAA
  CCAGGGTCCT GAGACGCTGC GGAGTCAGGT TAGTAGGCAG TGTCAGGAGA TTGACTTGCA TGATCTTTTC GAGGGCGTGA GGGAGGTTCA GATGGTACTT

       8310       8320       8330       8340       8350       8360       8370       8380       8390       8400
  CTAGAGGTGC CCAGGCAACC ACCTCTACAG CTACCGAACG TCCCAAGGCA CGGGGAACCC GCGATGGTGG CACGGGAACA AAAAGGAAAA CCCGCCGCCA
  GATCTCCACG GGTCCGTTGG TGGAGATGTC GATGGCTTGC AGGGTTCCGT GCCCCTTGGG CGCTACCACC GTGCCCTTGT TTTTCCTTTT GGGCGGCGGT

            start URF 9  stop pTP
       8410       8420       8430       8440       8450       8460       8470       8480       8490       8500
  CCGAGACAAC GAAGAACGTA CAAATCTTCG CCACAGCTCC CGCCCGTGGC CCGCCGTCCC CGCCGAGCCC TGGGCCGCCG TACCGACCGT CACCATGCAG
  GGCTCTGTTG CTTCTTGCAT GTTTAGAAGC GGTGTCGAGG GCGGGCACCG GGCGGCAGGG GCGGCTCGGG ACCCGGCGGC ATGGCTGGCA GTGGTACGTC

       8510       8520       8530       8540       8550       8560       8570       8580       8590       8600
  CCGTGGCGCG CGCCCATCCA AGACCATGAC GCGGGACTCT TCTGAGCGTA CGCGCTGCTG CGCCGCCAAC TGTAGGACCT AGACTGCGGA GACCCACTTT
  GGCACCGCGC GCGGGTAGGT TCTGGTACTG CGCCCTGAGA AGACTCGCAT GCGCGACGAC GCGGCGGTTG ACATCCTGGA TCTGACGCCT CTGGGTGAAA
                                                  start URF 5
       8610       8620       8630       8640       8650       8660       8670       8680       8690       8700
  CGATGGCCGG GGCACTCGAA CTTGGACTTT CTCTCAAGTT GTCTTAGTTA GAGCCATAGC AACTGCCGCC GAACGGATTC CTAAAGAACG TGCAGCGGTC
  GCTACCGGCC CCGTGAGCTT GAACCTGAAA GAGAGTTCAA CAGAATCAAT CTCGGTATCG TTGACGGCGG CTTGCCTAAG GATTTCTTGC ACGTCGCCAG

       8710       8720       8730       8740       8750       8760       8770       8780       8790       8800
  TCAACAGGAC CATCCGCTAG AGCCGGTACT TGACGAGCTA GAGAAGGAGA ACCTCTAGAG GCGCCGGGCG AGAGAGCTGC CACCGGCGCT CCAGCAACCT
  AGTTGTCCTG GTAGGCGATC TCGGCCATGA ACTGCTCGAT CTCTTCCTCT TGGAGATCTC CGCGGCCCGC TCTCTCGACG GTGGCCGCGA GGTCGTTGGA

       8810       8820       8830       8840       8850       8860       8870       8880       8890       8900
  CTACGCGGGT TACTCAACTC TCTTTCGTAA GTACGGGCGG AGCAAGGTCT GCGCCGACAT CTGGTGTCGG GGGTGCCCTA GAGAGCGCGC GTACTGGTGG
  GATGCGCCCA ATGAGTTGAG AGAAAGCATT CATGCCCGCC TCGTTCCAGA CGCGGCTGTA GACCACAGCC CCCACGGGAT CTCTCGCGCG CATGACCACC

       8910       8920       8930       8940       8950       8960       8970       8980       8990       9000
  ACCCGCTCCA ACTCGAGGTG CACCGCCCAC TTCTGGCGTA TCAACGTATC CGCGACCTTT TCCATCAACT CACACCACCG CTACACGAGC CACTGCTTCT
I TGGGCGAGGT TGAGCTCCAC GTGGCGGGTG AAGACCGCAT AGTTGCATAG GCGCTGGAAA AGGTAGTTGA GTGTGGTGGC GATGTGCTCG GTGACGAAGA
                                                  stop URF 5

       9010       9020       9030       9040       9050       9060       9070       9080       9090       9100
  TTATGTACTA GGTAGCAGAG TCGCCGTAGA GCGACTGTAG CGGGTCGCGA AGGTTCGCGA GGTACCGGAG CATCTTCAAG TGCCGTTTCA ACTTTTTGAC
  AATACATGAT CCATCGTCTC AGCGGCATCT CGCTGACATC GCCCAGCGCT TCCAAGCGCT CCATGGCCTC GTAGAAGTTC ACGGCAAAGT TGAAAAACTG

       9110       9120       9130       9140       9150       9160       9170       9180       9190       9200
  CCTCAATGCG CGCCTGTGCC AGTTGAGGAG AAGGTCTTCT GCCTATTCAA GCCGCTACCA CCACGCGTGG AGCGCGAGCT TTCGAGGATC CTAAAGAAGG
  GGAGTTACGC GCGGACACGG TCAACTCCTC TTCCAGAAGA CGGATAAGTT CGGCGATGGT GGTGCGCACC TCGCGCTCGA AAGCTCCTAG GATTTCTTCC

       9210       9220       9230       9240       9250       9260       9270       9280       9290       9300
  AGTTAGAGAA GAAGAAGGTG ATTGTAGAGA AGGAGAAGTC CACCCCGACG TCCTCCTCCC CCTTGCGCCG CTGCGGCCGC CGCGTGCCCG TCTGCCAGCT
  TCAATCTCTT CTTCTTCCAC TAACATCTCT TCCTCTTCAG GTGGGGCTGC AGGAGGAGGG GGAACGCGGC GACGCCGGCG GCGCACGGGC AGACGGTCGA

       9310       9320       9330       9340       9350       9360       9370       9380       9390       9400
  ACTTAGAAAG TTACTGGAGA GGCGCCGCCG CCGCGTATCA GAGCCACTGC CGTGCTGGCA AGAGGGACCC AGAGTCTCAC TTCTGCGGAG GCGCGTAGAG
  TGAATCTTTC AATGACCTCT CCGCGGCGGC GGCGCATAGT CTCGGTGACG GCACGACCGT TCTCCCTGGG TCTCAGAGTG AAGACGCCTC CGCGCATCTC

       9410       9420       9430       9440       9450       9460       9470       9480       9490       9500
  GGACTTCACC ACTGACCCTC CGAGAGGCAA CCCGTCCCTG TGGCGCGACT AATACGTAAA ATAGTTAACG GGGCATCCAT GAGGCGCGTT CCTGGACTAA
  CCTGAAGTGG TGACTGGGAG GCTCTCCGTT GGGCAGGGAC ACCGCGCTGA TTATGCATTT TATCAATTGC CCCGTAGGTA CTCCGCGCAA GGACCTGATT
                                                                             splice ↑ 3rd leader
       9510       9520       9530       9540       9550       9560       9570       9580       9590       9600
  CAGAGTTCTA GGTGCCCTAG ACTTTTGGAA AGCTGCTTTC GCAGATTGGT CAGCGTTAGC GTTCCATCCG ACTCGTGACA AAGAACGCCC GCCCCCGCCG
  GTCTCAAGAT CCACGGGATC TGAAAACCTT TCGACGAAAG CGTCTAACCA GTCGCAATCG CAAGGTAGGC TGAGCACTGT TTCTTGCGGG CGGGGGCGGC
                                                              splice ↑ 3rd leader
       9610       9620       9630       9640       9650       9660       9670       9680       9690       9700
  ATCTGCGAGC CAGCCCCAAG AGAGAAAGAA GAGGAAGGGG GAGAACGCTC CCACTCTGCT ACGACGACCA CTACTTTAAT TTTATCCGTC AAAATTCTGC
  TAGACGCTCG GTCGGGGTTC TCTCTTTCTT CTCCTTCCCC CTCTTGCGAG GGTGAGACGA TGCTGCTGGT GATGAAATTA AAATAGGCAG TTTTAAGACG

       9710       9720       9730       9740       9750       9760       9770       9780       9790       9800
  CGCCTACCAC CGCTCCTCGT GGTTCAGAAA CCCAGGCCGA ACAACCTACG CGTCCGCTAC TCGGTAGGGG GTTCGTAGTA GGACTGTAGA CCGGTCTAGA
  GCGGATGGTG GCGAGGAGCA CCAAGTCTTT GGGTCCGGCT TGTTGGATGC GCAGGCGATG AGCCATCCCC CAAGCATCAT CCTGACATCT GGCCAGATCT

       9810       9820       9830       9840       9850       9860       9870       9880       9890       9900
  AATATCATCA GAACGTACTC AGCAAGGTGC CCGTGAAGAA GAAGCGGGCG GGACGGTACG TACGCTCACT AGGGCTTGGG CGCGTACCCG ACCTGTTCAC
  TTATAGTAGT CTTGCATGAG TCGTTCCACG GGCACTTCTT CTTCGCCCGC CCTGCCATGC ATGCGAGTGA TCCCGAACCC GCGCATGGGC TGGACAAGTG

       9910       9920       9930       9940       9950       9960       9970       9980       9990      10000
  GGTCGAGGCG ATGTTGGGAA AGCCGCTCCT ACCGAACGAC GTGGACCCAC TCCCACCGAA CTTTCAGCAG TTTCAGGTGC TTCGCCACCA TCCGGGGCCA
J CCAGCTCCGC TACAACCCTT TCGGCGAGGA TGGCTTGCTG CACCTGGGTG AGGGTGGCTT GAAAGTCGTC AAAGTCCACG AAGCGGTGGT AGGCCCCGGT
```

FIGURE 26.1 (*Continued*)

```
     10010      10020      10030      10040      10050      10060      10070      10080      10090      10100
CAACTAACAC ATCCTCGTCA ACCGGTACTG ACTGGTCAAC TGACAGACCA CGGGTCCCGC GTGCTCGAGC CACATGAAGT CCGCGCTCAT ACGCGCGCAC
GTTGATTGTG TAGGAGCAGT TGGCCATGAC TGACCAGTTG ACTGTCTGGT GCCCAGGGCG CACGAGCTCG GTGTACTTCA GGCGCGAGTA TGCGCGCGTG

     10110      10120      10130      10140      10150      10160      10170      10180      10190      10200
AGTTTCTACA TTAGCAACGT CCACGCGTGG TCCATGACCA TCGGCTACTC TTTCACACCG CTACCGACCG CCATGTCCCC GGTAGCGAGA CATCGGCCCC
TCAAAGATGT AATCGTTGCA GGTGCGCACC AGGTACTGGT AGCCGATGAG AAAGTGTGGC GATGGCTGGC GGTACAGGGG CCATCGCTCT GTAGCCGGGG

     10210      10220      10230      10240      10250      10260      10270      10280      10290      10300
GAGGCCCCCG CTCCAGAAGG TCGTACTCCG CCACCATCGG CATCTACATG GACCTGTAGG TCCACTATGG CCTCCGCCAC CACCTACGTG CACCCTTGAG
CTCCGGGGGC GAGGTCTTCC AGCATGAGGC GGTGGTAGCC GTAGATGTAC CTGGACATCC AGGTGATACC GGAGGCGGTG GTGGATGCAC GTGGGAACTC

     10310      10320      10330      10340      10350      10360      10370      10380      10390      10400
CGCGTGCGCC AAGGTCTACA ACGCGTCGCC GTACTTCATC AAGTACCATC CGTGCCAGAC CGGTCACTCC GCGCGCGTCA GTAACTGCGA GACATCTGTG
GCGCACGCGG TTCCAGATGT TGCGCAGCGG CATGAAGTAG TTCATGGTAG GCACGGTCTG GCCAGTGAGG CGCGCGCAGT CATTGACGCT CTGTAGACAC
                                 start URF 6

     10410      10420      10430      10440      10450      10460      10470      10480      10490      10500
CCTCTTTTGC TTTCGCTACT CGCCGAGCTG AGGCACCGGA CCCCCTTGCA CCTGCCCAAC CCAGCGCCAC ATGGGGCCAA GCTCAGGTTT CGATTCGTTA
GGAGAAAACG AAAGCGATGA GCGGCTCGAC TCCGTGGCCT GGGGGAACGT GGACGGGTTG GGTCGCGGTG TACCCCGGTT CGAGTCCAAA GCTAAGCAAT

     10510      10520      10530      10540      10550      10560      10570      10580      10590      10600
GTGTGAGCCT AGCCGGCCTC GGCGCCGATT GCACCATAAC CGATAGGGCA GAGCTGGGTC GGCTGCTTAT AGGTCCCATG CCTCATCTCA GCAAAAACGA
CACACTCGGA TCGGCCGGAG CCGCGGCTAA CGTGGTATTG GCTATCCCGT CTCGACCCAG CCGACGAATA TCCAGGGTAC GGAGTAGAGT CGTTTTTGCT

                      start pTP
     10610      10620      10630      10640      10650      10660      10670      10680      10690      10700
CGAAAAAAGG ACCTGCACAC GGTAACGGTG CAGTTCGAAA TGTTGCGAGT CAAGAGCCCG GCACTCACCG AGCGCGGGCA TCAGACCTCT TAGTCAGCGG
GCTTTTTTCC TGGACGTGTG CCATTGCCAC GTCAAGCTTT ACAACGCTCA GTTCTCGGGC CGTGAGTGGC TCGCGCCCGT AGTCTGGAGA ATCAGTCGCC
                                                                                  stop URF 6

     10710      10720      10730      10740      10750      10760      10770      10780      10790      10800
TCCCAACGCA ACGCCATACG GGGGCCAAGC TCGGATTCGC GCCGAGCATA GCCGGCCAAA GGCGCTGTTC GCTCCCATAC CGTCGGGTCA GTAAAGGTTC
AGGGTTGCGT TGCGGTATGC CCCCGGTTCG AGCCTAAGCG CGGCTCGTAT CGGCCGGTTT CCGCGACAAG CGAGGGTATG GCAGCCCAGT CATTTCCAAG

     10810      10820      10830      10840      10850      10860      10870      10880      10890      10900
TGGGGCGGTC GGCTGAAGAG GTCAAATGCC CTCGCTCGGG AAAAAAAAAA AACAAAAAGA GCGGGTCTAC GTAGGTCACG ACGCTGTCTA CGCGGGGGTC
ACCCCGCCAG CCGACTTCTC CAGTTTACGG GAGCGAGCCC TTTTTTTTTT TTGTTTTTCT CGCCCAGATG CATCCAGTGC TGCGACAGAT GCGCCCCCAG
                                                                  start 52, 55 kD proteins

     10910      10920      10930      10940      10950
 GTTGTCCGGG GAAGAGTCGT TGTCGGTGTT TTCCGAGAAG AACGAGGA
KCAACAGGCCC CTTCTCAGCA ACAGCCACAA AAGGCTCTTC TTGCTCCT
```

FIGURE 26.1 (*Continued*)

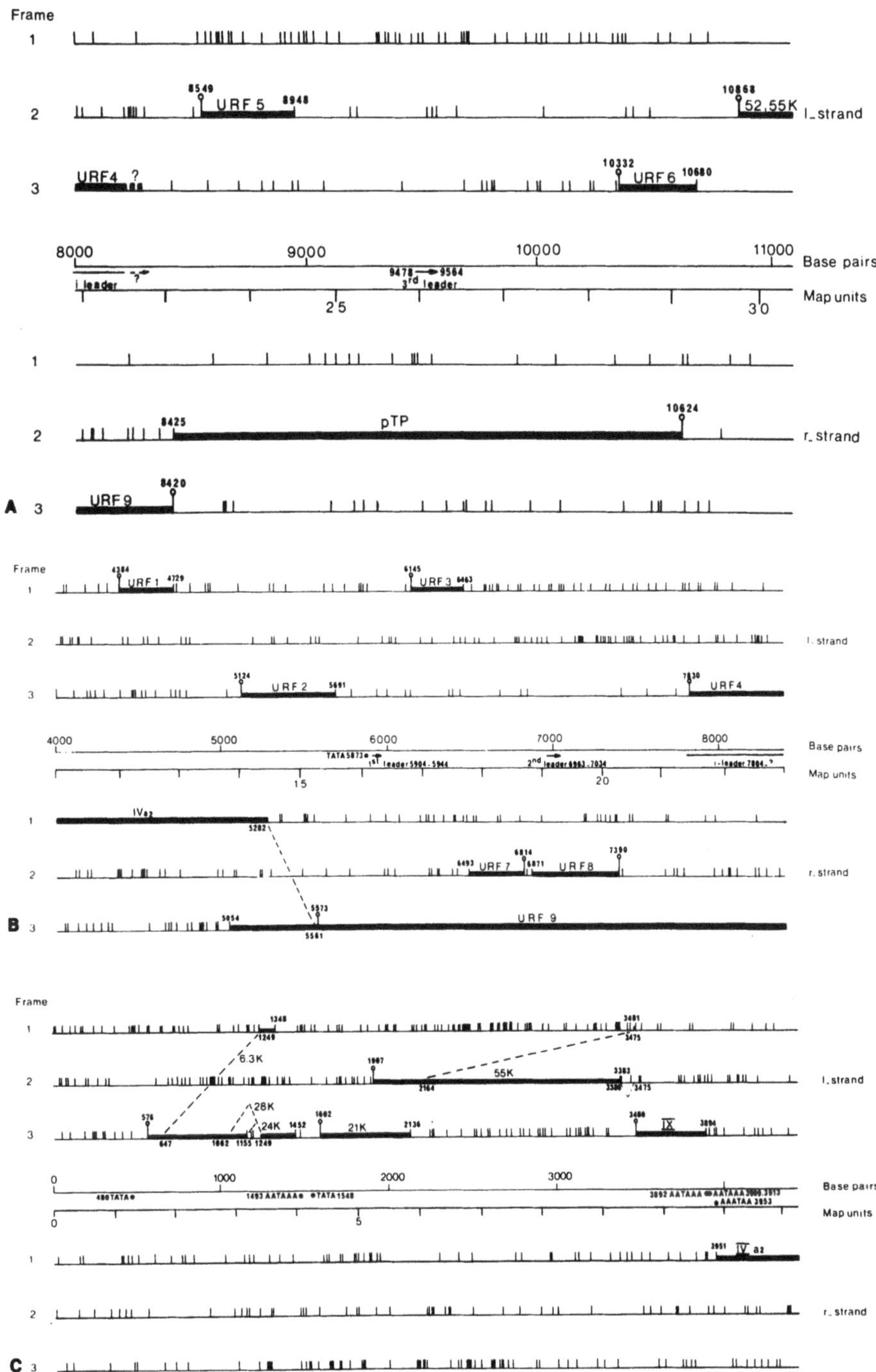

FIGURE 26.2A–C. Structural organization of a region between coordinates 0 and 31.7 on the Ad7 genome. This map is derived from the nucleotide sequence in Fig. 26.1. For details, see Fig. 3 (Section VII).

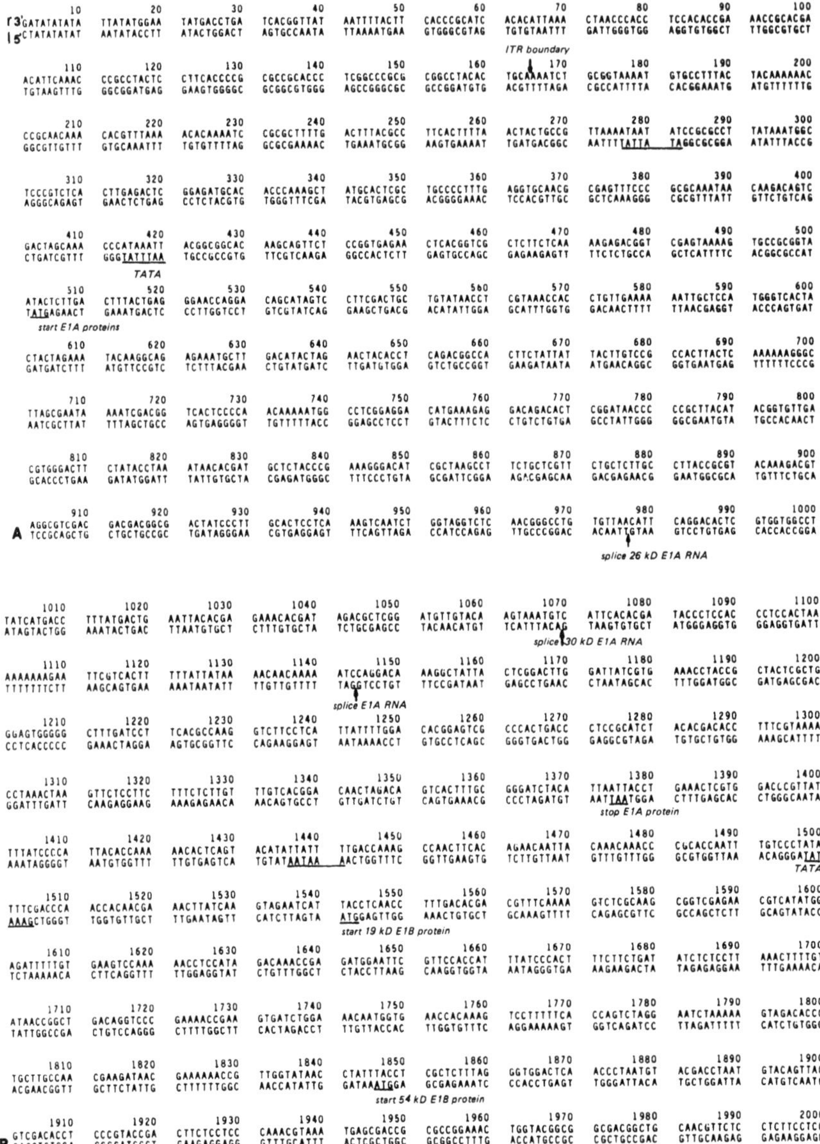

FIGURE 27.1A–D. Nucleotide sequence of a region between coordinates 0.0 and 11.5 on the Ad12 genome. This sequence and strategic signals were established by Fujinaga *et al.* (1979), Sugisaka *et al.* (1980), Kimura *et al.* (1981), and Bos *et al.* (1981).

```
      2010       2020       2030       2040       2050       2060       2070       2080       2090       2100
TCCTCGCCTT GGGACGCCAC CACCTCTTCA TTTGTACCTT GTTGTCCACG TTCTTCCGGT ACATGAACTG AGACCGCTTC CCGGATCAAC GCGTCTACTA
AGGAGCGGAA CCCTGCGGTG GTGGAGAAGT AAACATGGAA CAACAGGTGC AAGAAGGCCA TGTACTTGAC TCTGGCGAAG GGCCTAGTTG CGCAGATGAT
                        stop 19 kD E1B protein               splice 19 kD E1B RNA

      2110       2120       2130       2140       2150       2160       2170       2180       2190       2200
TCTCTATTCG TCCTTTTTTT TCTTTCAAAT TTCCTTCGAC GACAAGAATC ATCCGATTGA CAATTAGACT ACAGGGCGGG CGCAAACCTT TGACATATAA
AGAGATAAGC AGGAAAAAAA AGAAAGTTTA AAGGAAGCTG CTGTTCTTAG TAGGCTAACT GTTAATCTGA TGTCCCGCCC GCGTTTGGAA ACTGTATATT

      2210       2220       2230       2240       2250       2260       2270       2280       2290       2300
CCGTCCTCAA CGTCCTACTT AAAGTCGCCC CACTATACGT AAATGTCATG TTTATGTCAA AACTTGTTAA TTTTTGGGTG ACCAATCTCG GTACCCTCCT
GGCAGGAGTT GCAGGATGAA TTTCAGCGGG GTGATATGCA TTTACAGTAC AAATACAGTT TTGAACAATT AAAAACCCAC TGGTTAGAGC CATGGGAGGA

      2310       2320       2330       2340       2350       2360       2370       2380       2390       2400
ATACCTCACA CGATAATTTC GAAAACGATT TAACCGGAAT GCAGGACTAA CATCGATGTC TTAATGATTT TGTCATTGGT AATGAAGTAC GCGGATATAA
TATGGAGTGT GCTATTAAAG CTTTTGCTAA ATTGGCCTTA CGTCCTGATT GTAGCTACAG AATTACTAAA ACAGTAACCA TTACTTCATG CGCCTATATT

      2410       2420       2430       2440       2450       2460       2470       2480       2490       2500
TATCCATTGC CCCGTTATCA ACTCCATCTA TGTTCGCTGT CTCAACGAAA ATCTACAGCT TACGTCCCAT ACCCGGGTCC CCACCACCCA AACCTACCTT
ATAGGTAACG GGGCAATAGT TGAGGTAGAT ACAAGCGACA GAGTTGCTTT TAGATGTCGA ATGCAGGGTA TGGGCCCAGG GGTGGTGGGT TTGGATGGAA

      2510       2520       2530       2540       2550       2560       2570       2580       2590       2600
AATGTAAATA TTTACAATCC AAACGACCTC TATTCAAATT TCCGTAATAC AAGCTTCGAT TATGGACAGA ACAGAACGTA CCACAAATGA AAGAATTGAA
TTACATTTAT AAATGTTAGG TTTGCTGGAG ATAAGTTTAA AGGCATTATG TTCGAAGCTA ATACCTGTCT TGTCTTGCAT GGTGTTTACT TTCTTAACTT

      2610       2620       2630       2640       2650       2660       2670       2680       2690       2700
ATCATTGTAA ACACATCTCA GAACCTTATT CCAAAGACGA TCCCCGACAT GAAAAATACC TACAACCTTC CCAAACCACC CATCTGGTTT TTCATTTGAC
TAGTAACATT TGTGTAGAGT CTTGGAATAA GGTTTCTGCT AGGGGCTGTA CTTTTTATGG ATGTTGGAAG GGTTTGGTGG GTAGACCAAA AAGTAAACTG

      2710       2720       2730       2740       2750       2760       2770       2780       2790       2800
AGACATTTTT TCACAAACAA ACTTTTTACA CATGAACGAA ATTAACATCT CCCCCTACGT GTATAATCCG TATTACGTCG AAGTCTTTTA CGGACAAAAC
TCTGTAAAAA AGTGTTTGTT TGAAAAATGT GTACTTGCTT TAATTGTAGA GGGGGATGCA CATATTAGGC ATAATGCAGC TTCAGAAAAT GCCTGTTTTG

      2810       2820       2830       2840       2850       2860       2870       2880       2890       2900
ATAATAACTT CCCTTACCGA TAAAATTTCG TATTATACCA AACACCCCAC AGACTAGTTT GATACGCTGC AAAACAATGG ACACGACTAC CTTTAACAGT
TATTATTGAA GGGAATGGCT ATTTTAAAGC ATAATATGGT TTGTGGGGTG TCTGATCAAA CTATGCGACG TTTTGTTACC TGTGCTGATG GAAATTGTCA

      2910       2920       2930       2940       2950       2960       2970       2980       2990       3000
 ATGGAATTTT TGACAAGTAT AACACTCGGT GTCATCTGTA ACAACCGGAC ATACACTAGT ATTGTACAAA TACGCGACAT GGTATGTAAA TCCGAATTCC
C TACCTTAAAA ACTGTTCATA TTGTGAGCCA CAGTAGACAT TGTTGGCCTG TATGTGATCA TAACATGTTT ATGCGCTGTA CCATACATTT AGGCTTAAGG

      3010       3020       3030       3040       3050       3060       3070       3080       3090       3100
GCCCCATACA AATCTGGAAG GGTTACATTG AAGTCGGTGA GTTTGTAATA CGACCTTGGA CTTCACAAAA GATCTCACAC AAATTTACCC CATAAACTAA
CGGGGTATGT TTAGACCTTC CCAATGTAAC TTCAGCCACT CAAACATTAT GCTGGAACCT GAAGTGTTTT CTAGAGTGTG TTTAAATGGG GTATTTGATT

      3110       3120       3130       3140       3150       3160       3170       3180       3190       3200
ATAGACACCT TAATACATTC CAATATTCTA TATTACTACT ATGAGCTGTA GCAACGGCTG TCACACTCAC ACCATCGTCA GTAGATCTTG AAGCAGGGTA
TATCTGTGGA ATTATGTAAG GTTATAAGAT ATAATGATGA TACTCGACAT CGTTGCCGAC AGTGTGAGTG TGGTAGCAGT CATCTAGAAC TTCGTCCCAT

      3210       3220       3230       3240       3250       3260       3270       3280       3290       3300
ACACGATTTA CATTGACTCC TCGACTCTTC ACTGGTGGAA TGGGACAGAA CGGACGCCTG ACTGATACTC AGTTCACTAC TTCTGCTGTT GACTCCATTC
TGTGCTAAAT GTAACTGAGG AGCTGAGAAG TGACCACCTT ACCCTGTCTT GCCTGCGGAC TGACTATGAG TCAAGTGATG AAGACGACAA CTGAGGTAAG
                                                                          stop 54 kD E1B protein
                                                                                             splice E1B RNA

      3310       3320       3330       3340       3350       3360       3370       3380       3390       3400
ACCCACCTCG ATCCACCCTA ATATTTTCCG ACCTTCAGTT GATTTTTAAC AAAAACAAGA AAATTGTCGT GCTACTTGCC TTGATGAGTC TTGTTGCGAC
TGGGTGGAGC TAGGTGGGAT TATAAAAGGC TGGAAGTCAA CTAAAAATTG TTTTTGTTCT TTTAACAGCA CGATGAACGG AACTACTCAG AACAACGCTG
                      TATA                                       splice E1B RNA   start protein IX

      3410       3420       3430       3440       3450       3460       3470       3480       3490       3500
GCGAAAAACT ACCTCCCCAA AAATCGGGAA TAAACTGAAG GTCCAATGGT ATAACCCGGC CTCATGCAGT CTTACACCAT CCTAGATGTC ACCTGCCAGC
CGCTTTTTGA TGGAGGGGTT TTTAGCCCTT ATTTGACTTC CAGGTTACCA TATTGGGCCG GAGTACGTCA GAATGTGGTA GGATCTACAG TGGACGGTCG

      3510       3520       3530       3540       3550       3560       3570       3580       3590       3600
TGGACACCGT GGACGTTTAA GTAGTTGTAA TTGGATACGT TGATAACCTG GGAGCGGAAA CCTATGGCGG CGGCGGCGAC GTCGAAGGCG CCGGCGAAGA
ACCTGTGGCA CCTGCAAATT CATCAACATT AACCTATGCA ACTATTGGAC CCTCGCCTTT GGATACCGCC GCCGCCGCTG CAGCTTCCGC GGCCGCTTCT

      3610       3620       3630       3640       3650       3660       3670       3680       3690       3700
TGCCGAGCGT CATACCGTCG ACTAAAGTCG AAGATGTTAG TGAACCGAAG CTTACGACAC TGTGCGTGGC GTCAAGCTCT CCTGTAAGAC TGACAATACG
ACGGCTCGCA GTATGGCAGC TGATTTCAGC TTCTACAATC ACTTGGCTTC GAATGCTGTG ACACGCACCG CAGTTCGAGA GGACATTCTG ACTGTTATGC

      3710       3720       3730       3740       3750       3760       3770       3780       3790       3800
AACGGTTCGA ACTTTGAGAT TGACGAGTCG ACCTTCTCGA TAGCGTTTTC CAACTCCTTA ATCGACTACG ATGATGGGTA TGGGGTCGGG TTGGACATTG
TTGCCAAGCT TGAAACTCTA ACTGCTCAGC TGGAAGAGCT ATCGCAAAAG GTTGAGGAAT TAGCTGATGC TACTACCCAT ACCCCAGCCC AACCTGTAAC

      3810       3820       3830       3840       3850       3860       3870       3880       3890       3900
GGTTATTTCT TTTTTGAATT TAACTCTACC ACAATACTTA GAAATAACTA TGAACAAAAA AGACTGTACC ATTCGAGAAC TGGTGGCAAG GGATAGTAAT
CCAATAAAGA AAAAACTTAA ATTGAGATGG TGTTATGAAT CTTTATTGAT ACTTGTTTTT TCTGACATGG TAAGCTCTTG ACCACCGTTC CCTATCATTA
 stop protein IX

      3910       3920       3930       3940       3950       3960
 TCTTGTGCCA CTTACACAAG GTCATAAAAC ATTTCTACTC GGACATATAA TTCCATG
D AGAACACGGT GAATGTGTTC CAGTATTTTG TAAAGATGAG CCTGTATATT AAGGTAC
```

FIGURE 27.1 (*Continued*)

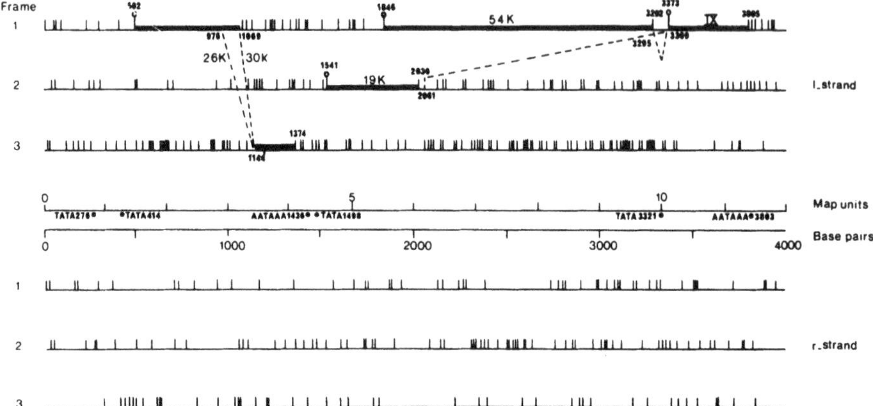

FIGURE 27.2. Structural organization of a region between coordinates 0.0 and 11.5 on the Ad12 genome. This map is derived from the nucleotide sequence in Fig. 27.1.

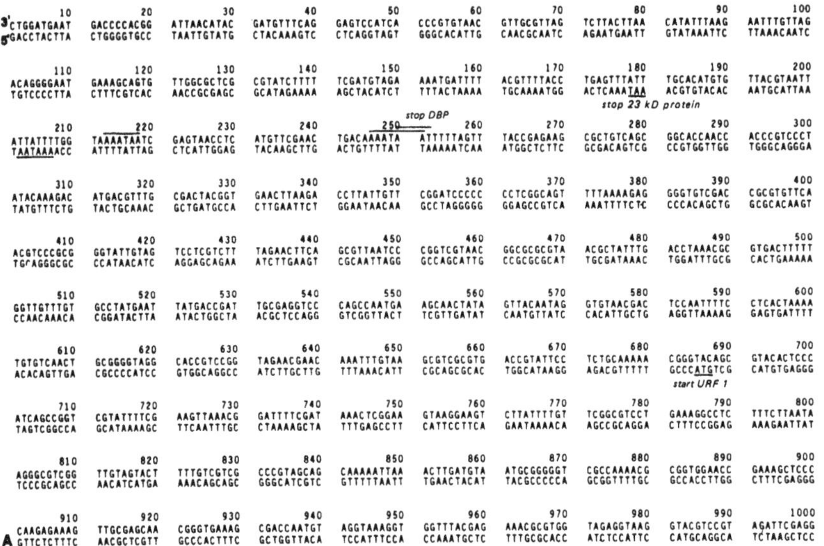

FIGURE 28.1A, B. Nucleotide sequence of a region between coordinates 61.5 and 67.0 on the Ad12 genome. This sequence was established by Kruijer *et al.* (1983).

```
      1010        1020        1030        1040        1050        1060        1070        1080        1090        1100
GGAAGCGCGA  GCCATGTGAA  TACGAGGGTG  TGCGTCGTTG  GCCACCCAAG  GGTCCTTAAG  ACAACCTGTG  GCCGTATTCG  AACGTATATA  GGAACGTTTT
CCTTCGCGCT  CGGTACACTT  ATGCTCCCAC  ACGCAGCAAC  CGGTGGGTTC  CCAGGAATTC  TGTTGGACAC  CGGCATAAGC  TTGCATATAT  CCTTGCAAAA

      1110        1120        1130        1140        1150        1160        1170        1180        1190        1200
TCGCAGGGTA  CTCGAGGACT  TTCCAAAAAA  CCCTGCTTTT  TCAGTCGACG  TTTGGCGCGA  AAAGAAGCAA  CTCGGTACAA  CACGTATAAA  AGAACATGTG
AGCGTCCCAT  GAGCTCCTGA  AAGGTTTTTT  CGGACGAAAA  AGTCAGCTGC  AAACCGCGCT  TTTCTTCGTT  GAGCCATGTT  GTGCATATTT  TCTTGTACAC

      1210        1220        1230        1240        1250        1260        1270        1280        1290        1300
CGACGGGACT  AGGCCGTTTT  TTGCTTTCCA  CCGCGCGAGC  AGCACTAGGT  GTACCATGAA  AAGGTAATCG  TATCGGTACC  GAAGGTACGG  AAAAAGGGTT
GCTGCCCTGA  TCCGGCAAAA  AACGAAAGGT  GGCGCGCTCG  TCGTGATCCA  CATGGTACTT  TTCCATTAGC  ATAGCCATGG  CTTCCATGCC  TTTTTCCCAA
                                                                        stop URF 1

      1310        1320        1330        1340        1350        1360        1370        1380        1390        1400
CGACTTTGAT  CCCCGACCGA  ACGGCCTAAC  GCTTGTTGTT  GTTGTAAGAA  AAGTAAAAGC  AGCGACAAAA  CTCGCCTTCG  GAAGTTTTGC  ACATGGACGG
GCTGAAACTA  GGGGCTGGCT  TGCCGGATTG  CGAACAACAA  CAACATTCTT  TTCATTTTCG  TCGCTGTTTT  GAGCGGAAGC  CTTCAAAACG  TGTACCTGCC

      1410        1420        1430        1440        1450        1460        1470        1480        1490        1500
ACCAAAGGTA  AAAAACTTTT  CTGACTCTTG  GCAGACGTAC  TACGTATTAC  GCCTGCCCGC  CGTACGACTT  TGGGTAATGA  GGATTTTGAC  GAGAACCACC
TGGTTTCCAT  TTTTTGAAAA  GACTGAGAAC  CGTCTGCATG  ATGCATAATG  CGGACGGGCG  GCATGCTGAA  ACCCATTACT  CCTAAAACTG  CTCTTGGTGG

      1510        1520        1530        1540        1550        1560        1570        1580        1590        1600
AAGACGGAGA  AGAAGAAGAC  GTGAGAGACC  CCTTTCTCCA  TAGCGTCGGT  ATCTAAAGAA  CTGAAAAAAG  AAACCTCCAT  TTCCGTGTCG  AAGGTCAAGA
TTCTGCCTCT  TCTTCTTCTG  CACTCTCTGG  GGAAAGAGGT  ATCGCAGCCA  TAGATTTCTT  GACTTTTTTC  TTTGGAGGTA  AAGGCACAGC  TTCCAGTTCT

      1610        1620        1630        1640        1650        1660        1670        1680        1690        1700
AGAAGCGAAA  GCCTTAGGTC  TTTCATAGAC  GGGTAAAAAC  CGCCGCCGCC  GACTCGCGAC  GCCAGACCCC  ACGCGAGGGA  GACACTCACG  ACTAACGACC
TCTTCGCTTT  CGGAATCCAG  AAAGTATCTG  CCCATTTTTG  GCGGCGGCGG  CTGAGCGCTG  CGGTCTGGGG  TGCGCTCCCT  CTGTGAGTGC  TGATTGCTGG

start DBP
      1710        1720        1730        1740        1750        1760        1770        1780        1790        1800
GGTAATAAAT  TAGGATCCGT  TTCTTTGTGT  ACTACCTAGA  CCTCGGTGTC  CTTTCGAATT  GGCGGGGGTG  GCGAGGGCGG  TAACCGCGAT  GCCGACAGTA
CCATTATTTA  ATCCTAGGCA  AAGAAACACA  TGATGGATCT  GGAGCCACAG  GAAAGCTTAA  CCGCCCCCAC  CGCTCCCGCC  ATTGGCGCTA  CGGCTGTCAT
                                    start 100 kD protein

      1810        1820        1830        1840        1850        1860        1870        1880        1890        1900
CCTCTTCCTG  TTTTCAGATG  AGTATGGGGT  TCTGCGTGGC  CAACTCGTCT  TGAACCCGAT  GCTCTGAGGG  GGGCTCCTTA  AACTTCCGAA  AGAAGTTTAG
GGAGAAGGAC  AAAAGTCTAC  TCATACCCCA  AGACGCACCG  GTTGAGCAGA  ACTTGGGCTA  CGAGACTCCC  CCCGAGGAAT  TTGAAGGCTT  TCTTCAAATC

      1910        1920        1930        1940
GTTTTCGTTG  GTTTACTCGT  TTTGCGACCC  GAGCTC
CAAAAGCAAC  CAAATGAGCA  AAACGCTGGG  CTCGAG
B
```

FIGURE 28.1 (*Continued*)

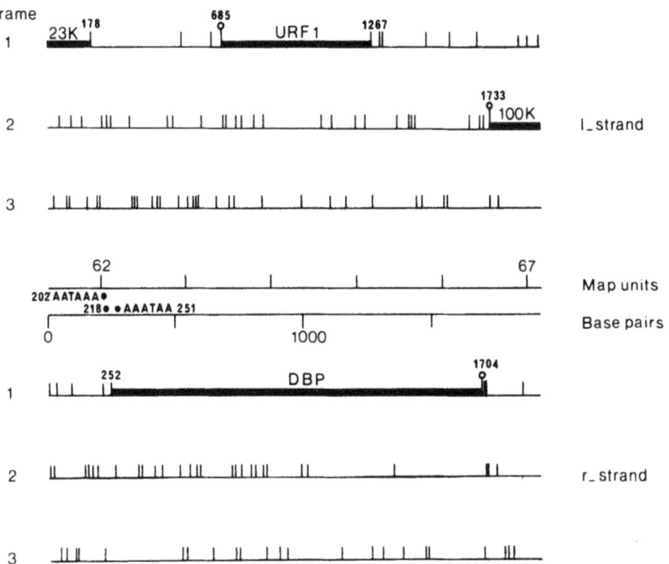

FIGURE 28.2. Structural organization of a region between coordinates 61.5 and 67.0 on the Ad12 genome. This map is derived from the nucleotide sequence in Fig. 28.1.

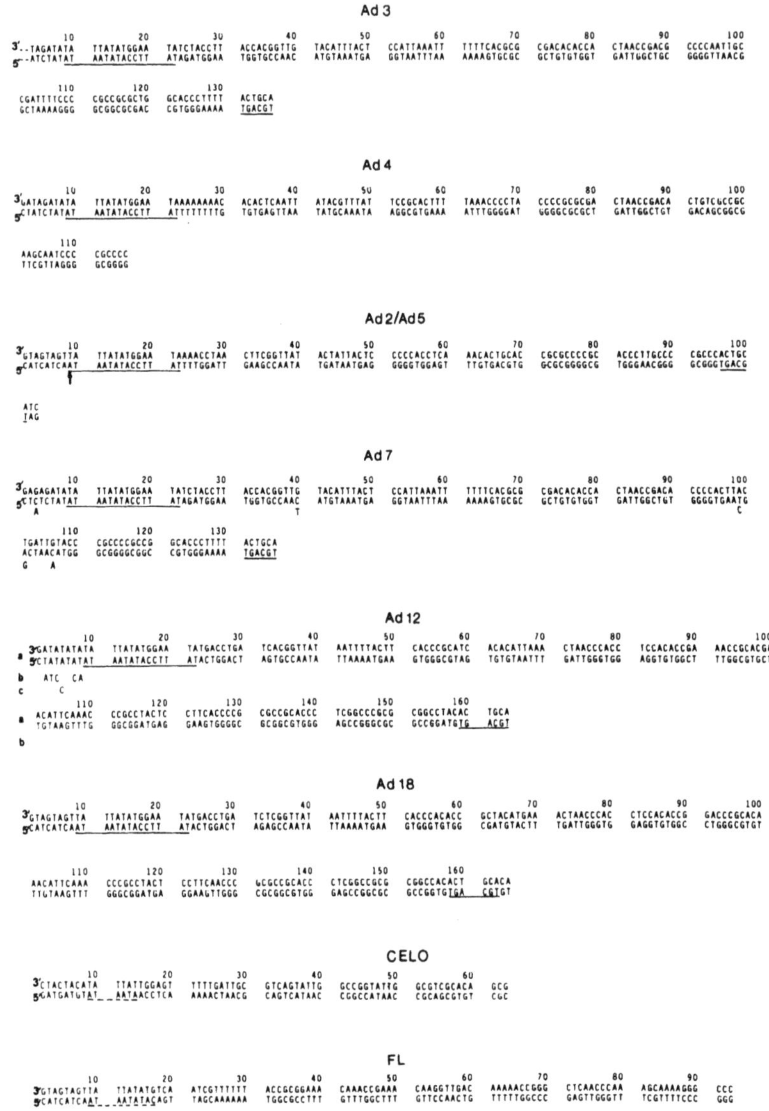

Ad 3

```
           10          20          30          40          50          60          70          80          90         100
3'..TAGATATA  TTATATGGAA  TATCTACCTT  ACCACGGTTG  TACATTTACT  CCATTAAATT  TTTTCACGCG  CGACACACCA  CTAACCGACG  CCCCAATTGC
5'..ATCTATAT  AATATACCTT  ATAGATGGAA  TGGTGCCAAC  ATGTAAATGA  GGTAATTTAA  AAAAGTGCGC  GCTGTGTGGT  GATTGGCTGC  GGGGTTAACG

       110         120         130
CGATTTTCCC  CGCCGCGCTG  GCACCCTTTT  ACTGCA
GCTAAAAGGG  GCGGCGCGAC  CGTGGGAAAA  TGACGT
```

Ad 4

```
         10          20          30          40          50          60          70          80          90         100
3'GATAGATATA  TTATATGGAA  TAAAAAAAAC  ACACTCAATT  ATACGTTTAT  TCCGCACTTT  TAAACCCCTA  CCCCGCGCGA  CTAACCGACA  CTGTCGCCGC
5'CTATCTATAT  AATATACCTT  ATTTTTTTTG  TGTGAGTTAA  TATGCAAATA  AGGCGTGAAA  ATTTGGGGAT  GGGGCGCGCT  GATTGGCTGT  GACAGCGGCG

       110
AAGCAATCCC  CGCCCC
TTCGTTAGGG  GCGGGG
```

Ad2/Ad5

```
         10          20          30          40          50          60          70          80          90         100
3'GTAGTAGTTA  TTATATGGAA  TAAAACCTAA  CTTCGGTTAT  ACTATTACTC  CCCCACCTCA  AACACTGCAC  CGCGCCCCGC  ACCCTTGCCC  CGCCCACTGC
5'CATCATCAAT  AATATACCTT  ATTTTGGATT  GAAGCCAATA  TGATAATGAG  GGGGTGGAGT  TTGTGACGTG  GCGCGGGGCG  TGGGAACGGG  GCGGGTGACG
          ↑

ATC
TAG
```

Ad 7

```
         10          20          30          40          50          60          70          80          90         100
3'GAGAGATATA  TTATATGGAA  TATCTACCTT  ACCACGGTTG  TACATTTACT  CCATTAAATT  TTTTCACGCG  CGACACACCA  CTAACCGACA  CCCCACTTAC
5'CTCTCTATAT  AATATACCTT  ATAGATGGAA  TGGTGCCAAC  ATGTAAATGA  GGTAATTTAA  AAAAGTGCGC  GCTGTGTGGT  GATTGGCTGT  GGGGTGAATG
   A                                         T                                                                       C

       110         120         130
TGATTGTACC  CGCCCCGCCG  GCACCCTTTT  ACTGCA
ACTAACATGG  GCGGGGCGGC  CGTGGGAAAA  TGACGT
G   A
```

Ad 12

```
           10          20          30          40          50          60          70          80          90         100
a 3'GATATATATA  TTATATGGAA  TATGACCTGA  TCACGGTTAT  AATTTTACTT  CACCCGCATC  ACACATTAAA  CTAACCCACC  TCCACACCGA  AACCGCACGA
  5'CTATATATAT  AATATACCTT  ATACTGGACT  AGTGCCAATA  TTAAAATGAA  GTGGGCGTAG  TGTGTAATTT  GATTGGGTGG  AGGTGTGGCT  TTGGCGTGCT
b    ATC  CA
c      C
         110         120         130         140         150         160
a  ACATTCAAAC  CCGCCTACTC  CTTCACCCCG  CGCCGCACCC  TCGGCCCGCG  CGGCCTACAC  TGCA
   TGTAAGTTTG  GGCGGATGAG  GAAGTGGGGC  GCGGCGTGGG  AGCCGGGCGC  GCCGGATGTG  ACGT
b
```

Ad 18

```
         10          20          30          40          50          60          70          80          90         100
3'GTAGTAGTTA  TTATATGGAA  TATGACCTGA  TCTCGGTTAT  AATTTTACTT  CACCCACACC  GCTACATGAA  ACTAACCCAC  CTCCACACCG  GACCCGCACA
5'CATCATCAAT  AATATACCTT  ATACTGGACT  AGAGCCAATA  TTAAAATGAA  GTGGGTGTGG  CGATGTACTT  TGATTGGGTG  GAGGTGTGGC  CTGGGCGTGT

       110         120         130         140         150         160
AACATTCAAA  CCCGCCTACT  CCTTCAACCC  GCGCCGCACC  CTCGGCCGCG  CGGCCACACT  GCACA
TTGTAAGTTT  GGGCGGATGA  GGAAGTTGGG  CGCGGCGTGG  GAGCCGGCGC  GCCGGTGTGA  CGTGT
```

CELO

```
         10          20          30          40          50          60
3'CTACTACATA  TTATTGGAGT  TTTTGATTGC  GTCAGTATTG  GCCGGTATTG  GCGTCGCACA  GCG
5'GATGATGTAT--AATAACCTCA  AAAACTAACG  CAGTCATAAC  CGGCCATAAC  CGCAGCGTGT  CGC
```

FL

```
         10          20          30          40          50          60          70          80          90
3'GTAGTAGTTA  TTATATGTCA  ATCGTTTTTT  ACCGCGGAAA  CAAACCGAAA  CAAGGTTGAC  AAAAACCGGG  CTCAACCCAA  AGCAAAAGGG  CCC
5'CATCATCAAT---AATATACAGT  TAGCAAAAAA  TGGCGCCTTT  GTTTGGCTTT  GTTCCAACTG  TTTTTGGCCC  GAGTTGGGTT  TCGTTTTCCC  GGG
```

FIGURE 29A, B. Nucleotide sequence of inverted terminal repetitions. The origins of the sequences are as follows: (A) Ad3: Tolun *et al.* (1979). Ad4: Tokunaga *et al.* (1982). Ad2/Ad5: The Ad2 sequence was determined by Shinagawa and Padmanabhan (1979) and the Ad5 sequence by Steenbergh *et al.* (1977). The two sequences are identical. Arrand and Roberts (1979) have analyzed an Ad2 strain that missed base pair 9 (↑). Ad7: These sequences were determined for strain Gomen by Dijkema and Dekker (1979) (a) and for strain Greider by Shinagawa and Padmanabhan (1980) (b). The differences between the sequences are indicated. (B) Ad12: Tolun *et al.* (1979) (a), Sugisaka *et al.* (1980) (a), Shinagawa and Padmanabhan (1980) (b), and Schwarz *et al.* (1982) (c). The differences between the sequences are indicated. Ad18: Garon *et al.* (1982). CELO: Aleström *et al.* (1982a). FL: Temple *et al.* (1981). In the human sequences, the conserved sequences 9–22 are underlined; the homologous regions in CELO and FL DNA are indicated by dashed underlines. The common sequence TGACGT discovered by Shinagawa and Padmanabhan (1980) is underlined.

REFERENCES

Akusjärvi, G., and Persson, H., 1981a, Gene and mRNA for precursor polypeptide VI from adenovirus type 2, *J. Virol.* **38**:469.

Akusjärvi, G., and Persson, H., 1981b, Control of RNA splicing and termination in the major late adenovirus transcription unit, *Nature (London)* **292**:420.

Akusjärvi, G., and Pettersson, U., 1978a, Sequence analysis of adenovirus DNA. I. Nucleotide sequence at the carboxy-terminal end of the gene for adenovirus type 2 hexon, *Virology* **91**:477.

Akusjärvi, G., and Pettersson, U., 1978b, Nucleotide sequence at the junction between the coding region of the adenovirus 2 hexon messenger RNA and its leader sequence, *Proc. Natl. Acad. Sci. U.S.A.* **75**:5822.

Akusjärvi, G., and Pettersson, U., 1979a, Sequence analysis of adenovirus DNA: Complete nucleotide sequence of the spliced 5'-non-coding region of adenovirus 2 hexon messenger RNA, *Cell* **16**:841.

Akusjärvi, G., and Pettersson, U., 1979b, Sequence analysis of adenovirus DNA. IV. The genomic sequences encoding the common tripartite leader of late adenovirus messenger RNA, *J. Mol. Biol.* **13**:143.

Akusjärvi, G., Mathews, M.B., Anderson, P., Vennström, B., and Pettersson, U., 1980, Structure of genes for virus-associated RNAI and RNAII of adenovirus type 2, *Proc. Natl. Acad. Sci. U.S.A.* **77**:2424.

Akusjärvi, G., Zabielsky, J., Perricaudet, M., and Pettersson, U., 1981, The sequence of the 3' non-coding region of the hexon mRNA discloses a novel adenovirus gene, *Nucleic Acids Res.* **9**:1.

Aleström, P., Akusjärvi, G., Perricaudet, M., Mathews, M.B., Klessig, D., and Pettersson, U., 1980, Sequence analysis of adenovirus DNA. VI. The nucleotide sequence of the gene for polypeptide IX from adenovirus type 2 and its colinear messenger RNA, *Cell* **19**:671.

Aleström, P., Stenlund, A., Li, P., and Pettersson, U., 1982a, A common sequence in the inverted terminal repetitions of human and avian adenoviruses, *Gene* **18**:193.

Aleström, P., Akusjärvi, G., Pettersson, M., and Pettersson, U., 1982b, DNA sequence analysis of the region encoding the terminal protein and the hypothetical N-gene product of adenovirus type 2, *J. Biol. Chem.* **257**:13492.

Anderson, C.W., and Lewis, J.B., 1980, Amino-terminal sequence of the adenovirus type 2 proteins: Hexon, fiber, component IX and the early protein 1B-15 K, *Virology* **104**:27.

Ariga, H., Shimojo, H., Hidaka, S., and Miura, M., 1979, Specific cleavage of the terminal protein from the adenovirus 5 DNA under the conditions of single-strand scission by nuclease S1, *FEBS Lett.* **107**:355.

Ariga, H., Klein, H., Levine, A.J., and Horwitz, M.S., 1980, A cleavage product of the adenovirus DNA binding protein is active for DNA replication *in vitro*, *Virology* **101**:307.

Arrand, J.R., and Roberts, R.J., 1979, The nucleotide sequences at the termini of adenovirus 2 DNA, *J. Mol. Biol.* **128**:577.

Arrand, J.R., Keller, W., and Roberts, R.J., 1975, Extent of terminal repetition in adenovirus 2 DNA, *Cold Spring Harbor Symp. Quant. Biol.* **39**:401.

Baker, C.C., and Ziff, E.B., 1980, Biogenesis, structures, and sites of encoding of the 5' termini of adenovirus-2 mRNAs, *Cold Spring Harbor Symp. Quant. Biol.* **44**:415.

Baker, C.C., and Ziff, E.B., 1981, Promoters and heterogeneous 5'-termini of the messenger RNAs of adenovirus-2, *J. Mol. Biol.* **149**:189.

Baker, C.C., Hérissé, J., Courtois, G., Galibert, F., and Ziff, E., 1979, Messenger RNA for the Ad2 DNA binding protein: DNA sequences encoding the first leader and heterogeneity at the mRNA 5' end, *Cell* **8**:569.

Bartok, K., Garon, C.F., Berry, K.W., and Rose, J.A., 1974, Specific fragmentation of adenovirus heteroduplex DNA molecules with single-strand specific nuclease of *Neurospora crassa*, *J. Mol. Biol.* **87**:437.

Bégin, M., and Weber, J., 1975, Genetic analysis of adenovirus type 2. I. Isolation and genetic characterization of temperature-sensitive mutants, *J. Virol.* **115**:1.

Berget, S.M., Moore, C., and Sharp, P.A., 1977, Spliced segments at the 5′ terminus of adenovirus 2 late mRNA, *Proc. Natl. Acad. Sci. U.S.A.* **74**:3171.

Berget, S.M., Berk, A., Harrison, T., and Sharp, P.A., 1978, Spliced segments at the 5′ termini of adenovirus 2 late mRNA: A role for heterogeneous nuclear RNA in mammalian cells, *Cold Spring Harbor Symp. Quant. Biol.* **42**:523.

Berk, A.J., and Sharp, P.A., 1977a, Sizing and mapping of early adenovirus mRNAs by gel electrophoresis of S1 endonuclease digested hybrids, *Cell* **12**:721.

Berk, A.J., and Sharp, P.A., 1977b, UV mapping of the adenovirus 2 early promoters, *Cell* **12**:45.

Berk, A.J., and Sharp, P.A., 1978, Structure of the adenovirus 2 early mRNAs, *Cell* **14**:695.

Berk, A.J., Lee, F., Harrison, T., Williams, J., and Sharp, P.A., 1979, Pre-early adenovirus 5 gene product regulates synthesis of early viral messenger RNA, *Cell* **17**:935.

Bernards, R., Houweling, A., Schrier, P.I., Bos, J.L., and van der Eb, A.J., 1982, Characterization of cells transformed by Ad5/Ad12 hybrid early region 1 plasmids, *Virology* **120**:422.

Bhatti, A.R., and Weber, J., 1979, Protease of adenovirus type 2, *J. Biol. Chem.* **254**:12265.

Binger, M.H., Flint, S.J., and Rekosh, D.M., 1982, Expression of the gene encoding the adenovirus DNA terminal protein precursor in productively infected and transformed cells, *J. Virol.* **42**:488.

Bos, J.L., Polder, L.J., Bernards, R., Schrier, P.I., van den Elsen, P., Van der Eb, A.J., and van Ormondt, H., 1981, The 2.2 kb mRNA of human Ad12 and Ad5 codes for two tumor antigens starting at different AUG triplets, *Cell* **27**:121.

Brackman, K.H., Green, M., Wold, W.S.M., Cartas, M., Matsuo, T., and Hashimoto, S., 1980, Identification and peptide mapping of human adenovirus type 2-induced early polypeptides isolated by two-dimensional gel electrophoresis, *J. Biol. Chem.* **255**:6772.

Broker, T.R., Chow, L.T., Dunn, A.R., Gelinas, R.E., Hassell, J.A., Klessig, D.F., Lewis, J.B., Roberts, R.J., and Zain, B.S., 1977, Adenovirus-2 messengers—An example of baroque molecular architecture, *Cold Spring Harbor Symp. Quant. Biol.* **42**:531.

Brown, D.T., Westphal, M., Burlingham, B.T., Winterhoff, U., and Doerfler, W., 1975, Structure and composition of the adenovirus type 2 core, *J. Virol.* **16**:366.

Burnett, J.P., and Harrington, J.A., 1968, Simian adenovirus SA7 DNA: Chemical, physical and biological studies, *Proc. Natl. Acad. Sci. U.S.A.* **60**:1023.

Carter, T.H., and Blanton, R.A., 1978, Possible role of the 72,000-dalton DNA-binding protein in regulation of adenovirus type 5 early gene expression, *J. Virol.* **25**:664.

Carter, T.H., and Ginsberg, H.S., 1976, Viral transcription in KB cells infected by temperature-sensitive early mutants of adenovirus type 5, *J. Virol.* **18**:156.

Carusi, E.A., 1977, Evidence for blocked 5′ termini in human adenovirus DNA, *Virology* **76**:380.

Celma, M.L., Pan, J., and Weissmann, S., 1977a, Studies of low molecular weight RNA from cells infected with adenovirus 2. I. The sequence at the 3′ end of VA-RNA 1, *J. Biol. Chem.* **252**:9032.

Celma, M.L., Pan, J., and Weissmann, S., 1977b, Studies of low molecular weight RNA from cells infected with adenovirus 2. II. Heterogeneity at the 5′ end of VA-RNA 1, *J. Biol. Chem.* **252**:9043.

Challberg, M.D., and Kelly, T.J., Jr., 1979a, Adenovirus DNA replication *in vitro*, *Proc. Natl. Acad. Sci. U.S.A.* **76**:655.

Challberg, M.D., and Kelly, T.J., Jr., 1979b, Adenovirus DNA replication *in vitro*: Origin and direction of daughter strand synthesis, *J. Mol. Biol.* **135**:999.

Challberg, M.D., and Kelly, T.J., Jr., 1981, Processing of the adenovirus terminal protein, *J. Virol.* **38**:272.

Challberg, M.D., Desiderio, S.V., and Kelly, T.J., Jr., 1980, Adenovirus DNA replication *in vitro*: Characterization of a protein covalently linked to nascent DNA strands, *Proc. Natl. Acad. Sci. U.S.A.* **77**:5105.

Chinnadurai, G., Chinnadurai, S., and Green, M., 1978, Enhanced infectivity of adenovirus DNA and a DNA–protein complex, *J. Virol.* **26**:195.

Chow, L.T., and Broker, T.R., 1978, The spliced structures of adenovirus-2 fiber message and the other mRNAs, *Cell* **15**:497.

Chow, L.T., Gelinas, R.E., Broker, T.R., and Roberts, R.J., 1977a, An amazing sequence arrangement at the 5′ ends of adenovirus 2 messenger RNA, *Cell* **12**:1.

Chow, L.T., Roberts, J.M., Lewis, J.B., and Broker, T.R., 1977b, A cytoplasmic RNA transcript map from adenovirus type 2, determined by electron microscopy of RNA:DNA hybrids, *Cell* **11**:819.

Chow, L.T., Broker, T.R., and Lewis, J.B., 1979a, Complex splicing patterns of RNAs from the adenovirus 2, *J. Mol. Biol.* **134**:265.

Chow, L., Lewis, J.B., and Broker, T., 1979b, RNA transcription and splicing at early and intermediate times after adenovirus-2 infection, *Cold Spring Harbor Symp. Quant. Biol.* **44**:401.

Coombs, D.H., and Pearson, G.D., 1978, Filter-binding assay for covalent DNA–protein complexes: Adenovirus DNA–terminal protein complex, *Proc. Natl. Acad. Sci. U.S.A.* **75**:5291.

Coombs, D.H., Robinson, A.J., Bodnar, J.W., Jones, C.J., and Pearson, G.D., 1978, Detection of DNA–protein complexes: The adenovirus DNA–terminal protein and HeLa DNA–protein complexes, *Cold Spring Harbor Symp. Quant. Biol.* **43**:741.

Corden, J., Engelking, H.M., and Pearson, G.D., 1976, Chromatin-like organization of the adenovirus chromosome, *Proc. Natl. Acad. Sci. U.S.A.* **73**:401.

Desiderio, S.V., and Kelly, T., 1981, Structure of the linkage between adenovirus DNA and the 55,000 molecular weight terminal protein, *J. Mol. Biol.* **145**:319.

Dijkema, R., and Dekker, B.M.M., 1979, The inverted terminal repetition of the DNA of the weakly oncogenic adenovirus type 7, *Gene* **8**:7.

Dijkema, R., Dekker, B.M.M., and van Ormondt, H., 1980a, The nucleotide sequence of the transforming BglII-H fragment of adenovirus type 7 DNA, *Gene* **9**:141.

Dijkema, R., Dekker, B.M.M., van Ormondt, H., Maat, J., and Boyer, H., 1980b, Gene organization of the transforming region of weakly oncogenic adenovirus type 7: The E1a region, *Gene* **12**:287.

Dijkema, R., Maat, J., Dekker, B.M.M., van Ormondt, H., and Boyer, W., 1981, The gene for polypeptide IX of adenovirus type 7, *Gene* **13**:375.

Dijkema, R., Dekker, B.M.M., and van Ormondt, H., 1982, Gene organization of the transforming region of adenovirus type 7, *Gene* **18**:143.

Dimitrov, D.H., Dubitchev, A.G., Naroditsky, B.S., Dreizin, R.S., and Tikchonenko, T.I., 1979, Physicochemical properties and restriction maps of simian adenovirus 38, *J. Gen. Virol.* **44**:69.

Doerfler, W., and Kleinschmidt, A., 1970, Denaturation pattern of the DNA of adenovirus type 2 as determined by electron microscopy, *J. Mol. Biol.* **50**:579.

Doerfler, W., Hellmann, W., and Kleinschmidt, A.K., 1972, The DNA of adenovirus type 12 and its denaturation pattern, *Virology* **47**:507.

Dunn, A.R., Mathews, M.B., Chow, L.T., Sambrook, J., and Keller, W., 1978, A supplementary adenoviral leader sequence and its role in messenger translation, *Cell* **15**:511.

Ellens, D.J., Sussenbach, J.S., and Jansz, H.S., 1974, Studies on the mechanism of replication of adenovirus DNA. III. Electron microscopy of replicating DNA, *Virology* **61**:427.

Engler, J., 1981, The nucleotide sequence of the polypeptide IX gene of human adenovirus type 3, *Gene* **13**:387.

Engler, J.A., and van Bree, M., 1982, The nucleotide sequence of the gene encoding protein IVa2 in human adenovirus type 7, *Gene* **19**:71.

Engler, J., Broker, T.R., and Chow, L.T., 1981, Sequences of human adenovirus Ad3 and Ad7 DNAs encoding the promoter and first leader segment of late RNAs, *Gene* **13**:133.

Engler, J.A., Hoppe, M.S., and van Bree, M.P., 1983, The nucleotide sequence of the genes encoded in the early region 2B in human adenovirus type 7, *Gene* **21**:147.

Enomoto, T., Lichy, J.H., Ikeda, J.E., and Hurwitz, J., 1981, Adenovirus DNA replication *in vitro*: Purification of the terminal protein in a functional form, *Proc. Natl. Acad. Sci. U.S.A.* **78**:6779.

Esche, H., Mathews, M.B., and Lewis, J.B., 1980, Proteins and messenger RNAs of the transforming region of wild-type and mutant adenoviruses, *J. Mol. Biol.* **142**:399.

Evans, R.M., Fraser, N., Ziff, E., Weber, J., Wilson, M., and Darnell, J.E., 1977, The initiation sites for RNA transcription in Ad2 DNA, *Cell* **12**:733.

Everitt, E., Sundquist, B., Petersson, O., and Philipson, L., 1973, Structural proteins of adenoviruses. X. Isolation and topography of low molecular weight antigens from the virion of adenovirus type 2, *Virology* **62**:130.

Forsblom, S., Rigler, R., Ehrenberg, M., Pettersson, U., and Philipson, L., 1976, Kinetic studies on the cleavage of adenovirus DNA by restriction endonuclease EcoRI, *Nucleic Acids Res.* **3**:3255.

Fraser, N., and Ziff, E., 1978, RNA structures near poly (A) of adenovirus-2 late messenger RNA's, *J. Mol. Biol.* **124**:27.

Fraser, N.W., Nevins, J.R., Ziff, E., and Darnell, J.E., 1979, The major late adenovirus type 2 transcription unit: Termination is downstream from the last poly (A) site, *J. Mol. Biol.* **129**:643.

Fraser, N.W., Baker, C.C., Moore, M.A., and Ziff, E.B., 1982, Poly(A) sites of adenovirus serotype 2 transcription units, *J. Mol. Biol.* **155**:207.

Freeman, A.E., Black, P.E., Vanderpool, E.A., Henry, P.H., Austin, J.B., and Huebner, R.J., 1967, Transformation of primary rat embryo cells by adenovirus type 2, *Proc. Natl. Acad. Sci. U.S.A.* **58**:1205.

Frost, E., and Williams, J., 1978, Mapping of temperature-sensitive and host-range mutants of adenovirus type 5 by marker rescue, *Virology* **91**:39.

Fujinaga, K., Sawada, Y., Uemizu, Y., Yamashita, T., Shimojo, H., Shiroki, K., Sugisaki, H., Sugimoto, K., and Takanami, M., 1979, Nucleotide sequences, integration, and transcription of the adenovirus-12 transforming genes, *Cold Spring Harbor Symp. Quant. Biol.* **44**:519.

Galibert, F., Hérissé, J., and Courtois, G., 1979, Nucleotide sequence of the EcoRI-F fragment of adenovirus 2 genome, *Gene* **6**:1.

Gallimore, P.H., 1974, Interactions of adenovirus type 2 with rat embryo cells: Permissiveness, transformation and *in vitro* characteristics of adenovirus transformed rat embryo cells, *J. Gen. Virol.* **25**:263.

Galos, R.S., Williams, J., Binger, M.H., and Flint, S.J., 1979, Location of additional early gene sequences in the adenoviral chromosome, *Cell* **17**:945.

Gambke, C., and Deppert, W., 1981, Late non-structural 100,000 and 33,000 dalton proteins of adenovirus type 2. II. Immunological and protein chemical analysis, *J. Virol.* **40**:594.

Garon, C.F., Berry, K.W., and Rose, J.A., 1972, A unique form of terminal redundancy in adenovirus DNA molecules, *Proc. Natl. Acad. Sci. U.S.A.* **69**:2391.

Garon, C.F., Berry, K.W., Hierholzer, J.C., and Rose, J.A., 1973, Mapping base sequence heterologies between genomes from different adenovirus serotypes, *Virology* **54**:414.

Garon, C.F., Berry, K.W., and Rose, J.A., 1975, Arrangement of sequences in the inverted terminal repetition of adenovirus 18 DNA, *Proc. Natl. Acad. Sci. U.S.A.* **72**:3039.

Garon, F.C., Parr, R.P., Padmanabhan, R., Roninson, I., Garrison, J.W., and Rose, J.A., 1982, Structural characterization of the adenovirus 18 inverted terminal repetition, *Virology* **121**:230.

Gingeras, T.R., Sciaky, D., Gelinas, R.E., Bing-Dong, J., Yen, C.E., Kelly, M.M., Bullock, P.A., Parsons, B.L., O'Neill, K.E., and Roberts, R.J., 1982, Nucleotide sequences from the adenovirus-2 genome, *J. Biol. Chem.* **257**:13475.

Ginsberg, H.S., Ensinger, M.J., Kauffman, R.S., Mayer, A.J., and Lundholm, U., 1974, Cell transformation: A study of regulation with types 5 and 12 adenovirus temperature-sensitive mutants, *Cold Spring Harbor Symp. Quant. Biol.* **44**:419.

Girardi, A.J., Hilleman, M.R., and Zwickey, R.E., 1964, Tests in hamsters for oncogenic quality of ordinary viruses including adenovirus type 7, *Proc. Soc. Exp. Biol. Med.* **115**:1141.

Goldbach, R.W., Evers, R.F., and Borst, P., 1978, Electrophoretic strand separation of long DNAs with poly-UG in agarose gels, *Nucleic Acids Res.* **5**:2743.

Goodhearst, C.R., 1971, DNA density of oncogenic and non-oncogenic simian adenoviruses, *Virology* **44**:645.

Graham, F.L., and van der Eb, A.J., 1973, A new technique for the assay of the infectivity of human adenovirus 5 DNA, *Virology* **52**:456.

Graham, F.L., van der Eb, A.J., and Heijneker, H.L., 1974a, Size and location of the transforming regions of human adenovirus type 5 DNA, *Nature* (*London*) **251**:687.

Graham, F.L., Abrahams, P.J., Mulder, C., Heijneker, H.L., Warnaar, S.O., de Vries, F.A.J., Fiers, W., and van der Eb, A.J., 1974b, Studies on *in vitro* transformation by DNA and DNA fragments of human adenoviruses and simian virus 40, *Cold Spring Harbor Symp. Quant. Biol.* **39**:637.

Green, M., 1970, Oncogenic viruses, *Annu. Rev. Biochem.* **39**:701.

Green, M., and Piña, M., 1963, Biochemical studies on adenovirus multiplication. IV. Isolation, purification and chemical analysis of adenovirus, *Virology* **20**:199.

Green, M., and Piña, M., 1964, Biochemical studies on adenovirus multiplication. VI. Properties of highly purified tumorigenic human adenoviruses and their DNA's, *Proc. Natl. Acad. Sci. U.S.A.* **51**:1251.

Green, M., Piña, M., Kimes, R., Wensink, P.C., Machattie, L.A., and Thomas, C.A., 1967, Adenovirus DNA. I. Molecular weight and conformation, *Proc. Natl. Acad. Sci. U.S.A.* **57**:1302.

Green, M., Parsons, J.T., Piña, M., Fujinaga, K., Caffier, H., and Landgraf-Leurs, I., 1970, Transcription of adenovirus genes in productively infected and transformed cells, *Cold Spring Harbor Symp. Quant. Biol.* **30**:803.

Green, M., Wold, W.S.M., Brackmann, K.H., and Cartas, M.A., 1979a, Identification of families of overlapping polypeptides coded by early "transforming" gene region 1 of human adenovirus type 2, *Virology* **97**:275.

Green, M., Mackey, J., Wold, W.S.M., and Rigden, P., 1979b, Thirty-one human adenovirus serotypes (Ad1–Ad31) form five groups (A–E) based on upon DNA genome homologies, *Virology* **93**:481.

Green, M., Brackmann, K., Wold, W.S.M., Cartas, M., Thornton, H., and Elder, J.H., 1979c, Conserved primary sequences of the DNA terminal proteins of five different human adenovirus groups, *Proc. Natl. Acad. Sci. U.S.A.* **76**:4380.

Green, M., Wold, W.S.M., Brackmann, K., and Cartas, M.A., 1979d, Studies on early proteins and transformation proteins in human adenoviruses, *Cold Spring Harbor Symp. Quant. Biol.* **44**:457.

Grodzicker, T., Williams, J., Sharp, P.A., and Sambrook, J., 1975, Physical mapping of temperature-sensitive mutations of adenoviruses, *Cold Spring Harbor Symp. Quant. Biol.* **39**:439.

Grodzicker, T., Anderson, C.W., Sambrook, J., and Mathews, M.B., 1977, The physical location of structural genes in adenovirus DNA, *Virology* **80**:111.

Harrison, T., Graham, F.L., and Williams, J., 1977, Host range mutants of adenovirus type 5 defective for growth in HeLa cells, *Virology* **77**:319.

Harter, M.L., and Lewis, J.B., 1978, Adenovirus type 2 early proteins synthesized *in vitro* and *in vivo*: Identification in infected cells of the 38,000- to 50,000-molecular-weight protein encoded by the left end of adenovirus type 2 genome, *J. Virol.* **26**:736.

Harter, M.L., Shanmugan, G., Wold, W.S.M., and Green, M., 1976, Detection of adenovirus type 2 induced early polypeptides using cycloheximide pretreatment to enhance viral protein synthesis, *J. Virol.* **19**:1976.

Hashimoto, S., Pursley, M.H., and Green, M., 1981, Nucleotide sequences and mapping of novel heterogeneous 5′-termini of adenovirus type 2 early region 4 mRNA, *Nucleic Acids Res.* **9**:1675.

Hérissé, J., and Galibert, F., 1981, Nucleotide sequence of the EcoRI E fragment of adenovirus-2 genome, *Nucleic Acids Res.* **9**:1229.

Hérissé, J., Courtois, G., and Galibert, F., 1980, Nucleotide sequence of the EcoRI D fragment of adenovirus 2 genome, *Nucleic Acids Res.* **8**:2173.

Hérissé, J., Rigolet, M., Dupont de Dinechin, S., and Galibert, F., 1981, Nucleotide sequence of adenovirus 2 fragment encoding for the carboxylic region of the fiber protein and the entire E4 region, *Nucleic Acids Res.* **9**:4023.

Hierholzer, J.C., 1973, Further subgrouping of the human adenoviruses by differential hemagglutination, *J. Infect. Dis.* **128**:541.

Horwitz, M.S., 1974, Location of the origin of DNA replication in adenovirus type 2, *J. Virol.* **13**:1046.

Horwitz, M.S., 1978, Temperature-sensitive replication of H5ts125 adenovirus DNA in the presence of cycloheximide, *J. Virol.* **11**:544.

Houweling, A., van den Elsen, P.J., and van der Eb, A.J., 1980, Partial transformation of primary rat cells by the leftmost 4.5% fragment of adenovirus 5 DNA, *Virology* **105**:537.

Huebner, R.J., Rowe, W.P., and Lane, W.T., 1962, Oncogenic effects in hamsters of human adenovirus types 12 and 18, *Proc. Natl. Acad. Sci. U.S.A.* **48**:2051.

Huebner, R.J., Casey, M.J., Chanock, R.M., and Schell, K., 1965, Tumors induced in hamsters by a strain of adenovirus type 3: Sharing of tumor antigens and neoantigens with those produced by adenovirus type 7 tumors, *Proc. Soc. Exp. Biol. Med.* **54**:381.

Jones, N., and Shenk, T., 1979a, An adenovirus type 5 early gene function regulates expression of other early viral genes, *Proc. Natl. Acad. Sci. U.S.A.* **76**:3665.

Jones, N., and Shenk, T., 1979b, Isolation of adenovirus type 5 host range deletion mutants defective for transformation of rat embryo cells, *Cell* **17**:683.

Jörnvall, H., Akusjärvi, G., Aleström, P., von Bahr-Lindstrom, H., Pettersson, U., Appella, E., Fowler, A.V., and Philipson, L., 1981a, The adenovirus hexon protein: The primary structure of the polypeptide and its correlation with the hexon gene, *J. Biol. Chem.* **256**:6181.

Jörnvall, H., Aleström, P., Akusjärvi, G., von Bahr-Lindstrom, H., Philipson, L., and Pettersson, U., 1981b, Order of the CNBr fragments in the adenovirus hexon protein, *J. Biol. Chem.* **256**:6204.

Keegstra, W.S., van Wielink, P.S., and Sussenbach, J.S., 1977, The visualization of a circular DNA–protein complex from adenovirions, *Virology* **76**:444.

Kilpatrick, B.A., Gelinas, R.E., Broker, T.R., and Chow, L.T., 1979, Comparison of late mRNA splicing among class B and class C adenoviruses, *J. Virol.* **30**:899.

Kimes, R., and Green, M., 1970, Adenovirus DNA. II. Separation of molecular halves of adenovirus type 2, *J. Mol. Biol.* **50**:203.

Kimura, T., Sawada, Y., Shinagawa, M., Shimizu, Y., Shiroki, K., Shimojo, H., Sugisaki, H., Takanami, M., Uemizu, Y., and Fujinaga, K., 1981, Nucleotide sequence of the transforming early region E1b of adenovirus 12 DNA: Structure and gene organization and comparison with those of adenovirus type 5 DNA, *Nucleic Acids Res.* **9**:6571.

Kitchingman, G.R., and Westphal, H., 1980, The structure of adenovirus 2 early nuclear and cytoplasmics RNAs. *J. Mol. Biol.* **137**:23.

Kitchingman, G.R., Lai, S.-P., and Westphal, H., 1977, Loop structures in hybrids of early RNA and the separated strands of adenovirus DNA, *Proc. Natl. Acad. Sci. U.S.A.* **74**:4329.

Klessig, D.F., and Grodzicker, T., 1979, Mutations that allow human Ad2 and Ad5 to express late genes in monkey cells map in the viral gene encoding the 72K DNA binding protein, *Cell* **17**:957.

Kruijer, W., van Schaik, F.M.A., and Sussenbach, J.S., 1980, Nucleotide sequence analysis of a region of adenovirus 5 DNA encoding a hitherto unidentified gene, *Nucleic Acids Res.* **8**:6033.

Kruijer, W., van Schaik, F.A.M., and Sussenbach, J.S., 1981, Structure and organization of the gene coding for the DNA binding protein of adenovirus type 5, *Nucleic Acids Res.* **9**:4439.

Kruijer, W., van Schaik, F.A.M., and Sussenbach, J.S., 1982, Nucleotide sequence of the gene encoding adenovirus type 2 DNA binding protein, *Nucleic Acids Res.* **10**:4493.

Kruijer, W., van Schaik, F.A.M., Speyer, J.G., and Sussenbach, J.S., 1983, Structure and function of denovirus DNA binding protein: comparison of the amino acid sequences of

Ad5 and Ad12 proteins derived from the nucleotide sequence of the corresponding genes, *Virology* **128:**140.

Kubinski, H., and Rose, J.A., 1967, Regions containing repeating base-pairs in DNA from some oncogenic and non-oncogenic animal viruses, *Proc. Natl. Acad. Sci. U.S.A.* **57:**1720.

Landgraf-Leurs, M., and Green, M., 1971, Adenovirus DNA. III. Separation of the complementary strands of adenovirus types 2, 7 and 12 DNA molecules, *J. Mol. Biol.* **60:**185.

Larsen, S.H., Margolskee, R.F., and Nathans, D., 1979, Alignment of the restriction map of mouse adenovirus FL and human adenovirus type 2, *Virology* **97:**406.

Larson, V.M., Girardi, A.J., Hilleman, M.R., and Zwickey, R.E., 1965, Studies on oncogenicity of adenovirus type 7 viruses in hamsters, *Proc. Soc. Exp. Biol. Med.* **118:**15.

Laver, W.G., 1970, Isolation of an arginine-rich protein from particles of adenovirus type 2, *Virology* **41:**488.

Laver, W.G., Suriano, J.R., and Green, M., 1967, Adenovirus proteins. II. N-terminal amino acid analysis, *J. Virol.* **1:**723.

Laver, W.G., Pereira, H.G., Russell, W.C., and Valentine, R.C., 1968, Isolation of an internal component of adenovirus type 5, *J. Mol. Biol.* **37:**379.

Laver, W.G., Younghusband, H.B., and Wrigley, N.G., 1971, Purification and properties of chick embryo lethal orphan virus (CELO), *Virology* **45:**598.

Lewis, J.B., and Mathews, M.B., 1980, Control of adenovirus early gene expression: A class of "immediate early" products, *Cell* **21:**303.

Lewis, J.B., Atkins, J.F., Anderson, C.W., Baum, P.R., and Gesteland, R.F., 1975, Mapping of late adenovirus genes by cell-free translation of RNA selected by hybridization to specific DNA fragments, *Proc. Natl. Acad. Sci. U.S.A.* **72:**1344.

Lewis, J.B., Atkins, J.F., Baum, P.R., Solem, R., Gesteland, R.B., and Anderson, C.W., 1976, Location and identification of the genes for adenovirus type 2 early polypeptides, *Cell* **17:**141.

Lewis, J.B., Anderson, C.W., and Atkins, J.F., 1977, Further mapping of late adenovirus genes by cell-free translation of RNA selected by hybridization to specific DNA fragments, *Cell* **12:**37.

Lewis, J.B., Esche, H., Smart, J.E., Stillman, B.W., Harter, M.L., and Mathews, M.B., 1979, Organization and expression of the left third of the genome of adenovirus, *Cold Spring Harbor Symp. Quant. Biol.* **44:**493.

Lichy, J.H., Field, J., Horwitz, M.S., and Hurwitz, J., 1982, Separation of the adenovirus terminal protein precursor from its associated DNA polymerase: Role of both proteins in the initiation of adenovirus DNA replication, *Proc. Natl. Acad. Sci. U.S.A.* **79:**5225.

Lupker, J.H., Davis, A., Jochemsen, H., and van der Eb, A.J., 1980, *In vitro* synthesis of adenovirus type 5 T antigens. I. Translation of early region 1-specific RNA from lytically infected cells, *J. Virol.* **37:**524.

Maat, J., and van Ormondt, H., 1979, The nucleotide sequence of the transforming HindIII-G fragment of adenovirus type 5 DNA, *Gene* **6:**75.

Maat, J., van Beveren, C.P., and van Ormondt, H., 1980, The nucleotide sequence of adenovirus type 5 early region E-1: The region between map positions 8.0 (HindIII site) and 11.8 (SmaI site), *Gene* **10:**27.

Maizel, J.V., White, D.O., and Scarff, M.D., 1968, The polypeptides of adenovirus. II. Soluble proteins, cores, top components and structure of the virion, *Virology* **36:**126.

Mathews, M.B., 1980, Binding of adenovirus VA RNA to mRNA: A possible role in splicing?, *Nature (London)* **285:**575.

Mathews, M.B., and Pettersson, U., 1978, The low molecular weight RNA's of adenovirus 2 infected cells, *J. Mol. Biol.* **119:**293.

Mathis, D.J., Elkaim, R., Kedinger, C., Sassone-Corsi, P., and Chambon, P., 1981, Specific *in vitro* initiation of transcription on the adenovirus type 2 early and late EII transcription units, *Proc. Natl. Acad. Sci. U.S.A.* **78:**7383.

Mautner, V., Williams, J., Sambrook, J., Sharp, P.A., and Grodzicker, T., 1975, The location of the genes coding for hexon and fibre proteins in adenovirus DNA, *Cell* **5:**93.

Mayer, A.J., and Ginsberg, H.S., 1977, Persistence of type 5 adenovirus DNA in cells transformed by the temperature-sensitive mutant H5ts125, *Proc. Natl. Acad. Sci. U.S.A.* **74**:785.

McAllister, R.M., Nicolson, M.O., Lewis, A.M., McPherson, I., and Huebner, R.J., 1969, Transformation of rat embryo cells by adenovirus type 1, *J. Gen. Virol.* **4**:29.

McBride, W.D., and Weiner, A., 1964, *In vitro* transformation of hamster kidney cells by human adenovirus type 12, *Proc. Soc. Exp. Biol. Med.* **115**:870.

McGrogan, M., and Raskas, H.J., 1978, Two regions of the adenovirus 2 genome specify families of late polysomal RNA's containing common sequences, *Proc. Natl. Acad. Sci. U.S.A.* **75**:625.

Miller, S.J., Ricciardi, R.P., Roberts, B.E., Paterson, B.M., and Mathews, M.B., 1980, Arrangement of messenger RNAs and protein coding sequences in the major late transcription unit of adenovirus 2, *J. Mol. Biol.* **142**:455.

Mirza, M.A., and Weber, J., 1982, Structure of adenovirus chromatin, *Biochim. Biophys. Acta* **696**:76.

Montell, C., Fisher, E.F., Caruthers, M.H., and Berk, A.J., 1982, Resolving the functions of overlapping viral genes by site-specific mutagenesis at a mRNA splice site, *Nature (London)* **295**:380.

Murray, V., and Holliday, R., 1979, Mechanism of RNA splicing for gene transcripts, *FEBS Lett.* **106**:5.

Naroditsky, B.S., Kolinina, T.I., Goldberg, E.Z., Borovik, A.S., Karamov, E.V., and Tikchonenko, T.I., 1980, Analysis of DNA from human adenovirus type 6 with restriction endonucleases HindIII, BglII, and BamHI, *Biochim. Biopys. Acta* **606**:214.

Nevins, J.R., 1981, Mechanism of activation of early viral transcription by the adenovirus E1A gene products, *Cell* **26**:213.

Nevins, J.R., and Darnell, J.E., 1978, Groups of adenovirus type 2 mRNA's derived from a large primary transcript: Probably nuclear origin and possible common 3'-ends, *J. Virol.* **25**:811.

Nevins, J.R., and Jensen-Winkler, J.J., 1980, Regulation of early adenovirus transcription: A protein product of early region 2 specifically represses region 4 transcription, *Proc. Natl. Acad. Sci. U.S.A.* **77**:1893.

Nevins, J.R., Blanchard, J.M., and Darnell, J.E., 1980, Transcription units of adenovirus type 2: Termination of transcription beyond the poly(A) addition site in early regions 2 and 4, *J. Mol. Biol.* **144**:377.

Nicolson, M.O., and McAllister, R.M., 1972, Infectivity of human adenovirus 1 DNA, *Virology* **48**:14.

Ohe, K., 1972, Virus-coded origin of a low molecular weight RNA from KB cells infected with adenovirus 2, *Virology* **47**:726.

Ohe, K., and Weissmann, S., 1970, Nucleotide sequence of an RNA from cells infected with adenovirus 2, *Science* **167**:879.

Ohen, K., and Weissman, S., 1971, The nucleotide sequence of a low molecular weight RNA from cells infected with adenovirus 2, *J. Biol. Chem.* **246**:6999.

Padmanabhan, R., and Green, M., 1976, Evidence for palindromic sequences near the termini of adenovirus 2 DNA, *Biochem. Biophys. Res. Commun.* **69**:860.

Padmanabhan, R., and Padmanabhan, R.V., 1977, Specific interaction of a protein(s) near the termini of adenovirus 2 DNA, *Biochem. Biophys. Res. Commun.* **80**:955.

Pan, J., Celma, M.L., and Weissman, S.M., 1977, Studies of low molecular weight RNA from cells infected with adenovirus 2. III. The sequence of the promoter for VA-RNA I. *J. Biol. Chem.* **252**:9047.

Patch, C.T., Lewis, A.M., and Levine, A.S., 1972, Evidence for a transcriptional control region of SV40 in the adenovirus 2–SV40 hybrid, Ad2+ND1, *Proc. Natl. Acad. Sci. U.S.A.* **69**:3375.

Pereira, M.S., Pereira, H.G., and Clarke, S.K., 1965, Human adenovirus type 31: A new serotype with oncogenic properties, *Lancet* **1**:21.

Perricaudet, M., Akusjärvi, G., Virtanen, A., and Pettersson, U., 1979, Structure of two spliced mRNA's from the transforming region of human subgroup C adenoviruses, *Nature (London)* **281**:694.

Perricaudet, M., le Moullec, J.M., and Pettersson, U., 1980, The predicted structure of two adenovirus T-antigens, *Proc. Natl. Acad. Sci. U.S.A.* **77**:3778.

Persson, H., and Philipson, L., 1982, Regulation of adenovirus gene expression, *Curr. Top. Microbiol. Immunol.* **97**:157.

Persson, H., Pettersson, U., and Mathews, M.B., 1978, Synthesis of a structural polypeptide in the absence of viral DNA replication, *Virology* **90**:67.

Persson, H., Mathisen, B., Philipson, L., and Pettersson, U., 1979a, A maturation protein in adenovirus morphogenesis, *Virology* **93**:198.

Persson, H., Signas, C., and Philipson, L., 1979b, Purification and characterization of an early glycoprotein from adenovirus type 2 infected cells, *J. Virol.* **29**:938.

Persson, H., Jansson, M., and Philipson, L., 1980a, Synthesis and genomic site for an adenovirus type 2 early protein, *J. Mol. Biol.* **136**:375.

Persson, H., Jörnvall, H., and Zabielski, J., 1980b, Multiple mRNA species for the precursor to an adenovirus-encoded glycoprotein: Identification and structure of the signal sequence, *Proc. Natl. Acad. Sci. U.S.A.* **77**:6349.

Pettersson, U., and Mathews, M.B., 1977, The gene and the messenger RNA for adenovirus polypeptide IX, *Cell* **12**:741.

Piña, M., and Green, M., 1965, Biochemical studies on adenovirus multiplication. IX. Chemical and base composition analysis of 28 human adenoviruses, *Proc. Natl. Acad. Sci. U.S.A.* **54**:547.

Ponomareva, T.I., Grodnitskaya, N.A., Goldberg, E.E., Chaplygina, N.M., Naroditsky, B.S., and Tikchonenko, T.I., 1979, Biological activity of intact and cleaved DNA of the simian adenovirus 7, *Nucleic Acids Res.* **5**:3119.

Prage, L., and Pettersson, U., 1971, Structural proteins of adenoviruses. VII. Purification and properties of an arginine-rich core protein from adenovirus type 2 and type 3, *Virology* **45**:364.

Prage, L., Pettersson, U., and Philipson, L., 1968, Internal base proteins in adenovirus, *Virology* **36**:508.

Prage, L., Pettersson, U., Hoglund, S., Lonberg-Holm, K., and Philipson, L., 1970, Structural proteins of adenoviruses. IV. Sequential degradation of the adenovirus type 2 virion, *Virology* **42**:341.

Price, R., and Penman, S., 1972, A distinct RNA polymerase activity, synthesizing 5.5S, 5S and 4S RNA in nuclei from adenovirus 2 infected HeLa cells, *J. Mol. Biol.* **70**:435.

Rekosh, D., 1981, Analysis of the DNA–terminal protein from different serotypes of human adenovirus, *J. Virol.* **40**:329.

Rekosh, D.M.K., Russell, W.C., Bellett, A.J.D., and Robinson, A.J., 1977, Identification of a protein linked to the ends of adenovirus DNA, *Cell* **11**:283.

Rijnders, A.W.M., van Maarschalkerweerd, M.W., Visser, L., Reemst, A.M.C.B., Sussenbach, J.S., and Rozijn, T.H., 1983, Expression of integrated viral DNA sequences outside the transforming region in eight adenovirus-transformed cell lines, *Biochim. Biophys. Acta* **739**:48.

Roberts, R.J., 1981, Restriction and modification enzymes and their recognition sequences, *Nucleic Acids Res.* **9**:r75.

Roberts, R.J., Arrand, J.R., and Keller, W., 1974, The length of the terminal repetition in adenovirus-2 DNA, *Proc. Natl. Acad. Sci. U.S.A.* **71**:3829.

Robinson, A.J., and Bellett, A.J.D., 1975a, A circular DNA–protein complex from adenovirus and its possible role in DNA replication, *Cold Spring Harbor Symp. Quant. Biol.* **39**:523.

Robinson, A.J., and Bellett, A.J.D., 1975b, Complementary strands of CELO virus DNA, *J. Virol.* **15**:458.

Robinson, A.J., Younghusband, H.B., and Bellett, A.J.D., 1973, A circular DNA–protein complex from adenoviruses, *Virology* **56**:54.

Roninson, I., and Padmanabhan, R., 1980, Studies on the nature of the linkage between the terminal protein and the adenovirus DNA, *Biochem. Biophys. Res. Commun.* **94**:398.

Rósen, L., 1960, A hemagglutination-inhibition technique for typing adenoviruses, *Am. J. Hyg.* **71**:120.

Ross, S., Flint, S.J., and Levine, A.J., 1980, Identification of the adenovirus early proteins and their genomic positions, *Virology* **100**:419.

Russell, W.C., McIntosh, K., and Skehel, J.J., 1971, The preparation and properties of adenovirus cores, *J. Gen. Virol.* **11**:35.

Sambrook, J., Williams, J.F., Sharp, P.A., and Grodzicker, T., 1975, Physical mapping of temperature-sensitive mutations of adenoviruses. *J. Mol. Biol.* **97**:369.

Schwarz, E., Reinke, C., Yamamoto, N., and Zur Hausen, H., 1982, Terminal rearrangements in the genome of adenovirus type 12 mutants adapted to growth in two human tumor cell lines, *Virology* **116**:284.

Seghal, P., Fraser, N., and Darnell, J., 1979, Early Ad2 transcription units: Only promoter-proximal RNA continues to be made in the presence of DRB, *Virology* **94**:185.

Sekikawa, K., Shiroki, K., Shimojo, H., Ojima, S., and Fujinaga, K., 1978, Transformation of a rat cell line by an adenovirus 7 DNA fragment *Virology* **88**:1.

Sharp, P.A., Gallimore, P.H., and Flint, S.J., 1975, Mapping of adenovirus 2 RNA sequences in lytically infected cells and transformed cell lines, *Cold Spring Harbor Symp. Quant. Biol.* **39**:457.

Sharp, P.A., Moore, C., and Haverty, J., 1976, The infectivity of adenovirus 5–DNA protein complex, *Virology* **75**:442.

Shaw, A.R., and Ziff, E.B., 1980, Transcripts from the adenovirus-2 late promoter yields a single family of coterminal mRNA's during early infection and five families at late times, *Cell* **22**:905.

Shinagawa, M., and Padmanabhan, R., 1979, Nucleotide sequences at the inverted terminal repetition of Ad2 DNA, *Biochem. Biophys. Res. Commun.* **87**:671.

Shinagawa, M., and Padmanabhan, R., 1980, Comparative sequence analysis of the inverted terminal repetitions from different adenoviruses, *Proc. Natl. Acad. Sci. U.S.A.* **77**:3831.

Shinagawa, M., Padmanabhan, R.V., and Padmanabhan, R., 1980, The nucleotide sequence of the right-hand terminal SmaI-K fragment of adenovirus type 2 DNA, *Gene* **9**:99.

Smart, J.E., and Stillman, B.W., 1982, Adenovirus terminal protein precursor: Partial amino acid sequence and the site of covalent linkage to virus DNA, *J. Biol. Chem.* **257**:13499.

Söderland, H., Pettersson, U., Vennström, B., Philipson, L., and Mathews, M.B., 1976, A new species of virus-coded low molecular weight RNA from cells infected with adenovirus type 2, *Cell* **7**:585.

Spector, D.J., McGrogan, M., and Raskas, H.J., 1978, Regulation of the appearance of cytoplasmic RNA's from region 1 of the adenovirus genome, *J. Mol. Biol.* **126**:395.

Spector, D.J., Crossland, L.D., Halbert, D.N., and Raskas, H.J., 1980a, A 28 K polypeptide is the translation product of 9 S RNA encoded by region 1A of adenovirus 2, *Virology* **102**:218.

Spector, D.J., Halbert, D.N., Crossland, L.D., and Raskas, H.J., 1980b, Expression of genes from the transforming region of adenovirus, *Cold Spring Harbor Symp. Quant. Biol.* **44**:437.

Steenbergh, P.H., and Sussenbach, J.S., 1979, The nucleotide sequence of the right-hand terminus of adenovirus type 5 DNA: Implications for the mechanism of DNA replication, *Gene* **6**:307.

Steenbergh, P.H., Maat, J., van Ormondt, H., and Sussenbach, J.S., 1977, The nucleotide sequence at the termini of adenovirus type 5 DNA, *Nucleic Acids Res.* **4**:4371.

Stenlund, A., Perricaudet, M., Tiollais, P., and Pettersson, U., 1980, Construction of restriction enzyme fragment libraries containing DNA from human adenovirus types 2 and 5, *Gene* **10**:47.

Stillman, B.W., Lewis, J.B., Chow, L.T., Mathews, M.B., and Smart, J.E., 1981, Identification of the gene and mRNA for the adenovirus terminal protein precursor, *Cell* **23**:497.

Sugisaka, H., Sugimoto, K., Takanami, M., Shiroki, K., Saito, I., Shimojo, H., Sawada, Y., Uemizu, Y., Uesigi, S.-I., and Fujinaga, K., 1980, Structure and gene organization in the transforming HindIII-G fragment of Ad12, *Cell* **20**:777.

Sussenbach, J.S., Ellens, D.J., and Jansz, H.S., 1973, Studies on the mechanism of adenovirus

DNA replication. II. The nature of single-stranded DNA in replicative intermediates, *J. Virol.* **12:**1131.

Tamanoi, F., and Stillman, B.W., 1982, Function of adenovirus terminal protein in the initiation of DNA replication, *Proc. Natl. Acad. Sci. U.S.A.* **79:**2221.

Tate, V.E., and Philipson, L., 1979, Parental adenovirus DNA accumulates in nucleosome-like structures in infected cells, *Nucleic Acids Res.* **6:**2769.

Temple, M., Antoine, G., Delius, H., Stahl, S., and Winnacker, E.L., 1981, Replication of mouse adenovirus strain FL DNA, *Virology* **109:**1.

Tibbetts, C., 1977, Physical organization of subgroup B human adenoviruses genomes, *J. Virol.* **24:**564.

Tibbetts, C., and Pettersson, U., 1974, Complementary strand-specific sequences from unique fragments of adenovirus type 2 DNA for hybridization mapping experiments, *J. Mol. Biol.* **88:**767.

Tibbetts, C., Johansson, K., and Philipson, L., 1973, Hydroxylapatite chromatography and formamide denaturation of adenovirus DNA, *J. Virol.* **12:**218.

Tibbetts, C., Pettersson, U., Johansson, K., and Philipson, L., 1974, Transcription of the adenovirus type 2 genome. I. Relationship of cytoplasmic RNA to the complementary strands of viral DNA, *J. Virol.* **13:**370.

Tokunaga, O., Shinagawa, M., and Padmanabhan, R., 1982, Physical mapping and sequence analysis at the inverted terminal repetition of adenovirus type 4, *Gene* **18:**329.

Tolun, A., Aleström, P., and Pettersson, U., 1979, Sequence of inverted terminal repetitions from different adenoviruses: Demonstration of conserved sequences and homology between SA7 and SV40 DNA, *Cell* **17:**705.

Tooze, J. (ed.), 1981, *The Molecular Biology of Tumor Viruses*, 2nd rev. ed., Part 10B, *DNA Tumor Viruses*, Cold Spring Harbor Press, Cold Spring Harbor, New York.

Trentin, J.J., Yabe, Y., and Taylor, G., 1962, The quest for human cancer viruses, *Science* **137:**835.

van Beveren, C.P., Maat, J., Dekker, B.M.M., and van Ormondt, H., 1981, The nucleotide sequence of the gene for protein IVa2 and of the 5′ leader segment of the major late mRNA's of adenovirus type 5, *Gene* **16:**179.

van den Elsen, P., de Pater, S., Houweling, A., Van der Veer, J., and van der Eb, A.J., 1982, The relationship between region E1A and E1B of human adenoviruses in cell transformation, *Gene* **18:**175.

van der Eb, A.J., and van Kesteren, L.W., 1966, Structure and molecular weight of the DNA of adenovirus type 5, *Biochim. Biopys. Acta* **129:**441.

van der Eb, A.J., van Kesteren, L.W., and van Bruggen, E.F.J., 1969, Structural properties of adenovirus DNAs, *Biochim. Biopys. Acta* **182:**530.

van der Eb, A.J., Mulder, C., Graham, F., and Houweling, A., 1977, Transformation with specific fragments of adenovirus DNA. I. Isolation of specific fragments with transforming activity of adenoviruses 2 and 5 DNA, *Gene* **2:**115.

van der Eb, A.J., van Ormondt, H., Schrier, P.I., Lupker, J.H., Jochemsen, H., van den Elsen, P.J., Deleys, R.J., Maat, J., van Beveren, C.P., Dijkema, R., and de Waard, A., 1979, Structure and function of the transforming genes of human adenoviruses and SV40, *Cold Spring Harbor Symp. Quant. Biol.* **44:**383.

van der Vliet, P.C., and Levine, A.J., 1973, DNA-binding proteins specific for cells infected by adenovirus, *Nature (London) New Biol.* **246:**170.

van der Vliet, P.C., and Sussenbach, J.S., 1975, An adenovirus type 5 gene function required for initiation of viral DNA replication, *Virology* **67:**415.

van der Vliet, P.C., Levine, A.J., Ensinger, M., and Ginsberg, H.S., 1975, Thermolabile DNA-binding proteins from cells infected with a temperature-sensitive mutant of adenovirus defective in viral DNA synthesis, *J. Virol.* **15:**348.

van der Vliet, P.C., Zandberg, J., and Jansz, H.S., 1977, Evidence for a function of the adenovirus DNA-binding protein in initiation of DNA synthesis as well as in elongation of nascent DNA chains, *Virology* **80:**98.

van Ormondt, H., Maat, J., de Waard, A., and van der Eb, A.J., 1978, The nucleotide sequence of the transforming HpaI-E fragment of adenovirus type 5 DNA, *Gene* **4:**309.

van Ormondt, H., Maat, J., and van Beveren, C.P., 1980a, The nucleotide sequence of the transforming region E-1 of adenovirus type 5 DNA, *Gene* **11**:299.

van Ormondt, H., Maat, J., and Dijkema, R., 1980b, Comparison of nucleotide sequences of the early E1A regions for subgroups A, B and C of human adenoviruses, *Gene* **12**:63.

van Wielink, P.S., 1978, A DNA–protein complex from adenovirus type 5 and its possible role in viral DNA replication, Academic thesis, State University of Utrecht, Utrecht.

Varsanyi, T.M., Winberg, G., and Wadell, G., 1977, DNA restriction site mapping of adenovirus type 16 with BamI, EcoRI, HpaI and SalI. *FEBS Lett.* **76**:151.

Vennström, B., Pettersson, U., and Philipson, L., 1978a, Initiation of transcription in nuclei isolated from adenovirus-infected cells, *Nucleic Acids Res.* **5**:205.

Vennström, B., Pettersson, U., and Philipson, L., 1978b, Two initiation sites for adenovirus 5.5S RNA, *Nucleic Acids Res.* **5**:195.

Virtanen, A., Pettersson, U., Le Moullec, J.M., Tiollais, P., and Perricaudet, M., 1982a, Different mRNAs from the transforming region of highly oncogenic and non-oncogenic human adenoviruses, *Nature (London)* **295**:705.

Virtanen, A., Aleström, P., Persson, H., Katze, M.G., and Pettersson, U., 1982b, An adenovirus agnogene, *Nucleic Acids Res.* **10**:2539.

Vlak, J.M., Rozijn, T.H., and Sussenbach, J.S., 1975, Studies on the mechanism of replication of adenovirus DNA. IV. Discontinuous DNA chain propagation, *Virology* **63**:168.

Wadell, G., 1978, Classification of human adenoviruses by SDS–polyacrylamide gel electrophoresis of structural polypeptides, *Intervirology* **11**:47.

Weber, J., Bégin, M., and Khitto, G., 1975, Genetic analysis of adenovirus type 2. II. Preliminary phenotypic characterization of temperature-sensitive mutants, *J. Virol.* **15**:1049.

Weinmann, R., Raskas, H.J., and Roeder, R.G., 1974, Role of DNA-dependent RNA polymerases II and III in transcription of the adenovirus genome late in productive infection, *Proc. Natl. Acad. Sci. U.S.A.* **71**:3436.

Weinmann, R., Brendler, T.G., Raskas, H.J., and Roeder, R.G., 1976, Low molecular weight viral RNA's transcribed by RNA polymerase III during adenovirus 2 infection, *Cell* **7**:5577.

Westphal, H., and Lai, S.-P., 1977, Quantitative electron microscopy of adenovirus RNA, *J. Mol. Biol.* **116**:525.

Westphal, H., Meyer, J., and Maizel, J.V., Jr., 1976, Mapping of adenovirus 2 mRNA by electron microscopy, *Proc. Natl. Acad. Sci. U.S.A.* **73**:2069.

Wilson, M.C., Fraser, N.W., and Darnell, J.E., 1979, Mapping of DNA initiation sites by high doses of ultraviolet irradiation: Evidence for three independent promoters within the left 11% of the Ad-2 genome, *Virology* **94**:15.

Winberg, G., and Hammarskjöld, M.L., 1980, Isolation of DNA from agarose gels using DEAE paper: Application to restriction site mapping of adenovirus type 16 DNA, *Nucleic Acids Res.* **8**:253.

Wolfson, J., and Dressler, D., 1972, Adenovirus DNA contains an inverted terminal redundancy, *Proc. Natl. Acad. Sci. U.S.A.* **69**:3054.

Wu, M., Roberts, R.J., and Davidson, N., 1977, Structure of the inverted terminal repetition of adenovirus type 2 DNA, *J. Virol.* **21**:766.

Younghusband, H.B., and Bellett, A.J.D., 1971, Mature form of DNA from chicken embryo lethal orphan virus, *J. Virol.* **8**:265.

Zain, B.S., and Roberts, R.J., 1979, Sequences from the beginning of the fiber messenger RNA of adenovirus-2, *J. Mol. Biol.* **131**:341.

Zain, S., Sambrook, J., Roberts, R.J., Keller, W., Fried, M., and Dunn, A.R., 1979a, Nucleotide sequence analysis of the leader segments in a cloned copy of adenovirus 2 fiber mRNA, *Cell* **16**:851.

Zain, S., Gingeras, T.R., Bullock, P., Wong, G., and Gelinas, R.E., 1979b, Determination and analysis of adenovirus-2 DNA sequences which may include signals for late mRNA processing, *J. Mol. Biol.* **135**:413.

Ziff, E.B., and Evans, R.M., 1978, Coincidence of the promoter and capped 5′ terminus of RNA from the adenovirus 2 major late transcription unit, *Cell* **155**:1463.

Ziff, E.B., and Fraser, N., 1978, Adenovirus type 2 late mRNA; Structural evidence for 3′-coterminal species. *J. Virol.* **25**:897.

CHAPTER 4

The Genetic System

C. S. H. YOUNG, T. SHENK, AND
HAROLD S. GINSBERG

I. INTRODUCTION

The use of genetics to study the interactions of adenoviruses with permissive and nonpermissive host cells has had two primary goals. One is to enumerate the genes and the functions encoded by them, a task that in recent years has depended more on the ability to map specific messenger RNAs (mRNAs) to particular regions of the genome than on classic genetic analysis. The other is to use mutants of various kinds in functional studies to determine the role of specific genes in such processes as the control of RNA transcription, DNA replication, host-cell transformation, and viral assembly. In these functional roles, adenovirus mutants have been of great practical benefit, as a reading of the following chapters in this volume will show. With the development of *in vitro* systems to study specific processes, mutants have been and will continue to be of great value in dissecting the steps involved and, importantly, in confirming the faithfulness of the reconstituted systems to the events detected *in vivo*.

The aims of this review are: (1) To show the development of the adenovirus genetic system since the mid-1970s; much of this development, of course, has depended on the technical advances of cloning and sequencing of specific nucleotide sequences, but equally it has depended on the ability to reconstruct viral genomes from sequences manipulated *in vitro* and of transfecting such genomes into suitable host cells by the

C. S. H. YOUNG AND HAROLD S. GINSBERG • Department of Microbiology, College of Physicians and Surgeons of Columbia University, New York, New York 10032. T. SHENK • Department of Microbiology, State University of New York at Stony Brook, Stony Brook, New York 11794.

methods developed by Graham and van der Eb (1973). (2) To put together a useful source for the various phenotypes and map locations for the many mutations that currently exist and to describe useful genetic techniques. (3) To point out areas in which genetic analysis has proved to be of great importance and in which further advances can be expected.

II. PRODUCTION OF VIRAL MUTANTS

A wide variety of mutagenic procedures have been applied to adenoviruses. Initially, investigators used physical and chemical mutagens that mainly induced point mutations. This approach has the advantage that it can be applied with little understanding of the physical organization of a viral chromosome, and it rapidly identifies and provides information about essential viral functions. More recently, a variety of mutagenesis schemes have been developed that are based on *in vitro* manipulation of the viral chromosome. Most of these schemes require a reasonably detailed understanding of the architecture of the viral chromosome. *In vitro* mutagenesis offers several advantages. First, any type of mutation can be generated (single base-pair changes, deletions, insertions, substitutions). Second the alterations can be targeted to specific locations, within either a gene of interest or a control region. Third, since the mutations can be introduced with high efficiency, there is no need to anticipate a phenotype in order to identify the mutation within a viral population.

A. Mutagens

Chemical and physical mutagens have been extensively utilized to increase the frequency of mutants within an adenovirus population. The mutagens used primarily induce single base-pair changes and fall into two categories: (1) those employed *in vitro* that induce alterations in viral DNA (e.g., nitrous acid, hydroxylamine, UV irradiation) and (2) those utilized during the viral growth cycle (e.g., nitrosoguanidine and 5-bromo-2′-deoxyuridine). Mutants are then identified by screening for an observable phenotype [e.g., temperature-sensitive (*ts*) and host-range (*hr*)] The use of mutagens for the production of adenovirus mutants has been reviewed in detail previously (Ginsberg and Young, 1977) and will not be discussed further here.

B. Screens for Preexisting Mutations

A variety of viral mutants have been identified on the basis of altered plaque morphology, virion density, or chromosomal structure. Takemori *et al.* (1968) isolated cytocidal adenovirus type 12 (Ad12) mutants on the basis of their altered plaque morphology. Werner and zur Hausen (1978) detected deletion and insertion mutations within an Ad12 viral stock serially passaged in Vero (African green monkey kidney) cells. From these stocks, it was possible to isolate by plaque purification several nondefective variants that carried small insertions within the right-end 12% of the viral chromosome. Deuring *et al.* (1981) isolated a defective Ad12 variant from a stock repeatedly passaged in human KB cells. The virions were less dense than those of wild-type (wt) virus and contained a DNA molecule in which a short segment from the left end of Ad12 DNA flanked both ends of a human DNA sequence. Hammerskjöld and Winberg (1980) identified and isolated by plaque purification viable Ad16 variants from multiple-passage stocks that contained left-end sequences duplicated and translocated to the right end of the viral chromosome. Groff and Daniell (1980) isolated from a multiple-passage Ad3 stock a viable variant carrying a deletion in early region 3, and Robinson and Tibbetts (personal communication), using a similar stock, have recently generated pure isolates of Ad3 variants carrying deletions and substitutions near the left end of the viral chromosome. Finally, Challberg and Ketner (1981) were able to propagate defective mutants containing large deletion mutations that had accumulated in a high-passage stock of Ad2. The mutants were complemented using a *ts* mutant, and it was possible to separate one of the mutants from the helper virus by equilibrium centrifugation.

In all these cases, mutant viruses accumulated to sufficiently high levels in adenovirus stocks so that they could be detected by physical methods and isolated without a biological selection procedure. The key to the accumulation of substantial numbers of variants appears to be serial passage of adenovirus stocks at a reasonably high multiplicity of infection. Presumably, this allows variants carrying mutations that lead to a selectively disadvantageous phenotype to be carried in the population as they arise. A further implication is that nonhomologous recombination events are common during the replicative cycle, since the fraction of the population that contains detectable changes is high.

C. Selection for Preexisting Mutations

The first preexisting mutant to be selected from an adenovirus stock was a heat-stable (*hs*) variant of Ad5 (Young and Williams, 1975). This variant was selected by several cycles of *in vitro* heat inactivation and growth.

More recently, preexisting mutants have been selected from viral stocks on the basis of resistance to restriction endonucleases. Mertz and Berg (1974a) first employed this type of selection to isolate variants lacking the unique *Hpa*II cleavage site on the simian virus 40 (SV40) chromosome. Jones and Shenk (1978, 1979a) extended this approach to Ad5 to isolate variants lacking either *Eco*RI or *Xba*I cleavage sites. Since Ad5 contains multiple cleavage sites for these enzymes, the viral DNA was cleaved and then re-ligated before it was used to infect cells. The rationale behind this procedure was that mutant DNAs lacking a cleavage site would generate fewer fragments on digestion with the endonuclease and would therefore have a higher probability than wt DNA of being ligated to generate an infectious DNA molecule. As originally described, this protocol is limited to enzymes that cleave DNA infrequently—probably four times or less, given the relatively poor infectivity of adenovirus DNA. Mathews and Grodzicker (1981) have applied this procedure using an Ad2–Ad5 recombinant virus containing a favorable combination of *BamHI* cleavage sites, and Rajagopalan and Chinnadurai (1981) have combined this procedure with an overlap recombination scheme (discussed in Section IV. B.2) to reduce effectively the number of cleavage sites to which the selection is applied.

D. Manipulation of Isolated Viral DNA segments

It is relatively easy to isolate a viral DNA segment and modify it *in vitro*. It has proved more difficult to rebuild the altered segment back into an intact viral chromosome. Two methods have been utilized. The first is to isolate a DNA segment using an enzyme that cleaves the viral chromosome infrequently. After modification, the segment can be reincorporated by ligation to fragments representing the remainder of the genome. The deletion (*dl*) mutant H5*dl*309 (Jones and Shenk, 1979a), which is cleaved only once at 4 map units (m.u.) by *Xba*I, has proved useful for this strategy. Carlock and Jones (1981a) inserted a short segment within the 0- to 4-m.-u. fragment and then rebuilt infectious virus by ligation of the altered segment to the 4- to 100-m.-u. segment. In a similar approach, Solnick (1981a) subjected the 0- to 4-m.-u. segment to chemical mutagenesis and then reincorporated it into an intact viral DNA. A second method for rebuilding a mutated segment into the viral chromosome is to utilize overlap recombination. Chinnadurai *et al.* (1979) demonstrated that transfection of cells with adenovirus DNA segments that overlap led to the efficient generation of intact viral DNAs resulting from recombination events within the overlap region. Ho *et al.* (1982) used overlap recombination to rebuild a 0- to 8-m.-u. *Hind*III-generated segment into the viral chromosome after *in vitro* chemical mutagenesis. The segment was mixed with the H5*dl*309-derived 4- to 100-m.-u. *Xba*I segment and used to infect cells. Mutations within the 4- to 8-m.-u. region of the *Hind*III-generated segment were incorporated into an intact chromosome when the recombination event occurred to the right of the mutation.

Stow (1981) demonstrated that it was possible to cleave a plasmid carrying a left-end segment of the viral chromosome at 4 m.u. with *Xba*I, ligate the linear plasmid to the H5*dl*309-derived 4- to 100-m.-u. *Xba*I segment, and regenerate an intact chromosome subsequent to infection. The precise left terminus of the viral chromosome was not required for infectivity so long as some portion of the terminal repeat remained. Further, the plasmid sequence could remain attached to the end of the viral sequences at the time of infection. Presumably, the intact, right-end terminal repeat was somehow copied onto the left end to regenerate a viral chromosome with normal termini. Stow (1981) went on to show that deletions constructed at a unique restriction endonuclease cleavage site within the 0- to 4-m.-u. cloned segment could be readily reincorporated into virus by this procedure. Subsequently, this approach has been used to introduce a variety of mutations into the left-end region of the viral chromosome, including deletions constructed by dropping fragments with a restriction endonuclease (Osborne *et al.*, 1982), deletions produced by a D-loop mutagenesis protocol (Hearing and Shenk, 1983), and site-specific, single-base-pair changes produced through the use of synthetic oligonucleotides (Montell *et al.*, 1982).

The same approach has been used to introduce *dl* mutations into the right-end region of the viral chromosome (Halbert, Cutt and Shenk, in prep.) Mutations produced in cloned segments derived from internal regions of the viral chromosome have also been reincorporated into infectious viruses using either overlap recombination (Kapoor and Chinnadurai, 1981) or three fragment ligations (Thimmappaya *et al.*, 1982). Thus, it is now possible to manipulate a cloned segment derived from any location on the viral genome and ultimately rebuild the altered region into an intact chromosome to study its effect. This, of course, makes it possible to employ any of the available mutagenesis procedures (reviewed in Shortle *et al.*, 1981) on the adenovirus chromosome.

E. Propagation of Mutants

A variety of schemes have been employed to propagate adenovirus mutants. In the simplest case, the mutant is viable and grows as well as the Wt virus. Plaque-morphology mutants (Takemori *et al.*, 1968) an *hs* variant (Young and Williams, 1975), and mutants lacking genes nonessential for growth in cell culture [e.g., region E3 (Kelly and Lewis, 1973; Jones and Shenk, 1978; Groff and Daniell, 1980; Kapoor *et al.*, 1981)] are examples of such mutants. Other variants have been isolated that are viable, but grow somewhat more poorly than their Wt parent [e.g., mutants unable to produce virus-associated RNAI (Thimmappaya *et al.*, 1982)].

Defective viral mutants of two types have been studied. Many conditionally defective variants have been isolated and characterized. These include *ts* (both heat- and cold-sensitive) and *hr* mutants. By far the most useful *hr* mutants are those that carry lesions in early region 1 (E1) (e.g.,

Harrison *et al.*, 1977; Jones and Shenk, 1979a). These mutants grow very poorly in normal human cells, but produce Wt yields when propagated in 293 cells [human embryonic kidney cells that contain and express the Ad5 E1 region (Graham *et al.*, 1977)]. Only one instance has been reported in which non-conditionally defective adenovirus mutants have been propagated and studied. Challberg and Ketner (1981) were able to propagate Ad2 variants carrying large deletions using a helper virus that contained a *ts* mutation at a locus outside the region deleted in the mutant. However, it seems unlikely that complementing, helper viruses will prove to be generally useful. Since adenovirus recombines efficiently, Wt recombinants are readily generated in mixed infections and can outgrow both the mutant under study and the helper virus. Further, it is also important to separate mutant from helper virus if the mutant's physiological defect is to be assessed. This then requires the two viral DNAs to be of quite different sizes and limits the types of mutations that can be effectively studied by this approach.

Clearly, additional methods for propagating defective viral mutants must be developed. Selectable markers have been utilized to cotransform human cells with adenovirus E1 genes that produce cell lines similar to 293 cells (Grodzicker and Klessig, 1980; Klessig *et al.*, 1982; Shiroki *et al.*, 1983; Babiss *et al.*, 1983b). The resultant transformants express E1 genes, and many can complement viruses with mutations in this region. Presumably, this approach can be used to generate cells that contain and express additional viral genes, producing valuable reagents for the propagation of new viral *hr* mutants. This topic is considered again in Section V.B.

Nonsense mutations may also play an important role in the future of adenovirus genetics. Laski *et al.* (1982) modified a *Xenopus* Tyr transfer RNA (tRNA) gene by site-specific mutagenesis to generate an amber tRNA gene. Infection of monkey cells with an SV40 recombinant carrying the amber tRNA gene induced a transient state permissive for an adenovirus–SV40 hybrid containing a nonsense mutation. Recently, this tRNA gene has been introduced into mouse cells and shown to suppress several nonsense mutations (Hudziak *et al.*, 1982). When human cells carrying this and other suppressor tRNA genes become available, it should be possible to complement nonsense mutations located within a variety of adenovirus genes.

III. ADENOVIRUS MUTANTS

A. Classes of Phenotypes

A previous review (Ginsberg and Young, 1977) outlined several classes of phenotypes into which adenovirus mutants could be, or had already been, classified. While no new phenotypic class has been observed, great advances in extending the usefulness of certain types have taken place, principally because of the development of new permissive cell lines containing expressed adenovirus gene sequences (Graham *et al.*, 1977) and the subsequent isolation of restricted host-range (*hr*) adenovirus mutants (Harrison *et al.*, 1977; Jones and Shenk, 1979a). In addition, the temperature-restricted category now includes so-called "cold-sensitive" mutations (Ho *et al.*, 1982; Babiss *et al.*, 1983a) and temperature-dependent *hr* mutations (Nicolas *et al.*, 1981). Table I summarizes the various classes, their respective gross phenotypic characteristics, and the known or suspected genetic changes responsible.

Since adenovirus is now known to code for a DNA polymerase (Stillman *et al.*, 1982a; Aleström *et al.*, 1982; Gingeras *et al.*, 1982) and probably for a protease (Weber, 1976; Kruijer *et al.*, 1980; Akusjärvi *et al.*, 1981), it is possible that specific drug-resistant mutants may be developed. This of course depends on the finding of suitable drugs and drug-resistant host cells either occurring naturally or following mutagenesis and selection.

B. Phenotypic Expression of the Mutants

As mentioned in Section I, details of the ways in which many of the adenovirus mutants have been used are given in other chapters of this volume. Here, only some highlights will be mentioned to reveal the advances made since the review of Ginsberg and Young (1977).

The continued characterization of selected temperature-sensitive (*ts*) mutations has been of great value in identifying the functions specified by the virus. Thus, for example, mutants belonging to complementation group N, in Williams's classification, were known to be defective in adenovirus DNA replication at the nonpermissive temperature (Wilkie *et al.*, 1973; Ginsberg *et al.*, 1974; van der Vliet and Sussenbach, 1975; Schutzbank, 1980). Subsequent marker-rescue data (Galos *et al.*, 1979) placed the mutations within a boundary of 18.5–22.0 m.u., and mRNA detection (Galos *et al.*, 1979) and translation *in vitro* (Stillman *et al.*, 1981) demonstrated early functions encoded by this region. *In vitro* complementation of DNA replication (Stillman *et al.*, 1982a) and sequence analysis (Aleström *et al.*, 1982; Gingeras *et al.*, 1982) suggested the presence of a gene encoding an adenovirus-specified DNA polymerase. A polymerase activity had been shown previously to copurify with the adenovirus terminal protein precursor as a higher-molecular-weight complex (Enomoto *et al.*, 1981; Lichy *et al.*, 1982). As mentioned in Section V.A.1, Galos *et al.* (1979) had proposed, on genetic grounds, that H5*ts*149 resided in a gene the product of which was enzymatic.

Other *ts* mutants that have strongly implicated adenovirus gene products in specific processes include H2*ts*1, the phenotype of which suggested that wild-type (Wt) virus encoded a protease necessary for the con-

TABLE I. Classes of Adenovirus Mutants

Phenotypic class	Gross phenotype	Genetic change	Example	References
1. Temperature-restricted				
a. Temperature-sensitive (*ts*)	Replication inhibited at 38–40°C, but not at 32°C	Missense base-pair change	H5*ts*125	Ensinger and Ginsberg (1972), Kruijer *et al.* (1981)
b. Host-range cold-sensitive (hr^{cs})	Replication inhibited in the restrictive host at 32°C, but less so at 37°C	Single base-pair deletion	H5*hr*1	Harrison *et al.* (1977), Ricciardi *et al.* (1981), Ho *et al.* (1982)
c. Host-range temperature-sensitive	Replication inhibited in the restrictive host at 39°C, but not at 32°C	Intragenic suppressor mutation caused by single base-pair change	H5r(*ts*107)202	Nicolas *et al.* (1981), Kruijer *et al.* (1983)
2. Host range[a]				
a. Restricted	i. Replication in human 293 cells; KB or HeLa cells less permissive	Deletions, insertions, or substitutions	H5*dl*312	Jones and Shenk (1979a)
	ii. Replication in HeLa or KB cells; hamster cells less permissive	Unknown	H5*hr*203	Minekawa *et al.* (1976)
b. Extended	Replication in monkey cells as extensive as in human cells	i. Single base-pair change in 72K DBP gene	H5*hr*404	Klessig and Grodzicker (1979), Kruijer *et al.* (1981)
		ii. Alterations to genome termini	Vero-adapted Ad12	Werner and zur Hausen (1978)

3. Plaque morphology				
a. Cytocidal	Degeneration of KB cell monolayer and formation of larger plaques	Unknown	H12*cyt*	Takemori *et al.* (1968), Takemori (1972)
b. Large plaque	Larger plaques than those of Wt on KB and HeLa cells	i. Single base-pair change in E1b	H2*lp*3	Chinnadurai *et al.* (1979), Chinnadurai (1983)
		ii. Deletion in E3	H5*sub*304	Volkert and Young (unpublished data)
4. Virion stability				
a. Heat-stable	Increased stability to *in vitro* thermal inactivation	Unknown	H5*hs*1	Young and Williams (1975)
b. Heat-labile	Decreased stability to *in vitro* thermal inactivation	i. Deletion of protein IX promoter	H5*dl*313	Jones and Shenk (1979a), Colby and Shenk (1981)
		ii. Substitution at *Eco*RI site in protein VIII gene	H5RI^r	Liu *et al.* (1984)

[a] The host range of adenovirus can be extended by the incorporation of early SV40 sequences into adenovirus E3. Restricted *hr* variants can then be isolated from these hybrids. The derivations and use of these mutants are reviewed in Tooze (1981).

version of precursor polypeptides pVI, pVII, and pVIII to virion polypeptides VI, VII, and VIII, respectively (Weber, 1976). Recent sequence data from the region of the genome in which H2*ts*1 lies show the presence of an open reading frame sufficient to code for a polypeptide with a molecular weight of 23,000 (23K) (Kruijer *et al.*, 1980; Akusjärvi *et al.*, 1981). Mutants that map within the 100K gene have revealed the crucial role the gene product plays in the assembly of trimeric hexon capsomers (Oosterom-Dragon and Ginsberg, 1981). Temperature-sensitive mutations in the 72K DNA-binding protein (DBP), revertants isolated from them, and extended *hr* mutations mapping within the DBP gene have revealed the multifunctional role of this product in adenovirus replication and transformation (summarized in Klessig and Quinlan, 1982). Since many of these mutations and reversions have been sequenced (Kruijer *et al.*, 1981, 1982, 1983) and the polypeptide is being characterized chemically and physically, a complete genetic and physiological description of the DBP can be expected.

Finally, the effects on viral assembly of *ts* mutations in late genes should be mentioned. Phenotypes ranging from the assembly of complete virions, albeit with uncleaved precursor polypeptides and a noninfectious character, to partial assembly and to complete absence of detectable particles have all been observed (reviewed by Philipson in Chapter 8).

The production of restricted *hr* mutants with DNA lesions varying from simple base-pair changes to extensive deletions and insertions in E1a and E1b is the development that has spurred perhaps the greatest interest. As mentioned in Section III.A, these mutants (Harrison *et al.*, 1977; Jones and Shenk, 1979a) fail to replicate efficiently on established human cell lines such as HeLa or KB, but do so on a line of human embryo kidney cells (293) transformed by sheared adenovirus DNA (Graham *et al.*, 1977) and expressing E1a and E1b sequences constitutively (Graham *et al.*, 1977; Aiello *et al.*, 1979). The *hr* mutants have a variety of phenotypes, depending on whether they map in E1a or E1b, on their position within these genes, and on the extent of the alteration. In summary, mutants in E1a can fail to transform nonpermissive rat cells, can cause a higher rate of "incomplete" transformation (Graham *et al.*, 1978; Babiss *et al.*, 1983a), and can cause a failure to initiate the remainder of the early transcription program efficiently (Jones and Shenk, 1979b; Berk *et al.*, 1979). Mutants in E1b fail to transform rat cells (Graham *et al.*, 1978) (although see Shiroki *et al.*, 1981; Mak and Mak, 1983), but do allow the transcriptional program to proceed. Mutations constructed to affect the 19K or 58K proteins of E1b fail to transform (Chinnadurai, 1983; Babiss and Ginsberg, 1984). The precise step at which the replicative cycle of the E1b mutants is affected is not clear yet, although it is known that viral DNA synthesis is normal, late viral polypeptides are underproduced, and the final viral yield is even more reduced (Lassam *et al.*, 1978). The detailed analysis of the defects in E1b expression may help to reveal how adenovirus transcription and translation are favored over the host in a permissive infection. A deletion in E1b extending past the promoter for virion polypeptide IX shows a marked thermolability (Colby and Shenk, 1981), demonstrating the importance of this polypeptide for virion structure.

Deletion mutants in regions other than E1a and E1b have revealed various phenotypes. Among some that might be mentioned are deletions of E3 that show that this region is nonessential for permissive infection in cultured cells (Kelly and Lewis, 1973; Kapoor *et al.*, 1981; Jones and Shenk, 1978). Interestingly, one deletion in E3, *sub*304 and its derivatives, displays a plaque-morphology phenotype on HeLa cells. Viruses with the deletion exhibit larger, clearer plaques than do Wt viruses (Volkert, Liu, and Ginsberg, unpublished findings). Deletions of one or another of the genes for the virus-associated RNAs (VA-RNAs) have revealed that VA-RNAII is dispensable, but that VA-RNAI is considerably more important. Deletions lacking the VA-RNAI gene translate mRNAs poorly and replicate to a titer some 50-fold lower than does Wt virus (Thimmappaya *et al.*, 1982).

Table II summarizes the current knowledge of the map positions, gross phenotypes, and functional analysis of mutants of group C adenoviruses. As this table shows, the detailed descriptions of many mutants are far from complete, and much remains to be understood. In particular, precise mapping data for many of the mutants collected after random mutagenesis would be of great value in pinpointing the polypeptides in which alterations have been made.

IV. MAPPING METHODS

One of the primary aims of genetic analysis is to map mutations relative to one another and to locate them precisely on the physical map of the genome. If the mutation that underlies a particular phenotype can be placed within a known coding region, and perhaps be traced to specific base-pair change(s), using sequencing techniques, functional analysis is placed on a sure footing, allowing conclusions as to the roles of specific polypeptides in the viral life cycle.

While it is clear that mapping is important for mutations isolated following random mutagenesis, it is also necessary to employ mapping techniques, and in particular sequencing methods, to confirm that site-directed mutagenesis has induced the changes expected from the specific technique and that subsequent DNA manipulations have not caused unwanted changes. Sequence analysis is also of great importance in determining the precise lesions present following region-directed mutagenesis, a technique that may cause several base-pair changes in the segment that was mutagenized. This principle is illustrated in the development of a new host-range (*hr*) mutant, H5*hr*440 (Solnick, 1981a). Nitrous acid mutagenesis of the 0- to 4-m.-u. segment caused changes in two adjacent

TABLE II. Mutants and Variants of the Subgroup C Adenoviruses[a]

Region[b]	Serotype[c]	Mutant[d]	Coordinates[e]	Polypeptide(s) mutated[f]	Phenotype[g]	Ref. Nos.[h]
E1a	5	*hr*1	1 bp Δ at 1055	51K is truncated to 28K	Host-range; reduced onset of transcription from E1b E2, E3, and E4; cold-sensitive for transformation in RE and CREF cells	1–6
		*hr*2	Lies between bp 448 and 1349	As for *hr*1	As for *hr*1	1, 2, 4, 5, 7
		*hr*3	—	—	As for *hr*1, but absolutely defective for transformation in RE cells	1, 4, 5
		*hr*4	—	—	As for *hr*3	1, 4, 5
		*hr*5	—	—	As for *hr*3	1, 4, 5
		*dl*311	Δbp 1282–1339	51K and 48K	Host-range; defective in DNA replication; somewhat reduced transformation frequency in RE cells	8, 3, 9, 10
		*dl*312	Δbp 448–1349	51K and 48K absent	Host-range; delayed onset of transcription from E1b through E4; defective for transformation in all rat cells tested	8, 3, 9, 10
		*in*500	Octanucleotide insert at bp 1010	51K is presumably truncated to 28K	Host-range; normal levels of E1b mRNA, but lower levels of E2a mRNA; transformation-defective in RE and BRK cells	11
		*dl*502	Δbp 1009–1340	51K and 48K	Host-range; unspliced E1a sequences found in cytoplasm; E1b, E2, E3, and E4 mRNAs also found	12
		*hr*440	2-bp change at bp 979 and 980	51K truncated, 48K absent	Host-range; normal levels of mRNA from E1b and E4; lower levels of E2a and E3; transformation-defective in BRK cells	13, 14
		*in*317	360-bp insert at *Xba*I site at m.u. 3.8	51K and 48K	Host range	8
		$hr^{cs}11$	Lies between 448 and 1349	51K or 48K or both	Cold-sensitive host-range; cold-sensitive for transformation in RE cells	5

	hr^{cs}12	As for *hr*^{cs}11	As for *hr*^{cs}11	As for *hr*^{cs}11	5
	*dl*101	Δbp 1008–1012	51K truncated to 30K (predicted)	Cold-sensitive host-range; cold-sensitive for transformation in CREF cells; E1b and E2a mRNA in lower amount; E3 and E4 mRNA not detectable	15
	in 106	16-bp insert at bp 1009	51K truncated to 24K (predicted)	As for *dl*101	15
	*dl*105	Δbp 1003–1071	51K reduced to 47K (predicted)	Host-range; E2a mRNA present; transformation-defective	15
2/5	4/1	7-bp Δ at *Sma*I site at 1010	51K	Host-range	16
	6/1	14-bp Δ at *Sma*I site at 1010	51K	Host-range	16
	*pm*975	1-bp change at bp 975 T→G	48K missing	No overt phenotype; 12 S mRNA missing owing to alteration in splice donor site; 13 S-encoded polypeptide unaltered because of degeneracy	17
	*dl*1501	Δbp −44 to −39 from cap site	—	No overt phenotype	18
	*dl*1502	Δbp −44 to −24 (includes TATA box)	—	No overt phenotype; E1a mRNA level 5-fold lower; E2a and E3 lower; E4 normal	18
	*dl*1503	Δbp −44 to +21	—	As for *dl*1502, except that E2a and E3 mRNA levels closer to normal	18
	*dl*1504	Δbp −44 to +62	First AUG triplet Δ for AU; second AUG sequence used with deletion of 14 a.a. from 51K and 48K (predicted)	As for *dl*1502, except that E4 mRNAs are decreased	18
2	*pm*1610	T→G at bp 1610	—	Polyadenylation signal of E1a mutated; some transcripts polyadenylated upstream from normal site, others proceed to E1b site	19

(*Continued*)

TABLE II. (*Continued*)

Region[b]	Serotype[c]	Mutant[d]	Coordinates[e]	Polypeptide(s) mutated[f]	Phenotype[g]	Ref. Nos.[h]
	5	310	Small change at *Xba*I site at m.u. 3.8	51K and 48K (predicted)	No overt phenotype	8
E1a + E1b	5	*dl*314	Δbp 1335–1679	51K and 48K (E1a)	Host-range; delayed onset of transcription from other early regions; transformation-defective in RE cells; does not complement E1a mutants	8, 3, 9, 20
		*sub*315	Δ3.7–8.7 m.u. Insertion of 4.1% foreign DNA	51K and 48K (E1a) 19K and 58K (E1b)	As for *dl*314; transformation-defective in CREF cells; complements E1a mutants	8, 20, 21
		*sub*316	Δ3.6–5.8 m.u. Insertion of 3.7% foreign DNA	As for *sub*315	As for *dl*314	8, 20
		*dl*434	Δ2.6–8.7 m.u.	As for *sub*315	Host-range	22
	2	*dl*201.1	Δ2.0–7.0 m.u. 0.3 m.u. repeated at right-hand terminus	As for *sub*315	Host-range	23
		*dl*201.2	As for *dl*201.1, but no terminal repeat	As for *sub*315	Host-range	23
E1b	5	*hr*6	Lies between m.u. 6.1 and 8.5	58K	Host-range; transformation-defective in all rat cells tested;	1, 7, 24–26
		*hr*7	Lies between m.u. 6.1 and 8.0	As for *hr*6	As for *hr*6; overproduction of 72K DBP	1, 7, 20, 24
		*hr*50	—	—	Transformation-defective	24, 26
		*hr*51	—	—	As for *hr*50	24, 26
		$hr^{cs}13$	Lies between m.u. 6.1 and 8.0	58K	Cold-sensitive host-range; defective for transformation in RE cells	5

		*dl*118	Δbp 2049–2591	58K and 19K	Host-range; all early mRNAs at normal levels; reduced synthesis of viral polypeptides; host protein synthesis not shut off; transformation-defective in CREF cells	27
		*dl*110	Δbp 2333–2804	58K	As for *dl*118	27
		*in*127	Insertion of 4 bp at bp 2808	As for *dl*110	As for *dl*118, but transformation decreased only 10-fold	27
		*dl*163	Δbp 3326–3333	As for *dl*110	As for *dl*176, but transformation in CREF cells decreased 5-fold	27
	2	*lp*3	1-bp change at 1718	Substitution of Val for Ala in N terminus of 19K	Large-plaque; 10-fold lower transformation frequency than wt in 3Y1 cells	28, 29
		*lp*1	—	—	Large-plaque; 3-fold lower transformation frequency in 3Y1 cells	28, 29
		*lp*2	—	—	As for *lp*1	28, 29
		*lp*5	Base-pair changes at 1954 and 2237	Substitution of Tyr for Asp in 19K; substitution of Leu for termination codon in 19K; substitution of Ile for Met in 58K	Large-plaque; transformation-defective in 3Y1 cells; absence of 19K polypeptide; presence of new 21K polypeptide	29
		*lp*4	—	—	Large-plaque; transformation frequency in 3Y1 10-fold lower	28, 29
E1a + E1b + IX	5	*dl*313	Δbp 1334–3639	51K and 48K (E1a) 19K and 58K (E1b) Polypeptide IX absent	Host range; transformation-defective in RE and CREF; some foci in 3Y1 and BRK cells at high m.o.i.; prelabeled host and newly synthesized viral DNA degraded; virion thermolabile	8, 3, 9, 21, 30–33
E2b	5	*ts*149	m.u. 18.5–22.0	Adenovirus-encoded DNA polymerase	Temperature-sensitive for DNA replication	34–38
		*ts*36[i]	As for *ts*149	As for *ts*149	As for *ts*149; transformation-deficient in RE cells	39, 36, 40, 41
		*ts*37[i]	—	—	As for *ts*36	39–41
		*ts*69[i]	m.u. 18.5–22.0	As for *ts*149	—	36

(*Continued*)

TABLE II. (Continued)

Region[b]	Serotype[c]	Mutant[d]	Coordinates[e]	Polypeptide(s) mutated[f]	Phenotype[g]	Ref. Nos.[h]
	2	*ts*105[i]	—	—	As for *ts*36, but transformation not tested	42
		*ts*114[i]	—	—	As for *ts*105	42
		52.11[i]	m.u. 15.5–22.0	—	Temperature-sensitive for DNA replication	43
		19[i]	m.u. 15.5–22.0	—	Temperature-sensitive for DNA replication	43
E2a	5	*ts*125	1-bp change at bp 1064	DBP; a.a. 413 Pro→Ser	Temperature-sensitive for DNA replication *in vivo* and *in vitro*; fails to shut off E4; transforms RE and CREF cells at higher frequency	44, 34, 35, 45–54
		*ts*107	As for *ts*125	As for *ts*125	As for *ts*125	55, 45
	2⁺ND1	*ts*23	1-bp change at bp 1457	DBP; a.a. 282 Leu→Phe	Temperature-sensitive for DNA replication	56
		*ts*63	m.u. 59.0–69.0	—		57
	2	*ts*201	—	—	Fails to complement H5*ts*125	58
		*hr*400	m.u. 62.9–65.6	DBP	Extended host range; grows on monkey cells as well as does Ad2⁺ND1; produces full complement of late polypeptides	59–61
		*hr*401	m.u. 63.6–70.7	—	As for *hr*400	60, 61
		*hr*402	As for *hr*401	—	As for *hr*400	60, 61
		*hr*403	As for *hr*401	—	As for *hr*400	60, 61
	5	*hr*404	1-bp change at bp 1913	DBP; a.a. 130 His→Tyr	As for *hr*400	60, 45, 61
	2⁺ND3	*hr*600	As for *hr*404	As for *hr*404	As for *hr*400	62, 63
		*hr*601	As for *hr*404	As for *hr*404	As for *hr*400	62, 63
		*hr*602	As for *hr*404	As for *hr*404	As for *hr*400	62, 63
		*hr*603	As for *hr*404	As for *hr*404	As for *hr*400	62, 63
	5[j]	r(*ts*125)13	1-bp change at bp 779	DBP; second site reversion at a.a. 508 His→Tyr	Plaques at 39°C with 10-fold lower efficiency than Wt; regulation of DBP synthesis not restored to Wt level	64–66
		r(*ts*125)28	As for 13	As for 13	—	64, 65

		r(*ts*125)20	1-bp change at bp 1262	DBP; second site reversion at a.a. 347 Ala→Pro	Transforms CREF cells with efficiency equal to *ts*125	64, 65, 67
		r(*ts*107)202	1-bp change at bp 1246	DBP; second site reversion at a.a. 352 Gly→Asp	Transforms rat 1 cells with efficiency equal to *ts*107, CREF cells at efficiency greater than *ts*125	64, 65, 67, 68
		r(*ts*107)177	As for 202	As for 202	Transforms CREF cells with efficiency greater than *ts*125	64, 65, 67
		r(*ts*107)164	As for 202	As for 202	—	64, 65
		r(*ts*107)197	As for 202	As for 202	—	64, 65
E3[k]	5	301	Lacks *Eco*RI site at m.u. 84.0	—	No overt phenotype	69
		*sub*304	Δ m.u. 83.2–85.1 Insertion of 1.2% foreign DNA	Uncertain	Larger, clearer plaques on HeLa cells	69, 70
		*dl*303	Δ m.u. 83.2–83.9	Uncertain	—	69
		*sub*305	Δ m.u. 83.2–85.1 Insertion of 4.8% foreign DNA	Uncertain	—	69
		*in*306	Insertion at 83.2 m.u. of 3.8% foreign DNA	Uncertain	—	69
		*in*307	Insertion at 83.2 m.u. of 4.7% foreign DNA	Uncertain	—	69
		*dl*324	Lacks *Xba*I-D fragment (m.u. 78.5–84.3)	E3 glycoproteins	Present in *dl*313 background and thus host range on HeLa cells	71
	2	Δp 305	Δ m.u. 83–85	Uncertain	—	Cited in 72
		*dl*801	Δ m.u. 78.5–83.5	E3 glycoproteins	No overt phenotype	73
E3 + L5	2	*dl*802	Δ m.u. 78.7–87.4	E3 glycoproteins and IV	Defective; replicates in presence of *ts* helper	73
		*dl*803	Δ m.u. 83.2–90.4	Perhaps E3 glycoproteins; IV	Defective; replicates in presence of helper	73

(*Continued*)

TABLE II. (Continued)

Region[b]	Serotype[c]	Mutant[d]	Coordinates[e]	Polypeptide(s) mutated[f]	Phenotype[g]	Ref. Nos.[h]
		*dl*804	Δ m.u. 77.8–86.7	E3 glycoproteins and perhaps IV	Defective?; isolated in presence of helper	73
		*dl*805	Δ m.u. 78.0–86.0	E3 glycoproteins	Presumably nondefective	73
E3, L5 + E4	2	*dl*807	Δ m.u. 82.5–95.0	E3 glycoproteins, IV, and E4 polypeptides	Defective; replicates in presence of helper; DNA replication normal; synthesis of several late polypeptides reduced	73
E4	5	*dl*341	Single base Δ at bp 7143	11K is predicted to be truncated to 6K	No overt phenotype	74
E4 + L5	2	*dl*808	Δ m.u. 92.0–97.1	Fiber perhaps truncated; E4 polypeptides absent	Defective?; isolated in presence of helper	73
VAI	5	*dl*330	29-bpΔ within VAI sequence, at +49 to +79 relative to transcription start site	None known in this region of the genome	No VA-RNAI present; virus replicates more slowly than the parent *dl*324; translation of late mRNAs decreased	71
		*dl*331	As for *dl*330	As for *dl*330	As for *dl*330, but the *dl*313 sequences have been replaced by Ad5 Wt sequences	71
		*dl*309	2-bpΔ at −22 and −23 from initiation site	As for *dl*330	No overt phenotype; VA-RNAI initiating with A residue is lost, that with G residue is enhanced	8, 75
	2/5	*var*1	Base-pair change at +66 (A→G) and at +72 (G→T) with respect to initiation site	As for *dl*330	No overt phenotype; base-pair changes result in an Ad2→Ad5 sequence change	76
		*var*2	As for *var*1	As for *var*1	As for *var*1	76
		*var*3	Insertion of GATC at +75	As for *var*1	No overt phenotype	76

	2	*dl*401	Small deletion at +75	As for *dl*330	No overt phenotype	77
	2/5	203	Small deletion at m.u. 29	As for *dl*330	No overt phenotype	78
		204	Small insertion at m.u. 29	As for 203	As for 203	78
VAII	5	*dl*328	17-bp deletion within VAII sequence at +56 to +74	None known in this region of the genome	No VA-RNAII present; no overt phenotype	71
Unknown	2	*ts*206	—	—	Fails to synthesize DNA at the nonpermissive temperature; complements mutants in E2a and E2b; mutant function expressed very early; no early mRNAs detected	58, 79
		*ts*214	—	—	Temperature-sensitive for DNA replication; complements *ts*206 and *ts*201; perhaps in E2b	58
		*ts*111	m.u. 0–29	—	Fails to synthesize DNA at the nonpermissive temperature; degrades prelabeled cellular DNA and any newly synthesized viral DNA; complements E2a and E2b mutants	42, 80, 81
	5	*ts*213	—	—	Fails to synthesize DNA at the nonpermissive temperature; unable to transform hamster NIL cells	82
		*ts*219	—	—	Fails to complement *ts*219; as for *ts*213	82
		*ts*212	—	—	Fails to synthesize DNA at the nonpermissive temperature; complements both *ts*213 and *ts*219; can transform NIL cells	82
		*ts*220	—	—	As for *ts*212	82

(Continued)

TABLE II. (*Continued*)

Region[b]	Serotype[c]	Mutant[d]	Coordinates[e]	Polypeptide(s) mutated[f]	Phenotype[g]	Ref. Nos.[h]
		*hr*203	—	—	Fails to replicate in hamster cells, but will transform them; deficient in DNA replication in hamster cells	82
L1	5	*ts*58	m.u. 31.5–36.7	IIIa or perhaps 52,55K	Accumulates intermediate-density particles; lowered amounts of IIIa antigen	83–85
		*ts*49	m.u. 31.5–37.3	As for *ts*58	No details reported	41, 84
		*ts*31	—	—	Fails to complement *ts*49	41
L1 or L2	2	*ts*112	m.u. 29.0–41.0	As for *ts*58, or III	Accumulates intermediate-density particles; lowered amounts of IIIa antigen; complements *ts*58	42, 80, 85, 86
		*ts*4	m.u. 29.0–42.0	As for *ts*112	As for *ts*112	87, 85, 88–90
L2 or L3	2	*ts*3	m.u. 41.0–56.9	Probably II	Temperature-sensitive for production of hexon, penton, and fiber antigens; fails to recombine with H5*ts*2, a mutant in hexon	87–89, 91, 92
		*ts*121[l]	m.u. 42.0–66.0	As for *ts*3	CNBr-fragments of hexon differ from Wt fragments; absence of hexon antigen	42, 80
L2 and L3	2	*dl*806	Δ m.u. 44.0–53.0	pVII, V, pVI, and II	Defective; replicates in presence of helper	73
L3	5	*ts*2[m]	m.u. 54.8–57.0	II	Abnormal distribution of hexon antigen; no particles formed	93, 39, 84
		*ts*147[n]	m.u. 45.9–70.7	II (from complementation data)	Hexon capsomer antigen remains in cytoplasm; no particles formed	34, 95, 97, 98
		*ts*40	m.u. 52.6–54.8	II	Complements *ts*2, presumably by intracistronic complementation	84, 96
L3	2	55	m.u. 58.5–59.5	II	Not reported	43
L3	2	*ts*1	1-bp change: C→T at bp 581; Pro→Leu	Probably in protease	Viral particles formed but pVI, pVII, pVIII, and pTP not cleaved	87, 99, 100
?	2	*ts*103	m.u. 58.5–66.0	II or E2a or protease	No viral particles produced; complements	42, 80

					mutants in hexon, E2a, and protease; lowered amounts of penton base and fiber antigens; possibility of mutation lying in "late-acting" domain of 72K binding protein	
	5	*ts*19	m.u. 63.6–68.0	72K or 100K	Accumulates incomplete particles; fails to induce interferon in CEF; complements mutants in 100K; "late-acting" domain of 72K?	93, 83, 84, 101, 102
L4	5	*ts*1[o]	m.u. 70.0–71.4	100K	Abnormal distribution of hexon antigen	93, 39, 84
					No particles formed	93, 39, 41, 96
		*ts*115[p]	m.u. 70–75.9	100K	Hexon monomers do not assemble into trimers; hexon and 100K polypeptides unstable at 39.5°C; no viral particles formed	44, 103, 104
		*ts*116	m.u. 59.5–70.0	100K	Same complementation group as *ts*115; phenotype similar	44, 103, 104
	2	*ts*118	m.u. 70.0–79.9	100K	Fails to complement H5*ts*1; 100K polypeptide thermolabile; no particles formed	42, 80
		*ts*48	m.u. 68.0–79.9	100K or 33K or pVIII	Fails to complement wide range of *ts* mutants from several complementation groups; no particles formed	105, 88
	2⁺ND1	*ts*4	m.u. 68.0–70.7	100K	Fails to complement H5*ts*1 and H5*ts*115	106, 57, 103, 104
		*ts*356	m.u. 58.5–70.7		Fails to complement *ts*4	106, 57
	5[q]					
	5	R1[r]	Missing *Eco*RI site at 75.9 m.u. Net insert of three extra base pairs	pVIII	Virion is thermolabile; poor yield at 39°C, but plaques visible, perhaps because of genetic background (RI[r] is derived from *sub*304)	69, 107

(*Continued*)

TABLE II. (Continued)

Region[b]	Serotype[c]	Mutant[d]	Coordinates[e]	Polypeptide(s) mutated[f]	Phenotype[g]	Ref. Nos.[h]
?	5	*ts*24	m.u. 77.9–81.0	?	A late mutant; virion is unstable to freezing and thawing	39, 41, 83, 84
L5	5	*ts*5[r]	m.u. 89.0–89.1	IV	Absence of fiber antigen; accumulation of empty particles	93, 39, 83, 84
		*ts*142	Maps very close to *ts*9	IV	Fiber antigen absent; fiber polypeptide not glycosylated, but 6 S fibers form; particles containing fragments of DNA from left-hand end of the genome are found	34, 108, 109
	2[s]	*ts*104	m.u. 80.6–89.7	IV	Deficient in fiber penton base and IIIa antigen; empty and intermediate particles formed	42, 80, 110
		*ts*125	m.u. 80.6–100.0	IV	Deficiency in fiber and IIIa antigens; some mature virions made	42, 80, 110
		*ts*136	m.u. 79.9–100.0	IV	Absence of fiber, penton base, and IIIa antigens; no particles formed	110, 80

[a] The authors welcome comments concerning organization and content. We would be particularly appreciative of corrections of errors of commission and omission. There are many *ts* mutations of late gene functions in Ad2 and Ad5 that have not been mapped and fully characterized. Readers interested in following up the isolation and preliminary characterization of these mutants are referred to Kathmann *et al.* (1976) and Minekawa *et al.* (1976).

[b] Regions are the transcriptional units currently classified in subgroup C adenovirus. Details are in Chow *et al.* (1977), Broker and Chow (1979), and Miller *et al.* (1980).

[c] Serotype refers to Ad2, Ad5, and Ad2^{+}ND1 or recombinants between Ad2 and Ad5 (2/5).

[d] Mutants are referred to by the nomenclature proposed by Ginsberg *et al.* (1973).

[e] In region E1, coordinates in base pairs (bp) are from the sequences published by van Ormondt *et al.* (1978), Bos *et al.* (1981), and Gingeras *et al.* (1982). In E2, coordinates are from the sequence published by Kruijer *et al.* (1981). In E4, coordinates are from Hérissé *et al.* (1981). Coordinates measured in map units (m.u.) refer to restriction sites, using the coordinates published in Tooze (1981), or are deductions from heteroduplex studies or from changes in mobility of specific fragments during gel electrophoresis, in the case of deletion and substitution mutations. Mapping techniques include intertypic recombination, marker rescue and deletion mapping, and, in a few cases, DNA sequence analysis. (Δ) Deletion.

[f] In most cases, assignment of mutations to specific polypeptide(s) is inferential. If a mutation maps to a specific region encoding a known polypeptide, this inference is strong. In some cases, aspects of gross phenotype and complementation data help to eliminate certain candidate polypeptides. (a.a.) Amino acid(s).

[g] This is necessarily a brief summary of often complex phenotypic characteristics. The class of mutant is usually deducible from its notation, e.g., *ts*, *hr*. (BRK) Baby rat kidney; (CEF) chicken embryo fibroblasts; (CREF) cloned rat embryo fibroblasts; (m.o.i.) multiplicity of infection; (RE) rat embryo.

[h] References (of which the first cited describes the first isolation and the others deal with genetic and phenotypic characterization): (1) Harrison *et al.* (1977); (2)

Ricciardi *et al.* (1981); (3) Esche *et al.* (1980); (4) Berk *et al.* (1979); (5) Ho *et al.* (1982); (6) Babiss *et al.* (1983a); (7) Galos *et al.* (1980); (8) Jones and Shenk (1979a); (9) Shenk *et al.* (1979); (10) Jones and Shenk (1979b); (11) Carlock and Jones (1981a); (12) Carlock and Jones (1981b); (13) Solnick (1981a); (14) Solnick and Anderson (1982); (15) Babiss *et al.* (1984); (16) Stow (1981); (17) Montell *et al.* (1982); (18) Osborne *et al.* (1982); (19) Montell *et al.* (1983); (20) Ross *et al.* (1980); (21) Babiss, Fisher, and Ginsberg (unpublished data); (22) Grodzicker and Klessig (1980); (23) Brusca and Chinnadurai (1981); (24) Graham *et al.* (1978); (25) Lassam *et al.* (1978); (26) Lassam *et al.* (1979); (27) Babiss and Ginsberg (1984); (28) Chinnadurai *et al.* (1979); (29) Chinnadurai (1983); (30) Shiroki *et al.* (1981); (31) Mak and Mak (1983); (32) Lai Fatt and Mak (1982); (33) Colby and Shenk (1981); (34) Ginsberg *et al.* (1974); (35) Carter and Ginsberg (1976); (36) Galos *et al.* (1979); (37) Schutzbank (1980); (38) Stillman *et al.* (1982a); (39) Russell *et al.* (1972); (40) Wilkie *et al.* (1973); (41) Williams *et al.* (1974); (42) Martin *et al.* (1978); (43) Chen and Ginsberg (unpublished data); (44) Ensinger and Ginsberg (1972); (45) Kruijer *et al.* (1981); (46) Horwitz (1978); (47) Kaplan *et al.* (1979); (48) van der Vliet and Sussenbach (1975); (49) van der Vliet *et al.* (1975); (50) van der Vliet *et al.* (1977); (51) Carter and Blanton (1978); (52) Nevins and Winkler (1980); (53) Babich and Nevins (1981); (54) Mayer and Ginsberg (1977); (55) Ginsberg and Young (1977); (56) Kruijer *et al.* (1982); (57) Grodzicker *et al.* (1977); (58) Kathmann *et al.* (1976); (59) Klessig (1977); (60) Klessig and Grodzicker (1979); (61) Klessig and Quinlan (1982); (62) Anderson (1981); (63) Anderson *et al.* (1983); (64) Nicolas *et al.* (1981); (65) Kruijer *et al.* (1983); (66) Nicolas *et al.* (1982); (67) Carter *et al.* (1982); (68) Logan *et al.* (1981); (69) Jones and Shenk (1978); (70) Volkert and Young (unpublished data); (71) Thimmappaya *et al.* (1982); (72) Rossini *et al.* (1981); (73) Challberg and Ketner (1981); (74) Sarnow *et al.* (1982); (75) Thimmappaya *et al.* (1979); (76) Mathews and Grodzicker (1981); (77) Kapoor and Chinnadurai (1981); (78) Rajagopalan and Chinnadurai (1981); (79) Winnacker (personal communication); (80) D'Halluin *et al.* (1982); (81) D'Halluin *et al.* (1979); (82) Minekawa *et al.* (1976); (83) Edvardsson *et al.* (1978); (84) Galos and Williams (1980); (85) Boudin *et al.* (1980); (86) D'Halluin *et al.* (1978); (87) Bégin and Weber (1975); (88) Hassell and Weber (1978); (89) Weber *et al.* (1975); (90) Khittoo and Weber (1977); (91) Weber *et al.* (1977); (92) Weber and Hassell (1979); (93) Williams *et al.* (1971); (94) Russell *et al.* (1974); (95) Luciw (1977); (96) Williams (personal communication); (97) Volkert and Young (1983); (98) Kauffman and Ginsberg (1975); (99) Weber (1976); (100) Yeh-Kai *et al.* (1983); (101) Ustaçelebi and Williams (1972); (102) Tarodi *et al.* (1979); (103) Oosterom-Dragon (1980); (104) Oosterom-Dragon and Ginsberg (1981); (105) Carstens *et al.* (1979); (106) Sambrook *et al.* (1975); (107) Liu *et al.* (1984); (108) Chee-Sheung (1978); (109) Chee-Sheung and Ginsberg (1982); (110) D'Halluin *et al.* (1980).

[i] All these *ts* mutants fall into a single complementation group and thus are expected to map in the DNA polymerase gene as does H5*ts*149.

[j] Several other revertants have been isolated and have one or more phenotypic characteristics that indicate a second site reversion (Nicolas *et al.*, 1981; Carter *et al.*, 1982).

[k] E3 is the site of variable insertion of the SV40 T-antigen gene and deletion of variable amounts of Ad2 sequences. Some ND derivatives have extended host range, and mutants with restricted host range have been isolated from them (for a review, see Tooze, 1981).

[l] Seven other mutants, H2*ts*102, 106, 108, 110, 117, 122, and 123, fall into overlapping complementation groups with H2*ts*121 and have absent or decreased amounts of hexon antigen (Martin *et al.*, 1978).

[m] Information is available on two other mutants, H5*ts*10 and 14; they and four others, H5*ts*42, 46, 52, and #465, fall into the same complementation group as H5*ts*2 (see footnote *h* references 93, 39, 41, and 94–96).

[n] Information is available on three other mutants, H5*ts*128, 135, and 138; all have defects in some aspect of hexon phenotype and show low or no complementation with H5*ts*147 (Luciw, 1977).

[o] Information is available for H5*ts*7, 16, and 30; (see footnote *h* references 93, 39, 41, and 96).

[p] Four other mutations of this complementation group are available (Ginsberg *et al.*, 1974).

[q] Several other *ts* mutations, in complementation groups other than that in which H5*ts*1 lies, map by marker rescue to the 100K gene. Several have been examined phenotypically and display abnormal hexon antigen localization. The mutants include H5*ts*3, 4, 12, 17, 18, 20, and 50 (see footnote *h* references 93, 39, 84, 96, and 102).

[r] Five other mutants, H5*ts*9, 13, 22, 39, and 59, fall into three overlapping complementation groups; those that have been mapped fall between m.u. 89.0 and 91.9. Where tested, the phenotype is similar to that of H5*ts*5 (see footnote *h* references 93, 39, 84, and 96).

[s] Two other mutants, H2*ts*115 and 116, fall into a set of overlapping complementation groups with H2*ts*104, 125, and 136 and almost certainly map in fiber (Martin *et al.*, 1978; D'Halluin *et al.*, 1980).

base pairs in the splice donor site for the E1a 12 S mRNA and a stop codon at that point in the 13 S mRNA. The reconstructed virus was replication- and transformation-defective (Solnick and Anderson, 1982). Other base-pair changes were present downstream, but were shown to be of little, if any, significance to the phenotype. Their removal by appropriate manipulation *in vitro* caused no change in phenotype. None of this complexity in base-pair alteration would have been revealed without sequence analysis.

Section IV.A and IV.B describe the strategies adopted for mapping mutations isolated following random mutagenesis. In Section IV.B also, the methods for confirmatory mapping of site- and region-directed mutagenesis are discussed.

A. Viral Crosses as Mapping Tools

1. Standard Two-Factor Crosses

A previous review presented data obtained in standard two-factor viral crosses using temperature-sensitive (*ts*) mutants of Ad5 (Ginsberg and Young, 1977). The genetic map that was presented has been confirmed in most particulars by subsequent physical mapping techniques. The predictive value of this simple genetic approach has improved as definite locations for certain markers have been obtained. Thus, Harrison *et al.*, (1977), knowing the approximate positions of *ts* mutations in hexon, 100K, and fiber genes, could suggest that their newly isolated *hr* mutations lay in the conventional left end of the genome. This prediction was subsequently confirmed by marker-rescue and sequencing methods (Galos *et al.*, 1980; Ricciardi *et al.*, 1981). Similarly, D'Halluin *et al.* (1982) could orient, indirectly, their extensive two-factor map of H2*ts* mutations with the corresponding Ad5 map, since cross-complementations had identified corresponding alleles in the fiber gene, located at the conventional right end. In general, the principle that close genetic linkage implies close physical location has been proved, and thus, as a first step, previously unmapped mutations can frequently be located with confidence by a set of two-factor crosses.

Two-factor data have another potential use in the construction of allele maps for a specific gene. In general, where alleles of a single complementation group have been crossed, the recombination frequencies have yielded linear unambiguous maps (Luciw, 1977; Chee-Sheung, 1978). An advantage of this technique over marker rescue is that to resolve closely spaced alleles by the latter method requires that a convenient restriction enzyme cleavage site exist between them, a situation that does not obtain in all cases.

2. Multifactor Crosses

The reliability of conventional genetic mapping increases with the number of genetic markers that can be scored among the progeny. Sets of three-factor crosses, in which an unselected third marker is scored among selected progeny, can yield an unambiguous order for the mutations. However, this technique, which has been widely used in organisms from *Drosophila* to phage, has proved to be of limited usefulness with adenoviruses because selected recombinants show a very high rate of supernumerary crossovers. In formal genetic parlance, the mutations have proved to be unlinked. Examples of this are to be found in papers employing *hr* (Williams *et al.*, 1975a; Sambrook *et al.*, 1975) and heat-stable (Young and Williams, 1975) unselected markers.

This high frequency of supernumerary crossovers, however, has proved to be a signal advantage in the technique of heterotypic recombination developed initially to map *ts* mutations in Ad5 and Ad2+ND1 viruses (Grodzicker *et al.*, 1974). The discovery that closely related adenovirus serotypes (e.g., Ad2 and Ad5) had distinguishing restriction endonuclease cleavage sites, and that they could recombine, allowed the use of these sites as genetic markers of precise physical location. Selected ts^+ recombinants could be scored for the presence of sites peculiar to one or the other parent, and the positions of intermolecular crossovers could be deduced. More important, for the purposes of mapping, regions of the selected ts^+ recombinant genome could be identified as coming from one or the other parent and therefore could not contain the *ts* lesion of that parent. With a set of ts^+ recombinants, all containing different supernumerary crossovers, the location of the *ts* mutations entering the crosses could be deduced, sometimes with great precision. Thus, Ad2+ND1*ts*4 was located between the H5-specific *Xho*I site at 70.0 m.u. and the Ad2+ND1-specific *Eco*RI site at 70.7 m.u. (Grodzicker *et al.*, 1977). Note that if supernumerary crossovers were rare, many more ts^+ isolates would have to be screened to obtain the same degree of precision of physical location. Heterotypic recombination has been employed by several groups to map *ts* mutations of Ad5 and Ad2 (results summarized in Table II).

3. Deletion Mapping

As mentioned previously, the development of cell lines constitutively expressing the E1 regions of the viral genome has allowed the isolation or construction of mutant viruses with nonconditional lesions in E1. Some of these mutants contain deletions and are suitable for "deletion mapping" by the techniques exemplified by those of Benzer (1961) in the *rII* locus of bacteriophage T_4. Briefly, point mutations in *rII* were crossed with a set of overlapping deletion mutations. If a point mutation lay within the bounds of a given deletion, no recombinants were formed.

Depending on the degree of overlap of the deletions employed, quite precise locations could be deduced for various point mutations.

The adenovirus E1 deletions have been characterized by restriction enzyme analysis and recently by direct sequence analysis (Shenk *et al.*, 1979; Colby and Shenk, 1981), so that their precise genomic locations are known. Using several of these deletions, Galos *et al.* (1980) were able to map two different *hr* mutations, one located in E1a, the other in E1b. The positions of the mutations were confirmed by marker rescue (Galos *et al.*, 1980). It is worth pointing out that other mutations isolated following random mutagenesis schemes and having phenotypes suggestive of early functions could also be mapped relative to this set of deletions. Examples might include H2*ts*111 and H2*ts*206, the map coordinates of which are not known with any certainty (D'Halluin *et al.*, 1982 and personal communication; Winnacker, personal communication). The extension of deletion mapping to other areas of the genome awaits new complementing cell lines capable of propagating the deletion-containing, defective viruses.

B. Mapping Methods Based on DNA-Mediated Transfection

1. Marker Rescue

The predictive value of two-factor mapping from viral crosses involves the assumptions that the outcome of recombination in adenovirus is similar to that of other organisms, so that the frequencies of recombination can be related to reproducible genetic distances, and that these distances in turn reflect physical distance. Similarly, the heterotypic method assumes that closely spaced double crossovers, placed between distinguishing restriction markers that would escape detection, are infrequent. This does not seem to have presented problems in practice, since the method has been internally consistent from cross to cross and within a particular set of ts^+ recombinants and the locations deduced for specific mutations have been confirmed subsequently in every instance. Deletion mapping has limitations based on the paucity of complementable deletions currently available from regions of the genome other than E1.

Thus, each technique has a residue of uncertainty, although this uncertainty can usually be resolved by using several techniques. Accordingly, a direct physical proof of location is desirable. For this reason, and following the precedent in $\phi \times 174$ and SV40, Arrand (1978) and Frost and Williams (1978) developed a system of "marker rescue" to define exact physical boundaries for several *ts* and *hr* mutations. In essence, this involves the cotransfection of appropriate cells with full-length mutant genomes and single purified restriction fragments corresponding to various regions of the genome. Provided the mutant phenotype arises from a single DNA lesion, one fragment from a set produced by a particular restriction enzyme should be capable of transferring the ts^+ allele to the recipient mutant genome.

The technique for accomplishing the mapping has varied from laboratory to laboratory. For instance, the recipient cell can be 293 or HeLa, the full-length DNA can be purified or prepared as a "DNA–protein complex" (Robinson *et al.*, 1973; Sharp *et al.*, 1976), and the assay itself can be performed as a direct selection under restrictive conditions or as a subsequent assay of yields prepared under either restrictive or permissive conditions. The choice depends on the reproducibility of direct selection assays in a particular laboratory. Yield analysis has the advantage that it can be repeated should an assay fail.

The source of the rescuing fragments can be from restricted, gel-purified viral DNA or from material cloned in a plasmid or phage vector. The latter sources have the advantage of being free of minor cross-contamination with other adenovirus-specific sequences. However, the use of cloned viral DNA sequences also has the disadvantages inherent in cloning in heterologous hosts, such as the possible loss of viral sequences.

In several cases, it has been noticed that when a mutation is located close to a restriction site being used as the boundary of a particular fragment, the rescuing efficiency declines (Oosterom-Dragon, 1980; Volkert, 1983). Furthermore, protecting the end of the fragment with heterologous bacterial DNA enhances the rescuing efficiency of the fragment (Volkert, 1983). These observations imply that there is an exonuclease activity degrading the wild-type (Wt) allele before it can enter into a rescuing reaction. In theory, this loss of terminally positioned sequences may limit the resolving power of marker rescue, unless the sequences are "buffered" by heterologous DNA. In practice, the technique has been sufficiently sensitive to place H5*ts*5 between an *Hpa*I site at 89.0 and an *Hin*dIII site at 89.1 m.u. (Galos and Williams, personal communication) (since Ad5 has not been sequenced in this region, the precise number of nucleotides that separate the two sites is not known, but it is probably less than 400, assuming a 0.5% maximum error in the estimates of the sites' positions).

Marker rescue continues to be the method of choice for mapping *ts* and *hr* mutations that have been isolated following random mutagenesis. The optimum strategy would appear to be that a set of fragments from an enzyme making few cuts in the genome is used to locate the approximate coordinates of a particular mutation. Location of the approximate mutation site is followed by detailed mapping using fragments, derived with different restriction enzymes, that overlap by a few hundred nucleotides in the region of a specific mutation. This procedure was used in the mapping of H5*ts*5 outlined above. With the ability to map mutations to small regions of the genome using marker rescue, strategies to detect the specific base-pair change(s) responsible for the phenotype (by cloning and DNA sequencing) are much clearer and involve less effort.

2. Overlap Recombination as a Mapping Tool

When cells are transfected with terminal overlapping DNA fragments, recombination within the overlap leads to the production of viable virus (Chinnadurai *et al.*, 1979; Rajagopalan and Chinnadurai, 1981; Kapoor and Chinnadurai, 1981; Ho *et al.*, 1982; Volkert and Young, 1983). With the development of detailed restriction maps, such overlaps can be placed from the left end 2.6 m.u. to at least 84.7 m.u. and can be of variable sizes. A list of overlaps currently available in some group C adenoviruses is given in Table III. As mentioned earlier, such overlaps are valuable for the reconstruction of complete genomes from segments manipulated *in vitro*. In addition, overlap recombination can be exploited as a mapping tool.

A marker contained within one or another terminal fragment but outside the region of overlap is always inherited, while markers within the overlap are inherited in a distance- and orientation-dependent manner (Volkert and Young, 1983). This is of particular value for closely spaced markers, the approximate positions of which are known, since it allows the order and physical distance separating the markers to be deduced (Volkert and Young, 1983; Nicolas *et al.*, 1983). Marker-rescue techniques do not address the question of physical distance, and determining order depends on the presence of a convenient restriction site between the markers. Using overlap recombination, it has been possible to segregate a *ts* mutation in the 72K DNA-binding protein (DBP), H5*ts*107, from an accompanying reversion mutation present in a ts^+ revertant. It has been possible to predict an order, left-*ts*107-reversion-right (Nicolas *et al.*, 1983), in complete agreement with recent sequencing data (Kruijer *et al.*, 1983). The sequence shows that no convenient six-base restriction enzyme site occurs between the two markers, so segregating the two markers by marker rescue would have been extremely difficult. Thus, overlap recombination should be of considerable value in ordering both alleles and closely spaced markers.

Provided the enzymes used cut a sufficient number of times outside the overlap to reduce infectivity to negligible proportions (some 10^4- or 10^5-fold reduction in transfection yields is a good benchmark), overlap techniques can be performed with unpurified cleavage mixtures of DNA–protein complexes.

3. Sequencing of Mutant Genomes

In the case of random mutagenesis, the site at which mutation has occurred to give rise to the mutant phenotype can only be inferred from the recombinational studies discussed and from complementation. To determine the exact site requires DNA sequence analysis by the methods developed by Maxam and Gilbert (1977) and Sanger *et al.* (1977). Furthermore, since random mutagenesis has the potential of generating si-

TABLE III. Terminal Fragments of Adenovirus Types 2 and 5 and Their Variants Suitable for Overlap Recombination in Strain Construction and Mapping

Fragments	Restriction endonuclease	Limitations[b]
Left-terminal fragments		
Rightward coordinate[a]		
2.8	*Sma*I	—
3.7	*Xba*I	—
4.3	*Hpa*I	—
4.8	*Sac*I	—
5.6	*Kpn*I	—
7.7	*Hind*III	—
9.1	*Bgl*II	—
10.0	*Sac*I	Δ of *Sac*I at 4.8
10.3	*Sph*I	—
10.8	*Sma*I	Δ of *Sma*I at 2.8
11.3	*Bcl*I	Ad2 only
15.4	*Sac*I	*dl*313 only
15.8	*Xho*I	—
17.1	*Hind*III	Δ of *Hind*III at 7.7
17.2	*Pvu*I	—
23.3	*Kpn*I	Δ of *Kpn*I at 5.6
24.4	*Bgl*II	*dl*1313 only
25.4	*Hpa*I	Δ of *Hpa*I at 4.3
25.9	*Sal*I	—
29.3	*Bam*HI	Ad2 only
33.5	*Bcl*I	Ad5 only
42.0	*Bam*HI	Ad5–Ad2 recombinant
58.5	*Eco*RI	Ad2 only
59.5	*Bam*HI	Ad5 only
76.0	*Eco*RI	Ad5 only
78.5	*Xba*I	*dl*313 – Ad5 recombinant
84.7	*Xba*I	310 – Ad2 recombinant
Right-terminal fragments		
Leftward coordinate		
2.6	*Cla*I	Ad5 only
3.7	*Xba*I	*dl*309
29.0	*Xba*I	*dl*309 – WT recombinant
41.9	*Bcl*I	Ad5 only
45.9	*Sal*I	—
51.2	*Cla*I	Ad2 only
59.5	*Bam*HI	—
65.2	*Pvu*I	—
66.0	*Xho*I	$Ad2^+$ND1 only
70.0	*Xho*I	$Ad2^+$ND1 – Ad5 recombinant
76.0	*Eco*RI	*sub*304 only
78.5	*Xba*I	*sub*304 only
82.8	*Xho*I	—
83.5	*Eco*RI	Ad5 only
84.0	*Sac*I	Ad5 only

[a] Coordinates are modified from Tooze (1981), and most are listed in the New England Biolabs catalog; others are taken from D'Halluin *et al.* (1983).

[b] Unless otherwise stated, the fragment is found in both Ad2 and Ad5. Variants that contain deletions (Δ) of restriction sites are not specified unless only a single variant contains such a deletion (e.g., *dl*313).

lent or modifying mutations in addition to the primary change, the ideal protocol is to sequence the region containing the Wt, the mutant allele, and a set of revertants with Wt phenotype. For adenoviruses, this has been performed to date only for the 72K DBP gene. Sussenbach and colleagues have determined the site of mutation in H5*ts*125, in H5*ts*107 (identical to H5*ts*125 down to the specific base-pair change), and in revertants with both same-site and second-site changes (Kruijer *et al.*, 1983). As mentioned above, one of the second-site changes can be segregated from the *ts* lesion by overlap recombination, revealing the *ts* phenotype.

Other mutations that have been sequenced include H5*hr*1, the location of which was deduced first from deletion and marker-rescue mapping (Galos *et al.*, 1980), then from the size of the polypeptide fragment specified by the mutant, which indicated a nonsense codon either in frame or generated by an upstream frameshift (Esche *et al.*, 1980), and finally by direct sequence analysis of the suspected region (Ricciardi *et al.*, 1981). As the latter authors point out, it is formally possible that some aspects of the phenotype may be determined by mutations upstream from the region sequenced. This now seems unlikely, since other deletions made by site-directed mutagenesis in the same region show a similar range of replicative and transformation phenotypes (Babiss *et al.*, 1984). As mentioned earlier, H5*hr*440, a mutant isolated following region-directed nitrous acid mutagenesis of the region 0–4 m.u., when sequenced, showed a number of base-pair changes, some of which had phenotypic consequences while others were silent.

V. GENETIC INTERACTIONS

A. Interactions between Genetically Distinguishable Viruses

Early in the studies of animal virus genetics, it was established that cells in culture could be infected with genetically differing viruses. These studies allowed the investigation of direct genetic interactions between genomes (i.e., recombination) and of functional interactions between virus-specified products (i.e., complementation and other *trans*-operating tests). Initially, complementation and recombination were exploited as methods of mutant classification, and indeed they continue to perform this function both for temperature-sensitive (*ts*) mutants (Martin *et al.*, 1978) and for host-range (*hr*) mutants (Harrison *et al.*, 1977; Ho *et al.*, 1982). Recently, genetic interactions have been used in more experimental ways: (1) to try to uncover whether or not a mutation lies in a catalytic polypeptide; (2) to determine the conservation of functions during the evolution of the human adenovirus group as a whole, in conjunction with sequence comparisons between serotypes; (3) to determine whether a phenotype derives from a mutated *cis*-acting element; (4) to detect and manipulate helper–defective coinfections; and (5) to discover the mechanisms of recombination within the infected cell. In addition, as new methods for manipulating genomes and for introducing genomes into cells have become available, it has been possible to examine interactions between whole and partial genomes and between incoming genomes and resident adenovirus genes. This section is concerned primarily with the experimental use of genetic interactions between viral genomes, rather than with the well-established classification methods summarized previously (Ginsberg and Young, 1977).

1. Dominance Interactions

Snustad (1968) established a genetic method for determining whether or not a nonsense mutation lay in a gene the product of which was required in catalytic or stoichiometric amounts. Briefly, the test entailed coinfecting a nonsuppressing host with a low, constant multiplicity of a wild-type (Wt) T_4 bacteriophage and varying amounts of the nonsense mutant to be tested. Burst size was measured, and, for a series of mutants known to lie in structural components of the virus, the yield of virus declined with increasing multiplicity of mutant input. This was to be expected for a product the absolute concentration of which was closely regulated to be in stoichiometric relations to those of other structural components. Conversely, mutations in genes encoding enzymatic functions showed little if any *trans* effects on the yield of Wt virus. Using these observations, Snustad suggested that T_4 gene 32 product was required in a noncatalytic role during T_4 DNA replication, a proposal subsequently verified (Alberts and Frey, 1970). The suggestion arose from its behavior in mixed infections, which, unlike that of other DNA replication mutants, showed a decline in Wt yields.

Several investigators have applied the same principles to adenovirus *ts* mutants (Galos *et al.*, 1979; Carstens *et al.*, 1979; Plaat and Weber, 1979; D'Halluin *et al.*, personal communication) and in general have found that such mutants have considerable *trans* effects, reducing yields of Wt some 2 orders of magnitude when the *ts*/Wt input ratio reaches 30:1. Of particular interest are the results with various *ts* mutants that affect DNA synthesis. Galos *et al.* (1979) showed clearly that H5*ts*125, defective in the 72K DNA-binding protein (DBP), behaved in a stoichiometric fashion, while H5*ts*149, in a region that at the time had no proven function, behaved catalytically. As mentioned in Section III.B, it has been shown that *ts*149 lies in an adenovirus-encoded DNA polymerase. In essence, the Snustad test showed the likelihood of a catalytic function in a DNA mutant by the lack of a *trans* dominant effect in coinfection with Wt virus. The principle of a catalytic type interaction in adenovirus *ts* mutants having been established, it would be profitable to screen other *ts* mutations of unknown function. Interestingly, mutations known to lie in the nonstructural 100K polypeptide have been shown to have strong dominant effects (Carstens *et al.*, 1979; Oosterom-Dragon, 1980; Williams, Young, and Austin, unpublished findings), suggesting a stoichiometric requirement for this gene product, consistent with its high concentration in the infected cell and its apparent function (Oosterom-Dragon and Ginsberg, 1981; Cepko and Sharp, 1982). H2*ts*111, a mutant defective in DNA replication, and mapping to the left of 29.0 m.u. by heterotypic recombination (D'Halluin *et al.*, 1982), has a strong dominant effect (D'Halluin *et al.*, personal communication). The exact location of this mutation is awaited with interest.

2. Complementation Tests between Viruses from Widely Diverged Serotypes

Cross-complementation between *ts* mutants from closely related serotypes has been observed repeatedly, even if the mutations lie in polypeptides of considerable amino acid divergence such as hexon and fiber. In some cases, it has been established that the capsids that emerge from such crosses showed phenotypic mixing (Mautner *et al.*, 1975), and in heterotypic recombinants of course, various serotype combinations of hexon, penton base, and fiber are viable. This suggests that at least within a subgroup, functional and structural equivalence exists among the various capsid polypeptides. Questions that arise from these observations are whether or not functional equivalence can cross subgroup boundaries and whether early functions are similarly catholic with regard to the genomes with which they interact. Early data showed that functions from Ad12 could substitute for several Ad5*ts* functions (Williams *et al.*, 1975b), while others were equivocal owing to considerable leakiness of the *ts* mutants under the conditions used. Subsequent work has tended to show almost complete interserotypic equivalence of early functions (Williams *et al.*, 1981; Rowe and Graham, 1981; Brusca and Chinnadurai, 1981), suggesting that despite considerable nucleic acid and amino acid sequence heterogeneity, polypeptides can interact with each other and with heterologous genomes. *In vitro* tests for DNA replication confirm that despite nucleic acid sequence differences, polypeptides from widely divergent serotypes can interact to allow replication (Stillman *et al.*, 1982b). However, both *in vitro* and *in vivo*, some cross-complementations seem to be less efficient than others (Stillman *et al.*, 1982b; McDonough and Rekosh, 1982).

3. Detection of Effects in Cis

In the classic application of complementation tests in *Drosophila* and in microorganisms, the pairs of mutants to be tested were crossed routinely both in *trans* and in *cis*. This latter test is normally a control that can be dispensed with, since most tests involve mutations in diffusible substances. In addition, constructing a set of strains that contain double *ts* mutations is time- and labor-consuming. However, for certain mutants, it is necessary to establish whether or not their effects are manifest in *cis*, and this can be established by crossing the mutant to Wt virus. Two examples the phenotypes of which do not automatically suggest either a *cis* or a *trans* effect are the extended monkey cell *hr* mutants *hr*400–404 isolated by Klessig (1977) and the splicing defect observed in the restricted *hr* mutant *hr*440 isolated by Solnick (1981a). In the former case, *hr*400 could cross-complement the replication of Ad7 in mixed infections (Klessig and Grodzicker, 1979), indicating that its effects could be mediated in *trans*. By contrast, *hr*440's defect in the processing of the 13S mRNA, occasioned by an aberrant 5′ donor splice junction, could act only in *cis*. In other words, the mutation lay in a *cis*-acting genetic element. Other such mutations have been isolated for the polyadenylation step of RNA processing (Berk *et al.*, personal communication).

4. Helper Virus and Defective Virus Mixed Infections

One of the most useful applications of the ability of viral functions to act in *trans* has been the development of complementing helper–defective mixed infection systems. These can be of the type in which both viruses mutually complement one another or of that in which one virus acts solely as a helper and is not itself defective. With the advent of DNA manipulation via cloning, much attention has been paid to the former type. For example, absolutely defective SV40 deletion mutants constructed *in vitro* can be complemented at the restrictive temperature by complementing *ts* mutants (Mertz and Berg, 1974b; reviewed by Berg, 1981). With adenoviruses, only one such system has been reported (Challberg and Ketner, 1981). Various *ts* mutants of Ad5 were used as helpers for the complementation of defective deletions encompassing parts of the L4 and L5, as well as the E3 and E4, regions. Some of the deletions were extensive enough to allow the separation of the helper and defective viruses by isopyknic centrifugation. This allowed direct proof of the defectiveness of the particular mutant, although several isolates could be surmised to contain defectives owing to the two-hit kinetics of plaquing of the virus in the original plaque. Other helper systems include those used for the propagation of novel vectors that express SV40 tumor (T) antigen under adenovirus early or late promoter control (Thummel *et al.*, 1981; Solnick, 1981b) or in which the late promoter replaces E1 sequences. Where the aim is to produce large quantities of a particular product, e.g., SV40 T antigen, the presence of the helper is no hindrance, but in many circumstances the helper must be removed. Fortunately, the packaging requirements for genome size are flexible in adenovirus, allowing genomes from at least as low as 88% to at least as high as 105% to be packaged (Thimmappaya *et al.*, 1982), and thus a judicious choice of helper-genome size can allow virions containing it to be purified away from the virus of interest by isopyknic centrifugation.

One of the problems of using helper systems in adenovirus studies is that recombination between helper and defective may be extensive, thus unlinking traits that are required to be together for purposes of purification or of expression. There are two possible ways of suppressing this recombination. One is to use a very distantly related serotype as helper. While cross-complementation is often excellent, recombination as measured by viable virus yield (Williams *et al.*, 1975b; Praszkier and Ginsberg, unpublished data) or by recombinant, intracellular viral DNA (Praszkier and Ginsberg, unpublished data) is negligible. An alternative may be to lower recombination by placing segments of heterologous DNA at the ends of molecules to inhibit recombination initiated terminally. Both physical and genetic data suggest that one recombination pathway in adenovirus is terminally initiated (Wolgemuth and Hsu, 1981; Munz and Young, submitted) and that regions of terminal heterology inhibit recombination internally (Munz and Young, submitted). There is no substantial evidence for adenovirus-specified functions essential only for recombination. If such are found, clearly they could be employed to circumvent the recombinational uncoupling of desirable traits.

5. Genetic Recombination Mechanisms

Recombination between genetically marked adenoviruses has been exploited as a mapping tool and also to construct new strains of viruses for functional studies and for the purpose of convenient genome manipulation. However, the study of adenovirus recombination also has potential for uncovering pathways of general, homology-based recombination in the cells of higher eukaryotes. The extent to which adenovirus encodes functions with roles in its own recombination is not known, but it seems likely that many of the polypeptides necessary for the initiation and resolution of recombination intermediates are supplied by the host. Thus, study of viral recombination is expected to reveal insight into cellular mechanisms.

a. Recombination in Viral Infection

Several lines of evidence suggest that adenovirus recombination can occur from some time in eclipse well into the late phase of viral replication and that individual lineages of DNA molecules can experience several rounds of recombination before being encapsidated. Hence, the kinetics of the production of Wt recombinants in *ts* × *ts* crosses show a significant increase in recombinant frequency from the end of eclipse until midexponential phase, and this increase is accompanied by a progressive rearrangement of unselected restriction endonuclease site differences (Young and Silverstein, 1980). Similarly, triparental crosses reveal the production of molecules with contributions from all three parents (Munz *et al.*, 1983), a necessary corollary to the hypothesis that multiple rounds of recombination can take place in the genome pool. Electron micrographs of adenovirus DNA molecules extracted from infected cells at midexponential phase, a period during which much recombination is presumed from the genetic data to be taking place, show that molecules consistent with recombination-intermediate structures are rare (Wolgemuth and Hsu, 1980, 1981). Given the high frequency of genetic recombinants in the final progeny yield, and the likelihood that recombination intermediates are not short-lived, the implication would be that the final recombinant frequency is achieved by the accumulation of many individual events over an extended period.

Adenovirus recombination is almost certainly homology-based. This is suggested by studies on intracellular molecules. Thus, electron micrographs of adenovirus DNA extracted from infected cells show typical "Holliday" structures with equal-length arms on opposite sides of the "chi," indicating base-pairing at equivalent position on each molecule (Wolgemuth and Hsu, 1980). Similarly, in restriction endonuclease analysis of intracellular DNA, diagnostic recombinant bands appear at high molarity (Young and Silverstein, 1980). While neither of these studies proves perfect, in register, base-pairing of the interacting parental molecules, the possibility that the final recombinant products are merely viable subsets of random nonhomologous exchanges seems remote. This conclusion is supported by the observation that recombination between heterologous serotypes is undetectable either within the pool of intracellular DNA [Ad2 × Ad7 (Praszkier and Ginsberg, unpublished findings)] or among viable progeny [Ad5 × Ad12 (Williams *et al.*, 1975b)]. Similarly, crossovers within the serotype-determining hexon genes of Ad2 and Ad5 are confined to regions of homology among viable recombinants (Boursnell and Mautner, 1981). As the authors point out, this last observation does not exclude the possibility of nonhomologous recombination leading to nonviable recombinant genomes that would escape detection and analysis.

The DNA structures that initiate recombination are not known, but several years ago it was pointed out that the single strands produced by adenovirus DNA replication could be important in the initiation reaction (Flint *et al.*, 1976). Direct physical evidence for single-strand "invasion" of duplex molecules has been provided by electron micrographs of intracellular adenovirus DNA (Wolgemuth and Hsu, 1981). Such molecules are not as common as the aforementioned Holliday structures, but the latter could be derived from the former by subsequent reciprocal strand exchange. The observation that the invasion by the donor strand is accompanied by terminal displacement of the strand of the same polarity in the recipient duplex suggests that recombination may be terminally initiated, although artifactual explanations, such as the greater stability of such forms in the extraction procedure, cannot be excluded. Recent genetic data (Munz and Young, submitted), however, strongly support the idea of terminal initiation as a major pathway in adenovirus recombination. Recombination between multiply marked adenovirus strains shows a marked bias toward either right or left termini. This is most clearly seen when a block of heterologous DNA is present in one parental terminus, for then a gradient of crossover frequencies is observed, with a maximum toward the other terminus. When both termini contain mutually heterologous sequences, recombination in internal regions is depressed, although not excluded.

The functions necessary to accomplish these reactions are not known. However, blocking early viral protein synthesis with anisomycin, a potent inhibitor, completely abolishes adenovirus recombination provided the drug is added no later than 5 h postinfection (Young and Silverstein, unpublished observations). Recombination was monitored by detecting diagnostic recombinant bands in the intracellular genome pool using Southern transfer. These data suggest a role for early viral function(s) or for labile host function(s). Attempts to determine the recombinational role of specific viral early functions are as yet inconclusive. Crosses at high-input multiplicities between suitably marked strains of *ts*125 show a marked decline in the production of diagnostic recombinant bands, with a long delay in their appearance, at the nonpermissive temperature of 40°C (Young and Silverstein, unpublished data). Two possibilities can be entertained: either the DBP has a role in recombination itself or DNA synthesis is a prerequisite for the major recombination pathway. The use of other early mutants to distinguish between these possibilities is complicated by the leakiness of most of the mutants available.

b. Recombination in DNA-Mediated Transfection

Sections IV.B.1 and IV.B.2 outlined the utility of marker rescue and overlap recombination as methods for the mapping of mutations and for strain construction. As first pointed out by Chinnadurai *et al.* (1979) for the latter phenomenon, both can be exploited to understand cellular recombination mechanisms. The genetic analysis of the events that take place in overlap recombination has shown that markers contained within the overlap are inherited in a manner dictated by their position within the overlap relative to the ends generated by restriction cleavage (Volkert and Young, 1983). In the case of one-point crosses, where a single *ts* marker of known physical location is present in the overlapping region, the inheritance of the marker and its Wt allele can be predicted from the simple geometry of the overlap, assuming that a single cross-over is necessary and sufficient to generate a full-length genome. In two-point crosses, where two *ts* markers are contained within the overlap, the generation of ts^+ recombinants is much more common in the orientation arranged so that a single crossover is sufficient to generate ts^+ than in the orientation requiring three crossovers. (It should be noted that there is a constraint on the detectable recombination events in this system, since only odd numbers of crossovers will lead to full-length genomes.). Thus, overlap behaves like a classic recombination system, and models based on recombination at the newly cleaved termini or on segregation from heteroduplex DNA the length of the overlap are unlikely. To investigate the nature of the events taking place in a single recombination, very low doses of overlapping fragment are employed, so that any cell giving rise to infectious virus has probably permitted a single left–right terminus interaction (Volkert, 1983). Markers within the overlap are scored, and the segregation suggests strongly that heteroduplex DNA is formed over intervals that included one or the other but not both markers.

The functions necessary to achieve overlap recombination are not known, but it is likely that cellular polypeptides are paramount. Circumstantial evidence suggests that adenovirus DNA replication is not a prerequisite. Overlapping fragments incapable of expressing the adenovirus-specified DNA polymerase are capable of recombining (Volkert, La Monica, and Young, unpublished observations). It also seems likely that cellular functions can be induced by prior uv irradiation of the host cell (293). Very small overlaps (0–4.5 m.u. × 3.8–100 m.u.), which yield few plaques on untreated cells, yield 20- to 100-fold more if the cells have been pretreated with UV at a dose of 5–15 J/m^2 some 16 hr before transfection (Babiss *et al.*, 1984; Babiss and Ginsberg, in preparation). There seems to be an induction period during which cellular protein synthesis is necessary. The nature of the functions induced by UV is of considerable interest. Moreover, the phenomenon of enhancement has been of considerable practical benefit in stimulating the production of plaques in strain construction using the overlap technique (Babiss and Ginsberg, unpublished observations).

Investigations into the nature of marker rescue are in their infancy. Cotransfer of unselected markers in selected marker rescues has been observed (Frost and Williams, 1978; Volkert and Young, 1983), but it has not been established conclusively that this results from concerted transfer of contiguous stretches of DNA from donor fragment to recipient duplex. Appropriate tests for genetic linkage, in which selected and unselected markers are presented either in *cis* or in *trans* on fragments of identical size, have yet to be attempted.

B. Genetic Interactions between Virus and Transformed Permissive Cells

The evidence that mammalian cells transformed by a variety of DNA tumor viruses contained and expressed viral genes led to the idea that permissive cells, once transformed, could act as novel hosts for defective viruses incapable of replicating in the untransformed parent (Benjamin, 1970). Similarly, the demonstration that permissive human embryo kidney cells could be transformed by sheared adenovirus DNA (Graham *et*

TABLE IV. Permissive Cell Lines Expressing Adenovirus Functions

Cell line	Derivation	Adenovirus sequences expressed	Properties	References
293	HEK cells transfected with sheared adenovirus DNA; visual selection of morphological transformant	E1a and E1b; no E1a or E1b 9 S mRNAs found	Complements mutants that map in E1a and E1b; efficient host for DNA transfection; useful plaquing line	Graham *et al.* (1977), Aiello *et al.* (1979)
A2 and A5	143tk$^-$ cells transfected with a gel-purified Ad2 *Eco*RI-A fragment and a plasmid containing HSV *tk* gene; selection for tk^+ and scoring for Ad sequences	E1a and E1b	Complements deletion mutants in E1a and E1b; subclones vary in complementing ability; selection for tk^- variants leads to loss of complementation ability	Grodzicker and Klessig (1980)
B1	Similar to A2, except that *Bam*HI-B fragment used	Similar to A2, but transcripts from 10.7 m.u. to at least 16.7 m.u. also observed	Similar to A2	Grodzicker and Klessig (1980)
B3, B15	Similar to B1	Not reported	Initial ability to complement lost rapidly	Grodzicker and Klessig (1980)
D10 and D27	143tk$^-$ cells transfected with a plasmid containing the HSV *tk* gene and adenovirus sequences from 0 to 7.9 m.u.	E1a sequences transcribed normally; E1b transcription initiated normally, but proceeds into flanking pBR322, *tk*, or host sequences	Complements deletion mutants in E1a *and E1b*	Klessig *et al.* (1982)

D16	Similar to D10	Not reported	Unstable for complementation	Klessig *et al.* (1982)
ABA40 and ABA46	KB cells transfected with plasmid containing *E. coli gpt* gene Ad12 *Bam*HI-A fragment; selection for gpt^+ and scoring for Ad sequences	E1a and E1b during early passage; ABA46 shows no evidence for E1b early mRNA	Complements E1a deletion mutant at early passage; Ad mRNA expression lost at high passage, even if cells maintained under selection	Shiroki *et al.* (1983)
E1A-5 and E1A-7	Similar to ABA40, except that Ad12 *Acc*I-H fragment present in plasmid	Similar to ABA40, but no E1b sequences present, since plasmid does not contain E1b sequences	Similar to ABA40, except that some E1a mRNA expression evident at late passage	Shiroki *et al.* (1983)
E1-1 and E1-7	Similar to ABA40, except that Ad12 *Eco*RI-C fragment present in plasmid	Similar to ABA40	Similar to ABA40, but complementation of E1b deletion mutants evident also; some E1a mRNA present at late passage	Shiroki *et al.* (1983)
	Five more lines, tested in part, showed complementation of an E1a deletion, and three had some E1b complementing ability.			
gpt^+6	KB cells transfected with plasmid *E. coli gpt* gene and Ad2 *Xho*I-C fragment; selection for gpt^+ and scoring for Ad sequences	E1a sequences transcribed, but mRNAs of larger size; E1b sequences transcribed, but one species of mRNA larger than normal also present	Complements mutants in E1a and E1b	Babiss *et al.* (1983b)
gpt^+8	Similar to gpt^+6	E1a mRNAs only	Complements E1a but not E1b mutants	Babiss *et al.* (1983b)
gpt^+16	Similar to gpt^+6	E1a and E1b mRNAs of correct size	Complements E1a and E1b mutants	Babiss *et al.* (1983b)
gpt^+18	Similar to gpt^+6	E1b mRNAs only (only E1b DNA sequences detectable)	Complements E1b but not E1a mutants; useful plaquing line for E1b mutants	Babiss *et al.* (1983b)

al., 1974) and that the transformed cell line expressed sequences corresponding to the E1 region of the genome (Graham *et al.*, 1977; Aiello *et al.*, 1979) allowed two groups of investigators to isolate mutants defective in E1a and E1b (Harrison *et al.*, 1977; Jones and Shenk, 1979a) by two different approaches (see earlier sections). This adenovirus-transformed cell line, called "293," has become the host of choice for many attempts to isolate mutants in E1 that have been constructed *in vitro* with defects in E1a or E1b. 293 cells also have excellent plaquing and transfection properties, important considerations when dealing with small quantities of valuable DNA.

Subsequent developments in the methods of transferring genes into cells have allowed the creation of new cell types suitable for specific functional and genetic studies. For example, it has been possible to cotransfer adenovirus genes into a human sarcoma cell line ($143tk^-$), which is deficient in thymidine kinase activity, using mixtures of herpes simplex virus (HSV) *tk* DNA and adenovirus DNA (Grodzicker and Klessig, 1980). Cotransfer is an efficient method (Wigler *et al.*, 1979), and 14% of selected tk^+ cells from one experiment were found to contain adenovirus genes expressing E1 functions. Later, the protocol was amended to use plasmid DNA in which HSV tk was covalently linked to adenovirus genes (Klessig *et al.*, 1982). Recently, the use of a dominant selectable gene, encoding xanthine guanine phosphoribosyltransferase (XGPRT) of *Escherichia coli*, has allowed the introduction, via linked transfer, of adenovirus genes into KB cells (Shiroki *et al.*, 1983; Babiss *et al.*, 1983b) and into 293 cells (Shenk *et al.*, unpublished observations).

It is probable that the potential of these and similar cell lines has not yet been fully exploited. While great strides have been made in the development of viral mutants in early region E1, other areas of the genome have not been fully expressible when integrated, and thus these areas have not proved to be amenable to mutant selection. Nevertheless, the cell types that contain restricted regions of the genome have great promise as tools for uncovering control circuits both between integrated regions and between them and incoming sequences delivered via viral infection or transfection of suitable plasmids. It remains to be seen whether adenovirus sequences present in cells, but unexpressed, can be transactivated by incoming genetic elements. The precedent set by bacteriophage λ in the uncovering of complex control circuits using combinations of integrated and extrachromosomal sequences (Thomas, 1971) suggests that similar advances may come from equivalent studies in adenovirus.

Table IV lists the permissive cell types available for such studies and their salient properties.

ACKNOWLEDGMENTS We should like to thank the many colleagues who supplied us with a wide variety of information characterizing numerous mutants. The task of completing the manuscript was greatly aided by the skill and patience of Nadirah Risbrook, to whom we extend our thanks.

REFERENCES

Aiello, L., Guilfoyle, R., Huebner, K., and Weinmann, R., 1979, Adenovirus 5 DNA sequences present and RNA sequences transcribed in transformed human embryo kidney cells (HEK-Ad-5 or 293), *Virology* **94**:460–469.

Akusjärvi, G., Zabielski, J., Perricaudet, M., and Pettersson, U., 1981, The sequence of the 3′ non-coding region of the hexon mRNA discloses a novel adenovirus gene, *Nucleic Acid. Res.* **9**:1–17.

Alberts, B.M., and Frey, L., 1970, T_4 bacteriophage gene 32: A structural protein in the replication and recombination of DNA, *Nature (London)* **227**:1313–1318.

Aleström, P., Akusjärvi, G., Pettersson, M., and Pettersson, U., 1982, DNA sequence analysis of the region encoding the terminal protein and the hypothetical N-gene product of adenovirus type 2, *J. Biol. Chem.* **257**:13492–13498.

Anderson, C.W., 1981, Spontaneous mutants of the adenovirus–simian virus 40 hybrid, $Ad2^+ND3$, that grow efficiently in monkey cells, *Virology* **111**:263–269.

Anderson, C.W., Hardy, M.M., Dunn, J.J., and Klessig, D.F., 1983. Independent, spontaneous mutants of adenovirus type 2-simian virus 40 hybrid $Ad2^+ND3$ that grow efficiently in monkey cells possess identical mutations in the adenovirus type 2 DNA-binding protein gene, *J. Virol.* **48**:31–39.

Arrand, J.E., 1978, Mapping of adenovirus type 5 temperature-sensitive mutations by marker rescue in enhanced double DNA infections, *J. Gen. Virol.* **41**:573–586.

Babich, A., and Nevins, J.R., 1981, The stability of early adenovirus mRNA is controlled by the viral 72kd DNA-binding protein, *Cell* **26**:371–379.

Babiss, L.E., and Ginsberg, H.S., 1984, Adenovirus type 5 early region 1b gene product is required for efficient shut-off of host protein synthesis, *J. Virol.* (in press).

Babiss, L.E., Ginsberg, H.S., and Fisher, P.B., 1983a, Cold-sensitive expression of transformation by a host range mutant of type 5 adenovirus, *Proc. Natl. Acad. Sci. U.S.A.* **80**:1352–1356.

Babiss, L.E., Young, C.S.H., Fisher, P.B., and Ginsberg, H.S., 1983b, Expression of adenovirus E1a and E1b gene products and the *Escherichia coli* XGPRT gene in KB cells, *J. Virol.* **46**:454–465.

Babiss, L.E., Fisher, P.B., and Ginsberg, H.S., 1984, Deletion and insertion mutations in early region 1a of type 5 adenovirus producing cold-sensitive or defective phenotypes for transformation, *J. Virol.* (in press).

Bégin, M., and Weber, J., 1975, Genetic analysis of adenovirus type 2. I. Isolation and genetic characterization of temperature-sensitive mutants, *J. Virol.* **15**:1–7.

Benjamin, T.L., 1970, Host range mutants of polyoma virus, *Proc. Natl. Acad. Sci. U.S.A.* **67**:394–399.

Benzer, S., 1961, On the topography of the genetic fine structure, *Proc. Natl. Acad. Sci. U.S.A.* **47**:403–415.

Berg, P., 1981, Dissections and reconstructions of genes and chromosomes, *Science* **213**:296–303.

Berk, A.J., Lee, F., Harrison, T., Williams, J., and Sharp, P.A., 1979, Pre-early adenovirus 5 gene product regulates synthesis of early viral messenger RNAs, *Cell* **17**:935–944.

Bos, J.L., Polder, L.J., Bernards, R., Schrier, P.I., van den Elsen, P.J., van der Eb, A.J., and van Ormondt, H., 1981, The 2.2 kb E1B mRNA of human Ad12 and Ad5 codes for two tumor antigens starting at different AUG triplets, *Cell* **27**:121–131.

Boudin, M.-L., D'Halluin, J.C., Cousin, C., and Boulanger, P., 1980, Human adenovirus type 2 protein IIIA: Maturation and encapsidation, *Virology* **101**:144–156.

Boursnell, M.E.G., and Mautner, V., 1981, Recombination in adenovirus: Cross-over sites in intertypic recombinants are located in regions of homology, *Virology* **112**:198–209.

Broker, T.R., and Chow,L.T., 1979, Alternative RNA splicing patterns and the clustered transcription and splicing signals of human adenovirus 2, in: *ICN–UCLA Symposia on Molecular and Cellular Biology*, Vol. XIV R. Axel, T. Maniatis, and C.F. Fox (eds.), Academic Press, pp. 611–635.

Brusca, J.S., and Chinnadurai, G., 1981, Transforming genes among three different oncogenic subgroups of human adenoviruses have similar replicative functions, *J. Virol.* **39**:300–305.

Carlock, L.R., and Jones, N.C., 1981a, Transformation-defective mutant of adenovirus type 5 containing a single altered E1a mRNA species, *J. Virol.* **40**:657–664.

Carlock, L., and Jones, N.C., 1981b, Synthesis of an unspliced cytoplasmic message by an adenovirus 5 deletion mutant, *Nature (London)* **294**:572–574.

Carstens, E.B., Magnan, J., and Weber, J., 1979, A dominant temperature-sensitive assembly mutant of adenovirus 2, *Can. J. Microbiol.* **25**:646–649.

Carter, T.H., and Blanton, R.A., 1978, Possible role of the 72,000-dalton DNA binding protein in regulation of adenovirus type 5 early gene expression, *J. Virol.* **25**:664–674.

Carter, T.H., and Ginsberg, H.S., 1976, Viral transcription in KB cells infected by temperature-sensitive "early" mutants of adenovirus type 5, *J. Virol.* **18**:156–166.

Carter, T.H., Nicolas, J.-C., Young, C.S.H., and Fisher, P.B., 1982, Multiple transformation phenotypes among revertants of temperature-sensitive mutants in the type 5 adenovirus DNA-binding protein, *Virology* **117**:519–521.

Cepko, C.L., and Sharp, P.A., 1982, Assembly of adenovirus major capsid protein is mediated by a nonvirion protein, *Cell* **31**:407–415.

Challberg, S.S., and Ketner, G., 1981, Deletion mutants of adenovirus 2: Isolation and initial characterization of virus carrying mutations near the right end of the viral genome, *Virology* **114**:196–209.

Chee-Sheung, C.C., 1978, Characterization of a type 5 adenovirus temperature-sensitive fiber mutant and its functional role, Ph.D. thesis, Columbia University, New York.

Chee-Sheung, C.C., and Ginsberg, H.S., 1982, Characterization of a temperature-sensitive fiber mutant of type 5 adenovirus and effect of the mutation on virion assembly, *J. Virol.* **42**:932–950.

Chinnadurai, G., 1983, Adenovirus 2 lp^+ locus codes for a 19kd tumor antigen that plays an essential role in cell transformation, *Cell* **33**:759–766.

Chinnadurai, G., Chinnadurai, S., and Brusca, J., 1979, Physical mapping of a large plaque mutation of adenovirus type 2, *J. Virol.* **32**:623–628.

Chow, L.T., Roberts, J.M., Lewis, J.B., and Broker, T.R., 1977, A map of cytoplasmic RNA transcripts from lytic adenovirus type 2, determined by electron microscopy of RNA:DNA hybrids, *Cell* **11**:819–836.

Colby, W.W., and Shenk, T., 1981, Adenovirus type 5 virions can be assembled *in vivo* in the absence of detectable polypeptide IX, *J. Virol.* **39**:977–980.

Deuring, R., Klotz, G., and Doerfler, W., 1981, An unusual symmetric recombinant between adenovirus type 12 DNA and human cell DNA, *Proc. Natl. Acad. Sci. U.S.A.* **78**:3142–3146.

D'Halluin, J.C., Milleville, M., Boulanger, P.A., and Martin, G.R., 1978, Temperature-sensitive mutant of adenovirus type 2 blocked in virion assembly: Accumulation of light intermediate particles, *J. Virol.* **26**:344–356.

D'Halluin, J.-C., Allart, C., Cousin, C., Boulanger, P.A., and Martin, G.R., 1979, Adenovirus early function required for protection of viral and cellular DNA, *J. Virol.* **32**:61–71.

D'Halluin, J.-C., Milleville, M., Martin, G.R., and Boulanger, P., 1980, Morphogenesis of human adenovirus type 2 studied with fiber and fiber and penton base-defective temperature-sensitive mutants, *J. Virol.* **33**:88–99.

D'Halluin, J.-C., Cousin, C., and Boulanger, P., 1982, Physical mapping of adenovirus type 2 temperature-sensitive mutations by restriction endonuclease analysis of interserotypic recombinants, *J. Virol.* **41**:401–413.

D'Halluin, J.-C., Milleville, M., and Boulanger, P.A., 1983, Restriction maps of human adenovirus types 2, 5 and 3 for *Bcl*I, *Cla*I, *Pvu*I and *Sph*I, *Gene* **21**:165–169.

Edvardsson, B., Ustaçelebi, S., Williams, J., and Philipson, L., 1978, Assembly intermediates among adenovirus type 5 temperature-sensitive mutants, *J. Virol.* **25**:641–651.

Enomoto, T., Lichy, J.H., Ikeda, J.E., and Hurwitz, J., 1981, Adenovirus DNA replication *in vitro*: Purification of the terminal protein in a functional form, *Proc. Natl. Acad. Sci. U.S.A.* **78**:6779–6783.

Ensinger, M.J., and Ginsberg, H.S., 1972, Selection and preliminary characterization of temperature-sensitive mutants of type 5 adenovirus, *J. Virol.* **10**:328–339.

Esche, H., Mathews, M.B., and Lewis, J.B., 1980, Proteins and messenger RNAs of the transforming region of wild-type and mutant adenoviruses, *J. Mol. Biol.* **142**:399–417.

Flint, S.J., Berget, S.M., and Sharp, P.A., 1976, Characterization of single-stranded viral DNA sequences present during replication of adenovirus types 2 and 5, *Cell* **9**:559–571.

Frost, E., and Williams, J., 1978, Mapping temperature-sensitive and host-range mutations of adenovirus type 5 by marker rescue, *Virology* **91**:39–50.

Galos, R., and Williams, J.F., 1980, Summary of locations of *ts* mutations, Distributed at EMBO workshop on adenoviruses, Peebles, Scotland, June 1980.

Galos, R.S., Williams, J., Binger, M.H., and Flint, S.J., 1979, Location of additional early gene sequences in the adenoviral chromosome, *Cell* **17**:945–956.

Galos, R.S., Williams, J., Shenk, T., and Jones, N., 1980, Physical location of host range mutations of adenovirus type 5: Deletion and marker rescue mapping, *Virology* **104**:510–513.

Gingeras, T.R., Sciaky, D., Gelinas, R.E., Bing-Dong, J., Yen, C.E., Kelly, M.M., Bullock, P.A., Parsons, B.L., O'Neill, K.E., and Roberts, R.J., 1982, Nucleotide sequences from the adenovirus 2 genome, *J. Biol. Chem.* **257**:13,475–13,491.

Ginsberg, H.S., and Young, C.S.H., 1977, The genetics of adenoviruses, in: *Comprehensive Virology*, Vol. 9 (H. Fraenkel-Conrat and R.R. Wagner, eds.), pp. 27–88, Plenum Press, New York.

Ginsberg, H.S., Williams, J.F., Doerfler, W.H., and Shimojo, H., 1973, Proposed nomenclature for mutants of adenoviruses, *J. Virol.* **12**:663–664.

Ginsberg, H.S., Ensinger, M.J., Kauffman, R.S., Mayer, A.J., and Lundholm, U., 1974, Cell transformation: A study of regulation with types 5 and 12 adenovirus temperature-sensitive mutants, *Cold Spring Harbor Symp. Quant. Biol.* **39**:419–426.

Graham, F.L., and van der Eb, A.J., 1973, A new technique for the assay of infectivity of human adenovirus 5 DNA, *Virology* **52**:456–467.

Graham, F.L., Abrahams, P.J., Mulder, C., Heijneker, H.L., Warnaar, S.O., De Vries, F.A.J., Fiers, W., and van der Eb, A.J., 1974, Studies on *in vitro* transformation by DNA and DNA fragments of human adenoviruses and simian virus 40, *Cold Spring Harbor Symp. Quant. Biol.* **39**:637–650.

Graham, F.L., Smiley, J., Russell, W.C., and Nairn, R., 1977, Characterization of a human cell line transformed by DNA from human adenovirus type 5, *J. Gen. Virol.* **36**:59–72.

Graham, F.L., Harrison, T., and Williams, J., 1978, Defective transforming capacity of adenovirus type 5 host range mutants, *Virology* **86**:10–21.

Grodzicker, T., and Klessig, D.F., 1980, Expression of unselected adenovirus genes in human cells co-transformed with the HSV-1 tk gene and adenovirus 2 DNA, *Cell* **21**:453–463.

Grodzicker, T., Williams, J., Sharp, P., and Sambrook, J., 1974, Physical mapping of temperature-sensitive mutations of adenoviruses, *Cold Spring Harbor Symp. Quant. Biol.* **39**:439–446.

Grodzicker, T., Anderson, C., Sambrook, J., and Mathews, M.B., 1977, The physical locations of structural genes in adenovirus DNA, *Virology* **80**:111–126.

Groff, D.E., and Daniell, E., 1980, Isolation and characterization of a viable deletion mutant of adenovirus type 3, *Virology* **106**:191–194.

Hammarskjöld, M.-L., and Winberg, G., 1980, Encapsidation of adenovirus 16 is directed by a small DNA sequence at the left end of the genome, *Cell* **20**:787–795.

Harrison, T.J., Graham, F.L., and Williams, J.F., 1977, Host range mutants of adenovirus 5, defective for growth in HeLa cells, *Virology* **77**:319–329.

Hassell, J.A., and Weber, J., 1978, Genetic analysis of adenovirus type 2. VIII. Physical locations of temperature-sensitive mutations, *J. Virol.* **28**:671–678.

Hearing, P., and Shenk, T., 1983, Functional analysis of the nucleotide sequence surrounding the cap site for adenovirus type 5 region E1a messenger RNAs ,*J. Mol. Biol.* **167**:809–822.

Hérissé, J., Rigolet, M., Dupont de Dinechin, S., and Galibert, F., 1981, Nucleotide sequence of adenovirus 2 DNA fragment encoding for the carboxylic region of the fiber protein and the entire E4 region, *Nucleic Acids Res.* **9**:4023–4042.

Ho, Y.-S., Galos, R., and Williams, J., 1982, Isolation of type 5 adenovirus mutants with a cold-sensitive host range phenotype: Genetic evidence of an adenovirus transformation maintenance function, *Virology* **122**:109–124.

Horwitz, M.S., 1978, Temperature-sensitive replication of H5*ts*125 adenovirus DNA *in vitro*, *Proc. Natl. Acad. Sci. U.S.A.* **75**:4291–4295.

Hudziak, R.M., Laski, F.A., RajBhandary, U.L., Sharp, P.A., and Capecchi, M.R., 1982, Establishment of mammalian cell lines containing multiple nonsense mutations and functional suppressor tRNA genes, *Cell* **31**:137–146.

Jones, N., and Shenk, T., 1978, Isolation of deletion and substitution mutants of adenovirus type 5, *Cell* **13**:181–188.

Jones, N., and Shenk, T., 1979a, Isolation of adenovirus type 5 host range deletion mutants defective for transformation of rat embryo cells, *Cell* **17**:683–689.

Jones, N.C., and Shenk, T., 1979b, An adenovirus type 5 early gene function regulates expression of other early viral genes, *Proc. Natl. Acad. Sci. U.S.A.* **76**:3665–3669.

Kaplan, L.M., Ariga, H., Hurwitz, J., and Horwitz, M.S., 1979, Complementation of the temperature-sensitive defect in H5*ts*125 adenovirus DNA replication *in vitro*, *Proc. Natl. Acad. Sci. U.S.A.* **76**:5534–5538.

Kapoor, Q.S., and Chinnadurai, G., 1981, Method for introducing site-specific mutations into adenovirus 2 genome: Construction of a small deletion mutant in VA-RNA$_{I}$ gene, *Proc. Natl. Acad. Sci. U.S.A.* **78**:2184–2188.

Kapoor, Q.S., Wold, W.S.M., and Chinnadurai, G., 1981, A nonessential glycoprotein is coded by early region E3 of adenovirus type 7, *Virology* **112**:780–784.

Kathmann, P., Schick, J., Winnacker, E.-L., and Doerfler, W., 1976, Isolation and characterization of temperature-sensitive mutants of adenovirus type 2, *J. Virol.* **19**:43–53.

Kauffman, R.S., and Ginsberg, H.S., 1975, Characterization of a temperature-sensitive hexon transport mutant of type 5 adenovirus, *J. Virol.* **19**:643–658.

Kelly, T.J., Jr., and Lewis, A.M., Jr., 1973, Use of non-defective adenovirus–simian virus 40 hybrids for mapping the simian virus 40 genome, *J. Virol.* **12**:643–652.

Khittoo, G., and Weber, J., 1977, Genetic analysis of adenovirus type 2. VI. A temperature-sensitive mutant defective for DNA encapsidation, *Virology* **81**:126–137.

Klessig, D.F., 1977, Isolation of a variant of human adenovirus serotype 2 that multiplies efficiently on monkey cells, *J. Virol.* **21**:1243–1246.

Klessig, D.F., and Grodzicker, T., 1979, Mutations that allow human Ad2 and Ad5 to express late genes in monkey cells map in the viral gene encoding the 72K DNA binding protein, *Cell* **17**:957–966.

Klessig, D.F., and Quinlan, M.P., 1982, Genetic evidence for separate functional domains on the human adenovirus specified 72kd, DNA binding protein, *J. Mol. Appl. Genet.* **1**:263–272.

Klessig, D.F., Quinlan, M.P., and Grodzicker, T., 1982, Proteins containing only half of the coding information of early region 1b of adenovirus are functional in human cells transformed with the herpes simplex virus type 1 thymidine kinase gene and adenovirus type 2 DNA, *J. Virol.* **41**:423–434.

Kruijer, W., van Schaik, F.M.A., and Sussenbach, J.C., 1980, Nucleotide sequence analysis of a region of adenovirus 5 DNA encoding a hitherto unidentified gene, *Nucleic Acids Res.* **8**:6033–6042.

Kruijer, W., van Schaik, F.M.A., and Sussenbach, J.S., 1981, Structure and organization of the gene coding for the DNA binding protein of adenovirus type 5, *Nucleic Acids Res.* **9**:4439–4457.

Kruijer, W., van Schaik, F.M.A., and Sussenbach, J.S., 1982, Nucleotide sequence of the gene encoding adenovirus type 2 DNA binding protein, *Nucleic Acids Res.* **10**:4493–4500.

Kruijer, W., Nicolas, J.-C., van Schaik, F.M.A., and Sussenbach, J.S., 1983, Structure and function of DNA binding proteins from revertants of adenovirus type 5 mutants with a temperature-sensitive DNA replication, *Virology* **124**:425–433.

Lai Fatt, R.B., and Mak, S., 1982, Mapping of an adenovirus function involved in the inhibition of DNA degradation, *J. Virol.* **42**:969–977.

Laski, F.A., Belagaje, R., RajBhandary, U.L., and Sharp, P.A., 1982, An amber suppressor tRNA gene derived by site-specific mutagenesis: Cloning and function in mammalian cells, *Proc. Natl. Acad. Sci. U.S.A.* **79**:5813–5817.

Lassam, N.J., Bayley, S.T., and Graham, F.L., 1978, Synthesis of DNA, late polypeptides and infectious virus by host-range mutants of adenovirus 5 in nonpermissive cells, *Virology* **87**:463–467.

Lassam, N.J., Bayley, S.T., and Graham, F.L., 1979, Tumor antigens of human Ad5 in transformed cells and in cells infected with transformation-defective host-range mutants, *Cell* **18**:781–791.

Lichy, J.H., Field, J., Horwitz, M.S., and Hurwitz, J., 1982, Separation of the adenovirus terminal protein precursor from its associated DNA polymerase: Role of both proteins in the initiation of adenovirus DNA replication, *Proc. Natl. Acad. Sci. U.S.A.* **79**:5225–5229.

Liu, G.Q., Babiss, L.E., Volkert, F.C., Young, C.S.H., and Ginsberg, H.S., 1984, A thermolabile mutant of type 5 adenovirus has a substitution mutation in protein VIII, *J. Virol.* (submitted).

Logan, J., Nicolas, J.-C., Topp, W.C., Girard, M. Shenk, T., and Levine, A.J., 1981, Transformation by adenovirus early region 2A temperature-sensitive mutants and their revertants, *Virology* **115**:419–422.

Luciw, P.A., 1977, Temperature-sensitive mutants of type 5 adenovirus affecting the hexon, Ph.D. Thesis, University of Pennsylvania, Philadelphia.

Mak, I., and Mak, S., 1983, Transformation of rat cells by *cyt* mutants of adenovirus type 12 and mutants of adenovirus type 5, *J. Virol.* **45**:1107–1117.

Martin, G.R., Warocquier, R., Cousin, C., D'Halluin, J.-C., and Boulanger, P.A., 1978, Isolation and phenotypic characterization of human adenovirus type 2 temperature-sensitive mutants, *J. Gen. Virol.* **41**:303–314.

Mathews, M.B., and Grodzicker, T., 1981, Virus-associated RNAs of naturally occurring strains and variants of group C adenoviruses, *J. Virol.* **38**:849–862.

Mautner, V., Williams, J., Sambrook, J., Sharp, P.A., and Grodzicker, T., 1975, The location of the genes coding for hexon and fiber proteins in adenovirus DNA, *Cell* **5**:93–99.

Maxam, A.M., and Gilbert, W., 1977, A new method for sequencing DNA, *Proc. Natl. Acad. Sci. U.S.A.* **74**:560–564.

Mayer, A.J., and Ginsberg, H.S., 1977, Persistence of type 5 adenovirus DNA in cells transformed by temperature-sensitive mutant, H5*ts*125, *Proc. Natl. Acad. Sci. U.S.A.* **74**:785–788.

McDonough, J.S., and Rekosh, D.M., 1982, Differential complementation of adenovirus type 5 temperature-sensitive early mutants by adenovirus types 3 and 12, *Virology* **120**:383–398.

Mertz, J.E., and Berg, P., 1974a, Viable deletion mutants of simian virus 40: Selective isolation by means of a restriction endonuclease from *Hemophilus parainfluenzae*, *Proc. Natl. Acad. Sci. U.S.A.* **71**:4879–4883.

Mertz, J.E., and Berg, P. 1974b, Defective simian virus 40 genomes: Isolation and growth of individual clones, *Virology* **62**:112–124.

Miller, J.S., Ricciardi, R.P., Roberts, B.E., Paterson, B.M., and Mathews, M.B., 1980, Arrangement of messenger RNAs and protein coding sequences in the major late transcription unit of adenovirus 2, *J. Mol. Biol.* **142**:455–488.

Minekawa, Y., Ishibashi, M., Yasue, H., and Takahashi, M., 1976, Characterization of host-range and temperature-sensitive mutants of adenovirus type 5 with particular regard to transformation of a hamster embryo cell line (Nil), *Virology* **71**:97–110.

Montell, C., Fisher, E.F., Caruthers, M.H., and Berk, A.J., 1982, Resolving the functions of overlapping viral genes by site-specific mutagenesis at a mRNA splice site, *Nature (London)* **295**:380–384.

Montell, C., Fisher, E.F., Caruthers, M.H., and Berk, A.J., 1983, Inhibition of RNA cleavage but not polyadenylation by a point mutation in mRNA 3′ consensus sequence AAUAAA, *Nature* **305:**600–605.

Munz, P.L., and Young, C.S.H., 1984, Polarity in adenovirus recombination, *Virology* (submitted).

Munz, P.L., Young, C., and Young, C.S.H., 1983, The genetic analysis of adenovirus recombination in triparental and superinfection crosses, *Virology* **126:**576–586.

Nevins, J.R., and Winkler, J.J., 1980, Regulation of early adenovirus transcription: A protein product of early region 2 specifically represses region 4 transcription, *Proc. Natl. Acad. Sci. U.S.A.* **77:**1893–1897.

Nicolas, J.-C., Suarez, F., Levine, A.J., and Girard, M., 1981, Temperature-independent revertants of adenovirus H5*ts*125 and H5*ts*107 mutants in the DNA binding protein: Isolation of a new class of host-range temperature conditional revertants, *Virology* **108:**521–524.

Nicolas, J.-C., Ingrand, D., Sarnow, P., and Levine, A.J., 1982, A mutation in the adenovirus type 5 DNA binding protein that fails to autoregulate the production of the DNA binding protein, *Virology* **122:**481–485.

Nicolas, J.-C., Young, C.S.H., Suarez, F., Girard, M., and Levine, A.J., 1983, Detection, rescue and mapping of mutations in the adenovirus DNA binding protein gene, *Proc. Natl. Acad. Sci. U.S.A.* **80:**1674–1677.

Oosterom-Dragon, E.A., 1980, Studies of the 100K protein of type 5 adenovirus using wild-type and temperature-sensitive mutants, Ph.D. thesis, Albert Einstein College of Medicine, Yeshiva University, The Bronx, New York.

Oosterom-Dragon, E.A., and Ginsberg, H.S., 1981, Characterization of two temperature-sensitive mutants of type 5 adenovirus with mutations in the 100,000-dalton protein gene, *J. Virol.* **40:**491–500.

Osborne, T.F., Gaynor, R.B., and Berk, A.J., 1982, The TATA homology and the mRNA 5′ untranslated sequence are not required for expression of essential adenovirus E1a functions, *Cell* **29:**139–148.

Plaat, D., and Weber, J., 1979, Intercistronic complementation between adenovirus 2 temperature-sensitive mutants, *Virology* **98:**55–62.

Rajagopalan, S., and Chinnadurai, G., 1981, Viable variants in VA-RNA$_I$ gene of an Ad2–Ad5 recombinant, *Virology* **112:**564–571.

Ricciardi, R.P., Jones, R.L., Cepko, C.L., Sharp, P.A., and Roberts, B.E., 1981, Expression of early adenovirus genes requires a viral encoded acidic polypeptide, *Proc. Natl. Acad. Sci. U.S.A.* **78:**6121–6125.

Robinson, A.J., Younghusband, H.B., and Bellett, A.J.D., 1973, A circular DNA–protein complex from adenoviruses, *Virology* **56:**54–69.

Ross, S.R., Levine, A.J., Galos, R.S., Williams, J., and Shenk, T., 1980, Early viral proteins in HeLa cells infected with adenovirus type 5 host-range mutants, *Virology* **103:**475–492.

Rossini, M., Jonak, G.J., and Baserga, R., 1981, Identification of adenovirus 2 early genes required for induction of cellular DNA synthesis in resting hamster cells, *J. Virol.* **38:**982–986.

Rowe, D.T., and Graham, F.L., 1981, Complementation of adenovirus type 5 host-range mutants by adenovirus type 12 in co-infected HeLa and BHK-21 cells, *J. Virol.* **38:**191–197.

Russell, W.C., Newman, C., and Williams, J.F., 1972, Characterization of temperature-sensitive mutants of adenovirus type 5-serology, *J. Gen. Virol.* **17:**265–279.

Russell, W.C., Skehel, J.J., and Williams, J.F., 1974, Characterization of temperature-sensitive mutants of adenovirus type 5: Synthesis of polypeptides in infected cells, *J. Gen. Virol.* **24:**247–259.

Sambrook, J., Williams, J., Sharp, P.A., and Grodzicker, T., 1975, Physical mapping of temperature-sensitive mutations of adenoviruses, *J. Mol. Biol.* **97:**369–390.

Sanger, F., Nicklen, S., and Coulson, A.R., 1977, DNA sequencing with chain-terminating inhibitors, *Proc. Natl. Acad. Sci. U.S.A.* **74**:5463–5467.

Sarnow, P., Hearing, P., Anderson, C.W., Reich, N., and Levine, A.J., 1982, Identification and characterization of an immunologically conserved adenovirus early region 11,000M_r protein and its association with the nuclear matrix, *J. Mol. Biol.* **162**:565–583.

Schutzbank, T.E., 1980, Use of DNA-minus temperature-sensitive mutants, H5*ts*125 and H5*ts*149, to study DNA replication, transformation and transcription, Ph.D. Thesis, Columbia University, New York.

Sharp, P.A. Moore, C., and Haverty, J.L., 1976, The infectivity of adenovirus 5 DNA–protein complex, *Virology* **75**:442–456.

Shenk, T., Jones, N., Colby, W., and Fowlkes, D., 1979, Functional analysis of adenovirus 5 host-range deletion mutants defective for transformation of rat embryo cells, *Cold Spring Harbor Symp. Quant. Biol.* **44**:367–375.

Shiroki, K., Maruyama, K., Saito, I., Fukui, Y., and Shimojo, H., 1981, Incomplete transformation of rat cells by a deletion mutant of adenovirus type 5, *J. Virol.* **38**:1048–1054.

Shiroki, K., Saito, I., Maruyama, K., Fukui, Y., Imatani, Y., Oda, K.-I., and Shimojo, H., 1983, Expression of adenovirus type 12 early region 1 in KB cells transformed by recombinants containing the gene, *J. Virol.* **45**:1074–1082.

Shortle, D., DiMaio, D., and Nathans, D., 1981, Directed mutagenesis, *Annu. Rev. Genet.* **15**:265–294.

Snustad, D.P., 1968, Dominance interactions in *Escherichia coli* cells mixedly infected with bacteriophage T4D wild type and amber mutants and their possible implications as to type of gene-product function: Catalytic vs. stoichiometric, *Virology* **35**:550–563.

Solnick, D., 1981a, An adenovirus mutant defective in splicing RNA from early region 1A, *Nature (London)* **291**:508–510.

Solnick, D., 1981b, Construction of an adenovirus–SV40 recombinant producing SV40 T antigen from an adenovirus late promoter, *Cell* **24**:135–143.

Solnick, D., and Anderson, M.A., 1982, Transformation-deficient adenovirus mutant defective in expression of region E1a but not region E1b, *J. Virol.* **42**:106–113.

Stillman, B.W., Lewis, J.B., Chow, L.T., Mathews, M.B., and Smart, J.E., 1981, Identification of the gene and mRNA for the adenovirus terminal protein precursor, *Cell* **23**:497–508.

Stillman, B.W., Tamanoi, F., and Mathews, M.B., 1982a, Purification of an adenovirus-coded DNA polymerase that is required for initiation of DNA replication, *Cell* **31**:613–623.

Stillman, B.W., Topp, W.C., and Engler, J.A., 1982b, Conserved sequences at the origin of adenovirus DNA replication, *J. Virol.* **44**:530–537.

Stow, N.D., 1981, Cloning of a DNA fragment from the left-hand terminus of the adenovirus type 2 genome and its use in site-directed mutagenesis, *J. Virol.* **37**:171–180.

Takemori, N., 1972, Genetic studies with tumorigenic adenoviruses. III. Recombination in adenovirus type 12, *Virology* **47**:157–167.

Takemori, N., Riggs, J.L., and Aldrich, C., 1968, Genetic studies with tumorigenic adenoviruses. I. Isolation of cytocidal (cyt) mutants of adenovirus type 12, *Virology* **36**:575–586.

Tarodi, B., Blair, G.E., Rekosh, D.M.K., and Russell, W.C., 1979, Characterization of two temperature-sensitive mutants of adenovirus type 5, *J. Gen. Virol.* **43**:531–540.

Thimmappaya, B., Jones, N., and Shenk, T., 1979, A mutation which alters initiation of transcription by RNA polymerase III on the Ad5 chromosome, *Cell* **18**:947–954.

Thimmappaya, B., Weinberger, C., Schneider, R.J., and Shenk, T., 1982, Adenovirus VAI RNA is required for efficient translation of viral mRNAs at late times after infection, *Cell* **31**:543–551.

Thomas, R., 1971, Control circuits, in: *The Bacteriophage Lambda* (A.D. Hershey, ed.), pp. 211–220, Cold Spring Harbor Laboratory, Cold Spring Harbor, New York.

Thummel, C., Tjian, R., and Grodzicker, T., 1981, Expression of SV40 T antigen under control of adenovirus promoters, *Cell* **23**:825–836.

Tooze, J. (ed.), 1981, *The Molecular Biology of Tumor Viruses* Part B, *DNA Tumor Viruses*, Cold Spring Harbor Press, Cold Spring Harbor, New York.

Ustaçelebi, S., and Williams, J.F., 1972, Temperature-sensitive mutants of adenovirus defective in interferon induction at non-permissive temperature, *Nature* (*London*) **235**:52–53.

van der Vliet, P.C., and Sussenbach, J.S., 1975, An adenovirus type 5 gene function required for initiation of viral DNA replication, *Virology* **67**:415–426.

van der Vliet, P.C., Levine, A.J., Ensinger, M.J., and Ginsberg, H.S., 1975, Thermolabile DNA binding proteins from cells infected with a temperature-sensitive mutant of adenovirus defective in viral DNA synthesis, *J. Virol.* **15**:348–354.

van der Vliet, P.C., Zandberg, J., and Jansz, H.S., 1977, Evidence for a function of the adenovirus DNA-binding protein in initiation of DNA synthesis as well as elongation of nascent DNA chains, *Virology* **80**:98–110.

van Ormondt, H., Maat, J., DeWaard, A., and van der Eb, A.J., 1978, The nucleotide sequence of the transforming HpaI E fragment of adenovirus type 5 DNA, *Gene* **4**:309–328.

Volkert, F.C., 1983, Genetic information transfer in transfection with adenovirus DNA, Ph.D. Thesis, Columbia University, New York.

Volkert, F.C., and Young, C.S.H., 1983, The genetic analysis of recombination using adenovirus overlapping terminal DNA fragments, *Virology* **125**:175–193.

Weber, J., 1976, Genetic analysis of adenovirus type 2. III. Temperature sensitivity of processing of viral proteins, *J. Virol.* **17**:462–471.

Weber, J., and Hassell, J.A., 1979, Genetic analysis of adenovirus type 2. IX. The physical locations of structural genes, *J. Gen. Virol.* **44**:639–655.

Weber, J., Bégin, M., and Khittoo, G., 1975, Genetic analysis of adenovirus type 2. II. Preliminary phenotypic characterization of temperature-sensitive mutants, *J. Virol.* **15**:1049–1056.

Weber, J., Bégin, M., and Carstens, E.B., 1977, Genetic analysis of adenovirus type 2. IV. Coordinate regulation of polypeptides 80K, IIIA and V, *Virology* **76**:709–724.

Werner, G., and zur Hausen, H., 1978, Deletions and insertions in adenovirus type 12 DNA after viral replication in Vero cells, *Virology* **86**:66–77.

Wigler, M., Sweet, R., Sim, G.K., Wold, B., Pellicer, A., Lacy, E., Maniatis, T., Silverstein, S., and Axel, R., 1979, Transformation of mammalian cells with genes from procaryotes and eucaryotes, *Cell.* **16**:777–785.

Wilkie, N.M., Ustaçelebi, S., and Williams, J.F., 1973, Characterization of temperature-sensitive mutants of adenovirus type 5: Nucleic acid synthesis, *Virology* **51**:499–503.

Williams, J.F., Gharpure, M., Ustaçelebi, S., and McDonald, S., 1971, Isolation of temperature-sensitive mutants of adenovirus type 5, *J. Gen. Virol.* **11**:95–101.

Williams, J.F., Young, C.S.H., and Austin, P.E., 1974, Genetic analysis of human adenovirus type 5 in permissive and nonpermissive cells, *Cold Spring Harbor Symp. Quant. Biol.* **39**:427–437.

Williams, J., Grodzicker, T., Sharp, P., and Sambrook, J., 1975a, Adenovirus recombination: Physical mapping of crossover events, *Cell* **4**:113–119.

Williams, J., Young, H., and Austin, P., 1975b, Complementation of human adenovirus type 5 *ts* mutants by human adenovirus type 12, *J. Virol.* **15**:675–678.

Williams, J., Ho, Y.-S., and Galos, R., 1981, Evidence for functional relatedness of products encoded by the transforming sequences of human adenovirus types 5 and 12, *Virology* **110**:208–212.

Wolgemuth, D.J., and Hsu, M.-T., 1980, Visualization of genetic recombination intermediates of human adenovirus type 2 DNA from infected HeLa cells, *Nature* (*London*) **287**:168–171.

Wolgemuth, D.J., and Hsu, M.-T., 1981, DNA replication-mediated recombination of molecules of adenovirus 2 DNA, *Proc. Natl. Acad. Sci. U.S.A.* **78**:5076–5080.

Yeh-Kai, L., Akusjärvi, G., Aleström, P., Pettersson, U., Tremblay, M., and Weber, J., 1983, Genetic identification of an endoproteinase encoded by the adenovirus genome, *J. Mol. Biol.* **167**:217–222.

Young, C.S.H., and Silverstein, S.J., 1980, The kinetics of adenovirus recombination in homotypic and heterotypic genetic crosses, *Virology* **101**:503–515.

Young, C.S.H., and Williams, J.F., 1975, Heat-stable variant of human adenovirus type 5: Characterization and use in three-factor crosses, *J. Virol.* **15**:1168–1175.

CHAPTER 5

Adenovirus Transcription

PHILLIP A. SHARP

I. INTRODUCTION

The emphasis in this chapter is on the regulation and kinetics of adenovirus transcription. Recent references have been selected to provide readers with starting points to explore the background literature. A more complete treatment of earlier work on the subject can be found in Tooze (1980). Two excellent recent reviews on the topic are Flint (1982) and Ziff (1980).

The study of adenovirus transcription and of viral RNA structure has contributed greatly to the understanding of messenger RNA (mRNA) synthesis in mammalian cells. Similar to the case of cellular genes, viral DNA is transcribed and viral RNA is processed in the nucleus. Mature mRNA is subsequently transported to the cytoplasm for translation. Viral DNA is transcribed by both cellular RNA polymerases II and III; most RNA processing and transport are probably also carried out by cellular activities. However, both the rate of transcription of viral promoters and RNA processing events are influence by viral-encoded products. Thus, the study of adenovirus transcription requires understanding of how a rapidly replicating virus subverts and controls cellular pathways for the regulated synthesis of its RNAs.

The vast majority of the work on adenovirus transcription has focused on the replication cycle of human adenovirus type 2 (Ad2) and Ad5. These serotype subgroup C viruses have a relatively brief replication cycle and grow to high titers, making them easier to study. It is likely that most adenoviruses express a similar set of RNAs and regulate their expression by equivalent processes.

PHILLIP A. SHARP • Center for Cancer Research and Department of Biology, Massachusetts Institute of Technology, Cambridge, Massachusetts 02139.

A productive cycle of adenovirus is notably punctuated by the onset of viral DNA replication. The period before replication is referred to as the *early* and the period postreplication as the *late* stage. Viral gene products active in stimulation of the cell in preparation for viral DNA replication should be expressed in the early stage. Virion components and assembly factors would be expected to be predominantly late products. As discussed in greater detail later, the replication cycle of adenovirus can be further divided into a continuous series of stages, each stage being a segment of the cycle wherein a single gene or a subgroup of genes is optimally expressed. It is convenient to partition the replication cycle into four stages, i.e., pre-early, early, intermediate, and late. These stages have a temporal sequence of 0–2, 2–6, 6–12, and 12–36 hr, respectively, after infection at moderate multiplicity of infection. Regulation during the later three stages results from the action of other viral genes. In a few cases, the viral gene responsible for some aspect of regulation has been identified.

Regulation of viral gene expression at the level of mRNA occurs by altering the balance of transcription and processing, i.e., synthesis and degradation. Changes in both rate of synthesis and half-lives of mRNAs have been documented as major steps in controlling the level of viral mRNAs. In fact, more detail is known about the kinetics of adenovirus RNA synthesis and degradation than about those of any other mammalian gene system. Promoter strengths have been investigated with pulses of less than a minute. The rates of turnover of RNAs have been analyzed. Intermediates in RNA splicing have been isolated and characterized. This information is an important part of the basis of our understanding of regulation of the replication cycle of adenovirus.

Adenovirus of humans typically produces a semipermissive infection of nonhuman cells (see Chapter 11). The most extensively studied semipermissive interaction is that of adenovirus subgroup C infection of monkey cells. This incomplete virus–cell interaction results in alterations of normal levels of late viral mRNAs and thus provides insight into regulation of viral transcription. Transformation of cells can result from infection of semipermissive cells (see Chapter 11). Transformed cells contained integrated viral DNA sequences and express viral mRNAs. As expected, only subsets of early mRNAs are synthesized in transformed cells. These mRNAs are probably regulated in a manner consistent with their expression during the replication cycle.

II. NINE TRANSCRIPTION UNITS

The sequence content of Ad2 mRNAs has been mapped by electron microscopy (EM), nuclease S1 mapping, and complementary DNA (cDNA) cloning and sequencing. In general, the mRNAs have been extracted from productively infected cells either early (6–8 hr) or late (18 hr) postinfection. These mRNAs have been further assigned to different stages on the basis of studies using either metabolic labeling or the phenotypes of mutants. In many cases, the molecular weight of the protein encoded by a given mRNA is known; this further identifies the mRNA. This information forms the foundation for considering adenovirus transcription and is summarized by transcription unit in Fig. 1.

The fundamental element of gene expression in mammalian cells is the transcription unit, sequences spanning the genome from the site of initiation of transcription to the site of termination of transcription (Berget *et al.*, 1977; Wilson *et al.*, 1979b). Since RNA splicing seems to be restricted to intramolecular processing, sequences within such a precursor can be processed to yield mRNA. In fact, many adenovirus transcription units are processed to multiple mRNAs either by polyadenylation at different sites or by excision of different sets of intervening sequences. This arrangement allows groups of mRNAs to be coordinately regulated at the level of transcription and to be differentially regulated at a posttranscriptional stage. Transcription units as defined above will also overlap if RNA polymerase II terminates at different positions. Thus, regulation of transcription can occur by termination as well as initiation.

All adenovirus mRNAs are modified by capping and polyadenylation (Philipson *et al.*, 1971). Both type 1 and 2 caps are found on viral mRNA, the former being much more abundant (Moss and Koczot, 1976; Sommer *et al.*, 1976). Internal methylation on adenine has been observed for both cellular and viral RNA. Both populations contain about one 6-methyl adenine per 1000 bases (Chen-Kiang *et al.*, 1979). The sites of modification of cellular mRNAs mainly have the sequence ${}^{A}_{G}m^6AC$ (Revel and Groner, 1978). The position of particular methylated adenine residues in viral RNA has not been determined, nor is the significance of methylation of RNA understood. Although the methylation of viral RNA probably occurs on the nascent chain and methylated sequences are conserved during processing, it is not clear that the modification affects posttranscriptional events (Chen-Kiang *et al.*, 1979).

III. TEMPORAL NATURE OF TRANSCRIPTION FROM ADENOVIRUS PROMOTERS

The multiplication cycle of adenovirus is initiated when viral DNA enters the nucleus about ½ hr after absorption of virus. Subsequently, transcription of various viral promoters commences in a defined sequence. Table 1 lists estimates of the relative rate of transcription from viral promoters at various times postinfection (p.i.). These relative rates of transcription have been determined by pulse-labeling with [^{3}H]uridine (Nevins *et al.*, 1979). RNA synthesis from the E1a promoter was detected by 1 hr p.i., while transcription from the other early promoter sites [E1b,

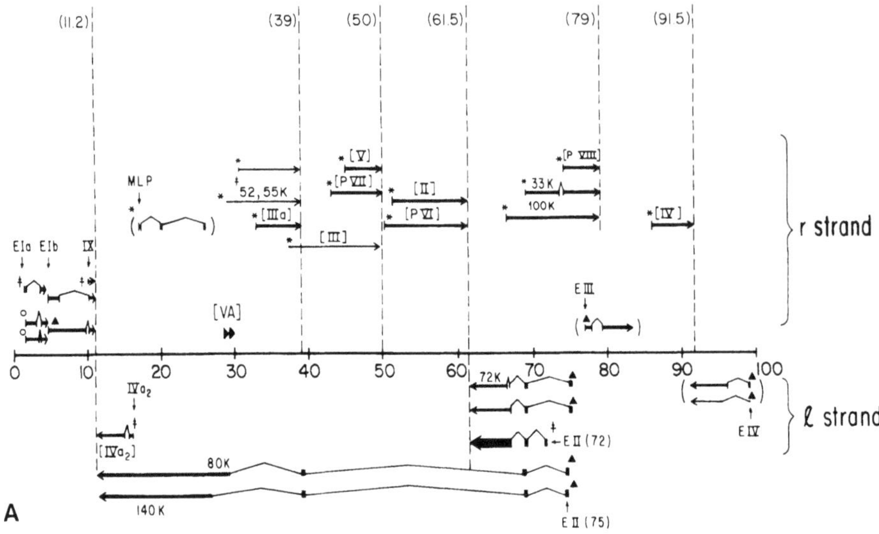

FIGURE 1. Transcription map of Ad2. The spliced structure and map coordinates of most of the mRNAs synthesized from the nine transcription units are indicated. The 35,000 base pairs of adenovirus DNA are divided into 100 units. Arrows drawn above and below the central line represent mRNAs from the r- and l-strands, respectively. Lines forming these arrows span sequences found in the RNA; lines joined by a caret indicate sequences spliced together. (A). The positions of the nine initiation sites for RNA polymerase II, E1a (EIa), E1b (EIb), E2(75) [EII(75)], E2(72) [EII(72)], E3 (EIII), E4 (EIV), IVa2, IX, and MLP, are indicated by arrows. In the late transcription unit, mRNAs are grouped by families that have coterminal 3′ termini mapping at L1–39, L2–50, L3–61.5, L4–79, and L5–91.5. Each mRNA in these families has the tripartite leader spliced to its 5′ terminus (*). The late polypeptide translated from each mRNA is given above the arrow. The other viral initiation sites are divided into pre-early (○), early (▲), and intermediate (‡) classes (see the text). The molecular weight of the polypeptide encoded by the E2(75) mRNAs is shown. The double arrowheads at 30 map units represent the abundant small RNA polymerase III products VA-RNAI and II. (B). More detailed structures for the mRNAs from the early transcription units E1a, E1b, E3 and E4 are shown. Below each schematic mRNA are the map coordinates of the DNA sequences. The sedimentation coefficients of mRNAs from E1a and E1b are listed along with the polypeptide encoded by each species. In the case of E3 and E4 mRNAs, polypeptides have not been assigned to unique mRNA species. The stage at which the mRNA is optimally expressed is indicated at the right. Details of the map coordinates for most of the viral mRNAs can be found in Berk and Sharp (1978) and Chow *et al.* (1979, 1980).

E2(75), E3,and E4] was detectable by 1.5–2.0 hr p.i. The rate of transcription from E1a, E3 and E4 sites reached maximum for the early phase by 3–4 hr p.i. and then declined slightly, while the E1b and E2(75) sites underwent a similar maximum slightly later (6–7 hr). With the onset of DNA replication, transcription from all the early sites increased 3- to 10-fold, probably reflecting the increase in number of DNA templates (Shaw and Ziff, 1982). During this intermediate period of infection, 8–12 hr p.i., the IX and IV_{a2} promoters attain maximal activity (Wilson *et al.*, 1979b). Transcription from the major late promoter (MLP) and the second promoter for the E2 region [E2(72)] increases during this stage (Chow *et al.*,

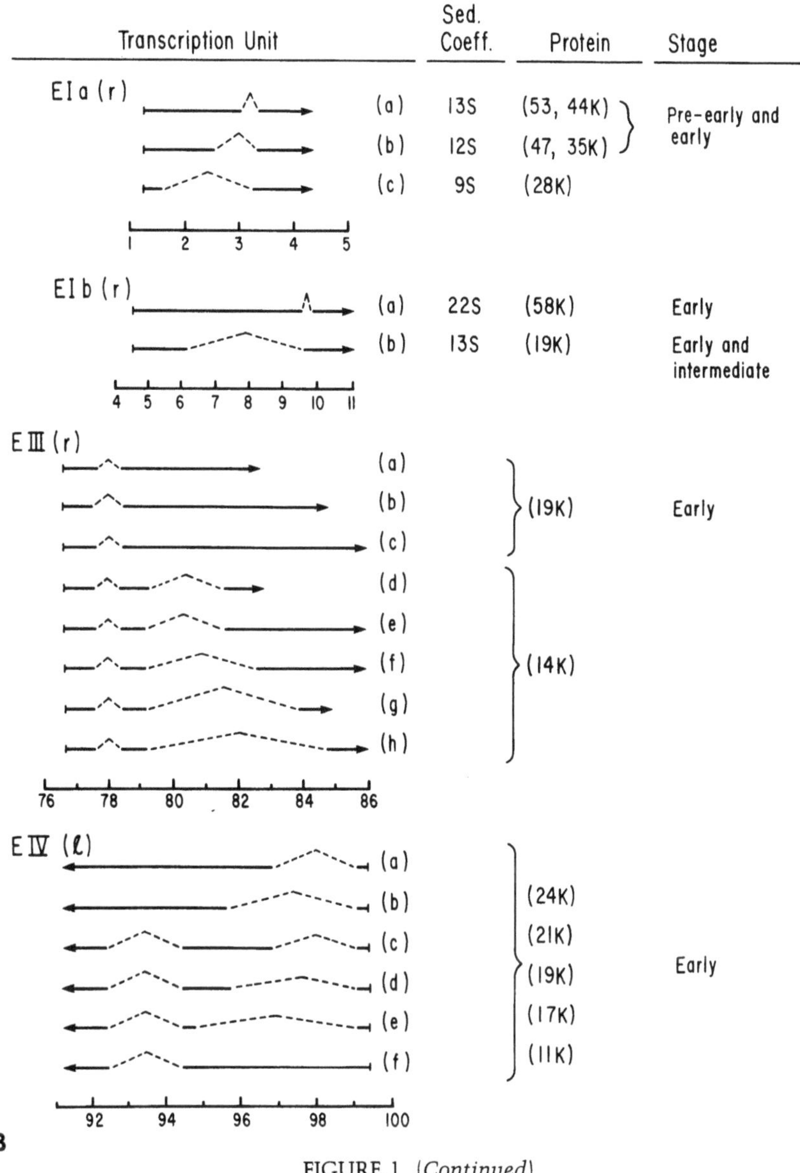

FIGURE 1 (*Continued*)

1979), and MLP attains a maximum level at 18 hr. This high level of transcription activity, about 20–30% of total RNA synthesis in the cell, then remains constant for at least 10 hr (Lucas and Ginsberg, 1971).

The absolute strength of viral promoters is difficult to estimate. The abundance of E1a mRNAs in transformed cells, containing 1–3 copies per cell of viral DNA, is 50–100 copies of viral mRNA per cell. Early during infection, mRNAs from early regions reach a steady-state level of approximately 500–1000 copies per cell, a level typical of an intermediate-

TABLE I. Rate of Transcription *in Vivo* from Viral Initiation Sites[a]

Transcription unit		Time p.i. (hr)	Relative rate at time	Fraction of Wt level (5 hr)[b]	
				*hr*1	*dl*312
E1a	(1.4)	3–5	0.5 of E3[c]	0.3	—
E1b	(4.7)	7–9	0.57 of E3[c]	0.2	0.09
E2	(75)	7	0.44 of E3[c]	0.07	0.04
E3	(76.6)	3	1.00[c]	0.03	0.01
E4	(99.1)	3	0.65 of E3[c]	0.01	0.01
IVa_2	(15.9)	4	0.1 of E3[d]	ND	ND
IX	(9.8)	3–4	<0.03 of E1b[e]	ND	ND
IX	(9.8)	15	0.3 of E1b[e]	—	—
MLP	(16.4)	5	0.22 of E3[f]	<0.03	ND
MLP	(16.4)	18	11 greater than E1b[g]	—	—

[a] Experiments employing either short pulses of [^{3}H]uridine or either UV attenuation or treatment with 5,6-dichloro-1-β-D-ribofuranosyl-benzimidazole to restrict transcription to promoter proximal regions have been used to estimate transcription rates of viral promoters at different times postinfection (p.i.). Only relative rates are listed, since these are internally controlled for DNA template number and specific activity of labeling. Transcription rates in HeLa cells infected with the E1a mutant Ad5*hr*1 or Ad5*dl*312 are compared to parallel infections with wild-type (Wt) Ad5 in the last two columns.

[b] Comparison made by 5-min pulses with [^{3}H]uridine 5 hr p.i. (Nevins, 1981). (ND) Not determined.

[c] From Table 1 of Nevins *et al.* (1979).

[d] Estimated from comparison to late site on other strand (Sehgal *et al.*, 1979).

[e] Estimated from UV effect on accumulation of short RNA chains after 10-min pulse of [^{3}H]uridine (Wilson *et al.*, 1979a).

[f] From comparison to E2 under similar labeling conditions (Nevins and Wilson, 1981).

[g] Estimated from data from 2-min pulse labels (Fraser *et al.*, 1979b).

abundance cellular mRNA (Flint and Sharp, 1976). At 18 hr p.i., a late mRNA typically reaches a steady-state level of 2000–5000 copies per cell, which increases to 20,000–50,000 copies per cell by 32 hr p.i. (Flint and Sharp, 1976). This high level of late transcription does not necessarily suggest that the late promoter is a strong promoter, since the cell contains 1000–100,000 copies of viral DNA. In fact, EM visualization of late transcription complexes suggests that a typical active viral DNA molecule has only four nascent chains distributed over 29,000 base pairs (bp) (Wolgemuth and Hsu, 1981; Beyer *et al.*, 1981). This should be compared to the situation in an erythroblast, where polymerases are densely packed on globin genes.

IV. VIRAL CHROMATIN STRUCTURE

Transcriptional activity has been shown to be correlated with distinct types of chromatin structures around cellular promoters. This subject has not been extensively studied for adenovirus promoters. However, it is likely that the nature of the viral DNA–protein template changes during the course of the cycle. Viral DNA can be detected in a nucleosome conformation shortly after infection; digestion of nuclei with deoxyribonuclease 1 releases viral DNA in a 200-bp ladder, which is typical of cellular chromatin (Sergeant *et al.*, 1979; Tate and Philipson, 1979; Daniell *et al.*, 1981). Similarly, integrated viral DNA in transformed cells is transcribed from a nucleosome-organized state (Larsen and Weintraub, 1982); thus, early viral promoters can be recognized in chromatin formed by nucleosomes. During the late stage of infection, viral DNA is organized in the basic proteins V and VII (Kedinger *et al.*, 1978). When observed by EM, this DNA–protein complex does not have nucleosome structure, nor does it yield a 200-bp ladder after partial nuclease digestion. Transcription complexes with nascent RNA chains prepared from late infected cells also do not have either a nucleosome structure or a compaction of typical chromatin (Wolgemuth and Hsu, 1981; Beyer *et al.*, 1981). The stage during the multiplication cycle at which transcription templates shift between these two types of nucleoprotein complexes is not known; however, it is possible that this shift has consequences in gene regulation.

V. SEQUENCE OF PROMOTER SITES

The biochemistry of cap formation suggests that only nucleotides with tri- or diphosphates can be utilized as substrate (Shatkin, 1976). Thus, it is assumed that the penultimate nucleotide of a cap demarks a transcription-initiation site. Ziff and colleagues have made extensive use of RNA fingerprinting methodology to map the penultimate nucleotides for mRNAs from all nine viral promoter sites (Ziff and Evans, 1978; Baker and Ziff, 1981). The map locations and sequences of these promoter sites are shown in Fig. 2. Most of the RNAs are initiated at either one or a few clustered sites (underlined). This degree of microheterogeneity is not uncommon to other viral and cellular initiation sites. In this regard, the E4 promoter is interesting; initiation occurs for about half the RNAs with the distal adenine, the other half of the RNAs being initiated with uridine at any of the five or six adjacent sites for Ad2 or Ad5, respectively. All the promoter sites have distinct TATA consensus (Goldberg, 1979) sequences between -34 and -26 with the exception of E2(75) and the IVa_2 promoters. This suggests a lack of strict correlation between the presence of a TATA sequence and transcriptional activity, since the E2(75) is one of the strongest promoters and the IVa_2 promoter is one of the weakest.

In vitro systems have recently been developed that yield specific initiation by RNA polymerase II on exogenously added DNA (Weil *et al.*, 1979; Manley *et al.*, 1980). Both systems were developed using well-defined adenovirus promoter sites as templates. The systems consist of extracts of mammalian cells that either contain saturating amounts of RNA polymerase II or are supplemented with purified enzyme. Specific transcription is typically measured by following incorporation of labeled nucleoside triphosphate into an RNA produced by initiation at the promoter site and transcription to the end of a fragment. In addition, the

"TATA" BOX / CAPS	Promoter (Coordinate)	Relative Efficiency: Mock	Relative Efficiency: Late
GTG[TATTTATA]CCCGGTGAGTTCCTCAAGAGGCCACTCTTGAGTG	Ad5 EIa (14)	02	003
GGG[TATATAAT]GCGCCGTGGGCTAATCTTGGTTACATCTGACCTC	Ad5 EIb (47)	0.35	005
GAA[TATATAA]GGTGGGGGTCTTATGTAGTTTTGTATCTGTTTTGC	Ad5 Protein IX (98)	035	06
GGC[TATAAAA]GGGGGTGGGGGCGCGTTCGTCCTCACTCTCTTCCG	Ad2 Major Late (164)	10	10
TCCTTCGTGCTGGCCTGGACGCGAGCCTTCGTCTCAGAGTGGTCC	Ad2 IVa_2 (15.9)	<0005	<0005
TAGTCCTTAAGAGTCAGCGCGCAGTATTTGCTGAAGAGAGCCTCC	Ad2 EII (75)	004	0005
AGGTACAAATTTGCGAAGGTAAGCCGACGTCCACAGCCCCGGAGT	Ad2 EIIa Late (72)	<0005	<0005
—— Not Sequenced ——	Ad5 EII (75)	002	0005
GGG[TATAA]CTCACCTGAAAATCAGAGGGCGAGGTATTCAGCTCAA	Ad2 EIII (76.6)	03	005
—— Not Sequenced ——	Ad5 EIII (76.6)	03	005
TCC[TATATATA]CTCGCTCTGCACTTGGCCCTTTTTTACACTGTGA	Ad5 EIV (99.1)	03	005

FIGURE 2. Sequence and initiation sites for adenovirus transcription units. The 5′ sequences of mRNAs of the nine transcription units of Ad2 or Ad5 are indicated by underlined nucleotides (Baker and Ziff, 1981). Sequences upstream from these initiation sites are shown with prominent TATA sequences boxed. The map coordinates of each site are given. The relative efficiency of DNA templates spanning each site in stimulation of *in vitro* transcription in whole cell extracts prepared from either uninfected (Mock) or late infected (Late) extracts is listed (Fire *et al.*, 1981).

specificity of initiation on the nine viral promoters has also been examined by mapping the 5′ terminus of the *in vitro* product (Fire *et al.*, 1981; Lee and Roeder, 1981). Transcription extracts prepared from uninfected human cells specifically recognize all the viral promoters except IVa_2 and EII(72) (see Fig. 2 for relative activity). The E2(75) promoter is also poorly utilized. Interestingly, a sequence similar to the consensus TATA is found at −59 N from the E2(75) region. In a cytoplasmic extract supplemented with purified RNA polymerase II, this site is recognized for initiation at an equal efficiency to that of the predominant *in vivo* site (Mathis *et al.*, 1981). Sensitive mapping of early *in vivo* mRNA also detects initiation at this upstream site, but at a much lower relative level. Thus, both *in vitro* systems efficiently recognize a common subset of viral promoters, those with the consensus TATA sequence, but also clearly recognize promoter sites that do not contain this sequence.

The IVa_2, E2(72), and E2(75) promoters are expressed at optimal rates during the intermediate stage of infection. Their low *in vitro* activity could suggest that these promoters are transcribed only when stimulated by a viral product. However, the IX promoter is also active during this stage, and it is an effective template for *in vitro* synthesis. Cell extracts from late infected cells show enhanced activity on major late promoter (MLP) and IX promoter (Fire *et al.*, 1981) (Fig. 2). Whether this change reflects the effect of a particular factor awaits further work. However, it is important to note that neither E2(75), E2(72), nor IVa_2 promoter is activated by a late extract (Fig. 2). Thus, the hypothetical factor necessary for their optimal transcription was either not solubilized or not active in this extract. The effect of deletion of sequences around the MLP (Hu and Manley, 1981; Corden *et al.*, 1980) and E3 promoter sites on their *in vitro* activity has been studied (Lee *et al.*, 1982). Alterations in sequences ranging from around the cap site to −60 change the efficiency of transcription. An enzyme of 500,000 daltons such as RNA polymerase II could easily interact with this length of DNA. Atypical of other promoters, deletions around the cap site of E3 reduced transcription 10-fold, suggesting a strong interaction in this region.

Recombinant DNA techniques have also been used to construct deletions in the vicinity of the E1a initiation site (Osborne *et al.*, 1982; Herring and Shenk, 1982, personal communication). These deletions have removed either the sequences spanning the cap site region or TATA region or sequences upstream of the TATA sequences. Viral mutants containing each of these deletions have been constructed and their *in vivo* phenotypes studied. In general, deletion of the cap site of EIa and even of sequences encompassing the first AUG does not impair expression of mRNA or biological activity from this region. Deletion of the TATA and flanking sequences results in a heterogeneous pattern of initiation, with many transcripts now initiated at sequences −150 to −300 N upstream. Low-abundance mRNAs initiated at these same sites can be detected during wild-type (Wt) virus infection, particularly at late times. This suggests that sequences far upstream from the normal initiation site are acting to direct initiation by RNA polymerase II in this vicinity. In Wt virus, the polymerase prefers to initiate with the TATA-specified reaction. However, in the absence of these sequences, the polymerase initiates at other sites. It is unclear how RNA polymerase II is directed to initiate in a particular region by sequences mapping at a distance. Similar roles for upstream sequences in other viral and cellular genes have been described previously (Grosschedl and Birnstiel, 1980; Benoist and Chambon, 1981; Gluzman *et al.*, 1980).

VI. E1a TRANSCRIPTION UNIT

Adenovirus genes responsible for maintenance of transformation are contained within the 0- to 11.0-map-unit (m.u.) sequences (see Chapter 4). Thus, this region has been intensively studied. The sequence of the three mRNAs expressed from E1a has been determined by analysis of recombinant clones of cDNAs (Perricaudet *et al.*, 1979, 1980a,b; Bos *et al.*, 1981; Virtanen and Pettersson, 1983). As in the case of the early mRNAs from simian virus 40 (Leibowitz and Weissman, 1979), the three mRNAs, (a), (b) and (c), are produced by splicing alternative 5′ SS to a common 3′ SS (see Fig. 1). All three mRNAs have common 5′ and 3′ termini and are probably processed from a common nuclear precursor. Nuclear precursors from the E1a region have been detected (Yang *et al.*,

1981). Little is known about the pathway of processing for the three mRNAs and what balances their relative production. Mutations have been isolated in the 5′ SS of both mRNAs (a) and (b) that block their synthesis (Solnick, 1981; Montell *et al.*, 1982). Mutants that fail to synthesize mRNA (a) still produce mRNAs (b) and (c). Similarly, mutants defective for mRNA (b) synthesized both mRNAs (a) and (c). Thus, neither of the longer mRNAs is an essential intermediate in splicing. Mutants constructed from cDNA copies of mRNA (a) where the corresponding intervening sequences have been precisely deleted, are transcribed and processed to aberrently spliced smaller RNAs. Thus, there is no sequential splicing between the larger and smaller mRNAs from E1a. (Svenson, Pattersson, and Akusjärvi, 1983).

Yang *et al.* (1981) have reported that spliced mRNA from this region can be synthesized in nuclei isolated from the early stage. Interestingly, antiserum that reacts with small nuclear particles containing the U1 RNA (small nuclear ribonucleoprotein) partially inhibits this splicing. This could be interpreted to suggest that U1-RNA-containing particles are involved in RNA splicing.

The first detectable viral mRNA synthesized postinfection is from E1a (Nevins *et al.*, 1979). Between 0.5 and 6 hr p.i., (a) (13 S) and (b) (12 S) mRNAs accumulate in the cytoplasm while the (c) (9 S) species accumulates at 1/20th of their level (Spector *et al.*, 1978). After 8–12 hr, the (c) (9 S) mRNA appears in the cytoplasm at a level equal to that of the two larger mRNAs. This shift in relative steady-state levels of these mRNAs may reflect changes in their cytoplasmic half-lives. At early times, mRNAs from E1a have half-lives of 15–20 min (Wilson *et al.*, 1979b); an increase in the half-life of the 9 S mRNA relative to that of the two larger mRNAs at late times could explain the shift in abundance. Similar preferential stabilization of shorter mRNAs at late times has been documented for the E1b region (Wilson and Darnell, 1981). However, the shift in abundance of the three mRNAs from E1a could also reflect some modification of RNA splicing.

Rat cells can be morphologically transformed with a short (0–4.5 μm) fragment. The two larger mRNAs, (a) and (b), are synthesized in these cells (Houweling *et al.*, 1980). Thus, transcription, processing, and translation of these mRNAs do not require other viral factors. Many other virus-transformed cells contain only the E1a and E1b transcription units. Expression of viral mRNA from these two units mirrors that found in early infected cells (Goldenberg and Raskas, 1979).

VII. E1b TRANSCRIPTION UNIT

Two mRNAs, (a) (22 S) and (b) (13 S), are processed from the E1b transcription unit. These mRNAs appear in the cytoplasm about 1.5–2 hr p.i., shortly after those from E1a (Nevins *et al.*, 1979). Transcription of E1b is stimulated by a protein encoded by E1a (see Section X). Both the E1b mRNAs have common 5′ and 3′ termini and are generated by differential splicing of a precursor (Berk and Sharp, 1978; Weber *et al.*, 1980). Mutants that terminate translation of protein from the 22 S mRNA do not affect the ratio of the two mRNAs (Blanchard *et al.*, 1978; Goldenberg and Raskas, 1979). About equal numbers of (a) (22 S) and (b) (13 S) mRNAs are synthesized from E1b at periods before 6–8 hr p.i. (Spector *et al.*, 1978). By 20 hr p.i., this ratio has changed so that mRNA (b) is 20-fold more abundant than mRNA (a) (Spector *et al.*, 1978). This shift seems to be partially due to a relative change in the half-lives of these mRNAs. Both mRNAs turn over with a half-life of 15–20 min in the early period. In the late period, the shorter mRNA is preferentially stabilized by a 5- to 10-fold increase in half-life (Wilson and Darnell, 1981). It is generally observed for other early regions that shorter mRNAs are preferentially accumulated at late times. Perhaps a similar preferential stability partially accounts for their increase. It is interesting to note that the measured relative change in half-lives of the two mRNAs from E1b at late times does not totally account for the observed change in cytoplasmic abundance. It is also possible that conditions in the late stage favor the splicing reaction that generates mRNA (b).

VIII. E2(75) TRANSCRIPTION UNIT

The E2(75) transcription unit has many features in common with the late unit. All the mRNAs are initiated at a single site, and the particular mRNA ultimately generated from a transcript is determined by post-transcriptional processing. At least five mRNAs are synthesized, two from the E2(75)-1 polyadenylation site and three from the E2(75)-2 polyadenylation site (see Fig. 1). Members of these two subfamilies of mRNAs have common 3′ termini, but vary in terms of their splicing pattern (Chow *et al.*, 1980; Stillman *et al.*, 1981). The more distal polyadenylation site generates about 1/10th the level of mRNAs generated by the proximal site. Nothing is known about the mechanism that balances synthesis of these two mRNA families.

Transcription initiation at the E2(75) promoter site is stimulated by an E1a factor; mRNAs from this region appear in the cytoplasm 2–2.5 hr p.i. (Nevins *et al.*, 1979). The site is optimally active during an intermediate period when the E2(75)-1 mRNAs attain levels comparable to those of the late mRNAs. Thus, regulation of this promoter is probably more complex than that of the other early units.

The kinetics and structure of nuclear precursors of the E2(75)-1 family have been well studied (Goldenberg and Raskas, 1979; Berk and Sharp, 1978; Weber *et al.*, 1980). After 10 min of labeling with [^{3}H]uridine, three polyadenylate-positive [poly(A)$^+$] nuclear RNA species can be resolved by gel electrophoresis: a 28 S colinear transcript from 75 to 61 m.u., a 23

S RNA with the 5′-proximal set of intervening sequences removed, and 20 S mature-size mRNA. Around 15 min after addition of label, a 20 S size mRNA begins to emerge in the cytoplasm. The 28 and 23 S nuclear precursors have a half-life of 5–7 min (Weber *et al.*, 1980). The mature 20 S mRNA seems to have a half-life greater than 3 hr. Pulse–chase experiments using actinomycin D suggest that nearly 100% of the 28 and 23 S precursors is processed to 20 S mRNA. These data suggest that RNA splicing of the E2(75)-1 precursors removes first the 5′-proximal intervening sequences and then the internal intervening sequences, the mature 20 S mRNA is formed in the nucleus and then transported to the cytoplasm, and the process is reasonably efficient.

The ease of detection of nuclear precursors from the E2(75)-1 region has led two groups, Blanchard *et al.* (1978) and Goldenberg and Raskas (1980), to use this region to demonstrate the validity of the RNA splicing hypothesis (Berget *et al.*, 1977). Nuclear precursors were prelabeled by pulsing with [^{3}H]uridine for 10 min, and nuclei were fractionated from cytoplasm. *In vitro* incubation of these prelabeled nuclei in the presence of a cytoplasmic extract from uninfected cells gave nearly quantitative conversion of the precursors to 20 S poly(A)$^+$ RNA. A component in the cytoplasmic extract enhanced the rate of splicing by 5-fold, but the chemical nature of this factor was not determined.

IX. LATE TRANSCRIPTION UNIT

Synthesis of mRNA by initiation at the late promoter had been thought to be restricted to the postreplicative part of the replication cycle. However, exhaustive EM mapping of cytoplasmic mRNA from the early stage of infection detected two mRNAs with polyadenylated 3′ termini at the L1 site (see Fig. 1), the poly(A) site of the first family of late mRNAs (Chow *et al.*, 1979; Kitchingham and Westphal, 1980). The relative level of early mRNAs from the L1 site is approximately 1/10th that of mRNAs from the other early regions, a level consistent with the transcription initiation at the late promoter at early times (see Table I). Both T1 fingerprint analysis of labeled RNA and the sensitivity of *in vivo* transcription to UV irradiation have shown that these early L1 mRNAs are initiated at the major late promoter (MLP) site (Nevins and Wilson, 1981; Shaw and Ziff, 1982). EM of RNA–DNA hybrids detected three leaders on half the early mRNAs and four leaders on the other half (Chow *et al.*, 1979). All mRNAs have the tripartite leader of 16.5, 19.5, and 26.5 m.u., while the latter half has a short fourth leader (440 N) specified by sequences near 22.0 m.u. (Virtanen *et al.*, 1982; Falvey and Ziff, 1983). The fourth leader form may either be a mature mRNA, a partially spliced precursor that leaked to the cytoplasmic fraction during extraction, or an aberrant end product. *In vitro* translation of cytoplasmic RNA containing the fourth leader (22.0 m.u.) yields a polypeptide with a molecular weight of 14,000 (14K), suggesting that this form might be a functional mRNA (Akusjärvi and Persson, 1981). Cytoplasmic mRNAs were not detected during the early stage from the L2 and L3 polyadenylation sites, suggesting that synthesis of early mRNAs from this unit is controlled by postinitiation events (Akusjärvi and Persson, 1981).

Metabolism of RNA from the MLP at early times (5 hr) has been examined by pulse-labeling and by following the accumulation of nuclear and cytoplasmic RNA during continuous labeling (Nevins and Wilson, 1981). At early times, transcription from the MLP terminates beyond L2 and gives fairly uniform transcription of L1 and L2 regions. This termination is somewhat surprising, since during late times, transcription continues to position at 99 m.u. (Fraser *et al.*, 1979a). The efficiency of polyadenylation at the L1 site seems to be 2- to 3-fold more than that at L2 and L3, and the transport of L1 mRNAs is also 2- to 3-fold more efficient (Nevins and Wilson, 1981). These two effects may account for the 10-fold greater levels of cytoplasmic L1 mRNAs relative to mRNAs from L2 and L3.

During the late stage of infection, two mRNAs of 3.8 kilobases (kb) and 2.3 kb, respectively, are processed from the L1 family (Akusjärvi and Persson, 1980; Nevins and Wilson, 1981). These two mRNAs differ by the positions where the tripartite leader is spliced in L1 sequences. Interestingly, at early times, neither of these late mRNAs is synthesized and only a 4.0-kb mRNA is detected (Nevins and Wilson, 1981). The 5′ SS of the body of this 4.0-kb mRNA is only 11 nucleotides downstream from the termination site of the virus-associated (VA)-RNAII gene, and the mRNA probably encodes a 52,55K polypeptide (Akusjärvi and Persson, 1981; Nevins and Wilson, 1981). This shift in spliced structure of mRNAs from L1 between early and late stages could be the consequence of either regulation of RNA splicing or mRNA half-lives. That splicing is regulated is suggested by the observation that at early times, [^{3}H]uridine labeling of cells for 40 min only resulted in incorporation into a 4.0-kb RNA (Nevins and Wilson, 1981). This is the best-documented example in any system of gene regulation at the level of RNA splicing. The biochemical factors responsible for this regulation are unknown.

Virion components and abundant nonvirion proteins are encoded by late mRNAs. These mRNAs can be assigned to one of five families, the members of each family having common 3′ termini that map at either 39 (L1), 50 (L2), 61.5 (L3), 78 (L4), or 94 m.u. (L5) (Chow *et al.*, 1979; Meyer *et al.*, 1977; Fraser and Ziff, 1978). Each member of a family has the tripartite leader spliced to the body of the mRNA (Berget *et al.*, 1977; Chow *et al.*, 1977). An early indication of the common initiation site for all late mRNAs was the recognition that a unique capped T1 ribonuclease undecanucleotide formed the 5′ terminus of most late mRNAs (Gelinas and Roberts, 1977). The sequences of the tripartite leader joined to the mRNA for polypeptide II (hexon) and polypeptide IV (fiber) have been determined, as well as the sequence around the 5′ SS of the body of these

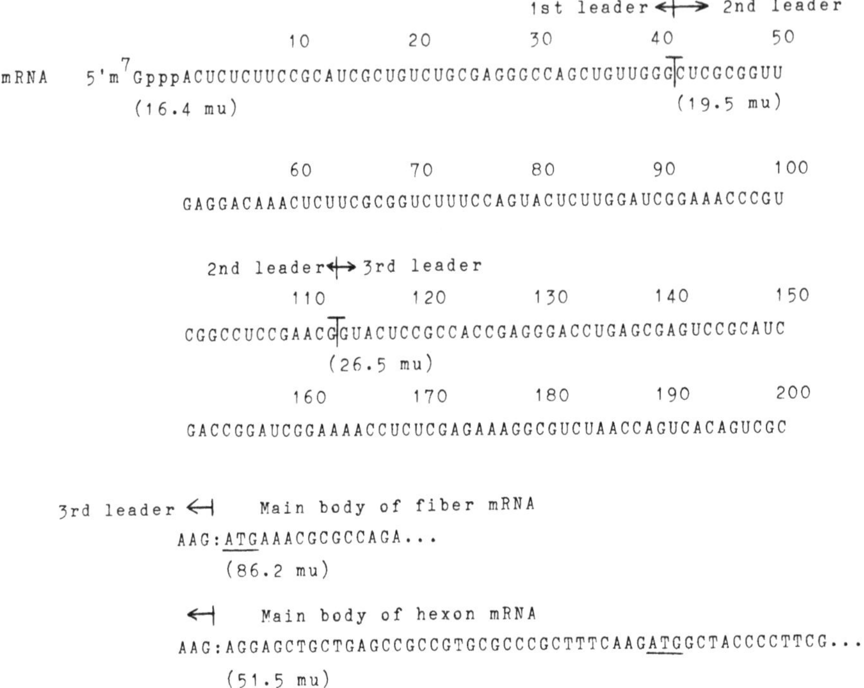

FIGURE 3. Sequence of the tripartite leader of Ad2 late mRNAs. Sequences of the three leaders attached to hexon and fiber mRNAs are shown. The map coordinates of the leaders are indicated. In the case of fiber mRNA, the third leader is spliced directly to the initiation codon (Zain and Roberts, 1979). In the case of hexon, the third leader is spliced 35 N upstream (Akusjärvi and Pettersson, 1979).

mRNAs (Akusjärvi and Pettersson, 1979; Zain *et al.*, 1979; Zain and Roberts, 1979). As shown in Fig. 3, the tripartite leader is joined immediately adjacent to the AUG and 35 N upstream from the AUG for initiation of polypeptides IV and II, respectively. In both cases, the initiation codon is the first AUG in the mRNA. Most late mRNAs are probably monocistronic mRNAs with initiation at the first AUG.

Although the abundant form of each late mRNA has the standard tripartite leader, variant forms of almost any late mRNA can be detected. These variant forms frequently have a fourth (440 N) leader mapping at 22.00 m.u. spliced between the second and third leader of the standard set (Chow *et al.*, 1979; Falvey and Ziff, 1982). Other less frequent forms have an additional leader inserted between the standard third leader and the body of the mRNA. The structure of mRNAs for fiber is particularly complex. At an appreciable frequency, various combinations of leader sequences, i.e., x, y, or z, mapping at 77, 79, and 85 m.u., respectively, are inserted between the standard third leader and the body of fiber mRNA (Chow *et al.*, 1979; Zain *et al.*, 1979; Uhlen *et al.*, 1982). Many of the splice sites on these three leaders seem to map near splice sites utilized in the synthesis of mRNA from the E3 region. Perhaps these internal

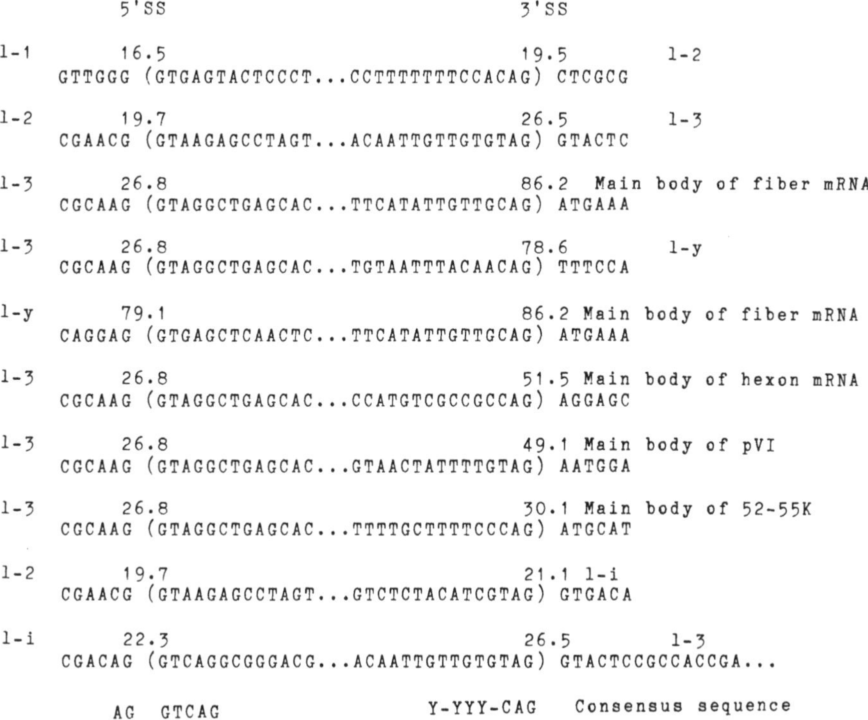

FIGURE 4. Sequences at the splice sites of late Ad2 mRNAs. Sequences listed in brackets are the splice sites of: the three late leaders, l-1, l-2, and l-3, the ancillary y leader on fiber mRNA (Zain *et al.*, 1979), the main body of fiber mRNA (Zain and Roberts, 1979), the main body of hexon mRNA (Akusjärvi and Pettersson, 1979), the main body of pVI mRNA (Akusjärvi and Persson, 1981), the main body of the 52,55K mRNA (Akusjärvi and Persson, 1981), and the i leader (Virtanen *et al.*, 1982). These have been aligned by assuming that the dinucleotide sequences GT and AG are at the 5′ and 3′ boundary of the introns (Breathnach *et al.*, 1978). Consensus sequences tabulated from 38 examples of splice sites in cellular and viral genes are shown (Sharp, 1981).

splice-site sequences interfere with the splicing of fiber mRNA. No role or function has been deduced for the late mRNAs with additional leaders. Given the overproduction of late polypeptides (see below) and perhaps also late mRNAs, they may be a nonessential by-product of the process.

The sequences at the splice sites of the late leaders of Ad2 have been determined by comparison of cDNA and genomic sequences. Somewhat surprisingly, these sequences seem to be typical of 5′ and 3′ SS and fit the consensus sequences compiled by comparison of many viral and cellular introns (Fig. 4). Thus, there is little justification for proposing that there is a unique splicing system for the processing of late mRNAs or that this process utilizes signals atypical of other splicing reactions. For example, it was proposed that the abundant small viral RNAs, VA-RNAI and II, might be involved in splicing; this no longer seems necessary. In addition, mutants that have deletions of the VA-RNAI gene still synthesize late mRNAs, but seem to have a reduced efficiency of translation (Thimmappaya *et al.*, 1982).

Only about 10% of most virion components synthesized during the late stage are actually assembled into viral particles (Horwitz *et al.*, 1969; White *et al.*, 1969). Thus, a tight balance of the relative amounts of different late mRNAs is probably not essential. No temperature-sensitive mutant is known to alter the balance of late mRNAs, and it is possible that the processing of different mRNAs from a single family or from different families is not regulated. In this scheme, nonconditional virus mutations that too drastically altered the relative abundance of late mRNAs would be nonviable. Alternatively, the balance of late mRNAs may be autoregulated. The relative abundance of late mRNAs could be significantly affected by changes in their cytoplasmic half-lives; at 18 hr p.i., mRNAs for hexon appear to have a short half-life of 30–40 min (J. Nevins, personal communication).

As mentioned previously, transcription from the late promoter of adenovirus counts for around 20–30% of the total RNA polymerase II activity at late times (Lucas and Ginsberg, 1971). This is probably an underestimate of the initiation events at the MLP, since 4/5ths of the initiated RNA chains terminate before elongation over 500 nucleotides from the promoter (Evans *et al.*, 1979; Fraser *et al.*, 1979b). These termination products are probably capped but not polyadenylated. An adenosine analogue, 5,6-dichloro-1-β-D-ribofuranosyl-benzimidazole, which inhibits most mRNA synthesis in mammalian cells, enhances this premature termination (Fraser *et al.*, 1979b; Sehgal *et al.*, 1979). It is unlikely that regulation of termination is responsible for the low level of synthesis of the late mRNAs at early times, since short termination RNAs are not detected during this stage (Fraser *et al.*, 1979b). Premature termination has been related to gene regulation by the classic studies of Yanofsky on attenuation of the Trp operon in bacteria (Bertrand *et al.*, 1975). However, control of attenuation in the Trp operon requires coupled translation and transcription, a process that clearly does not occur in a eukaryotic cell.

Experiments analyzing the processing of precursors from the late transcription unit are of two types: analysis of the structure of nuclear RNA (which does not reveal the fate of an RNA) or analysis of the metabolism of RNA sequences (which frequently does not resolve the contributions of different RNAs). In fact, the pulse-labeling of long late nuclear RNA gave an early glimpse of the extensive processing of late mRNA (Wall *et al.*, 1972; Bachenheimer and Darnell, 1975). EM and hybridization–nuclease digestion analysis of nuclear RNAs extracted 18 hr p.i. suggest that the processing of late precursors is complex (Berget and Sharp, 1979). Most of the nuclear RNA has the intervening sequence between the first and second leader excised. Subsequent labeling experiments have shown that this splice occurs rapidly, probably while the RNA chain is still nascent (Keohavong *et al.*, 1982). Nuclear RNAs with a bipartite leader of 1 and 2 spliced to sequences at either 21.6, 24.8, 25.6,

26.0, 26.2, 26.4, or 26.9 m.u. are common (Berget and Sharp, 1979). The latter splicing reaction represents the joining of the second and third leader, while the others are either intermediate splices in this reaction or accumulated aberrant forms. Their high abundance is most consistent with a role as intermediates. Intermediate splices in the excision of a set of intervening sequences have been detected in other systems (Avvedimento *et al.*, 1980). The structure of nuclear intermediates with the late tripartite leaders spliced to sites beyond 26.9 m.u. has not been well investigated. It is possible that the tripartite leader is joined directly from the 26.9 m.u. site to a 5′ SS, forming a mature mRNA; alternatively, there could be many intermediates.

Polyadenylated nuclear RNA longer than mRNA can be readily detected at the L1 (38.5), L2 (50.0), L3 (61.5), and L4 (78.3) sites (Berget and Sharp, 1979; Manley *et al.*, 1979). Few long nuclear RNAs accumulate spanning the L4–L5 (92.5) region, suggesting that few intact RNAs extend beyond L4 (Manley *et al.*, 1979). This could be the result either of formation of several intermediate splices with 5′ SS beyond L4 or of rapid processing of transcripts extending beyond L4. These data are consistent with the typical nuclear intermediate having a partial matured 5′ leader set and polyadenylated 3′ terminus. The tripartite leader is then spliced to a 5′ SS to generate a mature mRNA that is transported to the cytoplasm. Longer mRNAs from a family are apparently not further spliced to shorter mRNAs in the cytoplasm (Nevins and Darnell, 1978b).

The morass of nuclear RNA structures present at late times makes it impossible to follow the fate of RNA sequences by pulse-labeling protocols. However, comparison of the rates of accumulation of nuclear and cytoplasmic RNA from regions L1 through L5 under conditions of uniform labeling has proved informative. The initial rate of accumulation of nuclear RNA should be proportional to the rate of transcription in a region, while the initial rate of accumulation of cytoplasmic RNA should reflect the efficiency at which these nuclear RNA sequences are processed to the cytoplasm. Such experiments suggest that: (1) only about 1/6th of total nuclear viral RNA sequences are processed to the cytoplasm (consistent with the ratio of sizes of nuclear RNAs and mRNAs), (2) the average half-life of nuclear viral RNA sequences is 18 min, (3) between 70 and 100% of the nuclear RNA sequences polyadenylated at any particular late site are ultimately transported to the cytoplasm, and (4) 10, 23, 28, 21 and 22%, respectively, of the nuclear RNA sequences immediately upstream from L1, L2, L3, L4, and L5 are processed to cytoplasmic RNAs (Nevins and Darnell, 1978a,b). These results suggest that cleavage and polyadenylation of the nuclear precursor is a critical step in determining which mRNA is processed from a given precursor. However, the experiments are limited because they measure only an average property of many RNA species and have sizable errors due to the necessity of quantitation of hybridization.

The rate of *in vivo* transcription by RNA polymerase II is 50 N/sec; thus, 10 min is required for polymerase to transverse the 29,000-N late unit. Determination of the lag time between the incorporation of [^{3}H]uridine into nuclear RNA and poly(A)$^+$ nuclear RNA complementary to sequences near a polyadenylation site gives an estimate of the distance polymerase moves before cleavage and polyadenylation. The measured lag is 1 ± 1/2 min, suggesting that RNAs are typically polyadenylated before the polymerase transverses an additional 3000 N (Nevins and Darnell, 1978a). The nascent RNA polymerase II does not seem to terminate following cleavage and polyadenylation, since labeling pulses as short as 1–2 min detect uniform transcription along the total late unit (Nevins and Darnell, 1978a). A termination site for late transcription has been roughly mapped near 99 m.u. RNA sequences from the r-strand between 91.9 m.u., the poly(A) site of fiber mRNA, and 99 m.u. are confined to the nucleus and rapidly degraded (Fraser *et al.*, 1979a).

Synthesis of a polyadenylated 3′ terminus could result from either termination of transcription or cleavage of a precursor. That the latter is the case is shown by the *in vitro* cleavage and polyadenylation of long nuclear precursors in either isolated nuclei (Manley *et al.*, 1979, 1982) or nascent RNAs on transcription complexes (Chen-Kiang *et al.*, 1982). This cleavage and polyadenylation reaction occurs even when transcription is blocked by α-amanitin or actinomycin D (Manley *et al.*, 1982). Three late polyadenylation sites, L2, L3 and L4, have been shown to have the customary sequence AAUAAA immediately upstream of the site of attachment of poly(A) (Fraser *et al.*, 1982). Most mammalian mRNAs have a similar sequence in their 3′ untranslated sequences (Proudfoot and Brownlee, 1976), suggesting that these mRNAs are also polyadenylated after cleavage. The longest mRNA in the L2 family extends beyond the 3′ terminus of the preceding family. Thus, these cytoplasmic mRNAs contain sequences that specify the L2 polyadenylation reaction. Either such mRNAs escape from the nucleus without complete processing or an RNA is immune from further cleavage once it is cleaved and polyadenylated.

X. REGULATION BY E1a PROTEIN

Mutants with defects in either E1a or E1b or both regions have been isolated by complementation with the 293-31 cell line. This line was established by transformation of a primary human kidney cell with a sheared fragment of Ad5 DNA and expresses E1a and E1b products from integrated viral sequences (Harrison *et al.*, 1977). Ad5 mutants with deletions of E1a and E1b are fully permissive in 293-31 cells (Jones and Shenk, 1979a; Solnick, 1981). Mutations in the E1b unit have not been shown to affect viral transcription, but do have interesting phenotypes. Both point and deletion mutants of the E1a region are defective for synthesis of early mRNAs from units other than E1a (Berk *et al.*, 1979; Jones and Shenk, 1979a).

Point mutations in the E1a region are typified by the host-range mutant Ad5*hr*1, which has been shown to be a frameshift mutant due to an insertion of an additional base into sequences transcribed only into mRNA (a) (Ricciardi *et al.*, 1981). Thus, the phenotype of Ad5*hr*1 is the consequence of the absence of the 50K polypeptide encoded by mRNA (a). Montell *et al.* (1982) have specifically mutated the signals for the 5′ SS of E1a mRNA (b) and shown that the loss of this mRNA and protein product does not affect productive growth in culture. Jones and Shenk (1979a) isolated a number of deletions in the E1a region; Ad5*dl*312 is the prototype mutant, with deletion of sequences from 2.8 to 4.0 m.u. Viral stocks of Ad5*dl*312, produced by infection of 293-31 cells, have no detectable contamination by wild-type (Wt) virus. Infection of HeLa cells with either Ad5*hr*1 or Ad5*dl*312 at 10 plaque-forming units (PFU)/cell results in almost no detectable mRNA from the other early regions or from the late unit [less than 0.025 Wt levels. In the Ad5*hr*1 case, E1a mRNAs (a) and (b) are synthesized at 1/3rd Wt levels, proving that the mutants are not defective for uncoating (Berk *et al.*, 1979) (*hr* mutants do efficiently complement other viral mutants, also suggesting that uncoating is not defective). Prolonged incubation of *hr*-mutant-infected cells results in synthesis of high levels of mRNA from early regions (Gaynor *et al.*, 1982). Similarly, infection of HeLa cells at high plaque-forming units/cell ratios with Ad5*dl*312 (100–500 PFU/cell) yields Wt levels of mRNA from all transcription units and a Wt yield of progeny virus (Shenk *et al.*, 1979). Thus, the E1a product is not required for transcription when high copy numbers of template are present, but stimulates transcription about 50-fold at low copy numbers. Following transcription rates using short pulses of [^{3}H]uridine also shows that the E1a protein (either directly or indirectly) stimulates transcription of both other early and late promoters (Nevins, 1981). There are many examples of stimulation of transcription by factors in mammalian cells: the most commonly recognized type of factors are hormones and their receptors. It will be of interest to determine whether the E1a protein acts in a process similar to that of different hormones.

The E1a protein, being the first required factor after infection, is the only viral product that must be solely expressed by cellular factors. The critical role of the E1a protein in controlling further development of the replication cycle offers the possibility that some virus–cell type interactions could result in blocking the cycle at this point, leading to a quiescent state. Humans infected with adenovirus are known to harbor virus for long periods of time; however, it is not clear whether this is due to a continuous low level of replication or to a latent state that can be activated (Davis *et al.*, 1980). Measurements of rates of transcription (Table 1) following infection with these mutants suggest that higher levels of some early mRNAs should have been observed. This difference between transcription rates and steady state measurements may reflect variations in virus stocks and cell conditions or instability of viral RNA in mutant infected cells.

Although transcription from eight promoters is stimulated by E1a expression, this protein may not interact directly with any of the promoter sites. First, the E1a protein could act indirectly by modifying host-cell factors. Alternatively, the E1a factor could directly control another viral gene the activity of which is required for further transcription of various promoters. Obviously, combinations of these modes of action are possible. An interesting observation in this regard is the induction of the cellular heat-shock genes following adenovirus infection. Adenovirus mutants with defects in E1a fail to induce the heat-shock genes (Nevins, 1982). Evidence that the E1a protein may regulate viral genes in a complex fashion comes from the phenotype of the Ad5*hr*440 mutant, which has an amber mutation terminating the E1a (50K) protein (Solnick, 1981). This mutant synthesizes Wt levels of E1b and E4 mRNAs, but no detectable E2 or E3 mRNAs. Thus, either a small amount of leakage by this mutant is adequate to stimulate the transcription of these two promoters or the gene-regulation activity of the E1a protein is encoded by two parts, only one of which is destroyed by the amber mutation (the former seems more likely). A useful observation, but one that is difficult to interpret, also relates transcription of the E1a, E1b and E2 promoters. Infection of cells by either *hr* mutants or Wt Ad5 followed by prolonged incubation under conditions in which DNA synthesis is blocked with cytosine arabinoside results in 50-fold overproduction of E1a, E1b, and E2 mRNAs (Gaynor *et al.*, 1982). The other early mRNAs reach only Wt levels. This overproduction may reflect the absence of a compensating suppressor activity that controls transcription from these three sites under normal conditions.

XI. REGULATION BY THE 72K DNA-BINDING PROTEIN OF E2

Temperature-sensitive (*ts*) mutations in the E2 gene that codes for the 72K DNA-binding protein (DBP) have permitted definition of some of the activities of this protein in regulation of transcription. The best-studied mutant is Ad5*ts*125, which produces a DBP that is thermolabile for binding to single-stranded DNA (Ensinger and Ginsberg, 1972; van der Vliet and Levine, 1973). This protein persists in Ad5*ts*125-infected cells at the nonpermissive temperature, but does not participate in DNA replication (van der Vliet *et al.*, 1975). Cells infected with Ad5*ts*125 and cultured at the nonpermissive temperature overproduce (3- to 5-fold) mRNAs from all early regions (Blanton and Carter, 1979). Part of this increase is probably due to an increase in half-lives of early mRNAs, suggesting that the DBP is partially responsible for the rapid turnover of early mRNAs (Babich and Nevins, 1981). However, the DBP also regulates transcription of an early promoter. During infection by Ad5 at high multiplicities of infection, the rate of transcription (measured by short pulses of [^{3}H]uridine) from E2(75) and E4 promoters decreases 4-fold about 6–8 hr p.i. (Nevins *et al.*, 1979). [A slightly less pronounced decrease has been observed for all other early transcription units following infection at lower multiplicities of infection (Shaw and Ziff, 1982).] The decrease in rate of transcription from the E4 region is probably due to a repressor activity of the DBP, since the decrease is not observed following infections at the nonpermissive temperature with Ad5*ts*125 (Nevins and Winkler, 1980). The same series of experiments showed that the DBP is not directly responsible for suppression of the E2(75) promoter. Interestingly, inhibition of protein synthesis by cycloheximide following infection by Wt Ad2 prevents the repression of transcription from both E4 and E2(75) promoters (Nevins *et al.*, 1979), again indicating that another viral or cellular protein may inhibit transcription from E2(75).

XII. EFFECT OF INHIBITION OF PROTEIN SYNTHESIS ON ADENOVIRUS TRANSCRIPTION

Inhibitors of either protein or DNA synthesis have frequently been employed to block viral DNA replication and collect cells in the early stage. It was soon recognized that inhibition of protein synthesis by addition of cycloheximide an hour after infection results in a 10-fold overproduction of many early RNAs (Parsons and Green, 1971; Craig and Raskas, 1974). The increase in early mRNAs is typically greater for E4 (~20-fold), but is substantial for the other early regions (~5- to 10-fold) (Eggerding and Raskas, 1978; Wilson *et al.*, 1979b). Only in the case of E2(75) and E4 is the increase due to an enhanced rate of transcription (about 4-fold). During a normal infection, transcription from these two promoters is suppressed 4–6 hours p.i., this suppression being blocked if cycloheximide is added (Nevins and Winkler, 1980). The general increase in the level of early mRNAs probably results from an increase in the half-life of early mRNAs. During normal infection, the half-lives of early mRNAs are approximately 20 min; this increases about 4-fold after inhibition of protein synthesis (Wilson *et al.*, 1979b). Not all mRNAs from a transcription unit are equivalently enhanced; for example, the 22 S mRNA (a) from the E1b region decreases in amount, while the 13 S mRNA (b) increases 10-fold, in the presence of cycloheximide (Eggerding and Raskas, 1978). It should be pointed out that most cellular mRNAs have normal turnover times postinfection (Phillipson *et al.*, 1974). Thus, the short half-life of viral mRNAs is reasonably novel, and the mechanism of their stabilization by inhibition of protein synthesis or inactivation of the DBP is unknown.

Inhibition of protein synthesis before addition of adenovirus should block infection at the stage of the first essential protein activity. This has proved hard to define. The best-controlled experiments have used anisomycin as a stringent and rapidly reversible inhibitor. Addition of anisomycin (100 μM) 2 hr preinfection blocked synthesis of mRNA from early regions, but still gave detectable mRNAs from around the 14- to 28-m.u. region (Lewis and Mathews, 1980). Addition of the drug 1/2 hr preinfection resulted in synthesis of mRNAs from E1a, E1b, and E4 (Shaw and Ziff, 1982) or, in another study, from all early regions (Persson *et al.*, 1981). (In the latter experiments, evidence was obtained for a translationally regulatory factor in the E1a or E1b region.) Interpretation of experiments using protein-synthesis inhibitors is difficult because (1) general cell physiology and viability could change during prolonged treatment and alter the normal mechanisms of viral gene regulation and (2) the stability of viral RNAs could increase or decrease, making changes in transcription control difficult to detect. Others have interpreted the synthesis of viral mRNAs in the presence of inhibitors of protein translation to mean that the E1a protein stimulates transcription by inactivating an unstable repressor of transcription from other viral promoters (Nevins, 1981). An alternative explanation would be that prolonged inhibition of protein synthesis modifies cell physiology so as to stimulate RNA polymerase II recognition of early promoters in the absence of expression of an E1a protein, an event known to occur at high DNA templates per cell. This nonphysiological recognition in conjunction with stabilization of RNA sequences could account for the observed synthesis of early mRNAs. The correct interpretation will follow from elucidation of the mode of action of the EIa protein.

XIII. VIRUS-ASSOCIATED RNAI and RNAII

All adenoviruses contain genes for short (~160 N) RNAs transcribed by RNA polymerase III. Ad2 or Ad5 has two such genes near position 30 on the r-strand (Ohe and Weissman, 1970; Akusjärvi *et al.*, 1980; Mathews, 1975). The sequences of these genes are shown in Fig. 5, as are hypothetical secondary structures for the two RNAs (Akusjärvi *et al.*, 1980). The virus-associated (VA)-RNAI gene is transcribed by initiation with either A [VA1(A)] or three nucleotides downstream at G [VA1(G)]. The VA-RNAII gene generates a single RNA initiated with a G. Neither VA-RNAI nor VA-RNAII is posttranscriptionally processed. An extensive set of deletions has been used to map the *in vitro* promoter sequences of the VA-RNAI gene (Fowlkes and Shenk, 1980). Like all other RNA polymerase III genes, the sequences essential for recognition are encoded within the gene between positions +9 and +72. Alterations of sequences upstream from the VA-RNAI gene can significantly affect the level of transcription either *in vitro* or *in vivo* (Fowlkes and Shenk, 1980). It is not clear how these upstream sequences are recognized.

The VA-RNAI and II genes compete for transcription factors in both *in vitro* reaction mixes and infected cells (Fowlkes and Shenk, 1980; Thimmappaya *et al.*, 1982). About 40-fold more VA-RNAI than VA-RNAII is synthesized during a normal infection. Inactivation of VA-RNAI by deletion of internal promoter sequences reduces its potency in suppressing *in vitro* transcription of VA-RNAII genes. The inverse is also observed. Similarly, a virus containing a mutation in the VA-RNAI gene that inactivates its transcription stimulates levels of VA-RNAII higher than wild type (Wt) levels (Thimmappaya *et al.*, 1982). This competition suggests that expression of the VA genes may be limited by the availability of a required factor. Since adenovirus DNA replicates to 100,000 copies per cell, this might not be surprising. However, the limiting factor cannot be required for synthesis of cellular transfer RNAs, another RNA polymerase III product, since their rate of transcription is not noticeably depressed during the late phase.

VA-RNAs accumulate to massive levels during the late stage, some 10^5–10^6 per cell. The high levels of accumulation probably result from the stability of the RNA. Some and perhaps all of these RNAs exist as ribonucleoprotein particles partially formed by a cellular protein antigen recognized by the anti-La class of lupus sera (Lerner *et al.*, 1981a,b). Similar-size small RNAs associated with the same cellular antigen are encoded by the Epstein–Barr virus (Rosa *et al.*, 1981). In addition, the cellular small cytoplasmic 7 S RNA is associated with the same antigen.

By constructing viral mutants with deletions of internal promoter sequences of either VA-RNAI or II genes, Thimmappaya *et al.* (1982) were able to determine the phenotype of viruses defective for either VA-RNAI or VA-RNAII synthesis. Infection with a VA-RNAI$^+$/VA-RNAII$^-$ mutant gave Wt yields of virus. Thus, the VA-RNAII gene is either not essential for virus encoding VA-RNAI, normally nonfunctional, or essentialy only under certain conditions. The converse VA-RNAI$^-$/VA-RNAII$^+$ mutant was defective for efficient translation of late viral mRNAs. Although the mutant-infected cells contained Wt levels of late mRNAs, only 1/20th Wt levels of late proteins were synthesized. These mRNAs were capped, polyadenylated, and spliced. In addition, late mRNAs from VA-RNAI$^-$/VA-RNAII$^+$ were efficiently translated in an *in vitro* system. This led to the proposal that VA-RNAI is essential for efficient *in vivo* initiation of translation of late mRNAs. How VA-RNAI functions in this capacity is not clear. As yet, mutants defective for synthesis of both VA-RNAI and VA-RNAII have not been isolated. Only with such mutants will it be possible to determine the full range of the VA-RNA genes.

XIV. SUMMARY

The structure of adenovirus mRNAs and the kinetics of their synthesis have been thoroughly studied. In addition, the nucleotide se-

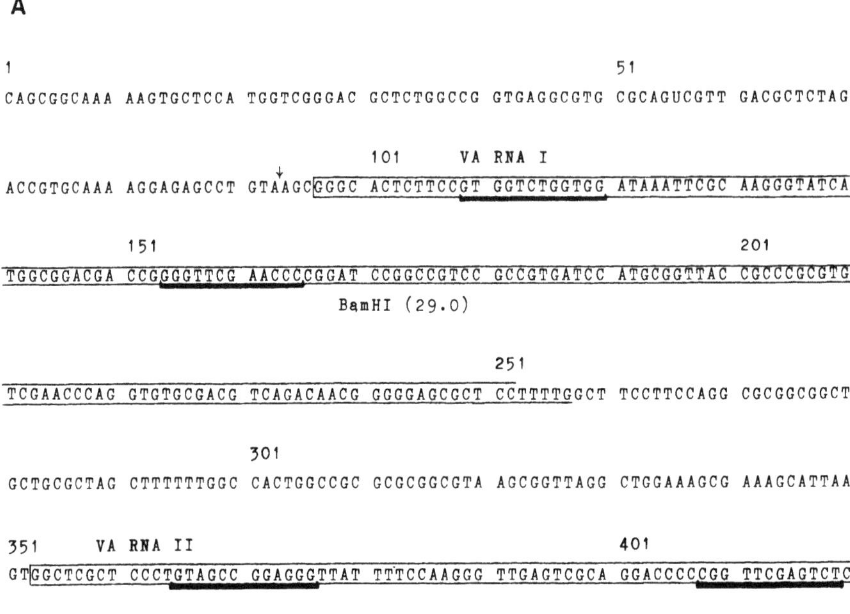

FIGURE 5. Sequence of VA-RNA genes of Ad2. (A) Nucleotide sequence of the region of Ad2 that encodes the small RNA polymerase III genes VA-RNAI and VA-RNAII. This r-strand sequence brackets the *Bam*HI cleavage site at 29.0 m.u. Sequences transcribed as VA-RNAI and VA-RNAII are boxed. The second initiation site for VA-RNAI is indicated by an arrow. The 3′ termini of both RNAs are heterogeneous, terminating in the T tracts. Sequences within the genes related to the two RNA polymerase III consensus sequences GTGG(Py)NN(Pu)GTGG and GGGTTCGAANCC are bold-underlined (Fowlkes and Shenk, 1980). (B) Hypothetical secondary structures for VA-RNAI and VA-RNAII drawn on the basis of the thermodynamics of base-pairing. The calculated free energy of each structure is indicated. Both RNAs probably have extensive base-pairing. (A, B) Adapted from Akusjärvi *et al.* (1980).

quences of all nine transcription initiation sites and many of the poly-adenylation sites have been determined. Thus, the basic descriptive structural data for understanding the regulation of viral mRNA synthesis are nearing completion.

Several viral mutants that are defective for regulation of mRNA synthesis have been identified. These range from mutations in the E1a protein that fail to stimulate transcription from many other promoters to mutations in the 72K DNA-binding protein (DBP) that fail to repress transcription from at least one promoter site. Undoubtedly, many other viral gene products yet to be identified also have a major role in the regulation of mRNA synthesis. Some of these other gene products may

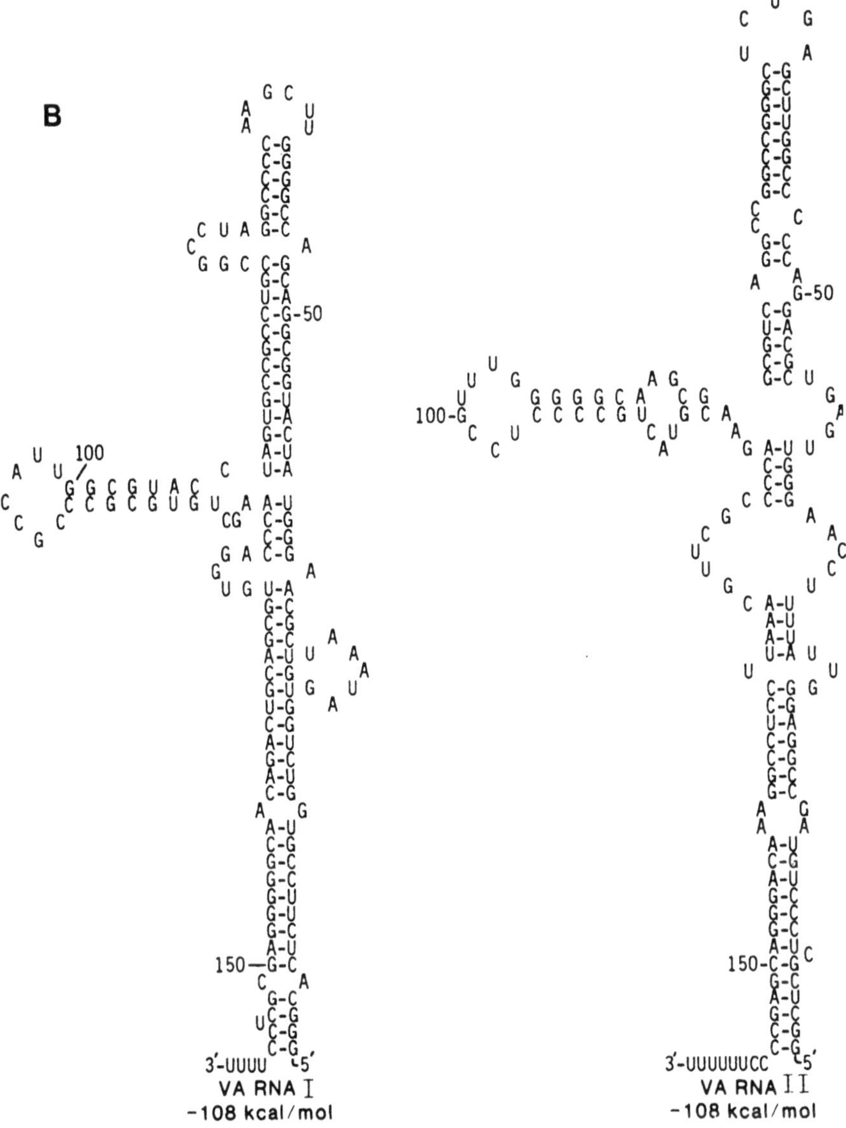

FIGURE 5 (*Continued*)

affect mRNA levels at stages of RNA splicing, polyadenylation, and RNA stability.

The most important step of viral gene regulation is initiation of transcription. Between different stages of the replication cycle, the ratio of transcription from two viral promoters can vary by more than 50-fold. The most striking example of this is the relative level of transcription from the major late promoter (MLP) and an early promoter such as E1b. Before DNA replication commences, transcription from E1b exceeds that from MLP by 5-fold. After the onset of DNA replication, transcription from the MLP exceeds that from E1b by 10-fold. How viral gene regulation is affected by viral or cellular factors is of interest. Simple schemes can be devised from current observations, but these must be incomplete. Some of the steps that are probably reponsible for regulation of transcription are:

1. The E1a (50K) protein either directly or indirectly stimulates initiation from E1b, E4, E2(75), E3, and MLP promoters. This stimulation probably occurs within the first 2 hr of infection.
2. About 6 hr p.i., the DBP of E2 represses transcription from E4 and may also directly or indirectly repress transcription or decrease mRNA stability from other early regions. In some fashion, the levels of all early mRNAs seem to be autoregulated so that infection with multiplicities of infection from 2 to 100 yield the same level of mRNA by 6 hr p.i. This autoregulation is probably not mediated solely through the DBP.
3. Some factor, probably a protein, represses transcription from E2(75) by 6 hr p.i.
4. Some factor(s) begin(s) to stimulate transcription from IX, E2(72), and MLP promoters by 8 hr p.i. This enhanced transcription could be partially due to an increase in viral DNA copies per cell as replication commences.
5. Some factor must either repress transcription from E1a, E1b, E3, and E4 promoters or inhibit the E1a protein stimulation activity at late times. It is undoubtedly the case that the high rate of transcription from the MLP at late times is at least partially dependent on DNA replication through amplification of template. However, something must depress transcription from the stronger early promoters.

Although characterization of regulation of viral mRNAs *in vivo* is still incomplete, primarily due to the lack of appropriate mutants, the study of regulation of adenovirus transcription is turning to *in vitro* biochemical methods. *In vitro* systems that actively transcribe viral promoters are being developed and characterized. The activity of these RNA polymerase II systems in terms of the fraction of DNA template transcribed is low. In addition, factors necessary for transcription have yet to be purified, but this is in progress. It is to be hoped that in the near future, it will be possible to use these and other *in vitro* systems to purify and define the molecular components responsible for adenovirus mRNA synthesis.

ACKNOWLEDGMENTS. This work was supported by National Science Foundation Grant PCM-7823230 (currently PCM-8200309), National Institutes of Health Grant PO1-CA26717 to P.A.S., and partially by Center for Cancer Biology at MIT (Core) Grant NIH-PO1-CA14051.

REFERENCES

Akusjärvi, G., and Persson, H., 1980, Controls of RNA splicing and termination in the major late adenovirus transcription unit, *Nature (London)* **292**:420.

Akusjärvi, G., and Persson, H., 1981, Gene and mRNA for precursor polypeptide VI from adenovirus type 2, *J. Virol.* **38**:469.

Akusjärvi, G., and Pettersson, U., 1979, Sequence analysis of adenovirus DNA: Complete nucleotide sequence of the spliced 5′ noncoding region of adenovirus 2 hexon messenger RNA, *Cell* **16**:841.

Akusjärvi, G., Mathews, M.B., Anderson, P., Vennstrom, B., and Pettersson, U., 1980, Structure of genes for virus-associated RNA_I and RNA_{II} of adenovirus type 3, *Proc. Natl. Acad. Sci. U.S.A.* **77**:2424.

Avvedimento, V.E., Vogeli, G., Yamada, Y., Maizel, J.V., Jr., Pastan, I., and de Crombrugghe, B., 1980, Correlation between splicing sites within an intron and their sequence complementarity with U1 RNA, *Cell* **21**:689.

Babich, A., and Nevins, J.R., 1981, The stability of early adenovirus mRNA is controlled by the viral 72 kd DNA-binding protein, *Cell* **26**:371.

Bachenheimer, S., and Darnell, J.E., 1975, Adenovirus-2 mRNA is transcribed as part of a high-molecular-weight precursor RNA, *Proc. Natl. Acad. Sci. U.S.A.* **72**:4445.

Baker, C.C., and Ziff, E.B., 1981, Promoters and heterogeneous 5′ termini of the messenger RNA's of adenovirus serotype 2, *J. Mol. Biol.* **149**:189.

Benoist, C., and Chambon, P., 1981, *In vivo* sequence requirements of the SV40 early promoter region, *Nature (London)* **290**:304.

Berget, S.M., and Sharp, P.A., 1979, Structure of late adenovirus 2 heterogeneous nuclear RNA, *J. Mol. Biol.* **129**:547.

Berget, S.M., Moore, C., and Sharp, P.A., 1977, Spliced segments at the 5′ terminus of adenovirus 2 late mRNA, *Proc. Natl. Acad. Sci. U.S.A.* **74**:3171.

Berk, A.J., and Sharp, P.A., 1978, Structure of adenovirus 2 early mRNAs, *Cell* **14**:695.

Berk, A.J., Lee, F., Harrison, T., Williams, J., and Sharp, P.A., 1979, Pre-early adenovirus 5 gene products regulate transcription and processing of early viral messenger RNAs, *Cell* **17**:935.

Bertrand, K., Korn, L., Lee, F., Platt, T., Squires, C.L., Squires, C., and Yanofsky, C., 1975, New features of the regulation of the tryptophan operon, *Science* **189**:22.

Beyer, A.L., Bouton, A.H., Hodge, L.D., and Miller, O.L., Jr., 1981, Visualization of the major late R strand transcription unit of adenovirus serotype 2, *J. Mol. Biol.* **147**:269–295.

Blanchard, J.M., Weber, J., Jelinek, W., and Darnell, J.E., 1978, *In vitro* RNA–RNA splicing in adenovirus 2 mRNA formation, *Proc. Natl. Acad. Sci. U.S.A.* **75**:5344.

Blanton, R.A., and Carter, T.H., 1979, Autoregulation of adenovirus type 5 early gene expression. III. Transcription studies in isolated nuclei, *J. Virol.* **29**:458.

Bos, J.L., Polder, L.J., Bernards, R., Schrier, P.I., van der Elsen, P.J., van der Eb, A.J., and van Ormondt, H., 1981, The 2.2 kb EIb mRNA of human Ad12 and Ad5 codes for two tumor antigens starting at different AUG triplets, *Cell* **27**:121.

Breathnach, R., Benoist, C., O'Hare, K., Gannon, F., and Chambon, P., 1978, Ovalbumin gene: Evidence for a leader sequence in mRNA and DNA sequences at the exon–intron boundaries, *Proc. Natl. Acad. Sci. U.S.A.* **75**:4853.

Chen-Kiang, S., Nevins, J.R., and Darnell, J.E., Jr., 1979, *N*-6 methyl-adenosine in adenovirus type 2 nuclear RNA is conserved in the formation of messenger RNA, *J. Mol. Biol.* **135**:733.

Chen-Kiang, S., Wolgemuth, D.J., Hsu, M.-T., and Darnell, J.E., 1982, Transcription and accurate polyadenylation *in vitro* of RNA from the major late adenovirus 2 transcription unit, *Cell* **28**:575.

Chow, L., Gelinas, R., Broker, T., and Roberts, R., 1977, An amazing sequence arrangement at the 5′ ends of adenovirus 2 messenger RNA, *Cell* **12**:1.

Chow, L.T., Broker, T.R., and Lewis, J.B., 1979, Complex splicing patterns of RNAs from the early regions of adenovirus-2, *J. Mol. Biol.* **134**:265.

Chow, L.T., Lewis, J.B., and Broker, T.R., 1980, RNA transcription and splicing at early and intermediate times after adenovirus-2 infection, *Cold Spring Harbor Symp. Quant. Biol.* **44**:401.

Corden, J., Wasylyk, B., Buchwalder, A., Sassone-Corsi, P., Kedinger, C., and Chambon, P., 1980, Promoter sequences of eukaryotic protein-coding genes, *Science* **209**:1406.

Craig, E.A., and Raskas, H.J., 1974, Effect of cycloheximide on RNA metabolism early in productive infection with adenovirus 2, *J. Virol.* **14**:26.

Daniell, E., Groff, D.E., and Fedor, M.J., 1981, Adenovirus chromatin structure at different stages of infection, *Mol. Cell. Biol.* **1**:1094.

Davis, B.D., Dulbecco, R., Eisen, H.N., and Ginsberg, H.S., 1980, *Microbiology*, 3rd ed., Chapter 54, Harper & Row, New York.

Eggerding, F., and Raskas, H.J., 1978, Effect of protein synthesis inhibitors on viral mRNA's synthesized early in adenovirus type 2 infection, *J. Virol.* **25**:453.

Ensinger, M.J., and Ginsberg, H.S., 1972, Selective and preliminary characterization of temperature-sensitive mutants of type 5 adenovirus, *J. Virol.* **10**:328.

Evans, R., Weber, J., Ziff, E., and Darnell, J.E., 1979, Premature termination during adenovirus transcription, *Nature* (*London*) **278**:367.

Falvey, E., and Ziff, E., 1983, Sequence arrangement and protein coding capacity of the adenovirus type 2 *i* leader, *J. Virol.* **45**:185–191.

Fire, A., Baker, C.C., Manley, J.L., Ziff, E.B., and Sharp, P.A., 1981, *In vitro* transcription of adenovirus, *J. Virol.* **40**:703.

Fitzgerald, M., and Shenk, T., 1981, The sequence 5′-AAUAAA-3′ forms part of the recognition site for polyadenylation of late SV40 mRNAs, *Cell* **24**:251.

Flint, S.J., 1982, Expression of adenoviral genetic information in productively infected cells, *B.B.A. Reviews on Cancer* **651**:175–208.

Flint, S.J., and Sharp, P.A., 1976, Adenovirus transcription. V. Quantitation of viral RNA sequences in adenovirus 2 infected and transformed cells, *J. Mol. Biol.* **106**:749.

Fowlkes, D.M., and Shenk, T., 1980, Transcription control regions of the adenovirus VAI RNA gene, *Cell* **22**:405.

Fraser, N.W., and Ziff, E., 1978, Adenovirus-2 late messenger RNAs: Mapping and comparison structure near poly(A), *J. Mol. Biol.* **124**:27.

Fraser, N.W., Nevins, J.R., Ziff, E., and Darnell, J.E., 1979a, The major late adenovirus type-2 transcription unit: Termination is downstream from the last poly(A) site, *J. Mol. Biol.* **129**:643.

Fraser, N.W., Sehgal, P., and Darnell, J.E., 1979b, Multiple discrete sites for premature RNA chain termination late in adenovirus-2 infection: Enhancement by 5,6-dichloro-1-β-D-ribofuranosylbenzimidazole, *Proc. Natl. Acad. Sci. U.S.A.* **76**:2571.

Fraser, N.W., Baker, C.C., Moore, M.A., and Ziff, E.B., 1982, Poly(A) sites of adenovirus serotype 2 transcription units, *J. Mol. Biol.* **155**:207.

Gaynor, R., Tsukamoto, A., Montell, C., and Berk, A.J., 1982, Enhanced expression of adenovirus transforming proteins, *J. Virol.* **44**:276–285.

Gelinas, R.E., and Roberts, R., 1977, One predominant 5′-undecanucleotide in adenovirus 2 late messenger RNAs, *Cell* **11**:533.

Gluzman, Y., Sambrook, J., and Frisque, R.J., 1980, Expression of early genes of origin-defective mutants of simian virus 40, *Proc. Natl. Acad. Sci. U.S.A.* **77**:1432.

Goldberg, M.L., 1979, Sequence analysis of Drosophila histone genes, Ph.D. thesis, Stanford University.

Goldenberg, C.J., and Raskas, H.J., 1979, Splicing patterns of nuclear precursors to the mRNA for adenovirus 2 DNA binding protein, *Cell* **16**:131.

Goldenberg, C.J., and Raskas, H.J., 1980, *In vitro* processing of intervening sequences in the precursors of messenger ribonucleic acid for adenovirus 2 deoxyribonucleic acid binding protein, *Biochemistry* **19**:2719.

Grosschedl, R., and Birnstiel, M.L., 1980, Identification of regulatory sequences in the prelude sequences of an H2A histon gene by the study of specific deletion mutants *in vivo*, *Proc. Natl. Acad. Sci. U.S.A.* **77**:1432.

Harrison, T., Graham, F., and Williams, J., 1977, Host range mutants of adenovirus type 5 defective for growth in HeLa cells, *Virology* **77**:319.

Horwitz, M.S., Scharff, M.D., and Maizel, J.V., Jr., 1969, Synthesis and assembly of adenovirus 2. I. Polypeptide synthesis, assembly of capsomers and morphogenesis of the virion, *Virology* **39**:682.

Houweling, A., Elsen, P.J., and van der Eb, A.J., 1980, Partial transformation of primary rat cells by the left most 4.5% fragment of adenovirus 5 DNA, *Virology* **105**:537–550.

Hu, S., and Manley, J.L., 1981, DNA sequence required for the initiation of transcription *in vitro* from the major late promoter of adenovirus 2, *Proc. Natl. Acad. Sci. U.S.A.* **78**:820.

Jones, N., and Shenk, T., 1979a, Isolation of adenovirus type 5 host range deletion mutants defective for transformation of rat embryo cells, *Cell* **17**:783.

Jones, N., and Shenk, T., 1979b, An adenovirus type 5 early gene function regulates expression of other early viral genes, *Proc. Natl. Acad. Sci. U.S.A.* **76**:3665.

Kedinger, C., Brison, O., Perrin, F., and Wilhelm, J., 1978, Structural analysis of viral replication intermediates isolated from adenovirus type 2-infected HeLa cell nuclei, *J. Virol.* **26**:364.

Keohavong, P., Gattoni, R., LeMoullec, J.M., Jacob, M., and Stevenin, J., 1982, The orderly splicing of the first three leaders of the adenovirus-2 major late transcripts, *Nucleic Acids Res.* **10**:1215.

Kitchingham, G.R., and Westphal, H., 1980, The structure of adenovirus 2 early nuclear and cytoplasmic RNAs, *J. Mol. Biol.* **137**:23.

Larsen, A., and Weintraub, H., 1982, An altered DNA conformation detected by S1 nuclease occurs at specific regions in active chick globin chromatin, *Cell* **29**:609–622.

Leibowitz, P., and Weissman, S.M., 1979, Organization and transcription of the simian virus 40 genome, *Curr. Top. Microbiol. Immunol.* **87**:44.

Lee, D.C., and Roeder, R.G., 1981, Transcription of adenovirus type 2 genes in a cell-free system: Apparent heterogeneity of initiation at some promoters, *Mol. Cell. Biol.* **1**:635.

Lee, D.C., Roeder, R.G., and Wold, W.S.M., 1982, DNA sequences affecting specific initiation of transcription *in vitro* from the EIII promoter of adenovirus 2, *Proc. Natl. Acad. Sci. U.S.A.* **79**:41.

Lerner, M.R., Andrews, N.C., Miller, G., and Steitz, J.A., 1981a, Two small RNAs encoded by Epstein–Barr virus and complexed with protein are precipitated by antibodies from patient with systemic lupus erythematosus, *Proc. Natl. Acad. Sci. U.S.A.* **78**:805.

Lerner, M.R., Boyle, J.A., Hardin, J.A., and Steitz, J.A., 1981b, Two novel classes of small ribonucleoproteins detected by antibodies associated with lupus erythematosus, *Science* **211**:400.

Lewis, J., and Mathews, M.B., 1980, Control of adenovirus early gene expression: A class of immediate early products, *Cell* **21**:303.

Lucas, J.J., and Ginsberg, H.S., 1971, Synthesis of virus-specific RNA in KB cells infected with type 2 adenovirus, *J. Virol.* **8**:203.

Manley, J.L., Sharp, P.A., and Gefter, M.L., 1979, RNA synthesis in isolated nuclei: Identification and comparison of Ad2 encoded transcripts synthesized *in vitro* and *in vivo*, *J. Mol. Biol.* **135**:171.

Manley, J.L., Fire, A., Cano, A., Sharp, P.A., and Gefter, M.L., 1980, RNA-dependent transcription of adenovirus genes in a soluble whole-cell extract, *Proc. Natl. Acad. Sci. U.S.A.* **77**:3855.

Manley, J.L., Sharp, P.A., and Gefter, M.L., 1982, RNA synthesis in isolated nuclei: Processing of adenovirus 2 late mRNA precursor, *J. Mol. Biol.* **159**:581.

Mathews, M.B., 1975, Genes for VA-RNA in adenovirus 2, *Cell* **6**:223.

Mathis, D.J., Elkaim, R., Kedenjer, C., Sassone-Corsi, P., and Chambon, P., 1981, Specific *in vitro* initiation of transcription on the adenovirus type 2 early and late EII transcription units, *Proc. Natl. Acad. Sci. U.S.A.* **78**:7383.

Meyer, J., Neuwald, P.D., Lai, S.-P., Maizel, J.V., and Westphal, H., 1977, Electron microscopy of late adenovirus type 2 mRNA hybridized to double-stranded viral DNA, *J. Virol.* **21**:1010.

Montell, C., Fisher, E.F., Caruthers, M.H., and Berk, A.J., 1982, Resolving the functions of overlapping viral genes by site-specific mutagenesis at a mRNA splice site, *Nature (London)* **295**:380.

Moss, B., and Koczot, F., 1976, Sequence of methylated nucleotides at the 5'-terminus of adenovirus-specific RNA, *J. Virol.* **17**:385.

Nevins, J.R., 1981, Mechanism of activation of early viral transcription by the adenovirus E1a gene product, *Cell* **26**:213.

Nevins, J.R., 1982, Induction of the synthesis of a 70,000 dalton mammalian heat shock protein by the adenovirus E1a gene product, *Cell* **29**:913.

Nevins, J.R., and Darnell, J.E., 1978a, Steps in the processing of Ad2 mRNA: Poly(A)$^+$ nuclear sequences are conserved and poly(A) addition precedes splicing, *Cell* **15**:1477.

Nevins, J.R., and Darnell, J.E., 1978b, Groups of adenovirus type 2 mRNA's derived from a large primary transcript: Probable nuclear origin and possible common 3' ends, *J. Virol.* **25**:811.

Nevins, J.R., and Wilson, M.C., 1981, Regulation of adenovirus-2 gene expression at the level of transcriptional termination and RNA processing, *Nature (London)* **290**:113.

Nevins, J.R., and Winkler, J.J., 1980, Regulation of early adenovirus transcription: A protein product of early region 2 specifically represses region 4 transcription, *Proc. Natl. Acad. Sci. U.S.A.* **77**:1893.

Nevins, J.R., Ginsberg, H.S., Blanchard, J.M., Wilson, M.C., and Darnell, J.E., 1979, Regulation of the primary expression of the early adenovirus transcription units. *J. Virol.* **32**:727.

Ohe, K., and Weissman, S.M., 1970, Nucleotide sequence of an RNA from cells infected with adenovirus 2, *Science* **167**:879.

Osborne, T.F., Gaynor, R.B., and Berk, A.J., 1982, The TATA homology and the mRNA 5' untranslated sequence are not required for expression of essential adenovirus E1A functions, *Cell* **29**:139.

Parsons, J.T., and Green, M., 1971, Resolution of early virus-specific RNA species in adenovirus infected and transformed cells, *Virology* **45**:154.

Perricaudet, M., Akusjärvi, G., Virtanen, A., and Pettersson, U., 1979, Structure of two spliced mRNAs from the transforming region of human subgroup C adenoviruses, *Nature (London)* **281**:694.

Perricaudet, M., Le Moullec, J.-M., and Pettersson, U., 1980a, Predicted structure of two adenovirus tumor antigens, *Proc. Natl. Acad. Sci. U.S.A.* **77**:3778.

Perricaudet, M., Le Moullec, J.-M., Tiollais, P., and Pettersson, U., 1980b, Structure of two adenovirus type 12 transforming polypeptides and their evolutionary implications, *Nature (London)* **288**:174.

Persson, H., Monstein, H.J., Akusjärvi, G., and Philipson, L., 1981, Adenovirus early gene products may control viral mRNA accumulation and translation *in vivo*, *Cell* **23**:485.

Philipson, L., Wall, R., Glickman, G., and Darnell, J.E., 1971, Addition of polyadenylate sequences to virus-specific RNA during adenovirus replication, *Proc. Natl. Acad. Sci. U.S.A.* **68**:2806.

Philipson, L., Pettersson, U., Lindberg, U., Tibbets, C., Vennström, B., and Persson, T., 1974, RNA synthesis and processing in adenovirus-infected cells, *Cold Spring Harbor Symp. Quant. Biol.* **39**:447.

Proudfoot, N.J., and Brownlee, G.G., 1974, Sequence at the 3' end of globin mRNA shows homology with immunoglobulin light chain mRNA, *Nature (London)* **252**:359–362.

Revel, M., and Groner, Y., 1978, Post-transcriptional and translational controls of gene expression in eukaryotes, *Annu. Rev. Biochem.* **47**:1079.

Ricciardi, R.P., Jones, R.L., Cepko, C.L., Sharp, P.A., and Roberts, B.E., 1981, Expression of early adenovirus genes requires a viral encoded acidic polypeptide, *Proc. Natl. Acad. Sci. U.S.A.* **78**:6121.

Rosa, M.D., Gottlieb, E., Lerner, M.R., and Steitz, J.A., 1981, Striking similarities are exhibited by two small Epstein–Barr virus encoded ribonucleic acids and the adenovirus-associated ribonucleic acid VAI and VAII, *Mol. Cell. Biol.* **1**:785.

Sehgal, P., Fraser, N., and Darnell, J., 1979, Early Ad-2 transcription units: Only promoter-proximal RNA continues to be made in the presence of DRB, *Virology* **94**:185.

Sergeant, A., Tigges, M.A., and Raskas, H.J., 1979, Nucleosome-like structural subunits of intranuclear parental adenovirus type 2 DNA, *J. Virol.* **29**:888.

Sharp, P.A., 1981, Speculations on RNA splicing (minireview), *Cell* **23**:643.

Shatkin, A.J., 1976, Capping of eukaryotic RNAs, *Cell* **9**:645.

Shaw, A.R., and Ziff, E.B., 1980, Transcripts from the adenovirus-2 major late promoter yield a single family of 3′ coterminal mRNAs during early infection and five families at late times, *Cell* **22**:905.

Shaw, A.R., and Ziff, E.B., 1982, Selective inhibition of adenovirus type 2 early region II and III transcription by an anisomycin block of protein synthesis, *Mol. Cell. Biol.* **2**:789.

Shenk, T., Jones, N., Colby, W., and Fowlkes, D., 1979, Functional analysis of adenovirus-5 host-range deletion mutants defective for transformation of rat embryo cells, *Cold Spring Harbor Symp. Quant. Biol.* **44**:367.

Solnick, D., 1981, An adenovirus mutant defective in splicing RNA from early region 1A, *Nature (London)* **291**:508.

Sommer, S., Salditt-Georgieff, M., Bachenheimer, S., Darnell, J.E., Furuichi, Y., Morgan, M., and Shatkin, A.J., 1976, The methylation of adenovirus-specific nuclear and cytoplasmic RNA, *Nucleic Acids Res.* **3**:749.

Spector, D.J., McGrogan, M., and Raskas, H.J., 1978, Regulation of the appearance of cytoplasmic RNAs from region 1 of the adenovirus 2 genome, *J. Mol. Biol.* **126**:395.

Stillman, B., Lewis, J., Chow, L., Mathews, M., and Smart, J., 1981, Identification of the gene and mRNA for the adenovirus terminal protein precursor, *Cell* **23**:497.

Svensson, C., Pettersson U., and Akusjärvi, G., 1983, Splicing of adenovirus 2 early region 1A mRNAs is nonsequential, *J. Mol. Biol.* **165**:475–495.

Tate, V.E., and Philipson, L., 1979, Parental adenovirus DNA accumulates in nucleosome-like structures in infected cells, *Nucleic Acids Res.* **6**:2769.

Thimmappaya, B., Weinberger, C., Schneider, R.J., and Shenk, T., 1982, Adenovirus VAI RNA is required for efficient translation of viral mRNAs at late times after infection, *Cell* **31**:543–551.

Tooze, J. (ed.), 1980, *DNA Tumor Viruses*, 2nd rev. ed., Cold Spring Harbor Press, Cold Spring Harbor, New York.

Uhlen, M., Svensson, C., Josephson, S., Alestrom P., Chattopadhyay, S., Pettersson, U., and Philipson, U., 1982, Leader arrangement in the adenovirus fiber mRNA, *The EMBO Journal* **1**:249–254.

van der Vliet, P.C., and Levine, A.J., 1973, DNA-binding proteins specific for cells infected by adenovirus, *Nature (London) New Biol.* **246**:170.

van der Vliet, P.C., Levine, A.J., Ensinger, M., and Ginsberg, H.S., 1975, Thermolabile DNA binding proteins from cells infected with a termperature-sensitive mutant of adenovirus defective in viral DNA synthesis, *J. Virol.* **15**:348.

Virtanen, A., and Pettersson, U., 1983, The molecular structure of the 9 S mRNA from early region 1A of adenovirus serotype 2, *J. Mol. Biol.* **165**:496–499.

Virtanen, A., Alestrom, P., Persson, H., Katze, M.G., and Pettersson, U., 1982, An adenovirus agno gene, *Nucleic Acids Res.* **10**:2539.

Wall, R., Philipson, L., and Darnell, J.E., 1972, Processing of adenovirus specific nuclear RNA during virus replication, *Virology* **50**:27.

Weber, J., Blanchard, J.M., Ginsberg, H., and Darnell, J.E., 1980, Polyadenylic acid addition and splicing events in early adenovirus mRNA formation, *J. Virol.* **33**:286.

White, D.L., Scharff, M.D., and Maizel, J.U., 1969, The polypeptides of adenovirus. III. Synthesis in infected cells, *Virology* **38**:395.

Weil, P.A., Luse, D.S., Segall, J., and Roeder, R.G., 1979, Selective and accurate initiation of transcription at the Ad2 major late promoter in a soluble system dependent on purified RNA polymerase II and DNA, *Cell* **18**:469.

Wilson, M.C., and Darnell, J.E., Jr., 1981, Control of messenger RNA concentration by differential cytoplasmic half-life: Adenovirus messenger RNAs from transcription units 1A and 1B, *J. Mol. Biol.* **148**:231.

Wilson, M.C., Fraser, N.W., and Darnell, J.E., 1979a, Mapping of RNA initiation sites by high doses of UV irradiation: Evidence for three independent promoters within the left 11% of the Ad-2 genome, *Virology* **94**:175.

Wilson, M.C., Nevins, J.R., Blanchard, J.-M., Ginsberg, H.S., and Darnell, J.E., 1979b, Metabolism of mRNA from the transforming region of adenovirus 2, *Cold Spring Harbor Symp. Quant. Biol.* **44**:447.

Wolgemuth, D.J., and Hsu, M.-T., 1981, Visualization of nascent RNA transcripts and simultaneous transcription and replication in viral nucleoprotein complexes from adenovirus 2-infected HeLa cells, *J. Mol. Biol.* **147**:247.

Yang, V.W., Lerner, M.R., Steitz, J.A., and Flint, S.J., 1981, A small nuclear ribonucleoprotein is required for splicing of adenovirus early RNA sequences, *Proc. Natl. Acad. Sci. U.S.A.* **78**:1371.

Zain, S.J., and Roberts, R.J., 1979, Sequences from the beginning of the fiber messenger RNA of adenovirus-2, *J. Mol. Biol.* **131**:341.

Zain, S., Sambrook, J., Roberts, R.J., Keller, W., Fried, M., and Dunn, A.R., 1979, Nucleotide sequence analysis of the leader segments in a cloned copy of adenovirus 2 fiber mRNA, *Cell* **16**:851.

Ziff, E.B., 1980, Transcription and RNA processing by DNA tumor virus, *Nature* (*London*) **287**:491–499.

Ziff, E.B., and Evans, R., 1978, Coincidence of the promoter and capped 5′ terminus of RNA from the adenovirus 2 major late transcription unit, *Cell* **15**:1463.

CHAPTER 6

Structural and Nonstructural Adenovirus Proteins

Ulf Pettersson

I. GENERAL CONSIDERATIONS

A. Classification

The adenovirus family comprises a large number of serotypes (species) of both human and animal origin. The human serotypes have been studied most intensively, and this review considers almost exclusively proteins that are produced as the result of a human adenovirus infection.

The human adenoviruses have been subdivided into a number of subgroups based on immunological, biological, and biochemical characteristics. Table I outlines the classification of human adenoviruses.

B. Polypeptide Composition of the Virion

It was originally thought that the adenovirus particle is composed of two kinds of proteins, hexons and pentons, in addition to the core. However, when the adenovirus proteins were analyzed by sodium dodecyl sulfate–polyacrylamide gel electrophoresis (SDS-PAGE) (Maizel *et al.*, 1968a; Maizel, 1971), an unexpected complexity was revealed, and it is today well established that the particle contains at least nine unique polypeptides (polypeptides II–IX and IIIa). In addition, several minor bands have been observed in the electropherograms, some of which may represent unique viral peptides present in only a few copies per virus particle.

ULF PETTERSSON • Department of Medical Genetics, The Biomedical Centre, University of Uppsala, S-751 23 Uppsala, Sweden.

TABLE I. Classification of Human Adenoviruses[a]

Subgroup[b,c]	Serotypes[c]	GC in DNA (%)	HA subgroup[d]	Length of fibers (nm)	Oncogenicity
A	12, 13, 31	47–49	IV	28–31	High
B	3, 7, 11, 14, 16, 21, 34, 35	50–52	I	9–11	Weak
C	1, 2, 5, 6	57–59	III	23–31	None[e]
D	8–10, 13, 15, 17, 19, 20, 22–30, 32, 33, 36, 37	57–60	II	12–13	None[e]
E	4	57	III	17	None

[a] Adapted from Wigand *et al.* (1982).
[b] Two additional subgroups, F and G, have recently been described (Wadell *et al.*, 1980; Wadell, personal communication). Few parameters, however, have so far been established for these subgroups.
[c] Subgroup should be *subgenus* and serotype should be *species* according to the current nomenclature (see Wigand *et al.*, 1982).
[d] Agglutination of erythrocytes by the hemagglutination subgroup members is as follows: I—monkey erthyrocytes, complete; II—rat erythrocytes, complete; III and IV—rat erythrocytes, incomplete.
[e] Almost all nononcogenic serotypes have been shown to transform rodent cells *in vitro.*

Other components such as polypeptides X–XII are likely to represent cleavage products that remain when certain capsid polypeptides mature during virion assembly. Figure I shows a schematic model of the adenovirus particle and also indicates the polypeptides that are known to be present in the virion.

II. MAJOR CAPSID PROTEINS

Soon after their discovery, it was observed that the adenoviruses produce large quantities of soluble viral antigens during their multiplication (Huebner *et al.*, 1954; Hilleman *et al.*, 1955). Separation by immunoelectrophoresis (Pereira *et al.*, 1959) or diethylaminoethyl (DEAE)-chromatography (Klemperer and Pereira, 1959; Philipson, 1960; Wilcox and Ginsberg, 1961; Haruna *et al.*, 1961) resolved them into three distinct components that were later shown to be related to the major capsid proteins by both antigenicity and structure (Wilcox and Ginsberg, 1963a; Wilcox *et al.*, 1963; Valentine and Pereira, 1965; Norrby, 1966a). Following the proposed nomenclature of Ginsberg *et al.* (1966), these are known as the hexons, pentons, and fibers of the adenovirus capsid. During the late phase of adenovirus replication, host-cell protein synthesis is heavily suppressed (Ginsberg *et al.*, 1967) and the infected cells are almost exclusively engaged in the synthesis of vast quantities of viral structural proteins and viral DNA. The nascent polypeptides assemble into multimeric structural proteins during or shortly after their synthesis (Horwitz

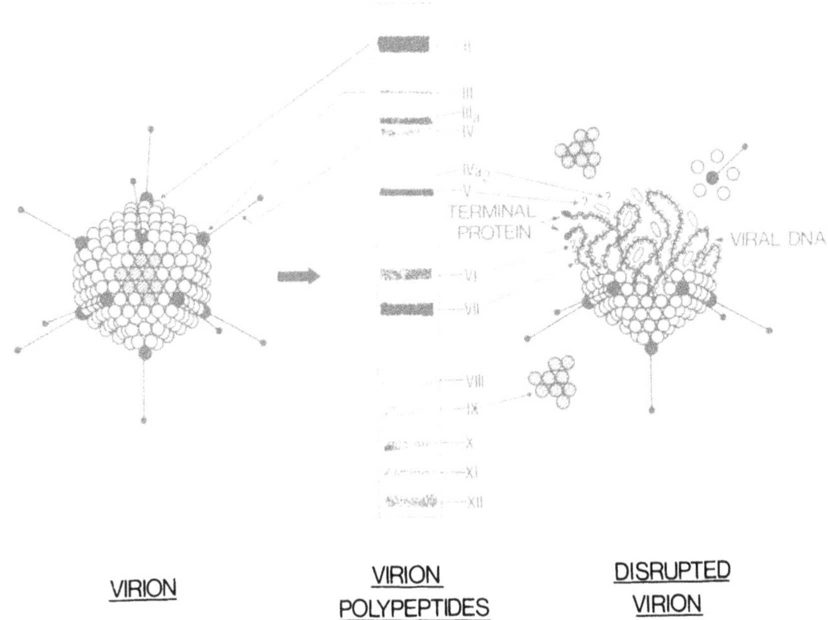

FIGURE 1. Schematic model illustrating the architecture of the adenovirus particle. The SDS-PAGE pattern is also shown, indicating all virion polypeptides known so far. The tentative locations of the different polypeptides in the virus capsid are also illustrated. Reprinted from Russell and Precious (1982) with permission.

et al., 1969; Velicer and Ginsberg, 1970), but only 1–5% of the fibers and penton bases and 20–30% of the hexons are used for assembly of virus particles (White *et al.*, 1969; Everitt *et al.*, 1971). The structural proteins of the adenovirion have attracted considerable attention, since they, unlike the structural proteins of most other animal viruses, are soluble under nondenaturing conditions and are available in sufficient quantities for detailed immunological and biochemical studies. The antigens that are present in the excess pool have been used as the source of material for purification of adenovirus capsid components, and several methods have been described for their production and purification, most of which are modifications of the original method of Green and Piña (1963).

A. HEXON

Several methods for purification of adenovirus hexons from infected cells have been reported in the literature. Köhler (1965) used methanol precipitation at pH 4.0 and DEAE chromatography, whereas other investigators (Valentine and Pereira, 1965; Hollinshead *et al.*, 1967; Russell *et al.*, 1967; Kjellén and Pereira, 1968; Maizel *et al.*, 1968b) used DEAE chromatography, either alone or in combination with other chromato-

TABLE II. Physical and Chemical Properties of Adenovirus Type 2 Proteins[a]

Polypeptide	Capsid component[b]	Molecular weight (K) Native	Molecular weight (K) Subunit[c]	Subunit composition	Sedimentation coefficient	Isoelectric point	Stokes's radius (nm)	Frictional coefficient (f/f_0)
III + IV	Penton (12)	365[d]	85 + 62	Trimer + trimer	10.5	Not determined	5.6	1.00
II	Hexon (240)	324[e]	108	Trimer	12.9	5.40–6.10	5.2	1.06
III	Penton base (12)	246–256[d]	85	Trimer	9.5	5.85	6.6	1.57
IV	Fiber (12)	156–207	62	Trimer	6.0	5.30–6.60	7.1	1.90
V	Core (180)	47	48	Monomer	3.5	6.70	3.4	1.42
VI	Hexon-associated (450)	50–72	24	Dimer or trimer	3.8	5.81	4.7	1.45
VII	Core (1070)	18–83	18.5	Monomer–tetramer	2.3	>7.50	3.3	1.14
VIII	Hexon-associated	13	13	Monomer	1.6	5.20	2.0	1.20
IX	Hexon nonamers (280)	23	14	Dimer	2.3	5.90	2.4	1.29
IIIa	Vertex region (60)	65	66	Monomer	6.0	5.95–6.08	2.7	1.02
100K	Nonstructural	94	100	Monomer	6.2	5.50–6.15	3.5	1.17
72K	Nonstructural	75	72	Monomer	3.5	6.70–7.10	4.5	1.61

[a] Modified from Lemay and Boulanger (1980).
[b] Numbers in parentheses are numbers of protein molecules per virus particle.
[c] Determined by SDS-PAGE.
[d] Data from Devaux *et al.* (1982).
[e] Data from Jörnvall *et al.* (1981).

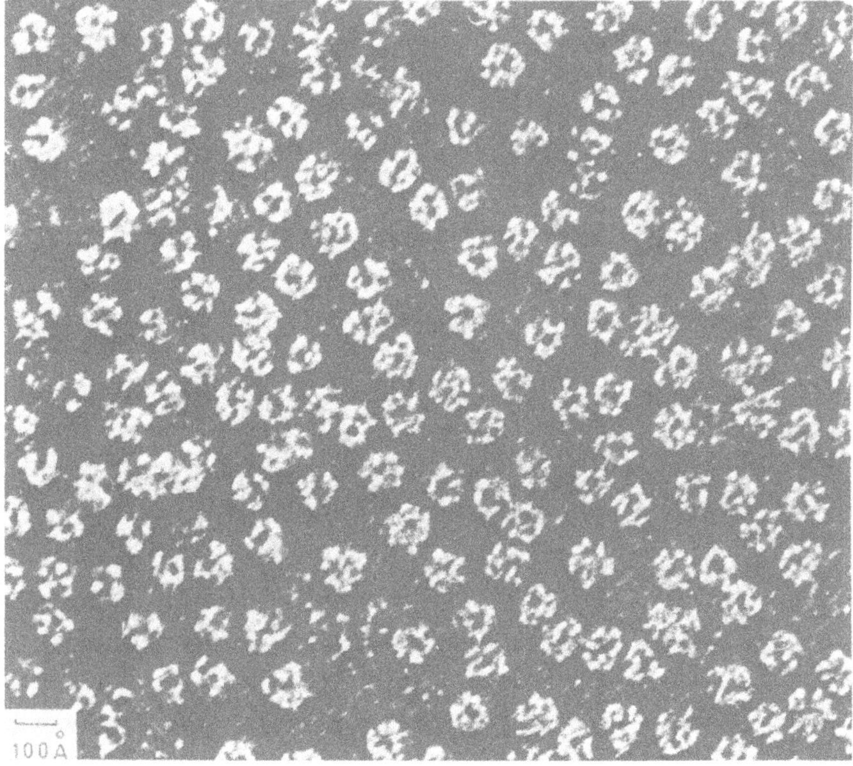

FIGURE 2. Electron micrograph of hexons.

graphic (Levine and Ginsberg, 1967) or electrophoretic methods (Pettersson *et al.*, 1967; Boulanger *et al.*, 1969). A convenient method for large-scale purification was introduced by Boulanger and Puvion (1973), giving high yields of purified proteins representing all the three major soluble antigens. After the discovery that hexons are easy to crystallize, methods based on crystallization have been introduced as a useful tool for large-scale purification of hexons. With the use of the latter procedure in combination with DEAE chromatography, sufficient amounts of purified hexon have been obtained to allow the determination of the complete amino acid sequence of the adenovirus type 2 (Ad2) hexon polypeptide (Jörnvall *et al.*, 1981).

1. Morphology

Using electron microscopy (EM) as the analytical tool, Horne *et al.* (1959) described hexons as solid spheres as did Valentine and Pereira (1965) later in their classic paper, which described the basic structure of adenovirus particles. Wilcox and Ginsberg (1963a) described hexons as hollow polygonal rods, and Pettersson *et al.* (1967) reported a cylindrical

1
ATPSMMPQWS YMHISGQDAS EYLSPGLVQF ARATETYFSL NNKFRNPTVA

51
PTHDVTTDRS QRLTLRFIPV DREDTAYSYK ARFTLAVGDN RVLDMASTYF

101
DIRGVLDRGP TFLPYSGTAY NALAPKGAPN SCEWEQTEDS GRAVAEDEEE
*

151
EDEDEEQEEE EQNARDQATK KTHVYAQAPL SGETITKSGL QIGSDNAETQ
*

201
AKPVYADPSY QPEPQIGESQ WNEADANAAG GRVLKKTTPM KPCYGSYARP

251
TNPFGGQSVL VPDEKGVPLP KVDLQLFSNT TSLNDRQGNA TKPKVVLYSE
*

301
DVNMETPDTH LSYKPGKGDE NSKAMLGQQS MPNRPNYIAF RDNFIGLMYY

351
NSTGNMGVLA GQASQLNAVV DLQDRNTELS YQLLLDDIGD DWRYFSMWNQ

401
AVDSYDPDVR IIENHGTEDE LPNYCFPLGG IGVTDTYQAI KANGNGSGDN

451
GDTTWTKDET FATRNEIGVG NNFAMEINLS ANLWRNFLYG NQALYLEDTA

501
KYNPTNVEIS DNPNTYDYMN KRVVAPGLVD CYINLGARWS LDYMDNVNPF

551
NHHRNAGLRY RSMLLGNGRY VPFHIQVPQK FFAIKNLLLL LGYYGTEWNF

601
RKDVNMVLQS SLGNDLRVDG ASIKFDSICL YATFFPMAHN TASTLEAMLR

651
NDTNDQSFND YLSAANMLYP IPANATNVPI SIPSRNWAAF RGWAFTRLKT

701
KETPSLGSGY NPYYTYSGSI PYLDGTFYLN HTFKKVAITF DSSVWWPGND

751
RLLTPNEFEI KRSVDGEGYN VAQCNMTKDS FLVQMLANYN IGYQGFYIPE

801
SYKDRMYSFF RNFQPMSRQV VDDTKYKEYQ QVGILHQHNN SGFVGYLAPT

851
MREGQAYPAN VPYPLIGKTA VDSITQKKFL CDRTLWRIPF SSNFMSMGAL

901
TDLGQNLLYA NSAHALDMTF EVDPMDEPTL LYVLFEVFDV VRVHQPHRGV

951
IETVYLRTPF SAGNATT

structure having a diameter of 8–10 nm and a central hole measuring 2.5 nm in diameter (Fig. 2). In a more recent study, Berger *et al.* (1978) described hexons as being pyramids with a triangular top and hexagonal base. A hole appears to extend from the apex to the base of the pyramid. The dimensions of the native Ad2 hexons have also been studied by low-angle X-ray diffraction (Tejg-Jensen *et al.*, 1972), and the results revealed cylinderlike objects with a height of 12.5 nm and a diameter of 8–8.5 nm. The radius of gyration has been reported to be 43–49 Å in different studies (Devaux *et al.*, 1982; Berger *et al.*, 1978). It is hoped that crystallographic studies of the hexon that have been initiated (Burnett *et al.*, 1979) will soon provide a definite answer as to the precise morphology of adenovirus hexons. On the basis of EM data, several three-dimensional models have been constructed (Nermut 1975, 1980; Nermut and Perkins, 1979); these models are discussed in Chapter 2.

2. Physicochemical Properties

Hexons from Ad2 sediment at 12.9 S (Table II) (Franklin *et al.*, 1971a), and their molecular weight is 324,000 (324K) (Jörnvall *et al.*, 1981). Several different estimates for the molecular weight of the hexon ranging between 180 and 400K have in the past been reported by different investigators (Valentine and Pereira, 1965; Wasmuth and Tytell, 1966; Köhler, 1965; Pettersson *et al.*, 1967; Grütter and Franklin, 1974; Devaux *et al.*, 1982). It is today established, however, that the Ad2 hexon consists of three identical polypeptide chains (Horwitz *et al.*, 1970; Cornick *et al.*, 1973; Boulanger and Puvion, 1974; Jörnvall *et al.*, 1974a), each with a molecular weight of 108K (Jörnvall *et al.*, 1981). The precise molecular weight of the hexon polypeptide is derived from the amino acid sequence that has been determined for Ad2 hexons.

Ad2 hexons consist of 967 amino acid residues (Jörnvall *et al.*, 1981; Aleström *et al.*, 1984), and the total composition is shown in Table III. Hexons are rich in dicarboxylic amino acids, and no carbohydrates have been detected (Pettersson *et al.*, 1967; Boulanger *et al.*, 1969). Seven CyS/Cys residues are present in the hexon, none of which appears to be engaged in disulfide-bond formation (Jörnvall *et al.*, 1974a; Grütter and Franklin, 1974). Only one Cys residue, however, is accessible in the native molecule (residue 629 in Fig. 3), presumably due to the conformational organization of native hexons (Grütter and Franklin, 1974; Jörnvall and Philipson, 1980).

It was originally reported by Laver *et al.* (1967) that proteins in the outer adenovirus capsid lack free N termini, and it was subsequently

←

FIGURE 3. Amino acid sequence of hexon from Ad2. A highly acidic region is underlined. (*) Sites at which trypsin cleaves the native hexon molecule. Data from Jörnvall *et al.* (1981) and Aleström *et al.* (1984).

TABLE III. Amino Acid Compositions of Different Structural and Nonstructural Adenovirus Type 2 Proteins[a]

		Hexon			Fiber												
Amino acid	Virion[b]	Type 2[c]	Type 3[d]	Penton base[e]	[f]	[g]	V[h]	VI[i]	VII[j]	VIII[k]	IX[l]	IIIa[m]	100K[n]	DBP[o]	IV_{a2}[p]	33K[q]	19K glycoprotein[r]
Lys	4.4	3.9	4.0	6.8	5.8	6.0	6.8	5.5	2.9	1.8	1.4	2.3	3.5	6.6	4.4	6.2	12.7
His	1.6	1.6	1.7	1.9	0.9	0.8	1.1	0.5	2.3	2.4	NF	1.9	2.4	3.0	4.2	0.4	1.4
Arg	7.9	4.7	4.9	5.3	2.0	1.5	13.0	7.4	23.6	5.5	5.8	5.2	7.6	7.0	7.8	7.5	2.8
Asp	11.8	14.6	14.6	10.4	13.1	13.1	5.4	7.8	5.7	7.0	8.6	10.2	9.3	8.8	10.5	7.1	7.7
Thr	6.9	6.9	8.3	6.1	11.7	11.1	8.9	6.0	6.9	4.8	8.6	5.8	4.4	4.2	4.4	8.4	9.9
Ser	6.7	6.3	6.8	6.0	10.7	11.2	4.9	9.7	3.4	20.0	15.8	10.4	5.3	8.3	5.3	7.9	3.5
Glu	9.0	9.4	8.3	12.1	6.0	6.4	11.7	11.1	3.4	13.1	6.5	11.8	14.8	12.3	9.6	14.5	9.9
Pro	7.2	5.9	6.1	5.8	5.8	5.2	8.9	11.1	6.9	5.1	5.8	8.3	6.7	8.1	8.7	10.6	4.2
Gly	7.8	6.8	6.4	7.3	8.0	8.5	5.9	8.3	6.3	20.0	4.3	10.8	6.6	4.8	5.8	2.2	2.8
Ala	9.0	7.2	6.8	7.9	6.3	6.5	8.4	6.9	19.0	9.1	16.5	8.5	8.9	8.7	7.3	15.0	5.6
Val	6.1	5.7	6.0	6.1	5.7	5.3	8.7	8.3	9.8	3.0	7.2	5.3	5.3	6.2	4.2	3.1	4.2
Met	2.3	2.9	3.0	2.0	2.1	1.5	1.6	1.4	1.7	0.5	1.4	1.2	1.2	2.8	3.6	0.9	3.5
Ile	3.4	3.5	3.9	4.2	5.5	5.0	2.4	3.2	2.3	2.3	2.2	3.5	3.7	2.8	4.7	1.8	5.6

Leu	7.4	8.0	7.3	9.2	9.6	9.9	7.9	7.8	2.3	2.7	11.5	7.4	11.2	7.9	11.1	8.8	7.7
Tyr	4.4	5.9	5.5	3.2	2.9	2.8	3.3	0.9	2.3	1.3	1.4	3.4	2.6	1.1	3.1	1.8	6.3
Phe	3.8	4.6	5.0	3.8	2.9	2.9	0.8	1.8	0.0	0.7	2.1	4.0	3.2	3.6	2.4	1.8	5.6
Trp	1.2	1.4	ND	0.9	0.7	1.6	0.5	1.8	1.1	ND	0.7	ND	0.7	1.1	1.1	1.3	1.4
CyS	0.3	0.7	0.5	1.0	0.5	0.7	0.0	0.5	0.0	ND	NF	ND	2.9	2.5	1.6	0.9	4.9

[a] (NF) Not found; (ND) not determined.
[b] Data from Polasa and Green (1967).
[c] Data from the amino acid sequence of Ad2 hexons (Jörnvall *et al.*, 1981).
[d] Data from Pettersson (1971).
[e] Data from Boudin *et al.*, (1979).
[f] Data from the predicted amino acid sequence of the Ad2 fiber (Hérissé *et al.*, 1981).
[g] Data from Pettersson *et al.* (1968).
[h] Data from Aleström *et al.*, (in prep).
[i] Data from Akusjärvi and Persson (1981).
[j] Data from Aleström *et al.*, (in prep).
[k] Data from Everitt and Philipson (1974).
[l] Data from the predicted amino acid sequence of polypeptide IX from Ad2 (Aleström *et al.*, 1980).
[m] Data from Lemay *et al.* (1980).
[n] Data from the predicted amino acid sequence (Galibert *et al.*, 1979; Kruijer *et al.*, 1981). The composition corresponds to a hypothetical hybrid protein consisting of N-terminal sequences from Ad5 and C-terminal sequences from Ad2.
[o] Data from the predicted sequence of Ad2 DNA-binding protein (DBP) (Kruijer *et al.*, 1982).
[p] Data from van Beveren *et al.* (1981) and Gingeras *et al.* (1982).
[q] Data from Oosterom-Dragon and Anderson (1983).
[r] Data from Hérissé *et al.* (1981).

established that hexons indeed contain an acetylated N-terminal amino acid (Laver, 1970; Pettersson, 1971). The N terminus of Ad2 hexons has the sequence acetyl-Ala-Thr-Pro-Ser (Jörnvall *et al.*, 1974b).

3. Primary Structure

The 967-residue-long sequence of the Ad2 hexon polypeptide is shown in Fig. 3. The sequence exhibits some unusual features; Pro residues show an extraordinary distribution in the sense that many are associated in pairs along the polypeptide chain separated by 1–5 other amino acid residues. In fact, 32 of the 57 Pro residues in the hexon polypeptide occur in such combinations, with 14 Pro residues in exactly the relative numerical positions 1 and 5. The Pro residues are furthermore unevenly distributed along the hexon polypeptide chain (Jörnvall and Philipson, 1980); they are frequent in the first third of the sequence (27 Pro of the first 335 residues), uncommon in the middle third (10 Pro of the middle 333 residues), and again common in the C-terminal third (20 prolines of the last 298 residues). Clusters of hydrophobic residues are essentially absent from the hexon sequence. In only one place are there more than 3 aliphatic hydrophobic residues in a row; the penta-Leu sequence starting at position 587 is located in a 20-residue-long segment in which additional hydrophobic and other neutral residues are overrepresented. Charged residues, on the other hand, are common. The sequence contains 207 such residues (21%), with an excess of 11 acidic residues over the sum of the basic ones. The previously determined amino acid composition shows an apparently greater excess of acidic residues, since dicarboxylic residues are frequent, but the majority of them are amidated (123 amides and 109 free carboxylic groups). Residues with positive and negative charges are intermingled in the primary structure, except in a highly acidic region that includes 15 dicarboxylic residues (positions 146–161) and a more widespread region, with 9 charged residues (positions 552–580). The former, very acidic, region appears, as expected, to be exposed on the surface of the protein in its native conformation, since it can be eliminated by limited tryptic digestion.

Some additional residues also appear in specific combinations, in particular Met and Thr residues. For instance, 13 Met residues are preceded by Asn, Pro, or Ser, giving 4–5 identical depeptide sequences. Similarly, in the case of Thr, 17 of 67 such residues are followed by Phe, His, or Lys. Repeats of 3 and 4 residues are also common in the hexon sequence, and 84 such repeats affecting a total of 542 residues have been identified. These findings suggest a sequence pattern of limited variability, although no regions with long repeated sequences are apparent. Consequently, the many short regions with similar or identical amino acid combinations probably do not reflect remnants of possible ancestral duplications. They indicate instead strict structural requirements on the molecule, limiting the sequence patterns and suggesting the presence of different types of conformations at the level of secondary or supersecondary structures. The importance of conformational rather than ancestral relationships to explain the repeats is also supported by the fact that Pro is frequently involved in short repeats, rather than Gly, which would be more typical for distant evolutionary connections.

Finally, the amino acid patterns in different parts of the molecule also explain the accessibility of some residues to labeling and proteolysis (Jörnvall and Philipson, 1980). Combined, all sequence data therefore suggest that three portions of the hexon polypeptide chain may be distinguished, with different conformations in the middle as compared to the two end portions and with elements regularly repeated in all regions. It will be interesting to see whether these conclusions are confirmed by high-resolution crystallographic work.

Although the complete amino acid sequence is not yet known for hexons from any serotype other than Ad2, it has been established that the amino acid compositions for hexons from other serotypes, i.e., Ad5 and Ad3, show only insignificant differences, suggesting that the hexon is a well-conserved entity in adenovirus evolution. Stinski and Ginsberg (1975) have compared the peptides that are generated after CNBr cleavage of hexons from Ad2, Ad5, and Ad3. They found that approximately two thirds of the CNBr fragments from Ad2 and Ad5 have identical isoelectric points, whereas Ad3 hexons had fewer CNBr fragments in common with Ad2 hexons. Von Bahr-Lindström *et al.* (1982) have recently determined part of the sequence for the hexon polypeptide from Ad5. In this study, 59% of the sequence (529 residues) was determined, and the results showed, as expected, that Ad2 and Ad5 hexons are extremely similar. Only 23 residues (4%) were found to be different, in most cases as the result of single base-pair mutations. The molecular ends of the hexon polypeptide were found to be particularly well conserved. At least two deletions appear to have occurred, however, in the Ad5 sequence, rendering Ad5 hexons 5–7 K (approximately 60 residues) smaller. These deletions include the three trypsin-sensitive sites located between residues 140 and 290, which makes the native Ad5 hexon resistant to the action of low trypsin concentrations (see Section II.A.5).

Boursnell and Mautner (1981) have used an indirect approach to identify heterologous regions in the hexon polypeptide; i.e., they have identified unique restriction enzyme cleavage sites in the genes for Ad2 and Ad5 hexons. Their results confirm that the two ends of the hexon molecule are very well conserved and suggest that the heterologous sites are clustered in the region of the Ad2 hexon polypeptide where the trypsin-sensitive sites are located, i.e., between residues 140 and 290. This region is thus likely to contain type-specific antigenic determinants.

4. Three-Dimensional Structure

The adenovirus hexon protein was the first animal virus protein to be crystallized. According to the original method described by Pereira *et al.* (1968), purified hexons were dialyzed against 0.8 M KH_2PO_4. Under these conditions, a precipitate is formed that is gradually converted into tetrahedral crystals. Franklin *et al.* (1971a) and Cornick *et al.* (1971) have used slightly different conditions and thereby obtained pyramidal crystals. Structural studies of both crystal types have revealed that they belong to the cubic system and have the space group $P2_13$. The length of the cubic cell is 149.9 Å, and each unit cell contains four hexons. Since there are three asymmetrical units in the cell, there must be three crystallographic asymmetrical units per hexon. The cylinder axis of the hexon is parallel to the axis of 3-fold symmetry. Consequently, there are three structural units symmetrically arranged around the cylinder axis of the hexon, which has also been confirmed by EM (Franklin *et al.*, 1971b; Crowther and Franklin, 1972). One dyad that is perpendicular and another that is parallel to the 3-fold axis have been observed (Franklin *et al.*, 1971b). The structural implications of these dyads are questionable, however, since it is now established that the hexon protein is a trimer. Neither does the amino acid sequence reveal along the hexon polypeptide any homologous regions of significant length to account for these observations (Jörnvall *et al.*, 1981). The three-dimensional (3D) structure of the hexon is discussed in detail in Chapter 2.

5. Immunological Properties

It was discovered very early in the history of adenovirus research that almost all adenoviruses contain a common antigenic specificity known as α (Pereira *et al.*, 1959; Wilcox and Ginsberg, 1963a,b). Hexons from avian and certain bovine adenoviruses seem to be the exceptions to this rule (Bürki *et al.*, 1979; Wigand *et al.*, 1982). More detailed immunological studies (Wilcox and Ginsberg, 1963a,b; Köhler, 1965; Pettersson, 1971) have revealed an additional type-specific component in the hexon designated ϵ (Table IV). With the use of more sensitive techniques, a more complex arrangement of the antigenic determinants in the hexon molecule was demonstrated (Norrby 1969a,b; Norrby and Wadell, 1969). Type, group, and intrasubgroup and intersubgroup specificities were revealed in these studies by cross-absorption and complement-fixation assays. Serotype 12, which appears to have a distant relationship to other human adenoviruses, has hexons that differ with regard to charge and that share few antigenic determinants with hexons from other serotypes (Norrby and Wadell, 1969). Hexons from Ad4, which is classified as the only member of subgroup E, are most closely related to hexons from subgroup B and in particular to hexons from Ad16 (Norrby and Wadell, 1969). Intermediate adenovirus strains that contain hexon specificities from one parental strain and fiber specificities from another parental strain have been described (Norrby, 1969c). These strains are presumably naturally occurring recombinants between two serotypes.

TABLE IV. Antigens Associated with the Major Structural Proteins[a]

Protein	Corresponding polypeptide	Designation	Antigens	
			Specificity	Remarks
Hexon	II	α	Group	Oriented toward the inside of the virion
		?	Inter- and intrasubgroup	—
		ϵ	Type	Available at the surface of the virion
Penton base	III	β	Group	Carries toxin activity
			Inter- and intrasubgroup	—
Fiber	IV	γ	Type	Reacts with hemagglutination-inhibition antibody
		?	Intersubgroup	Shared between members of subgroups C and D
		δ	Intrasubgroup	Located at the proximal part of the fiber; present only in subgroups A, C, D, and E
Major core protein	VII	—	Group and type	Only Ad2 and Ad3 have been examined
Minor virion polypeptides	IIIa	—	Group	—
	IX	—	Group and type	Only Ad2 and Ad5 have been examined

[a] Adapted from Philipson *et al.* (1975).

Type-specific hexon determinants have been demonstrated on the surface of the virion (Norrby *et al.*, 1969a; Willcox and Mautner, 1976). The group-specific determinant (α), in contrast, does not appear to be exposed on the surface, since only homologous antisera give rise to a visible antibody interaction (Norrby *et al.*, 1969a). Adenoviruses in subgroup C seem to have few determinants exposed on the outside of the particle, since homologous hexon antibodies fail to aggregate the virions as judged by EM (Wadell, 1972) and sucrose-gradient centrifugation (Prage *et al.*, 1970). An interesting observation is the finding that type-specific antibodies directed against the ϵ determinant of the hexon interfere with the capacity of adenoviruses from subgroups A and B to agglutinate erythrocytes. This inhibition is presumably caused by the attachment of antibodies to hexons in the peripentonal region, which thereby prevent the attachment of adenoviruses with short fibers, i.e., subgroup B and D viruses (see Table I), to the red-cell receptor because of steric hindrance (Norrby, 1969b; Norrby and Wadell, 1969). Subgroup C virions, in contrast, having long fibers (Table I), are less inhibited by homologous hexon antibodies in their interaction with erythrocytes.

The topography of the antigenic determinants on the hexon protein has been studied by the use of partial proteolysis. Treatment of purified Ad2 hexons with trypsin introduces breaks in the polypeptide without changing the antigenic specificity (Pettersson, 1971). The sites in the native hexon molecule that are sensitive to mild trypsin treatment have been identified in the Ad2 hexon sequence and are located between the following residues: 142–143, 165–166, 286–287. These three sites are the only ones that are recognized in the native molecule, suggesting that the region between residues 140 and 290 is exposed on the surface of the hexon (von Bahr-Lindström *et al.*, 1982). Limited digestion with chymotrypsin, papain, or subtilisin, in contrast, removes more than 50% of the hexon polypeptide, and parts of the group-specific determinant α are destroyed under these conditions (Pettersson, 1971). The type-specific determinant (ϵ) remains unaltered after limited proteolysis of native hexon and can be removed only by drastic proteolytic treatment of hexons, suggesting that this part of the hexon polypeptide has an unusually compact structure.

The use of monoclonal antibodies provides yet another way to study the antigenic determinants of the hexon. Cepko *et al.* (1981) have prepared group-reacting monoclonal antibodies that precipitate trimeric (native) hexons as well as hexon nonamers.* They fail, however, to precipitate intact virions and unassembled hexon polypeptide chains. The results thus demonstrate two properties of the hexon antigenic determinants; the native hexon molecule is recognized by different antibodies than are

Aggregates of nine hexons (nonamers) are released from the facets of the adenovirus capsid when it is disintegrated by certain agents [also see Pereira and Wrigley (1974) and Chapter 2].

unassembled hexon polypeptides. This has also been demonstrated by Ginsberg and co-workers (Ginsberg *et al.*, 1974; Stinski and Ginsberg, 1974; Ginsberg, 1979), who found that hexons that have been treated with strong denaturing agents induce a different set of antibodies than does native hexon. The results of Cepko *et al.* (1981) also show that certain antigenic determinants of the hexon are unavailable in the intact virus particle, which confirms previous results obtained by EM (Norrby *et al.*, 1969a). A collection of monoclonal antibodies against Ad5 hexons has been described by Russell *et al.* (1981). These antibodies include five apparently different epitopes against the hexon, confirming the immunological complexity of the hexon molecule, previously demonstrated in cross-adsorption studies. One monoclonal antibody against Ad5 hexons failed to react with Ad2 hexons but did react with Ad1 and Ad6 hexons, revealing a specificity that had been overlooked in previous studies. It seems likely that the use of monoclonal antibodies will reveal a large number of antigenic determinants that exhibit widely different reactivities with hexons from different serotypes.

Two classes of hexon molecules, a major and a minor, more slowly migrating population, have been identified after purification of native hexons by PAGE (Pettersson, 1971; Boulanger *et al.*, 1978a). Analysis of the slowly migrating hexons by SDS-PAGE suggests that they are composed of a slightly shorter polypeptide, lacking approximately 50 residues at the N-terminal end (Boulanger *et al.*, 1978a). The reported presence of an unblocked N-terminal amino acid suggests, moreover, that they arise through proteolytic trimming (Boulanger *et al.*, 1978a). Immunological studies have failed to reveal any distinct differences between the two classes, although the slow hexons are more reactive against type-specific antibodies (Boulanger *et al.*, 1978a). Antisera against the slowly migrating population of Ad2 hexons also appear to neutralize virus infectivity more efficiently than antisera against the major population (Pettersson, 1971). The slowly migrating hexons are, however, absent from virions, and their biological function therefore remains to be determined.

Hexons have been reported by some investigators to carry the antigen that is responsible for adenovirus neutralization (Wilcox and Ginsberg, 1963b; Kjellén and Pereira, 1968; Pereira and Laver, 1970; Norrby *et al.*, 1969a; Haase and Pereira, 1972), whereas other investigators (Pettersson *et al.*, 1968) have claimed that highly purified preparations of Ad2 hexons induce insignificant titers of neutralizing antibody. These conflicting results could be interpreted in two ways. One possibility is that hexons, when extensively purified, change their conformation and thereby fail to induce antibodies with proper neutralizing activity. Alternatively, less pure hexon preparations might contain small amounts of a contaminating protein that is thus true neutralizing antigen. The finding of Stinski and Ginsberg (1975) that hexons, when free in solution, have different antigenic properties than hexons in the virus particle lends support to the former hypothesis. Also, the observation that the slowly migrating hexons, which appear to have lost a short segment, are more efficient in inducing neutralizing antibodies suggests that the conformation rather than the primary sequence accounts for the difference.

6. Physiological Functions

No specific nonstructural function has yet been assigned to the hexon, although DNA-binding studies have revealed that hexons have weak DNA-binding properties. It has been suggested on the basis of this observation that hexons can bind to DNA and thereby inhibit DNA polymerase and DNA-dependent RNA polymerase during the lytic infection (Levine and Ginsberg, 1968). The validity of this hypothesis is unclear, however, since mutants that fail to synthesize structural proteins under nonpermissive conditions still have the capacity to inhibit host-cell macromolecular synthesis (Wilkie *et al.*, 1973).

The hexon polypeptide carries a reactive Ser residue as revealed by its affinity for diisopropylfluorophosphate (Devaux and Boulanger, 1980). However, no enzymatic activity seems to be connected with this property, and it has been proposed that the reactive residue plays its role during polypeptide assembly (Devaux and Boulanger, 1980).

B. Fiber

Several different methods have been described for the purification of fibers from the soluble antigen pool in cells infected with Ad2 or Ad5. Most purification schemes include one or more cycles of DEAE chromatography in combination with other separation methods such as isoelectric focusing, hydroxylapatite chromatography, or gel-exclusion chromatography. In this way, homogeneous preparations of fibers have been obtained, suitable for biochemical studies. Fibers are present in infected cells at a much lower concentration than hexons, which has hampered the biochemical studies of this protein.

1. Morphology

The fiber is a highly assymetrical structure [$f/f_0 = 1.90$ (Table II)], consisting of a rod with a terminal knob (Fig. 4). The diameter of the rod is estimated to be 2 nm and the diameter of the knob 4 nm. The length of the fiber is a characteristic property for members of different adenovirus subgroups (see Table I). Subgroup B adenoviruses have the shortest fibers, 10 nm, and subgroup C the longest, 23–31 nm (Norrby, 1969b). Ad4, which constitutes its own subgroup (E), has fibers that are 17 nm long. Avian adenoviruses, except the agent of the so called egg drop syndrome, have been found to have a peculiar morphology, with two fibers of different lengths extending from different sites on the penton base (Laver

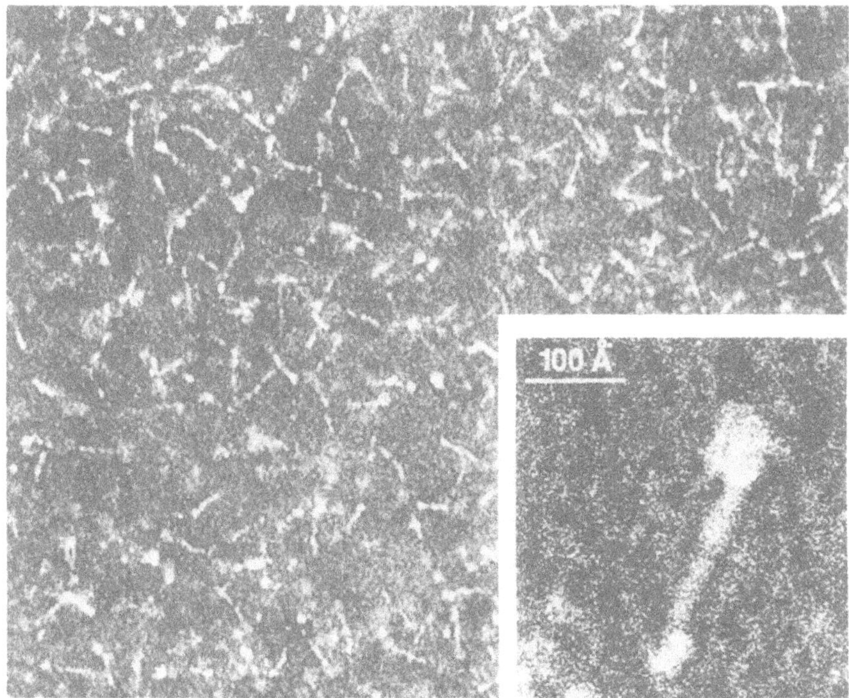

FIGURE 4. Electron micrograph of fibers.

et al., 1971; Gelderblom and Maichle-Lauppe, 1982). Extracts from adenovirus-infected cells contain dimers of fibers that are stable after several purification steps. It is not yet known, however, whether these dimers exist under physiological conditions or whether they contain an extra component that links the two units (Wadell and Norrby, 1969a; Norrby *et al.*, 1969b). Fibers from Ad5 were crystallized as early as 1971 (Mautner and Pereira, 1971), but no detailed 3D studies were reported, presumably because the crystals are too small to allow a detailed structural examination. Conformational studies suggest that the native fiber is mainly in a β-pleated sheet structure (Boulanger and Loucheux, 1972).

2. Physicochemical Properties

Fibers from Ad2 and Ad5 sediment at about 6 S (see Table II) (Levine and Ginsberg, 1967; Pettersson *et al.*, 1968; Wadell *et al.*, 1969; Devaux *et al.*, 1982), and molecular-weight determinations based on sedimentation equilibrium centrifugation or neutron scattering show that fibers from Ad2 and Ad5 have a molecular weight of 156–207K (Sundquist *et al.*, 1973; Dorsett and Ginsberg, 1975; Devaux *et al.*, 1982). The radius of gyration is 75 Å (Devaux *et al.*, 1982). The polypeptide composition of the fiber has been examined by SDS-PAGE, and the results reveal that fibers are composed of polypeptide chains with an estimated molecular weight of about 62K (Pettersson *et al.*, 1968; Maizel *et al.*, 1968b; Sundquist *et al.*, 1973; Dorsett and Ginsberg, 1975). The molecular weight of the fiber protein thus suggests that it is a trimeric structure, although the estimate is not entirely unambiguous due to the fluctuation among the values that have been reported for the molecular weight of the native fiber protein. As to the identity of the three polypeptide subunits of the fiber, the results are to some extent conflicting. SDS-PAGE resolves a single polypeptide band (Maizel *et al.*, 1968b; Sundquist *et al.*, 1973). Dorsett and Ginsberg (1975), on the other hand, have provided evidence for the existence of two types of fiber polypeptides using two-dimensional (2D) fingerprint analysis and polypeptide separation in urea-containing polyacrylamide gels. Their results suggest that the trimeric fiber consists of two copies of one polypeptide and one copy of another. A single type of polypeptide chain in the highly assymetrical fiber structure would be unexpected on a theoretical basis, since it is difficult to envisage how identical polypeptide subunits could be assembled to form the fiber structure. On the other hand, it appears clear that the adenovirus genome contains a single fiber gene that encodes a polypeptide chain with a predicted molecular weight of 62K. It has been established, moreover, that purified trimeric fiber protein has the same amino acid composition (Pettersson *et al.*, 1968; Dorsett and Ginsberg, 1975) as can be predicted from the DNA sequence of the fiber gene (see Table III). Thus, if two populations of fiber polypeptides exist, their difference is likely to be caused by posttranslational modification of the polypeptide chain.

Fibers contain low amounts of Arg and are rich in hydroxyamino acids as compared to hexons (Table III) (Pettersson *et al.*, 1968; Dorsett and Ginsberg, 1975). Ishibashi and Maizel (1974) originally discovered that the fiber is a glycoprotein containing two residues of *N*-acetyl glycosamine in each polypeptide chain. The functional significance of the carbohydrate moiety remains unknown. A mutant of Ad5 that fails to glycosylate fibers at the nonpermissive temperature has been isolated (Ginsberg and Young, 1977; Cheng Chee-Sheung and Ginsberg, 1982). This mutant produces fibers with physical properties similar to those of fibers produced by wild-type virus, although they do not seem to be immunologically reactive (Cheng Chee-Sheung and Ginsberg, 1982). It can be predicted from the nucleotide sequence (Hérissé *et al.*, 1981) that the Ad2 fiber polypeptide consists of 581 amino acid residues (Fig. 5) with a molecular weight of 61.9K, a value that is in almost perfect agreement with the previous size estimates by SDS-PAGE and ultracentrifugation. The predicted amino acid sequence for the Ad2 fiber does not reveal any striking features. Three Cys residues have been identified in the sequence, but it is not known whether any of them are engaged in disulfide bond formation. Nine potential glycosylation sites are present in the sequence, but it remains to be established which of these are actually glycosylated.

```
1
KRARPSEDTF NPVYPYDTET GPPTVPFLTP PFVSPNGFQE SPPGVLSLRV

51
SEPLDTSHGM LALKMGSGLT LDKAGNLTSQ NVTTVTQPLK KTKSNISLDT

101
SAPLTITSGA LTVATTAPLI VTSGALSVQS QAPLTVQDSK LSIATKGPIT

151
VSDGKLALQT SAPLSGSDSD TLTVTASPPL TTATGSLGIN MEDPIYVNNG

201
KIGIKISGPL QVAQNSDTLT VVTGPGVTVE QNSLRTKVAG AIGYDSSNNM

251
EIKTGGGMRI NNNLLILDVD YPFDAQTKLR LKLGQGPLYI NASHNLDINY

301
NRGLYLFNAS NNTKKLEVSI KKSSGLNFDN TAIAINAGKG LEFDTNTSES

351
PDINPIKTKI GSGIDYNENG AMITKLGAGL SFDNSGAITI GNKNDDKLTL

401
WTTPDPSPNC RIHSDNDCKF TLVLTKCGSQ VLATVAALAV SGDLSSMTGT

451
VASVSIFLRF DQNGVLMENS SLKKHYWNFR NGNSTNANPY TNAVGFMPNL

501
LAYPKTQSQT AKNNIVSQVY LHGDKTKPMI LTITLNGTSE STETSEVSTY

549
SMSFTWSWES GKYTTETFAT NSYTFSYIAQ E
```

FIGURE 5. Amino acid sequence of Ad2 fiber. Data from Hérissé and Galibert (1981) and Hérissé *et al.* (1981).

Green *et al.* (1983) have recently proposed a new model for the fiber based on the predicted amino acid sequence of the fiber polypeptide. According to this model the fiber protein contains a repeating motif of 15 residues. These repeats extend from residue 43 to residue 400 of the 581 residue long fiber polypeptide. The model predicts that each 15-residue segment contains two short β-strands and two β-bends, one of which incorporates an extra residue in a β-bulge of the Gx type. The 44 strands together form a long (210 Å) narrow amphipathic β-sheet which could be stabilized by dimer formation to give the shaft of the fiber. The knob may arise from a dimer of the C-terminal 180 residue segment predicted to be an 8-10 stranded β-sandwich. According to this model the fiber protein would be a dimeric unit consisting of two fiber polypeptides each with a molecular weight of 62K.

Nearly all studies on adenovirus fibers have so far been carried out with fibers from subgroup C members, and unfortunately no data are yet available concerning the structure of fibers from other subgroups.

3. Immunological Properties

The fiber contains one type-specific determinant, γ, that resides in the knob of the fiber molecule as revealed by EM (Norrby *et al.*, 1969a). Serotypes that belong to subgroup B have the shortest fibers (see Table I) and contain exclusively the type-specific γ determinant. For serotypes with longer fibers, a positive correlation has been found between the length of the fiber and its antigenic complexity (see Table I). The long fibers present in virions from subgroups A, C, and D have in their rodlike portion an additional antigenic determinant, δ, that is shared among fibers from members within the same subgroup. The long fibers of subgroup C members have in addition inter-subgroup-specific determinants (Norrby, 1968, 1969b; Pettersson *et al.*, 1968). The type-specific determinant, γ, can be demonstrated in hemagglutination-inhibition and immunodiffusion assays (Pereira and Figueiredo, 1962; Valentine and Pereira, 1965; Norrby, 1966b; Pettersson *et al.*, 1968) and binds at least two immunoglobulin G (IgG) molecules (Norrby *et al.*, 1969a). Antibodies against the γ specificity prevent the attachment of virions to red-cell receptors (Pereira and Figueiredo, 1962) and also prevent the attachment of fibers to KB- and HeLa-cell receptors (Philipson *et al.*, 1968). It is believed that the δ determinants are located near the junction between the fiber and the penton base, since they are masked in the intact penton structure (Pettersson and Höglund, 1969; Wadell and Norrby, 1969b).

4. Physiological Functions

It is assumed that the fiber recognizes specific receptors on the plasma membrane during the first stages of a productive infection and that it thereby mediates the early contact between viruses and cells. Evidence in favor of this assumption is available from studies on adenovirus infection of KB and HeLa cells (Levine and Ginsberg, 1967; Philipson *et al.*, 1968). Purified fiber in quantities 10 to 100-fold higher than the number of virus receptors (around 10^4 per cell) can inhibit the attachment of virus to the cell surface (Philipson *et al.*, 1968). With the use of highly labeled fiber preparations, it has been demonstrated that the fiber binds to the cell and that attachment can be prevented by antibodies against fiber (Philipson *et al.*, 1968). After penetration of the plasma membrane, the virus seems to lose the fiber together with other proteins in the vertex region (Sussenbach, 1967). Experiments with ^{125}I-labeled virions have shown that intracellular virus lacks the fiber together with other polypeptides associated with the peripentonal region (unpublished data).

High concentrations of fiber have been found to inhibit the replication of adenoviruses as well as polio and vaccinia viruses, and fibers have thus been thought to be identical with the *adenovirus interfering principle*, first described by Pereira (1960). Fibers, like hexons, bind to DNA at a low ionic strength, and the binding inhibits the activity of

DNA and RNA polymerases *in vitro*. It has therefore been concluded that purified fiber may be responsible for the inhibition of macromolecular synthesis in uninfected cells after a latency period of 24–28 hr (Levine and Ginsberg, 1967, 1968). DNA and RNA as well as protein synthesis is reduced when cell cultures are treated with fiber at concentrations of 50 $\mu g/10^6$ cells. It is unclear, however, whether the fiber alone is responsible for the inhibition of host-cell macromolecular synthesis during viral infection, since the inhibition during the lytic cycle is observed prior to the synthesis of fiber protein (Russell *et al.*, 1967). It has also been shown that adenovirus mutants that are defective with regard to synthesis of fibers still inhibit host-cell DNA synthesis at the nonpermissive temperature (Wilkie *et al.*, 1973). The effect of the fiber on host macromolecular synthesis during a lytic infection is thus unclear despite the clear-cut results that have been observed under *in vitro* conditions.

No enzymatic effect has so far been linked to the fiber. Purified fibers from Ad1 and Ad2 are associated with an activity known as the *erythrocyte-receptor-modifying (ERM) factor* that inactivates receptors on human erythrocytes (Kasel *et al.*, 1961; Kasel and Huber, 1964). More detailed studies indicate, however, that this effect is due to the binding of excess fibers to erythrocyte receptors, rather than to enzymatic modification of the receptor itself (Wadell, 1969).

C. Penton

The pentons, which are composed of two structural units, the base and the fiber (Fig. 6), are more labile than the other major capsid components. It has therefore been more difficult to purify pentons in large quantities, and they have been subjected to much less thorough studies than hexons and fibers. Pentons from Ad2 and Ad5 have been purified by procedures similar to those used for hexon purification, i.e., DEAE chromatography either alone or in combination with hydroxyapatite chromatography, gel-exclusion chromatography, or electrophoresis (Valentine and Pereira, 1965; Maizel *et al.*, 1968b; Pettersson and Höglund, 1969). In this way, it has been possible to obtain sufficient amounts of homogeneous pentons for biochemical studies.

1. Morphology

EM studies have shown that penton bases have a morphology very similar to that of isolated hexons, and pentons were originally described as spheres with an average diameter of 8 nm (Valentine and Perera, 1965). Some investigators have observed a conical morphology when using different procedures for preparing the specimens (Laver *et al.*, 1969; Wadell *et al.*, 1969), whereas others have observed globular or polygonal structures sometimes exhibiting a pentagonal outline (Pettersson and Hög-

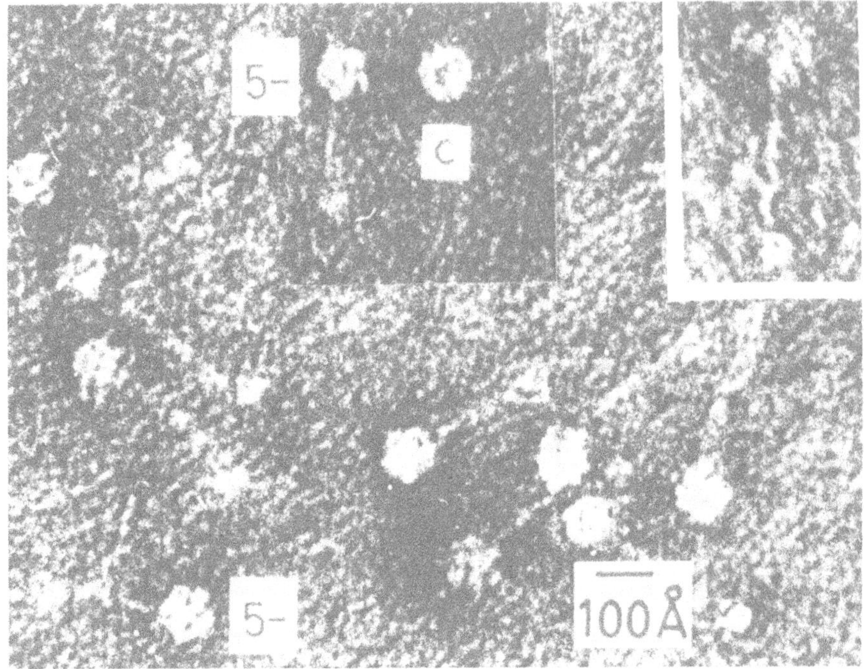

FIGURE 6. Electron micrograph of pentons. Penton bases with a pentagonal outline are indicated (5-) as well as one with a central cavity (C).

lund, 1969; Boudin *et al.*, 1979). As in hexons, central holes were sometimes seen in the penton bases, although not as consistently as in hexon preparations (Pettersson and Höglund, 1969; Boudin *et al.*, 1979). When attached to the penton base, the fiber is indistinguishable from the free fibers present in infected cells. Pentons, like fibers, have sometimes been found to appear as aggregates when isolated from infected cells (Norrby, 1966a; Wadell *et al.*, 1969), and two types of aggregates have been recognized. Dimers of pentons are present in cells infected with certain members of subgroups B and C; other members of subgroups B and D as well as of E produce aggregates consisting of 12 pentons in a starlike arrangement, known as dodecons [see Fig. 11 (Section IV.A)]. Although dodecons from some different serotypes contain pentons of identical size, they have been found to sediment at widely different rates (50–100 S) (Norrby, 1969b). This observation suggests that the pentons are oriented around a central core structure of unknown properties. A polypeptide with an approximate molecular weight of 37K has been found to be associated with purified dodecons, but it is not yet known whether it represents a virus-encoded protein (Winberg and Hammarskjöld, personal communication).

2. Physicochemical Properties

Pentons from Ad2 and Ad5 sediment at approximately 10.5–11 S (Table II) (Pettersson and Höglund, 1969; Wadell *et al.*, 1969). Widely different estimates have been reported in the literature for the molecular weight of pentons. Pettersson and Höglund (1969) arrived at a value of 400K for Ad2 pentons by analytical ultracentrifugation, whereas Wadell (1970) suggested 485–505K for Ad5 pentons using ultracentrifugation and gel-exclusion chromatography. EM has yielded considerably lower estimates (Valentine and Pereira, 1965), probably due to shrinkage of the specimen during fixation. Devaux *et al.* (1982) have recently applied more precise methods for sizing of native adenovirus capsid proteins and arrived at a value of 365K for the complete penton structure. This is likely to be a reliable estimate for Ad2 pentons, since several methods were applied in this study (Devaux *et al.*, 1982). Analysis by SDS-PAGE of virions and purified pentons has revealed that the penton base is composed of identical polypeptide chains, each having a molecular weight of approximately 85K (Anderson *et al.*, 1973; Everitt *et al.*, 1973; Devaux *et al.*, 1982). Molecular-weight determinations using neutron-scattering techniques have given a molecular weight of 246K for the base (Devaux *et al.*, 1982), which therefore suggests that it is a trimeric structure. This is an unexpected finding because of the pentagonal symmetry at the corners of the adenovirus icosahedron, and the results may suggest that polypeptide IIIa, rather than the base, is connected to the peripentonal hexons (see Section III.A). The fiber is joined to the penton base by noncovalent bonds, and it can be dissociated into its two constituents by treatment with guanidine hydrochloride (Norrby and Skaaret, 1967), formamide (Neurath *et al.*, 1968), pyridine (Pettersson and Höglund, 1969), or deoxycholate (DOC) (Boudin *et al.*, 1979). Penton bases that have been dissociated from the fiber have been purified and shown to retain their antigenicity and their ability to induce cytopathic changes (see Section II.C.4) when added to cell monolayers (Pettersson and Höglund, 1969). Boudin *et al.* (1979) and Winters *et al.* (1970) have shown, moreover, that free vertex capsomers exist in adenovirus-infected cells, and they have been purified to homogeneity directly from the soluble antigen pool. Cells infected with temperature-sensitive mutants, producing defective fiber polypeptides, overproduce free penton bases, which has provided a convenient source for purification (Boudin *et al.*, 1979). The free bases sediment at 9.5 S and appear to have a blocked N terminus, like the other major capsid components (Boudin *et al.*, 1979). Penton bases and fibers are able to assemble *in vitro* into pentons that are indistinguishable from their naturally occurring counterparts (Boudin and Boulanger, 1982). The reaction is reversible with a dissociation constant $K_D = 2 \times 10^{-7}$M (fiber). Evidence has been provided that the C-terminal end of the fiber is involved in the attachment to the penton base (Boudin and

Boulanger, 1982). Boudin and Boulanger (1981) have furthermore made the interesting observation that the addition of antibodies against the fiber, particularly those directed against the δ determinant, appear to cause dissociation of the penton into its two constituents, suggesting that the antibodies induce conformational changes in a critical portion of the fiber.

The amino acid composition of pentons from Ad2 has been determined and reveals no unusual features (see Table III). Pentons resemble hexons in their overall composition, although significant differences are observed in particular with regard to Tyr and the hydroxyamino acids (Table III). The penton bases are considerably more sensitive to the action of proteolytic enzymes as compared to most other adenovirus capsid proteins. The base is completely destroyed at high enzyme concentrations, liberating free fibers, whereas the antigenic as well as the morphological properties remain unchanged at low concentrations of trypsin (Pettersson and Höglund, 1969; Wadell and Norrby, 1969b).

3. Immunological Properties

The penton base carries a weak antigenic determinant, known as β, that is common to pentons from all adenovirus serotypes. In addition, inter- and intrasubgroup-specific determinants have been revealed (Wadell and Norrby, 1969b), and these can be demonstrated in hemagglutination-enhancement assays, whereas immunodiffusion reveals primarily the intrasubgroup-specific antigen. Pentons from Ad12, which is more distantly related to other human adenovirus serotypes, appear to carry fewer determinants on their surface than pentons from other serotypes (Wadell and Norrby, 1969b).

4. Physiological Functions

Adenovirus-infected cells show cytopathic changes during the course of the lytic infection. These changes usually manifest themselves as a granular appearance of the cytoplasm, a rounded cell morphology, and detachment of the cells from the surface. Pereira and Kelly (1957) originally demonstrated that two different kinds of cytopathic effects can be distinguished. The first type is apparent 8–24 hr after infection with high concentrations of crude virus, and the second type appears 7–20 days after infection and in conjunction with production of progeny virus. It was later shown (Everett and Ginsberg, 1958; Pereira, 1958; Rowe *et al.*, 1958) that soluble proteins from the infected cells could induce the early type of cytopathic changes. When the structural components of the adenovirion were identified, this early cytopathic effect was linked to the pentons (Valentine and Pereira, 1965). A penton concentration of 0.1 $\mu g/10^6$ cells is required to induce a clearly detectable effect. The molecular mechanism behind the early cytopathic effect remains enigmatic, although the phenomenon has now been known for more than two decades. The morphological changes do not seem to be accompanied by perturbation of the macromolecular synthesis, and the effect is fully reversible after washing with fresh medium (Pereira, 1958; Rowe *et al.*, 1958). The cytopathic effect of the penton is destroyed after mild proteolytic degradation, and low concentrations of trypsin destroy this effect without interfering with the antigenic properties of the penton base (Pettersson and Höglund, 1969; Wadell and Norrby, 1969b). The cytopathic effect, induced with the free pentons, is neutralized by antipenton sera, but not by sera against fibers (Pettersson and Höglund, 1969; Wadell and Norrby, 1969b), which suggests that the penton base alone is responsible for the phenomenon. In accord with this hypothesis, it has been found that isolated penton bases are able to induce early cytopathic changes (Pettersson and Höglund, 1969; Winters *et al.*, 1970; Boudin *et al.*, 1979).

An endonuclease activity associated with pentons (Burlingham *et al.*, 1971) has been demonstrated in cells infected with both Ad2 and Ad12. The endonuclease has also been detected in preparations of purified virions, and when highly purified structural components were tested for activity, only pentons were found to be active (Burlingham *et al.*, 1971). Hexons and fibers apparently lack the activity, even at high concentrations. The endonuclease appears to be specific for DNA, and it cleaves double-stranded DNA (dsDNA) 20 times faster than denatured DNA, with a preference for GC regions (Burlingham *et al.*, 1971). Other investigators (Marusyk *et al.*, 1975) have reported that the activity primarily causes single-stranded scissions. More recent observations suggest, however, that the endonuclease can be disassociated from pentons by treatment with high concentrations of salt. The most likely interpretation of the results is that the endonuclease is a cellular component that remains tightly associated with pentons even after several purification steps (Reif *et al.*, 1977a,b). The biological significance of the associated endonuclease remains unknown, although the activity has been demonstrated in several independent studies (Burlingham *et al.*, 1971; Marusyk *et al.*, 1975; Cajean-Feroldi *et al.*, 1977).

D. Major Core Proteins

The adenoviruses, unlike the papovaviruses, encode their own basic core proteins. Two major polypeptides are associated with adenovirus cores, polypeptides V and VII, as defined by SDS-PAGE of complete virions (Maizel *et al.*, 1968a; Laver *et al.*, 1968; Russell *et al.*, 1971; Everitt *et al.*, 1973; Anderson *et al.*, 1973). Polypeptide VII is the major core protein and accounts for about 10% of the protein mass of the virion. There are approximately 1070 copies in each particle, and the VII/DNA weight ratio is approximately 1 : 1 (Corden *et al.*, 1976; Sung *et al.*, 1977). Polypeptide VII has been extensively purified from both Ad2 and Ad3

1
SILISPSNNT GWGLRFPSKM FGG*AKKRSDQ HPVRVRGHYR APWGAHKRGR

51
TGRTTVDDAI DAVVEEARNY TPTPPPVSTV DAAIQTVVRG ARRYAKMKRR

101
RRRVARRHRR RPGTAAQRAA AALLNRARRT GRRAAMRAAR RLAAGIVTVP

151
PRSRRRAAAA AAAAISAMTQ GRRGNVYWVR DSVSGLRVPV RTRPPRN

FIGURE 7. Amino acid sequence of polypeptide pVII from Ad2. The position where the precursor is cleaved is indicated (*). (Data from Aleström *et al.*, in preparation.)

cores (Laver, 1970; Prage and Pettersson, 1971; Everitt *et al.*, 1973). The purification procedures include extraction at an acidic pH in combination with PAGE. Separation of the core proteins can also be achieved by phosphocellulose chromatography and gel-exclusion chromatography on Sephadex G-75 (Sung *et al.*, 1977). Anderson *et al.* (1973) discovered that the major adenovirus core protein is synthesized as a precursor, polypeptide pVII, that is cleaved during virion maturation into the mature core protein, which has an estimated molecular weight of 19.4K. The proteolytic cleavage of the precursor, which has a molecular weight of 22K, results in the loss of one Met-containing tryptic peptide (Anderson *et al.*, 1973). The molecular weight of the native protein VII has been estimated to be 19K by equilibrium centrifugation (Prage and Pettersson, 1971), and SDS-PAGE (Maizel, 1971; Everitt *et al.*, 1973; Anderson *et al.*, 1973) gives an almost identical value, suggesting that the protein exists as a monomeric unit. Under certain conditions, however, the protein can aggregate into a tetramer (Lemay and Boulanger, 1980). The gene for polypeptide VII and its precursor has recently been sequenced (Aleström *et al.*, in prep.) and the primary structure of the protein can thus be predicted (Fig. 7). The sequence shows that protein VII contains both Tyr and Try residues and it has Ala as the N-terminal amino acid (Prage and Pettersson, 1971; Sung *et al.*, 1977). The amino acid composition (see Table III) shows, furthermore, that the protein is very basic, containing approximately 24% Arg. With regard to its amino acid composition it resembles the Arg-rich histones in having a high Arg and Ala content (Prage *et al.*, 1968), but differs from these histones in containing Try and lower amounts of Lys. The Arg residues are relatively evenly distributed along the polypeptide chain (Laver, 1970) except in a few positions where clusters occur (Fig. 7). The middle portion of the protein, in particular contains long regions with basic residues. A cluster of Ala residues is also found near the C-terminal end of the VII-polypeptide (Fig. 7). It has been calculated that approximately 50% of the phosphate groups in the DNA

```
1
SKRKIKEEML QVIAPEIYGP PKKEEQDYKP RKLKRVKKKK KDDDDDELDD

51
EVELLHATAP RRRVQWKGRR VRRVLRPGTT VVFTPGERST RTYKRVYDEV

101
YGDEDLLEQA NERLGEFAYG KRHKDMLALP LDEGNPTPSL KPVTLQQVLP

151
TLAPSEEKRG LKRESGDLAP TVQLMVPKRQ RLEDVLEKMT VEPGLEPEVR

201
VRPIKQVAPG LGVQTVDVQI PTTSSTSIAT ATEGMETQTS PVASAVADAA

251
VQAAAAAASK TSTEVQTDPW MFRVSAPRRP RRSRKYGTAS ALLPEYALHP

301
SIAPTPGYRG YTYRPRRRAT TRRRTTTGTR RRRRRRQPVL APISVRRVAR

351
EGGRTLVLPT ARYHPSIV
```

FIGURE 8. Amino acid sequence of polypeptide V from Ad2. (Data from Aleström *et al.*, in preparation.)

could be neutralized by the basic residues present in protein VII (Prage and Pettersson, 1971). The purified protein interacts with adenovirus DNA in a highly specific manner, forming a compact structure in which the DNA appears to be organized in the same way as in cores isolated from virus particles (Black and Center, 1979). The complex between DNA and polypeptide VII is extremely stable; treatment by 3 M NaCl, 1% sarcosyl, or 5% DOC leaves the complex intact. The major core protein, like other basic proteins, is weakly antigenic and special care must be taken to reveal precipitation lines by immunodiffusion (Prage and Pettersson, 1971). Core proteins from Ad2 and Ad3 have been compared immunologically and found to share immunological properties, but antigenic differences are also apparent (see Table IV).

Surprisingly little is known about polypeptide V; it is moderately basic (see Table IV) and has an estimated molecular weight of 41.7K, and there are about 180 molecules of polypeptide V per virus particle. Polypeptide V, unlike polypeptide VII, seems to mature without proteolytic trimming, and it appears to be less tightly bound to the DNA than protein VII (Brown *et al.*, 1975) The gene for polypeptide VII has recently been sequenced (Aleström *et al.*, in prep.) and its predicted amino acid sequence is shown in Fig. 8. The N-terminal end of polypeptide V is very rich in charged residues. As many as 21 out of 26 residues in a region which starts with amino acid number 22 are acidic or basic.

III. MINOR POLYPEPTIDES IN THE ADENOVIRUS CAPSID

When the polypeptide composition of the adenovirus particle was analyzed by SDS-PAGE, a remarkable complexity was revealed. In addition to the polypeptide constituents of the well-known capsid components hexons, pentons, fibers, and cores, several other polypeptides were revealed (Maizel *et al.*, 1968 a,b). One of these is polypeptide IIIa.

A. Protein IIIa

This polypeptide is a minor component of the virion, and there are approximately 60 copies of polypeptide IIIa in the virus particle (Boudin *et al.*, 1980; Devaux *et al.*, 1982). When the adenovirus capsid is disintegrated by mild methods that exclusively release components from the vertices of the adenovirus icosahedron, polypeptide IIIa is released together with pentons and a limited number of hexons (Everitt *et al.*, 1973). It was therefore concluded that polypeptide IIIa is associated with the vertex region, and the stoichiometry implies the existence of 5 polypeptide IIIa molecules for each penton structure. Antibodies against polypeptide IIIa aggregate virions, indicating that the antigenic determinants are exposed on the surface of the virion (Everitt *et al.*, 1975). Analysis of the virion topography by external iodination has failed, however, to reveal polypeptide IIIa as a labeled component. These results are to some extent contradictory, but could be explained if polypeptide IIIa has its Tyr residues located on the interior side of the virus particle. Polypeptide IIIa appears to be produced in excess during the infectious cycle, and it has been purified both from this source and from virions (Lemay *et al.*, 1980). The purification procedure includes extraction in the presence of urea, as well as chromatography on (QAE)–Sephadex or hydroxyapatite. Homogeneous preparations of polypeptide IIIa have been obtained in this way to allow biochemical and immunological studies. Studies on the topography of the adenovirus particle using cross-linkers have shown that polypeptide IIIa may have contact with the core polypeptides V and VII inside the adenovirus particle (Everitt *et al.*, 1975). This observation reinforces the idea that polypeptide IIIa, although accessible on the surface, might extend through the capsid and have contacts with components in the interior of the virion. Boudin *et al.* (1980) have reported that polypeptide IIIa, like most internal adenovirus capsid proteins, suffers proteolytic cleavage during virion maturation. Immature virus particles appear to contain a slightly larger form of polypeptide IIIa (pIIIa) that has an estimated molecular weight of 67K. During virion maturation, the precursor appears to be cleaved to an entity with a molecular weight of 66K. It has been suggested from recent results (Devaux *et al.*, 1982) that polypeptide IIIa plays an important role in maintaining the adenovirus capsid structure and also during virion assembly, since temperature-sensitive mutants that are defective with regard to polypeptide IIIa fail to assemble virions properly.

1. Physicochemical Properties

Polypeptide IIIa sediments at 6.0 S, and the native molecule has an estimated molecular weight of 65.4K, which is in good agreement with the value of 66K determined by SDS-PAGE. The native protein IIIa thus appears to exist as a monomer. The Stokes's radius was found to be 2.7 nm and the frictional coefficient f/f_0 was estimated to be 1.0, suggesting that the protein has a spherical shape [see Table II (Section II.A.2)] (Lemay *et al.*, 1980). Electrophoretic analysis of protein IIIa suggests that it has an isoelectric point around 6.0 (Table II), and the amino acid composition of protein IIIa [see Table III (Section II.A.2)] reveals that it, like hexons and fibers, is rich in dicarboxylic amino acids (Lemay *et al.*, 1980). The contents of Ser and Gly are high, whereas the basic amino acids are underrepresented. The comparatively high isoelectric point obviously suggests that most of the acidic amino acid residues are amidified. The mature protein IIIa appears to have Gly as the N-terminal amino acid, whereas the N terminus of the precursor is blocked by acetylation (Lemay *et al.*, 1980).

Polypeptide IIIa is a phosphoprotein, and a protein kinase activity that is associated with adenovirus particles (see Section IV.B) appears to preferentially phosphorylate this polypeptide (Akusjärvi *et al.*, 1978; Blair and Russell, 1978).

2. Immunological Properties

Antisera prepared against purified polypeptide IIIa precipitate similar polypeptides from cells infected with Ad2, Ad3, and Ad5, and no type-specific antigenic determinants have been detected on protein IIIa by immunoprecipitation. Although it was originally claimed that antibodies against polypeptide IIIa have neutralizing activity (Everitt and Philipson, 1974), more recent data (Boudin *et al.*, 1980) suggest that antibodies against protein IIIa do not play a major role in adenovirus neutralization.

B. Protein VI and Its Precursor Polypeptide

The minor adenovirus capsid components VI, VIII, IX, and X–XII have been subject to much less detailed studies than the other adenovirus capsid components, primarily because they are available only in small amounts in infected cells as well as in virions. This has obviously hampered their isolation, purification, and biochemical characterization. Analysis of adenovirus particles by SDS-PAGE reveals the presence of a minor polypeptide with an estimated molecular weight of 24K, which,

```
1
EDINFASLAP RHGSRPFMGN WQDIGTSNMS GG*AFSWGSL WSGIKNFGST
50
IKNYGSKAWN SSTGQMLRDK LKEQNFQQKV VDGLASGISG VVDLANQAVQ
100
NKINSKLDPR PPVEEPPPAV ETVSPEGRGE KRPRPDREET LVTQIDEPPS
150
YEEALKQGLP TTRPIAPMAT GVLGQHTPVT LDLPPPADTQ QKPVLPGPSA
200
VVVTRPSRAS LRRAASGPRS MRPVASGNWQ STLNSIVGLG VQSLKRRRCF
```

FIGURE 9. Amino acid sequence of polypeptide pVI from Ad2. (*) Position at which the precursor is cleaved. Data from Akusjärvi and Persson (1981).

due to its relative mobility, is designated polypeptide VI (Maizel *et al.*, 1968a; Everitt *et al.*, 1973; Anderson *et al.*, 1973). Quantitative estimates suggest that there are about 450 molecules of polypeptide VI per virion. Subsequent studies have revealed that this polypeptide is synthesized as a precursor, polypeptide pVI, with a molecular weight of 27K. The precursor polypeptide is cleaved to the mature polypeptide VI during virion maturation. The mature polypeptide VI can be extracted from virions by low concentrations of urea at an acidic pH, and the protein has been purified to homogeneity by a multistep procedure involving gel and ion-exchange chromatography. The purified protein exists as a dimer or a trimer with a molecular weight of 50–72K (Everitt and Philipson, 1974; Lemay and Boulanger, 1980). The location in the capsid of the polypeptide VI has not yet been determined unambigously. A stepwise degradation procedure was used by Everitt *et al.* (1973) to investigate the topography of the virion, and results from this study suggest that polypeptide VI is likely to be associated with the facets of the adenovirus icosahedron. The basis for this assumption is that polypeptide VI is released together with the hexons when the capsid is broken by repeated cycles of freezing and thawing. Enzymatic labeling of the exterior of the capsid demonstrates, however, that polypeptide VI cannot be labeled from the outside, suggesting that it is located on the inside of the capsid (Everitt *et al.*, 1975). Cross-linking experiments have also revealed an association between polypeptides VI and V in the virion, supporting the notion that it is located on the interior side of the capsid (Everitt *et al.*, 1975). Recent studies by Russell and Precious (1982) indicate that polypeptide VI is a DNA-binding protein. This observation, together with the results discussed above, points toward a more intimate relationship between protein VI and the core than with the outer capsid. In fact, it cannot be excluded at present that polypeptide VI is a core protein that is less tightly bound to the DNA than the other core proteins, polypeptides VII and V.

The nucleotide sequence of the gene for polypeptide pVI has been determined, and it is consequently possible to predict the precise amino acid sequence of polypeptides pVI and VI (Fig. 9). Polypeptide pVI consists of 249 amino acids and has a molecular weight of 27.0K (Akusjärvi and Persson, 1981). During maturation of polypeptide VI, 32 amino acids are removed from the N-terminal end of the precursor (Akusjärvi and Persson, 1981). The cleavage takes place between amino acid residues 32 (Gly) and 33 (Ala). The overall amino acid composition of polypeptide VI reveals, furthermore, high contents of Pro, Ser, and dicarboxylic residues [see Table III (Section II.A.2)]. It has been noted that the carboxy-terminal end of polypeptide pVI resembles a transmembrane protein in having a long stretch of uncharged, predominantly hydrophobic, residues followed by four basic amino acids. The significance of this structure is, however, not known at present.

C. Protein VIII

Polypeptide VIII has a molecular weight of 13K as estimated by SDS-PAGE (Maizel *et al.*, 1968a; Anderson *et al.*, 1973; Everitt *et al.*, 1973). It appears to be synthesized as a considerably larger precursor with an estimated molecular weight of 27K. Thus, the proteolytic cleavage that takes place during virion assembly eliminates a considerably larger fragment from the pVIII precursor than from the other polypeptide precursors. Nothing is yet known as to the fate of the parts of the precursor that are eliminated during cleavage. Polypeptide VIII has been extracted from virions with urea in high salt and purified to homogeneity by preparative PAGE and gel-exclusion chromatography (Everitt and Philipson, 1974). The protein appears to exist as a monomer, but little is known about its biochemical properties. By the application of methods for sequential degradation, it has been found that polypeptide VIII, when present in the virion, is associated with hexons, although less firmly than polypeptide IX. The hexon nonamers (see Chapter 2) that can be isolated in an intact form from the facets of adenovirions appear to lack this polypeptide.

The nucleotide sequence of the gene for the precursor polypeptide VIII has been established by Galibert and co-workers (Galibert *et al.*, 1979; Hérissé *et al.*, 1980). The nucleotide sequence predicts that polypeptide pVII consists of 227 amino acids, with a calculated molecular weight of 24.7K. The predicted amino acid sequence reveals some unusual features; the polypeptide is unusually rich in Pro and Arg, whereas Cys residues are absent from the sequence. Remarkably high contents of Gly and Ser have been noted in purified protein VIII [see Table III (Section II.A.2)]. Although it is not known where in the sequence the cleavage takes place to generate the mature polypeptide VIII, it has been established that the cleavage of pVIII leads to a drastic alteration in the shape of the polypeptide (Lemay and Boulanger, 1980).

D. Protein IX

Polypeptide IX is a minor constituent of the adenovirus particle (Maizel *et al.*, 1968a; Everitt *et al.*, 1973; Anderson *et al.*, 1973). Unlike many

```
1
SANSFDGSIV SSYLTTRMPP WAGVRQNVMG SSIDGRPVLP ANSTTLTYET
51
VSGTPLETAA SAAASAAAAT ARGIVTDFAF LSPLASSAAS RSSARDDKLT
101
ALLAQLDSLT RELNVVSQQL LDLRQQVSAL KASSPPNAV
```

FIGURE 10. Amino acid sequence of polypeptide IX from Ad2. Data from Aleström *et al.* (1980).

other minor adenovirus polypeptides, protein IX does not mature through proteolytic trimming. The native polypeptide IX has been purified from virus particles or hexon nonamers by urea extraction at low pH followed by ion-exchange chromatography on QAE–Sephadex (Everitt and Philipson, 1974; Boulanger *et al.*, 1979). Large amounts of polypeptide IX are produced at late times after infection, and it appears that more protein IX is synthesized than is needed for encapsidation. The gene for polypeptide IX has been sequenced from several serotypes, and the complete amino acid sequence can thus be predicted for polypeptide IX from Ad2 (Aleström *et al.*, 1980; Anderson and Lewis, 1980), Ad3 (Engler, 1981), Ad5 (Maat *et al.*, 1980), Ad7 (Dijkema *et al.*, 1981), and Ad12 (Bos *et al.*, 1981). Polypeptide IX from Ad2 comprises 139 amino acids (Fig. 10) and has a predicted molecular weight of 14.3K. The N-terminal amino acid is acetylated, and the amino acid composition reveals high contents of Ser, Ala, and Leu [see Table III (Section II.A.2)]. The Ser and Ala residues are clustered in some regions of the polypeptide. There is a single Trp residue in the sequence, but no His or Cys residues. From its sequence, it has been proposed that polypeptide IX can be divided into C-terminal and N-terminal domains that are joined by an Ala-rich spacer (Engler, 1981). A comparison of the amino acid sequences of polypeptide IX, from different serotypes reveals that adenovirus members of the same subgroup possess protein IX molecules with similar sequences. Substantial homologies are also found when polypeptide IX, from the more distantly related serotypes are compared, particularly in the N-terminal domain (Engler, 1981). Antibodies have been raised against the purified polypeptide IX, and these have been used to study the immunological properties of the polypeptide. The results (Boulanger *et al.*, 1979) show that protein IX preparations from different serotypes cross-react, although type-specific components appear to be present as well (Boulanger *et al.*, 1979).

In the virus particle, polypeptide IX has been found to be associated with those hexons that constitute the facets of the adenovirus icosahedron (Maizel *et al.*, 1968b; Everitt *et al.*, 1973). When the adenovirus particle is disrupted by certain agents, so-called "nonamers" of hexons are formed that presumably originate from the facets of the icosahedron (also see Chapter 2). Analysis of these nonamers by SDS-PAGE has revealed the presence of polypeptide IX in addition to the hexon polypeptide. The bond between polypeptide IX and the hexon appears to be rather firm, and enzymatic treatment of the hexon nonamers leaves polypeptide IX antigenically intact. Boulanger *et al.* (1979) have estimated that there are approximately 280 molecules of polypeptide IX per virion, and from the stoichiometry they suggest that polypeptide IX is located at the "corner-to-edge" contacts between the hexon trimers in the icosahedron. Antibodies against polypeptide IX cause aggregation of virus particles, which suggests that the antigenic determinants of the polypeptide are exposed on the outside of the capsid.

The gene for polypeptide IX differs from other genes for adenovirus structural proteins in being expressed at intermediate as well as late times after infection, even before the onset of viral DNA replication (Persson *et al.*, 1978a). It has also been found that the gene for polypeptide IX is located in another transcription unit than the genes for the other structural proteins in the adenovirus particle, and the messenger RNA (mRNA) for polypeptide IX is also unique in not requiring splicing for its maturation (Aleström *et al.*, 1980). These findings may suggest that polypeptide IX has other biological functions besides being a structural protein in the virus particle. Speaking against this notion, however, is the observation of Colby and Shenk (1981) that adenovirus mutants that lack the gene for polypeptide IX are still viable. Mutant virions are, however, less heat-stable than wild-type virions, reinforcing the idea that polypeptide IX is involved in the maintenance of the adenovirus capsid structure, possibly by strengthening the bonds between the hexon capsomers.

E. Polypeptides X–XII

Maizel *et al.* (1968a) reported in their original study of adenovirus polypeptides that the adenovirus particle contains a polypeptide, designated X, that migrates with the dye front when subjected to SDS-PAGE. Anderson *et al.* (1973) were later able to resolve this component into three polypeptide components that range in molecular weight from 5 to 6.5K. Nothing is known about their function; experimental evidence suggests that one component may be associated with the virus core. It remains to be established, however, that they are unique translation products, and it is perhaps more likely that at least some of them represent degradation products resulting from the proteolytic cleavage of other capsid proteins.

Hosokawa and Sung (1976) have described a virion polypeptide, designated μ, that is associated with adenovirus cores. Polypeptide μ was estimated to have a molecular weight of 4–5K and may be identical to one of polypeptides X–XII. It has been estimated that approximately 125 copies of the μ polypeptide are present in the adenovirus particle, and a biochemical analysis has revealed a very unusual amino acid composition; the basic residues account for 69% of the amino acids, with 54%

Arg. With available techniques for *in vitro* translation, it should be possible to decide in the near future whether polypeptides X–XII, including μ, are encoded by unique mRNAs.

IV. PHYSIOLOGICAL EFFECTS OF THE STRUCTURAL PROTEINS

The adenovirus particle contains more polypeptides than most other viruses of comparable size. More than 50% of the coding capacity of the viral DNA is required to encode the virion polypeptides. It is thus conceivable that some virion proteins have specific functions in addition to being structural proteins. Despite intensive efforts, however, it has been difficult to assign specific functions to the known structural proteins.

A. Hemagglutination

Hemagglutination by adenoviruses was first demonstrated by Rosén (1958), and it has subsequently been shown that all human serotypes can agglutinate erythrocytes (Rosén, 1960; Rosén *et al.*, 1962; Bauer and Wigand, 1963; Schmidt *et al.*, 1965; Norrby, 1969b). When the structural proteins of the virion were isolated and morphologically defined, it became possible to show that the type-specific determinant γ of the fiber is the component that interacts with the red-cell surface (Pereira and Figueiredo, 1962; Valentine and Pereira, 1965; Norrby, 1966b; Pettersson *et al.*, 1968; Norrby *et al.*, 1969a). Specific receptors seem to be present on the erythrocyte membrane, and an adenovirus receptor from monkey red cells has been solubilized and partly purified (Neurath *et al.*, 1969). Since intact virions carry several vertex projections, they can establish a bridge between the erythrocytes and give a complete hemagglutination pattern. Dodecons, like dimers of pentons and fibers, can in the same way give rise to complete hemagglutination (Fig. 11). In contrast, pentons or fibers alone can only establish a monovalent link with the cell, and agglutination can be detected only when antibodies are used to bridge the receptor-associated structural units (Fig. 11) (Pereira and Figueiredo, 1962; Norrby, 1966b, 1969b). Heterotypic antibodies must be used to demonstrate hemagglutination by penton and fiber monomers, since the type-specific determinant of the fiber is responsible for the attachment of the erythrocyte receptor (Pereira and Figueiredo, 1962; Norrby and Skaaret, 1967). This method of revealing adenovirus hemagglutination by heterotypic antisera has been called the *hemagglutination-enhancement* reaction (Fig. 11). The short fibers of subgroup I viruses fail to hemagglutinate, since they contain only a type-specific determinant (Norrby and Skaaret, 1967) and are therefore unable to form divalent aggregates after interaction with heterotypic antibodies. Antibodies directed against

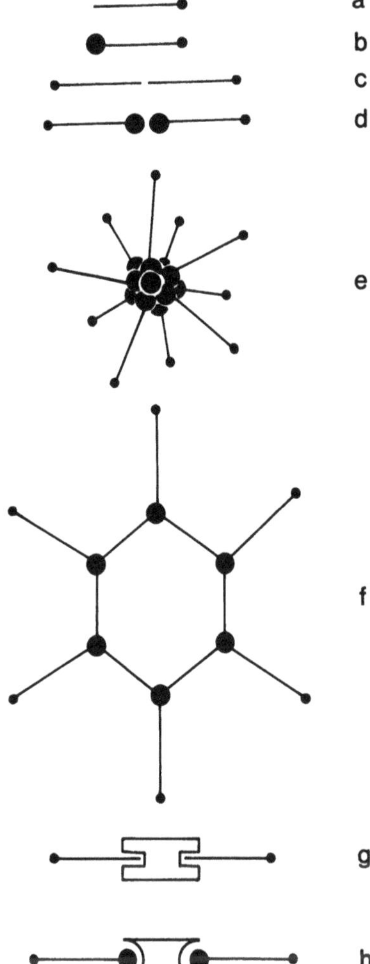

FIGURE 11. Adenovirus hemagglutinins. Fibers (a) or pentons (b) do not cause hemagglutination, since they are monovalent. They bind to the receptors on the erythrocyte membrane but do not agglutinate, since no multivalent aggregates can be created. Dimers of fibers (c), dimers of pentons (d), dodecons (e), or virions (f), on the other hand, can cause agglutination. Fibers or pentons are converted into hemagglutinins by the use of antibodies directed against the proximal part of the fiber (g) or the base of the penton (h).

the γ specificity of the fiber can be assayed by *hemagglutination-inhibition* tests. Specific antibody-consumption tests have been designed to assay for the group- and inter- and intrasubgroup-specific determinants on penton bases and fibers. The penton base may thus be assayed by *hemagglutination-enhancement consumption* (HEC) tests (Norrby and Skaaret, 1967), and free fibers from subgroup I viruses, which lack subgroup specificity, can be assayed by the *hemagglutination-inhibition consumption* (HIC) test (Norrby and Skaaret, 1967). With the aid of these techniques, Norrby and Wadell (Norrby, 1968, 1969b; Wadell, 1970; Wadell and Norrby, 1969b) have elucidated the detailed immunological relationships between the capsid components from many different adenovirus serotypes. On the basis of the ability of adenoviruses to agglutinate different red cells, Rosén (1960) proposed a subgroup classification for human adenoviruses [see Table I (Section I.A)]. Subgroup I viruses agglutinate monkey erythrocytes, whereas subgroup II viruses agglutinate rat erythrocytes. Subgroup III viruses also agglutinate rat erythrocytes, but with a partial pattern, this partial pattern being one in which a fraction of the erythrocytes have sedimented without being agglutinated and therefore form a ring in a less dense lattice. The partial agglutination is probably due to a competition between mono- and the multivalent virus component for the receptors on the erythrocytes (Wadell, 1969). Members of subgroup IV, which include the highly oncogenic adenoviruses, were originally thought to lack hemagglutination capacity, but were later shown to agglutinate rat erythrocytes with a partial pattern (Schmidt *et al.*, 1965; Norrby *et al.*, 1969b). Further subdivisions based on the types of erythrocytes that the different serotypes within one subgroup agglutinate have been suggested (Norrby, 1968, 1969b).

B. Protein Kinase Activity Associated with Adenovirus Particles

A protein kinase activity is released from highly purified adenovirus particles on disintegration (Akusjärvi *et al.*, 1978; Blair and Russell, 1978). The kinase has been solubilized and partially purified from Ad2 and Ad5 virions, and the results show that the enzymatic activity does not copurify with any of the known virion polypeptides (Akusjärvi *et al.*, 1978). Therefore, it remains to be established that it is a viral enzyme. The γ-phosphate of ATP is transferred by the enzyme to Ser and Thr residues. The activity preferentially phosphorylates polypeptide IIIa when endogenous reactions are carried out, although small amounts of label are also incorporated into polypeptides V, VI, VII, and X. Casein as well as phosvitin can be used as exogenous acceptor proteins by the virion-associated kinase, which does not seem to require cyclic nucleotides for its activity (Akusjärvi *et al.*, 1978; Blair and Russell, 1978).

C. The Viral Protease

It was discovered in 1973 (Anderson *et al.*, 1973) that the major core protein polypeptide VII suffers proteolytic cleavage during virion maturation. It has subsequently been demonstrated that most internal adenovirus polypeptides, including the terminal protein that is covalently linked to the virus DNA, are processed in connection with viral assembly. Weber (1976) has isolated a temperature-sensitive (*ts*) mutant of Ad2, H2*ts*1, that fails to cleave the polypeptide precursors. This observation suggests that a virus-encoded function is responsible for the proteolytic activity. Hassell and Weber (1978) were later able to map the mutation in H2*ts*1 to a region located between map positions 57 and 69. When the nucleotide sequence of this region was determined (Kruijer *et al.*, 1980; Akusjärvi *et al.*, 1981), a hitherto undetected gene was disclosed that is located on the immediate 3′ side of the hexon gene. This hypothetical gene encodes a 23K polypeptide that is likely to be the viral protease. Recent sequence studies of the H2*ts*1 mutant (Liang *et al.*, 1983) have indeed identified a mutation in this region of the viral genome, adding additional strength to the hypothesis that the 23K protein is the protease. The protease is present in only minute amounts in infected cells, which has hampered its biochemical characterization. Bhatti and Weber (1979) have been able to detect the activity in nuclei from infected cells and in extracts from immature virus particles. They report that the protease is chymotrypsinlike neutral protease. From the known cleavage site present in the precursor for polypeptide VI, it appears that the protease cleaves Gly–Ala bonds (Akusjärvi and Persson, 1981).

V. ADENOVIRUS PROTEINS SYNTHESIZED AT INTERMEDIATE TIMES AFTER INFECTION

It now seems to be well established that the adenovirus replication cycle goes through several phases. A certain class of adenovirus polypeptides is produced in very low amounts before the onset of viral DNA replication, whereas their synthesis increases drastically late after infection. The kinetics of their appearance thus resemble those of the quasi-late proteins that are synthesized during the replication of bacteriophage T4. Polypeptides IX and IVa_2 appear to belong to this particular class of adenovirus proteins. Polypeptide IX is a structural component of the virion, and its properties are described in Section III.D. Polypeptide IVa_2 was originally recognized as a minor component of the virion (Maizel *et al.*, 1968a; Everitt *et al.*, 1973; Anderson *et al.*, 1973). Analysis of its molar representation reveals, however, that only a few copies are present per virus particle, which raised doubts concerning its functional significance. Polypeptide IVa_2 has not yet been purified from either infected cells or disintegrated virions. There is consequently little information as to its structural and functional properties. The molecular weight of polypeptide IVa_2 has been estimated to be 50K by SDS-PAGE (Everitt *et al.*, 1973; Anderson *et al.*, 1973). Persson *et al.* (1979b) have observed that immature adenovirus particles contain a 50K polypeptide that disappears during maturation of the virion and thus behaves like a scaffolding protein. The 50K polypeptide present in the assembly intermediates appears to be related and possibly identical to the IVa_2 polypeptide, and on the basis of this observation, it has been proposed that it serves as a maturation protein during adenovirus morphogenesis. The presence of polypeptide IVa_2 in mature virions may thus be an artifact due to the presence of small amounts of immature virions, even in highly purified virus preparations. Like many other virus proteins, polypeptide IVa_2 appears to be phosphorylated (D'Halluin *et al.*, 1978), and it also appears to be a DNA-binding protein (Russell and Precious, 1982). A possible function for polypeptide IVa_2 may therefore be to interact with the virus DNA when it enters the preformed capsid. Polypeptide IVa_2 expression is controlled by a separate promoter, which lends additional support to the hypothesis that polypeptide IVa_2 has a unique function in the adenovirus life cycle. The gene that encodes polypeptide IVa_2 has been sequenced in Ad2, Ad5, and Ad7 (Dijkema *et al.*, 1981; Gingeras *et al.*, 1982; Engler and van Bree, 1982), and it is consequently possible to predict the primary sequence for polypeptide IVa_2 from different serotypes. A surprising conclusion that can be drawn from the nucleotide sequence is that polypeptide IVa_2 is encoded by two separate exons, one of which specifies the first four N-terminal amino acids. The sequence of the 448-amino-acid-long polypeptide IVa_2 (predicted molecular weight 51K) from Ad2 is shown in Fig. 12. A comparison between the sequences of polypeptide IVa_2 from Ad5 and of that from Ad7 reveals that the N-terminal portion is poorly conserved (54% homology for the first 82 amino acids), whereas the C-terminal portions exhibit a very high degree of homology [88% homology for the C-terminal 366 amino acids (Engler and van Bree, 1982)]. The primary sequences of polypeptide IVa_2 from Ad2 and of that from Ad5 appear to be identical (Gingeras *et al.*, 1982).

VI. LATE NONSTRUCTURAL PROTEINS

The mRNAs that are controlled by the major late promoter (MLP) at map position 16.5 encode eight distinct polypeptides that are all present in the virion. In addition to the capsid components, it has been observed that a prominent polypeptide, designated 100K because of its estimated molecular weight, is controlled by the MLP. The mRNA for the 100K polypeptide has, like the other mRNAs in the MLP transcription unit, a tripartite leader attached to its 5′ end. The presence of an adenovirus-specific 100K polypeptide was first noticed when extracts from adenovirus-infected cells at late times after infection were analyzed by SDS-PAGE (Russell and Skehel, 1972; Anderson *et al.*, 1973; Everitt *et al.*, 1973) and when adenovirus mRNA was translated *in vitro* (Anderson *et al.*, 1974).

A. 100K POLYPEPTIDE

This polypeptide is present in large quantities late after infection, and its mRNA is one of the most abundant viral messages in the infected cell late after adenovirus infection (Flint and Sharp, 1976). Analysis of ^{32}P-labeled extracts has demonstrated, further, that the 100K polypeptide is a phosphoprotein. The amino acid sequence that has been predicted for the 100K protein reveals that the protein is unusually Leu-rich. The 100K protein has been purified by Oosterom-Dragon and Ginsberg (1980) by a procedure that includes ion-exchange chromatography and removal

1
ETRGRRPAAL OHQQDQPQAH PGQRAARSAP LHRDPDYADE DPAPVERHDP

51
GPSGRAPTTA VQRKPPQPAK RGDMLDRDAV EQVTELWDRL ELLGQTLKSM

101
PTADGLKPLK NFASLQELLS LGGERLLADL VRENMRVRDM LNEVAPLLRD

151
DGSCSSLNYQ LHPVIGVIYG PTGCGKSQLL RNLLSSQLIS PTPETVFFIA

201
PQVDMIPPSE LKAWEMQICE GNYAPGPDGT IIPQSGTLRP RFVKMAYDDL

251
ILEHNYDVSD PRNIFAQAAA RGPIAIIMDE CMENLGGHKG VSKFFHAFPS

301
KLHDKFPKCT GYTVLVVLHN MNPRRDMAGN IANLKIQSKM HLISPRMHPS

351
QLNRFVNTYT KGLPLATSLL LKDIFRHHAQ RSCYDWIIYN TTPQHEALQW

401
CYLHPRDGLM PMYLNIQSHL YHVLEKIHRT LNDRDRWSRA YRARKTPK

FIGURE 12. Amino acid sequence of polypeptide IVa_2. Data from van Beveren *et al.* (1982) and Gingeras *et al.* (1982).

of contaminating hexon protein by affinity chromatography using immobilized antiserum against purified virions. A surprising observation is that antibodies prepared against purified 100K protein precipitate hexon polypeptides as well as the 100K polypeptide from infected cell extracts, and it appears that the reactivity against hexon polypeptides cannot be removed by adsorption (Oosterom-Dragon and Ginsberg, 1980). From genetic studies (Williams *et al.*, 1974; Ginsberg *et al.*, 1974; Kauffman and Ginsberg, 1976), it was proposed that the 100K protein functions in conjunction with the assembly of trimeric hexon. This hypothesis is based on the recognition of two separate complementation groups of adenovirus temperature-sensitive mutants that both fail to produce functionally assembled hexon molecules. Mapping of these mutants has revealed that they indeed represent two separate genes, one encoding the hexon polypeptide and the other encoding the 100K polypeptide. Mutants that fail to produce functional 100K protein at the nonpermissive temperature accumulate in the cytoplasm hexon polypeptides that fail to assemble into trimeric structures. It has therefore been proposed that the 100K protein is required for proper assembly of hexon trimers. This notion is also supported by the fact that antisera prepared against highly purified 100K protein coprecipitate the hexon polypeptide from infected-cell extracts, a finding that suggests a close physical association between the 100K and hexon polypeptides. It has been demonstrated, moreover, that antibodies against the 100K protein prevent the assembly of trimeric hexon in cell extracts (Oosterom-Dragon and Ginsberg, 1981). It has been reported that late after infection, the 100K polypeptide accumulates in the infected-cell nucleus, although the functional significance of this observation remains to be elucidated (Gambke and Deppert, 1981a).

A polypeptide with an estimated molecular weight of 100K has been found to be associated with mRNA in adenovirus-infected cells (Lindberg and Sundquist, 1974; Tasseron-de Jong *et al.*, 1979). It is therefore conceivable that the 100K polypeptide becomes associated with the nascent hexon polypeptide already present on the polyribosome. The binding of the 100K protein to polyribosomes in adenovirus-infected cells could in addition, however, serve yet unknown functions.

Monoclonal antibodies have been prepared against the 100K polypeptide from Ad5 (Russell *et al.*, 1981). These antibodies react with the corresponding polypeptide in cells infected with Ad2, but not with that in cells infected with Ad3. These results thus suggest that the 100K polypeptide reacts in a subgroup-specific manner.

B. 33K Phosphoprotein

Several reports have described a 33K adenovirus-specific phosphoprotein that is present late after infection (Russell and Blair, 1977; Levinson *et al.*, 1977; Axelrod, 1978). It was originally thought that this phosphoprotein is related to the 100K polypeptide, possibly being a cleavage or degradation product (Axelrod, 1978). This conclusion was based on similarities between the tryptic peptide maps that were generated from the two polypeptides. Subsequent studies have failed to confirm these results, however, and it is now established that the 33K polypeptide is unique, although it is encoded in the same region of the adenovirus genome as the 100K polypeptide. The first clue that indicated that the 33 and 100K polypeptides have unrelated sequences came from the immunological studies; Gambke and Deppert (1981b) showed that the two proteins induce antibodies that fail to cross-react. *In vitro* translation of a 33K polypeptide was first reported by Lewis *et al.* (1975), whereas Ricciardi *et al.* (1979) and Miller *et al.* (1980) demonstrated that the 33K polypeptide is an independently encoded adenovirus product. A recent study performed by Oosterom-Dragon and Anderson (1982) has clarified the precise relationship between the 33 and 100K proteins. This study demonstrates that the 33K polypeptide is encoded by an mRNA that completely overlaps with the mRNA for the 100K protein. The coding regions for the two polypeptides also overlap to some extent, but the proteins still have completely unrelated sequences, since different translational reading frames are utilized. The results show, moreover, that an internal splice in the mRNA for the 33K protein gives rise to a frameshift whereby the N- and C-terminal parts of the 33K polypeptide are encoded by different translational reading frames. The DNA sequence allows the complete amino acid sequence of the 33K polypeptide to be predicted; the protein consists of 227 amino acids with 45% nonpolar and 29% charged residues. The contents of Ala, Pro, Glu and Leu are high, and the hydroxyamino acids together constitute about 16% of the total number of residues [see Table III (Section II.A.2)] Unfortunately, nothing is yet known about the function of the 33K polypeptide, although studies on its intracellular location show that it is a nuclear protein, being associated with different nuclear substructures than the 100K polypeptide (Gambke and Deppert, 1981a).

VII. STRUCTURE AND FUNCTION OF ADENOVIRUS PROTEINS SYNTHESIZED EARLY AFTER INFECTION

The proteins that are encoded by the early genes are in some respects more interesting than the late structural proteins, since they are likely to carry catalytic and regulatory functions. Consequently, by studying their structure and function, it may be possible to learn important facts about the principal mechanisms behind eukaryotic gene regulation. Bacteriophages have in the past been instrumental in enabling molecular biologists to reach the current, very sophisticated level of understanding concerning prokaryotic gene regulation. The adenoviruses resemble bacteriophage λ with regard to genome size and genetic complexity. It thus appears likely that the adenoviruses will provide an important research tool to unravel the basic mechanisms for eukaryotic gene regulation. Our current knowledge about early adenovirus proteins is very limited at present, however, the reason being that they are produced in small quantities and with few exceptions constitute only a minute fraction of the total amount of protein in the infected cell. Very few early proteins have been purified to date, and we therefore know very little about their properties, particularly with regard to their functions. Approximately 50% of the Ad2 genome is transcribed into mRNA early after infection (Tibbetts *et al.*, 1974; Chow *et al.*, 1979; Stillman *et al.*, 1981). It was discovered long ago that there exist on the adenovirus genome four separate early regions, which have been designated E1, E2, E3, and E4 (Sharp *et al.*, 1974; Philipson *et al.*, 1974; Pettersson *et al.*, 1976; Chow *et al.*, 1977). Region E1 contains two transcription units that subdivide it into regions E1a and E1b. It was furthermore observed that two sets of proteins are controlled by the E2 promoter(s), and region E2 has therefore been subdivided into E2a and E2b.

A. Polypeptides Encoded by Region E1a

The proteins that are encoded by region E1 have attracted a great deal of attention, since they play a key role in adenovirus transformation (Chapter 9). The E1 proteins are often referred to as tumor (T) antigens because of their postulated role in transformation. Several attempts have been made to purify adenovirus T antigens (Gilead and Ginsberg, 1968; Biron and Raska, 1977; Levinson and Levine, 1977; Green *et al.*, 1979), but only one E1 protein has so far been purified to homogeneity. From genetic studies of the transforming region and from structural studies of the E1a mRNAs, certain predictions can be made about the adenovirus T antigens, with regard to both function and primary structure. Region E1a encodes a set of proteins that are specified by overlapping DNA sequences. There exist two large E1a mRNAs, a 13 and a 12 S mRNA, that both encode a set of related polypeptides ranging in molecular weight between 35 and 53K (Lewis *et al.*, 1976, 1980; Harter and Lewis, 1978; Esche *et al.*, 1980; Halbert *et al.*, 1979; Jochemsen *et al.*, 1981; Lupker *et al.*, 1981). The 13 S mRNA appears to encode a 53 and a 41K polypeptide, whereas the 12 S mRNA encodes polypeptides with molecular weights of 35 and 47K (Halbert *et al.*, 1979). Each polypeptide set can be resolved into additional components by 2D gel electrophoresis. It is not known at present how each mRNA can specify two or more polypeptides that differ slightly in molecular weight. The phenomenon is most likely the result of posttranslational modification events the nature of which is not yet understood. From the DNA sequence, it has been predicted that the products of the 12 and the 13 S mRNAs are completely overlapping (Fig. 13), but that the protein product or products of the larger

```
1
RHIICHGGVI TEEMAASLLD QLIEEVLADN LPPPSHFEPP TLHELYDLDV

51
TAPEDPNEEA VSQIFPESVM LAVOEGIDLF TFPPAPGSPE PPHLSRQPEQ

101
PEORALGPVS MPNLVPEVID LTCHEAGFPP SDDEDEEGEE FVLDYVGHPG

151
HGCRSCHYHR RNTGDPDIMC SLCYMRTCGM FVYSPVSEPE PEPEPEPEPA

201
RPTRRPKLVP AILRRPTSPV SRECNSSTDS CDSGPSNTPP EIHPVVPLCP

251
IKPVAVRVGG RRQAVECIED LLNESGOPLD LSCKRPRP
```

FIGURE 13. Amino acid sequence of the protein product of the 13 S mRNA from region E1a of Ad2. The nucleotides that are absent from the product of the 12 S mRNA are underlined. Data from Gingeras *et al.* (1982) and Perricaudet *et al.* (1979).

mRNA presumably contain an extra functional domain (Perricaudet *et al.*, 1979; Smart *et al.*, 1981). The predicted amino acid sequence of the largest E1a protein shows some unusual features; the protein is exceptionally rich in Pro (16%), and a region consisting of alternating Glu and Pro residues is found in its C-terminal part (residues 188–199 in Fig. 13). None of the 32–53K polypeptides has yet been purified, and no specific functions have been assigned to them through a biochemical analysis of purified proteins. Genetic studies have shed light, however, on some possible functions; mutants that fail to produce a functional 13 S mRNA have lost their transforming properties (Jones and Shenk, 1979a). It has also been shown that a product of this mRNA enhances the expression of other early regions (Berk *et al.*, 1979; Jones and Shenk, 1979b) as discussed in Chapter 5. It is unclear from the genetic studies whether the product of the 12 S mRNA, which is closely related to that of the 13 S mRNA, probably has an indispensable biological function, since mutants that fail to produce this mRNA are still viable (Montell *et al.*, 1982). There is, unfortunately, no information as to the mechanism by which the gene product or products of the 12 and 13 S mRNAs exert their interesting functions, although preliminary evidence suggests that they may interact with cellular regulatory protein(s) (Katze *et al.*, 1981). A third protein from region E1a is encoded by a small 9 S mRNA. This mRNA is produced in large quantities late after infection, and it has been reported that its product is a 28K polypeptide (Spector *et al.*, 1980; Esche *et al.*, 1980). This is a remarkably large size for a product of such a small mRNA and may suggest that the protein is heavily modified. No information concerning the function of the 28K protein has yet been published.

B. Polypeptides Encoded by Region E1b

Three unique polypeptides have so far been assigned to region E1b. Sequence studies of the spliced mRNAs from this region have revealed the existence of a 13 and a 22 S mRNA that encode products with predicted molecular weights of 21 and 55K, respectively, in cells infected or transformed by Ad2 (Halbert *et al.*, 1979; Esche *et al.*, 1980; Lupker *et al.*, 1981). Polypeptide IX is also encoded by region E1b, but its properties are discussed in connection with the structural proteins of the particle (see Section III.D). Concerning the function of the E1b proteins, it has been postulated that they play a critical role in transformation, since cells that carry region E1a alone are immortalized, but lack many of the phenotypic properties that are associated with transformation (van der Eb *et al.*, 1979). The relative significance of the polypeptides encoded by the transforming region of adenoviruses is discussed in detail in Chapter 9.

1. E1b/21K Polypeptide

The 21K* protein encoded by region E1b has been purified to homogeneity (Persson *et al.*, 1982). The purification procedure involves phosphocellulose, hydroxyapatite, DEAE–cellulose, and aminohexyl-Sepharose chromatography. The resulting product is homogeneous as judged by SDS-PAGE, and monospecific antiserum, raised against the purified protein, immunoprecipitates the E1b/21K protein from a variety of virus-transformed cell lines of human, rat, and hamster origin. The E1b/21K protein is associated with the membrane fraction in both productively infected and virus-transformed cell lines and can be released only after detergent treatment (Persson *et al.*, 1982). A major part of the E1b/21K protein is protected in membrane vesicles after treatment with trypsin of proteinase K, suggesting an extensive membrane interaction. *In vitro* translation of early viral mRNA in a cell-free system, supplemented with rough microsomes, has shown that the E1b/21K protein is also associated with membranes *in vitro*. No specific signal polypeptide has been identified in the known sequence of the 21K polypeptide (Fig. 14), and the region of the protein that is responsible for the interaction

* The molecular weight of this polypeptide is sometimes given as 15K. The DNA sequence predicts, however, a molecular weight of 21K.

```
1
EAWECLEDFS AVRNLLEQSS NSTSWFWRFL WGSSQAKLVC RIKEDYKWEF EELLKSCGEL

51
FDSLNLGHQA LFQEKVIKTL DFSTPGRAAA AVAFLSFIKD KWSEETHLSG GYLLDFLAMH

101
LWRAVVRHKN RLLLLSSVRP AIIPTEEQQQ EEARRRRRQE QSPWNPRAGL DPRE
```

FIGURE 14. Amino acid sequence of the E1b 21K polypeptide. Data from Gingeras *et al.* (1982).

with the membranes is not yet identified, although the predicted amino acid sequence (Fig. 12) reveals two internal hydrophobic segments (Persson *et al.*, 1982; Bos *et al.*, 1981). The protein does not incorporate [^{3}H]mannose or [^{3}H]glucosamine, suggesting that it is not a glycoprotein. The failure of the protein to bind to lectin columns lends additional support to this notion (Persson *et al.*, 1982). The predicted amino acid sequence (Fig. 12) shows that the protein has an unusually high content of Trp (4.6%) and that the C-terminal part is rich in charged residues.

Green *et al.* (1982) have purified a 20K polypeptide from region E1b. This protein is probably different from the 21K polypeptide described above, since it is related to the large E1b protein (55K).

2. Protein Kinase Activity Encoded by Region E1

Following the original observation of Collett and Erikson (1978) that the transforming genes of avian sarcoma viruses are associated with protein kinase activity, it was reported that a virus-induced protein kinase can also be detected in adenovirus-infected and -transformed cells (Lassam *et al.*, 1979; Branton *et al.*, 1981). This kinase activity was demonstrated by immunoprecipitation using sera from tumor-bearing animals, and, as is the case for the (*src*) protein, the adenovirus-induced kinase phosphorylates the heavy chain of IgG when present in the immunoprecipitate. It can, however, also phosphorylate exogenous acceptor proteins such as histones. From its kinetics of appearance, it has been concluded that the adenovirus-induced kinase is an early protein, most likely encoded by region E1 (Branton *et al.*, 1981). The latter conclusion is based on the fact that the antiserum that was used for kinase detection was prepared against cells expressing only the E1 region. It should be pointed out, however, that it has not yet been established which of the E1 proteins is associated with the kinase activity, although studies using host-range mutants (see Chapter 7) suggest that the kinase activity is low or absent in cells infected with mutants carrying lesions in the E1b region. The kinases that are induced by the nononcogenic serotype 5 and the highly oncogenic serotype 12 are immunologically different, as would be expected, since the T antigens of these two adenovirus serotypes fail to cross-react (Branton *et al.*, 1981). Unlike the product of the *src* gene, the adenovirus protein kinase seems to preferentially phosphorylate Thr and Ser residues, rather than Tyr (Branton *et al.*, 1981). It is not yet known whether the purified E1b/21K protein described in Section VII.3.1 carries kinase activity, although it is a possible candidate.

C. Polypeptides Encoded by Region E2

Region E2 is expressed from two promoters located at map positions 72 and 75. The latter promoter is preferentially used for the synthesis of E2 mRNAs before the onset of viral DNA replication, whereas the promoter located at map position 72 is activated late after infection (Chow *et al.*, 1979). One or more different leader segments are joined to a single mRNA body by splicing, generating an overlapping set of E2a mRNAs. It is not yet known why different promoters are used in region E2 before and after the onset of viral DNA replication. The different leader structures that are joined to the E2a mRNA bodies at early and late times after infection do not give rise to structurally different polypeptides, since the initiator AUG triplet is located within the common body of the E2a mRNAs. It was originally thought that all the E2 mRNAs had a common polyadenylation site located at map position 61. It was recently found, however, that there are transcribed from the E2 promoters, by differential splicing, an additional set of mRNAs that have their polyadenylation site located close to map position 11. Early region E2 has therefore been subdivided into region E2a, with mRNA bodies mapping between positions 61 and 67, and region E2b, with mRNA bodies located between positions 28.5 and 11 (Stillman *et al.*, 1981; Chow *et al.*, 1979). The E2 region has the properties of an operon for proteins involved in DNA replication, and proteins that function during initiation as well as DNA chain elongation have been assigned to this transcription unit.

1. DNA-Binding Protein

Although several E2a mRNAs exist that differ with regard to their leader sequences, only one protein product has been assigned to region E2a, namely, the 72K DNA-binding protein (DBP). The DBP was discovered in 1973 by van der Vliet and Levine (1973), who used DNA–cellulose affinity chromatography to fractionate cell extracts from uninfected and adenovirus-infected cells. It is by far the best-characterized nonstructural adenovirus protein. Originally, two types of DBP were described with estimated molecular weights of 44 and 72K (van der Vliet and Levine, 1973). Subsequent studies have shown, however, that the proteins are intimately related and that the 44K polypeptide can be generated from the 72K polypeptide by proteolytic cleavage, using low concentrations of chymotrypsin. Chymotryptic treatment gives rise to both a 26 and a 44K polypeptide, suggesting that the protein consists of two defined structural domains (Klein *et al.*, 1979; Linné and Philipson, 1980; Schechter *et al.*, 1981). The DPB has been purified from cells infected with Ad2, Ad5, and Ad12 using procedures that include DNA–cellulose chromatography, DEAE or phosphocellulose chromatography, selective precipitation, and gel-exclusion chromatography (Linné *et al.*, 1977; Sugawara *et al.*, 1977; Klein *et al.*, 1979; Linné and Philipson, 1980; Schechter *et al.*, 1981). The DBP has a highly assymmetrical shape in solution (van der Vliet *et al.*, 1978) and self-associates very easily, particularly at a low ionic strength (Schechter *et al.*, 1981). It is a phosphoprotein (Russell and Blair, 1977; Jeng *et al.*, 1977; Axelrod, 1978), and it has been reported that there are 4–7 phosphate groups per polypeptide chain (Linné and Philipson, 1980), whereas other studies have revealed as many as 11 tryptic phosphopeptides (Klein *et al.*, 1979). The phosphate groups are linked to Ser as well as Thr residues (Russell and Blair, 1977; Jeng *et al.*, 1977; Levinson *et al.*, 1977; Axelrod, 1978). It has been shown, moreover, that all phosphate groups are recovered in the 26K polypeptide after chymotryptic cleavage of the DBP. The 26K polypeptide itself does not bind to DNA, and it is therefore not surprising that the presence or absence of phosphate groups has little or no influence on the DNA-binding properties of the protein. This has been established by treating purified DBP with alkaline phosphatase, which removes 90–95% of the phosphate groups (Klein *et al.*, 1979; Linné and Philipson, 1980). The molecular weight of the DBP decreases to approximately 70K after treatment with alkaline phosphatase. The purified DBP has been resolved into as many as 15 subspecies using isoelectric focusing. The heterogeneity can be ascribed in part to the presence of different amounts of phosphate in different polypeptide species. Due to the limited number of phosphate groups present, however, the results may suggest additional heterogeneity in the polypeptide part of the DBP. The cause of this heterogeneity is, however, not known at present (Linné and Philipson, 1980).

It was possible to assign a specific function to the 72K protein shortly after its discovery, thanks to the availability of the temperature-sensitive (*ts*) H5*ts*125 mutant. This mutant, which is unable to replicate its DNA at the nonpermissive temperature, was first isolated by Ensinger and Ginsberg (1972) and was later utilized by van der Vliet *et al.* (1975) to demonstrate that it encodes a 72K protein that is temperature-sensitive in its DNA-binding properties. The 45K subspecies of the DBP, which is derived from the C-terminal part of the protein, binds to DNA (Klein *et al.*, 1979), and it has been possible to demonstrate that the 45K polypeptide fragment is biologically active *in vitro*, since it can complement the H5*ts*125 mutant at the nonpermissive temperatures in an *in vitro* replication system (Ariga *et al.*, 1980). Antibodies against purified DBP precipitate the 45K polypeptide as well as the 72K DBP. The genes for DBP from Ad2 and Ad5 were recently sequenced, and the complete amino acid

```
1
ASREEEQRET TPERGRGAAR PPTMEDVSSP SPSPPPPRAP PKKRLRRRLE

51
SEDEEDSSQD ALVPRTPSPR PSTSTADLAI ASKKKKKRPS PKPERPPSPE

101
VIVDSEEERE DVALQMVGFS NPPVLIKHGK GGKRTVRRLN EDDPVARGMR

151
TQEEKEESSE AESESTVINP LSLPIVSAWE KGMEAARALM DKYHVDNDLK

201
ANFKLLPDQV EALAAVCKTW LNEEHRGLQL TFTSNKTFVT MMGRFLQAYL

251
QSFAEVTYKH HEPTGCALWL HRCAEIEGEL KCLHGSIMIN KEHVIEMDVT

301
SENGQRALKE QSSKAKIVKN RWGRNVVQIS NTDARCCVHD AACPANQFSG

351
KSCGMFFSEG AKAQVAFKQI KAFMQALYPN AQTGHGHLLM PLRCECNSKP

401
GHAPFLGRQL PKLTPFALSN AEDLDADLIS DKSVLASVHH PALIVFQCCN

451
PVYRNSRAQG GGPNCDFKIS APDLLNALVM VRSLWSENFT ELPRMVVPEF

501
KWSTKHQYRN VSLPVAHSDA RQNPFDF
```

FIGURE 15. Amino acid sequence of the 72K DBP from Ad2. Data from Kruijer *et al.* (1982).

sequence can consequently be predicted; the predicted polypeptide contains 527 amino acid residues (Fig. 15) and has a molecular weight of 59K; its amino acid composition is shown in Table III (Section II.A.2). The DBP is rich in dicarboxylic amino acid residues, many of which are present in the amide form, giving the protein a neutral character. The N-terminal part of the protein is rich in Pro and contains clusters of basic and acidic residues (Kruijer *et al.*, 1981). A comparison between the Ad2 and the Ad5 DBP shows that the protein is extremely well conserved, particularly in its C-terminal part. Not a single residue among the C-terminal 354 residues has been altered, whereas 9 of 175 residues are different in the N-terminal portion.

It was originally thought that the 72K DBP binds specifically to single-stranded DNA in a cooperative fashion (van der Vliet and Levine,

1973; van der Vliet *et al.*, 1978). More recent studies (Fowlkes *et al.*, 1979; Schechter *et al.*, 1981) have shown that the protein also interacts with dsDNA. The interaction between DBP and natural or synthetic polynucleotides has been examined, and polynucleotide-melting techniques have revealed a tight, yet dissociable, binding to the helix structure of dsDNA. A specific binding to dsDNA termini with a dissociation constant of 1×10^{-9} to 2×10^{-9} M has been demonstrated at high DNA/protein ratios. The ability of the DBP to interact with heat-denatured DNA was confirmed in this study (Fowlkes *et al.*, 1979), but the binding to dsDNA termini was found to be more specific on a molar basis. The DBP can apparently recognize both even and staggered ends of dsDNA molecules (Fowlkes *et al.*, 1979).

Several different functions have been assigned to the 72K DBP. One function is defined by the defect in the *ts* mutant H5*ts*125 (Ensinger and Ginsberg, 1972; van der Vliet *et al.*, 1975). The precise role of the DBP in replication is not yet defined in biochemical terms, although it seems likely that the protein binds to the single-stranded tails that are exposed during adenovirus DNA replication, perhaps thereby protecting them from nuclease action. The DBP seems, however, to be multifunctional protein, since several different properties have been attributed to it by different investigators. Temperature-shift experiments indicate, for instance, that the DBP is involved in both initiation and elongation, although it has not yet been possible to demonstrate the former role *in vitro* (Horwitz, 1978; van der Vliet and Sussenbach, 1975; van der Vliet *et al.*, 1977). At least one function that has been linked to the DBP can be complemented by the A gene product of simian virus (SV40), since coinfection with SV40 complements the replication-defective H5*ts*125 mutant (Rabek *et al.*, 1981). It is also clear that the DBP regulates its own expression and also the expression of other early proteins. This function has also been established by the use of the H5*ts*125 mutant; Carter and Blanton (1978a,b) have shown that cells infected with H5*ts*125 produce considerably higher levels of early virus mRNA at the nonpermissive temperature. The results have been interpreted to mean that the DBP represses its own synthesis during a normal wild-type infection and also exerts a similar influence on the expression of other early proteins. Babich and Nevins (1981) have shown that the increase in early protein synthesis in H5*ts*125-infected cells is caused by a prolonged half-life of the message, rather than by a direct effect on transcription.

An increase in the transformation frequency has been observed, moreover, after infection of rat cells with the H5*ts*125 mutant at the nonpermissive temperature, and the mutation also seems to influence the amount of virus DNA sequences that become integrated (Mayer and Ginsberg, 1977).

Other results suggest that the DBP influences the host range of the virus, possibly by interacting with host-cell functions. It has long been known that human adenoviruses replicate inefficiently in monkey cells

(Rabson *et al.*, 1964), probably because the mRNA for the fiber protein is spliced incorrectly (Klessig and Chow, 1980; Klessig and Andersson, 1975; Klessig and Grodzicker, 1979). Several investigators have studied the basis for the host-range restriction and found that mutants can be obtained that grow efficiently in both human and monkey cells (Klessig, 1977). These mutants have been mapped to the 26K domain of the DBP (Klessig and Grodzicker, 1979; Kruijer *et al.*, 1981). Nicholas *et al.* (1981) have in addition reported that revertants of the H5*ts*125 mutant exhibit certain remarkable host-range properties, again suggesting that the DBP is a multifunctional protein that influences the replication of the virus in a very complex manner.

Sequence studies of adenovirus mutants with lesions in the DBP have been very informative. Kruijer *et al.* (1981, 1982) have studied three different DBP mutants, and the results show that mutants that are defective for DNA replication map in the C-terminal 45K domain, whereas those that affect the host range of the virus map in the N-terminal portion of the protein. These results reinforce from both a functional and a structural viewpoint the idea that the DBP contains two domains.

2. E2b Proteins

Stillman *et al.* (1981) discovered that the l-strand of the region of the Ad2 genome which is located between map positions 11 and 30 encodes three polypeptides with estimated molecular weights of 87, 105, and 75K. This conclusion is based on the finding that mRNAs for these three protein products hybridize specifically to this part of the adenovirus genome. Subsequent studies have shown that the 87K polypeptide is the precursor for the so-called "terminal protein" that is covalently bound to the termini of adenovirus DNA (Gingeras *et al.*, 1982; Aleström *et al.*, 1982; Smart and Stillman, 1982). This protein plays a very important role during initiation of viral DNA replication and is discussed in detail in Chapter 3. The 105K polypeptide is also involved in DNA replication, and compelling evidence is accumulating to suggest that this polypeptide is a viral DNA polymerase that is responsible for the replication of adenovirus DNA (Chapter 3). The nature of the 75K polypeptide is not firmly established. It is possible that it is in fact the DBP, simply being translated from a rare mRNA species that extends with its 3′ noncoding regions all the way to the polyadenylation site at map position 11.

D. Polypeptides Encoded by Region E3

Early region E3 is located between map coordinates 76 and 86, and an analysis of the transcripts from this region has identified at least eight mRNA species that are all partly overlapping (Chow *et al.*, 1979; Kitchingman and Westphal, 1980; Persson *et al.*, 1980). It has been shown, for instance, that three mRNAs, which differ only in their 3′ noncoding regions, encode the same polypeptide, namely, the 19K glycoprotein. However, at least five additional mRNAs have been described, most of which are expected to encode unique polypeptides. In light of the large number of mRNA species assigned to region E3, it is surprising that only two unique E3 polypeptides have been identified so far, namely, the E3/14K polypeptide and the 19K glycoprotein. A 14.5K polypeptide has also been assigned to the E3 region, but it is unclear whether the 14 and 14.5K polypeptides are structurally related or not (Persson *et al.*, 1980). By DNA sequence studies of region E3 (Hérissé and Galibert, 1981; Hérissé *et al.*, 1981), several open translational reading frames have been identified that in theory could encode polypeptides larger than 10K. It is therefore likely that additional proteins will be assigned to the E3 region in the future.

Region E3 appears to be dispensable for virus replication, since viruses that lack all or most of the E3 region are still viable and able to replicate in tissue-culture cells. An interesting biological property has, however, been assigned to the E3/19K glycoprotein. Two E3 proteins have been purified to homogeneity, i.e., the 19K glycoprotein and 14K polypeptide.

1. E3/19K Glycoprotein

An Ad2 early glycoprotein with an apparent molecular weight of 19K has been purified from infected cells (Jeng *et al.*, 1978; Persson *et al.*, 1979a). The purification procedure involves detergent solubilization of the membrane fraction from infected cells, affinity chromatography on a lectin column, and DEAE–Sephadex chromatography. The 19K polypeptide accounts for at least 90% of the purified material (Persson *et al.*, 1979a) and contains glucosamine as revealed by the incorporation of [^{3}H]glucosamine during synthesis in infected cells and the amino acid analysis of purified 19K glycoprotein. Immunoprecipitation with a monospecific antiserum has shown that synthesis of the 19K glycoprotein begins 2 hr after infection, reaching a maximal rate at 4 hr after infection, whereas its synthesis is reduced later after infection. The protein has been found to be present in nearly all transformed cell lines that contain the E3 region.

The 19K glycoprotein is synthesized as an 18K unglycosylated precursor polypeptide that is around 1.5K larger than the polypeptide moiety of the mature glycosylated E3/19K protein. The amino acid sequences of the 19K glycoprotein and its precursor have been deduced from the DNA sequence of region E3 (Fig. 16). The results show that a 16-amino-acid-long signal sequence is removed from the N terminus when the 19K glycoprotein matures. The signal sequence is, like other signal sequences, very rich in hydrophobic amino acids and ends with the sequence Ala-Ala-Lys. There are two potential sites for carbohydrate attachment in the protein based on the presence of the sequence Asn-Val-Thr, which is

```
1
EAWECLEDFS AVRNLLEQSS NSTSWFWRFL WGSSQAKLVC RIKEDYKWEF

51
EELLKSCGEL FDSLNLGHQA LFQEKVIKTL DFSTPGRAAA AVAFLSFIKD

101
KWSEETHLSG GYLLDFLAMH LWRAVVRHKN RLLLLSSVRP AIIPTEEQQQ

151
EEARRRRRQE QSPWNPRAGL DPRE
```

FIGURE 16. Amino acid sequence of the 19K glycoprotein. Two hypothetical glycosylation sites (Asn-Val-Thr) are underlined. (*) Position at which the signal sequence is cleaved. Data from Hérissé *et al.* (1981) and Persson *et al.* (1980).

known to be a typical site for glycosylation. The C terminus of the protein contains a 23-amino-acid-long region of uncharged, predominantly hydrophobic amino acids, followed by a hydrophilic region of 15 amino acids. This structure resembles the parts of the histocompatibility antigens and glycophorin that span the cell membrane and have hydrophilic C termini on the cytoplasmic side of the membrane. Both potential glycosylation sites in the 19K glycoprotein are located on the exterior side of the membrane according to the proposed model of the protein (Persson *et al.*, 1980).

The 19K glycoprotein is also very interesting from a functional point of view. It has been observed that it is linked to the class I major histocompatibility antigens in transformed cells and also during productive infection (Kvist *et al.*, 1978; Signäs *et al.*, 1982). This conclusion is based on the observation that antisera against either the 19K glycoprotein or the heavy chain of the class I histocompatibility antigens precipitate a complex between the two components. Complexes that are precipitated contain reduced amounts of β_2-microglobulin, which may suggest that the 19K glycoprotein induces conformational changes in the class I antigens so that they bind less β_2-microglobulin. It has also been possible to demonstrate an *in vitro* association between the two components by performing cell-free translation. By immobolizing the 19K glycoprotein on a solid matrix, it is possible to bind and release the heavy chain of human and mouse class I antigens (Signäs *et al.*, 1982). It has also been proposed that the interaction between the two components takes place on the cytoplasmic side of the cell membrane. Taken together, these pieces of available experimental evidence point toward an interaction between the two components in the living cell, and it has been proposed that this interaction forms the basis for the recognition of adenovirus-infected cells by cytotoxic T cells.

2. E3/14K Polypeptide

A polypeptide encoded by region E3 with an apparent molecular weight of 14K has been purified to homogeneity by fractionation of cytoplasmic extracts, using precipitation at low pH and DEAE, phosphocellulose, and hydroxyapatite chromatography (Persson *et al.*, 1978b). The purified protein has a sedimentation coefficient of 5.5 S in its native state, corresponding to a molecular weight of around 95K. This suggests that the protein exists as a multimeric unit containing 6–7 polypeptide chains. The amino acid composition reveals a high content of charged amino acids [see Table III (Section II.A.2)]. Pulse–chase experiments have shown that the E3/14K polypeptide is a primary translational product, not suffering posttranslational modification. Immunoprecipitation with monospecific antisera has furthermore shown that the protein is exclusively localized in the cytoplasm of infected cells. Synthesis of the protein begins 2 hr after infection, the rate becoming maximal 4–6 hr postinfection. The E3/14K protein is completely unrelated to the 19K glycoprotein in its primary sequence, and the two proteins are encoded by nonoverlapping regions of the Ad2 genome.

E. Polypeptides Encoded by Region E4

Region E4 has been studied in less detail than the other early regions of the adenovirus genome. A large number of overlapping mRNAs have been identified by EM and the nuclease-S1 analysis. Eight polypeptides in the molecular-weight range between 11 and 35K have been attributed to region E4 (Matsuo *et al.*, 1982). The 11K protein is highly conserved in different adenovirus serotypes. It has been purified to homogeneity (Sarnow *et al.*, 1982) and has been shown to be tightly bound to the nuclear matrix. A frameshift mutant in which the size of the 11K protein is reduced to 6K has been found to be viable (Sarnow *et al.*, 1982). Other mutants with lesions in region E4 that are defective in their expression of certain late proteins have been isolated (Challberg and Ketner, 1981), suggesting that region E4 may be involved in the regulation of certain groups of genes.

VIII. PROSPECTIVES FOR FUTURE STUDIES OF ADENOVIRUS PROTEINS

Soon after the discovery of adenoviruses it became apparent that they provide a very useful research tool for structural studies of viral proteins. The adenoviruses have a number of advantages when compared to other animal viruses in this respect; they can be produced in large quantities, they have capsid proteins that are soluble under native conditions, and they have a complex and interesting morphology. In the past, numerous reports have appeared describing the biochemical and immunological properties of different adenovirus components, and even 3D studies of the hexon proteins have been initiated. However, our present knowledge is very limited when it comes to the details of the capsid architecture and particularly the function of structural as well as nonstructural proteins. The model of the adenovirus particle illustrated in Fig. 1 (Section I.3), although presented in its basic aspects a decade ago, must be regarded as tentative. The assignments of different polypeptides to different parts of the capsid are based primarily on evidence obtained by sequential disintegration of the virus particle. With available methods, it should be possible to refine the picture considerably. Monoclonal antibodies have already been produced against a number of capsid components (Cepko *et al.*, 1981; Russell *et al.*, 1981). It should thus be possible to identify also the minor components in the capsid by observing the interaction between tagged antibodies and adenovirus particles in the EM. By the use of improved reversible cross-linkers, it should also be possible to study the neighbor relationship in the virus particle and thereby further refine the picture. The core, finally, deserves more attention, since it provides a unique opportunity to study the interaction between DNA-binding proteins and a well-defined segment of DNA. Surprisingly little is known about the relative significance of the different protein components of the core and their possible role in gene regulation.

The adenoviruses obviously provide one of the best model systems to study both structural and nonstructural proteins, and the assignment of specific functions to the early adenovirus proteins will be particularly exciting.

ACKNOWLEDGMENTS The author is greatly indebted to Marianne Gustafson, Ingegärd Schiller, Jeanette Backman, and Christina Pelletieri for outstanding secretarial help. Dr. W. C. Russell kindly provided Fig. 1, and Hannu Ukkonen made the rest of the artwork. The electronmicrographs were prepared by Dr. Stefan Hoglund. Valuable comments on the manuscript by Drs. C. Welch, Z. Dinter, T. Linné, and G. Akusjärvi are also gratefully acknowledged. A. Virtanen has kindly helped to compile the amino acid sequences. Work in the author's laboratory has been supported by grants from the Swedish Medical Research Council, the Swedish Cancer Society, and the National Swedish Board for Technical Development.

REFERENCES

The references in this list do not represent an exhaustive review of the literature. The author apologizes for his prejudices in selecting the papers to be quoted and for any omission owing to oversight.

Akusjärvi, G., and Persson, H., 1981, Gene and mRNA for precursor polypeptide VI from adenovirus type 2, *J. Virology* **38**:469.

Akusjärvi, G., Philipson, L., and Pettersson, U., 1978, A protein kinase associated with adenovirus type 2, *Virology* **87**:276.

Akusjärvi, G., Zabielski, J., Perricaudet, M., and Pettersson, U., 1981, The sequence of the 3′ non-coding region of the hexon mRNA discloses a novel adenovirus gene, *Nucleic Acids Res.* **9**:1.

Aleström, P., Akusjärvi, G., Perricaudet, M., Mathews, M. B. Klessig, D.F., and Pettersson, U., 1980, The gene for polypeptide IX of adenovirus type 2 and its unspliced messenger RNA, *Cell* **19**:671.

Aleström, P., Akusjärvi, G., Pettersson, M., and Pettersson, U., 1982, DNA sequence analysis of the region encoding the terminal protein and the hypothetical N-gene product of adenovirus type 2, *J. Biol. Chem.* **257**:13492.

Aleström, P., Akusjärvi, G., Pettersson, M., Liang, Y.K., and Pettersson, U., 1984, The adenovirus 2 hexon gene, *J. Biol. Chem.* (submitted).

Anderson, C.W., and Lewis, J. B., 1980, Amino-terminal sequence of adenovirus type 2 proteins: Hexon, fiber, component IX, and early protein 1B-15K, *Virology* **104**:27.

Anderson, C. W., Baum, P.R., and Gesteland, R. F., 1973, Processing of adenovirus 2-induced proteins, *J. Virol.* **12**:241.

Anderson, C.W., Lewis, J. B., Atkins, J.F., and Gesteland, R.F., 1974, Cell-free synthesis of adenovirus 2-specific proteins programmed by fractionated mRNA: A comparison of polypeptide product size and messenger RNA lengths, *Proc. Natl. Acad. Sci. U.S.A.* **71**:2756.

Ariga, H., Klein, H., Levine, A.J., and Horwitz, M.S., 1980, A cleavage product of the adenovirus DNA binding protein is active in DNA replication *in vitro*, *Virology* **101**:307.

Axelrod, N., 1978, Phosphoproteins of adenovirus type 2, *Virology* **87**:366.

Babich, A., and Nevins, J.R., 1981, The stability of early adenovirus mRNA is controlled by the viral 72K DNA binding protein, *Cell* **26**:371.

Bauer, H., and Wigand, R., 1963, Eigenschaften der Adenovirus-Hämagglutinine, *Z. Hyg. Infektions* **149**:96.

Berger, J., Burnett, R.M., and Franklin, R.M., 1978, Small angle X-ray scattering studies on adenovirus type 2 hexon, *Biochem. Biophys. Acta* **535**:233.

Berk, A.J., Lee, F., Harrison, T., Williams, J., and Sharp, P.A., 1979, Pre-early adenovirus 5 gene product regulates synthesis of early viral messenger RNAs, *Cell* **17**:935.

Bhatti, A.R., and Weber, J., 1979, Protease of adenovirus type 2: Partial characterization, *Virology* **96**:478.

Biron, K.K., and Raska, K., Jr., 1977, Purification of adenovirus type 12 tumor antigen from transformed hamster cells, *Virology* **76**:516.

Black, B.C., and Center, M.S., 1979, DNA-binding properties of the major core protein of adenovirus 2, *Nucleic Acids Res.* **6**:2339.

Blair, G.E., and Russell, W.C., 1978, Identification of a protein kinase activity associated with human adenoviruses, *Virology* **86**:157.

Bos, J.L., Polder, L.J., Bernards, R., Schrier, P.I., van den Elsen, P.J., van der Eb, A.J., and van Ormondt, H., 1981, The 2.2kb E1b mRNA of human ad12 and ad5 codes for two tumor antigens starting at different AUG triplets, *Cell* **27**:121.

Boudin, M.-L., and Boulanger, P., 1981, Antibody-triggered dissociation of adenovirus penton capsomer, *Virology* **113**:781.

Boudin, M.-L., and Boulanger, P., 1982, Assembly of adenovirus penton base and fiber, *Virology* **116**:589.

Boudin, M.-L., Moncany, M., D'Halluin, J.C., and Boulanger, P., 1979, Isolation and characterization of adenovirus type 2 vertex capsomer (penton base), *Virology* **92**:125.

Boudin, M.-L., D'Halluin, J.C., Cousin, C., and Boulanger, P., 1980, Human adenovirus type 2 protein IIIa. II. Maturation and encapsidation, *Virology* **101**:144.

Boulanger, P.A., and Loucheux, M.H., 1972, Conformational study of adenovirus type 2 hexon and fiber antigens, *Biochem. Biophys. Res. Commun.* **47**:194.

Boulanger, P.A., and Puvion, F., 1973, Large-scale preparation of soluble adenovirus hexon, penton and fiber antigens in highly purified form, *Eur. J. Biochem.* **39**:37.

Boulanger, P.A., and Puvion, G., 1974, Adenovirus assembly: Cross-linking of adenovirus type 2 hexons *in vitro*, *Eur. J. Biochem.* **43**:465.

Boulanger, P.A., Flamencourt, P., and Biserte, G., 1969, Isolation and comparative chemical study of structure proteins of the adenovirus 2 and 5: Hexon and fiber antigens, *Eur. J. Biochem.* **10:**116.

Boulanger, P. Devaux, C., and Lemay, P., 1978a, Isolation and characterization of a slow-migrating class of adenovirus type 2 hexons, *Virology* **84:**456.

Boulanger, P., Devaux, C., and Loucheaux-Lefebvre, M.H., 1978b, Comparative optical properties of free and assembled hexon capsomeres of human adenovirus type 2, *FEBS Lett.* **85:**52.

Boulanger, P., Lemay, P., Blair, E.G., and Russell, W.C., 1979, Characterization of adenovirus protein IX, *J. Gen. Virol.* **44:**783.

Boursnell, M.E.G., and Mautner, V., 1981, Recombination in adenovirus: Crossover sites in intertypic recombinants are located in regions of homology, *Virology* **112:**198.

Branton, P.E., Lassam, N. J., Downey, J.F., Yee, S.-P., Graham, F.L., Mak, S., and Bayley, S.T., 1981, Protein kinase activity immunoprecipitated from adenovirus-infected cells by sera from tumor-bearing hamsters, *J. Virol.* **37:**601.

Brown, D.T., Westphal, M., Burlingham, B.T., Winterhoff, U., and Doerfler, W., 1975, Structure and composition of the adenovirus type 2 core, *J. Virol.* **16:**366.

Bürki, F., Hinaidy, B., and Böckmann, J., 1979, Bovine adenoviruses. III. Subclassification into two subgroups by means of immunodiffusion-tests, *Microbiologica* **2:**65.

Burlingham, B.T., Doerfler, W., Pettersson, U., and Philipson, L., 1971, Adenovirus endonuclease: Association with the penton of adenovirus type 2, *J. Mol. Biol.* **60:**45.

Burnett, R.M., Grütter, MG., Markovic, Z., and White, J.L., 1979, The molecular envelope of adenovirus type 2 hexon and its interaction in the viral capsid, *J. Supramol. Struct. Suppl.* **3:**92.

Cajean-Feroldi, C., Chardonnet, Y., and Chantepie-Auray, J., 1977, Deoxyribonuclease activity associated with adenovirus 5 and 7, *Eur. J. Biochem.* **74:**457.

Carter, T.H., and Blanton, R.A., 1978a, Possible role of the 72,000-dalton DNA-binding protein in regulation of adenovirus type 5 early gene expression, *J. Virol.* **25:**664.

Carter, T.H., and Blanton, R.A., 1978b, Autoregulation of adenovirus type 5 early gene expression. II. Effect of temperature-sensitive early mutations on virus RNA accumulation, *J. Virol.* **28:**450.

Cepko, C.L., Changelian, P.A., and Sharp, P.A., 1981, Immunoprecipitation with two-dimensional pools as a hybridoma screening technique: Production and characterization of monoclonal antibodies against adenovirus 2 proteins, *Virology* **110:**385.

Challberg, S.S., and Ketner, G., 1981, Deletion mutants of adenovirus 2: Isolation and initial characterization of virus carrying mutations near the right end of the viral genome, *Virology* **114:**196.

Cheng Chee-Sheung, C., and Ginsberg, H.S., 1982, Characterization of a temperature-sensitive fiber mutant of type 5 adenovirus and effect of the mutation on virion assembly, *J. Virol.* **42:**932.

Chow, L.T., Roberts, J.M., Lewis, J.B., and Broker, T.R., 1977, A map of cytoplasmic RNA transcripts from lytic adenovirus type 2, determined by electron microscopy of RNA:DNA hybrids, *Cell* **11:**819.

Chow, L.T., Lewis, J.B., and Broker, T.R., 1979, RNA transcription and splicing at early and intermediate times after adenovirus-2 infection, *Cold Spring Harbor Symp. Quant. Biol.* **44:**401.

Colby, W.W., and Shenk, T., 1981, Adenovirus type 5 virions can be assembled *in vivo* in the absence of detectable polypeptide IX, *J. Virol.* **39:**977.

Collet, M.S., and Erikson, R.L., 1978, Protein kinase activity associated with the avian sarcoma virus src gene product, *Proc. Natl. Acad. Sci. U.S.A.* **75:**2021.

Corden, J., Engelking, H.M., and Pearson, G.D., 1976, Chromatin-like organization of the adenovirus chromosome, *Proc. Natl. Acad. Sci. U.S.A.* **73:**401.

Cornick, G. Sigler, P.B., and Ginsberg, H.S., 1971, Characterization of crystals of adenovirus type 5 hexon, *J. Mol. Biol.* **57:**397.

Cornick, G., Sigler, P.B., and Ginsberg, H.S., 1973, Mass of protein in the asymmetric unit of hexon crystals—a new method, *J. Mol. Biol.* **73**:533–537.

Crowther, R.A., and Franklin, R.M., 1972, The structure of the groups of nine hexons from adenovirus, *J. Mol. Biol.* **68**:181.

Devaux, C., and Boulanger, P., 1980, Reactive serine in human adenovirus hexon polypeptide, *Virology* **102**:94.

Devaux, C., Zulauf, M., Boulanger, P., and Jacrot, B., 1982, Molecular weight of adenovirus serotype 2 capsomers: A new characterization, *J. Mol. Biol.* **156**:927.

D'Halluin, J.-C., Martin, G.R., Torpier, G., and Boulanger, P.A., 1978, Adenovirus type 2 assembly analysed by reversible cross-linking of labile intermediates, *J. Virol.* **26**:357.

Dijkema, R., Maat, J., Dekker, B.M.M., van Ormondt, H., and Boyer, H.W., 1981, The gene for polypeptide IX of human adenovirus type 7, *Gene* **13**:375.

Dorsett, P.H., and Ginsberg, H.S., 1975, Characterization of type 5 adenovirus fiber protein, *J. Virol.* **15**:208.

Engler, J.A., 1981, The nucleotide sequence of the polypeptide IX gene of human adenovirus type 3, *Gene* **13**:387.

Engler, J.A., and van Bree, M., 1982, The nucleotide sequence of the gene encoding protein IVa2 in human adenovirus type 7, *Gene* **19**:71.

Ensinger, M.J., and Ginsberg, H.S., 1972, Selection and preliminary characterization of temperature-sensitive mutants of type 5 adenovirus, *J. Virol.* **10**:328.

Esche, H., Mathews, M.B., and Lewis, J.B., 1980, Proteins and messenger RNAs of the transforming region of wild-type and mutant adenoviruses, *J. Mol. Biol.* **142**:399.

Everett, S.F., and Ginsberg, H.S., 1958, A toxin-like material separable from type 5 adenovirus particles, *Virology* **6**:770.

Everitt, E., and Philipson, L., 1974, Structural proteins of adenoviruses. XI. Purification of three low molecular weight virion proteins of adenovirus type 2 and their synthesis during productive infection, *Virology* **62**:253.

Everitt, E., Sundquist, B., and Philipson, L., 1971, Mechanism of arginine requirement for adenovirus synthesis. I. Synthesis of structural proteins, *J. Virol.* **8**:742.

Everitt, E., Sundquist, B., Pettersson, U., and Philipson, L., 1973, Structural proteins of adenoviruses. X. Isolation and topography of low molecular weight antigens from the virion of adenovirus type 2, *Virology* **52**:130.

Everitt, E., Lutter, L., and Philipson, L., 1975, Structural proteins of adenoviruses. XII. Location and neighbor relationships among proteins of adenovirus type 2 as revealed by enzymatic iodination, immunoprecipitation and chemical cross-linking, *Virology* **67**:197.

Flint, S.J., and Sharp, P.A., 1976, Adenovirus transcription. V. Quantitation of viral RNA sequences in adenovirus 2 infected and transformed cells, *J. Mol. Biol.* **106**:749.

Fowkles, D.M., Lord, S.T., Linné, T., Pettersson, U., and Philipson, L., 1979, Interaction between the adenovirus DNA-binding protein and double-stranded DNA, *J. Mol. Biol.* **132**:163.

Franklin, R.M., Pettersson, U., Åkervall, K., Strandberg, B., and Philipson, L., 1971a, Structural proteins of adenoviruses. V. On the size and structure of the adenovirus type 2 hexon, *J. Mol. Biol.* **57**:383.

Franklin, R.M., Harrison, S.C., Pettersson, U., Brändén, C.I., Werner, P.E., and Philipson, L., 1971b, Structural studies on the adenovirus hexon, *Cold Spring. Harbor Symp. Quant. Biol.* **36**:503.

Galibert, F., Hérissé, J., and Courtois, G., 1979, Nucleotide sequence of the *Eco* RI F fragment of adenovirus 2 genome, *Gene* **6**:1.

Gambke, C., and Deppert, W., 1981a, Late nonstructural 100,000- and 33,000-dalton proteins of adenovirus type 2. I. Subcellular localization during the course of infection. *J. Virol.* **40**:585.

Gambke, C., and Deppert, W., 1981b, Late nonstructural 100,000- and 33,000-dalton proteins of adenovirus type 2. II. Immunological and protein chemical analysis, *J. Virol.* **40**:594.

Gelderblom, H., and Maichle-Lauppe, I., 1982, The fibers of fowl adenoviruses, *Arch. Virol.* **72:**289.

Gilead, Z., and Ginsberg, H.S., 1968, Characterization of the tumor like (T) antigen induced by type 12 adenovirus. I. Purification of the antigen from infected KB cells and a hamster tumor cell line, *J. Virol.* **2:**7.

Gingeras, T.R., Sciaky, D., Gelinas, R.E., Bing-Dong, J., Yen, C.E., Kelly, M.M., Bullock, P.A., Parsons, B.L., O'Neill, K.E., and Roberts, R. J., 1982, Nucleotide sequences from the adenovirus-2 genome, *J. Biol. Chem.* **257:**13,475–13,491.

Ginsberg, H.S., 1979, Adenovirus structural proteins, in: *Comprehensive Virology*, Vol. 13 (H. Fraenkel-Conrat and R.R. Wagner, eds.), pp. 409–457, Plenum Press, New York.

Ginsberg, H.S., and Young, C.S.H., 1977, Genetics of adenovirus, in: *Comprehensive Virology*, Vol. 9 (H. Fraenkel-Conrat and R.R. Wagner, eds.), pp. 27–86, Plenum Press, New York.

Ginsberg, H.S., Pereira, H.G., Valentine, R.C., and Wilcox, W.C., 1966, A proposed terminology for the adenovirus antigens and virion morphological subunits, *Virology* **28:**782.

Ginsberg, H.S., Bello, L.J., and Levine, A.J., 1967, Control of biosynthesis of host macromolecules in cells infected with adenovirus, in: *The Molecular Biology of Viruses* (J.S. Colter and W. Paranchych, eds.), pp. 547–572, Academic Press, New York.

Ginsberg, H.S., Ensinger, M.J., Kaufman, R.S., Mayer, A.J., and Lundholm, U., 1974, Cell transformation: A study of regulation with types 5 and 12 adenovirus temperature-sensitive mutants, *Cold Spring Harbor Symp. Quant. Biol.* **39:**419.

Green, M., and Piña, M., 1963, Biochemical studies on the adenovirus multiplication. IV. Isolation, purification and chemical analysis of adenovirus, *Virology* **20:**199.

Green, M., Wold, W.S.M., Brackman, K., and Cartas, M.A., 1979, Studies on early proteins and transformation proteins of human adenoviruses, *Cold Spring Harbor Symp. Quant. Biol.* **44:**457.

Green, M., Brackmann, K.H., Cartas, M.A., and Matsuo, T., 1982, Identification and purification of a protein encoded by the human adenovirus type 2 transforming region, *J. Virol.* **42:**30.

Green, N.M., Wrigley, N.G., Russell, W.C., Martin, S.R., and McLachlan, A.D., 1983, Evidence for a repeating cross-β sheet structure in the adenovirus fibre, *The EMBO Journal* **2:**1357.

Grütter, M., and Franklin, R.M., 1974, Studies on the molecular weight of the adenovirus type 2 hexon and its subunit, *J. Mol. Biol.* **89:**163.

Haase, A.T., and Pereira, H.G., 1972, The purification of adenovirus neutralising activity: Adenovirus type 5 hexon immunoadsorbent, *J. Immunol.* **108:**633.

Halbert, D.N., Spector, D.J., and Raskas, H.J., 1979, *In vitro* translation products specified by the transforming region of adenovirus type 2, *J. Virol.* **31:**621.

Harter, M.L., and Lewis, J.B., 1978, Adenovirus type 2 early proteins synthesized *in vitro* and *in vivo*: Identification in infected cells of the 38,000- to 50,000-molecular-weight protein encoded by the left end of the adenovirus type 2 genome, *J. Virol.* **26:**736.

Haruna, J., Yaosi, H., Kono, R., and Watanabe, I., 1961, Separation of adenovirus by chromatography on DEAE–cellulose, *Virology* **13:**264.

Hassell, J.A., and Weber, J., 1978, Genetic analyses of adenovirus type 2. VIII. Physical locations of temperature sensitive mutations, *J. Virol.* **28:**671.

Hérissé, J., and Galibert, F., 1981, Nucleotide sequence of the EcoRI E fragment of adenovirus-2 genome, *Nucleic Acids Res.* **9:**1229.

Hérissé, J., Courtois, G., and Galibert, F., 1980, Nucleotide sequence of the Eco RI D fragment of adenovirus 2 genome *Nucleic Acids Res.* **8:**2173.

Hérissé, J., Rigolet, M., Dupont de Dinechin, S., and Galibert, F., 1981, Nucleotide sequence of adenovirus 2 DNA fragment encoding for the carboxylic region of the fiber protein and the entire E4 region, *Nucleic Acids Res.* **9:**4023.

Hilleman, R.J., Werner, J.H., and Stewart, M.T., 1955, Grouping and occurrence of RI (prototype RI-67) viruses, *Proc. Soc. Exp. Biol. Med.* **90:**555.

Hollinshead, A.C., Bunnag, B., and Alford, T.C., 1967, Relationship between the subunits and "T" antigens of adenovirus type 2, *Nature (London)* **215**:397.

Horne, R.W., Brenner, S., Waterson, A.P., and Wildy, P., 1959, The icosahedral form of an adenovirus, *J. Mol. Biol.* **1**:84.

Horwitz, M.S., 1978, Temperature-sensitive replication of H5ts125 adenovirus DNA *in vitro*, *Proc. Natl. Acad. Sci. U.S.A.* **75**:4291.

Horwitz, M.S., Scharff, M.D., and Maizel, J.V., Jr., 1969, Synthesis and assembly of adenovirus 2. I. Polypeptide synthesis, assembly of capsomers, and morphogenesis of the virion, *Virology* **39**:682.

Horwitz, M.S., Maizel, J.V., and Scharff, M.D., 1970, Molecular weight of adenovirus type 2 hexon polypeptide, *J. Virol.* **6**:569.

Hosokawa, K., and Sung, M.T., 1976, Isolation and characterization of an extremely basic protein from adenovirus type 5, *J. Virol.* **17**:924.

Huebner, R.J., Rowe, W.P., Ward, T.G., Parrot, R.H., and Bell, J.A., 1954, Adenoidal–pharyngeal–conjunctival agents: A newly recognized group of common viruses of the respiratory system, *N. Engl. J. Med.* **251**:1077.

Ishibashi, M., and Maizel, J.V., 1974, The polypeptides of adenovirus. VI. The early and late glycopeptides, *Virology* **58**:345.

Jeng, Y.H., Wold, W.S.M., Sugawara, K., Gilead, Z., and Green, M., 1977, Adenovirus type 2 coded single-stranded DNA binding protein: *In vivo* phosphorylation and modification, *J. Virol.* **22**:402.

Jeng, Y.H., Wold, W.S.M., Sugawara, K., and Green, M., 1978, Evidence for an adenovirus type 2-coded early glycoprotein, *J. Virol.* **28**:314.

Jochemsen, H., Hertoghs, J.J.L., Lupker, J.H., Davis, H., and van der Eb, A.J., 1981, *In vitro* synthesis of adenovirus type 5 T-antigens. II. Translation of virus-specific RNA from cells transformed by fragments of adenovirus type 5 DNA, *J. Virol.* **37**:530.

Jones, N., and Shenk, T., 1979a, Isolation of adenovirus type 5 host range deletion mutants defective for transformation of rat embryo cells, *Cell* **17**:683.

Jones, N., and Shenk, T., 1979b, An adenovirus type 5 early gene function regulates expression of other early viral genes, *Proc. Natl. Acad. Sci. U.S.A.* **76**:3665.

Jörnvall, H., and Philipson, L., 1980, Limited proteolysis and a reactive cysteine residue define accessible regions in the native conformation of the adenovirus hexon protein *Eur. J. Biochem.* **104**:237.

Jörnvall, H., Petterson, U., and Philipson, L., 1974a, Structural studies of adenovirus type 2 hexon protein, *Eur. J. Biochem.* **48**:179.

Jörnvall, H., Ohlsson, H., and Philipson, L., 1974b, An acetylated N-terminus of adenovirus 2 hexon protein, *Biochem. Biophys. Res. Commun.* **56**:304.

Jörnvall, H., Akusjärvi, G., Aleström, P., van Bahr-Lindström, H., Pettersson, U., Appella, E., Fowler, A.V., and Philipson, L., 1981, The adenovirus hexon protein: The primary structure of the polypeptide and its correlation with the hexon gene, *J. Biol. Chem.* **256**:6181.

Kasel, J.A., and Huber, M., 1964, Relationship of the adenovirus erythrocyte-receptor modifying factor to the type specific complement fixing antigen, *Proc. Soc. Exp. Biol. Med.* **116**:16.

Kasel, J.A., Rowe, W.P., and Nemes, J.L., 1961, Further characterization of the adenovirus erythrocyte receptor modifying factor, *J. Exp. Med.* **114**:717.

Katze, M.G., Persson, H., and Philipson, L., 1981, Control of adenovirus early gene expression: A posttranscriptional control mediated by both viral and cellular gene products, *Mol. Cell. Biol.* **1**:807.

Kauffman, R.S., and Ginsberg, H.S., 1976, Characterization of a temperature-sensitive, hexon transport mutant of type 5 adenovirus, *J. Virol.* **19**:643.

Kitchingman, G.R., and Westphal, H., 1980, The structure of adenovirus 2 early nuclear and cytoplasmic RNAs, *J. Mol. Biol.* **137**:23.

Kjellén, L., and Pereira, H.G., 1968, Role of adenovirus antigens in the induction of virus neutralizing antibodies, *J. Gen. Virol.* **2**:177.

Klein, H., Maltzman, W., and Levine, A.J., 1979, Structure–function relationships of the adenovirus DNA-binding protein, *J. Biol. Chem.* **254:**11051.

Klemperer, H.G., and Pereira, H.G., 1959, Study of adenovirus antigens fractionated by chromatography on DEAE–cellulose, *Virology* **9:**536.

Klessig, D.F., 1977, Isolation of a variant of human adenovirus serotype 2 that multiplies efficiently on monkey cells, *J. Virol.* **21:**1243.

Klessig, D.F., and Anderson, C.W., 1975, Block to multiplication of adenovirus serotype 2 in monkey cells, *J. Virol.* **16:**1650.

Klessig, D.F., and Chow, L.T., 1980, Incomplete splicing and deficient accumulation of the fiber messenger RNA in monkey cells infected by human adenovirus type 2, *J. Mol. Biol.* **139:**221.

Klessig, D.F., and Grodzicker, T., 1979, Mutations that allow human ad2 and ad5 to express late genes in monkey cells map in the viral gene encoding the 72K DNA binding protein, *Cell* **17:**957.

Köhler, K., 1965, Reinigung und Charakterisierung zweier Proteine des Adenovirus-type 2, *Z. Naturforsch.* **20b:**747.

Kruijer, W., van Schaik, F.M.A., and Sussenbach, J.S., 1980. Nucleotide sequence analysis of a region of adenovirus 5 DNA encoding a hitherto unidentified gene, *Nucleic Acids Res.* **8:**6033.

Kruijer, W., van Schaik, F.M.A., and Sussenbach, J.S., 1981, Structure and organization of the gene coding for the DNA binding protein of adenovirus type 5, *Nucleic Acids Res.* **9:**4439.

Kruijer, W., Van Schaik, F.M.A., and Sussenbach, J.S., 1982, Nucleotide sequence of the gene encoding adenovirus type 2 DNA binding protein, *Nucleic Acids Res.* **10:**4493.

Kvist, S., Östberg, L., Persson, H., Philipson, L., and Pettersson, P.A., 1978, Molecular association between transplantation antigens and a cell surface antigen in an adenovirus-transformed cell line, *Proc. Natl. Acad. Sci. U.S.A.* **75:**5674.

Lassam, N.J., Bayley, S.T., and Graham, F.L., 1979, Tumor antigens of human ad5 in transformed cells and in cells infected with transformation-defective host range mutants, *Cell* **18:**781.

Laver, W.G., 1970, Isolation of an arginine-rich protein from particles of adenovirus type 2, *Virology* **41:**488.

Laver, W.G., Suriano, R., and Green, M., 1967, Adenovirus proteins. II. N-terminal amino acid analysis, *J. Virol.* **1:**723.

Laver, W.G., Pereira, H.G., Russell, W.C., and Valentine, R., 1968, Isolation of an internal component from adenovirus type 5, *J. Mol. Biol.* **37:**379.

Laver, W.G., Wrigley, N.G., and Pereira, H.G., 1969, Removal of pentons from particles of adenovirus type 2, *Virology* **39:**599.

Laver, W.G., Younghusband, H.B., and Wrigley, N.G., 1971, Purification and properties of chick embryo lethal orphan virus (an avian adenovirus), *Virology* **45:**598.

Lemay, P., and Boulanger, P., 1980, Physiochemical characteristics of structural and nonstructural proteins of human adenovirus 2, *Ann. Virol.* **131:**259.

Lemay, P., Boudin, M.-L., Milleville, M., and Boulanger, P., 1980, Human adenovirus type 2 protein IIIa. I. Purification and characterization, *Virology* **101:**131.

Levine, A.J., and Ginsberg, H.S., 1967, Mechanism by which fiber antigen inhibits multiplication of type 5 adenovirus, *J. Virol.* **1:**747.

Leving, A.J., and Ginsberg, H.S., 1968, Role of adenovirus structural proteins in the cessation of host-cell biosynthetic functions, *J. Virol.* **2:**430.

Levinson, A.D., and Levine, A.J., 1977, The isolation and identification of the adenovirus group C tumor antigens. *Virology* **76:**1.

Levinson, A.D., Postel, E.H., and Levine, A.J., 1977, *In vivo* and *in vitro* phosphorylation of the adenovirus 5 single-strand specific DNA-binding protein, *Virology* **79:**144.

Lewis, J.B., Atkins, J.F., Anderson, C.W., Baum, P.K., and Gestland, R.K., 1975, Mapping of late adenovirus genes by cell-free translation of RNA selected by hybridization to specific DNA fragments, *Proc. Natl. Acad. Sci. U.S.A.* **72:**1344.

Lewis, J.B., Atkins, J.F., Baum, P.R., Solem, R., Gesteland, R.B., and Anderson, C.W., 1976, Location and identification of the genes for adenovirus type 2 early polypeptides, *Cell* **7**:141.

Lewis, J.B., Esche, H., Smart, J.E., Stillman, B.W. Harter, M.L., and Mathews, M.B., 1980, Organization and expression of the left third of the genome of adenovirus, *Cold Spring Harbor Symp. Quant. Biol.* **44**:493.

Liang, Y.-K., Akusjärvi, G., Aleström, P., Pettersson, U., Tremblay, M., and Weber, J., 1983, Genetic identification of an endoproteinase encoded by the adenovirus genome, *J. Mol. Viol.* **57**:217–222.

Lindberg, U., and Sundquist, B., 1974, Isolation of messenger ribonucleoproteins from mammalian cells, *J. Mol. Biol.* **86**:451.

Linné, T., and Philipson, L., 1980, The nature of the phosphate moiety of the adenovirus type 2 DNA-binding protein, *Eur. J. Biochem.* **103**:259.

Linné, T., Jörnvall, H., and Philipson, L., 1977, Purification and characterization of the phosphorylated DNA-binding protein from adenovirus type 2 infected cells, *Eur. J. Biochem.* **76**:481.

Lischwe, M.A., and Sung, M.T., 1977, A histone-like protein from adenovirus chromatin, *Nature (London)* **267**:552.

Lupker, J.H., Davis, A., Jochemsen, H., and van der Eb, A.J., 1981, *In vitro* synthesis of adenovirus type 5 T antigens. I. Translation of early region 1-specific RNA from lytically infected cells, *J. Virol.* **37**:524.

Maat, J., van Beveren, C.P., and van Ormondt, H., 1980, The nucleotide sequence of adenovirus type 5 early region E1: The region between map positions 8.0 (HindIII site) and 11.8 (SmaI site), *Gene* **10**:27.

Maizel, J.V., Jr., 1971, Polyacrylamide gel electrophoresis of viral proteins, in: *Methods of Virology*, Vol. V (K. Maramosch and H. Koprowski, eds.), pp. 179–246, Academic Press, New York.

Maizel, J.V., Jr., White, D.O., and Scharff, M.D., 1968a, The polypeptides of adenovirus. I. Evidence for multiple protein components of the virion and a comparison of types 2, 7A and 12, *Virology* **36**:115.

Maizel, J.V., Jr., White, D.O., and Scharff, M.D., 1968b, The polypeptides of adenovirus. II. Soluble proteins, cores, top components and the structure of the virion, *Virology* **36**:126.

Marusyk, R.G., Morgan, A.R., and Wadell, G., 1975, Association of endonuclease activity with serotypes belonging to the three subgroups of human adenoviruses, *J. Virol.* **16**:456.

Matsuo, T., Hashimoto, S., Wold, W.S.M., and Green, M., 1982, Identification of adenovirus type 2 early region E4 polypeptides by *in vitro* translation and tryptic peptide map analysis, *J. Virol.* **41**:334.

Mautner, V., and Pereira, H.G., 1971, Crystallization of a second adenovirus protein (the fiber), *Nature (London)* **230**:456.

Mayer, A.J., and Ginsberg, H.S., 1977, Persistence of type 5 adenovirus DNA in cells transformed by a temperature sensitive mutant, H5ts125, *Proc. Natl. Acad. Sci. U.S.A.* **74**:785.

Miller, J.S., Ricciardi, R.P., Roberts, B.E., Paterson, B.M., and Mathews, M.B., 1980, Arrangement of messenger RNAs and protein coding sequences in the major late transcription unit of adenovirus 2, *J. Mol. Biol.* **142**:455.

Montell, C., Fisher, E.F., Caruthers, M.H., and Berk, A.J., 1982, Resolving the functions of overlapping viral genes by site-specific mutagenesis at a mRNA splice site, *Nature (London)* **295**:380.

Nermut, M.V., 1975, Fine structure of adenovirus type 5. I. Virus capsid, *Virology* **65**:480.

Nermut, M.V., 1980, The architecture of adenoviruses: Recent views and problems, *Arch. Virol.* **64**:175.

Nermut, M.V., and Perkins, W.J., 1979, Consideration of the three dimensional structure of the adenovirus hexon from electron microscopy and computer modelling, *Micron* **10**:247.

Neurath, A.R., Rubin, B.A., and Stasny, J.T., 1968, Cleavage by formamide or intercapsomer bonds in adenovirus type 4 and 7 virions and hemagglutinins, *J. Virol.* **2:**1086.

Neurath, A.R., Hartzell, R.W., and Rubin, B.A., 1969, Solubilization and some properties of erythrocyte receptors for adenovirus type 7 hemagglutinin, *Nature (London)* **221:**1069.

Nicolas, J.C., Suarez, F., Levine, A.J., and Girard, M., 1981, Temperature-independent revertants of adenovirus H5ts125 and H5ts107 mutants in the DNA binding protein: Isolation of a new class of host range temperature conditional revertants, *Virology* **108:**521.

Norrby, E., 1966a, The relationship between the soluble antigens and the virion of adenovirus type 3. I. Morphological characteristics, *Virology* **28:**236.

Norrby, E., 1966b, The relationship between the soluble antigens and the virion of adenovirus type 3. II. Identification and characterization of an incomplete hemagglutinin, *Virology* **30:**608.

Norrby, E., 1968, Biological significance of structural adenovirus components, *Curr. Top. Microbiol. Immunol.* **43:**1.

Norrby, E., 1969a, The relationship between the soluble antigens and the virion of adenovirus type 3. IV. Immunological characteristics, *Virology* **37:**565.

Norrby, E., 1969b, The structural and functional diversity of adenovirus capsid components, *J. Gen. Virol.* **5:**221.

Norrby, E., 1969c, Capsid mosaics of intermediate strains of human adenoviruses, *J. Virol.* **4:**657.

Norrby, E., and Skaaret, P., 1967, The relationship between the soluble antigens and the virion of adenovirus type 3. III. Immunological identification of fiber antigen and isolated vertex capsomer antigen, *Virology* **32:**489.

Norrby, E., and Wadell, G., 1967, Soluble components of adenovirus type 4, *Virology* **31:**592.

Norrby, E., and Wadell, G., 1969, Immunological relationships between hexons of certain human adenoviruses, *J. Virol.* **4:**663.

Norrby, E., Marusyk, H.L., and Hammarskjöld, M.-L., 1969a, The relationship between the soluble antigens and the virion of adenovirus type 3. V. Identification of antigen specificities available at the surface of virions, *Virology* **38:**477.

Norrby, E., Wadell, G., and Marusyk, H., 1969b, Fiber-associated incomplete and complete hemagglutinins of adenovirus type 6, *Arch. Gesamte Virusforsch.* **28:**239.

Oosterom-Dragon, E., and Anderson, C.W., 1983, Polypeptide structure and encoding location of the serotype 2 adenovirus late, non-structural 33K protein, *J. Virol.* **45:**25.

Oosterom-Dragon, E.A., and Ginsberg, H.S., 1980, Purification and preliminary immunological characterization of the type 5 adenovirus late, nonstructural 100,000 dalton protein, *J. Virol.* **33:**1203.

Oosterom-Dragon, E.A., and Ginsberg, H.S., 1981, Characterization of two temperature-sensitive mutants of type 5 adenovirus with mutations in the 100,000-dalton protein gene. *J. Virol.* **40:**491.

Pereira, H.G., 1958, A protein factor responsible for the early cytopathic effect of adenoviruses, *Virology* **6:**601.

Pereira, H.G., 1960, A virus inhibitor produced in HeLa cells infected with adenoviruses, *Virology* **11:**590.

Pereira, H.G., and Figueiredo, M.V.T. de, 1962, Mechanism of hemagglutination by adenovirus types 1,2,4,5, and 6, *Virology* **18:**1.

Pereira, H.G., and Kelly, B., 1957, Dose response curves of toxic and infective actions of adenovirus in HeLa cell cultures, *J. Gen. Microbiol.* **17:**517.

Pereira, H.G., and Laver, W.G., 1970, Comparison of adenovirus types 2 and 5 hexons by immunological and biochemical techniques, *J. Gen. Virol.* **9:**163.

Pereira, H.G., and Wrigley, N.G., 1974, *In vitro* reconstitution, hexon bonding, and handedness of incomplete adenovirus capsid, *J. Mol. Biol.* **85:**617.

Pereira, H.G., Allison, A.C., and Farthing, C., 1959, Study of adenovirus antigens by immunoelectrophoresis, *Nature (London)* **183:**895.

Pereira, H.G., Valentine, R.C., and Russell, W.C., 1968, Crystallization of an adenovirus protein, *Nature (London)* **219**:946.

Perricaudet, M., Akusjärvi, G., Virtanen, A., and Pettersson, U., 1979, Structure of two spliced mRNAs from the transforming region of human subgroup C adenoviruses, *Nature (London)* **281**:694.

Persson, H., Pettersson, U., and Mathews, M.B., 1978a, Synthesis of a structural adenovirus polypeptide in the absence of viral DNA replication, *Virology* **90**:67.

Persson, H., Öberg, B., and Philipson, L., 1978b, Purification and characterization of an early protein (E14K) from adenovirus type 2-infected cells, *J. Virol.* **28**:119.

Persson, H., Signäs, C., and Philipson, L., 1979a, Purification and characterization of an early glycoprotein from adenovirus type 2-infected cells, *J. Virol.* **29**:938.

Persson, H., Mathisen, B., Philipson, L., and Pettersson, U., 1979b, A maturation protein in adenovirus morphogenesis, *Virology* **93**:198.

Persson, H., Jörnvall, H., and Zabielski, J., 1980, Multiple mRNA species for the precursor to an adenovirus-encoded glycoprotein: Identification and structure of the signal sequence, *Proc. Natl. Acad. Sci. U.S.A.* **77**:6349.

Persson, H., Katze, M.G., and Philipson, L., 1982, An adenovirus tumor antigen associated with membranes *in vivo* and *in vitro*, *J. Virol.* (in press).

Pettersson, U., 1971, Structural proteins of adenoviruses. VI. On the antigenic determinants of the hexon, *Virology* **43**:123.

Pettersson, U., and Höglund, S., 1969, Structural proteins of adenoviruses. III. Purification and characterization of the adenovirus type 2 penton antigen, *Virology* **39**:90.

Pettersson, U., Philipson, L., and Höglund, S., 1967, Structural proteins of adenoviruses. I. Purification and characterization of adenovirus type 2 hexon antigen, *Virology* **33**:575.

Pettersson, U., Philipson, L., and Höglund, S., 1968, Structural proteins of adenoviruses. II. Purification and characterization of adenovirus type 2 fiber antigen, *Virology* **35**:204.

Pettersson, U. Tibbetts, C., and Philipson, L., 1976, Hybridization maps of early and late mRNA sequences on the adenovirus type 2 genome, *J. Mol. Biol.* **101**:479.

Philipson, L., 1960, Separation on DEAE–cellulose of components associated with adenovirus reproduction, *Virology* **10**:459.

Philipson, L., Lonberg-Holm, K., and Pettersson, U., 1968, Virus receptor interaction in an adenovirus system, *J. Virol.* **2**:1064.

Philipson, L., Pettersson, U., Lindberg, U., Tibbetts, C., Vennström, B., and Persson, T., 1974, RNA synthesis and processing in adenovirus infected cells, *Cold Spring Harbor Symp. Quant. Biol.* **39**:447.

Polasa, H., and Green, M., 1967, Adenovirus proteins. I. Amino acid composition of oncogenic and nononcogenic human adenovirus, *Virology* **31**:565.

Prage, L., and Pettersson, U., 1971, Structural proteins of adenoviruses. VII. Purification and properties of an arginine-rich core protein from adenovirus type 2 and 3, *Virology* **45**:364.

Prage, L., Pettersson, U., and Philipson, L., 1968, Internal basic proteins in adenovirus, *Virology* **36**:508.

Prage, L., Pettersson, U., Höglund, S., Lonberg-Holm, K., and Philipson, L., 1970, Structural proteins of adenoviruses. IV. Sequential degradation of the adenovirus type 2 virion, *Virology* **42**:341.

Rabek, J.P., Zakian, V.A., and Levine, A.J., 1981, The SV40 A gene product suppresses the adenovirus H5ts125 defect in DNA replication, *Virology* **109**:290.

Rabson, A.S., O'Conor, G.T., Berezesky, I.K., and Paul, F.J., 1964, Enhancement of adenovirus growth in African Green monkey kidney cell cultures by SV40, *Proc. Soc. Exp. Biol. Med.* **116**:187.

Reif, U.K., Winterhoff, U., and Doerfler, W., 1977a, Characterization of the pH 4.0 endonuclease from adenovirus-type 2-infected cells, *Eur. J. Biochem.* **73**:327.

Reif, U., Winterhoff, U., Lundholm, U., Philipson, L., and Doerfler, W., 1977b, Purification of an endonuclease from adenovirus-infected KB cells, *Eur. J. Biochem.* **73**:313.

Ricciardi, R.P., Miller, J.S., and Roberts, B.E., 1979, Purification and mapping of specific mRNAs by hybridization–selection and cell-free translation, *Proc. Natl. Acad. Sci. U.S.A.* **76**:4927.

Rosén, L., 1958, Hemagglutination of adenoviruses, *Virology* **5**:574.

Rosén, L., 1960, A hemagglutination-inhibition technique for typing adenoviruses, *Am. J. Hyg.* **71**:120.

Rosén, L., Hovis, J.F., and Bell, J.A., 1962, Further observation on typing adenovirus and a description of two possible additional serotypes, *Proc. Soc. Exp. Biol. Med.* **110**:710.

Rowe, W.P., Hartley, J.W., Roizmann, B., and Levey, H.B., 1958, Characterization of a factor formed in the course of adenovirus infection of tissue cultures causing detachment of cells from glass, *J. Exp. Med.* **108**:713.

Russell, W.C., and Blair, G.E., 1977, Polypeptide phosphorylation in adenovirus infected cells, *J. Gen. Virol.* **34**:19.

Russell, W.C., and Precious, B., 1982, Nucleic acid binding properties of adenovirus structural polypeptides, *J. Gen. Virol.* **56**:393.

Russell, W. C., and Skehel, J.J., 1972, The polypeptides of adenovirus-infected cells, *J. Gen. Virol.* **15**:45.

Russell, W.C., Valentine, R.C., and Pereira, H.G., 1967, The effect of heat on the anatomy of the adenovirus, *J. Gen. Virol.* **1**:509.

Russell, W.C., Laver, W.G., and Sanderson, P.J., 1968, Internal components of adenovirus, *Nature (London)* **219**:1127.

Russell, W.C., McIntosh, K., and Skehel, J.J., 1971, The preparation and properties of adenovirus cores, *J. Gen. Virol.* **11**:35.

Russell, W.C., Patel, G., Precious, B., Sharp, I., and Gardner, P.S., 1981, Monoclonal antibodies against adenovirus type 5: Preparation and preliminary characterization, *J. Gen. Virol.* **56**:393.

Sarnow, P., Hearing, P., Anderson, C.W., Reich, N., and Levine, A.J., 1982, Identification and characterization of an immunologically conserved adenovirus early region E4-11Kd protein and its association with the nuclear matrix, *J. Mol. Biol.* **162**:565.

Schechter, N.M., Davies, W., and Anderson, C.W., 1981, Adenovirus coded deoxyribonucleic acid binding protein: Isolation, physical properties, and effects of proteolytic digestion, *Biochemistry* **19**:2802.

Schmidt, N.J., King, C.J., and Lennette, E.H., 1965, Hemagglutination and hemagglutination-inhibition with adenovirus type 12, *Proc. Soc. Exp. Biol. Med.* **118**:208.

Sharp, P.A., Gallimore, P.H., and Flint, S.J., 1974, Mapping of adenovirus 2 RNA sequences in lytically infected cells and transformed cell lines, *Cold Spring Harbor Symp. Quant. Biol.* **39**:457.

Signäs, C., Katze, M.G., Persson, H., and Philipson, L., 1982, An adenovirus glycoprotein binds heavy chains of class I transplantation antigens from man and mouse, *Nature (London)* **299**:175.

Smart, J.E., and Stillman, B.W., 1982, Adenovirus terminal protein precursor: Partial amino acid sequence and the site of covalent linkage to virus DNA, *J. Biol. Chem.* **257**:13,499.

Smart, J.E., Lewis, J.B., Mathews, M.B., Harter, M.L., and Anderson, C.W., 1981, Adenovirus type 2 early proteins: Assignment of the early region 1A proteins synthesized *in vivo* and *in vitro* to specific mRNAs, *Virology* **112**:703.

Spector, D.J., Crossland, L.D., Halbert, D.N., and Raskas, H.J., 1980, A 28K polypeptide is the translation product of 9S RNA encoded by region 1A of adenovirus 2, *Virology* **102**:218.

Stillman, B.W., Lewis, J.B., Chow, L.T., Mathews, M.B., and Smart, J.E., 1981, Identification of the gene and mRNA for the adenovirus terminal protein precursor, *Cell* **23**:497.

Stinski, M.F., and Ginsberg, H.S., 1974, Antibody to the type 5 adenovirus hexon polypeptide: Detection of nascent polypeptides in cytoplasm of infected KB cells, *Intervirology* **4**:226.

Stinski, M.F., and Ginsberg, H.S., 1975, Hexon polypeptides of type 2, 3, and 5 adenoviruses and their relationship to hexon structure, *J. Virol.* **15**:898.

Sugawara, K., Gilead, Z., and Green, M., 1977, Purification and molecular characterization of adenovirus type 2 DNA-binding protein, *J. Virol.* **21**:338.

Sundquist, B., Pettersson, U., and Philipson, L., 1973, Structural proteins of adenoviruses. IX. Molecular weight and subunit composition of adenovirus type 2 fiber, *Virology* **51**:252.

Sung, M.T., Lischwe, M.A., Richards, J.C., and Hosokawa, K., 1977, Adenovirus chromation. I. Isolation and characterization of the major core protein VII and precursor pro-VII, *J. Biol. Chem.* **252**:4981.

Sussenbach, J.S., 1967, Early events in the infection process of adenovirus type 5 in Hela cells, *Virology* **33**:567.

Tasseron-de Jong, J.G., Brouwer, J., Rietveld, K., Zoetemelk, C.E.M., and Bosch, L., 1979, Messenger ribonucleoprotein complexes in human KB cells infected with adenovirus type 5 contain tightly bound viral-coded "100k" proteins, *Eur. J. Biochem.* **100**:271.

Tejg-Jensen, B., Furugrin, B., Kindquist, I., and Philipson, L., 1972, A small angle X-ray study of adenovirus type 2 hexon, *Monastsh. Chem.* **103**:1730.

Tibbetts, C., Pettersson, U., Johansson, K., and Philipson, L., 1974, Relationship of messenger RNA from productively infected cells to the complementary strands of adenovirus type 2 DNA, *J. Virol.* **13**:370.

Valentine, R.C., and Pereira, H.G., 1965, Antigens and structure of the adenovirus, *J. Mol. Biol.* **13**:13.

van Beveren, C.P., Maat, J., Dekker, B.M.M., and van Ormondt, H., 1981, The nucleotide sequence of the gene for protein IVa2 and of the 5′ leader segment of the major late mRNAs of adenovirus type 5, *Gene* **16**:179.

van der Eb, A.J., van Ormondt, H., Schrier, P.I., Lupker, J.H., Jochemsen, H., van den Elsen, P.J., de Leys, R.J., Maat, J., van Beveren, C.P., Dijkema, R., and de Waard, A., 1979, Structure and function of the transforming genes of human adenoviruses and SV40, *Cold Spring Harbor Symp. Quant. Biol.* **44**:383.

van der Vliet, P.C., and Levine, A.J., 1973, DNA-binding proteins specific for cells infected by adenovirus, *Nature* (*London*) **246**:170.

van der Vliet, P.C., and Sussenbach, J.S., 1975, An adenovirus type 5 gene function required for initiation of viral DNA replication, *Virology* **64**:415.

van der Vliet, P.C., Levine, A.J., Ensinger, M.S., and Ginsberg, H.S., 1975, Thermolabile DNA binding proteins from cells infected with a temperature-sensitive mutant of adenovirus defective in viral DNA synthesis, *J. Virol.* **15**:348.

van der Vliet, P.C., Zandberg, J., and Jansz, H.S., 1977, Evidence for a function of the adenovirus DNA-binding protein in initiation of DNA synthesis as well as in elongation of nascent DNA chains, *Virology* **80**:98.

van der Vliet, P.C., Keegstra, W., and Jansz, H.S., 1978, Complex formation between the adenovirus type 5 DNA-binding protein and single-stranded DNA, *Eur. J. Biochem.* **86**:389.

Velicer, L.F., and Ginsberg, H.S., 1970, Synthesis, transport, and morphogenesis of type 5 adenovirus capsid proteins, *J. Virol.* **5**:338.

Von Bahr-Lindström, H., Jörnvall, H., Althin, S., and Philipson, L., 1982, Structural differences between hexons from adenovirus types 2 and 5: Correlation with differences in size and immunological properties, *Virology* **118**:353.

Wadell, G., 1969, Hemagglutination with adenovirus serotypes belonging to Rosén's subgroup II and III, *Proc. Soc. Exp. Biol. Med.* **132**:413.

Wadell, G., 1970, Structural and biological properties of capsid components of human adenoviruses, Ph.d. thesis, Karolinska Institutet, Stockholm.

Wadell, G., 1972, Sensitization and neutralization of adenovirus by specific sera against capsid subunits, *J. Immunol.* **108**:622.

Wadell, G., and Norrby, E., 1969a, The soluble hemagglutinins of adenoviruses belonging to Rosén's subgroup III. I. The rapidly sedimenting hemagglutinin, *Arch. Gesamte Virusforsch.* **26**:33.

Wadell, G., and Norrby, E., 1969b, Immunological and other biological characteristics of pentons of human adenoviruses, *J. Virol.* **4**:671.

Wadell, G., Norrby, E., and Skaaret, P., 1969, The soluble hemagglutinins of adenoviruses belonging to Rosén's subgroup III. I. The rapidly sedimenting hemaglutinin, *Arch. Gesamte. Virusforsch.* **26**:32.

Wadell, G., Hammarskjöld, M.-L., Winberg, G., Varsanyi, T.W., and Sundell, G., 1980, Genetic variability of adenoviruses, *Ann. N. Y. Acad. Sci.* **354**:16.

Wasmuth, E.H., and Tytell, A.A., 1966, Physical studies with adenovirus hexon antigens, *Life Sci.* **6**:1063.

Weber, J., 1976, Genetic analysis of adenovirus type 2. III. Temperature-sensitivity of processing or viral proteins, *J. Virol.* **17**:462.

White, D.O., Scharff, M.D., and Maizel, J.V., Jr., 1969, The polypeptides of adenoviruses. III. Synthesis in infected cells, *Virology* **38**:395.

Wigand, R., Bartha, A., Dreizin, R.S., Esche, H., Ginsberg, H.S., Green, M., Hierholzer, J.C., Kalter, S.S., McFerran, J.B., Pettersson, U., Russell, W.C., and Wadell, G., 1982, Adenoviridae, second report, *Intervirology* **18**:169.

Wilcox, W.C., and Ginsberg, H.S., 1961, Purification and immunological characterization of types 4 and 5 adenovirus soluble antigens, *Proc. Natl. Acad. Sci. U.S.A.* **47**:512.

Wilcox, W.C., and Ginsberg, H.S., 1963a, Structure of type 5 adenovirus. I. Antigenic relationship of virus structural proteins to virus specific soluble antigens from infected cells, *J. Exp. Med.* **118**:295.

Wilcox, W.C., and Ginsberg, H.S., 1963b, Production of specific neutralizing antibody with soluble antigens of type 5 adenovirus, *Proc. Soc. Exp. Biol. Med.* **114**:37.

Wilcox, W.C., Ginsberg, H.S., and Andersson, T.F., 1963, Structure of type 5 adenovirus. II. Fine structure of virus subunits: Morphological relationship of structural subunits to virus specific soluble antigens from infected cells, *J. Exp. Med.* **118**:307.

Wilkie, N.M., Ustaçelebi, S., and Williams, J.F., 1973, Characterization of temperature-sensitive mutants of adenovirus type 5: Nucleic acid synthesis, *Virology* **51**:499.

Willcox, N., and Mautner, V., 1976, Antigenic determinants of adenovirus capsids. I. Measurement of antibody cross-reactivity, *J. Immunol.* **116**:19.

Williams, J.F., Young, C.S.H., and Austin, P.E., 1974, Genetic analysis of human adenovirus type 5 in permissive and non-permissive cells, *Cold Spring Harbor Symp. Quant. Biol.* **39**:427.

Winters, W.D., Brownstone, A., and Pereira, H.G., 1970, Separation of adenovirus penton base antigens by preparative gel electrophoresis, *J. Gen. Virol.* **9**:105.

CHAPTER 7

Adenovirus DNA Replication

THOMAS J. KELLY, JR.

I. INTRODUCTION

As documented throughout this book, the adenoviruses have proved to be useful models for probing the basic mechanisms of macromolecular synthesis in eukaryotic cells. Much of the work on adenovirus DNA replication has been motivated by the idea that analysis of the replication of simple viral chromosomes may provide insights useful for understanding the complex mechanisms involved in the replication of eukaryotic chromosomes. Progress in this area has been rapid in recent years. Many of the general features of the adenovirus DNA replication pathway are now reasonably well understood from analysis of viral DNA synthesis *in vivo*. In addition, a cell-free replication system dependent on exogenous adenovirus DNA templates has been developed. This system has provided a means to explore the molecular mechanisms involved in initiation and chain elongation and to purify and characterize replication proteins. This chapter summarizes the available information on adenovirus DNA replication with particular emphasis on current developments. Additional information may be found in several recent reviews of the subject (Winnacker, 1978; Flint and Broker, 1980; Kelly, 1982; Challberg and Kelly, 1982).

II. STRUCTURE OF THE TERMINI OF ADENOVIRUS GENOMES

As described in greater detail below, the origins of adenovirus DNA replication are known to lie at the ends of the viral genome. For this

THOMAS J. KELLY, JR. • Department of Molecular Biology and Genetics, The Johns Hopkins University School of Medicine, Baltimore, Maryland 21205.

```
                               20                  40                  60                  80

Ad2   5'    CATCATCAAT AATATACCTT ATTTTGGATT GAAGCCAATA TGATAATGAG GGGGTGGAGT TTGTGACGTG GCGCGGGGCG TGGCAACGGG

Ad3   5'      ATCTATAT AATATACCTT ATAGATGGAA TGGTGCCAAC ATGTAAATGA GGTAATTTAA AAAAGTGCGC GCTGTGTGGT GATTGGCTGC

Ad4   5' (C)TCTCTCTAT AATATACCTT ATTTTTTTTG TGTGAGTTAA TATGCAAATA AGGCGTGAAA ATTTGGGGAT GGGGCGCGCT GATTGGCTGT

Ad5   5'    CATCATCAAT AATATACCTT ATTTTGGATT GAAGCCAATA TGATAATGAG GGGGTGGAGT TTGTGACGTG GCGCGGGGCG TGGGAACGGG

Ad7   5'    CTCTCTATAT AATATACCTT ATAGATGGAA TGGTGCCAAC ATGTAAATGA GGTAATTTAA AAAAGTGCGC GCTGTGTGGT GATTGGCTGT

Ad9   5' (C)TATCTATAT AATATACCCC ACAAAGTAAA CAAAAGTTAA TATGCAAATG AGCTTTTGAA TTTTAACGGT TTCGGGGCGG AGCCAACGCT

Ad10  5' (C)TTCATCAAT AATATACCCC ACAAAGTAAA CAAAAGTTAA TATGCAAATG AGCTTTTGAA TTTTAACGGT TTCGGGGCGG AGCCAACGCT

Ad12  5'    CATCATCAAT AATATACCTT ATACTGGACT AGTGCCAATA TTAAAATGAA GTGGGCGTAG TGTGTAATTT GATTGGGTGG AGGTGTGGCT

Ad18  5'    CATCATCAAT AATATACCTT ATACTGGACT AGAGCCAATA TTAAAATGAA GTGGGTGTGG CGATGTACTT TGATTGGGTG GAGGTGTGGC

Ad31  5' (C)ATCATCAAT AATATACCTT ACACTGGACT TGAGCCAATA TTAAAATGAA GTGGGCGGAG TGAATAGTTA ATTGACCGTA GGCGTGGTTT

SA7   5'        ATCAAT AATATACCTT ATTTGGGAAC GGTGCCAATA TGCTAATGAG GTGGGCGGAG TTTGGTGACG TATGCGGAAA TGGGCGGAGT

FL    5'    CATCATCAAT AATATACAGT TAGCAAAAAA TGGCGCCTTT GTTTGGCTTT GTTCCAACTG TTTTTGGCCC GAGTTGGGTT TCGTTTTCCC

CELO  5'    GATGATGTAT AATAACCTCA AAAACTAACG CAGTCATAAC CGGCCATAAC CGCA


            100                 120                 140                 160

Ad2   GCGGGTGACG TAG    3'

Ad3   GGGGTTAACG GCTAAAAGGG GCGGCCCGAC CGTGGGAAAA TGACGT

Ad4   GACAGCGGCG TTCGTTAGGG GCGGGG

Ad5   GCGGGTGACG TAG

Ad7   GGGGTGAATG ACTAACATGG GCGGGGCGGC CGTGGGAAAA TGACGT    3'

Ad9   GATTGGACGA GAGAAGACGA TGCAAATGAC GTCACGACGC ACGGCTAACG GTCGCCGCGG AGGCGGGC

Ad10  GATTGGACGA GAGAAGACGA TGCAAATGAC GTCACGACGC ACGGC    3'

Ad12  TTGGCGTGCT TGTAAGTTTG GGCGGATGAG GAAGTGGGGC GCGGCGTGGG AGCCGGCGCG CCGGTGTGAC GT    3'

Ad18  CTGGGCGTGT TTGTAAGTTT GGGCGGATGA GGAAGTTGGG CGCGGCGTGG GAGCCGGCGC GCCGGTGTGA CGTGT

Ad31  GCAAGTTTGC CGAAGCCGGA TGTGACGCGT GTGGGAGCCG GGCGCGCCGG ATGTGACG

SA7   TAGGGGCGGG GTTTGGCGGT AGGCGTGGCT

FL    GGG
```

reason, the detailed structure of the molecular termini has been the subject of considerable study. Two novel structural features have been identified: (1) the inverted terminal repetition (ITR) and (2) the terminal protein (TP).

A. Terminal Nucleotide Sequences

The existence of the ITR was initially suggested by the observation that the single strands of human adenovirus DNAs are capable of forming circles when incubated under renaturation conditions (Garon *et al.*, 1972; Wolfson and Dressler, 1972). The single-stranded circles so formed could be converted to linear molecules by digestion with exonuclease III or by heating to temperatures approaching the melting temperature of native DNA. These findings were interpreted as indicating the presence of complementary nucleotide sequences at or near the two ends of the viral genome. Similar findings were subsequently reported for avian adenovirus [chick embryo lethal orphan (CELO)], mouse adenovirus, and bovine adenovirus (Robinson and Bellett, 1975; Niicjama *et al.*, 1975; Larsen and Nathans, 1977).

The nucleotide sequences at the termini of the genomes of a number of adenoviruses have been determined and are shown in Fig. 1 (Steenbergh *et al.*, 1977; Arrand and Roberts, 1979; Dijkema and Dekker, 1979; Tolun *et al.*, 1979; Shinagawa and Padmanabhan, 1979, 1980; Sugisaki *et al.*, 1980; Temple *et al.*, 1981; Aleström *et al.*, 1982b; Garon *et al.*, 1982; Stillman *et al.*, 1982b). In the case of the human adenoviruses, the following general points may be made: (1) the ITRs extend to the very ends of the viral genome; (2) the nucleotide sequences at the two ends of the genome are precisely the same; (3) the sizes of the ITRs are different in different adenoviruses, ranging from 103 base pairs (bp) [adenovirus type 2 (Ad2) and Ad5] to 163 bp (Ad12); (4) the 5′-terminal nucleotide is dCMP; (5) the first 50 bp in each inverted repeat sequence are very rich in adenine and thymine (AT) residues, while the remaining base pairs show a strong bias toward guanine and cytosine (GC) residues; and (6) the sequences of the ITRs of different adenoviruses show a strong degree of homology, especially in the first 50 bp.

FIGURE 1. Nucleotide sequences of the ITRs of various adenoviruses. The sequence data were compiled from the following sources: (Ad2) Arrand and Roberts (1979), Shinagawa and Padmanabhan (1979); (Ad5) Steenbergh *et al.* (1977); (Ad3) Tolun *et al.* (1979); (Ad4, Ad9, Ad10, Ad31) Stillman *et al.* (1982b); [Ad7 (Greider strain)] Shinagawa and Padmanabhan (1980) [the inverted terminal repetition of the Gomen strain of Ad7 differs at five positions (Dijkema and Dekker, 1979)]; (Ad12) Tolun *et al.* (1979), Shinagawa and Padmanabhan (1980), Sugisaki *et al.* (1980); (Ad18) Garon *et al.* (1982); [SA7 (simian adenovirus)] Tolun *et al.* (1979); [FL (mouse adenovirus)] Temple *et al.* (1981); [CELO (avian adenovirus)] Aleström *et al.* (1982b). Partial sequences for the ITRs of Ad10 and SA7 are shown. The sequence common to all human adenoviruses is underlined.

The most striking region of homology is a sequence of 10 bp (nucleotides 9–18) that is perfectly conserved in all human adenoviruses examined to date (see Fig. 1) (Stillman *et al.*, 1982b). All or part of this sequence (ATAATATACC) is also present in the ITRs of nonhuman adenoviruses. The conservation of this sequence in the face of significant divergence elsewhere in the genomes of these viruses implies that it has an important biological function. Indeed, recent evidence suggests that it may play a role in initiation of viral DNA replication.

The GC-rich portion of the ITR (distal to 50 bp) shows little sequence conservation and is also quite variable in length. However, two features of this region appear to be common to the various adenoviruses studied so far. The first is the occurrence of short segments that are homologous to sequences found in or near the origins of replication of papovaviruses (for discussions, see Stillman *et al.*, 1982b; Temple *et al.*, 1981), and the second is the presence of the sequence TGACG (or a homologue) at or near the junction of the ITR with the remainder of the genome. The function of these conserved sequences is not known.

B. Terminal Protein

A second novel structural feature of the termini of adenovirus DNA was discovered in 1973 by Bellett and colleagues (Robinson *et al.*, 1973; Robinson and Bellett, 1974). These workers found that viral genomes released from purified virions by disruption in 4 M guanidine-HCl readily formed duplex circles or oligomers on removal of the guanidine by dialysis. The circular and oligomeric forms were rapidly converted to unit-length linear molecules by treatment with proteases, suggesting that their ends were held together by protein. These findings were subsequently confirmed and extended by a number of laboratories (Brown *et al.*, 1975; Sharp *et al.*, 1976; Carusi, 1977; Keegstra *et al.*, 1977; Padmanabhan and Padmanabhan, 1977; Rekosh *et al.*, 1977; Coombs *et al.*, 1979). It has been established that the 5′ end of each adenovirus DNA strand is bound to a single molecule of protein and that the protein–DNA linkage is resistant to treatment with a variety of chaotropic agents that would be expected to disrupt noncovalent bonds. In the case of the human adenoviruses, the molecular weight of the terminal protein is approximately 55,000 (55K) (Rekosh *et al.*, 1977; Green *et al.*, 1979; Rekosh, 1981). The protein–DNA linkage consists of a phosphodiester bond between the β-OH of a serine residue in the TP and the 5′-OH of the terminal deoxycytidine residue in the DNA (Desiderio and Kelly, 1981).

The circular and oligomeric structures originally observed by Bellett and co-workers appear to be formed as a result of interactions between TP molecules. The biological significance, if any, of these structures remains unclear. It is possible that end-to-end joining of adenovirus DNA molecules plays some important role in viral replication. However, since the interactions between TP molecules appear to have no fixed stoichiometry, the circles and oligomers may well be the product of nonspecific protein-protein interactions and thus may have no functional role *in vivo.*

Interestingly, the presence of the TP greatly increases the infectivity of adenovirus DNA (Sharp *et al.*, 1976). The basis for this effect is not understood at present. On one hand, the TP could simply enhance the uptake of viral DNA or provide protection against cellular exonucleases; on the other hand, it may serve to increase the efficiency of some important early step in viral development. One possibility, suggested by *in vitro* studies, is that the TP may facilitate initiation of viral DNA replication (see Section VI).

In infected cells, the adenovirus TP is synthesized in the form of a larger (molecular weight 80K) precursor TP (pTP) that is subsequently cleaved to produce the mature 55K species (Challberg *et al.*, 1980; Stillman *et al.*, 1981; Challberg and Kelly, 1981; Smart and Stillman, 1982). Analysis by hybrid-selected translation techniques has demonstrated that the 80K pTP is encoded by the viral genome (Stillman *et al.*, 1981; Smart and Stillman, 1982). The 5′ termini of nascent adenovirus DNA strands synthesized *in vitro* or *in vivo* are attached exclusively to the pTP (Challberg *et al.*, 1980; Challberg and Kelly, 1981; Stillman *et al.*, 1981; Lichy *et al.*, 1981; Enomoto *et al.*, 1981; Tamanoi and Stillman, 1982). The processing event that results in conversion of the 80K pTP to the 55K TP appears to occur at a late step in virion assembly (Challberg and Kelly, 1981; Stillman *et al.*, 1981). As described in detail in Section VI, there is now a large body of evidence that indicates that the 80K TP plays a central role in initiation of viral DNA replication.

III. ADENOVIRUS DNA REPLICATION *IN VIVO*

A. Overview

Studies of the biochemistry of the lytic cycle of adenoviruses have largely focused on the events that occur on infection of human cells (HeLa or KB) with Ad2 and Ad5. Adenovirus virions attach to specific cellular receptors via fibers that project from each vertex of the icosahedral particle (Philipson *et al.*, 1968). The mechanism by which virions enter the cell has not yet been firmly established, but the available electron-microscopic (EM) data suggest that entry occurs either by endocytosis (Chardonnet and Dales, 1970a,b) or by direct penetration of the cytoplasmic membrane (Lonberg-Holm and Philipson, 1969; Morgan *et al.*, 1969). Removal of the viral capsid proteins occurs largely in the cytoplasm of the infected cell (Philipson *et al.*, 1968; Lonberg-Holm and Philipson, 1969; Morgan *et al.*, 1969; Sussenbach, 1967), and virion cores enter the nucleus (Morgan *et al.*, 1969; Chardonnet and Dales, 1972). The final stages of uncoating occur in the nucleus, yielding adenovirus genomes that appear to be almost free of virion proteins (Lonberg-Holm and Philipson, 1969). EM autoradiographic experiments suggest that the replication of adenovirus DNA takes place at sites within the nucleoplasm and is not associated with the nuclear membrane (Martinez-Paloma and Granboulan, 1967; Shiroki *et al.*, 1974; Simmons *et al.*, 1974; Vlak *et al.*, 1975b).

A one-step growth curve for adenovirus can be approximated by infecting HeLa or KB cells in suspension culture at a high multiplicity of infection (50–100 plaque-forming units/cell) (Green *et al.*, 1970). Under these conditions, replication of viral DNA is first detected at about 6–8 hr after infection. Viral DNA synthesis proceeds by a semiconservative mechanism (Pearson and Hanawalt, 1971; Bellett and Younghusband, 1972; van der Vliet and Sussenbach, 1972; van der Eb, 1973) and reaches a maximal rate about 19 hr after infection (Green *et al.*, 1970). Approximately 10^5–10^6 progeny viral genomes are produced by 24 hr, but only about 20% are ultimately packaged to form progeny viral particles (Green *et al.*, 1970; Green, 1962). Infection with adenovirus produces a profound inhibition of cellular DNA replication (Ginsberg *et al.*, 1967; Piña and Green, 1969; Mantyjarvi and Russell, 1969). The inhibition is first detectable at 6–8 hr after infection, coincident with the onset of viral DNA replication, and is virtually complete by 12 hr after infection.

B. Structure of Replicative Intermediates

Viral replicative intermediates were first identified by pulse-labeling methods (Pearson and Hanawalt, 1971; Bellett and Younghusband, 1972; van der Vliet and Sussenbach, 1972; van der Eb, 1973; Burlingham and Doerfler, 1971; Sussenbach and van der Vliet, 1972; Robin *et al.*, 1973; Pearson, 1975). When adenovirus-infected cultures were exposed to a short pulse of [^{3}H]thymidine, a significant fraction of the labeled DNA was observed to sediment more rapidly (31–60 S) than mature viral DNA (31 S) in neutral sucrose gradients. Pulse–chase experiments established that this rapidly sedimenting species represents a precursor to mature viral DNA. Sedimentation of replicative intermediates under denaturing conditions revealed a broad distribution of nascent viral DNA strands with lengths less than or equal to the length of mature viral DNA strands (Pearson and Hanawalt, 1971; Bellett and Younghusband, 1972; van der Vliet and Sussenbach, 1972; van der Eb, 1973; Pearson, 1975; Horwitz, 1971). The buoyant density of replicative intermediates was observed to be significantly greater than that of mature viral DNA (Pearson and Hanawalt, 1971; Bellett and Younghusband, 1972; Sussenbach and van der Vliet, 1972; Robin *et al.*, 1973; Pearson, 1975; Sussenbach *et al.*, 1972). The density difference could be abolished by treatment of replicative intermediates with nuclease S1, but not by treatment with ribonuclease, indicating that the intermediates contain extensive regions of single-stranded DNA (ssDNA). A similar conclusion was reached by examination of the chromatographic behavior of replicative intermediates on benzoylated–naphthoylated diethylaminoethyl–cellulose (van der Eb, 1973; Sussenbach *et al.*, 1972).

The basic structural features of adenovirus replicative intermediates have been defined by several different kinds of analysis. Understanding of the structure of the intermediates has led in turn to a reasonably clear picture of the pathways involved in adenovirus DNA replication. One of the most revealing methods of analysis has been direct visualization of replicative intermediates in the EM (Sussenbach *et al.*, 1972, 1973; Ellens *et al.*, 1974; Lechner and Kelly, 1977). Two basic types of replicative intermediates have been observed (Fig. 2): Y-shaped molecules composed of a unit-length, double-stranded segment with one or more single-stranded branches (type I) and unbranched, unit-length, linear molecules containing both double- and single-stranded segments (type II). In a population of type I replicative intermediates, the lengths of the single-stranded branches are variable, but within a given molecule, each single-stranded branch is equal in length to one of the duplex arms defined by the branch point. During the period of maximal DNA replication, type I replicative intermediates with multiple single-stranded branches are relatively common, accounting for as much as 25% of the total. The vast majority of type II replicative intermediates contain only one double-stranded and one single-stranded region, and the position of the double-to-single-stranded transition point is variable from one molecule to another. A small fraction of replicative intermediates have the characteristics of both type I and type II replicative intermediates. These type I–II molecules contain a unit-length linear DNA segment with one double-to-single-strand transition point, like type II molecules; however, they also contain one or more single-stranded branches emerging from the duplex segment, like type I molecules (see Fig. 2).

The EM observations described above were interpreted in terms of a model in which initiation of replication on duplex molecules occurs at or near an end and proceeds by a displacement mechanism that produces type I replicative intermediates (Ellens *et al.*, 1974; Lechner and Kelly, 1977). Type II molecules were interpreted as replicative intermediates in which complementary strand synthesis occurs on ssDNA templates displaced from type I intermediates. The structure of type II molecules suggested that initiation of synthesis on displaced strands also takes place at or near an end. To determine whether initiation of adenovirus DNA replication occurs at a unique end, partially denatured replicative intermediates were analyzed. Ad2 (and Ad5) DNA molecules contain clusters of AT-rich regions at map positions 0–0.15, 0.5–0.6, and 0.8–1.0 (Doerfler and Kleinschmidt, 1970). Under suitable partial denaturation conditions, these regions are melted and the right-to-left polarity of the molecules can be determined by inspection. Analysis of the partially denatured replicative intermediates showed conclusively that the initiation events that produce both type I and type II molecules occur at approximately the same frequency at either end of the viral genome (Lechner and Kelly, 1977). Consistent with this finding was the observation that a significant fraction of multiply branched type I molecules contained replication forks moving in opposite directions.

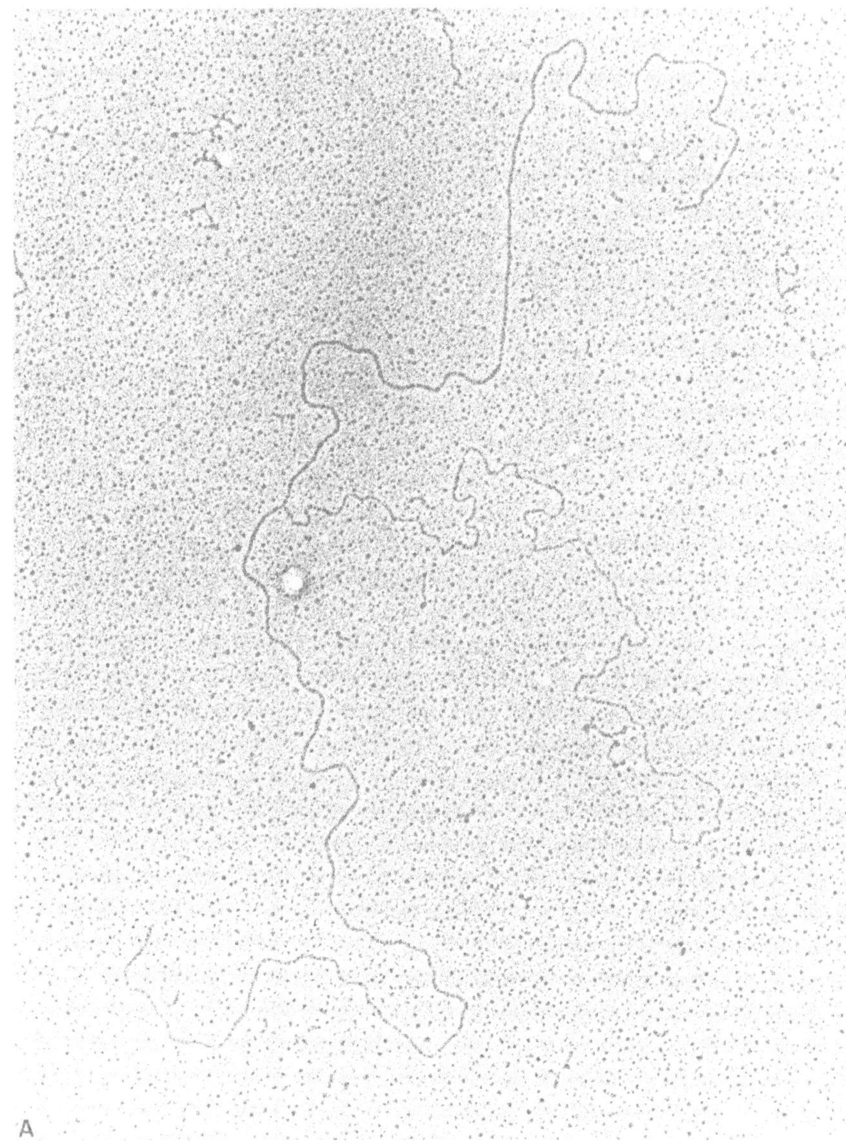

FIGURE 2. Electron micrographs of adenovirus replicative intermediates. (A) Type I replicative intermediate with one single-stranded branch; (B) type I replicative intermediate with two single-stranded branches; (C) type II replicative intermediate; (D) type I–II replicative intermediate. [(C, D)→] Transitions from double- to single-stranded DNA. (D) Scale bar: 1 μm.

The overall polarity of daughter-strand synthesis was also analyzed by EM (Lechner and Kelly, 1977). For this purpose, the enzyme terminal transferase (Bollum, 1974) was used to label the 3′-OH termini of replicative intermediates. Following incubation with terminal transferase and thymidine triphosphate, polydeoxythymidylic acid [poly(dT)] "tails"

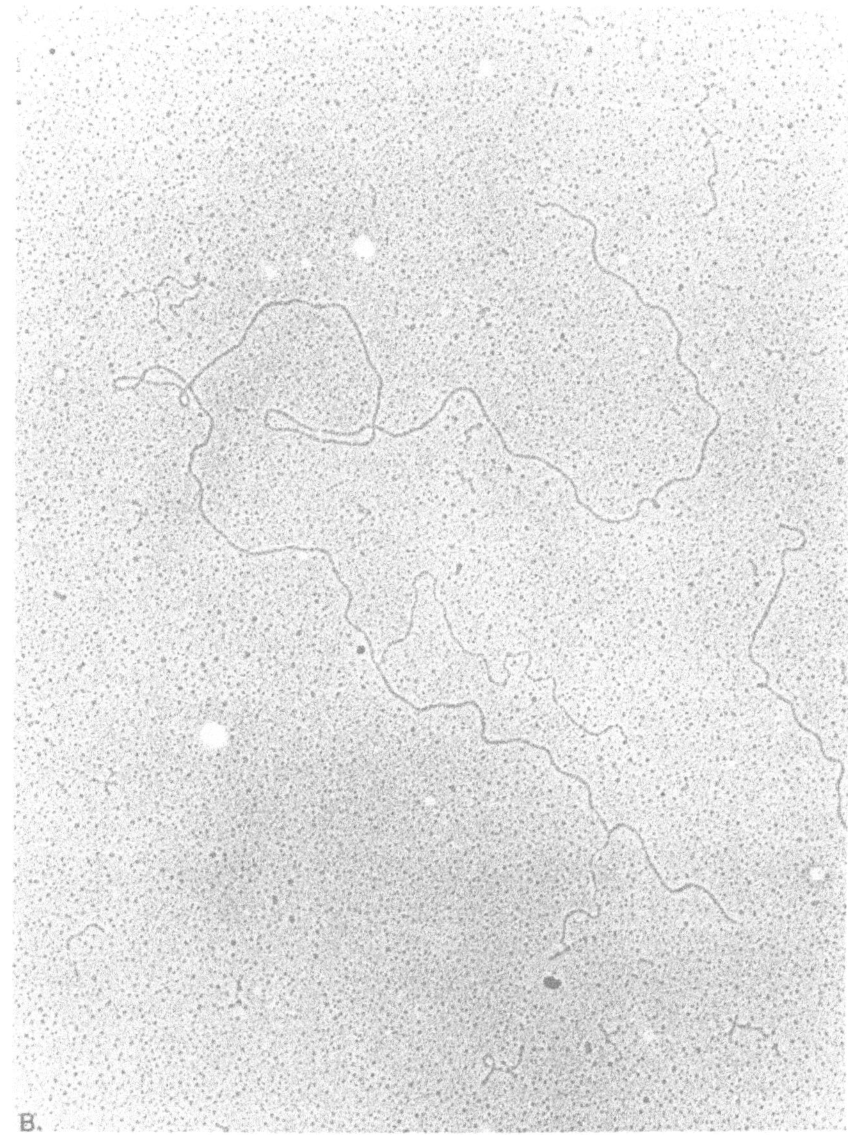

FIGURE 2 (*Continued*)

were observed at the branch points of type I molecules and at the double-to-single-stranded transition points in type II molecules. This result indicated that the nascent daughter strands in adenoviral replicating molecules grow from their 5′ termini toward their 3′ termini and that each nascent strand contains only one free 3′-OH group, which is located at the growing point.

The basic conclusions of the EM analysis of adenovirus replication intermediates were confirmed and, in some instances, anticipated by in-

FIGURE 2 (*Continued*)

dependent studies using other physical methods. The locations of the termination points of DNA replication were determined by analysis of the distribution of incorporated radioactivity in newly synthesized genomes, according to the method originally developed in the simian virus 40 system by Danna and Nathans (1972). In this method, infected cells are exposed to a short pulse of [^{3}H]thymidine, and the progeny viral genomes completed during the pulse are isolated and separated from other

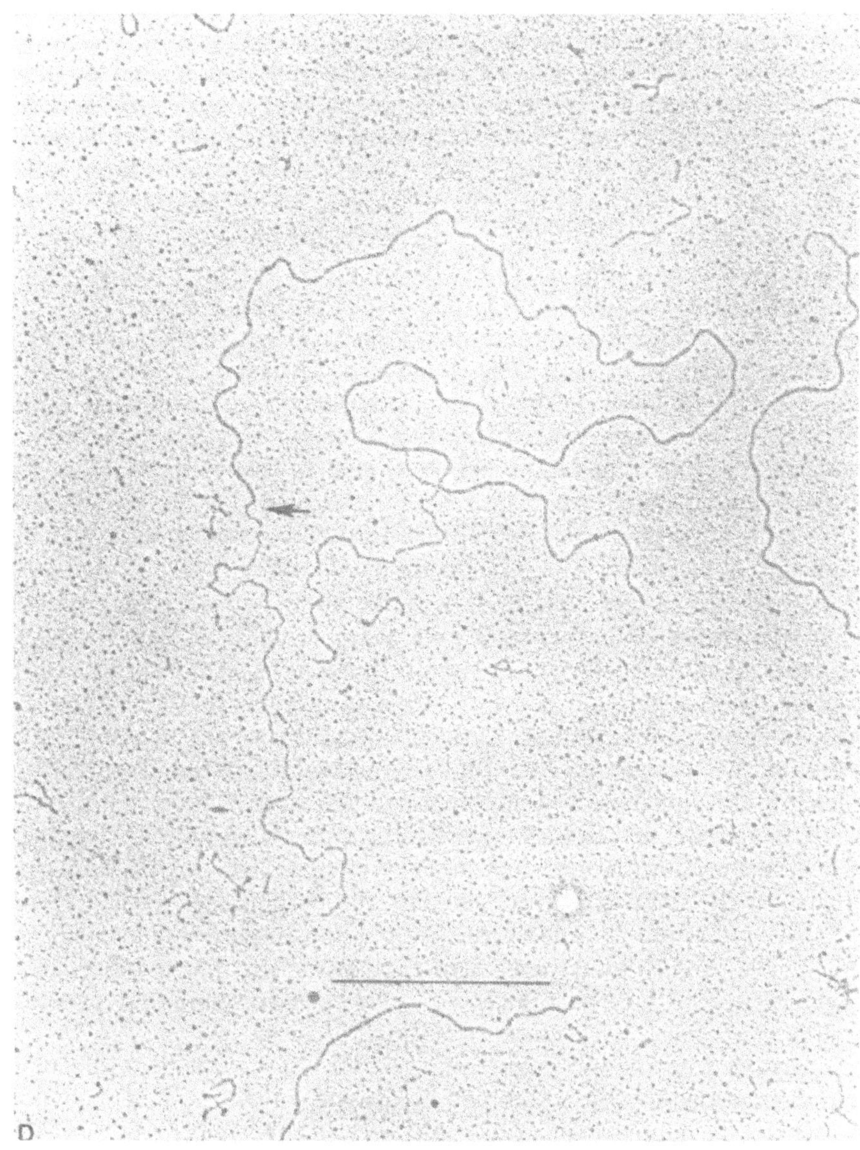

FIGURE 2 (*Continued*)

labeled species. Such progeny genomes are preferentially labeled in regions adjacent to the termini of replication. When newly synthesized adenovirus genomes were analyzed in this manner, restriction fragments at the two molecular ends were found to have the highest specific radioactivities (Schilling *et al.*, 1975; Tolun and Pettersson, 1975; Sussenbach and Kuijk, 1977; Ariga and Shimojo, 1977). Moreover, the radioactivity incorporated into the terminal restriction fragments was distributed asymmetrically between the two complementary strands. Essentially all the radioactivity incorporated into restriction fragments derived from the right end of the genome was recovered in the leftward-transcribed strand (l-strand), whereas radioactivity in restriction fragments derived from the left end of the genome was confined to the rightward transcribed strand (r-strand) (Sussenback and Kuijk, 1977; Weingartner *et al.*, 1976; Horwitz, 1976). These results demonstrated quite clearly that the synthesis of each of the two adenovirus DNA strands terminates at its 3′ end. While the data were not sufficient to define the precise locations of the origins of DNA replication, they suggested that the synthesis of each daughter strand begins near its 5′ end.

To localize the origins more accurately, attempts were made to synchronize the onset of viral DNA replication (Ariga and Shimojo, 1977; Sussenbach and Kuijk, 1978). In one such study (Sussenbach and Kuijk, 1978) cells were infected with H5*ts*125, a temperature-sensitive mutant defective in DNA replication (see Section V.A), and maintained at the restrictive temperature for 16 hr. Under these conditions, little or no viral DNA replication could be detected (Ensinger and Ginsberg, 1972; van der Vliet and Sussenbach, 1975; Ginsberg *et al.*, 1975). The infected cells were then shifted to the permissive temperature, and the newly synthesized adenovirus DNA was labeled for a short time with [^{3}H]thymidine. Analysis of the distribution of radioactivity in the labeled DNA after cleavage with restriction enzymes revealed that the specific radioactivities of restriction fragments derived from the two molecular termini were significantly greater than those of internal restriction fragments. In addition, the distribution of incorporated radioactivity between the two DNA strands of the terminal restriction fragments was just the opposite of that observed in the termination experiments described above. In particular, at the right end of the genome, the r-strand was labeled preferentially, and at the left end of the genome, the l-strand was labeled preferentially. These findings suggested that during the first round of DNA replication, both molecular ends can serve as origins, and that synthesis begins at or near the 5′ end of each daughter strand.

Further information concerning the structure of adenovirus replication intermediates was obtained by analysis of the ssDNA sequences present in the pool of replicating molecules. In this type of study, ssDNA from infected cells was assayed for its content of l and r sequences by hybridization to each of the two complementary strands of the adenovirus genome (Tolun and Pettersson, 1975; Lavelle *et al.*, 1975; Flint *et al.*, 1976; Sussenbach *et al.*, 1976). This analysis revealed that sequences derived from both strands of the genome are represented in the pool of intracellular ssDNA in roughly equal amounts. A careful study of the relative frequency of ssDNA sequences derived from different regions of the genome showed that while all regions are represented in the ssDNA fraction, the distribution is far from uniform (Flint *et al.*, 1976). Specifically, at the right end of the genome, r-strand sequences predominate over l-strand sequences, whereas at the left end of the genome, l-strand sequences predominate over r-strand sequences. The simplest interpre-

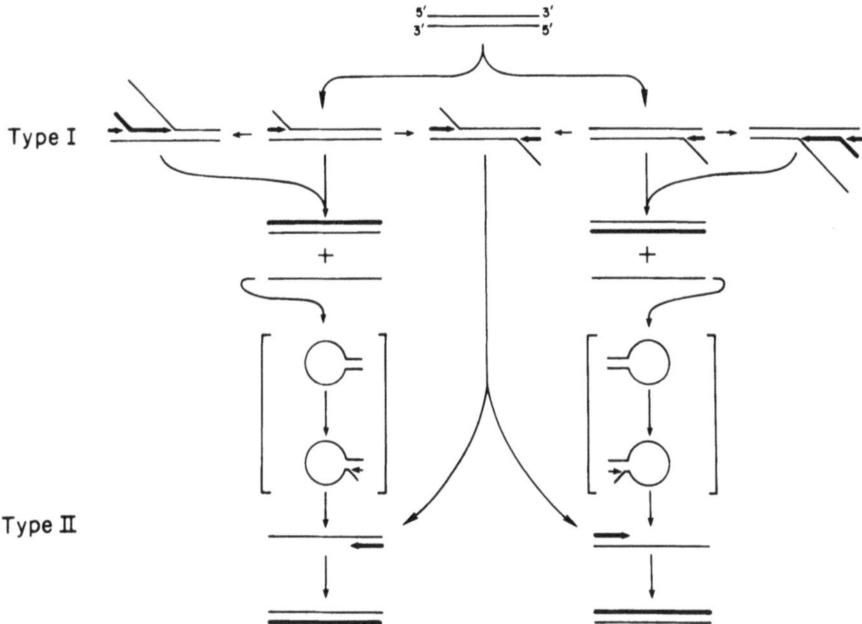

FIGURE 3. Model for adenovirus DNA replication. Newly synthesized DNA strands are depicted by bold lines, parental DNA strands by light lines. Arrows indicate the direction of chain elongation. The bracketed circular "panhandle" forms are theoretical intermediates. Adapted from Lechner and Kelly (1977).

tation of these results is that both adenovirus DNA strands are displaced during DNA replication.

C. Model for Adenovirus DNA Replication

The information derived from EM and biochemical studies provides strong support for the replication model illustrated in Fig. 3 (Lechner and Kelly, 1977). In this model, replication can be initiated at either end of the double-stranded parental genome. Following each initiation event, a daughter strand is synthesized in the 5′-to-3′ direction with concomitant displacement of the parental strand of the same polarity, producing a type I replicative intermediate. Thus, when initiation occurs at the right molecular end, the parental l-strand serves as template and the parental r-strand is displaced; conversely, when initiation occurs at the left molecular end, the parental r-strand is copied and the parental l-strand is displaced. The frequencies of initiations at the two ends of the viral genome are approximately the same, and multiple initiation events on the same replicating molecule are common.

The model in Fig. 3 indicates two possible pathways for the synthesis of the viral DNA strands complementary to the displaced parental strands. Both pathways were deduced from EM studies of adenovirus DNA replication. In the major pathway, displacement synthesis proceeds to completion, resulting in the formation of a daughter duplex and the release of a parental single strand. The latter then serves as a template for the synthesis of a complementary strand. Complementary-strand synthesis is initiated at the 3′ end of the released single strand and proceeds in the 5′-to-3′ direction, generating a type II replicative intermediate. The minor pathway is a consequence of the fact that displacement synthesis can be initiated at either end of the genome. Thus, there is a low, but finate, probability that before completion of a round of displacement synthesis initiated at one end of the genome, a second initiation will occur at the opposite end. In such doubly initiated molecules, both parental strands serve as templates for daughter-strand synthesis. When the two oppositely moving displacement forks meet, the parental strands will no longer be held together by base-pairing and will separate, yielding two type II molecules (fork annihilation).

Although not shown in Fig. 3, the proposed model for adenovirus DNA replication also conveniently accounts for type I–II replicative intermediates. Such molecules could arise by initiation of displacement synthesis at the duplex end of a type II replicative intermediate or by fork annihilation in a type I intermediate that contains more than two branches.

All initiation events that occur during adenovirus DNA replication take place at the 3′ terminus of the template DNA strand. Since the nucleotide sequences at the 3′ termini of the l- and r-strands of adenovirus DNA are identical, it seems probable that all initiation events proceed by the same molecular mechanism. Of course, the pathway of viral DNA replication as currently understood would appear to demand two different sorts of initiations, those that occur at duplex ends and those that occur at single-stranded ends. However, since the nature of the actual substrate for initiation is not known in either case, the difference may be more apparent than real. For example, if the duplex parental genome is partially unwound prior to initiation, the exposed 3′ end of one of the strands would be indistinguishable from the 3′ end of a single strand released at the completion of a round of displacement synthesis. Another possibility, first suggested by Sambrook (in Daniell, 1976), is that the single strands released by displacement synthesis form circles by hybridization of their self-complementary terminal sequences (Fig. 3). The sequence of nucleotide pairs in the duplex "panhandle" formed by this cyclization process is identical to that at the right and left ends of the parental viral genome and could presumably be operated on by the same initiation mechanism. Since little is yet known about the molecular events involved in initiation, schemes such as these remain speculative. They are mentioned here simply to emphasize the potential equivalence of all initiation events in adenovirus DNA replication. Further discussion of the initiation problem is presented in Section VI.

D. Mechanism of DNA Chain Elongation

It is apparent from the model discussed in Section III.C that there is no requirement, at least in principle, for a discontinuous mode of DNA synthesis during any phase of adenovirus DNA replication. All nascent daughter strands grow from their 5′ termini toward their 3′ termini and thus can be elongated without the need for the synthesis and subsequent joining of Okazaki fragments. There is, in fact, considerable experimental support for the idea that adenovirus DNA replication proceeds by a completely continuous mechanism. The more recent EM studies of adenovirus replicative intermediates (Lechner and Kelly, 1977; Kedinger *et al.*, 1978) indicate that internal initiations on displaced single strands seldom, if ever, occur. Furthermore, the poly(dT) "tailing" experiments described in Section III.B suggest that replicative intermediates contain no free 3′-OH groups other than those at the growing end of nascent daughter strands (Lechner and Kelly, 1977). On the other hand, analysis of the size distribution of pulse-labeled nascent chains has given somewhat conflicting data. In a number of studies (van der Eb, 1973; Horwitz, 1971; Brison *et al.*, 1977; Yamashita *et al.*, 1977), nascent DNA strands labeled *in vitro* or *in vivo* for short periods of time (less than 2 min) were found to sediment under alkaline conditions as a continuum, ranging in size from a few hundred to 35,000 nucleotides. By suitable pulse–chase experiments, it was demonstrated that the shorter chains were elongated to full length during longer incubations. These results are entirely consistent with the hypothesis of completely continuous replication. In other studies, however, a discrete class of short (6–12 S in alkaline sucrose gradients) nascent chains were the primary species observed after pulse labeling (Bellett and Younghusband, 1972; Pearson, 1975; Vlak *et al.*, 1975a; Winnacker, 1975). These short chains also appeared to be intermediates in the synthesis of full-length viral strands on the basis of their behavior in pulse–chase experiments. Restriction endonuclease analysis of pulse-labeled viral DNA demonstrated that the 6–12 S chains were derived from all regions of the viral genome, and hybridization studies revealed the presence of sequences from both the l- and the r-strand. The short chains accumulated when infected cells were exposed to hydroxyurea or high concentrations of thymidine, uracil, or deoxycytidine. Their size in isolated nuclei was strongly dependent on the ratio of ribonucleoside triphosphates to deoxyribonucleoside triphosphates. Although the observation of these short DNA chains was originally interpreted as evidence of discontinuous DNA synthesis during adenovirus DNA replication, their real significance is not yet clear. Studies in both prokaryotic (Tye *et al.*, 1977) and eukaryotic (Brynelf *et al.*, 1978; Wist *et al.*, 1978) systems suggest that transient short chains can be generated by excision-repair mechanisms that operate to remove abnormal bases (e.g., uracil, hypoxanthine, 3-methyladenine) from DNA. For example, both bacteria and mammalian cells contain a uracil-DNA-glycosylase activity that removes uracil residues that arise either from misincorporation of dUTP during replication or from spontaneous deamination of cytidine (Lindahl, 1974; Sekiguchi *et al.*, 1976). The resulting apyrimidinic site represents a substrate for an excision process mediated by an apyrimidinic endonuclease and an exonuclease activity. The broken DNA strand is subsequently repaired by polymerase and ligase activities, thus giving the appearance of transient intermediates in replication. Experimental support for the hypothesis that excision-repair processes may be responsible for the short nascent strands observed during adenovirus DNA replication has come from analysis of the incorporation of uracil into viral DNA (Ariga and Shimojo, 1979). When Ad12-infected cells were exposed to [^{3}H]uridine for brief periods, the short chains were preferentially labeled. Moreover, when the labeled short chains were isolated from alkaline sucrose gradients and digested to completion with deoxyribonuclease I and venom phosphodiesterase, 90% of the incorporated radioactivity was recovered as dUTP. While these data do not directly prove that excision repair is the source of the shirt chains, they are strongly suggestive that this is the case. Thus, the weight of the evidence available at present favors the idea that adenovirus DNA synthesis is a completely continuous process.

IV. ADENOVIRUS DNA REPLICATION *IN VITRO*

By analogy with recent experience in prokaryotic systems, it seems likely that further progress in understanding the mechanism of adenovirus DNA replication at the biochemical level will require analysis of *in vitro* replication systems. The major goals of such analysis are to define the essential components of the replication reaction and to determine the function of each component.

Two basic types of *in vitro* systems capable of synthesizing adenovirus DNA have been developed. In the first type, endogenous DNA synthesis occurs on viral replicative intermediates in isolated nuclei or subnuclear replication complexes (van der Vliet and Sussenbach, 1972; Brison *et al.*, 1977; Winnacker, 1975; Kaplan *et al.*, 1977; Yamashita *et al.*, 1975). In such systems, initiation of DNA replication is very infrequent, if it occurs at all. Studies of endogenous viral DNA synthesis have so far proved most valuable in analyzing the structure of replicative intermediates and in defining some of the requirements for DNA chain elongation. More recently, a soluble enzyme system capable of replicating exogenously added adenovirus DNA has been described (Challberg and Kelly, 1979a,b). The system carries out both the initiation and the elongation of viral DNA chains by mechanisms that appear to resemble closely those observed *in vivo*. The requirement of the system for an exogenous template provides an approach to defining the structural features of the template that are important in replication. In addition, the system seems well suited to fractionation–reconstitution experiments of the type that have proved so powerful in the analysis of prokaryotic DNA replication.

A. Replication on Endogenous DNA Templates

The first *in vitro* studies of adenovirus DNA replication were performed with isolated nuclei from infected cells. Isolated nuclei are permeable to deoxyribonucleoside triphosphates (dNTPs), and under the appropriate conditions, such dNTPs are incorporated into growing DNA chains (van der Vliet and Sussenbach, 1972; Sussenbach and van der Vliet, 1972; Winnacker, 1975). At 37°C, incorporation is linear for 30–60 min and then levels off. DNA synthesis requires all four dNTPs and a divalent cation. It is stimulated slightly by ATP. The synthetic product consists predominantly of unit-length, double-stranded adenovirus DNA molecules that contain uninterrupted single strands. The nature of the DNA synthesis reaction that occurs in isolated nuclei is revealed most clearly by density-shift experiments in which 5-bromo-2′-deoxyuridine triphosphate is substituted for dTTP (van der Vliet and Sussenbach, 1972). Such experiments show that the reaction product is heterogeneous in buoyant density, with greater than 95% of the molecules falling in the range between light (LL) and hybrid (HL) densities. Moreover, by prelabeling the infected cells prior to the isolation of nuclei, it is possible to demonstrate that the DNA strands synthesized *in vitro* are covalently linked to DNA strands that were being synthesized *in vivo* at the time of cell lysis. These findings indicate that DNA replication in isolated nuclei represents a continuation of synthesis on replicative intermediates formed *in vivo*. Thus, nascent strands initiated *in vivo* can be completed *in vitro*, but *de novo* initiation of new DNA chains apparently does not occur.

The ability to incorporate a density label into growing DNA chains in isolated nuclei provides a convenient method for isolating and purifying replicative intermediates. Much of the early work on the structure of replicative intermediates (described in Section III.B) utilized this approach (Sussenbach *et al.*, 1972, 1973; Ellens *et al.*, 1974).

As a logical extension of studies with nuclear replication systems, a number of methods have been developed to liberate viral replicative intermediates and associated proteins from isolated nuclei. Under suitable conditions, such "replication complexes" will support continued DNA synthesis *in vitro*. The first attempts along these lines involved the isolation of replicative intermediates in association with nuclear membrane components by the so-called "M-banding" technique or by sucrose-density-gradient sedimentation of nuclear sonicates (Frenkel, 1978; Yamashita *et al.*, 1975; Shiroki *et al.*, 1974). Membrane complexes of this type have not proved very useful for the biochemical analysis of adenovirus DNA replication. Considerable fragmentation of the viral DNA occurs during incubation *in vitro*. Furthermore, only rather short DNA chains (6–9 S) are synthesized. Thus, the *in vitro* reaction bears little resemblance to viral DNA replication *in vivo*. The association of replicative intermediates with membranes in this type of preparation may well be artifactual. The available EM evidence strongly suggests that adenovirus DNA replication takes place in the nucleoplasm and is not associated with the nuclear membrane *in vivo* (Martinez-Paloma and Granboulan, 1967; Simmons *et al.*, 1974; Vlak *et al.*, 1975b).

Alternative approaches to the isolation of subnuclear replication complexes have been developed. These involve incubation of isolated nuclei in the presence of 0.2–0.3 M ammonium sulfate (Brison *et al.*, 1977; Kaplan *et al.*, 1977; Shaw *et al.*, 1979) or 1 mg/ml sodium heparin (Yamashita *et al.*, 1977). Under these conditions, replication complexes are released from the nuclei in a soluble form and can be separated from cellular chromatin by centrifugation methods or gel filtration. The replication complexes support the synthesis of adenovirus DNA at a rate comparable to the *in vivo* rate for short periods of time (30–60 min). The synthetic reaction requires the four dNTPs and Mg^{2+} and is stimulated to a modest extent by ATP. As in the nuclear systems, the observed DNA synthesis represents the elongation of nascent chains initiated *in vivo*. During a 30-min incubation *in vitro*, a significant fraction of the nascent strands are completed, yielding full-length, double-stranded viral DNA molecules. The *in vitro* elongation reaction clearly resembles that observed *in vivo*. Synthesis takes place on branched (type I) and linear (type II) replication intermediates identical to those present in intact infected cells, and replication terminates at the ends of the template DNA strands (Kaplan *et al.*, 1977; Horwitz *et al.*, 1979). The analysis of replication complexes represents a potentially useful approach to defining mechanisms involved in adenovirus DNA chain elongation. However, it has not yet been possible to fractionate such systems and reconstitute the replication reaction, and this inability has limited their usefulness.

B. Replication on Exogenous DNA Templates

The analysis of adenovirus DNA replication *in vitro* was taken a step further with the discovery that nuclear extracts from adenovirus-infected cells are capable of carrying out the replication of exogenously added adenovirus DNA (Challberg and Kelly, 1979a,b). The extracts are normally prepared from cells infected in the presence of hydroxyurea to prevent the accumulation of endogenous replicative intermediates. Efficient replication *in vitro* is observed when the extracts are incubated with adenovirus DNA containing the covalently bound terminal protein. The replication reaction requires the four dNTPs and Mg^{2+}. It is stimulated 2- to 3-fold by ATP, but is not further stimulated by the addition of the other three ribonucleoside triphosphates.

Several lines of evidence indicate that the DNA synthesis observed in this *in vitro* system closely resembles adenovirus DNA replication *in vivo* (Challberg and Kelly, 1979a,b; Challberg *et al.*, 1980; Stillman, 1981; Kaplan *et al.*, 1979; Reiter *et al.*, 1980). First, daughter-strand synthesis is initiated at either terminus of the genome and proceeds exclusively in the 5′-to-3′ direction. Second, replication takes place by a displacement mechanism on type I replicative intermediates identical in structure to those observed *in vivo*. Third, the rate of DNA chain elongation *in vitro* (about 1200 nucleotides/min) is quite comparable to the *in vivo* rate. Finally, the DNA synthesized *in vitro* consists of long adenovirus DNA strands that are hydrogen-bonded, but not covalently linked, to the input

DNA template. A significant fraction of the newly synthesized DNA strands approximate full length.

The *in vitro* system that is dependent on exogenous template has proved valuable for the elucidation of molecular mechanisms involved in initiation and elongation of viral DNA chains and for the purification and characterization of replication proteins. The major conclusions derived from the analysis of adenovirus DNA replication *in vitro* are described in detail in Sections V and VI.

V. REPLICATION PROTEINS

A. Adenovirus DNA-Binding Protein

The observation that adenovirus replication intermediates contain extensive stretches of ssDNA led to a search for ssDNA-binding proteins specific for virus-infected cells. In 1973, van der Vliet and Levine (1973) reported the isolation from type 5 Ad-infected cells of two such proteins, with polypeptide molecular weights of 72 and 48K, respectively. Both proteins were quantitatively retained on a column of ssDNA–cellulose and could be eluted only at relatively high ionic strengths (0.5–1 M NaCl). Neither protein was detectable in extracts from uninfected cells or in purified virions. Subsequent studies have revealed that the 48K species is closely related in structure to the 72K species; the two proteins share common antigenic determinants and have similar tryptic peptide maps (Rosenwirth *et al.*, 1976; Linné *et al.*, 1977). These findings, together with the observation that the amount of the 48K protein varies widely from preparation to preparation, suggest that the 48K protein is derived from the 72K species by proteolytic cleavage during isolation (Rosenwirth *et al.*, 1976). Similar DNA-binding proteins have been detected in cells infected with Ad2 and Ad12 (Levine *et al.*, 1975; Rosenwirth *et al.*, 1975).

In Ad2- or Ad5-infected cells, the 72K DNA-binding protein (DBP) is first detected in about 4 hr after infection, prior to the onset of viral DNA replication, and is synthesized continuously until late in infection (van der Vliet and Levine, 1973; Gilead *et al.*, 1976; Chin and Maizel, 1976; Beltz and Flint, 1979). Newly synthesized DBP accumulates in the nucleus of the infected cell. It is the most abundant early viral protein, eventually reaching a level of about 5×10^6 molecules/cell (Schechter *et al.*, 1980). EM studies indicate that the single-stranded regions of adenovirus replicative intermediates are completely coated by DBP (Kedinger *et al.*, 1978).

Evidence that the adenovirus DBP is encoded by the viral genome was initially obtained in studies of H5*ts*125, a temperature-sensitive mutant of Ad5 that is defective in viral DNA replication (van der Vliet *et al.*, 1975). When cells were infected with H5*ts*125 at the restrictive temperature, less than 5% of the normal wild-type (wt) level of DBP protein was produced. Furthermore, DBP isolated from H5*ts*125-infected cells, grown at the permissive temperature, was found to be thermolabile for binding to ssDNA. These findings strongly suggested that the locus of the H5*ts*125 mutation is the structural gene for the adenovirus DBP. Direct evidence that the DBP is virus-coded was obtained by *in vitro* translation of Ad2 messenger RNA (mRNA) selected by hybridization to specific restriction fragments (Lewis *et al.*, 1976). These studies localized the coding sequences for the DBP roughly between adenovirus map positions 58.5 and 70.7 (early region E2). Messenger RNA encoding DBP first appears in infected cells at about 2–3 hr after infection. Transcription is initiated at approximately 75 map units (m.v.) and proceeds right to left on the viral genome (Evans *et al.*, 1977; Seghal *et al.*, 1979; Baker *et al.*, 1979). The major messenger species has a short 5′ leader sequence at map position 75.2, an internal leader at map position 68.8 (missing in a fraction of DBP mRNAs), and a main body from map positions 66.5 to 61.6 (Kitchingman *et al.*, 1977; Berk and Sharp, 1978; Chow *et al.*, 1979; Kruijer *et al.*, 1981). Late in infection, the majority of the mRNA molecules that encode the DBP have 5′ leaders mapping at 72 m.u., suggesting that different promoters may be utilized at early and late times during infection (Chow *et al.*, 1979, 1980).

The nucleotide sequence of the gene that codes for the Ad5 DBP has been determined (Kruijer *et al.*, 1981). The amino acid sequence of the DBP is encoded entirely by a long open reading frame of 1587 nucleotides in the main body of DBP mRNA. On the basis of the nucleotide sequence analysis, it can be inferred that the DBP contains 529 amino acid residues and has a molecular weight (before posttranslational modification) of 59,049. The site of mutation of H5*ts*125 lies within the carboxy-terminal part of the DBP at amino acid residue 413. Similar sequence analysis of the Ad2 DBP has revealed only 9 amino acid differences from the Ad5 protein, all of which reside in the amino-terminal portion of the molecule.

The Ad2 and Ad5 DBPs have been purified essentially to homogeneity by several different laboratories (van der Vliet *et al.*, 1975, 1978; Linné *et al.*, 1977; Sugawara *et al.*, 1977; Schechter *et al.*, 1980). At low protein concentration and high ionic strength, the DBP behaves as a monomer with a molecular weight closely approximating the polypeptide molecular weight of 72K (Sugawara *et al.*, 1977; van der Vliet *et al.*, 1978). The monomer has a high frictional coefficient ($f/f_0 = 1.8$), suggesting that it is a markedly anisometric protein. At high protein concentrations, DBP molecules self-associate, forming a variety of multimeric forms (Schechter *et al.*, 1980; Sugawara *et al.*, 1977; van der Vliet *et al.*, 1978). A stable trimeric form has been observed at high ionic strength (Schechter *et al.*, 1980). Highly purified DBP displays extensive charge heterogeneity; approximately 9–11 different charged species are detectable by isoelectric focusing (Klein *et al.*, 1979). The majority of this charge heterogeneity is the result of posttranslational phosphorylation of Ser and Thr residues (Linné *et al.*, 1977; Jeng *et al.*, 1977; Levinson *et al.*, 1977; Axelrod, 1978; Klein *et al.*, 1979; Linné and Philipson, 1980). Differences in the degree of phosphorylation appear to have only slight effects on the DNA-binding properties of the DBP (Klein *et al.*, 1979).

Mild chymotrypsin treatment of the 72K DBP generates two polypeptide fragments of 26 and 44K (Klein *et al.*, 1979; Linné and Philipson, 1980). The 44K fragment is derived from the carboxy-terminal portion of the DBP and contains the locus of the H5*ts*125 mutation. It seems possible that the 44K fragment possesses all the activities required for viral DNA replication, since it retains the ability to bind to ssDNA (Klein *et al.*, 1979; Linné and Philipson, 1980) and can complement the replication defect in an *in vitro* DNA synthesis system derived from H5*ts*125-infected cells (Ariga *et al.*, 1980). The 26K fragment is derived from the amino-terminal portion of the DBP and contains nearly all the phosphoryl groups of the molecule (Klein *et al.*, 1979; Linné and Philipson, 1980). It does not appear to bind to ssDNA. The functional significance of the amino-terminal domain of the DBP is not yet clear, although there is genetic evidence that suggests that it may play some role during the expression of the late genes of adenovirus (Klessig and Grodzicker, 1979; Kruijer *et al.*, 1981).

The binding of the adenovirus DBP to DNA has been studied by sedimentation and filter-binding techniques (van der Vliet and Levine, 1973, van der Vliet *et al.*, 1978; Fowlkes *et al.*, 1979; Schechter *et al.*, 1980). The protein binds cooperatively and tightly to ssDNA from a variety of sources. At saturation, approximately 1 molecule of DBP is bound for every 7 nucleotides (van der Vliet *et al.*, 1978). Hydrodynamic measurements and EM observations indicate that the protein holds the DNA chain in an extended configuration. Single-stranded DNA bound to DBP is partially resistant to hydrolysis by a variety of nucleases (van der Vliet *et al.*, 1978; Nass and Frenkel, 1980). Thus, the DNA-binding properties of adenovirus DBP are similar in some respects to those of prokaryotic ssDBPs, such as bacteriophage T_4 gene 32 protein. However, attempts to demonstrate that the adenovirus DBP lowers the melting temperature of DNA have so far been unsuccessful (van der Vleit *et al.*, 1978; Fowlkes *et al.*, 1979), and the effect of the protein on the kinetics of renaturation has not been determined. Interestingly, careful filter-binding studies have demonstrated that the adenovirus DBP also binds to a limited extent to double-stranded DNA (dsDNA) (Fowlkes *et al.*, 1979; Schechter *et al.*, 1980). The binding apparently occurs exclusively at the ends of linear duplex molecules, since no binding is observed with circular dsDNA.

Evidence that the adenovirus DBP is involved in viral DNA replication has been obtained by analysis of the properties of the mutant H5*ts*125. As indicated above, this mutant produces a thermolabile DBP in infected cells. At the restrictive temperature, no detectable viral DNA replication occurs in H5*ts*125-infected cells (Ensinger and Ginsberg, 1972; van der Vliet and Sussenbach, 1975). Moreover, when H5*ts*125-infected cultures are maintained at the permissive temperature until after the onset of viral DNA replication and then shifted to the nonpermissive temperature, the rate of viral DNA synthesis declines to very low levels within 1 hr. This finding is consistent with the idea that the DBP is directly involved in the replication process.

Attempts to define the precise role(s) of the DBP in adenovirus DNA replication have not yielded completely unambiguous answers. Early studies indicated that the apparent rate of viral DNA chain elongation in cells infected with H5*ts*125 is normal after a shift from the permissive to the nonpermissive temperature, even though the overall rate of DNA synthesis is reduced by more than 80% (van der Vliet and Sussenbach, 1975). This fact, together with the observation that replicative intermediates do not accumulate after the shift up, was interpreted as indicated that the DBP is required for the initiation of adenovirus DNA chains, but not for DNA chain elongation. However, a number of other explanations of the data are possible, including the possibility that the DBP has increased thermal stability when it is bound to DNA. More recently, the role of the DBP in replication has been investigated in several different *in vitro* systems. The first such studies employed isolated nuclei or subnuclear replication complexes (see Section IV.A). These systems are capable of elongating nascent adenovirus DNA chains present in endogenous replication intermediates, but do not initiate the synthesis of new chains. Under the usual assay conditions, DNA synthesis in nuclei or replication complexes derived from H5*ts*125-infected cells is not significantly more thermolabile than that observed in similar preparations from Wt-infected cells (van der Vliet *et al.*, 1977; Horwitz, 1978). However, if replication complexes from H5*ts*125-infected cells are preincubated for 1 hr at 38°C in the absence of DNA synthesis, they are irreversibly inactivated. Inactivation does not occur with replication complexes derived from cells infected with Wt virus and can be prevented in the case of H5*ts*125 complexes by inclusion of Wt DBP in the preincubation mixture (Horwitz, 1978). While these observations are consistent with a requirement for DBP in chain elongation, it is not clear whether the requirement is direct or indirect. It is possible, for example, that the DBP is needed simply to prevent some change in the structure of the viral replicative intermediates that would otherwise occur during prolonged incubation in the absence of DNA replication (Horwitz, 1978). The effect of anti-DBP antibody on viral DNA replication has also been examined in nuclear and subnuclear replication systems. Antibody specifically inhibits chain elongation *in vitro*, but the inhibition is only partial (40–50%), even at high antibody concentrations (Kedinger *et al.*, 1978; van der Vliet *et al.*, 1977). Analysis of DNA replication in the *in vitro* system that is dependent on exogenous viral DNA template (see Section IV.B) has provided some new insights into the role of the DBP. *In vitro* complementation studies with extracts from H5*ts*125-infected cells as well as reconstitution experiments with purified replication proteins indicate that the DBP is not required for the formation of terminal protein (TP)–nucleotide complexes, the proposed first step in adenovirus DNA replication (see Section VI), but is required for subsequent steps in the replication process (Enomoto *et al.*, 1981; Challberg *et al.*, 1982; Tamanoi and Stillman, 1982; Stillman *et al.*, 1982a; Ostrove *et al.*, 1983). Thus, the weight of current evidence favors the idea that the DBP functions during the chain-elongation step. By analogy with prokaryotic systems, the DBP could act to promote unwinding of the helix (or prevent re-winding) or to facilitate the polymerization reaction by holding template strands in a favorable configuration. In addition, it could play a role in protecting single-stranded regions from nuclease attack. Further work will be required to determine which, if any, of these possibilities is correct.

B. Adenovirus Terminal Protein

As indicated in Section II.B, the 5′ ends of mature adenovirus DNA strands are covalently linked to a 55K TP via a phosphodiester bond to a Ser residue. In infected cells, the TP appears to be synthesized in the form of a larger precursor TP (pTP) that is subsequently cleaved to produce the mature 55K species. The precursor was first identified during studies of the structure of adenovirus DNA replicated *in vitro* (Challberg *et al.*, 1980). These studies demonstrated that the 5′ ends of nascent viral DNA strands are covalently linked to a protein with a molecular weight of 80K. (Estimates of the molecular weight of the pTP vary from 80 to 87K depending on the molecular-weight standards employed.) It was further shown that the 80K protein shares common peptides with the 55K protein (Challberg *et al.*, 1980; Stillman *et al.*, 1981) and that both proteins are attached to DNA by the same linkage (Challberg *et al.*, 1980). Subsequent studies showed that the termini of nascent DNA strands replicated *in vivo* are also attached to the 80K protein (Challberg and Kelly, 1981). Pulse–chase experiments were consistent with the hypothesis that the 80K protein is, in fact, the precursor to the 55K protein.

Analysis of a mutant of Ad2 (H2*ts*1) suggested that processing of the pTP to the mature TP occurs as a late step in virion assembly (Challberg and Kelly, 1981; Stillman *et al.*, 1981). At the restrictive temperature, the H2*ts*1 mutant produces viral particles that resemble Wt virions morphologically, but are noninfectious. Analysis of the polypeptides of these noninfectious particles has demonstrated that the H2*ts*1 mutant does not process the precursors of several virion structural proteins (pVI, pVII, pVIII) to their mature forms during assembly (Weber, 1976; Mirza and Weber, 1977). Under the same conditions, the mutant also fails to process the pTP (Challberg and Kelly, 1981; Stillman *et al.*, 1981). Thus, the virion DNA strands are attached to the 80K protein instead of to the 55K protein. Since normal amounts of viral DNA are produced in H2*ts*1-infected cells, these findings also demonstrate that processing of the 80K protein to the 55K protein is not required for continued adenovirus DNA replication. Additional support for this view has come from the finding that adenovirus chromosomes containing attached 80K TP (obtained from H2*ts*1 virions) are fully competent as templates for adenovirus DNA replication *in vitro* (Challberg and Kelly, 1981; Stillman *et al.*, 1981).

Evidence that the adenovirus pTP is encoded by the viral genome has been obtained by hybrid-selected translation techniques (Stillman *et al.*, 1981). Adenovirus mRNA complementary to the l-strand of the viral genome between map positions 11.2 and 31.5 directs the synthesis *in vitro* of an 80K protein the peptide map of which is identical to that of the 80K TP obtained from H2*ts*1 chromosomes. Detailed EM heteroduplex analysis has revealed a family of mRNAs from this region, which has been designated early region E2b (Stillman *et al.*, 1981). The E2b mRNAs share a common 5′ terminus at map position 75 and internal leader sequences at map positions 39 and 68.5; their main bodies extend from map positions 30, 26, or 23 to map position 11.1. Nucleotide sequence analysis of the region between map positions 30 and 11.1 has revealed the existence of three large open translational reading frames (Gingeras *et al.*, 1982; Aleström *et al.*, 1982a). The majority of the coding sequences for the pTP have been localized to one of these open frames (map positions 28.9–23.5) by aligning partial amino acid sequences with the amino acid sequence predicted from the DNA sequence (Smart and Stillman, 1982). Thus, it seems likely that the pTP is translated from the mRNA species the main body of which extends from map positions 30 to 11.1. The mature form of TP (55K protein) that is found in virions derives from the carboxy-terminal portion of the pTP (Smart and Stillman, 1982).

The development of the *in vitro* adenovirus DNA replication system that is dependent on exogenous templates has provided a convenient assay that can be used for the purification of replication proteins. Using a modification of this system designed to detect virus-specific replication proteins, Enomoto *et al.* (1981) purified the 80K pTP in functional form. Purification of the pTP has also been accomplished using *in vitro* complementation assays that make use of viral mutants defective in viral DNA replication (see Section VI) (Ostrove *et al.*, 1983; Stillman *et al.*, 1982a).

Current ideas concerning the function of the TP in viral DNA replication are focused on its possible roles in the initiation reaction. Detailed discussion of this topic is deferred to Section VI.

C. Adenovirus N Protein

The involvement of a third adenovirus-coded replication protein is implied by the existence of a second group of *ts* mutants defective in viral DNA replication, the group N mutants. Members of this group, which include H5*ts*36 and H5*ts*149, complement mutants (such as H5*ts*125) the lesions of which lie in the DBP gene (Ginsberg *et al.*, 1975; Williams *et al.*, 1975). At the restrictive temperature, no detectable viral DNA synthesis occurs in cells infected with group N mutants (Wilkie *et al.*, 1973; Ginsberg *et al.*, 1975). When mutant-infected cultures are maintained at the permissive temperature until after the onset of viral DNA replication

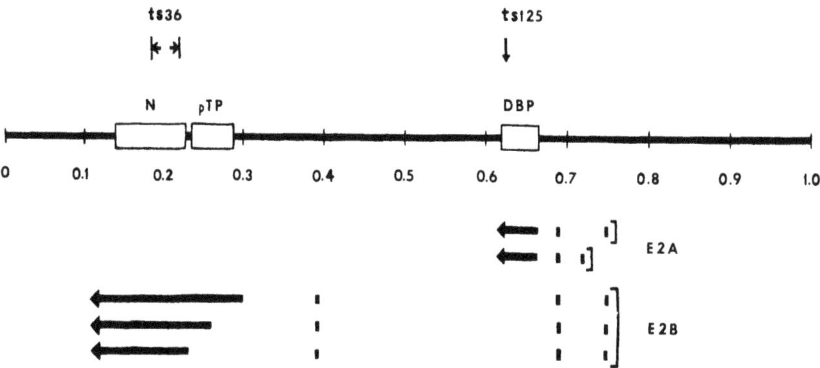

FIGURE 4. Map of adenovirus genes involved in viral DNA replication. The map (top line) shows the locations of the coding sequences for the adenovirus DBP, the 80 K pTP, and the N protein (Kruijer *et al.*, 1981; Stillman *et al.*, 1981; Smart and Stillman, 1982; Gingeras *et al.*, 1982; Aleström *et al.*, 1982a). The boundaries of the N gene are not known with certainty. The loci of N group mutations (e.g., H5*ts*36) map within an open reading frame between map positions 24.1 and 14.2. The first AUG in this open reading frame maps at position 22.9 (Galos *et al.*, 1979; Gingeras *et al.*, 1982; Aleström *et al.*, 1982a). The structures of the major mRNA species derived from early regions E2a and E2b are shown in the lower portion of the figure (Chow *et al.*, 1979, 1980; Stillman *et al.*, 1981). The arrows indicate the direction of transcription, and the vertical brackets indicate the location of promoter sites. Gaps between promoter sites and 3′ termini (arrowheads) represent intervening sequences removed from primary transcripts by splicing.

and then shifted to the restrictive temperature, DNA synthesis declines to low levels after a lag of about 6 hr (van der Vliet and Sussenbach, 1975; Ginsberg *et al.*, 1975). This rather slow shutoff of viral DNA synthesis has hampered attempts to define the nature of the replication defect by *in vivo* studies.

The group N mutations fall within early region E2b, between map positions 18.5 and 22.0 (Galos *et al.*, 1979). This segment of the viral genome lies outside the boundaries of the gene that encodes the 80K pTP and within a large open reading frame (map positions 22.9–14.2) with a coding capacity of about 120K of protein (Smart and Stillman, 1982; Gingeras *et al.*, 1982; Aleström *et al.*, 1982a). While the product of the N gene has not yet been identified with certainty (but see below), it seems likely that its synthesis is directed by one or more members of the E2b family of mRNAs described in Section V.B. Thus, it appears that all three virus-coded proteins known to be involved in viral DNA replication (the DBP, the TP, and the N protein) are products of the same transcriptional unit, sharing a common promotor at map position 75. This may allow for coordinate regulation of the synthesis of these proteins during the viral multiplication cycle. Figure 4 summarizes the currently available information on the organization and expression of the adenovirus replication genes.

The best candidate for the product of the N gene is a polypeptide with an apparent molecular weight of 140K that was observed to copurify with the 80K pTP during fractionation of infected-cell extracts (Enomoto *et al.*, 1981). The sedimentation properties of the 80 and 140K proteins in glycerol gradients suggest that the two proteins may form a bimolecular complex (Enomoto *et al.*, 1981). Highly purified fractions containing the two proteins were subsequently shown to complement the replication defect in nuclear extracts derived from cells infected with group N mutants (Stillman *et al.*, 1982a; Ostrove *et al.*, 1983; Friefeld *et al.*, 1983). It has also been reported that the complementing activity in partially purified fractions cosediments with the 140K protein in glycerol gradients containing urea (Stillman *et al.*, 1982a; Friefeld *et al.*, 1983). These observations together with the fact that mRNA selected by hybridization to cloned DNA fragments from the EZb region directs the *in vitro* synthesis of a polypeptide of approximately the correct size (Stillman *et al.*, 1982a) are clearly suggestive that the N gene encodes the 140K protein; however, at this writing, direct evidence on this point is lacking. Recently, the 80K pTP and 140K protein have been separated by sedimentation in glycerol gradients containing urea and 0.5 M NaCl (Lichy *et al.*, 1982). The 140K protein was shown to be required for adenovirus DNA replication *in vitro* and to contain DNA polymerase activity detectable with activated DNA templates (Enomoto *et al.*, 1981; Lichy *et al.*, 1982). The 140K polymerase is distinct from the known eukaryotic DNA polymerases, α, β, and γ, in its chromatographic properties. It resembles DNA polymerase α in its template specificity and its sensitivity to *N*-ethylmaleimide, NaCl, and cytosine arabinosyl CTP; however, it is insensitive to aphidicolin, a potent inhibitor of DNA polymerase α, and is more sensitive than DNA polymerase α to ddTTP. The 140K polymerase appears to be directly involved in the initiation of adenovirus DNA replication as discussed in Section VI.

D. Cellular Replication Proteins

Fractionation–reconstitution experiments with the *in vitro* adenovirus DNA replication system have implicated cellular proteins in the replication process. The reconstitution of the complete *in vitro* replication reaction has been shown to require factors present in uninfected cells in addition to purified components (80K pTP, DBP, 140K protein) derived from infected cells (Enomoto *et al.*, 1981; Lichy *et al.*, 1982; Stillman *et al.*, 1982a; Ostrove *et al.*, 1983). In addition, the formation of a covalent complex between the pTP and dCMP, the proposed initial step in viral DNA replication (see Section VI), is greatly stimulated by a cellular factor. Little is yet known about the nature of these cellular components or their role in adenovirus DNA replication. Using stimulation of synthesis of pTP–dCMP complexes as an assay, one cellular factor, termed nuclear factor I, has been partially purified (Nagata *et al.*, 1982). The major polypeptide present in the most highly purified fraction has an apparent molecular weight of 47K by sodium dodecyl sulfate–polyacrylamide gel electrophoresis (SDS-PAGE). The protein has no detectable nuclease, RNA or DNA polymerase, ATPase, or topoisomerase I activity.

A number of studies have addressed the question of whether adenovirus DNA in infected cells is complexed with cellular histones. Analysis of the fate of ^{32}P-labeled adenovirus has shown that after uncoating takes place, at least a fraction of the parental viral DNA molecules become organized into nucleoprotein structures that have some of the properties of cellular chromatin (Tate and Philipson, 1969; Sergeant *et al.*, 1979). In particular, micrococcal nuclease digestion yields a characteristic pattern of viral nucleoprotein complexes containing DNA segments that are integral multiples of 165–185bp. Although it has been suggested that this nucleosomelike pattern is a consequence of the association of histones with the viral DNA, this has not yet been directly demonstrated. It is also not known whether the parental molecules that are actually biologically active (i.e., those that participate in transcription or replication) have a chromatinlike structure. Indeed, EM visualization of adenovirus DNA present in preparations of replication complexes reveals that duplex viral DNA has a relatively smooth contour, in contrast to cellular chromatin, in which the typical nucleosomal pattern is evident (Kedinger *et al.*, 1978; Beyer *et al.*, 1981). In addition, direct analysis by SDS-PAGE indicates that histones are not a prominent constituent of purified replication complexes (Kedinger *et al.*, 1978). These studies would appear to rule out the idea that the bulk of intracellular adenovirus DNA is packaged in nucleosomes. However, they leave open the question of whether replicating viral DNA is complexed with nonhistone cellular proteins or with viral proteins.

VI. INITIATION OF ADENOVIRUS DNA REPLICATION

All the DNA polymerases that have been carefully studied to date are capable of extending DNA chains in the 5′-to-3′ direction from a preexisting primer, but are incapable of starting DNA chains *de novo*. Thus, the question of how new DNA chains are initiated reduces to the question of how the necessary primers are generated. In the case of adenovirus, a good deal of evidence has accumulated in support of a novel initiation mechanism in which DNA synthesis is primed by protein (Rekosh *et al.*, 1977). Figure 5 shows a model based on this general idea (Challberg *et al.*, 1980). In the model, the primary initiation event is the formation of an ester linkage between the α-phosphoryl group of dCTP and the β-OH of a serine residue in the 80K precursor terminal protein (pTP). It is envisioned that this event takes place as part of a concerted reaction in which a free pTP molecule associates with the terminus of the parental genome. Such an association is presumably mediated by terminal nucleotide sequences (Tamanoi and Stillman, 1982) and may also be facilitated by interactions with the TP attached to parental DNA (Challberg and Kelly, 1979a,b). As a result of these interactions, the 3′-

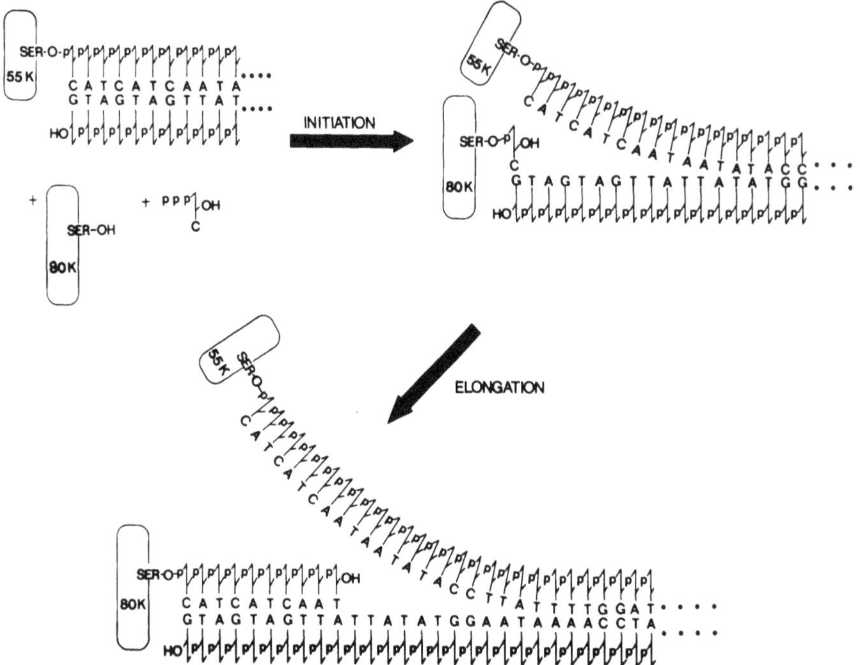

FIGURE 5. Model for the initiation of adenovirus DNA replication. (55K) TP of molecular weight 55,000; (80K) pTP of molecular weight 80,000. From Challberg *et al.* (1980).

OH of the dCMP residue, which becomes covalently attached to the pTP, is positioned in such a way that it can serve as primer for subsequent chain elongation.

Experimental support for the protein priming model has been obtained by analysis of viral DNA replication in the *in vitro* system that is dependent on exogenous templates. It has been demonstrated that the 5′ ends of nascent adenovirus DNA synthesized *in vitro* are attached to the pTP by a phosphodiester bond between the β-OH of a Ser residue in the protein and the 5′-OH of the terminal Cyt residue in the DNA (Challberg *et al.*, 1980). Less direct studies indicate that nascent viral DNA strands synthesized *in vivo* are also covalently linked to pTP (Challberg and Kelly, 1981; Coombs *et al.*, 1979; Girard *et al.*, 1977; Horwitz *et al.*, 1979; Kelly and Lechner, 1979; Stillman and Bellett, 1979). The synthesis of a covalent complex between pTP and dCMP, the predicted intermediate in the initiation reaction, has been detected *in vitro* (Lichy *et al.*, 1981; Pincus *et al.*, 1981; Challberg *et al.*, 1982; Tamanoi and Stillman, 1982). Formation of this complex occurs in the presence of nuclear extract from adenovirus-infected cells, adenovirus DNA containing the 5′-linked terminal protein, $MgCl_2$, ATP, and dCTP. Other deoxyribonucleoside triphosphates (dNTPs) will not substitute for dCTP. Most, if not all, of the pTP–dCMP complex that is synthesized is bound by noncovalent bonds to the input DNA template. The results of pulse–chase experiments suggest that pTP–dCMP complexes can be incorporated into newly synthesized adenovirus DNA on addition of the other three dNTPs.

Recent progress in the purification of replication proteins has provided insights into the requirements for the protein–nucleotide joining reaction (Enomoto *et al.*, 1981; Lichy *et al.*, 1982; Nagata *et al.*, 1982; Stillman *et al.*, 1982a; Ostrove *et al.*, 1983). Optimal synthesis of pTP–dCMP complexes requires a purified fraction containing the 80K pTP and the 140K polymerase described in Section V.C and, in addition, a factor present in uninfected cells (Enomoto *et al.*, 1981; Lichy *et al.*, 1982; Nagata *et al.*, 1982; Stillman *et al.*, 1982a; Ostrove *et al.*, 1983). Partial separation of the pTP and the 140K polymerase has been achieved by glycerol-gradient sedimentation in the presence of urea, and the results of mixing experiments indicate that both components participate in the protein–nucleotide joining reaction (Lichy *et al.*, 1982). These data are consistent with the hypothesis that the 140K polymerase catalyzes the transfer of dCMP to the pTP, although there is, as yet, no direct evidence on this point (Lichy *et al.*, 1982). As indicated in Section V.D, the cellular factor involved in the reaction (nuclear factor I) has been purified free of nuclease and polymerase activities (Nagata *et al.*, 1982). The role of this 47K protein in the reaction is not known.

Some progress has been made in understanding the template requirements for the protein–nucleotide joining reaction. When crude nuclear extracts are employed as the source of protein factors, the only template that effectively supports formation of pTP–dCMP complexes is adenovirus DNA with the covalently linked TP (Lichy *et al.*, 1981; Pincus *et al.*, 1981; Challberg *et al.*, 1982; Tamanoi and Stillman, 1982). Removal of the TP renders the DNA inactive in the reaction. Identical results have been obtained with systems reconstituted from purified proteins (Enomoto *et al.*, 1981; Stillman *et al.*, 1982a; Ostrove *et al.*, 1983). These data suggest that the TP molecules attached to parental strands may play a role in the initiation process. Interestingly, a partially purified fraction prepared from crude infected-cell extract by batch elution at 200 mM NaCl from denatured DNA cellulose does not show the requirement for TP on the parental strands of duplex DNA templates (Tamanoi and Stillman, 1982). With this fraction, deproteinized adenovirus DNA as well as plasmids containing terminal adenovirus DNA sequences support the protein–nucleotide joining reaction. In the latter case, the plasmid DNA must be linearized by restriction enzyme cleavage prior to the reaction to expose the adenovirus origin sequence. Neither linearized plasmid DNA with an internal origin nor circular plasmid DNA supports the formation of pTP–dCMP complexes. While the mechanism of this reaction is not yet understood, it provides direct evidence that interactions with specific adenovirus nucleotide sequences is an obligatory feature of the initiation of adenovirus DNA replication. Subsequent studies with plasmid DNAs containing mutations in the origin region indicate that the terminal 18 nucleotide pairs of the adenovirus genome are sufficient to direct formation of pTP–dCMP complexes (Challberg and Rawlins, unpublished findings; Tamanoi and Stillman, unpublished findings). It is not yet known which protein factor or factors are responsible for recognition of the adenovirus origin; however, it seems reasonable to suggest that this function may be a property of the pTP.

It has also been observed that a variety of ssDNA molecules support formation of pTP–dCMP complexes, as well as a limited degree of DNA synthesis. Active templates include denatured adenovirus DNA, denatured plasmids containing adenovirus origin sequences, single-stranded φX174, M13, f1, or G4 DNAs, and synthetic oligonucleotides containing the 3′-terminal 15 nucleotides of adenovirus DNA (Ikeda *et al.*, 1982; Tamanoi and Stillman, 1982; Tamanoi and Stillman, unpublished findings; Rawlins and Kelly, unpublished findings). It has been suggested that the reaction is dependent on the presence in these DNAs of specific nucleotide sequences related to the adenovirus origin (Tamanoi and Stillman, 1982; Ikeda *et al.*, 1982). However, the recent finding that single-stranded oligonucleotides that are essentially unrelated to the adenovirus origin can support the formation of pTP-dCMP complexes indicates that the sequence requirements for the reaction are not very stringent (Rawlins and Kelly, unpublished findings). Further work will be required to assess the significance of these observations with regard to the mechanism of initiation of adenovirus DNA replication.

Addendum

A number of significant studies of adenovirus DNA replication have been completed since this review was written. Although space limitations preclude a comprehensive account of these studies, the following is a brief survey of some of them.

A protein that stimulates elongation of nascent adenovirus DNA chains has been purified from nuclear extracts of uninfected HeLa cells (Nagata *et al.*, 1983). The protein, designated nuclear factor II, stimulates the synthesis of full length viral DNA strands in the presence of highly purified 80 kd terminal protein precursor (pTP), 140 kd Ad polymerase, 72 kd Ad DNA binding protein (DBP) and nuclear factor I. Purified preparations of nuclear factor II contain a DNA topoisomerase activity, and type I DNA topoisomerases from HeLa cells or calf thymus can substitute for nuclear factor II in the *in vitro* adenovirus DNA replication system. Gel filtration experiments indicate that the native molecular weight of nuclear factor II is 25–45 kd, which is considerably smaller than the known type I topoisomerases of eukaryotic cells.

It has been demonstrated that nuclear factor I, a cellular protein that is required for efficient initiation of adenovirus DNA replication, is a sequence-specific DNA binding protein (Nagata *et al.*, 1983; Rawlings *et al.*, 1984; Guggenheimer *et al.*, 1984). The protein binds to a sequence near the terminus of the viral genome with high affinity ($K_d = 2 \times 10^{-11}$). Under similar experimental conditions the affinity of nuclear factor I for DNA molecules lacking adenovirus origin sequences is at least two orders of magnitude lower. DNAse I footprinting experiments indicate that the binding site lies in the region between nucleotides 17 and 48 of the Ad 5 genome.

The sequence requirements for initiation of DNA replication have been explored further by studying plasmid templates containing cloned DNA segments derived from the termini of the viral genome (Tamanoi and Stillman, 1983; van Bergen *et al.*, 1983; Challberg and Rawlings, 1984; Lally *et al.*, 1984; Rawlings *et al.*, 1984; Guggenheimer *et al.*, 1984). Recent studies with templates carrying deletions or base substitutions in the viral terminal sequences indicate that the adenovirus origin of replication consists of two functionally distinct regions (Rawlings *et al.*, 1984; Guggenheimer *et al.*, 1984). The first 18 base pairs of the viral genome are sufficient to support a limited degree of initiation. This minimal origin of replication may contain sequences required for the binding of the pTP-Ad polymerase complex (Rijnders *et al.*, 1983). The presence of sequence elements in the region between nucleotides 19 and 48 greatly enhances the efficiency of the initiation reaction. This region of the origin contains the binding site for nuclear factor I as described above. Binding of the protein appears to be responsible for the enhancement of initiation since mutations that abolish binding simultaneously reduce the efficiency of formation of pTP-dCMP complexes *in vitro*.

REFERENCES

Aleström, P., Akusjärvi, G., Pettersson, M., and Pettersson, U., 1982a, DNA sequence analysis of the region encoding the terminal protein and the hypothetical N-gene product of adenovirus type 2, *J. Biol. Chem.* **257**:13492.

Aleström, P., Stenlund, A., Li, P., Bellett, A., and Pettersson, U., 1982b, A common sequence in the inverted terminal repetitions of human and avian adenoviruses, *Gene* **18**:193.

Ariga, H., and Shimojo, H., 1977, Initiation and termination sites of adenovirus 12 DNA replication, *Virology* **78**:415.

Ariga, H., and Shimojo, H., 1979, Incorporation of uracil into the growing strand of adenovirus 12 DNA, *Biochem. Biophys. Res. Commun.* **87**:588.

Ariga, H., Klein, H., Levine, A.J., and Horwitz, M.S., 1980, A cleavage product of the adenovirus DNA binding protein is active in DNA replication *in vitro*, *Virology* **101**:307.

Arrand, J.R., and Roberts, R.J., 1979, The nucleotide sequences at the termini of adenovirus 2 DNA, *J. Mol. Biol.* **128**:577.

Axelrod, N., 1978, Phosphoproteins of adenovirus 2, *Virology* **87**:366.

Baker, C.C., Hérissé, J., Courtois, G., Galibert, F., and Ziff, E., 1979, Messenger RNA for the adenovirus-2 DNA binding protein: DNA sequences encoding the first leader and heterogeneity at the mRNA 5′ end, *Cell* **18**:569.

Bellett, A.J.D., and Younghusband, H.B., 1972, Replication of the DNA of chick embryo lethal orphan virus, *J. Mol. Biol.* **72**:691.

Beltz, G.A., and Flint, S.J., 1979, Inhibition of HeLa cell protein synthesis during adenovirus infection: Restriction of cellular messenger RNA sequences to the nucleus, *J. Mol. Biol.* **131**:353.

Berk, A.J., and Sharp, P., 1978, Structure of the adenovirus 2 early mRNAs, *Cell* **14**:695.

Beyer, A.L., Bouton, A.H., Hodge, L.D., and Miller, O.L., Jr., 1981, Visualization of the major late r strand transcription unit of adenovirus serotype 2, *J. Mol. Biol.* **147**:269.

Bollum, F.J., 1974, Terminal deoxynucleotidyl transferase, in: *The Enzymes*, 3rd ed., Vol. 10 (P.D. Boyer, ed.), pp. 145–171, Academic Press, New York.

Brison, O., Kedinger, C., and Wilhelm, J., 1977, Enzymatic properties of viral replication complexes isolated from adenovirus type 2 infected HeLa cell nuclei, *J. Virol.* **24**:423.

Brown, D.T., Westphal, M., Burlingham, B.T., and Doerfler, W., 1975, Structure and composition of the adenovirus type 2 core, *J. Virol.* **16**:366.

Brynelf, K., Eliasson, R., and Reichard, P., 1978, Formation of Okazaki fragments in polyoma DNA synthesis caused by misincorporation of uracil, *Cell* **13**:573.

Burlingham, B.T., and Doerfler, W., 1971, Three size classes of intracellular adenovirus DNA, *J. Virol.* **7**:707.

Carusi, E.A., 1977, Evidence for blocked 5′ termini in human adenovirus DNA, *Virology* **76**:380.

Challberg, M., and Kelly, T.J., Jr., 1979a, Adenovirus DNA replication *in vitro*, *Proc. Natl. Acad. Sci. U.S.A.* **76**:655.

Challberg, M.D., and Kelly, T.J., Jr., 1979b, Adenovirus DNA replication *in vitro*: Origin and direction of daughter strand synthesis, *J. Mol. Biol.* **135**:999.

Challberg, M.D., and Kelly, T.J., Jr., 1981, Processing of the adenovirus terminal protein, *J. Virol.* **38**:272.

Challberg, M.D., and Kelly, T.J., Jr., 1982, Eukaryotic DNA replication: Viral and plasmid model systems, *Annu. Rev. Biochem.* **51**:901–934.

Challberg, M.D., and Rawlins, D.R., 1984, Template requirements for the initiation of adenovirus DNA replication, *Proc. Natl. Acad. Sci. U.S.A.* **81**:100.

Challberg, M.D., Desiderio, S.V., and Kelly, T.J., Jr., 1980, Adenovirus DNA replication *in vitro*: Characterization of a protein covalently linked to nascent DNA strands, *Proc. Natl. Acad. Sci. U.S.A.* **77**:5105.

Challberg, M.D., Ostrove, J., and Kelly, T.J., Jr., 1982, Initiation of adenovirus DNA replication: Detection of covalent complexes between nucleotide and the 80 kd terminal protein, *J. Virol.* **41**:365.

Chardonnet, Y., and Dales, S., 1970a, Early events in the interaction between adenoviruses and HeLa cells. I. Penetration of type 5 and intracellular release of the DNA genome, *Virology* **40**:462.

Chardonnet, Y., and Dales, S., 1970b, Early events in the interaction between adenoviruses and HeLa cells. II. Comparative observations on the penetration of types 1, 5, 7, and 12, *Virology* **40**:478.

Chardonnet, Y., and Dales, S., 1972, Early events in the interaction of adenoviruses with HeLa cells. III. Relationship between an ATPase activity in nuclear envelopes and transfer of core material: A hypothesis, *Virology* **48**:342.

Chin, W.W., and Maizel, J.V., 1976, The polypeptides of adenovirus. VII. Further studies of early polypeptides *in vivo* and localization of E2 and E2A to the cell plasma membrane, *Virology* **7**:518.

Chow, L.T., Broker, T.R., and Lewis, J.B., 1979, Complex splicing patterns of RNAs from the early regions of adenovirus 2, *J. Mol. Biol.* **134**:265.

Chow, L.T., Lewis, J.B., and Broker, T.R., 1980, RNA transcription and splicing at early and intermediate times after adenovirus-2 infection, *Cold Spring Harbor Symp. Quant. Biol.* **44**:401.

Coombs, D.H., Robinson, A.J., Bodnar, J.W., Jones, C.J., and Pearson, G., 1979, Detection of covalent DNA–Protein complexes: The adenovirus DNA–terminal protein complex and HeLa DNA–protein complexes, *Cold Spring Harbor Symp. Quant. Biol.* **43**:741.

Daniell, E., 1976, Genome structure of incomplete particles of adenovirus, *J. Virol.* **19**:685.

Danna, K.J., and Nathans, D., 1972, Bidirectional replication of simian virus 40 DNA, *Proc Natl. Acad. Sci. U.S.A.* **69**:3097.

Desiderio, S.V., and Kelly, T.J., Jr., 1981, The structure of the linkage between adenovirus DNA and the 55,000 dalton terminal protein, *J. Mol. Biol.* **145**:319.

Dijkema, R., and Dekker, B.M.M., 1979, The inverted terminal repetition of the DNA of weakly oncogenic adenovirus type 7, *Gene* **8**:7.

Doerfler, W., and Kleinschmidt, A., 1970, Denaturation pattern of the DNA of adenovirus type 2 as determined by electron microscopy, *J. Mol. Biol.* **50**:579.

Ellens, D.J., Sussenbach, J.S., and Jansz, H.S., 1974, Studies on the mechanism of replication of adenovirus DNA. III. Electron microscopy of replicating DNA, *Virology* **61**:427.

Enomoto, T., Lichy, J.H., Ikeda, J.-E., and Hurwitz, J., 1981, Adenovirus DNA replication *in vitro*: Purification of the terminal protein in a functional form, *Proc. Natl. Acad. Sci. U.S.A.* **78**:6779.

Ensinger, M.J., and Ginsberg, H.S., 1972, Selection and preliminary characterization of temperature-sensitive mutants of type 5 adenovirus, *J. Virol.* **10**:328.

Evans, R.M., Fraser, N., Ziff, E., Weber, J., Wilson, M., and Darnell, J.E., 1977, The initiation sites for RNA transcription in Ad2 DNA, *Cell* **12**:733.

Flint, S.J., and Broker, T.R., 1980, Lytic infection by adenoviruses, in: *The Molecular Biology of Tumor Viruses*, Part 2, *DNA Tumor Viruses*, 2nd rev. ed. (J. Tooze, ed.), pp. 443–546, Cold Spring Harbor Press, Cold Spring Harbor, New York.

Flint, S.J., Berget, S.M., and Sharp, P.A., 1976, Characterization of single-stranded viral DNA sequences present during replication of adenovirus types 2 and 5, *Cell* **9**:559.

Fowlkes, D.M., Lord, S.T., Linné, T., Pettersson, U., and Phillipson, L., 1979, Interactions between the adenovirus DNA-binding protein and double-stranded DNA, *J. Mol. Biol.* **132**:163.

Friefeld, B.R., Lichy J.H., Hurwitz, J. and Horwitz, M.S., 1983, Evidence for an altered adenovirus DNA polymerase in cells infected with the mutant H5ts 149, *Proc. Natl. Acad. Sci. U.S.A.* **80**:1589.

Frenkel, G.D., 1978, Adenovirus DNA synthesis *in vitro* in an isolated complex, *J. Virol.* **25**:459.

Galos, R., Williams, J., Binger, M.H., and Flint, S.J., 1979, Location of additional early gene sequences in the adenoviral chromosome, *Cell* **17**:945.

Garon, C.F., Berry, K.W., and Rose, J.A., 1972, A unique form of terminal redundancy in adenovirus DNA molecules, *Proc. Natl. Acad. Sci. U.S.A.* **69**:2391.

Garon, C.F., Parr, R.P., Padmanabhan, R., Roninson, I., Garrison, J.W., Rose, J.A., 1982, Structural characterization of the adenovirus-18 inverted terminal repetition, *Virology* **121**:230.

Gilead, Z., Sugawara, K., Shanmugan, G., and Green, M., 1976, Synthesis of the adenovirus-coded DNA binding protein in infected cells, *J. Virol.* **18**:454.

Gingeras, T.R., Sciaky, D., Gelinas, R.E., Bing-Dong, J., Yen, C.E., Kelly, M.M., Bullock, P.A., Parsons, B.L., O'Neill, K.E., and Roberts, R.J., 1982, Nucleotide sequences from the adenovirus-2 genome, *J. Biol. Chem.* **257**:13,475.

Ginsberg, H.S., Bello, L.J., and Levine, A.J., 1967, Control of biosynthesis of host macromolecules in cells infected with adenoviruses, in: *The Molecular Biology of Viruses* (J.S. Colter and W. Paranchych, eds.), pp. 547–572, Academic Press, New York.

Ginsberg, H.S., Ensinger, M.J., Kauffman, R.S., Mayer, A.J., and Lundholm, U., 1975, Cell transformation: A study of regulation with type 5 and type 12 adenovirus temperature-sensitive mutants, *Cold Spring Harbor Symp. Quant. Biol.* **39**:419.

Girard, M., Bouche, J.P., Marty, L., Revel, B., and Berthelot, N., 1977, Circular Adenovirus DNA–protein complexes from infected HeLa cell nuclei, *Virology* **83**:34.

Green, M., 1962, Studies on the biosynthesis of viral DNA. IV. Isolation, purification and chemical analysis of adenovirus, *Cold Spring Harbor Symp. Quant. Biol.* **27**:219.

Green, M., Parsons, J.T., Piña, M., Fujinaga, K., Caffier, H., and Landgraf-Leurs, I., 1970, Transcription of adenovirus genes in productively infected and in transformed cells, *Cold Spring Harbor Symp. Quant. Biol.* **35**:803.

Green, M., Brackman, K., Wold, W.S.M., Cartas, M., Thorton, H., and Elder, J.H., 1979, Conserved primary sequences of the DNA terminal proteins of five different human adenovirus groups, *Proc. Natl. Acad. Sci. U.S.A.* **76**:4380.

Guggenheimer, R.A., Stillman, B.W., Nagata, K., Tamanoi, F., and Hurwitz, J., 1984, DNA sequences required for the *in vitro* replication of adenovirus DNA, *Proc. Natl. Acad. Sci. U.S.A.* in press.

Horwitz, M.S., 1971, Intermediates in the replication of type 2 adenovirus DNA, *Virology* **8**:675.

Horwitz, M.S., 1976, Bidirectional replication of adenovirus type 2 DNA, *J. Virol.* **18**:307.

Horwitz, M.S., 1978, Temperature-sensitive replication of H5ts125 adenovirus DNA *in vitro*, *Proc. Natl. Acad. Sci. U.S.A.* **75**:4291.

Horwitz, M.S., Kaplan, L.M., Abboud, M., Moritato, J., Chow, L.T., and Broker, T.R., 1979, Adenovirus DNA replication in soluble extracts of infected cell nuclei, *Cold Spring Harbor Symp. Quant. Biol.* **43**:769.

Ikeda, J.-E., Enomoto, T., and Hurwitz, J., 1982, Adenoviral protein-primed initiation of DNA chains *in vitro*, *Proc. Natl. Acad. Sci. U.S.A.* **79**:2442.

Jeng, Y.H., Wold, S.M., Sugawara, K., Gilead, Z., and Green, M., 1977, Adenovirus type 2 coded single-stranded DNA binding protein: *In vivo* phosphorylation and modification, *J. Virol.* **22**:402.

Kaplan, L.M., Kleinman, R.E., and Horwitz, M.S., 1977, Replication of adenovirus type 2 DNA *in vitro*, *Proc. Natl. Acad. Sci. U.S.A.* **74**:4425.

Kaplan, L.M., Hiroyoshi, A., Hurwitz, J., and Horwitz, M.S., 1979, Complementation of the temperature-sensitive defect in H5ts125 adenovirus DNA replication *in vitro*, *Proc. Natl. Acad. Sci. U.S.A.* **76**:5534.

Kedinger, C., Brison, O., Perrin, F., and Wilhelm, J., 1978, Structural analysis of replicative intermediates from adenovirus type 2 infected HeLa cell nuclei, *J. Virol.* **26**:364.

Keegstra, C.S., van Wielink, P.S., and Sussenbach, J.S., 1977, The visualization of a circular DNA–protein complex from adenovirions, *Virology* **76**:444.

Kelly, T.J., Jr., 1982, Organization and replication of adenovirus DNA, in: *Organization and Replication of Viral DNA* (A.S. Kaplan, ed.), pp. 115–146, CRC Press, Boca Raton, Florida.

Kelly, T.J., Jr., and Lechner, R.L., 1979, The structure of replicating adenovirus DNA molecules: Characterization of DNA–protein complexes from infected cells, *Cold Spring Harbor Symp. Quant. Biol.* **43**:721.

Kitchingman, G.R., Lai, S.P., and Westphal, H., 1977, Loop structures in hybrids of early RNA and the separated strands of adenovirus DNA, *Proc. Natl. Acad. Sci. U.S.A.* **74**:4392.

Klein, H., Malzman, W., and Levine, A.J., 1979, Structure–function relationships of the adenovirus DNA-binding protein, *J. Biol. Chem.* **254**:11051.

Klessig, D.F., and Grodzicker, T., 1979, Mutations that allow human Ad2 and Ad5 to express late genes in monkey cells map in the viral gene encoding the 72K DNA binding protein, *Cell* **17**:957.

Kruijer, W., van Schaik, F.M.A., and Sussenbach, J.S., 1981, Structure and organization of the gene coding for the DNA binding protein of adenovirus type 5, *Nucleic Acids Res.* **9**:4439.

Lally, C., Dörper, T., Gröger, W., Antoine, G., and Winnacker, E-L., 1984, A size analysis of the adenovirus replicon, *The EMBO Journal* **3**:333.

Larsen, S.H., and Nathans, D., 1977, Mouse adenovirus: Growth of plaque-purified F-L virus in cell lines and characterization of viral DNA, *Virology* **82**:182.

Lavelle, G., Patch, C., Khoury, G., and Rose, J., 1975, Isolation and partial characterization of single-stranded adenoviral DNA produced during synthesis of adenovirus type 2 DNA, *J. Virol.* **16**:775.

Lechner, R.L., and Kelly, T.J., Jr., 1977, The structure of replicating adenovirus 2 DNA molecules, *Cell* **12**:1007.

Levine, A.J., van der Vliet, P.C., Rosenwirth, B., Rabek, J., Frenkel, G., and Ensinger, M., 1975, Adenovirus-infected cell-specific DNA binding proteins, *Cold Spring Harbor Symp. Quant. Biol.* **39**:559.

Levinson, A.D., Postel, E.H., and Levine, A.J., 1977, *In vivo* and *in vitro* phosphorylation of the adenovirus type 5 single-strand-specific DNA-binding protein, *Virology* **79**:144.

Lewis, J.B., Atkins, J.F., Baum, P.R., Solem, R., Gesteland, R.F., and Anderson, C.W., 1976, Location and identification of the genes for adenovirus type 2 early polypeptides, *Cell* **7**:141.

Lichy, J.H., Horwitz, M.S., and Hurwitz, J., 1981, Formation of a covalent complex between the 80,000-dalton adenovirus terminal protein and 5′-dCMP *in vitro*, *Proc. Natl. Acad. Sci. U.S.A.* **78**:2678.

Lichy, J.H., Field, J., Horwitz, M.S., and Hurwitz, J., 1982, Separation of the adenovirus terminal protein precursor from its associated DNA polymerase: Role of both proteins in the initiation of adenovirus DNA replication, *Proc. Natl. Acad. Sci. U.S.A.* **79**:5225.

Lindahl, T., 1974, An *N*-glycosidase from *E. coli* that releases free uracil from DNA containing deaminated cytosine residue, *Proc. Natl. Acad. Sci. U.S.A.* **71**:3649.

Linné, T., and Philipson, L., 1980, Further characterization of the phosphate moiety of the adenovirus type 2 DNA-binding protein, *Eur. J. Biochem.* **103**:259.

Linné, T., Jornvall, H., and Philipson, L., 1977, Purification and characterization of the phosphorylated DNA-binding protein from adenovirus-type 2-infected cells, *Eur. J. Biochem.* **76**:481.

Lonberg-Holm, K., and Philipson, L., 1969, Early events of virus–cell interactions in an adenovirus system, *J. Virol.* **4**:323.

Mantyjarvi, R., and Russell, W.C., 1969, Studies on DNA synthesis in cells infected with adenovirus, *J. Gen. Virol.* **5**:339.

Martinez-Paloma, A., and Granboulan, N., 1967, Electron microscopy of adenovirus 12 replication. I. High resolution autoradiography of infected KB cells labelled with tritiated thymidine, *J. Virol.* **1**:1010.

Mirza, M.A.A., and Weber, J., 1977, Genetic analysis of adenovirus type 2. VII. Cleavage modified affinity for DNA of internal virion protein, *Virology* **80**:83.

Morgan, G., Rosenkranz, H.S., and Medris, B., 1969, Structure and development of viruses as observed in the electron microscope. X. Entry and uncoating of adenovirus, *J. Virol.* **4**:777.

Nagata, K., Guggenheimer, R.A., and Hurwitz, J., 1983, Specific binding of a cellular DNA replication protein to the origin of replication of adenovirus DNA, *Proc. Natl. Acad. Sci. U.S.A.* **80**:6177.

Nagata, K., Guggenheimer, R.A., Enomoto, T., Lichy, J.H., and Hurwitz, J., 1982, Adenovirus DNA replication *in vitro*: Identification of a host factor that stimulates synthesis of the preterminal protein–dCMP complex, *Proc. Natl. Acad. Sci. U.S.A.* **79**:6438.

Nass, K., and Frenkel, G., 1980, The adenovirus-specific DNA-binding protein inhibits the hydrolysis of DNA by DNase *in vitro*, *J. Virol.* **35**:314.

Niicjama, Y., Igarashi, K., Tsukamoto, K., Kurokawa, T., and Sugino, Y., 1975, Biochemical studies on bovine adenovirus type 3. I. Purification and properties, *J. Virol.* **16**:621.

Ostrove, J.M., Rosenfeld, P., Williams, J., and Kelly, T.J., Jr., 1983, *In vitro* complementation as an assay for purification of adenovirus DNA replication proteins, *Proc. Natl. Acad. Sci. U.S.A.* **80**:935.

Padmanabhan, R., and Padmanabhan, R.V., 1977, Specific interaction of a protein(s) near the termini of adenovirus 2 DNA, *Biochem. Biophys. Res. Commun.* **80**:955.

Pearson, G.D., 1975, Intermediates in adenovirus type 2 replication, *J. Virol.* **16**:17.

Pearson, G.D., and Hanawalt, P.C., 1971, Isolation of DNA replication complexes from uninfected and adenovirus-infected HeLa cells, *J. Mol. Biol.* **62**:65.

Philipson, L., Lonberg-Holm, K., and Pettersson, U., 1968, Virus–receptor interaction in an adenovirus system, *J. Virol.* **2**:1064.

Piña, M., and Green, M., 1969, Biochemical studies on adenovirus multiplication. XIV. Macromolecule and enzyme synthesis in cells replicating oncogenic and non-oncogenic human adenoviruses, *Virology* **38**:573.

Pincus, S., Robertson, W., and Rekosh, D., 1981, Characterization of the effect of aphidicolin on adenovirus DNA replication: Evidence in support of a protein primer model of initiation, *Nucleic Acids Res.* **9**:4919.

Rawlins, D.R., Rosenfeld, P.J., Wides, R.J., Challberg, M.D., and Kelly, T.J., Jr., 1984, Structure and function of the adenovirus origin of replication, *Cell*, in press.

Reiter, T., Futterer, J., Weingartner, B., and Winnacker, E.-L., 1980, Initiation of adenovirus DNA replication, *J. Virol.* **35**:662.

Rekosh, D., 1981, Analysis of the DNA–terminal protein from different serotypes of human adenovirus, *J. Virol.* **40**:329.

Rekosh, D.M.K., Russell, W.C., Bellett, A.J.D., and Robinson, A.J., 1977, Identification of a protein linked to the ends of adenovirus DNA, *Cell* **11**:283.

Rijnders, A.W.M., van Bergen, B.G.M., van der Vliet, P.C., and Sussenbach, J.S., 1983, Specific binding of the adenovirus terminal protein precursor-DNA polymerase complex to the origin of DNA replication, *Nucl. Acids Res.* **11**:8777.

Robin, J.D., Bourgaux-Ramoizy, D., and Bourgaux, P., 1973, Single-stranded regions in replicating DNA of adenovirus type 2, *J. Gen. Virol.* **20**:233.

Robinson, A.J., and Bellett, A.J.D., 1974, A circular DNA–protein complex from adenoviruses and its possible role in DNA replication, *Cold Spring Harbor Symp. Quant. Biol.* **39**:523.

Robinson, A.J., and Bellett, A.J.D., 1975, Complementary strands of CELO virus DNA, *J. Virol.* **15**:458.

Robinson, A.J., Younghusband, H.B., and Bellett, A.J.D., 1973, A circular DNA–protein complex from adenovirus, *Virology* **56**:54.

Rosenwirth, B., Shiroki, K., Levine, A.J., and Shimojo, H., 1975, Isolation and characterization of adenovirus type 12 DNA binding proteins, *Virology* **67**:14.

Rosenwirth, B., Anderson, C.W., and Levine, A.J., 1976, Tryptic fingerprint analysis of adenovirus types 2, 5 and 12 DNA-binding proteins, *Virology* **69**:617.

Schechter, N.M., Davies, W., and Anderson, C.W., 1980, Adenovirus coded DNA binding protein: Isolation, physical properties and effects of proteolytic digestion, *Biochemistry* **19**:2802.

Schilling, R., Weingartner, B., and Winnacker, E.-L., 1975, Adenovirus type 2 DNA replication. II. Termini of DNA replication, *J. Virol.* **16**:767.

Seghal, P.B., Fraser, N.W., and Darnell, J.E., 1979, Early Ad2 transcription units: Only promoter proximal RNA continues to be made in the presence of DRB, *Virology* **94**:185.

Sekiguchi, M., Hayakawa, H., Makino, F., Tanaka, K., and Okada, Y., 1976, A human enzyme that liberates uracil from DNA, *Biochem. Biophys. Res. Commun.* **73**:293.

Sergeant, A., Tigges, M.A., and Raskas, H.J., 1979, Nucleosome-like structural subunits of intranuclear parental adenovirus type 2 DNA, *J. Virol.* **29**:888.

Sharp, P.A., Moore, C., and Harverty, J., 1976, The infectivity of adenovirus 5–DNA protein complex, *Virology* **75**:442.

Shaw, C.H., Russell, W.C., and Rekosh, D.M.K., 1979, Association of adenovirus early protein with a nuclear fraction that synthesizes DNA *in vitro*, *Virology* **92**:436.

Shinagawa, M., and Padmanabhan, R., 1979, Nucleotide sequences at the inverted terminal repetition of Ad2 DNA, *Biochem. Biophys. Res. Commun.* **87**:671.

Shinagawa, M., and Padmanabhan, R., 1980, Comparative sequence analysis of the inverted terminal repetitions from different adenoviruses, *Proc. Natl. Acad. Sci. U.S.A.* **77**:3831.

Shiroki, K., Shimojo, H., and Yamaguchi, K., 1974, The viral DNA replication complex of adenovirus 12, *Virology* **60**:192.

Simmons, T., Heywood, P., and Hodge, L.D., 1974, Intranuclear site of replication of adenovirus DNA, *J. Mol. Biol.* **89**:423.

Smart, J.E., and Stillman, B.W., 1982, Adenovirus terminal protein precursor, *J. Biol. Chem.* **257**:13,499.

Steenbergh, P.H., Maat, H.J., van Ormondt, H., and Sussenbach, J.S., 1977, The nucleotide sequence at the termini of adenovirus type 5 DNA, *Nucleic Acids Res.* **4**:4371.

Stillman, B.W., 1981, Adenovirus DNA replication *in vitro*: A protein linked to the 5′ end of nascent DNA strands, *J. Virol.* **37**:139.

Stillman, B.W., and Bellett, A.J.D., 1979, An adenovirus protein associated with the ends of replicating DNA molecules, *Virology* **93**:69.

Stillman, B.W., Lewis, J.B., Chow, L.T., Mathews, M.B., and Smart, J.E., 1981, Identification of the gene and mRNA for the adenovirus terminal protein precursor, *Cell* **23**:497.

Stillman, B.W., Tamanoi, F., and Mathews, M.B., 1982a, Purification of an adenovirus-coded DNA polymerase that is required for initiation of DNA replication, *Cell* **31**:613.

Stillman, B.W., Topp, W.C., and Engler, J.A., 1982b, Conserved sequences at the origin of adenovirus DNA replication, *J. Virol.* **44**:530.

Sugawara, K., Gilead, Z., and Green, M., 1977, Purification and molecular characterization of adenovirus type 2 DNA-binding protein, *J. Virol.* **21**:338.

Sugisaki, H., Sugimoto, K., Takanami, M., Shiroki, K., Saito, I., Shimojo, H., Sawada, Y., Uemizu, Y., Uesugi, S.-I., and Fujinaga, K., 1980, Structure and gene organization in the transforming Hind III-G fragment of Ad12, *Cell* **20**:777.

Sussenbach, J.S., 1967, Early events in the infection process of adenovirus type 5 in HeLa cells, *Virology* **33**:567.

Sussenbach, J.S., and Kuijk, M.G., 1977, Studies on the mechanism of replication of adenovirus DNA. V. The location of termini of replication, *Virology* **77**:149.
Sussenbach, J.S., and Kuijk, M.G., 1978, The mechanism of replication of adenovirus DNA. VI. Localization of the origins of the displacement synthesis, *Virology* **84**:509.
Sussenbach, J.S., and van der Vliet, P.C., 1972, Characterization of adenovirus DNA in KB cells infected with adenovirus type 12, *Virology* **49**:224.
Sussenbach, J.S., van der Vliet, P.C., Ellens, D.J., and Jansz, H. S., 1972, Linear intermediates in the replication of adenovirus DNA, *Nature (London) New Biol.* **239**:47.
Sussenbach, J.S., Ellens, D.J., and Jansz, H.S., 1973, Studies on the mechanism of replication of adenovirus DNA. I. The nature of single-stranded DNA in replicative intermediates, *J. Virol.* **12**:1131.
Sussenbach, J.S., Tolun, A., and Pettersson, U., 1976, On the nature of the single-stranded DNA in replicating adenovirus type 5 DNA, *J. Virol.* **20**:532.
Tamanoi, F. and Stillman, B.W., 1982, Function of adenovirus terminal protein in the initiation of DNA replication, *Proc. Natl. Acad. Sci. U.S.A.* **79**:2221.
Tamanoi, F. and Stillman, B.W., 1983, Initiation of adenovirus DNA replication *in vitro* requires a specific DNA sequence, *Proc. Natl. Acad. Sci. U.S.A.* **80**:6446.
Tate, V.E., and Philipson, L., 1979, Parental adenovirus DNA accumulates in nucleosome-like structures in infected cells, *Nucleic Acids Res.* **6**:2769.
Temple, M., Antoine, G., Delius, H., Stahl, S., and Winnacker, E.-L., 1981, Replication of mouse adenovirus strain FL DNA, *Virology* **109**:1.
Tolun, A., and Pettersson, U., 1975, Termination sites for adenovirus type 2 DNA replication, *J. Virol.* **16**:759.
Tolun, A., Aleström, P., and Pettersson, U., 1979, Sequences of inverted terminal repetitions from different adenoviruses: Demonstration of conserved sequences and homology between SA7 termini and SV40 DNA, *Cell* **17**:705.
Tye, B.-K., Nyman, P.-O., Lehman, J.R., Hochhauser, S., and Weiss, B., 1977, Transient Accumulation of Okazaki fragments as a result of uracil incorporation into nascent DNA, *Proc. Natl. Acad. Sci. U.S.A.* **74**:154.
van Bergen, B.G.M., van der Ley, P.A., van Driel, W., van Mansfeld, A.D.M., and van der Vliet, P.C., 1983, Replication of origin containing adenovirus DNA fragments that do not carry the terminal protein, *Nucl. Acids Res.* **11**:1975.
van der Eb, A.J., 1973, Intermediates in type 5 adenovirus DNA replication, *Virology* **51**:11.
van der Vliet, P.C., and Levine, A.J., 1973, DNA-binding proteins specific for cells infected by adenovirus, *Nature (London) New Biol.* **246**:170.
van der Vliet, P.C., and Sussenbach, J.S., 1972, The mechanism of adenovirus DNA synthesis in isolated nuclei, *Eur. J. Biochem.* **30**:584.
van der Vliet, P.C., and Sussenbach, J.S., 1975, An adenovirus type 5 gene function required for initiation of viral DNA replication, *Virology* **67**:415.
van der Vliet, P.C., Levine, A.J., Ensinger, M., and Ginsberg, H.S., 1975, Thermo-labile DNA-binding proteins from cells infected with a temperature-sensitive mutant of adenovirus defective in viral DNA synthesis, *J. Virol.* **15**:348.
van der Vliet, P.C., Landberg, J., and Jansz, H.S., 1977, Evidence for a function of the adenovirus DNA-binding protein in initiation of DNA synthesis as well as in elongation on nascent DNA chains, *Virology* **80**:98.
van der Vliet, P., Keegstra, N., and Jansz, H., 1978, Complex formation between the adenovirus type 5 DNA binding protein and single-stranded DNA, *Eur. J. Biochem.* **86**:389.
Vlak, J.M., Rozjin, T.H., and Sussenbach, J.S., 1975a, Studies on the mechanism of replication of adenovirus DNA. IV. Discontinuous DNA chain propagation, *Virology* **63**:168.
Vlak, J.M., Rozijn, T.H., and Spies, F., 1975b, Localization of adenovirus DNA replication in KB cells, *Virology* **65**:535.
Weber, J., 1976, Genetic analysis of adenovirus type 2. III. Temperature-sensitivity of processing of viral proteins, *J. Virol.* **17**:462.
Weingartner, B., Winnacker, E.-L., and Pettersson, U., 1976, Two complementary strand-specific termination sites for adenovirus DNA replication, *Cell* **9**:259.
Wilkie, N.M., Ustaçelebi, S., and Williams, J.F., 1973, Characterization of temperature-sensitive mutants of adenovirus type 5: Nucleic acid synthesis, *Virology* **51**:499.

Williams, J.F., Young, H., and Austin, P., 1975, Genetic analysis of human adenovirus type 5 in permissive and nonpermissive cells, *Cold Spring Harbor Symp. Quant. Biol.* **39:**427.

Winnacker, E.-L., 1975, Adenovirus type 2 DNA replication. I. Evidence for discontinuous replication, *J. Virol.* **15:**744.

Winnacker, E.-L., 1978, Adenovirus DNA: Structure and function of a novel replicon, *Cell* **14:**761.

Wist, E., Unhjem, O., and Krokan, H., 1978, Accumulation of small fragments of DNA in isolated HeLa cell nuclei due to transient incorporation of dUMP, *Biochim. Biophys. Acta* **520:**253.

Wolfson, J., and Dressler, D., 1972, Adenovirus DNA contains an inverted terminal repetition, *Proc. Natl. Acad. Sci. U.S.A.* **69:**3054.

Yamashita, T., Arens, M., and Green, M., 1975, Adenovirus DNA replication. III. Synthesis of viral DNA by a nuclear membrane fraction from infected KB cells, *J. Biol. Chem.* **250:**3273.

Yamashita, T., Arens, M., and Green, M., 1977, Adenovirus DNA replication: Isolation of a soluble replication system and analysis of the *in vitro* DNA product, *J. Biol. Chem.* **252:**7940.

CHAPTER 8

Adenovirus Assembly

Lennart Philipson

I. INTRODUCTION

The basic structure of all virions is a capsid enclosing the nucleic acid. Crick and Watson (1957) pointed out that the capsid structures must be built of multiple copies of one or a few protein subunits, since viruses do not have enough genetic information in their nucleic acids to code for a great variety of capsid proteins. There are two ways in which identical asymmetrical subunits can assemble to build stable regular capsids, either through helical assemblies or by forming closed shells. It was long thought that the capsid proteins, by a process of self-assembly, would transfer from an energetically unfavorable state into the capsid such that a structure of minimal free energy was obtained for the specific protein(s) in the virion. The shape of the protein molecules that serve as building blocks in the capsid may therefore determine the shape and size of the capsid.

In complete contrast to this perhaps oversimplified view, it has been shown for several complex bacteriophages that the assembly of virions is an elaborate and precise process. All virion proteins and other late proteins required for assembly are synthesized simultaneously and accumulate in a precursor pool. From the pool, they are withdrawn by specific interactions with other protein molecules to form assembly intermediates that in turn are assembled into complete virions, a process that in many cases involves several enzymatic steps. An elaborate example is provided by the T-even phages; *in vivo* studies with conditional lethal mutants first showed the general pathway, but an *in vitro* assembly sys-

LENNART PHILIPSON • Department of Microbiology, The Biomedical Centre, University of Uppsala, S-751 23 Uppsala, Sweden. Dr. Philipson's present address is: European Molecular Biology Laboratory, D-6900 Heidelberg, Federal Republic of Germany.

tem was required to delineate the molecular details. Some 15 proteins are involved in head assembly and around 20 proteins for the tail (Edgar and Wood, 1966). With an *in vitro* complementation system, the exact pathway for assembly was identified, and the same approach has been used in several other phage systems to clarify the morphogenesis of these bacterial viruses.

With animal viruses, it has been more difficult to (1) identify all the proteins involved in the assembly pathway and (2) develop *in vitro* assembly systems. In mammalian cells, many of the intermediate steps in virus assembly appear to involve interaction with specific cellular structures including internal membranes, and the reconstruction of the system *in vitro* has therefore not been achieved or is extremely difficult. The introduction of temperature-sensitive and, more recently, precise deletion mutants for the animal viruses, together with new methods to characterize the proteins and nucleic acids associated with the assembly intermediates, have, however, given some hope for the future. Although progress in understanding the assembly pathway has been made with several enveloped and nonenveloped RNA viruses, the pathway for assembly of complex DNA viruses is still unresolved. Only during the last five years have all the tools become available to attack this problem for adenovirus, and this chapter attempts to summarize the progress in this field. Few if any reviews on this topic have appeared, but some results have been discussed in recent publications (Philipson, 1979; Flint and Broker, 1981).

II. PROTEIN NEIGHBORS IN THE VIRION

The structure of adenoviruses is described in detail in Chapters 2 and 6. The polypeptide composition of purified adenovirus particles has been studied extensively, and it is now established by genomic and antigenic analyses that at least 10 distinct virus-coded polypeptides (Maizel *et al.*, 1968; Everitt *et al.*, 1973; Everitt and Philipson, 1974; Miller *et al.*, 1980) are required to build the virion (Fig. 1). These polypeptides together account for around 80% of the coding capacity of the virus genome.

More important for an understanding of assembly is the information about the precise topography of the proteins in adenovirus particles, i.e., the identification of the protein neighbors in the virions, because this has implications for understanding the structure of the assembly intermediates. The internal relationships among the various polypeptides in the virion have been studied extensively, mainly by methods devised to liberate morphological substructures and analyze them by sodium dodecyl sulfate–polyacrylamide gel electrophoresis (SDS-PAGE). The virus is icosahedral with 252 capsomers; the 12 vertex capsomers that have local 5-fold symmetry have been called *pentons*, and the remaining 240

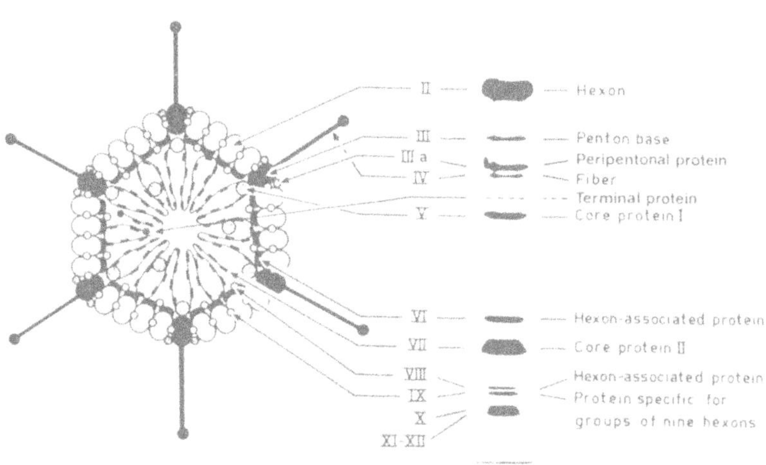

FIGURE 1. Tentative model for the location of the proteins in the adenovirus type 2 virion. The polypeptide compositions of the virion proteins are shown in a stained exponential 10–16% sodium dodecyl sulfate (SDS)–polyacrylamide gel, and the evidence for the location of the different protein moieties is summarized in the text. A covalently linked terminal protein is also indicated in the figure, but it cannot be seen in a protein gel, and the corresponding polypeptide band is therefore indicated with a dashed line. Modified from Everitt *et al.* (1975).

capsomers with local 6-fold symmetry have been called *hexons* (Ginsberg *et al.*, 1966; Valentine and Pereira, 1965).

Treatment of purified virions with heat (Russell *et al.*, 1967) or dialysis against low-ionic-strength buffers with a pH between 6.0 and 6.8 (Prage *et al.*, 1968, 1970; Laver *et al.*, 1969) removes the vertex regions, comprising the fiber, the penton base together with the 60 peripentonal hexons surrounding the 12 pentons at the vertices, and in addition protein IIIa (Everitt *et al.*, 1973) (Fig. 1). Capsids devoid of the vertex region are stable (Russell *et al.*, 1967) and obviously held together by intermolecular bonds between the hexons. The perforated capsids still contain virion DNA that is now accessible to deoxyribonuclease (DNase), in contrast to DNA in intact virions (Prage *et al.*, 1970). Treatment of purified virions with pyridine (Prage *et al.*, 1970) or deoxycholate (DOC) and heat (Russell *et al.*, 1971) disintegrates the virion even further. The pentons and the peripentonal hexons are released as capsomers, and the main body of the capsid is also disrupted, so that the majority of the hexons are liberated in groups of nine hexons (nonamers) that display local 3-fold symmetry (Crowther and Franklin, 1972). Capsomeric units of all hexons may be obtained by disrupting the particles with one of several different agents such as SDS (Smith *et al.*, 1965), formamide (Stansy *et al.*, 1968), or 8 M urea (Shortridge and Biddle, 1970). Polypeptides II and III of adenovirus type 2 (Ad2) particles, with molecular weights around 110,000 (110K) and

85K, respectively, form the two capsomers, the hexon and the penton base, by association of three identical polypeptides, giving the capsomers molecular weights of 330 and 255K, respectively (Grütter and Franklin, 1974; Devaux *et al.*, 1982).

Polypeptide IV (62K) is probably also present in three copies per fiber (Sundquist *et al.*, 1973b), but it is not yet certain whether all three polypeptides are identical (Dorsett and Ginsberg, 1975), although only one gene for polypeptide IV has been identified in the viral genome (Miller *et al.*, 1980). Recent structural studies imply, however, a dimer structure (Devauz *et al.*, 1983, Green *et al.*, 1983).

Hexons purified from infected cells contain only polypeptide II, whereas hexons obtained by dissociating virions with pyridine are bound to polypeptide VI (Everitt *et al.*, 1973). For each mole of hexon protein two moles of polypeptide VI (24K) can be recovered suggesting that a dimer of polypeptide VI is associated with each hexon trimer. Polypeptide VI cannot, however, be labeled from the outside by iodination suggesting that it is located on the inside of the hexon shell (Everitt *et al.*, 1975). Polypeptide IIIa is probably located in the peripentonal region since it is associated with peripentonal hexons in degraded virus preparations. In crosslinking experiments polypeptide IIIa is linked to peripentonal hexons, the penton base and the internal core polypeptide V. It therefore probably extends from the surface of the virion to its interior (Everitt *et al.*, 1975).

Since the penton base appears to be a trimeric unit (Devaux *et al.*, 1982), polypeptide IIIa which is probably present in 5 copies per vertex region, (Boudin *et al.*, 1980; Devaux *et al.*, 1982) may provide the symmetry required to build a quasi-icosahedral structure at the vertices of the adenoviruses. The location of polypeptide VIII in the virion is still not resolved even though this protein is associated with hexons after disruption of viral particles by freezing and thawing (Everitt *et al.*, 1973). It has been assigned a tentative position inside the hexon shell (Fig. 1) because of its basic isoelectric point.

Finally, polypeptide IX can be recovered in the nonameric hexon units released with pyridine or DOC. Around 15 copies of polypeptide IX have been recovered per nonameric hexon unit, and it has been suggested from considerations of symmetry that polypeptide IX is located at the corner-to-edge contacts between the hexons in the nonamers (Boulanger *et al.*, 1979). Polypeptide IX is not labeled from the outside by iodination, but its antigenic determinants are exposed on the exterior of the virons. It has therefore been suggested that this polypeptide plays a cementing role in the capsid structure (Everitt *et al.*, 1975). Recently, it was shown with a deletion mutant of Ad5 (H5*dl*313) that lacks the polypeptide IX gene that this protein is not essential for assembly of Ad5 virions (Colby and Shenk, 1981). The mutant virions, on the other hand, appear to be considerably more heat-labile than wild-type particles. Mutant virus also fails to produce hexon nonamers on disruption with pyridine. Thus, polypeptide IX might not be essential, but provides a stabilization function in the outer capsid.

Inside the capsid, the DNA is packed into a core structure. The proteins of the core have been identified and constitute around 20% of the protein content of the virion, there being around 1000 copies of polypeptide VII (18.5K) and around 200 copies of polypeptide V (48.5K), usually referred to as the major and the minor core proteins or core proteins II and I, respectively (Fig. 1). The adenovirus core is around 67 nm in diameter, and when the core is treated with ionic detergent, long filaments of DNA radiating out from the central body can be seen under the electron microscope (EM) (Brown *et al.*, 1975).

After disruption of viruses with the ionic detergent sarcosyl, the cores contain 8–10 discernible spherical subunits, with 2–4 units masked by the negative stain in the EM. It has therefore been suggested that each core might consist of 12 subunits, allowing for 1 subunit per vertex (Brown *et al.*, 1975). Only polypeptide VII is recovered from sarcosyl-generated cores, but after pyridine or DOC disintegration, both polypeptides V and VII are associated with the cores (Brown *et al.*, 1975; Prage *et al.*, 1970; Russell *et al.*, 1971). Cross-linking studies have shown that polypeptide V is in contact with the penton base and also the polypeptide VII–DNA complex, suggesting that polypeptide V is associated with each vertex, forming a link between the outer shell and the core (Everitt *et al.*, 1975; Nermut, 1979). Polypeptide VII is an extremely basic polypeptide, with around 23% Arg (Prage and Pettersson, 1971). Its N-terminal amino acid sequence is similar to those of some of the cellular histone proteins, especially H_4 (Lischwe and Sung, 1977). At the two ends of the linear adenovirus DNA in the particle, there is a terminal protein (55K) that is covalently linked to the DNA and required for initiation of DNA replication (Rekosh *et al.*, 1977; Challberg *et al.*, 1980). It has been suggested that adenovirus chromatin has a subunit structure similar to that of the nucleosomes in eukaryotic chromatin (Corden *et al.*, 1976), but this suggestion has not been confirmed. Subsequent analyses have shown that adenovirus DNA in the particles is protected only in a 150-base-pair (bp) core particle after digestion with micrococcal nuclease irrespective of the conditions of digestion (Tate and Philipson, 1979; Brown and Weber, 1980; Daniell *et al.*, 1981; Mirza and Weber, 1982). More detailed analysis of the adenovirus core has revealed that only polypeptide VII can be cross-linked to the DNA by UV irradiation (Sato and Hosokawa, 1981), and it has been suggested that viral protein VII forms a helical core and the DNA is wound around it with a compaction factor of 5–6 and a protected fragment of DNA of around 160 per 6 units of protein VII. Virion protein V may form a thin pelliculum around the core and has therefore been called a core shell (Nermut, 1979, 1980). If a nucleocapsid is formed by major core protein VII, it is a misnomer to refer to the virion structure as a capsid, and the whole concept of icosahedral capsids for nonenveloped viruses should probably be reconsidered (Nermut, 1980).

At early times in the infectious cycle, however, parental viral DNA appears in a chromatinlike nucleosome pattern, suggesting that the virus-coded proteins are replaced by cellular histones shortly after infection (Tate and Philipson, 1979; Sergeant *et al.*, 1979). The location of the small polypeptides X, XI, and XII has not yet been determined, but they may arise during protein cleavage at late times in the assembly process (Everitt *et al.*, 1975). The minor components, proteins IVa_1 and IVa_2, appear to be necessary for the assembly process; the trace amounts identified in purified virus may derive from an incompletely assembled fraction of the particles.

III. ASSEMBLY OF CAPSOMERS

During the late phase of adenovirus productive infection, host-cell protein synthesis is shut off and viral mesenger RNA (mRNA) is preferentially translated. Most of the viral polypeptides are rapidly released from polyribosomes and transported to the nucleus with a lag period of only 3–6 min (Horwitz *et al.*, 1969; Velicer and Ginsberg, 1970). During this short interval, the monomeric structural polypeptides of the hexon, the penton base, and the fiber assemble into trimeric proteins with sedimentation coefficients of 13, 10.5, and 6 S, respectively (Velicer and Ginsberg, 1970). The very first step in adenovirus assembly is therefore the formation of the major structural units of the capsid from the monomeric polypeptide chains. Where this occurs in the cell is not yet clear, but it probably takes place at or shortly after translation, since the monomeric polypeptide for all three major capsid units is insoluble, but freely soluble in the multimeric form. The penton base and the fiber are combined into the penton, which may be regarded as an intermediary step in capsomer assembly. Penton assembly occurs rapidly for around 25% of newly synthesized polypeptides of fiber and penton base, but it takes up to 24 hr before this process is completed (Horwitz *et al.*, 1969); penton assembly may therefore be a partially reversible process.

A. Hexon

The hexon capsomer contains three identical polypeptides of known amino acid sequence with a molecular weight of 108K (Jörnvall *et al.*, 1981) for Ad2, but with 5–8% variation in size among different serotypes. The trimeric unit sediments at 13 S and contains type-, subgroup-, and group-specific antigenic determinants (Wilcox *et al.*, 1963; Norrby and Wadell, 1969; Pettersson, 1971). The monomeric polypeptide is not recognized by antibodies raised against the trimer, however, but only by antibodies against fully denatured monomers (Stinski and Ginsberg,

1974). The product formed by *in vitro* translation of hexon mRNA can therefore be detected only with antimonomeric antibodies, which recognize only the nascent chains in infected cells (Stinski and Ginsberg, 1974; Öberg *et al.*, 1975). When polysomes from infected cells are used for *in vitro* translation, however, the trimer is formed (Wilhelm and Ginsberg, 1972; Persson *et al.*, 1977), suggesting that formation of a trimer requires some other adenovirus product.

The first genetic evidence that assembly of hexons requires one additional gene function was provided by Leibowitz and Horwitz (1975) using two temperature-sensitive (*ts*) mutants. Hexon polypeptides synthesized in cells infected with H5*ts*17 and H5*ts*20 at the nonpermissive temperature could assembly after a shift to permissive temperature if the cells were kept for only 15 min, but not if they were kept for 2 hr, at the nonpermissive temperature. Since no difference in size was observed among the hexons made in cells infected by H5*ts*17, H5*ts*20, and wild-type (wt) Ad2, the authors suggested that the product of an additional adenovirus gene is required for formation of the trimeric unit. It was later established by marker-rescue experiments that these two *ts* mutants are located in the nonstructural 100K protein gene (Frost and Williams, 1978), suggesting that this protein is necessary for assembly of the hexon polypeptide into trimeric units. These results were recently confirmed with two other *ts* mutants located in the 100K protein gene, H5*ts*115 and H5*ts*116 (Oosterom-Dragon and Ginsberg, 1981). A direct association between the 100K protein and nascent hexon polypeptides has also been shown using antiserum or monoclonal antibodies against the 100K protein or against denatured hexons (Oosterom-Dragon and Ginsberg, 1981; Cepko and Sharp, 1982). It is therefore tempting to suggest that the 100K protein, which is almost as abundant as the hexon, is required in stoichiometric amounts to polymerize the hexon polypeptide into the trimeric unit with its complicated structure of type-specific determinants on the outer surface and group-specific determinants on the inner (Wilcox and Ginsberg, 1963). The folding process may therefore require the 100K protein as a scaffolding protein. Once the final structure of the hexon has been achieved, the 100K protein may dissociate from the trimeric unit. This process may involve the reactive Ser within the nascent hexon polypeptide that can interact with diisopropylfluorophosphate (DFP), which binds only to the active Ser residues in proteases (Devaux and Boulanger, 1980). DFP obviously interacts only with the denatured polypeptide *in vitro* and not with the trimeric hexon. It appears justifiable to conclude that the assembly of the hexon polypeptides into a trimeric unit requires the 100K protein, but the molecular detail of this process remains to be established. The efficiency of hexon formation is, however, remarkable. The hexon accumulates in the nuclei within 5 min of completion of the chain in the polysomes, and as much as 10% of the total protein content of the cell is hexon protein at the end of the infectious cycle.

B. Penton

The penton capsomer consists of the penton base and the fiber unit. Each appears to consist of trimers of the penton base (III) and the fiber polypeptide (IV), respectively (Sundquist *et al.*, 1973b; Devaux *et al.*, 1982). A molecular weight of around 450K for the intact penton base has been claimed, suggesting a pentameric arrangement of the 85K subunit (Pettersson and Höglund, 1969; Boudin *et al.*, 1979). However, a careful structural analysis using neutron diffraction appears to establish that the penton base is a trimer of the 85K polypeptide (Devaux *et al.*, 1982). The genes for the penton base and the fiber polypeptides are located more than 10 kilobases apart on the viral genome, but both are located within the major late transcriptional unit in the L2 and L5 regions, respectively (see Chapter 3). Several *ts* mutants have been isolated that map in the fiber region of the genome, but no *ts* mutants have physically been mapped in the penton-base region, with the possible exception of H2*ts*3 (D'Halluin *et al.*, 1982).

The penton base and the fiber are separately synthesized on polyribosomes, and within minutes the two subunits are accumulating in the nucleus of the infected cell (Velicer and Ginsberg, 1970). Pulse–chase experiments suggest that the penton base is synthesized first and then assembled with the fiber to form complete pentons (Horwitz *et al.*, 1969; Velicer and Ginsberg, 1970). During productive infection with wt virus, an excess of fiber units is formed, and the formation of penton base is probably rate-limiting in the assembly process, since only minute amounts of free penton base can be found (Boudin *et al.*, 1979). Penton base accumulates, however, in cells infected with fiber-defective *ts* mutants at the nonpermissive temperature (Boudin *et al.*, 1979). An *in vitro* assembly system for pentons has recently been described (Boudin and Boulanger, 1982), using purified fiber from Ad2 Wt infection and penton base accumulated in cells infected with a fiber-defective *ts* mutant, H2*ts*125. Assembly as revealed by two-dimensional immunoelectrophoresis (Fig. 2) occurs over a broad range of pH (5.5–9.0) and at ionic strength from 0.05 to 1.0. The association between the penton base and the fiber is a reversible reaction with a dissociation constant of 2×10^{-7} M determined on the basis of fiber molarity. The fiber appears to combine with the penton base via its C terminus as shown with carboxypeptidase. These experiments also reveal that the last 20 amino acids of the fiber polypeptide are required for assembly. The formation of chimeric pentons by incubating the two units from different serotypes suggests that a group-specific area of the fiber is recognized by the penton base. This interpretation was corroborated by the formation of serotypic ts^+ recombinants *in vivo*. A change in antigenicity of the fiber moiety was also observed at complex formation, possibly reflecting a three-dimensional structural change. These results establish that the penton capsomer can form by

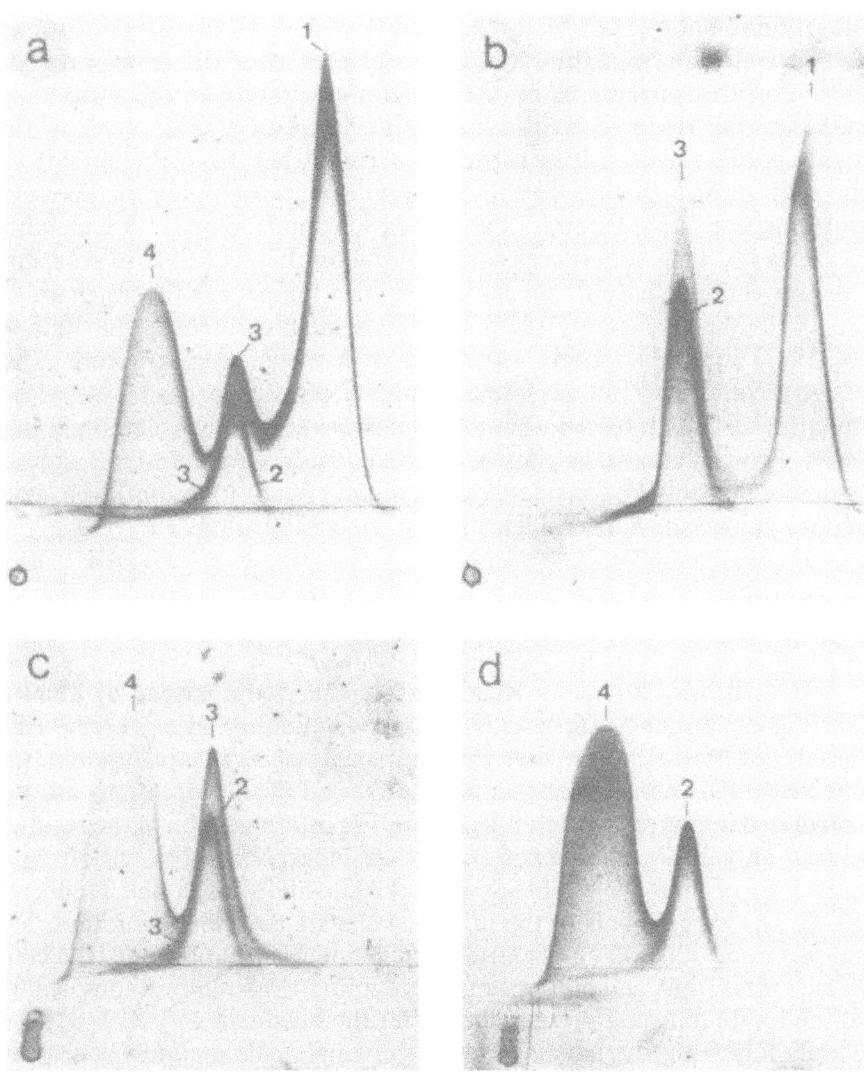

FIGURE 2. Two-dimensional immunoelectrophoretic analysis of penton assembled *in vitro*. (a) Partial assembly of penton base (1) and of fiber (4), forming the pentons (2, 3); line (2) corresponds to antigenic determinants of the fiber, line (3) to the antigenic determinants of the penton base in the penton complex. (b) Complete assembly of fiber into the penton unit in an excess of penton base. (c) Complete assembly of penton base into the penton unit in an excess of fiber. (a–c) Stained with Coomassie blue; (d) an autoradiograph of (c) revealing only the antigenic determinants of the labeled fiber. Reprinted from Boudin and Boulanger (1982).

self-assembly between the two trimeric structures, the fiber and the penton base. No additional proteins appear to be necessary for formation of the trimeric units.

Several important functions are associated with the penton capsomer. The fiber appears to be the recognition protein of the virus that identifies the receptor on the host cell (Philipson *et al.*, 1968), while highly purified penton base carries an activity that, from the outside of the cell, mimics the cytopathic changes during virus infection (Everett and Ginsberg, 1958; Pereira, 1958; Pettersson and Höglund, 1969; Boudin *et al.*, 1979). This activity appears to reside in the penton-base polypeptide. An endonuclease found associated with the penton (Burlingham *et al.*, 1971; Marusyk *et al.*, 1975) appears to be of cellular origin (Reif *et al.*, 1977) and only loosely associated with the penton structure. The activity could also not be found in highly purified penton base accumulating in cells infected with fiber-defective *ts* mutants (Boudin *et al.*, 1979). Any enzymatic activity associated with a viral capsomer should therefore probably be considered as a cellular contaminant until the gene and its product have been identified on the viral genome.

IV. FORMATION OF EMPTY CAPSIDS

The pathway for adenovirus assembly was first analyzed by kinetic labeling experiments (Horwitz *et al.*, 1969; Sundquist *et al.*, 1973a; Ishibashi and Maizel, 1974), but the large pools of structural proteins in infected cells, and the fact that only a fraction of the capsomers are incorporated into virus particles, hampered the interpretation of the results. In fact, only 20% of the hexon trimers are incorporated into virions and only 10% of viral DNA enters virus particles (Philipson and Lindberg, 1974). The excess pools in the infected cells are therefore enormous, because around 100,000 virus particles are produced per infected HeLa cell. In fact, viral DNA and proteins must accumulate in amounts comparable to the total DNA and protein content of the host cell.

By following the labeling kinetics of single polypeptides, it became clear that different polypeptides of the virion entered viral particles with different kinetics (Horwitz *et al.*, 1969; Sundquist *et al.*, 1973a; Ishibashi and Maizel, 1974). Core proteins V and VII appear in intact virions within 2–5 min, but hexons appear only 30–60 min later, suggesting a differential rate of insertion of the different building blocks into virions (Horwitz *et al.*, 1969; Sundquist *et al.*, 1973a).

An explanation for this difference was offered when the different subviral structures were studied in more detail. From 13 hr after infection, cells infected by Ad2 or Ad3 contain both virions and empty capsids that can be readily separated by CsCl equilibrium gradients. These particles increase in concentration in parallel, and both are synthesized at maximum rates 24 hr after infection. More careful analysis of the appearance of labeled hexon in empty capsid and mature virus showed that the hexon trimers appeared without a lag in empty capsids, but 60–80 min elapsed before mature virions were labeled (Sundquist *et al.*, 1973a). Pulse–chase experiments with cells infected by Ad2, which has a small pool of empty capsids, suggest in fact a precursor–product relationship between incomplete and mature particles (Sundquist *et al.*, 1973a; Ishibashi and Maizel, 1974). It was also established that the radioactivity in core proteins appears immediately in the complete virions (Sundquist *et al.*, 1973a). These results favor the idea that mature virions are formed by the insertion of viral DNA and core proteins into preassembled empty capsids.

A more careful analysis of the protein composition of empty capsids compared to full virions showed that the former contain no core protein V or VII (Maizel *et al.*, 1968; Prage *et al.*, 1972; Ishibashi and Maizel, 1974), and proteins VI, VIII, and X–XII are also missing. But several other polypeptides not present in mature virions were identified in the low-density particles (Prage *et al.*, 1972; Sundquist *et al.*, 1973a; Wadell *et al.*, 1973; Ishibashi and Maizel, 1974; Rosenwirth *et al.*, 1974; Winberg and Wadell, 1977; Edvardsson *et al.*, 1976; 1978). Some of these polypeptides have been identified as precursor polypeptides to virion proteins (Anderson *et al.*, 1973; Öberg *et al.*, 1975). They are larger than their virion counterparts and are designated pVI and pVIII, respectively. Other polypeptides are found only in empty capsids and may be scaffolding proteins, like the IVa_2, the 40K, and the 32–33K proteins (Edvardsson *et al.*, 1976; D'Halluin *et al.*, 1978a). They are rapidly incorporated into empty capsids, but do not persist as stable polypeptides in mature virions. In some cases, these polypeptides are proteolytically cleaved at a late stage in the assembly pathway.

When the results from studies on kinetics of labeling and protein composition are combined, it appears that when empty capsids are assembled, they contain precursor polypeptides and possibly scaffolding proteins; viral DNA and core proteins either together or separately are then inserted, and the final step in morphogenesis appears to involve proteolytic cleavage of precursor polypeptides. We do not understand in detail, however, the mechanism of assembly of empty capsids. No *in vitro* systems are available, but a number of attempts to generate capsidlike structures with different substructures of the virion have been successful. Hexon trimers purified from infected cells do not all have the same charge (Pettersson, 1971; Boulanger, 1975). Two classes can be separated by electrophoresis or ion-exchange chromatography, and they may be used to assemble nonamers of hexons, associated with the faces of the icosahedron or the peripentonal hexons at each vertex region. The nonamers of hexons purified from virions disrupted with pyridine or sodium DOC and heat can reaggregate at low pH. They can form dimers, rings of five, and in some cases icosahedral shells containing 20 nonamers similar to the perforated structures obtained after removal of the peripentonal regions (Pereira and Wrigley, 1974). These results suggest that the nonamers of hexons might self-assemble to form a cagelike capsid. The role in this process of other hexon-associated proteins such as VI and IX is unclear. Polypeptide IX appears to be non-essential for the assembly of the viral icosahedron, but it will stabilize the shell (Colby and Shenk, 1981). An early step in virus assembly may therefore involve the formation of an icosahedral lattice through the interaction of nonamers of hexons, perhaps by a self-assembly reaction. Proteins pVI and pVIII might be necessary *in vivo* together with the recently identified presumptive scaffolding proteins (Edvardsson *et al.*, 1976; D'Halluin *et al.*, 1978a). Since the penton units and protein IIIa are also associated with the empty capsids, the process *in vivo* involves several additional steps that cannot be reconstructed *in vitro*. It therefore appears unlikely that formation of empty capsids can be explained simply by a self-assembly mechanism. Formation of an empty capsid appears to be a prerequisite, however, for the insertion of the DNA into the virion.

V. INSERTION OF VIRAL DNA

Before discussing the mechanism of insertion of DNA into empty capsids, we should summarize the properties of the defective and incomplete virus particles recovered from adenovirus-infected cells. These particles may contain limiting amounts of either viral or cellular DNA, and they are often isolated after sonication of cells by virtue of their buoyant density in CsCl gradients, which is intermediate between that of infectious virions and that of empty capsids. The infectivity associated with these particles is orders of magnitude less than that of virions (Prage *et al.*, 1972; Wadell *et al.*, 1973; Burlingham *et al.*, 1974; Niiyama *et al.*, 1975) and is probably due to contamination with trace quantities of infectious virions. The defective particles of human group B adenoviruses do not interfere with infection by virions or alter the yield of progeny, in contrast to the case with several animal RNA viruses (Prage *et al.*, 1972; Daniell, 1976). Defective particles of human Ad12 and bovine Ad3, which contain nearly full lengths of viral DNA, seem to be prevalent only after high-multiplicity infection (Mak, 1971; Niiyama *et al.*, 1975; Igarishi *et al.*, 1975). These defective particles contain a deletion of the viral genome that causes the change in buoyant density. Some defective particles have the same potential as intact virus to transform cells in culture (Schaller and Yohn, 1974; Igarishi *et al.*, 1975), suggesting that the left-hand end of the adenovirus genome, which is necessary for transformation (Graham *et al.*, 1974; Gallimore *et al.*, 1974), is overrepresented.

A whole spectrum of defective particles ranging in density from empty capsid to intact virions has been observed among all adenovirus groups, but human adenovirus group B in particular appears to accumulate empty capsids and low-density defective particles at a much higher rate than the group A or group C viruses. The protein composition of the defective particles mimics in all cases those of empty capsids, and instead of fully processed virion polypeptide, precursor and scaffolding polypeptides are encountered (Sundquist *et al.*, 1973a; Ishibashi and Maizel, 1974; Edvardsson *et al.*, 1976). Defective particles that contain specific deletions of the adenovirus genome, on the other hand, have the protein composition of mature virions (Mak *et al.*, 1979; Niiyama *et al.*, 1975; van Roy *et al.*, 1979).

Conflicting reports have appeared on the type of DNA associated with defective particles. It has been claimed that they contain cellular DNA (Tija *et al.*, 1977; Hammarskjöld *et al.*, 1977; Khittoo and Weber, 1981), but precautions to remove unspecifically adsorbed DNA by DNase digestion prior to purification of the particles have not always been taken. Cellular DNA was claimed in one report to be covalently linked to adenovirus DNA in defective particles as revealed by a two-step hybridization procedure (Tija *et al.*, 1977), but it remains to be shown that such packaging is functionally significant. Since cellular DNA associates readily and nonspecifically with capsids *in vitro* (Tibbetts and Giam, 1979), it is necessary to be cautious in interpreting these results. Aberrant forms of viral DNA have in addition been detected in defective particles from group B viruses, mainly molecules with long panhandles (Daniell, 1976). They are probably generated by defective DNA replication.

Since only about 10% of the total viral DNA is packaged into virions, recognition for packaging must be a discriminatory event. Quantitative analysis suggests that 95% of defective particles contain adenovirus DNA preferentially from the left-hand end, and a pattern of insertion can be recognized (Tibbetts, 1977; Brown and Weber, 1982; Cheng Chee-Sheung and Ginsberg, 1982) (Fig. 3). Only around 5% or less of the particles contain aberrant forms of viral DNA. In human group B viruses, each empty capsid interacts with only one viral DNA molecule, starting insertion at the molecular left end of the conventional viral genomic map. Since human group B viruses accumulate defective particles of varying density, the DNA associated with different density classes of Ad7 incomplete particles was shown to be a linear duplex sequence from the left end of the Ad7 genome with a variable right terminus (Tibbetts, 1977).

The control for this experiment was provided by Daniell and Mullenbach (1978), who demonstrated that both ends of the genome were equally represented in the pool of subgenomic DNA molecules in the cells. This places the recognition signal for packaging and insertion of DNA in empty capsids in the left-hand end of the adenovirus genome. In an elegant study utilizing a plaque-purified (PP) and a multiple-passage (MP) strain of Ad16, Hammarskjöld and Winberg (1980) showed that the PP virus inserted its DNA with preference for the left-hand end, but the MP strain, containing reduplications of the left-hand end at the right-hand end of the genome, could insert either end into the capsid (Fig. 4). A reduplication of only 390 bp was sufficient to allow packaging of DNA from the right-hand end. A variant of the MP virus containing a redu-

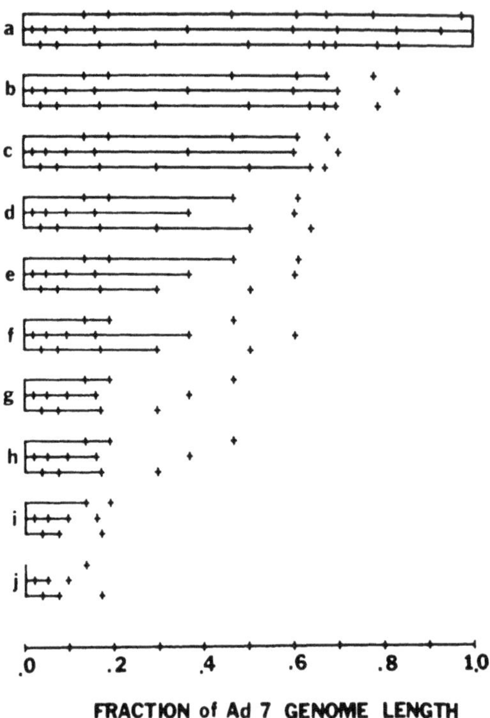

FIGURE 3. Physical maps of DNA associated with Ad7 incomplete particles. DNAs from different density classes of incomplete particles were digested with restriction endonucleases as denoted by the three bars in each set: *SmaI* (*top*), *BamI* (*middle*), and *Hind*III (*bottom*). Fragments recovered in single-copy yield from each DNA sample; (+) position for the next cleavage site of the respective enzymes on the viral DNA and at the same time the next-most-abundant cleavage fragment observed in the digest. The lengths of the DNA were linearly related to the buoyant densities of the particles. Reprinted from Tibbetts (1977).

plication, which was only 100 bp shorter, was not encapsidated from the opposite end. The conclusion appears obvious: there is a region between 290 and 390 nucleotides from the left end of the viral genome that is essential for the packaging of viral DNA. This finding is of considerable interest, especially since all adenovirus genomes contain an inverted terminal repetition (ITR) of around 150–200 bp that makes both ends identical (Garon *et al.*, 1972; Wolfson and Dressler, 1972). Placing the packaging signal internally to this sequence explains how the left-hand end can be preferred.

The recognition sequence for packaging is probably not transcribed, since the major E1a promoter closest to this region starts around 500 nucleotides from the left-hand end (Baker and Ziff, 1981). On the other hand, the DNA sequence reveals (van Ormondt *et al.*, 1980) that there

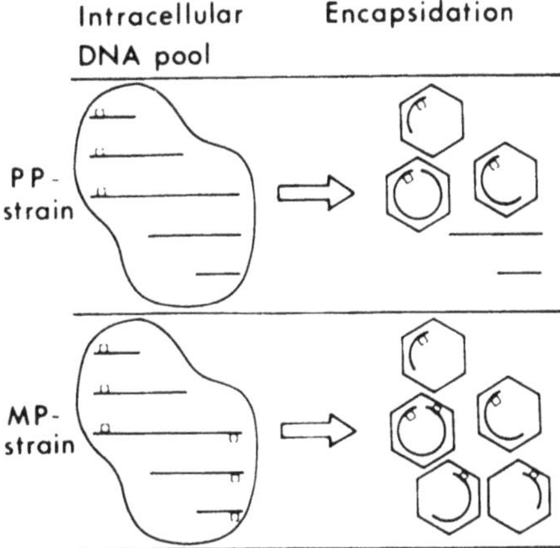

FIGURE 4. Schematic representation of the difference between a PP and an MP strain of Ad16 during packaging of viral DNA. DNA molecules in infected cells contain both ends in equal proportion, but on encapsidation, viral DNA molecules are recognized from the left-hand end as shown for the PP strain. The MP strain, which contains reduplications of the package-recognition sequence in the right-hand end of the genome, will therefore package both right-hand and left-hand ends of the DNA. Reprinted from Hammarskjöld and Winberg (1980).

is a TATA box and an AUG initiation site within the ITRs of the adenovirus DNA, although no mRNA from this position has been identified. The packaging signal acts only in *cis*, also suggesting that it does not code for a protein (Hammarskjöld and Winberg, 1980). The pool of replicating DNA in infected cells containing both ends in equal concentration, confirming that the recognition sequence, which may span from the left-hand end to nucleotide 390, is really a packaging signal. This signal in the left end of adenovirus DNA mimics the protein-recognition site in the λ phage DNA where a specific sequence close to the left end but outside the cohesive end sequence mediates the binding to proteins that are required to insert λ DNA into the head structure (Feiss *et al.*, 1979). In fact, the cosmids used as cloning vectors in conjunction with the λ *in vitro* packaging system probably also carry this piece of λ DNA (Collins and Hohn, 1978). The identification of this specific signal in adenovirus DNA is essential for designing adenovirus vectors for recombinant DNA in mammalian cells. It would be rewarding to clone this region of the viral DNA and identify the protein or proteins that specifically interact with it. These results also explain the failure to encapsidate cellular DNA into adenovirus empty capsids even if cellular DNA readily associates with empty capsids (Tibbetts and Giam, 1979). The tight binding of DNA to empty capsids may in fact explain why several authors have found cellular DNA in defective particles (Tija *et al.*, 1977; Khittoo and Weber, 1981).

We may conclude that adenovirus DNA is inserted into preformed capsids preferentially through the left end and that the recognition signal for the insertion is located within the first 400 nucleotides of the genome. Most defective particles observed are probably intermediates in assembly that have varying amounts of viral DNA associated with the capsid structure. Particles that contain DNA with small deletions in the genomes can, however, go through a complete assembly process.

VI. PROTEOLYTIC CLEAVAGE DURING MATURATION

Studies with one temperature-sensitive (*ts*) mutant, H2*ts*1, established the requirement for proteolytic cleavage in virion maturation (Weber, 1976). Cells infected with H2*ts*1 and maintained at the nonpermissive temperature accumulate virion structures, but none of the precursor polypeptides pVI, pVII, and pVIII is processed in the virion. This mutation obviously blocks the appearance of several posttranslational products, such as polypeptides VI–VIII and X–XIII, that normally accumulate in mature virions. The proteolytic cleavages necessary to generate the mature virion proteins are inhibited by temperature shift-up, but occur after temperature shift-down of cells infected with the mutant. The particles that accumulate at the nonpermissive temperature in cells infected by H2*ts*1 are not infectious, although they contain the DNA and the DNA retains infectivity. This single adenovirus *ts* mutant led to the identification of a separate step in the maturation pathway called *young virions* (Ishibashi and Maizel, 1974). Young virions must be exposed to an internal endopeptidase to convert them into infectious mature virions. The gene affected in H2*ts*1 has probably been identified by sequence analysis (Kruijer *et al.*, 1980; Akusjärvi *et al.*, 1981) following its localization by restriction endonuclease analysis of intertypic ts^+ recombinants (Hassell and Weber, 1978). It is located in the L3 region in the late transcription unit around coordinate 60 and beyond the coding sequence for the hexon. The endopeptidase appears to be around 23K in molecular weight, but the scarcity of the mRNA and the product has precluded its purification from infected cells. The enzymatic activity has, however, been detected in nuclei from infected cells and inside mature virions, but not in empty capsids (Bhatti and Weber, 1979). The mRNA for the endopeptidase is present at a concentration 1000-fold lower than the hexon mRNA, which contains in a nontranslatable form at its 3′ terminus the encoded sequences for the endopeptidase. The H2*ts*103 mutant may also be a protease mutant (D'Halluin *et al.*, 1980a, 1982).

Recently, it has been shown that several other viral proteins are processed by the endopeptidase during the final step in maturation. The terminal protein covalently linked to the DNA appears *in vivo* as an 80K species that is cleaved during maturation of the virion into the 55K species of infectious virions (Challberg *et al.*, 1980). In fact, the *in vitro* translation product of this protein, derived from the E2b region, is 87K, and in this form it accumulates in H2*ts*1 virions at the nonpermissive temperature (Lewis and Mathews, 1980; Stillman *et al.*, 1981). This terminal protein, which appears to have a role in initiation of DNA replication, is covalently linked to DNA in infected cells in the 80K form during the peak of DNA replication; it must, however, also be functional in DNA replication in its 55K form linked to the parental DNA.

Polypeptide IIIa appears to undergo proteolytic cleavage from a 67 to a 66K protein during the final steps of maturation of the virion (Boudin

TABLE I. Protease Cleavage Sites in Adenovirus Type 2 Precursor Proteins[a]

Protein	Molecular weight	Location of Gly-Ala residues
pVI	27,000	SerAsnMetSerGlyGly-AlaPheSerTrpGlySer
pVII	20,457	(1...Gly_3......20)- AlaLysLysArgSerAsp
pVIII	24,707	GlnMetThrAsnSerGly-AlaGlnLeuAlaGlyGly

[a] Modified from Tremblay *et al.* (1983). The sequences were obtained from pVI (Akusjärvi and Persson, 1981), pVII (Sung *et al.*, 1977), and pVIII (Hérissé *et al.*, 1980). The proteins are probably cleaved at the hyphenated Gly-Ala residues.

et al., 1980). To identify all proteolytic cleavages, it might be worthwhile to sequence the minor polypeptides X–XII, which probably constitute the cleaved-off portions of several of the processed precursors. Since most of the DNA sequence of the adenovirus genome is now available (Flint and Broker, 1981), it should be possible with this indirect approach to identify the sites on the genome at which proteolytic cleavage occurs during maturation. It can already be deduced directly from the DNA or protein sequences that the recognition signal for the viral endopeptidase probably involves a Gly–Ala bond that is cleaved in pVI, pVII, and pVIII as shown in Table I (Tremblay *et al.*, 1983).

VII. ARREST OF THE ASSEMBLY PATHWAY

Almost all temperature-sensitive (*ts*) mutants that affect proteins in the L2, L3, and L4 regions of the late transcription unit, except the H2*ts*1 mutated in the protease (L3 region), fail to accumulate empty capsids and assembly intermediates (Edvardsson *et al.*, 1978). Some mutants defective in assembling the capsomers, however, such as H2*ts*3, do process some precursor proteins (Weber *et al.*, 1977). Temperature-sensitive mutants in proteins from the L1 and L5 regions, on the other hand, form empty capsids, but do not process precursors (Khittoo and Weber, 1977; Edvardsson *et al.*, 1978). The L1 region encodes only polypeptide IIIa, and the L5 region appears to encode only the fiber (IV) (Miller *et al.*, 1980), of the structural proteins. Assembly of adenoviruses is thus impaired drastically in many *ts* mutants from several different areas of the genome.

Assembly is also severely inhibited if infected cells are maintained in a medium that lacks Arg (Bonifas and Schlesinger, 1959; Rouse and Schlesinger, 1967) or at high temperatures (Leibowitz and Horwitz, 1974). In the absence of Arg, all major structural proteins of the virion including the Arg-rich core protein are made, but in somewhat smaller amounts (Everitt *et al.*, 1971; Winters and Russell, 1971; Prage and Rouse, 1976). DNA replication is also essentially normal, but the yield of virus is 3 orders of magnitude lower than in the presence of Arg (Rouse and Schlesinger, 1967; Russell and Becker, 1968). In the absence of Ile, on the other hand, a similar reduction in synthesis for most of the viral proteins is observed, but a much higher yield of virus is obtained (Heilman and Rouse, 1980). The Arg-sensitive function, therefore, probably does not involve protein synthesis, but affects a subtle reaction during maturation for which Arg may be required in catalytic amounts.

At high temperature (42°C), the yield of infectious virus is decreased by 2 orders of magnitude, but viral DNA and mRNA are synthesized at faster rates than at 37°C even if the proportion between the polypeptides might vary. Although the formation of hexon trimers is clearly temperature-sensitive, there also appears to be a block in assembly at the higher temperature (Okubo and Raskas, 1971; Leibowitz and Horwitz, 1974). The dramatic effect of Arg starvation and a number of *ts* mutants on adenovirus assembly suggest that the process is a coordinated activity in which all the structural units must participate in the correct proportions to achieve assembly.

In fact, the only *ts* mutants that will accumulate assembly intermediates are those defective in fiber or the IIIa proteins, which can form empty capsids but not pass beyond this step. The changes of developing an *in vitro* system under these conditions are low, and more insight into the assembly pathway might be achieved only by identifying the events that are affected during defective assembly, e.g., during Arg deprivation, or by identifying the specific defects in the *ts* polypeptides. It is almost mandatory, however, that an *in vitro* system be developed if the details of the assembly pathway are to be understood.

VIII. INTERMEDIATE PARTICLES IN ASSEMBLY

The true adenovirus assembly intermediates are fragile, and the DNA in the nucleoprotein is easily sheared by sonication or even deproteinized by salt during extraction from the cells or during isodensity centrifugation in CsCl gradients (Edvardsson *et al.*, 1976). When the extruding nucleoprotein has been removed by shearing, the assembly intermediates are recovered in CsCl gradients at densities that differ depending on the amount of DNA that has been encapsidated (Tibbets, 1977) (Fig. 5). After fixation by reversible cross-linkers, it is possible to isolate the intermediates in CsCl gradients and to investigate their polypeptide composition after reversion of the cross-linkage (D'Halluin *et al.*, 1978a). By sucrose-gradient centrifugation, which preserves the intermediates better, two assembly intermediates sedimenting at 750 and around 600 S, respectively, can be isolated from cells infected with several adenovirus types. Following reversible cross-linking, the 750 S species, which accumulates in a sharp peak, has the properties of complete virions with a density of 1.34 g/cm^3 (D'Halluin *et al.*, 1978a). It contains viral DNA with a sedimentation coefficient of 34 S in alkaline sucrose gradients and a polypeptide complement that is similar to that identified in young virions,

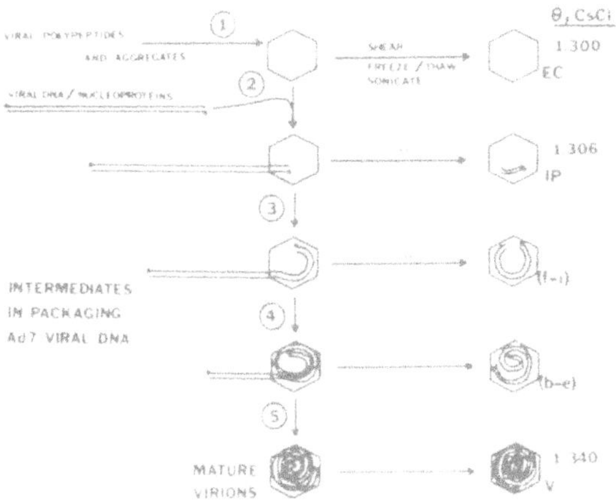

FIGURE 5. Packaging of adenovirus DNA. During interaction of viral DNA with empty capsids, the left-hand end of the viral genome is preferred, and the viral DNA or its nucleoproteins are inserted gradually into the empty capsids as shown in steps 1–5. When cells are broken by shearing, incomplete particles are observed corresponding to the different density classes encountered in CsCl gradients ranging in density from empty capsids at 1.30 g/cm^3 to virions at 1.34 g/cm^3. Incomplete particles are frequently observed among the subgroup B human adenoviruses, which yield large quantities of such particles. Whether the naked DNA or the nucleoproteins are inserted into the particles has not yet been resolved. A heavy intermediate (1.37 g/cm^3) particle has been observed (D'Halluin *et al.*, 1978a) when shearing forces are kept at a minimum, but on the other hand, high salt concentrations were then used to extract the virions from the nuclei of the cells. High concentrations of salt may dissociate some proteins from the DNA. Reprinted from Tibbetts (1977).

i.e., with all the precursor polypeptides in the uncleaved form. Both mature and young virions will sediment at this density, and they can therefore be differentiated only by SDS-PAGE (Ishibashi and Maizel, 1974). The 600 S peak of intermediates is heterogeneous on sucrose gradients and after cross-linking is resolved on CsCl gradients into two components with densities of 1.315 and 1.37 g/cm^3. The lighter class is associated with DNA that sediments at 11 S in alkaline gradients, and it contains precursor polypeptides pVI and pVIII, but none of the core polypeptides pV and pVII. The association of 11 S instead of full-length DNA with the light intermediates may be the result of DNA shearing. The heavier intermediates contain full-length viral DNA, but again no core polypeptides. The lighter particles contain 50, 40, and 32–33K proteins believed to be scaffolding proteins necessary for forming the empty capsids (D'Halluin *et al.*, 1978a; Edvardsson *et al.*, 1976). The 50K protein is the IVa_2 adenovirus gene product (Persson *et al.*, 1979), but the other two presumptive scaffolding proteins have not yet been shown to be virally coded. In kinetic labeling experiments, the light intermediate is detected before the heavy intermediate, and the label can also be chased partially into the heavier form (D'Halluin *et al.*, 1978a). These results suggest that DNA

FIGURE 6. EM structure of an assembly intermediate from Ad5-infected cells. The virus particles were extracted with 0.2 M ammonium sulfate, and the assembly intermediates were separated by sucrose-gradient centrifugation and analyzed in the EM. A single-stranded DNA (ss^d) covered with a coat of the E2 72K DNA-binding protein is shown on a viral DNA structure that appears to be associated with an empty capsid. Replication thus appears to proceed at one end of the DNA molecule, while encapsidation occurs at the other end, suggesting a coupled DNA replication–packaging process in the formation of adenoviruses. This result supports the notion that DNA is packaged into the virion before the core proteins. (PM2) marker phage DNA; (ds^d) double-stranded DNA. Rephotographed from Moncany *et al.* (1980).

can associate with empty capsids before it associates with the core proteins, which may complex with the encapsidated DNA only later. In a rather similar study, however, intermediates were formed that contained different amounts of DNA in all cases associated with the core proteins (Edvardsson *et al.*, 1976). The difference between these two experiments was that 0.3 M ammonium sulfate was used to extract the intermediates the DNA of which lacked core proteins, while when sonication was used, the core proteins were found associated with the DNA. It is therefore still not resolved whether the DNA enters preformed empty capsids together with or before the core proteins. The salt resistance of the bonds between protein VII and DNA has recently been investigated (Black and Center, 1979), suggesting that the initial interaction is salt-sensitive but that the complex resists 3 M NaCl. Similar studies on the interaction between the precursor pVII and DNA have not been performed, but such studies would help to establish whether the intermediates without core proteins are artifacts. The identification by EM of an intermediate in adenovirus assembly (Moncany *et al.*, 1980) that comprises a viral particle directly associated with replicating DNA (Fig. 6) provides some evidence for the direct insertion of DNA into the empty capsid. If so, it would explain the observation (D'Halluin *et al.*, 1980b) that novobiocin, which inhibits initiation but not prolongation of DNA replication, also blocks adenovirus DNA encapsidation. On the other hand, since only a very small fraction of the newly replicated viral DNA in the cells is packaged into virions (Green, 1962), it remains an open question whether or not the DNA is covered with core proteins at the time of entry.

The protein composition of the intermediates suggests that all the structural units are present, but it is difficult to determine quantitatively whether the full complement of the vertex proteins has been inserted (Edvardsson *et al.*, 1976). None of the precursor polypeptides pVI, pVII, pVIII, and pIIIa or the DNA terminal protein is processed in the intermediates, and they are still uncleaved in the young virions that accumulate in the nucleus of the infected cell (Ishibashi and Maizel, 1974; Edvardsson *et al.*, 1976; Weber, 1976; Boudin *et al.*, 1980; Stillman *et al.*, 1981).

Antigenic analysis of the intermediates during virus assembly suggests that several polypeptides in the vertex region (III, IIIa, and IV) are not recognized by antibodies to mature structural units until the final maturation involving proteolytic cleavage has occurred (D'Halluin *et al.*, 1980a). Attempts to verify the tentative assembly pathway shown in Fig. 7 have been disappointing (Khittoo and Weber, 1977; Edvardsson *et al.*, 1978; D'Halluin *et al.*, 1978b) for two reasons: first, we lack precise knowledge about the protein affected by the *ts* mutation in most cases, and some of the *ts* mutants might be double mutants; second, among several complementation groups of Ad5 *ts* mutants, only a few *ts* mutants accumulate intermediates but little or no young or mature virions at the nonpermissive temperature (Edvardsson *et al.*, 1978). Temperature-sensitive mutants from the fiber gene (Williams *et al.*, 1974) accumulated intermediates and a small amount of young and mature virions. In addition, H5*ts*19 and H5*ts*58 accumulated intermediates, but no young or mature virions. The H5*ts*19 may map in the region for the pVIII polypeptide, the L4 region, and H5*ts*58 has been unequivocally located in the region for the IIIa polypeptide, the L1 region, of the Ad5 genome (Frost and Williams, 1978). Shift-down experiments revealed, however, that the intermediates formed at the nonpermissive temperature are thermolabile, and only a fraction of them can be utilized for virus assembly after a temperature shift (Edvardsson *et al.*, 1978). Experiments with the Ad2 *ts* mutant H2*ts*112 have also provided evidence that a fraction of the label incorporated into the light intermediates at the nonpermissive temperature could be chased into the heavy intermediates and then into mature virions under conditions in which protein synthesis was inhibited by cycloheximide (D'Halluin *et al.*, 1978b). The H2*ts*112 mutant has recently also been placed in the polypeptide IIIa region of the virus genome by restriction endonuclease analysis of interserotypic recombinants (D'Halluin *et al.*, 1982). An arrest of assembly at the level of empty capsids has also been observed after infection at the nonpermissive temperature with the H2*ts*4 mutant, this probably also being located in the polypeptide IIIa gene region (Khittoo and Weber, 1977). Thus, only *ts* mutants in the protein IIIa region are stable enough to demonstrate partial assembly

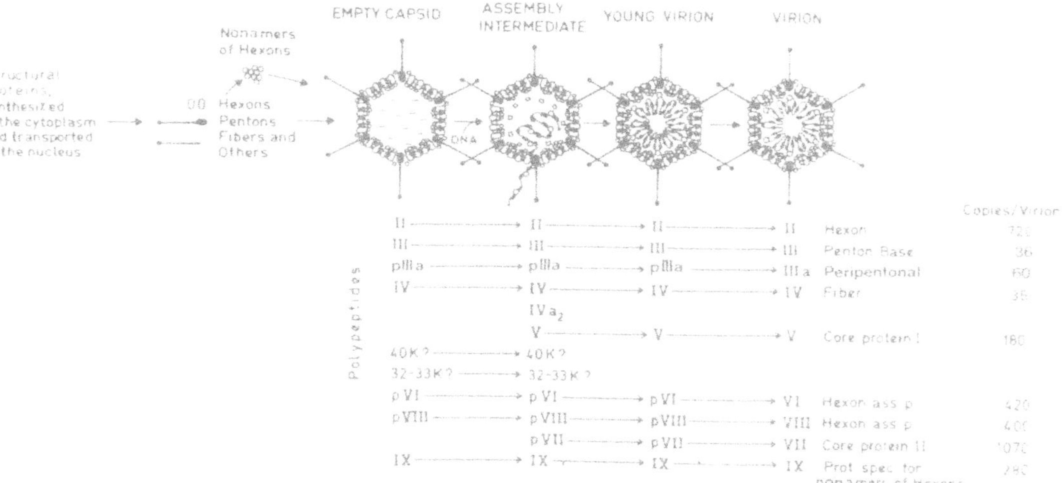

FIGURE 7. Tentative assembly pathway for adenoviruses. The structural proteins are synthesized in the cytoplasm and rapidly transported to the nucleus, probably in the form of a trimer for the main capsid units: hexon, penton base, and fiber. From nonamers of hexon, empty capsids are possibly formed by a process of self-assembly. The main assembly intermediates observed will contain varying portions of the DNA or varying amounts of core proteins, depending on the procedure used to extract the particles from the cell, especially the shear force and the salt concentration used to liberate the intermediates. It is not yet resolved whether the DNA is packaged before the core protein, although both a light and a heavy intermediate have been identified (D'Halluin *et al.*, 1978a). The next step is the young virions, in which several polypeptides are present in a precursor form, followed by the final step involving proteolytic cleavage of at least five polypeptides in the virion structure. The four stages of virus assembly are indicated at the top and the polypeptides present in each structure below; the numbers of copies of the polypeptides per virion are indicated at right below.

in shift-down experiments. More recently, D'Halluin *et al.* (1980a) have done shift-down experiments with two fiber-defective mutants that also suggest that a fraction of the light intermediates can chase into mature virions. In this case, however, it appears that the mutation in the fiber created a deletion of the polypeptide that reduced its molecular weight by 2000. The arrest of the H2*ts*1 mutant (Weber, 1976) at the stage of young virions and the possibility of chasing the young virions into mature virions after shift-down was essential, however, for the identification of one intermediate step in virus assembly.

In conclusion, because of the many polypeptides in the adenovirus virion, the multiple cmRNAs from each region, and the heterogeneous effects of *ts* mutations on the expression of viral proteins, the *ts* mutants have not been as useful as the corresponding mutants of bacterial viruses in clarifying the assembly pathway. On the other hand, when several more of the available *ts* mutations have been mapped precisely on the viral genome, we will probably be able to reinterpret the accumulated results in a more comprehensible manner.

IX. CONCLUSION

The assembly of adenoviruses is a complex process involving first the assembly of the individual polypeptide chains for the major capsomers into trimeric units, i.e., for the hexon, the fiber, and the penton base. From these, the capsid structure is then generated in a process involving four to five additional structural polypeptides and probably three scaffolding proteins, of which only one has been identified as a virus-gene product (IVa_2). The assembly pathway is difficult to study by pulse–chase kinetics because of the large pools of the structural subunits inside the nuclei of infected cells, where only 10–20% of the structural polypeptides assemble into virions. At least five but probably more of the virion polypeptides are present as precursor proteins (pVI, pVII, pVIII, pIIIa, and the DNA terminal protein), and several classes of particles accumulate in the cell, all appearing to be assembly intermediates on the pathway to mature virions. A tentative diagram of the assembly pathway is shown in Fig. 7. After the formation of the structural units and their aggregate substructures, the empty capsids are formed and appear to contain all the virion proteins except for the core proteins. In addition, two polypeptides, 40 and 32–33K, of viral or host origin are associated with the empty capsids and may be scaffolding proteins. The DNA is then inserted into the empty capsid via a specific recognition sequence located in the first 400 nucleotides at the left-hand end of the genome. Protein IVa_2, encoded on the l-strand of the adenovirus genome, may be inserted together with the DNA and may therefore recognize the packaging signal on the DNA. It is still unresolved, however, whether the DNA enters with or without the core proteins. A separate step for the insertion of the core proteins may therefore be envisaged. The young virions, the next stage in the assembly, have lost the scaffolding proteins and the IVa_2 polypeptide, but all the precursor polypeptides are present in an uncleaved form. After maturation, which primarily involves proteolytic cleavage, probably of a Gly–Ala bond, the virion is infectious and the cycle is complete. A more detailed knowledge of the locations on the genome of the temperature-sensitive mutations in the mutants able to form assembly intermediates, and a more precise mapping of all the proteins involved in the late transcription unit of the adenovirus genome, are required before we can design specific reconstruction experiments. An *in vitro* system for assembly is badly needed to dissect the molecular details of this pathway.

ACKNOWLEDGMENTS. I would like to thank Dr. John Tooze for invaluable help with the editing of this contribution and Mrs. Waltraud Ackerman and Ms. Nelly van der Jagt for expert secretarial assistance. I am also greatful to several colleagues and friends who have contributed comments and preprints.

REFERENCES

Akusjärvi, G., and Persson, H., 1981, Gene and mRNA for precursor polypeptide VI from adenovirus type 2, *J. Virol.* **38**:469.

Akusjärvi, G., Zabielski, J., Perricaudet, M., and Pettersson, U., 1981, The sequence of the 3′ non-coding region of the hexon mRNA discloses a novel adenovirus gene, *Nucleic Acids Res.* **9**:1.

Anderson, C.W., Baum, P.R., and Gesteland, R.F., 1973, Processing of adenovirus 2 induced proteins, *J. Virol.* **12**:241.

Baker, C., and Ziff, E., 1981, Promoters and heterogeneous 5′ termini of the messenger RNAs of adenoviruses serotype 2, *J. Mol. Biol.* **149**:189.

Bhatti, A.R., and Weber, J., 1979, Protease of adenovirus type 2, *J. Biol. Chem.* **254**:1265.

Black, B.C., and Center, M.S., 1979, DNA-binding properties of the major core protein of adenovirus 2, *Nucleic Acids Res.* **6**:2329.

Bonifas, V., and Schlesinger, R.W., 1959, Nutritional requirement for plaque production by adenovirus, *Fed. Proc. Fed. Am. Soc. Exp. Biol.* **18**:500.

Boudin, M.-L., and Boulanger, P., 1982, Assembly of adenovirus penton base and fiber, *Virology* **116**:589.

Boudin, M.-L., Moncany, M., D'Halluin, J.-C., and Boulanger, P.A., 1979, Isolation and characterization of adenovirus type 2 vertex capsomer (penton base), *Virology* **92**:125.

Boudin, M.-L., D'Halluin, J.-C., Cousin, C., and Boulanger, P., 1980, Human adenovirus type 2 protein IIIa. II. Maturation and encapsidation, *Virology* **101**:144.

Boulanger, P.A., 1975, Adenovirus assembly: Self-assembly of partially digested hexons, *J. Virol.* **16**:1678.

Boulanger, P., Lemay, P., Blair, G.E., and Russell, W.C., 1979, Characterization of adenovirus protein IX, *J. Gen. Virol.* **44**:783.

Brown, M., and Weber, J., 1980, Virion core-like organization of intranuclear adenovirus chromatin late in infection, *Virology* **107**:306.

Brown, M., and Weber, J., 1982, Discrete subgenomic DNA fragments in incomplete particles of adenovirus type 2, *J. Gen. Virol.* **62**:81.

Brown, D.T., Westphal, M., Burlingham, B.T., Winterhoff, U., and Doerfler, W., 1975, Structure and composition of the adenovirus type 2 core, *J. Vriol.* **16**:366.

Burlingham, B., Doerfler, W., Pettersson, U., and Philipson, L., 1971, Adenovirus endonuclease: Association with the penton of adenovirus type 2, *J. Mol. Biol.* **60**:45.

Burlingham, B.T., Brown, D.T., and Doerfler, W., 1974, Incomplete particles of adenovirus. I. Characteristics of the DNA associated with incomplete adenovirions of types 2 and 12, *Virology* **60**:419.

Cepko, C.L., and Sharp, P.A., 1982, Assembly of adenovirus major capsid protein is mediated by a non-virion protein, *Cell* **31**:407.

Challberg, M.D., Desiderio, S.V., and Kelly, T.J., Jr., 1980, Adenovirus DNA replication *in vitro*: Characterization of a protein covalently linked to nascent DNA strands, *Proc. Natl. Acad. Sci. U.S.A.* **77**:5105.

Cheng Chee-Sheung, C., and Ginsberg, H.S., 1982, Characterization of a temperature-sensitive fiber mutant of type 5 adenovirus and effect of the mutation on virion assembly, *J. Virol.* **42**:932.

Colby, W.W., and Shenk, T., 1981, Adenovirus type 5 virions can be assembled *in vivo* in the absence of detectable polypeptide IX, *J. Virol.* **39**:977.

Collins, J., and John, B., 1978, Cosmids: A type of plasmid gene-cloning vector that is packageable *in vitro* in bacteriophage lambda heads, *Proc. Natl. Acad. Sci. U.S.A.* **75**:4242.

Corden, J., Engelking, H.M., and Pearson, G.D., 1976, Chromatin-like organization of the adenovirus chromosome, *Proc. Natl. Acad. Sci. U.S.A.* **73**:401.

Crick, F.H.C., and Watson, J.D., 1957, The structure of small viruses, *Nature (London)* **177**:473.

Crowther, R.A., and Franklin, R.M., 1972, The structure of the groups of nine hexons from adenovirus, *J. Mol. Biol.* **68**:181.

Daniell, E., 1976, Genome structure of incomplete particles of adenovirus, *J. Virol.* **19**:685.

Daniell, E., and Mullenbach, T., 1978, Synthesis of defective viral DNA in HeLa cells infected with adenovirus type 3, *J. Virol.* **26**:61.

Daniell, E., Groff, D.E., and Fedor, M.J., 1981, Adenovirus chromatin structure at different stages of infection, *Mol. Cell. Biol.* **1**:1094.

Devaux, C., and Boulanger, P., 1980, Reactive serine in human adenovirus hexon polypeptide, *Virology* **102**:94.

Devaux, C., Zulauf, M., Boulanger, P., and Jacrot, B., 1982, Molecular weight of adenovirus serotype 2 capsomers: A new characterization, *J. Mol. Biol.* **156**:927.

Devaux, C., Berthet–Colominas, C., Timmins, P.A., and Jacrot, B., 1984, Crystal packing and stoichiometry of the fiber protein adenovirus type 2, *J. Mol. Biol.* (in press).

D'Halluin, J.-C., Martin, G.R., Torpier, G., and Boulanger, P.A., 1978a, Adenovirus type 2 assembly analysed by reversible cross-linking of labile intermediates, *J. Virol.* **26**:357.

D'Halluin, J.-C., Milleville, M., Boulanger, P.A., and Martin, G.R., 1978b, Temperature-sensitive mutant of adenovirus type 2 blocked in virion assembly: Accumulation of light intermediate particles, *J. Virol.* **26**:344.

D'Halluin, J.-C., Milleville, M., Martin, G.R., and Boulanger, P., 1980a, Morphogenesis of human adenovirus type 2 studied with fiber- and fiber and penton base-defective temperature-sensitive mutants, *J. Virol.* **33**:88.

D'Halluin, J.-C., Milleville, M., and Boulanger, P., 1980b, Effects of novobiocin on adenovirus DNA synthesis and encapsidation, *Nucleic Acids Res.* **8**:1625.

D'Halluin, J.-C., Cousin, C., and Boulanger, P., 1982, Physical mapping of adenovirus type 2 temperature-sensitive mutations by restriction endonuclease analysis of interserotypic recombinants, *J. Virol.* **41**:401.

Dorsett, P.H., and Ginsberg, H.S., 1975, Characterization of an adenovirus type 5 fiber protein, *J. Virol.* **15**:208.

Edgar, R.S., and Wood, W.B., 1966, Morphogenesis of bacteriophage T_4 in extracts of mutant-infected cells, *Proc. Natl. Acad. Sci. U.S.A.* **55**:498.

Edvardsson, B., Everitt, E., Jörnvall, H., Prage, L., and Philipson, L., 1976, Intermediates in adenovirus assembly, *J. Virol.* **19**:533.

Edvardsson, B., Ustaçelebi, S., Williams, J., and Philipson, L., 1978, Assembly intermediates among adenovirus type 5 temperature-sensitive mutants, *J. Virol.* **25**:641.

Everett, S.F., and Ginsberg, H.S., 1958, A toxin-like material separable from type 5 adenovirus particles, *Virology* **6**:770.

Everitt, E., and Philipson, L., 1974, Structural proteins of adenoviruses. XI. Purification of three low molecular weight virion proteins of adenovirus type 2 and their synthesis during productive infection, *Virology* **62**:253.

Everitt, E., Sundquist, B., and Philipson, L., 1971, Mechanism of arginine requirement for adenovirus synthesis. I. Synthesis of structural proteins, *J. Virol.* **8**:742.

Everitt, E., Sundquist, B., Pettersson, U., and Philipson, L., 1973, Structural proteins of adenoviruses. X. Isolation and topography of low molecular weight antigens from the virion of adenovirus type 2, *Virology* **62**:130.

Everitt, E., Lutter, L., and Philipson, L., 1975, Structural proteins of adenoviruses. XII. Location and neighbour relationship among proteins of adenovirus type 2 as revealed by enzymatic iodination, immunoprecipitation and chemical cross-linking, *Virology* **67**:197.

Feiss, M., Fisher, R.A., Siegele, D.A., Nichols, B.P., and Donelson, J.E., 1979, Packaging of the bacteriophage lambda chromosome: A role for the base sequences outside *cos*, *Virology* **92**:56.

Flint, S.J., and Broker, T.R., 1981, Lytic infection by adenoviruses, in: *The Molecular Biology of Tumor Viruses, DNA Tumor Viruses,* 2nd rev. ed. (J. Tooze, ed.), pp. 443–546, Cold Spring Harbor Press, Cold Spring Harbor, New York.

Frost, E., and Williams, J., 1978, Mapping temperature-sensitive and host-range mutations of adenovirus type 5 by marker rescue, *Virology* **91**:39.

Gallimore, P.H., Sharp, P.A., and Sambrook, J., 1974, Viral DNA in transformed cells. II. A study of the sequences of adenovirus 2 DNA in nine lines of transformed rat cells using specific fragments of the viral genome, *J. Mol. Biol.* **89**:49.

Garon, C.F., Gerry, K.W., and Rose, J.A., 1972, A unique form of terminal redundancy in adenovirus DNA molecules, *Proc. Natl. Acad. Sci. U.S.A.* **69**:2391.

Ginsberg, H.S., Pereira, H.G., Valentine, R.C., and Wilcox, W.C., 1966, A proposed terminology for the adenovirus antigens and virion morphological subunits, *Virology* **28**:782.

Graham, F.L., van der Eb, A.J., and Heijneker, H.L., 1974, Size and location of the transforming region of human adenovirus type 5 DNA, *Nature (London)* **251**:687.

Green, M., 1962, Studies on the biosynthesis of viral DNA. IV. Isolation, purification and chemical analysis of adenovirus, *Cold Spring Harbor Symp. Quant. Biol.* **27**:219.

Green, N.M., Wrigley, N.G., Russell, W.C., Martin, S.R., and McLachlan, A.D., 1983, Evidence for a repeating cross B-sheet structure in the adenovirus fibre, *EMBO Journal* **2**:1357.

Grütter, M., and Franklin, R.M., 1974, Studies on the molecular weight of the adenovirus type hexon and its subunits, *J. Mol. Biol.* **89**:163.

Hammarskjöld, M.-L., and Winberg, G., 1980, Encapsidation of adenovirus 16 DNA is directed by a small DNA sequence at the left end of the genome, *Cell* **20**:787.

Hammarskjöld, M.-L., and Winberg, G., Norrby, E., and Wadell, G., 1977, Isolation of incomplete adenovirus 16 particles containing viral and host cell DNA, *Virology* **82**:449.

Hassell, J.A., and Weber, J., 1978, Genetic analyses of adenovirus type 2. VIII. Physical locations of temperature sensitive mutations, *J. Virol.* **28**:671.

Heilman, C.A., and Rouse, H., 1980, Adenovirus type 2 polypeptide synthesis in arginine-deprived cells, *Virology* **105**:159.

Hérissé, J., Courtois, G., and Galibert, F., 1980, Nucleotide sequences of the EcoRI-D fragment of the adenovirus 2 genome, *Nucleic Acids Res.* **8**:2173.

Horwitz, M.S., Scharff, M.D., and Maizel, J.V., 1969, Synthesis and assembly of adenovirus 2. I. Polypeptide synthesis, assembly of capsomers and morphogenesis of the virion, *Virology* **39**:682.

Igarishi, K., Niiyama, Y., Tsukamoto, K., Kurokawa, T., and Sugino, Y., 1975, Biochemical studies on bovine adenovirus type 3. II. Incomplete virus, *J. Virol.* **16**:634.

Ishibashi, M., and Maizel, J.V., 1974, The polypeptides of adenovirus. V. Young virions, structural intermediates between top components and aged virions, *Virology* **57**:409.

Jörnvall, H., Akusjärvi, G., Aleström, P., von Bahr-Lindström, H., Petersson, U., Apella, E., Fowler, A., and Philipson, L., 1981, The adenovirus hexon protein: The primary structure of the polypeptide and its correlation with the hexon gene, *J. Biol. Chem.* **256**:6181.

Khittoo, G., and Weber, J., 1977, Genetic analysis of adenovirus type 2. VI. A temperature-sensitive mutant defective for DNA encapsidation, *Virology* **81**:126.

Khittoo, G., and Weber, J.M., 1981, The nature of the DNA associated with incomplete particles of adenovirus type 2, *J. Gen. Virol.* **54**:343.

Kruijer, W., Van Schiak, F.M.A., and Sussenbach, J.S., 1980, Nucleotide sequence analysis of a region of adenovirus 5 DNA encoding a hitherto unidentified gene, *Nuleic Acids Res.* **8**:6033.

Laver, W.G., Wrigley, N.G., and Pereira, H.G., 1969, Removal of pentons from particles of adenovirus type 2, *Virology* **38**:579.

Leibowitz, J., and Horwitz, M.S., 1974, Synthesis and assembly of adenovirus type 2 polypeptides. II. Reversible inhibition of hexon assembly at 42°, *Virology* **61**:129.

Leibowitz, J., and Horwitz, M.S., 1975, Synthesis and assembly of adenovirus polypeptides. III. Reversible inhibition of hexon assembly in adenovirus type 5 temperature-sensitive mutants, *Virology* **66**:10.

Lewis, J.B., and Mathews, M., 1980, Control of adenovirus early gene expression: A class of immediate early products, *Cell* **21**:303.

Lischwe, M.A., and Sung, M.T., 1977, A histone-like protein from adenovirus chromatin, *Nature (London)* **267**:552.

Maizel, J.V., Jr., White, D.O., and Scharff, M.D., 1968, The polypeptides of adenovirus. I. Evidence for multiple protein components in the virion and a comparison of types 2, 7, and 12, *Virology* **36**:115.

Mak, S., 1971, Defective virions in human adenovirus type 12, *J. Virol.* **7**:426.

Mak, I., Ezo, H., and Mak, S., 1979, Structure and function of adenovirus type 12 defective virions, *J. Virol.* **32**:240.

Marusyk, R.G., Morgan, A.R., and Wadell, G., 1975, Association of endonuclease activity with serotypes belonging to the three subgroups of human adenoviruses, *J. Virol.* **16**:456.

Miller, J.S., Ricciardi, R.P., Roberts, B.E., Patterson, B.M., and Mathews, M.B., 1980, Arrangement of messenger RNAs and protein coding sequences in the major late transcription unit of adenovirus 2, *J. Mol. biol.* **142**:455.

Mirza, M.A., and Weber, J., 1982, Structure of adenovirus chromatin, *Biochim. Biophys. Acta* **696**:76.

Moncany, M.L.J., Révet, B., and Girard, M., 1980, Characterization of a new adenovirus type 5 assembly intermediate, *J. Gen. Virol.* **50**:33.

Nermut, M.V., 1979, Structural elements in adenovirus cores: Evidence for a "core shell" and linear structures in "relaxed" cores, *Arch. Virol.* **62**:101.

Nermut, M.V., 1980, *The architecture of adenoviruses: Recent views and problems, Arch. Virol.* **64**:175.

Niiyama, Y., Igarishi, K., Tsukamoto, K., Kurokawa, T., and Sugino, Y., 1975, *Biochemical studies on bovine adenovirus type 3. I. Purification and properties, J. Virol.* **16**:621.

Norrby, E., and Wadell, G., 1969, Immunological relationships between hexons of certain human adenoviruses, *J. Virol.* **4**:663.

Öberg, B., Saborio, J., Persson, T., Everitt, E., and Philipson, L., 1975, Identification of the *in vitro* translation products of adenovirus mRNA by immuno-precipitation, *J. Virol.* **15**:199.

Okubo, C.K., and Raskas, H.J., 1971, thermo-sensitive events in the replication of adenovirus type 2 at 42°, *Virology* **46**:175.

Oosterom-Dragon, E.A., and Ginsberg, H.S., 1981, Characterization of two temperature-sensitive mutants of type 5 adenovirus with mutations in the 100,000-dalton protein gene, *J. Virol.* **40**:491.

Pereira, H.G., 1958, A protein factor responsible for the early cytopathic effect of adenoviruses, *Virology* **6**:601.

Pereira, H.G., and Wrigley, N.G., 1974, *In vitro* reconstitution, hexon-bonding and handedness of incomplete adenovirus capsids, *J. Mol. Biol.* **85:**617.

Persson, H., Öberg, B., and Philipson, L., 1977, *In vitro* translation with adenovirus polyribosomes, *J. Virol.* **21:**187.

Persson, H., Mathiesen, B., Philipson, L., and Pettersson, U., 1979, A maturation protein in adenovirus morphogenesis, *Virology* **93:**198.

Pettersson, U., 1971, Structural proteins of adenoviruses. VI. On the antigenic determinants of the hexon, *Virology* **43:**123.

Pettersson, U., and Höglund, S., 1969, Structural proteins of adenoviruses. III. Purification and characterization of the adenovirus type 2 penton antigen, *Virology* **39:**90.

Philipson, L., 1979, Adenovirus proteins and their messenger RNAs, *Adv. Virus Res.* **25:**357.

Philipson, L., and Lindberg, U., 1974, Reproduction of adenoviruses, in: *Comprehensive Virology* Vol. 3 (Fraenkel-Conrat and R.R. Wagner, eds.), p. 143, Plenum Press, New York.

Philipson, L., Lonberg-Holm, K., and Petterson, U., 1968, Virus receptor interaction in an adenovirus system, *J. Virol.* **2:**1064.

Prage, L., and Pettersson, U., 1971, Structural proteins of adenoviruses. VII. Purification and properties of an arginine-rich core protein from adenovirus type 2 and type 3, *Virology* **45:**364.

Prage, L., and Rouse, H.C., 1976, Effect of arginine starvation on macromolecular synthesis in infection with type 2 adenovirus. III. Immunofluorescence studies of the synthesis of the hexon and the major core antigen (AAP), *Virology* **69:**352.

Prage, L., Pettersson, U., and Philipson, L., 1968, Internal base proteins in adenovirus, *Virology* **36:**508.

Prage, L., Pettersson, U., Höglund, S., Lonberg-Holm, K., and Philipson, L., 1970, Structural proteins of adenoviruses. IV. Sequential degradation of the adenovirus type 2 virion, *Virology* **42:**341.

Prage, L., Höglund, S., and Philipson, L., 1972, Structural proteins of adenoviruses. III. Characterization of incomplete particles of adenovirus type 3, *Virology* **49:**745.

Reif, U.M., Winterhoff, U., Lundholm, U., Philipson, L., and Doerfler, W., 1977, Purification of an endonuclease from adenovirus-infected KB cells, *Eur. J. Biochem.* **73:**313.

Rekosh, D.M.K., Rusell, W.C., Bellett, A.J.D., and Robinson, A.J., 1977, Identification of a protein linked to the ends of adenovirus DNA, *Cell* **11:**283.

Rosenwirth, B., Tija, S., Westphal, M., and Doerfler, W., 1974, Incomplete particles of adenovirus. II. Kinetics of formation and polypeptide composition of adenovirus type 2, *Virology* **60:**431.

Rouse, H.C., and Schlesinger, R.W., 1967, An arginine-dependent step in the maturation of type 1 adenovirus, *Virology* **33:**513.

Russell, W.C., and Becker, Y., 1968, A maturation factor for adenovirus, *Virology* **35:**18.

Russell, W.C., Valentine, R.C., and Pereira, H.G., 1967, The effect of heat on the anatomy of the adenovirus, *J. Gen. Virol.* **1:**509.

Russell, W.C., McIntosh, K., and Skehel, J.J., 1971, The preparation and properties of adenovirus cores, *J. Gen. Virol.* **11:**35.

Sato, K., and Hosokawa, K., 1981, The structure of adenovirion chromatin revealed by ultraviolet light induced cross-linking, *Biochem. Biophys. Res. Commun.* **101:**1318.

Schaller, J.P., and Yohn, D.S., 1974, Transformation potentials of the non-infectious (defective) component in pools of adenoviruses type 12 and simian adenovirus 7, *J. Virol.* **14:**392.

Sergeant, A., Tigges, M.A., and Raskas, H.J., 1979, Nucleosomelike structural subunits of intranuclear parental adenovirus type 2DNA, *J. Virol.* **29:**888.

Shortridge, K.F., and Biddle, F., 1970, The proteins of adenovirus type 5, *Arch. Gesamte Virusforch.* **29:**1.

Smith, K.O., Gehle, W.D., and Trousdale, M.D., 1965, Architecture of the adenovirus capsid, *J. Bacteriol.* **90:**254.

Stansy, J.T., Neurath, A.R., and Rubin, B.A., 1968, Effect of foramide on the capsid morphology of adenovirus types 4 and 7, *J. Virol.* **2**:1429.
Stillman, B.W., Lewis, J.B., Chow, L.T., Mathews, M.B., and Smart, J.E., 1981, Identification of the gene and mRNA for the adenovirus terminal protein precursor, *Cell* **23**:497.
Stinski, M.F., and Ginsberg, H.S., 1974, Antibody to the type 5 adenovirus hexon polypeptide: Detection of nascent polypeptides in cytoplasm of infected KB cells, *Intervirology* **4**:226.
Sundquist, B., Everitt, E., Philipson, L., and Höglund, S., 1973a, Assembly of adenoviruses, *J. Virol.* **11**:449.
Sundquist, B., Pettersson, U., Thelander, L., and Philipson, L., 1973b, Structural proteins of adenoviruses. IX. Molecular weight and subunit composition of adenoviruses type 2 fibre, *Virology* **51**:252.
Sung, M.T., Lischwe, M.A., Richards, J.C., and Hosokawa, K., 1977, Adenovirus chromatin I: Isolation and characterization of the major core protein VII and precursor PRO-VII, *J. Biol. Chem.* **252**:4981.
Tate, V.E., and Philipson, L., 1979, Parental adenovirus DNA accumulates in nucleosome-like structures in infected cells, *Nucleic Acids Res.* **6**:2769.
Tibbetts, C., 1977, Viral DNA Sequences from incomplete particles of human adenovirus type 7, *Cell* **12**:243.
Tibbetts, C., and Giam, C.-Z., 1979, *In vitro* association of empty adenovirus capsids with double-stranded DNA, *J. Virol.* **32**:995.
Tija, S., Fanning, E., Schick, J., and Doerfler, W., 1977, Incomplete particles of adenovirus type 2. III. Viral and cellular DNA sequences in incomplete particles, *Virology* **76**:365.
Tremblay, M.L., Déry, C.V., Talbot, B.G., and Weber, J., 1983, *In vitro* cleavage specificity of the adenovirus type 2 proteinase *Biochim. Biophys. Acta* **743**:239.
Valentine, R.C., and Pereira, H.G., 1965, Antigens and structure of the adenovirus, *J. Mol. Biol.* **13**:13.
van Ormondt, H., Maat, J., and van Bevern, C.P., 1980, The nucleotide sequence of the transforming early region E1 of adenovirus type 5 DNA, *Gene* **11**:299.
van Roy, F., Engler, G., and Fiers, W., 1979, Isolation and characterization of a specific deletion mutant of human adenovirus type 2, *Virology* **96**:486.
Velicer, L.F., and Ginsberg, H.S., 1970, Synthesis, transport and morphogenesis of type 5 adenovirus capsid proteins, *J. Virol.* **5**:338.
Wadell, G., Hammarskjöld, M.L., and Varsanyi, T., 1973, Incomplete particles of adenovirus type 16, *J. Gen. Virol.* **20**:287.
Weber, J., 1976, Genetic analysis of adenovirus type 2. III. Temperature-sensitivity of processing or viral proteins, *J. Virol.* **17**:462.
Weber, J., Begin, M., and Carstens, E.B., 1977, Genetic analysis of adenovirus type 2. IV. Coordinate regulation of polypeptides 80k, IIIa and V, *Virology* **76**:709.
Wilcox, W.C., and Ginsberg, H.S., 1963, Structure of type 5 adenovirus. I. Antigenic relationship of virus structural proteins to virus specific soluble antigens from infected cells, *J. Exp. Med.* **118**:295.
Wilcox, W.C., Ginsberg, H.S., and Anderson, T.F., 1963, Structure of type 5 adenovirus. II. Fine structure of virus subunits, morphologic relationship of structural subunits to virus-specific soluble antigens from infected cells, *J. Exp. Med.* **118**:307.
Wilhelm, J.M., and Ginsberg, H.S., 1972, Synthesis *in vitro* of type 5 adenovirus capsid proteins, *J. Virol.* **9**:973.
Williams, J.F., Young, C.S., and Austin, P.E., 1974, Genetic analysis of human adenovirus type 5 in permissive and non-permissive cells, *Cold Spring Harbor Symp. Quant. Biol.* **39**:427.
Winberg, G., and Wadell, G., 1977, Structural polypeptides of adenovirus type 16 incomplete particles, *J. Virol.* **22**:389.
Winters, W.D., and Russell, W.C., 1971, Studies on the assembly of adenovirus *in vitro*, *J. Gen. Virol.* **10**:181.
Wolfson, J., and Dressler, D., 1972, Adenovirus 2 DNA contains an inverted terminal repetition, *Proc. Natl. Acad. Sci. U.S.A.* **69**:3054.

CHAPTER 9

Transformation by and Oncogenicity of Human Adenoviruses

F. L. GRAHAM

I. INTRODUCTION

Just slightly more than 20 years ago, human adenovirus type 12 (Ad12) gained the distinction of being the first human virus shown to have oncogenic properties when Trentin *et al.* (1962) reported the induction of tumors following injection of the virus into newborn hamsters. Their observations were quickly confirmed and extended to other human adenovirus serotypes and to other rodents (Huebner *et al.*, 1962, 1965; Rabson *et al.*, 1964; Yabe *et al.*, 1964; Pereira *et al.*, 1965), with the result that attention shifted from attempts to develop vaccines against adenovirus infections to attempts to understand the mechanisms by which adenoviruses induce tumors and to determine whether adenoviruses play any role in human malignancy. Although two decades later we still do not know why adenoviruses are oncogenic in rodents or whether they might cause certain tumors in man, adenoviruses have nevertheless become important tools for the study not only of malignant transformation but also of gene expression in mammalian cells generally.

A. Oncogenic and Nononcogenic Adenoviruses

Following the initial studies with Ad12, it soon became clear that not all adenovirus serotypes could induce tumors, and different serotypes

F. L. GRAHAM • Departments of Biology and Pathology, McMaster University, Hamilton, Ontario L85 4K1, Canada.

were classed as nononcogenic, weakly oncogenic, or highly oncogenic, depending on the efficiency with which they induced tumors and the latent period of the malignancy (Huebner, 1967; Trentin *et al.*, 1968; Green, 1970). On the basis of oncogenicity and a variety of additional criteria such as hemagglutination properties (Rosén, 1970), viral DNA guanine–cytosine (GC) content (Piña and Green, 1965), and, more recently, DNA sequence homology (Green *et al.*, 1979a), the human adenoviruses are now subdivided into five groups: highly oncogenic group A viruses (12, 18, 31), weakly oncogenic group B viruses (3, 7, 11, 14, 16, 21), and nononcogenic viruses of groups C (1, 2, 5, 6), D (8–10, 13, 15, 17, 19, 20, 22–30) and E (4). Both oncogenic and nononcogenic viruses are able to induce a morphological transformation of cells in culture, resulting in characteristic and distinctive foci of small, often polygonal cells with an ill-defined, scanty cytoplasm (McBride and Weiner, 1964; Freeman *et al.*, 1967a–c; Casto, 1968; McAllister and Macpherson, 1968; Gilden *et al.*, 1968; McAllister *et al.*, 1969a,b). Such foci can usually be expanded into lines of cells capable of dividing indefinitely, just as cells explanted from adenovirus-induced tumors can usually generate immortal cell lines. Morphologically, cells transformed in culture by adenoviruses are similar to cells of adenovirus-induced tumors, and in particular, rodent cells transformed by oncogenic adenoviruses are, like cells explanted from tumors and grown in cell culture, or like the virus itself, often capable of inducing tumors following injection into newborn or weanling animals (Kitamura *et al.*, 1964; Freeman *et al.*, 1967b,c; Strohl *et al.*, 1967; Casto, 1969).

The process of transformation of cells in culture should not, however, be taken as being completely identical to tumor induction in animals. Many cell lines transformed by tumor viruses are totally unable to induce tumors following inoculation into animals. In particular, rat cells transformed by group C adenoviruses such as Ad2 or Ad5 do not usually induce tumors in rats, unlike their Ad12-transformed counterparts. These considerations require a definition of what is meant by transformation. In this review, *transformation* is defined as any heritable phenotypic (usually selectable) alteration in mammalian cells resulting from infection with virus or viral DNA.

It is not known why some adenovirus serotypes are oncogenic whereas others are not. Tumorigenicity seems to correlate inversely with DNA GC content, the most highly oncogenic viruses having a low GC content (48–49%), not too dissimilar from that of mammalian-cell DNA, and the nononcogenic viruses a relatively high GC content (55–61%) (Piña and Green, 1965; Green, 1970). However, this apparent correlation is probably fortuitous, since only a small fraction of the adenovirus genome is required for transformation and, in any case, the transforming functions are located in the GC-rich half of the adenovirus DNA molecule (Gallimore *et al.*, 1974; Graham *et al.*, 1974a,b). In addition, certain simian adenoviruses are very highly oncogenic in rodents, but contain DNA with a very high GC content (Burnett and Harrington, 1968a; Piña and Green, 1968). Gallimore (1972) has shown that even rat cells transformed by nononcogenic adenovirus can induce tumors in rats provided the animals are immunosuppressed, whereas tumor induction by cells transformed by oncogenic viruses can occur in immunocompetent animals. These results indicate that the differences in oncogenicity among different adenoviruses or among cells transformed by them may be related to the ability of the animal host to mount an immune response against adenovirus-transformed cells. Recent studies that are beginning to shed light on the interactions between adenovirus-transformed cells and the host immune response are discussed in Section VI.

B. Persistence of Viral Information

Typically, adenovirus-induced tumors are free of virus, and attempts to rescue infectious virus from tumors or tumor cells by a variety of means have generally been unsuccessful (Huebner *et al.*, 1963; Landau *et al.*, 1966; Dunn *et al.*, 1973). Nevertheless, it became apparent very early that viral information must persist in tumor and transformed cells, since these cells were found to contain virus-specific antigens (Huebner *et al.*, 1963, 1964; Girardi *et al.*, 1964) and virus-specific RNA (Fujinaga and Green, 1966, 1967). Thus, adenoviruses, like simian virus 40 (SV40) and polyoma, appear to transform according to the now more or less standard pattern of viruses whose genes persist within the host cell, an indication that the continued presence and expression of viral information may be necessary for maintenance of the transformed phenotype. As discussed in the following sections, the viral genes that are involved in adenovirus transformation are now known, most if not all of their products have been identified, and a variety of useful mutants in the transforming genes have been isolated and characterized. It will become apparent, however, that much remains to be learned about the function of these gene products in transformation as well as in viral replication.

II. STRUCTURE OF INTEGRATED ADENOVIRUS DNA SEQUENCES IN VIRUS-INDUCED TUMORS AND IN TRANSFORMED CELLS

A. Quantitation of Viral DNA in Transformed Cells

The discovery in transformed cells of virus-specific antigens and messenger RNA (mRNA) led naturally to studies designed to detect the presence and to analyze the structure of viral DNA in tumors and transformed cells. Such studies were facilitated by the development of a variety of techniques for detecting specific DNA sequences including RNA–DNA or DNA–DNA hybridization (on filters or in solution) and, more recently, the particularly useful Southern (1975) blot technique for transferring DNA and RNA from agarose gels to nitrocellulose filters for hybridization analysis. Early studies with adenovirus-transformed cells (Green, 1970; Green *et al.*, 1970), in which labeled adenovirus complementary RNA was hybridized to cell DNA immobilized on nitrocellulose filters, suggested the presence of 20–60 viral genomes per transformed cell. These experiments, however, probably overestimated the amount of viral DNA present, since RNA–DNA hybrids may have been selectively lost from filters used in reconstruction experiments (Haas *et al.*, 1972). More reliable quantitation of the amount of viral DNA contained in transformed cells was obtained by measuring the reassociation kinetics (Britten and Kohne, 1968) of radioactive viral DNA in the presence of unlabeled transformed cell DNA, a technique that does not require reconstruction experiments to accurately measure DNA concentrations (Gelb *et al.*, 1971; Pettersson and Sambrook, 1973). This technique, in combination with the use of specific restriction enzyme fragments of viral DNA as probes, has permitted a detailed analysis of viral DNA in transformed cells. Studies with a number of independently isolated Ad2-transformed rat cells showed that all the lines contained detectable viral DNA sequences, but that not all regions of the viral genome were represented equally, and in fact none of the lines seemed to contain the entire genome (Sharp *et al.*, 1974; Gallimore *et al.*, 1974). The only sequences of viral DNA that were detected in every line were those derived from the left end, approximately 14%, of the genome, an observation that suggested that genes mapping in this region might be involved in transformation. These findings have been confirmed and extended by a number of workers (Flint *et al.*, 1976; Johansson *et al.*, 1977, 1978; Sambrook *et al.*, 1974), and with exceptions that are discussed below, it seems a general rule that rodent cells transformed by group C adenoviruses (Ad2 and Ad5) contain only part of the viral genome, though always at least the left approximately 12–14%.

It was also seen that the pattern of viral DNA was different in each line but was stable over many generations of cell growth in culture, suggesting that viral DNA might be stably integrated into the host-cell DNA. Other observations that suggested integration were the detection of viral sequences in the nuclear and chromosomal DNA of transformed cells (Green *et al.*, 1970) and the detection, in certain transformed cells, of viral RNA covalently linked to RNA sequences of cellular origin (Tsuei *et al.*, 1972; Wall *et al.*, 1973).

Studies on hamster cells transformed by an avian adenovirus [chick embryo lethal orphan (CELO)], in which cell DNA was fractionated by techniques that would separate it from free viral DNA or unintegrated viral DNA fragments, indicated that all the CELO DNA was probably joined covalently to high-molecular-weight cell DNA (Bellet, 1975); similar results were obtained for Ad12-transformed hamster cells (Green *et al.*, 1976, 1977; Groneberg *et al.*, 1977). So far, all indications are that adenovirus DNA is always integrated into chromosomal DNA of transformed cells.

B. Structure of Integrated Viral DNA

A more detailed picture of the structure of integrated adenovirus sequences in a large number of virus-transformed cells has been built up in recent years chiefly through the use of the Southern blot technique and, for a small but rapidly increasing number of cell lines, molecular cloning and sequencing. The following general conclusions can be drawn from blot analyses: first, the structure of integrated viral DNA varies greatly from one cell line to another, but covalent integration seems to be the rule; second, the left end of the genome is always present, confirming results of reassociation kinetics; third, many cell lines appear to have one or only a few inserts of viral DNA, but these are often repeated or amplified many times (especially in the case of inserts consisting of left-end sequences) along with flanking cellular sequences; and finally, there seems to be no specificity with respect to the integration sites in the cellular DNA as judged from restriction enzyme cleavage patterns (Sutter *et al.*, 1978; Sambrook *et al.*, 1979; Doerfler *et al.*, 1979; Stabel *et al.*, 1980; Dorsch-Hasler *et al.*, 1980; Vardimon and Doerfler, 1981; Green *et al.*, 1981; Visser *et al.*, 1981).

Though persistence of only segments of viral DNA in Ad2- and Ad5-transformed rodent cells seems to be the most common occurrence, there are special cases in which very often all or nearly all of the viral genome is found integrated. Certain host-range (*hr* group I) mutants of Ad5 (Harrison *et al.*, 1977) (discussed in more detail in Sections IV.A and V.G), though defective in one or more of the functions encoded by the left end of the genome (Frost and Williams, 1978; Galos *et al.*, 1980), are nevertheless able to induce a morphological transformation of primary baby rat kidney cells that can result in immortal cell lines (Graham *et al.*, 1978). Such lines frequently contain what appears to be the entire viral genome integrated as a structure that is colinear with virion DNA (Ruben *et al.*, 1982). Similar structures have been seen in cells transformed by an Ad5 mutant, H5*ts*125, that is temperature-sensitive for viral DNA replication as a result of a defect in the molecular-weight 72,000 (72K) DNA-binding protein (Dorsch-Hasler *et al.*, 1980). In addition, an unusual rat cell line [cloned rat embryo fibroblasts (CREF)] has been developed that can be efficiently transformed by Ad5, and in this system even transformation by wild-type (Wt) virus results in cell lines containing the entire viral genome (Fisher *et al.*, 1982). Ad12 tumor and transformed cells provide another example of frequent integration of the entire viral DNA (Green *et al.*, 1976; Ibelgaufts *et al.*, 1980; Kuhlmann *et al.*, 1982), though at least one study has been reported in which the majority of Ad12-induced tumors contained less than half the viral genome (Lee and Mak, 1977).

Thus, though it is still true that most adenovirus-transformed cells contain only a fraction of the viral genome, the exceptions to this rule are numerous. One feature that many of these exceptions may have in common is that in terms of cell–virus interactions, they represent relatively nonpermissive systems: either the virus is defective, as in the case of *hr* or *ts* mutants, or the cells are completely nonpermissive so that viral replication is blocked at an early step, as in the case of Ad12 infection of hamster cells (Doerfler, 1968, 1969; S. Mak, 1969, 1975). In contrast, as reported by Gallimore (1974), rat cells are semipermissive for Ad2 (and Ad5), with the result that limited viral replication occurs following virus infection and most of the infected cells are killed. In the case of H5*hr*1, for example, the difference between infections of rat cells by mutant and by Wt virus is striking, the latter being followed by extensive cytopathic effect (CPE) and destruction of most of the infected cells, whereas no CPE is visible in mutant-infected cultures (Ruben and Graham, unpublished observations). Infection of the readily transformable CREF cell line with Ad5 was also characterized by absence of any noticeable CPE (Ginsberg, personal communication). In this system, though neither H5*hr*1 nor Wt Ad5 was cytotoxic, higher transforming activity was still obtained with H5*hr*1, suggesting that the high frequency of focus formation by group I *hr* mutants may be due to other factors besides, or in addition to, differential cell killing.

Another frequently observed arrangement of viral DNA in adenovirus-transformed cells is one in which the left and right ends of the genome are joined together, usually, though not always, in a head-to-tail joint (Sambrook *et al.*, 1979; Stabel *et al.*, 1980; Visser *et al.*, 1981, 1982). The fact that both ends of the viral genome are often found in transformed cells, joined either to cellular DNA or to each other, suggests that the termini of the adenovirus DNA molecule may be specifically involved in integration, possibly through the formation of circular intermediates. Indeed, covalently closed circles of Ad5 have recently been detected in infected cells, though it is not yet clear whether these play a role in integration or in viral DNA replication or in both (Ruben *et al.*, 1983).

A simple model for adenovirus DNA integration that can account for most of the viral DNA structures found in transformed cells is shown in Fig. 1. In the most general terms, it is assumed that during the first stages of integration, both ends of the viral genome interact with cellular DNA in such a way that the viral genome is integrated as a linear structure, the termini or sequences near the termini being covalently linked to cellular DNA. In stringently nonpermissive cells, or in mutant-transformed cells, such a structure could be stable. However, in permissive cells, there would presumably be a strong selection, very early after infection, for rearrangements or deletions that would prevent expression of cytotoxic genes (assumed to be in the middle of the genome, e.g., late

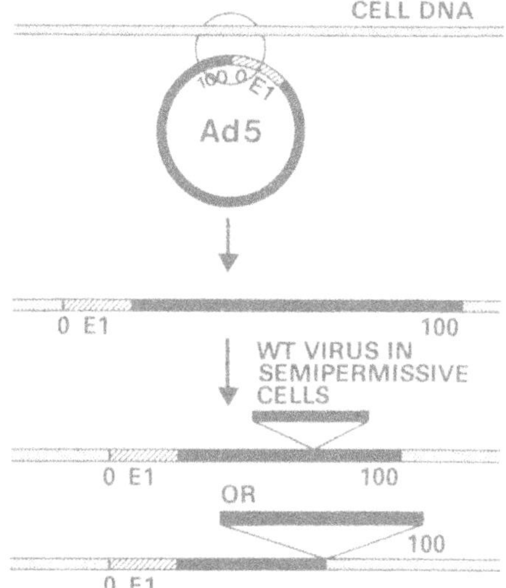

FIGURE 1. Model for the integration and subsequent rearrangement of Ad5 DNA during transformation. See the text for details.

genes) while retaining the left-end sequences that seem to be needed for maintenance of transformation. The possibility that rearrangement of viral DNA can occur after integration is suggested by the fact that selection for morphological revertants of Ad12-transformed cells frequently results in lines that have lost all or part of the viral DNA (Groneberg *et al.*, 1978; Eick *et al.*, 1980; Groneberg and Doerfler, 1979; Schirm and Doerfler, 1981). In addition, the structure and copy number of viral DNA in a number of adenovirus-transformed rat cells suggest that amplification of left-end sequences may have occurred after integration (Sambrook *et al.*, 1979; Vardimon and Doerfler, 1981). Of course, rearrangement of viral DNA is not the only mechanism by which expression of putative cytotoxic genes might be blocked. In particular, methylation has been implicated in adenovirus-gene inactivation in transformed cells in a number of studies (Sutter and Doerfler, 1979, 1980; Eick *et al.*, 1980; Kuhlmann and Doerfler, 1982). Alternatively, during the initial integration step, crossover events that flank the head-to-tail-joined termini of viral DNA and include the transforming genes might be possible. Even if such integration events were rare, since they could immediately excise most of the viral genome, they might result in a relatively large fraction of surviving transformed cells following viral infection.

This model can more or less account for the variety of different structures seen in transformed cells, but is difficult to test in the absence of any *in vitro* system for studying integration or of well-defined mutants defective in integration. However, if integration requires some sort of specific involvement of the termini of viral DNA, it may also require virus-coded functions. It may therefore be worthwhile to examine the effect of various mutants on integration, especially those mutants known to be defective in initiation of transformation. These are discussed in Section IV.

An alternative model is one in which transformation is frequently induced by defective virions containing only part of the viral genome. Incomplete particles have been reported for several adenovirus serotypes of both oncogenic and nononcogenic groups (S. Mak, 1971; Prage *et al.*, 1972; Burlingham *et al.*, 1974; Hammarskjöld *et al.*, 1977) and have been shown to contain either the left end of the genome (Daniell, 1976; Tibbetts, 1977; Hammarskjöld and Winberg, 1980; Khittoo and Weber, 1981; Brown and Weber, 1982) or both left and right ends (Hammarskjöld and Winberg, 1980; Daniell and Mullenbach, 1978; Rajagopalan and Chinnadurai, 1979). Thus, most of the observed forms of integrated viral DNA could result from transformation of cells by viral particles containing subgenomic DNA, though this model would not account for integrated sequences consisting of left and right ends joined head to tail.

C. Sequence Analysis of Integrated Viral DNA

Additional clues as to how integration occurs may come from studies on the fine structure of viral sequences in transformed cells. At the level of detail available from Southern blot analysis of transformed cells, adenovirus DNA seems to insert randomly in cell DNA during transformation. However, molecular cloning and sequencing of cell–virus junctions might yet provide evidence for a degree of specificity. There are still only a few examples of transformed cells from which cell–virus joint sequences have been cloned and analyzed in detail.

Sambrook *et al.* (1979) reported a rather detailed analysis of an Ad2-transformed rat cell line, F4. This cell line contains 16 copies of an insert consisting of the left 60% joined to the right 4% of the viral DNA, not, however, in a head-to-tail arrangement, but with the right-end sequences inverted with respect to the left 60% segment. All 16 copies in F4 cells seem to have an identical structure, a finding that Sambrook *et al.* (1979) have taken as evidence suggesting that amplification of a single insert along with flanking cellular sequences probably occurred after integration. Cloning and sequencing of the virus–virus joint segment and of one of the cell–virus junctions revealed that 61 nucleotides from the left end were deleted and nucleotide 62 was joined to nucleotide 1545 from the right end, while at the cell–virus junction, 2 base pairs (bp) from the right end of the viral DNA molecule had been lost. The DNA of cellular origin hybridized strongly to rat DNA, suggesting that integration into a region of repetitive cellular DNA may have occurred.

A fragment containing the junction between cell DNA and the left end of the viral genome has also been cloned from DNA of an AD12-induced hamster tumor line, CLAC3, and the region spanning the joint has been analyzed (Deuring *et al.*, 1981). It was found that the first 45 bp were deleted from the left end of the viral DNA molecule and that nucleotide 46 was joined to a stretch of cell DNA that contained scattered octa- to undecanucleotide stretches with homology to distant viral DNA sequences. Regions of dyad symmetry spanning the cell–virus junction were also seen, but neither the significance of these regions nor that of the patch homologies is known. Similar analyses have been carried out on the junction between the right end of Ad2 and cell DNA in an Ad2-transformed hamster cell line, HE5 (Gahlmann *et al.*, 1982; Gahlmann and Doerfler, 1983). In this case 10 and 8 nucleotides, respectively, were deleted from the left and right viral termini at the virus-cell junctions and the integrated viral DNA molecule contained an internal deletion extending from 35% to 82% (Gahlmann and Doerfler, 1983). Interestingly, cloning and sequencing of the unoccupied site in cell DNA showed that insertion of viral DNA had occurred without the generation of either deletions, rearrangements or duplications in cellular DNA sequences. Again short patch homologies between cell DNA and sequences at the end of viral DNA were noted.

The fact that transformed cells often contain viral DNA structures in which left- and right-end segments are joined was mentioned previously. One such structure, a head-to-tail joint, has been cloned from the DNA of an Ad5-transformed rat cell line, 5RK20 (Visser *et al.*, 1982). Sequence studies showed that the entire inverted terminal repetition (ITR) of the right end was deleted along with most of the left-end ITR, with the result that nucleotide 63 from the left end was joined to the nucleotide 112 bp from the right terminus. Visser *et al.* (1982) suggested that joining of left and right ends of viral DNA in this fashion most probably preceded integration.

Cloning and sequence analysis were also carried out by Westin *et al.* (1982) on a hamster cell line, 268.C31, transformed by sheared Ad5 DNA. This line has two inserts of viral DNA, one of which was cloned along with flanking DNA. Interestingly, one end of the viral DNA insert was joined to salmon sperm DNA, the DNA used as carrier in the transformation assay (Graham, unpublished results), while the right end was linked to hamster cell DNA. Ligation of transforming DNA to carrier DNA or other cotransfected DNA sequences in DNA-mediated transformation assays has been observed in other systems, and it is thought that such structures may persist for some time prior to integration into the cellular chromosome (Wigler *et al.*, 1980; Scangos *et al.*, 1981). It is not clear, however, whether the structure of integrated viral DNA in cells such as 268.C31 can shed light on the mechanism of integration of viral DNA during virus-mediated transformation, since this probably occurs through a rather different process from that which occurs during DNA-mediated transformation (see Rowe *et al.*, 1983a; Graham *et al.*, 1983).

From the limited number of cell lines analyzed so far, a few general conclusions can be tentatively put forward: First, integration appears to involve the loss of viral sequences at the termini, from a few to a few dozen base pairs or more. If this is generally true, it would explain why it is impossible to rescue infectious virus even from cells that apparently contain the entire Ad5 genome (Dorsch-Hasler *et al.*, 1980; Ruben *et al.*, 1982). Second, short patch homologies between cell DNA and viral DNA may play a role in determining the site of viral DNA integration and the mechanism by which integration occurs. It is interesting that a short (5-bp) homology was found between SV40 DNA and cell DNA at the site of integration of viral DNA in an SV40-transformed rat cell line (Stringer, 1982). Site specificity at the level of a few base pairs would still be consistent with apparently random insertion on a larger scale and might be analogous to transposition of certain bacterial transposable elements that may also involve short stretches of homology between bacterial chromosomal DNA and the ends of the element (Halling and Kleckner, 1982).

III. TRANSFORMATION BY ADENOVIRUS DNA

A. Identification of Transforming DNA Fragments

As discussed in Section II, studies on the viral DNA content of a variety of adenovirus-transformed cells have indicated that only the left-end sequences of the viral genome are required for maintenance of transformation. Independent evidence for this hypothesis came from studies on transformation by purified viral DNA and DNA fragments. In the case of every adenovirus serotype studied so far, only DNA fragments that contain the left approximately 8–10% of the viral genome are needed for transformation in DNA-mediated transfection assays.

The first evidence that purified adenovirus DNA could be biologically active was obtained by Burnett and Harrington (1968a,b), who demonstrated that simian adenovirus 7 (SA7) DNA was infectious and also capable of inducing tumors following injection into newborn or weanling animals. Subsequently, it was shown that type 7 simian adenovirus DNA (SA7) was also effective in inducing tumors (Mayne *et al.*, 1971). More extensive studies with Ad5 DNA, in which a DNA-mediated cell-transformation assay was utilized (Graham and van der Eb, 1973a,b), indicated that adenovirus DNA could be sheared to very small fragments, less than 10% of the size of the intact genome, and still retain the ability to transform primary rodent cells. Digestion of Ad5 DNA inward from the ends with exonuclease III combined with single-stranded-specific endonuclease S1 did not eliminate transforming activity until digestion had proceeded inward about 1%, suggesting that the extreme terminal nucleotides were not required for transformation by DNA (Graham *et al.*, 1974a,b). Separation of Ad5 DNA fragments obtained by shearing or restriction enzyme digestion, and assays of transforming activity, showed that the viral genes necessary and sufficient for transformation were located within the left-end approximately 10% of the Ad2 and Ad5 genomes

(Graham *et al.*, 1974a,b; van der Eb *et al.*, 1977). Similar results have been obtained with serotypes Ad3 and Ad7 (Sekikawa *et al.*, 1978; Dijkema *et al.*, 1979) and Ad12 (Shiroki *et al.*, 1977; Mak *et al.*, 1979).

Thus, two distinct approaches to the problem of identifying the viral genes involved in transformation had clearly implicated genes mapping at the left end of the genome. However, whereas all virus-transformed cells appear to retain at least 12–14% of the adenovirus genome (Gallimore *et al.*, 1974), encompassing all of early region 1 (E1), less of the viral genome, as detailed below, is needed to transform cells in a DNA-mediated transfection assay.

B. Organization of Early Region 1: The Transforming Region

As discussed in Chapter 5, E1 is one of five regions expressed early in productive viral replication; because of its importance in transformation, this region has been very intensively studied in recent years. E1 was one of the first regions to be sequenced for Ad5 (van Ormondt *et al.*, 1978, 1980; Maat and van Ormondt, 1979; Maat *et al.*, 1980; Bos *et al.*, 1981), Ad7 (Dijkema *et al.*, 1980), and Ad12 (Sugisaki *et al.*, 1980; Kimura *et al.*, 1981; Bos *et al.*, 1981); transcription patterns have been characterized in detail (see Chapter 5), complementary DNA (cDNA) has been cloned and partially sequenced for several E1 messages (Perricaudet *et al.*, 1979, 1980a,b), and the proteins encoded by this region have been largely identified and partially characterized (see Section V).

Some of the more salient features of E1 are illustrated in Fig. 2, which presents a compilation of data from a large number of studies involving the group C viruses Ad2 and Ad5 (see Tooze, 1981). The structure and organization of E1 for other adenovirus serotypes follow a very similar pattern. E1 contains two independently promoted sets of mRNAs defining E1a, between 1.4 and 4.5 map units (m.u.), and E1b from 4.8 to 11.2 m.u., plus a third transcription unit giving rise to the 9 S mRNA that encodes the virion structural protein IX. Within E1a, three different mRNA species have been detected (see Fig. 2). The 13 S and 12 S mRNAs are 5′- and 3′-coterminal, but differ in the way they are generated by splicing. The 9 S mRNA, like the polypeptide IX message, is expressed late and is not detected in transformed cells. Its product has so far been detected only by *in vitro* translation of selected Ad2 mRNA as a polypeptide with an apparent molecular weight of 28K (Spector *et al.*, 1980; Esche *et al.*, 1980). This is surprisingly large, since studies with Ad7 (Dijkema *et al.*, 1980) and recent work with Ad2 (Virtanen and Pettersson, 1983) indicate that the 9 S message splices into a different reading frame from that of the second exon of the 12 and 13 S mRNAs and is predicted to encode a polypeptide of molecular weight 6.3K for Ad7 or 6.1K for Ad2. The 12 and 13 S E1a messages should theoretically code for polypeptides 242 and 288 amino acids long for Ad2 and 243 and 289 amino acids long for Ad5

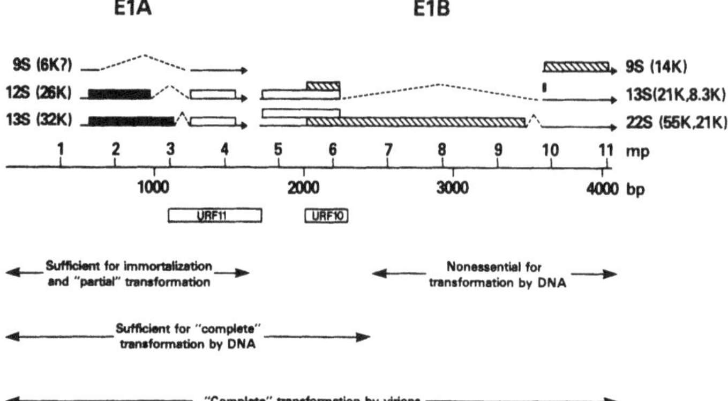

FIGURE 2. Early region 1 of group C adenoviruses Ad2 and Ad5 showing spliced mRNA species that have been detected and their polypeptide coding sequences. On the r-strand, open reading frames are indicated for reading frames 1 (▭), 2 (■), and 3 (▨). (URF11, URF10) Unidentified open reading frames on the l-strand starting with the first AUG initiator. Theoretical molecular weights for E1 proteins are shown in parentheses following the corresponding mRNAs. (←→) Sequences required for cell transformation under various conditions. See the text for a more detailed discussion and for references.

(26K and 32K, respectively). Except for an internal stretch of 46 amino acids, present in the larger protein but spliced out of the message for the smaller polypeptide, the 12 and 13 S mRNA translation products are predicted to be identical.

Within E1b, the 9 S mRNA, which is unspliced and extends from 9.9 to 11.2 m.u. (Aleström *et al.*, 1980), is made late in the replication cycle and appears not to be expressed in transformed cells. Besides this message, a 13 and a 22 S mRNA have been mapped to E1b, and both these messages are made early in infected cells and are usually expressed in transformed cells. The 22 S mRNA is thought to encode both a 55 and a 21K polypeptide, the latter also encoded by the 13 S mRNA. By utilizing the start codon at 2019, the 13 S mRNA could also encode an 8.3K protein predicted to be highly related to the 55K polypeptide.

Finally, two unassigned open reading frames exist on the leftward transcribed strand (l-strand) within E1; in the case of Ad2, they extend from an AUG codon at 2290 to a UAA stop codon at 2005 (URF10) and from an AUG at 1712 to a stop codon at 1196 (URF11) and could encode polypeptides of 14 and 23K, respectively (Gingeras *et al.*, 1982). Only recently has any evidence been obtained that either of these open reading frames may be utilized: Katze *et al.* (1982) detected a 20 S mRNA hybridizing to the l-strand of E1 that could be translated *in vitro* to give a polypeptide reported to be 11K (perhaps the product of URF10).

The actual detection and mapping of E1-encoded proteins have relied on a variety of techniques, including the use of various mutants in E1, the analysis of cells transformed by virus or by a variety of different DNA fragments, the use of inhibitors, and the use of *in vitro* translation of

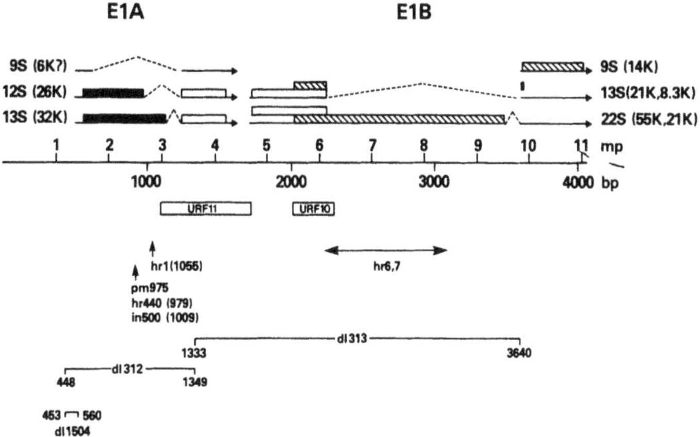

FIGURE 3. Map locations of a selection of E1 mutants. Horizontal bars below the map indicate either the region in which mutants have been mapped (*hr*6,7) or the viral sequences removed by deletion (*dl*312, 313, 1504).

mRNAs selected by hybridization to specific DNA fragments. A more detailed examination of E1 proteins is presented in Section V after further discussion of transformation by DNA fragments and by various recombinant plasmids containing adenovirus DNA and after description of results obtained with various transformation-defective mutants. It can be pointed out here, however, that the E1 proteins actually detected in infected and transformed cells present a picture that is considerably more complex than that depicted in Fig. 2, especially in the case of E1a products.

C. Characteristics of Cells Transformed by DNA Fragments

Which of all the different E1 mRNAs and polypeptides is required for cell transformation by viral DNA seems to depend to some extent on the selective conditions of the transformation assay. Transformation assays are generally carried out on nonpermissive or semipermissive cells (e.g., rat or hamster cells for Ad12, rat cells for Ad2 and Ad5) and usually make use of the fact that adenovirus-transformed cells grow preferentially in medium containing low concentrations of calcium ions (Freeman *et al.*, 1966). Cells transformed by fragments of viral DNA containing all of E1, such as *Xho*I-C (left 16%) of Ad5, or analogous fragments of Ad12, are indistinguishable from cells transformed by virus in terms of morphology, growth properties, and tumorigenicity (Graham *et al.*, 1974b; Shiroki *et al.*, 1977; Yano *et al.*, 1977; Mak *et al.*, 1979; van der Eb *et al.*, 1979; Graham, unpublished observations). In the case of slightly smaller fragments (see Fig. 3) such as *Bgl*II-D (left 9.5%) or *Hind*III-G (left 8%) of Ad5, or analogous fragments of Ad7 and Ad12 (i.e., fragments that contain E1a and part of E1b but lack the right end of E1b), there seems to be general agreement that the resulting transformed cells are morphologically very similar to cells transformed by virus or by larger DNA fragments, though differences in tumor antigens are seen with fluorescent-antibody staining (van der Eb *et al.*, 1977, 1979; Shiroki *et al.*, 1977; Sekikawa *et al.*, 1978; Jochemsen *et al.*, 1982). There is not, however, complete agreement as to whether cells transformed by such fragments (which are missing coding sequences for the carboxy terminus of the larger 22 S mRNA product) are able to induce tumors. Jochemsen *et al.* (1982) reported that rat cells transformed by the *Hind*III-G (left 7.2%) of Ad12 were nontumorigenic in nude mice, whereas cells transformed by a larger fragment, *Eco*RI-C (left 16%), were able to induce tumors. A similar observation was made in preliminary studies with Ad5 *Hind*III-G (left 8%)-transformed rat cells (van der Eb *et al.*, 1979). In contrast, Shiroki *et al.* (1977) reported that cells of the established rat line 3Y1 could be transformed by Ad12 *Hind*III-G to give anchorage-independent foci that were capable of inducing tumors following injection into newborn rats. Finally, it has been found that primary hamster kidney cells transformed by Ad5 *Hind*III-G are tumorigenic in hamsters (Rowe *et al.*, 1984a). One possible explanation for these conflicting observations, aside from differences due to the use of different cells and different tumorigenicity assays, is that tumorigenicity may vary quantitatively depending on how the cells are transformed. Studies on tumorigenicity much more extensive than those reported to date would be needed to identify anything other than very large variations in degree of tumorigenicity among cells transformed by different fragments.

Under certain assay conditions, even smaller DNA fragments containing only E1a, e.g. *Hpa*I-E of Ad5 (left 4.5%), or analogous fragments of Ad7 and Ad12, have been shown to induce a morphological transformation of rat cells resulting in "immortalization" (Shiroki *et al.*, 1979b; Dijkema *et al.*, 1979; Houweling *et al.*, 1980; McKinnon and Graham, unpublished data). There is general agreement that the resulting transformed cell lines are distinguishable in a number of properties from cell lines transformed by larger DNA fragments or by virus. Morphologically, Ad5 *Hpa*I-E transformants were somewhat fibroblastic rather than epithelioid and tended to grow slower and to stop dividing at lower saturation densities than "fully" transformed cells (Houweling *et al.*, 1980). Rat 3Y1 cells transformed by the *Bpa*I-H (left 4.5%) fragment of Ad12 were morphologically similar to cells transformed by larger DNA fragments, but otherwise displayed the same "intermediate" range of transformed properties seen in their Ad5 counterparts. In addition, Ad12 *Bpa*I-H transformants had reduced tumorigenicity in syngeneic rats (Shiroki *et al.*, 1979b).

An unusual property of E1a-fragment transformation of baby rat kidney cells is that no foci appear when the transformation assays utilize low-calcium-ion medium supplemented with horse serum, but foci are detected when selective media contain normal calcium ion concentrations and bovine serum (McKinnon and Graham, unpublished observations).

Additional studies utilizing viral DNA segments cloned in plasmids to transform cells have, by and large, confirmed the findings discussed above. McKinnon *et al.* (1982) constructed a series of plasmids containing the 5.6-kb bacterial transposable element, Tn5, inserted at various positions within Ad5 E1 sequences. DNA-mediated transformation assays with these insertion mutants indicated that mutations in the region of E1b to the right of the 5′ splice site for the 13 S mRNA had no effect on transforming activity, whereas insertions in E1a or the 5′ region of E1b generally reduced or eliminated transforming activity. In addition, deletion of all of E1b from E1-containing plasmids completely eliminated transforming activity when the selective media had low calcium ion concentrations and horse serum (McKinnon *et al.*, 1982 and unpublished observations).

Thus, both E1a and at least part of E1b are needed for full transformation of rodent cells by adenovirus DNA. The two regions need not be physically linked, however, since primary cells could be transformed by mixtures of fragments containing E1a and E1b of Ad5 separately, or even mixtures containing E1a of Ad5 and E1b of Ad12 (van den Elsen *et al.*, 1982; Bacchetti and Graham, unpublished). Preliminary studies carried out on cell lines established following transformation with different combinations of fragments suggested that the degree of tumorigenicity in nude mice was higher for cells carrying Ad12 E1b sequences than for cells with E1b derived from Ad5. Additional experiments using plasmids containing E1a of Ad5 (or Ad12) and E1b of Ad12 (or Ad5) essentially confirmed the results of tumorigenicity assays in nude mice, relatively high tumorigenicity being associated with E1b of Ad12, whichever serotype E1a was derived from (Bernards *et al.*, 1982; van der Eb *et al.*, 1983). However, only cells transformed by plasmids containing both E1a and E1b sequences from Ad12 were highly oncogenic in syngeneic rats (van der Eb *et al.*, 1983).

In interpreting results of studies on transformation by adenoviruses and adenovirus DNA, it is important to bear in mind that transforming activity depends very much on the cell type and assay conditions employed. In addition, tumorigenicity of transformed cells depends on the host animal being used, as well (presumably) as on expression of viral functions. Nevertheless, several general conclusions can be drawn from the studies outlined above. First, in DNA-mediated transformation assays, only E1a is needed to "immortalize" cells and to induce an "intermediate" transformed phenotype. In particular, cells transformed by E1a alone, whether from Ad5 or Ad12, are nontumorigenic or weakly tumorigenic. Acquisition of the full complement of transformed properties appears to require additional DNA sequences to the right of E1a; whether this indicates a need for the product of the 13 S E1b mRNA, for all or part of the 22 S mRNA product, or for some other protein remains to be determined. Finally, no viral functions coded by sequences outside E1 seem to be needed for "complete" transformation of rodent cells by adenovirus DNA.

Recently E1a has been shown to be able to "cooperate" with viral and cellular oncogenes such as polyoma middle T antigen and Harvey ras1 to induce oncogenic transformation of primary rat cells (Ruley, 1983; Land *et al.*, 1983). This is consistent with, and indeed might have been predicted by, the ability of E1a to immortalize, since alone, py mid T and Ha ras1 can transform established cell lines but not primary cells.

IV. TRANSFORMATION-DEFECTIVE MUTANTS

For reasons that are still not clear, early attempts to obtain temperature-sensitive (*ts*) mutants in functions encoded by the left end of adenovirus were relatively unsuccessful, although a few *ts* mutations that affect transformation were isolated and subsequently found to be in genes located to the right of E1. These are mutations in complementation group N (Williams *et al.*, 1971, 1974, 1979), which is now thought to affect a virus-coded DNA polymerase (Enomoto *et al.*, 1981; Lichy *et al.*, 1982; Stillman *et al.*, 1982) and maps between 19 and 24 m.u. (Galos *et al.*, 1979), and mutations in the 72K DNA-binding protein (DBP), the best-characterized being H5*ts*125 (Ensinger and Ginsberg, 1972; van der Vliet *et al.*, 1975; Ginsberg *et al.*, 1977). The effect that these mutations have on transformation, and speculations on the function of the corresponding proteins in transformation, are discussed later in this section.

A. E1 Mutants

To obtain mutants defective in E1 functions, a particularly successful approach has been to isolate or construct host-range (*hr*) mutants that can replicate in an Ad5-transformed human embryonic kidney cell line (293), but fail to replicate or, more accurately, replicate poorly in HeLa or KB cells. Since 293 cells contain and express E1 (Graham *et al.*, 1977; Aiello *et al.*, 1979), the functions encoded within this region effectively become nonessential for replication of Ad5, with the result that E1 sequences can be altered more or less at will.

The first series of mutants obtained through this approach were isolated from mutagenized virus stocks. Two complementation groups of *hr* mutants were identified, I and II, both distinct from groups defined by previously isolated *ts* mutants (Harrison *et al.*, 1977) and both altered in their ability to transform primary rat cells (Graham *et al.*, 1978). Host-range mutants of group I are now known to map in E1a (Frost and Williams, 1978) (see Fig. 3 for map locations of *hr* mutants and a few other E1 mutants), and one mutant, *hr*1, has been partially sequenced and shown to contain a deletion of nucleotide 1055 (Ricciardi *et al.*, 1981). This results in a frameshift that prematurely terminates translation of the 13 S E1a mRNA species; since the *hr*1 mutation is spliced out of the 12 and 9 S mRNAs, the translation products of these smaller messages should not be altered (unless *hr*1 contains other mutations as well). In HeLa cells, group I *hr* mutants fail to synthesize viral DNA or late viral proteins (Lassam *et al.*, 1979a; Ross *et al.*, 1980), defects that may be only indirectly caused by the *hr*1 defect, since E1a mutants have also been shown to be defective in expression of early regions other than E1a (N. Jones and Shenk, 1979b; Shenk *et al.*, 1979; Berk *et al.*, 1979a,b).

Group I mutants fail to transform rat embryo or rat embryo brain cells; however, as mentioned in Section II.B, they induce an abnormal but highly efficient (in terms of numbers of foci) transformation of baby rat kidney (BRK) cells (Graham *et al.*, 1978). The reason for this difference among different cell types is not known, but the suggestion was made that *hr*1 transformants may be less able than wild-type (Wt) transformants to divide under the conditions of somewhat higher cell density that exist in rat embryo or rat embryo brain cell cultures, compared to BRK cell cultures (Graham *et al.*, 1978).

Transformation of primary BRK cells by group I *hr* mutants is semiabortive in the sense that foci could not be individually isolated and expanded into cell lines (Graham *et al.*, 1978). Although mixed populations containing many transformed foci could be subcultured to generate immortal cell lines, these failed to express a fully transformed phenotype: compared to Wt-virus-transformed cells, the resulting cells were altered morphologically, did not form colonies in soft agar, and were much less tumorigenic in nude mice (Graham *et al.*, 1978; Ruben *et al.*, 1982). Since cells transformed by group I mutants expressed normal levels of E1b proteins (Lassam *et al.*, 1979a,b; Ruben *et al.*, 1982), these observations suggest that one or more E1a products may be required for maintenance of transformation.

Group II *hr* mutants have been mapped in E1b between 6.1 and 8.5 m.u., approximately within the sequences that are spliced out of the 13 S mRNA but retained within the 22 S message (Galos *et al.*, 1980). Consistent with this mapping, all group II *hr* mutants analyzed appear to be defective in the synthesis of the large protein encoded by the 22 S mRNA (Lassam *et al.*, 1979a,b; Ross *et al.*, 1980; Rowe *et al.*, 1983b), but express an apparently normal 13 S mRNA product (Lassam *et al.*, 1979a,b; Rowe *et al.*, 1983b). Group II mutants are able to synthesize viral DNA and even late viral proteins in HeLa cells (Lassam *et al.*, 1978), but are unable to transform any of the cell types (rat embryo, rat embryo brain, or BRK) tested to date (Graham *et al.*, 1978). This is somewhat surprising in light of the results discussed in Section III that suggested that insertions in or deletions of the right part of E1b had no effect on transformation in DNA-mediated assays. Recently, it has been found that DNA extracted from group II mutant virions was as efficient at transforming rat cells as was

DNA extracted from Wt virions (Rowe and Graham, 1983). Since the resulting transformants had a fully transformed phenotype including the ability to induce tumors, it has been suggested that the 22 S mRNA product is needed only for initiation of transformation by virions, but not for transformation by purified viral DNA or for maintenance of transformation (Rowe and Graham, 1983). Moreover, since a group II mutant has been shown to be incapable of rescuing adeno-associated virus (AAV) from the integrated state in AAV-transformed cells (Ostrove and Berns, 1980), it has been hypothesized (Rowe *et al.*, 1983a; Graham *et al.*, 1983) that the group II function may be involved in viral DNA integration. According to this hypothesis, during transformation by viral DNA (a very inefficient process requiring relatively large numbers of viral genomes per transformant), integration of viral DNA fragments would involve only cellular functions, thus bypassing defects such as deletions or *hr*II mutations in putative virus-coded integrative functions.

Recently, several cold-sensitive host-range (hr^{cs}) mutants have been isolated and partially characterized; in addition, previously isolated *hr* mutants (*hr*1 and *hr*6 of groups I and II, respectively) were found to be cold-sensitive in one or more phenotypic properties (Ho *et al.*, 1982; Babiss *et al.*, 1983). Evidence was obtained from transformation assays with these mutants for a role of both E1a and E1b functions in maintenance of transformation.

A series of E1 deletion mutants with properties somewhat similar to those of the *hr* mutants have been isolated by N. Jones and Shenk (1979a), also using 293 cells as a permissive host. These mutants were obtained through a very elegant series of constructions involving the isolation of a derivative of Ad5 (*dl*309) with a single *Xba*I cleavage site at 3.8% and the use of this virus to isolate deletion mutants in E1 that had lost this *Xba*I site and various flanking viral sequences. The most extensively characterized of these deletion mutants were *dl*312, with a deletion from 448 to 1349 bp (Shenk *et al.*, 1979) that removes essentially all of E1a, and *dl*313, which extends from nucleotide 1333 to 3640 and has consequently lost all of E1b and the carboxy-terminal region of E1a (Colby and Shenk, 1981). The *dl*312 mutant complemented *hr*II mutants and, like *hr*I mutants, was DNA$^-$ and defective in expression of other early regions of Ad5 (N. Jones and Shenk, 1979a,b; Galos *et al.*, 1980). The *dl*313 mutant could complement *hr*1 as well as *dl*312, suggesting that it expresses functional E1a products even though it lacks the coding sequences for the carboxy-terminal 69 amino acids of E1a. Unlike *hr*II mutants, *dl*313 was defective in viral DNA synthesis in nonpermissive cells (N. Jones and Shenk, 1979a), a property that could be due to the extension of the *dl*313 deletion into E1a or to the elimination of the coding sequences for the E1b 13 S mRNA product.

Both *dl*312 and *dl*313 were found by N. Jones and Shenk (1979a) to be unable to transform rat embryo or rat embryo brain cells, but recently Shiroki *et al.* (1981) have reported that *dl*313 could induce a "partial" transformation of rat 3Y1 cells. Similar findings were made by I. Mak and S. Mak (1983), using *dl*313 to infect primary rat kidney cells at high multiplicities of infection. In contrast, the group II mutant, *hr*6, did not transform even at high multiplicities of infection, a rather surprising observation, since the *dl*313 deletion must span the sequences that contain the *hr*6 mutation.

With the use of *dl*309, an insertion mutant, in 52, has been constructed that may prove useful in analyzing transformation events (Graham, unpublished observations). This mutant has a 2.2-kilobase (kb) segment of pBR322 (containing the β-lactamase gene and the origin of plasmid DNA replication) inserted into the *Xba*I site at 3.8% in *dl*309. As a consequence, DNA from cells infected with this mutant can be used to transform *Escherichia coli* to ampicillin resistance, a property that should allow the rescue from infected cells of short-lived or rare forms of viral DNA that might otherwise be difficult to study. This insertion mutant belongs to complementation group I in the sense that it could complement group II but not group I *hr* mutants, but unlike other group I mutants was unable to transform primary BRK cells and synthesized viral DNA in HeLa cells.

Solnick (1981) made use of *dl*309 to obtain mutations specifically in E1a of Ad5. This was done by isolating and mutagenizing the *Xba*I-E fragment (left 3.8%) of Ad5, then ligating it to the large fragment of *dl*309, infecting 293 cells, and screening for *hr* mutants. One E1a mutant, *hr*440, was characterized in detail and shown to contain mutations near the 5′ splice for the 12 S mRNA that appeared to prevent proper splicing of RNA transcripts to the 12 S species. These same mutations introduced an amber mutation into the coding sequences for the 13 S mRNA product. Further studies (Solnick and Anderson, 1982) showed that although *hr*440 was deficient in expression of E1 and E3, production of mRNA from E1b and E4 in *hr*440-infected cells was normal. Mutant *dl*309 was also used by Carlock and Jones (1981) to construct an E1a mutant containing an octanucleotide insert at the *Sma*I site, position 1009, in E1a of Ad5. This introduced a frameshift in the coding sequences of the 13 S mRNA and resulted in a truncated protein, while leaving the 12 S mRNA product unaffected. Like *hr*I mutants, the insertion (*in*) mutant, *in*500, was defective for viral DNA replication in nonpermissive cells. However, although expression of E2 and E4 was reduced, E1b and E3 expression were not impaired. Both *in*500 and *hr*440 were unable to transform rat cells (Carlock and Jones, 1981; Solnick and Anderson, 1982), again indicating that E1b expression alone is insufficient for transformation and that one or more E1a functions must also be directly involved.

Through the use of a variety of recombinant DNA techniques, it is now possible, at least in principle, to obtain virtually any type of mutant in adenovirus DNA sequences cloned in bacterial plasmids. The difficulty arises in the rescue of these mutations into full-length infectious viral DNA. One approach to this problem, applicable to mutations in E1a to the left of 3.8 m.u., was developed by Stow (1981), again using the Ad5 mutant *dl*309. He showed that if plasmid DNA containing the left-end 4.5% of Ad2 or Ad5 was cut with *Xba*I, ligated to the large 3.8–100% *dl*309 fragment, and transfected into 293 cells, mutations that had been generated in the cloned viral DNA sequences could be efficiently rescued into infectious virus progeny. With the use of this method, several mutants were isolated that contained small deletions at the *Sma*I site at 2.8 m.u. (Stow, 1981). The methods developed by Stow have also been used to rescue several Ad5 insertion mutants, derived from the Tn5 mutations (McKinnon *et al.*, 1982) described in Section III.C, into Ad5 DNA (Haj Ahmad and Graham, unpublished observations). Only mutants having net insertions of less than about 2 kb could be rescued with ease, suggesting that packaging (or other) constraints may severely limit the size of Ad5 DNA variants that can be replicated as infectious virus.

The Stow method of rescuing mutants has so far been limited to rescue of mutations in E1a to the left of the *Xba*I site at 3.8%. A method of rescuing mutants near the ends of adenovirus DNA but not necessarily restricted to the left 3.8% has been described by Dunsworth-Browne *et al.* (1980). They showed that hybridization of denatured restriction fragments to exonuclease-III-digested DNA–protein complex and transfection of 293 cells resulted in a small but significant fraction of progeny carrying sequences derived from the hybridizing restriction enzyme fragments. A third method of rescuing mutations made in viral DNA fragments or in viral DNA sequences cloned in plasmids involves simply mixing DNA fragments with full-length viral DNA and infecting cells. Recombination between fragments of viral DNA and full-length DNA is efficient enough that a significant fraction of the progeny will contain DNA sequences from the fragment. With the use of this technique in a marker-rescue format, several Ad2 and Ad5 mutants have been physically mapped on the viral genome (Frost and Williams, 1978; Arrand, 1978; Galos *et al.*, 1979, 1980). Used in a reverse marker-rescue format, the technique could in principle be used to obtain mutations in any region of the genome. Recently, Grodzicker and Klessig (1980) showed that unselected Ad2 DNA fragments could be introduced into human thymidine-kinase-negative (TK^-) cells by cotransformation with the herpes simplex 1 virus *tk* gene. This approach could lead to the development of cell lines that contain and express viral genes from regions of the adenovirus genome other than E1 and that could provide a system for the isolation of additional *hr* and *dl* mutants.

One of the best examples of the power and precision that can be brought to bear on the problem of obtaining useful mutations in the transforming genes of adenovirus is *pm*975, a mutant constructed by Montell *et al.* (1982). By applying techniques developed by Gillam *et al.* (1979) and Gillam and Smith (1980), in which synthetic oligonucleotides are used to introduce specific sequence changes into DNA, Montell and coworkers introduced a single T-to-G transversion at position 975 in E1a of Ad2, a mutation in the donor sequence for the splice in the 12 S mRNA. At the same time, the amino acid specified by the 13 S mRNA codon containing the transversion was unaltered due to the degeneracy of the genetic code. After rescuing the mutation from plasmid DNA into infectious virus using the Stow method, Montell and co-workers showed that no 12 S mRNA was made in *pm*975-infected HeLa cells, yet the mutant was phenotypically Wt for replication in HeLa cells. The transforming activity of *pm*975 was reduced 4- to 5-fold relative to Wt, and *pm*975 transformants gave only very small colonies in soft agar (C. Montell, personal communication). These results, combined with the data for *hr*1 presented above, suggest that the protein encoded by the 13 S mRNA may be the most important E1a product in viral replication, but that both 12 and 13 S mRNA products may play a role in transformation.

Another series of useful mutants, consisting of deletions around the 5′ end of E1a, have recently been constructed and partially characterized (Osborne *et al.*, 1982). It was found that removal of sequences upstream from the TATA signal at position 468 had no effect on E1a transcription. Removal of the TATA signal resulted in a 5- to 10-fold reduction in the rate of E1a transcription and shifted the 5′ end of the mRNAs to a minor set of species that were also detected in Wt infections. Analysis of other early mRNAs suggested that E2 and E3 were most dependent on expression of E1a, and E4 and E1b less dependent, in accord with results of other studies (Solnick and Anderson, 1982; Shaw and Ziff, 1982). Interestingly, neither removal of the TATA and entire untranslated leader sequence of E1a mRNA nor even deletion of sequences extending into the first AUG of E1a (mutant *dl* 1504) had any effect on the ability of the resulting mutants to replicate in HeLa cells or to transform rat cells (Osborne *et al.*, 1982). Evidently, not only are Wt levels of E1a mRNA 5- to 10-fold higher than needed for viral replication or transformation, but also the N-terminal 14 amino acids are nonessential for function of E1a proteins.

B. Mutants in E2a and E2b

As mentioned briefly at the beginning of this section, in addition to mutants that map in E1, two groups of Ad5 mutants have been isolated that are altered in their ability to transform rodent cells but that are now known to map well outside E1. Mutants represented by complementation group N, *ts*36 being the best characterized, are located between 18.5 and 22 m.u. (Galos *et al.*, 1979) within an l-strand transcription block, E2b that encodes the 80–87K protein precursor of the 55K terminal protein (TP) and a 140K protein now thought to be a DNA polymerase (Stillman *et al.*, 1981, 1982; Enomoto *et al.*, 1981; Lichy *et al.*, 1982). Williams *et al.* (1974, 1979) showed that N mutants *ts*36, 69, and 149 are temperature-sensitive for initiation of transformation but not for its maintenance once cells had been transformed and cultured for 2 days at the permissive temperature. This is consistent with the wealth of data discussed in Section III that indicate that sequences outside E1 are not needed for maintenance of transformation. Recently, the coding sequences of the 55K TP have been identified and shown not to overlap with the map location of the N-mutants (Smart and Stillman, 1982), and it now seems likely that the N group of mutants are temperature-sensitive in the virus-coded DNA polymerase (Stillman *et al.*, 1982). It is unclear why a DNA polymerase should be needed for initiation of transformation. Viral DNA replication seems not to be a requirement, since, as discussed below, the DNA$^-$ mutant *ts*125 can transform at both the permissive and the nonpermissive temperature, and Ad12 can transform hamster cells even though these cells are completely incapable of supporting Ad12 DNA replication (Doerfler, 1968, 1969; S. Mak, 1969). However, the 140K polymerase and the 80–87K TP are known to form a complex that is involved in initiation of viral DNA synthesis at the ends of the viral DNA molecule (Enomoto *et al.*, 1981; Lichy *et al.*, 1982; Stillman *et al.*, 1982), and one possibility is that this complex plays a role in the integration of the viral genome into cell DNA during transformation. This would be consistent with the model for integration discussed in Section III and with the fact that N mutants are defective only in initiation of transformation.

Results of studies on transformation by group N mutants using cloned rat embryo fibroblast (CREF) cells (Ginsberg, personal communication) differ from results obtained by Williams and co-workers using primary rat embryo cells. Ginsberg and co-workers found that *ts*36 transformed CREF cells with reduced efficiencies relative to Wt at both the permissive and the nonpermissive temperature, whereas *ts*149 was found not to be transformation-defective at either temperature. The reasons for this discrepancy are unclear at present, and consequently it is possible that the adenovirus-coded DNA polymerase does not play a role in transformation.

The other class of mutations mapping outside of E1 that are altered in their ability to transform are mutants in the 72K DBP encoded in E2a, *ts*125 being the best-characterized (Ensinger and Ginsberg, 1972; van der Vliet *et al.*, 1975; Ginsberg *et al.*, 1977). Rather than being transformation-defective, *ts*125 transforms more efficiently than Wt virus at both the permissive and the nonpermissive temperature (Ginsberg *et al.*, 1974; Logan *et al.*, 1981; Fisher *et al.*, 1982). This property has been shown to be unrelated to differences in cytotoxicity of the virus for infected cells (Fisher *et al.*, 1982) and hence must be a consequence of some less trivial effect on cell–virus interactions. One possible explanation for the enhanced transformation efficiency of *ts*125 may stem from effects that the DBP has on expression of other early regions, in particular of E1, and through autoregulation of E2 transcription, of E2b, the region in which *ts*36 maps. In cells infected with *ts*125 at the nonpermissive temperature, mRNA levels from E2 and other early regions are enhanced relative to Wt (Carter and Ginsberg, 1976; Carter and Blanton, 1978a,b; Blanton and

Carter, 1979; Babich and Nevins, 1981); since E1 and possibly E2b functions are required for transformation, enhanced expression of either region could result in enhanced efficiency of transformation. Logan *et al.* (1981) have indeed postulated that enhanced levels of expression of E2b (containing the *ts*36 function) in cells infected with *ts*125 or with certain of its revertants might account for enhanced transformation efficiencies. On the other hand, effects on E1a expression might be equally important. In particular, Braithwaite *et al.* (1983) have shown that E1a functions are required to induce cell-cycle abnormalities that might be associated with transformation, and Murray *et al.* (1982b) have shown that these effects are enhanced in *ts*125 infections. Models involving enhanced expression of either E1a or E2b can adequately explain enhanced transformation by *ts*125 at the nonpermissive temperature, but do not so readily account for increased transformation efficiencies at the permissive temperature. Perhaps the *ts*125 defect is also expressed to some extent at low temperatures.

C. Adenovirus Type 12 Mutants

Although by far the most detailed information is available for Ad2 and Ad5, a few transformation-defective mutants have also been isolated in other serotypes, chiefly Ad12. One temperature-sensitive Ad12 mutant, *ts*401, has been shown to be defective in a function required both for viral DNA replication and for maintenance of transformation (Ledinko, 1974, 1978a; Ledinko *et al.*, 1979). This mutant has not been mapped, but from its properties it would be predicted to map in E1, possibly E1a.

A series of interesting Ad12 mutants were isolated several years ago by Takemori *et al.* (1968). These mutants, called *cyt*, gave large, clear plaques on human embryonic kidney cells (in contrast to the small plaques obtained with the parental strain of Ad12) and were characterized by a more extensive and rapid cytopathic effect (in contrast to Wt-virus infection, which tends to leave infected cells intact even very late in the viral replication cycle). The *cyt* mutants were less tumorigenic than Wt Ad12 and failed to transform primary baby hamster kidney cells (Takemori *et al.*, 1968, 1969). The reason for cellular destruction by *cyt* mutants may now be understood in light of the finding that *cyt* mutants induce extensive degradation of both cellular and viral DNA in the course of the infection (Ezoe *et al.*, 1981). Thus, although virus yields are low, cell DNA degradation presumably results in early cell lysis, rapid release of virus, and the formation of large, clear plaques (due to lysed cells). For most mutants, the *cyt* defect is recessive and can be suppressed by coinfection with Ad5 Wt, *hr*I or *hr*II virus, but not by *dl*313 (Lai Fatt and Mak, 1982), suggesting that the associated defect is located in E1b but is not related to the product of the 22 S mRNA affected by group II *hr* mutations.

One *cyt* mutant, *cyt*61, has a dominant phenotype, is not defective for transformation, and may therefore map outside E1 (Lai Fatt and Mak, 1982; Mak and Mak, 1983). Interestingly, *dl*313 itself was found to have a *cyt*-like phenotype in the sense that cell and viral DNA were degraded in *dl*313-infected KB cells (Lai Fatt and Mak, 1982). It is not known why DNA is degraded during *cyt* and *dl*313 infections, but it has been postulated that the role of the *cyt* function is to suppress a virus-induced nuclease activity, consistent with the recessive nature of most *cyt* mutants (Ezoe *et al.*, 1981).

Besides the Ad5 mutant *dl*313, at least two other mutants have been isolated that may be analogous to Ad12 *cyt* mutants. One of these, H2*ts*111, is a recessive DNA^- mutant that results in extensive cell and viral DNA degradation at the nonpermissive temperature (Martin *et al.*, 1978; D'Halluin *et al.*, 1979). Another mutant, or series of mutants, isolated by Chinnadurai *et al.* (1979) give large plaques on KB cells. Both *ts*111 and the large-plaque mutants map in the left part of the Ad2 genome, *ts*111 to the left of 28% (D'Halluin *et al.*, 1982) and the Ad2 large-plaque mutant to the left of 41% (Chinnadurai *et al.*, 1979). More recently a selection of Ad2 lp mutants has been mapped precisely by marker rescue and sequencing and shown to affect the E1b 19K antigen (Chinnadurai, 1983). Some of these mutants were shown to be transformation-defective both in virus-mediated and DNA mediated assays, providing further evidence that 19K plays a key role in oncogenic transformation (Chinnadurai, 1983).

V. TUMOR ANTIGENS AND THEIR ROLE IN TRANSFORMATION

A. Maintenance of Transformation by E1 Proteins

Although any protein capable of eliciting an immune response in tumor-bearing animals can be considered a tumor (T) antigen, most of the emphasis in this section is on proteins encoded within the left end of the adenovirus genome. Much of the data discussed in the previous sections provides at least indirect evidence that E1 proteins are involved not only in the initiation of transformation but also in its maintenance. First, data from a large number of studies have shown that cells from adenovirus-induced tumors, and cells known to have been transformed by adenovirus, almost invariably retain adenovirus DNA and in particular retain sequences from the left end of the genome. Second, as mentioned in Section IV, the properties of cells transformed by host-range (*hr*) mutants and by cold-sensitive *hr* (hr^{cs}) mutants suggest that one or more E1 functions are active in maintaining at least some of the phenotypic characteristics of the transformed cell.

Additional evidence for a role of E1 functions in maintenance of transformation comes from studies on revertants of Ad12-transformed cells. Counterselection of transformed cells for variants that have lost certain phenotypic properties characteristic of the transformed state has been shown to result in cells that have lost or rearranged viral DNA sequences (Groneberg *et al.*, 1978; Groneberg and Doerfler, 1979; Eick *et al.*, 1980; Eick and Doerfler, 1982) or that have undergone changes in methylation patterns of viral DNA, resulting in cells that no longer express E1 proteins (Eick *et al.*, 1980).

A particularly instructive study was carried out by Kuhlmann *et al.* (1982), who showed that all of 39 Ad12-induced hamster tumors contained Ad12 DNA: in 30 tumors, what appeared to be the entire genome was found integrated in a colinear arrangement, while in the remaining 9, either the left end or the left and right ends of the viral DNA persisted. When cell lines were established from primary tumors, it was found that in some cases all viral DNA sequences were lost after several passages in culture, resulting in a shift from an epithelioid to a more fibroblastic morphology. Leaving aside the possibility that a subpopulation of non-adenovirus-transformed cells was selected out of the tumor-cell population, these results are still consistent with adenovirus gene expression being required for maintenance of at least the epithelioid morphology typical of adenovirus transformation. Kuhlmann *et al.* (1982) also made the unusual observation that one of these fibroblastic viral DNA lines was tumorigenic, suggesting the possibility that Ad12 may have acted as a "hit-and-run" carcinogen. In another study utilizing an established rat liver cell line, two ostensibly Ad12-transformed lines were obtained that were T-antigen-negative and had no detectable Ad12 DNA, again indicating a possible "hit-and-run" action (Paraskeva and Gallimore, 1980; Paraskeva *et al.*, 1982). However, since the vast majority of adenovirus-induced tumors and transformed cells have a characteristic morphology and retain and express viral DNA sequences, it seems unlikely that any but a small minority of oncogenic transformations by adenoviruses can occur via such a mechanism. Thus, a major goal in attempts to understand transformation by adenovirus has been the identification and characterization of proteins coded by the transforming genes of adenoviruses, i.e., the genes contained within E1.

The structure and organization of E1 were discussed briefly in Section III.B and illustrated in Fig. 2. As mentioned earlier, only the products encoded by the E1a 12 and 13 S mRNAs and the E1b 13 and 22 S mRNAs seem to be expressed early in infected cells or in transformed cells, and only these and proteins that may be encoded by leftward-transcribed-strand (l-strand) mRNAs are discussed in detail in this section. Theoretical molecular weights are for polypeptides encoded by E1 of Ad5 and differ only slightly from values predicted for Ad2 (Gingeras *et al.*, 1982).

B. Detection of E1 Antigens

Most of the proteins encoded by E1 are made in relatively small amounts prior to the onset of viral DNA synthesis and before cellular protein synthesis is inhibited. To detect virus-specific T antigens, therefore, a variety of enhancement techniques have been used. Their rate of synthesis can be enhanced relative to host-cell protein synthesis through the use of various metabolic inhibitors. Virus-specific proteins can be translated *in vitro* off mRNA selected by hybridization to viral DNA or specific viral DNA fragments, or T antigens can be partially purified by immune precipitation using specific antisera. The latter approach has the advantage of permitting analysis of products made *in vivo* in the absence of any metabolic inhibitors, and thus under "normal" conditions of viral replication, but depends on the availability of appropriate sera. Usually, radioactively labeled virus-specific proteins have been analyzed by sodium dodecyl sulfate–polyacrylamide gel electrophoresis (SDS-PAGE) and autoradiography, though recently high-performance liquid chromatography has been used to fractionate adenovirus T antigens (Green and Brackmann, 1982).

Interpretation of the literature describing adenovirus-specific tumor antigens is complicated by several factors: First, different laboratories use different PAGE conditions and different molecular-weight markers, with the result that molecular-weight assignments for a single protein can vary widely from one study to another. A particularly troublesome contributing factor in this variability arises from the fact that molecular-weight values assigned to Ad2 virion polypeptides, one of the more common set of markers, are in error by a substantial amount in the low-molecular-weight range—below approximately 30K (J. Downey and S. T. Bayley, personal communication; Green *et al.*, 1982). Second, different groups study different adenovirus serotypes, and though Ad2 and Ad5 E1 proteins fortunately are fairly similar, direct comparison with Ad12 antigens based on mobility on gels is virtually impossible. Third, the complexity and apparent molecular weight of many E1 products often bear little relationship to those predicted from the sequence data, and different groups not only assign different molecular weights to E1 proteins, but also detect different numbers of protein species. In the discussion that follows, E1b antigens are designated according to their apparent or reported molecular weights as determined by mobility on SDS-PAGE. Where it is deemed helpful and the identity of the protein under discussion is reasonably clear, observed values are followed in brackets by the theoretical molecular weight predicted from sequence data. In the case of E1a proteins, only apparent molecular weights are given, since there is no similarity between the predicted size and complexity of E1a species and those actually observed.

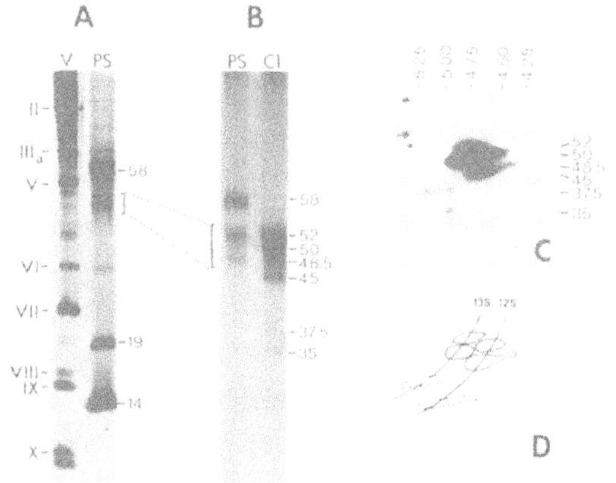

FIGURE 4. E1-encoded T antigens of Ad5. (A) Autoradiograph of a 14% polyacrylamide gel analysis of [^{35}S]methionine-labeled Ad5 T antigens immunoprecipitated with a polyvalent hamster antitumor serum (PS). The polyvalent serum was made by pooling antitumor sera with different major antigenic specificities. (V) ^{14}C-labeled virus marker. Structural proteins are indicated by roman numerals at left. E1a proteins migrating as an unresolved smear around 40–50K are indicated by the vertical bar at right. (B) Autoradiograph of a 10% polyacrylamide gel analysis of ^{32}P-labeled tumor antigens immunoprecipitated with polyvalent antitumor serum (PS) or a rabbit antiserum directed against a synthetic peptide corresponding to the predicted carboxy terminus of the E1a products (C1) (Yee *et al.*, 1983). (C) 2D gel analysis of ^{35}S-labeled E1a products immunoprecipitated with C1 serum. Proteins were separated on an isoelectric focusing tube gel (horizontal dimension) and then on an SDS–12% polyacrylamide slab gel (vertical dimension). Only that portion of the slab gel containing the E1a products is presented, and the approximate pH of the focusing gel is indicated above. (D) Schematic diagram of the gel presented in (C). The identification of different polypeptide species as products of E1a 13 or 12 S mRNAs was made through the use of various E1a mutants (see the text).

C. E1a Antigens

Until recently, the study of E1a was hampered by the lack of good sera capable of immunoprecipitating E1a products, since the major avidity of tumor sera was usually against E1b proteins. Consequently, many studies used *in vitro* translation of selected mRNA to identify E1a proteins or in some cases used two-dimensional (2D) gel electrophoresis to separate virally coded from host proteins. *In vitro* translation of E1a specific mRNA has yielded between two and six polypeptides with molecular weights usually reported to be in the range from 35 to 50K (see Table I and references cited therein). Two-dimensional electrophoresis of infected-cell proteins according to the technique of O'Farrell (1975) separates E1a proteins from most host-cell proteins and allows the detection of a cluster of highly acidic proteins that, from several studies and from a variety of different techniques, have been shown to consist of two related families. Each family, containing up to two major and one minor species, appears to be encoded by either the 12 S or the 13 S E1a message. Smart *et al.* (1981) showed this definitively by isolation and partial sequencing of peptides of E1a proteins purified from 2D gels and comparison of the amino acid sequences with those predicted by cDNA sequence data (Perricaudet *et al.*, 1979). Analyses of proteins made *in vivo* in cells

TABLE 1. Analysis of Ela Proteins of Subgroup C Adenoviruses

Method[a]	Serotype	Proteins	Messinger RNA assignment	Properties	References
In vitro translation of selected mRNA	Ad2	44, 50K	—	—	J. B. Lewis *et al.* (1976)
In vitro translation of selected mRNA, CHX enhancement *in vivo*, and 2D gels	Ad2	40–50K *in vitro*, 38–50K *in vivo*	—	Related acidic	Harter and Lewis (1978)
In vitro translation of selected and size-fractionated mRNA	Ad2	28K 35, 47K 41, 53K	9 S 12 S 13 S	Late, not seen *in vivo*; all 5 proteins highly related by peptide mapping	Halbert *et al.* (1979), Spector *et al.* (1980), Halbert and Raskas (1982)
In vivo and 2D gels	Ad2	40–50K	—	Four major related species of acidic proteins in two families	Green *et al.* (1979), Brackmann *et al.* (1980)
In vitro translation of selected and size-fractionated mRNA	Ad2	28K 42, 54K 48, 58K + two minor species	9 S 12 S 13 S	Late Related	J. B. Lewis *et al.* (1979)
Immune precipitation with tumor sera	Ad5	25K 34–42K 47, 52K	—	—	Schrier *et al.* (1979)

In vitro translation of selected and size-fractionated mRNA	Ad2, 5	28K	9 S	Late	Esche *et al.* (1980)
		42, 54K	12 S		
		48, 58K	13 S	26K in *hr*1 infections	
	Ad5	48K	12 S		Ricciardi *et al.* (1981)
		51K	13 S	28K in *hr*1 infections	
	Ad5	31–42K	—	—	Lupker *et al.* (1981)
In vitro translation 2D gels of *in vivo* products	Ad2	28K	9 S	Late, 28K not detected *in vivo*; mapped to specific mRNAs by partial sequencing of tryptic peptides	Smart *et al.* (1981)
		42, 54K *in vitro*, 38, 42, 46K *in vivo*	12 S		
		48, 58K *in vitro*, 42, 46, 50K *in vivo*	13 S		
ARA-C enhancement *in vivo* and 2D gels	Ad5	Two major proteins	12 S 13 S	Phosphoproteins, assigned to specific messages using mutants	Gaynor *et al.* (1982)
Immune precipitation with sera specific for Ela carboxy-terminal peptide	Ad5	35, 45, 50K 37.5, 48.5, 52K	12 S 13 S	Phosphoproteins, assignment made using mutants	Yee *et al.* (1983), Rowe *et al.* (1983b)

[a] (CHX) cyclohexamide; (ARA-C) cytosine arabinoside.

infected with a variety of E1 mutants, or *in vitro* off mRNA extracted from mutant infected cells, are consistent with the assignments made by Smart and colleagues (Esche *et al.*, 1980; Ricciardi *et al.*, 1981; Gaynor *et al.*, 1982; Yee *et al.*, 1983) (see the following discussion and Fig. 4).

A particularly useful antiserum (E1a.C1 serum) that is specific for the carboxy terminus of Ad5 E1a polypeptides has recently been prepared (Yee *et al.*, 1983). This antiserum was raised in rabbits against a synthetic peptide (Tyr-Gly-Lys-Arg-Pro-Arg-Pro-COOH) that mimics the last five amino acids predicted from sequence data for the products of the 12 and 13 S E1a messages. Studies with wild-type (Wt) Ad5 and E1a mutants have shown that E1a.C1 serum immunoprecipitates a total of at least four major and two minor proteins species from Ad5-infected cells [see Fig. 4], half of which are products of the 12 S mRNA, the other half, of the 13 S mRNA. In addition, a number of tumor sera have been obtained against transformed hamster cells. These sera are also relatively specific for E1a products (Rowe *et al.*, 1983b, 1984b), but they have the rather intriguing property of immunoprecipitating only a subset of the proteins recognized by E1a.C1 serum (see Fig. 4). Furthermore, studies with E1a mutants have shown that the tumor sera react only with the product of the 13 S mRNA, even though the transformed cells contained both 13 and 12 S products (Rowe *et al.*, 1983b, 1984b). The predicted polypeptide specified by the 13 S mRNA differs from the 12 S product only in an internal stretch of 46 amino acids removed from the 12 S protein by the larger intron of 12 S mRNA. As pointed out by Gingeras *et al.* (1982), this region contains 5 Cys residues, and "the proximity of so many potential intra- or inter-molecular cross-linking residues is reminiscent of the hinge region of the immunoglobulin polypeptides." From the studies with mutants discussed in Section IV, and the fact that tumor sera precipitate the 13 S but not the 12 S products, it is clear that this internal stretch of amino acids must have profound effects on both the function and the antigenicity of the 13 S mRNA products.

At present, it is not known why there are so many different species of proteins derived from only two E1a messages or why their apparent molecular weights in SDS-PAGE are so much greater than those predicted from DNA and RNA sequence data. Results with E1a.C1 serum indicate that all these species share a common carboxy terminus. In addition, studies with an Ad5 deletion mutant, *dl*1504 (Osborne *et al.*, 1982), suggest that all E1a polypeptide species share the same N terminus. This mutation removes the first AUG in the coding sequences for E1a proteins, apparently resulting in a shift in the translation start to the next AUG 14 codons downstream, and in *dl*1504-infected cells, there is a corresponding shift in the mobility of all the E1a proteins relative to Wt products (Rowe *et al.*, 1983b). It is therefore likely that the heterogeneity among different E1a proteins, and perhaps also their anomalous migration patterns on SDS-PAGE, are due to posttranslational modification. All the E1a species detected to date have been shown to be phosphorylated (Gaynor *et al.*, 1982; Yee *et al.*, 1983), but whether this explains their unusual gel-electrophoretic mobilities remains to be seen. It also remains to be determined how different E1a species within each family differ functionally.

D. E1b Antigens

The identification and characterization of E1b-coded proteins have been slightly more straightforward than for E1a, because E1b proteins are generally made in larger amounts during productive infection, because tumor sera against E1b antigens are more readily obtained, and finally because the patterns and mobilities on gels correspond rather closely to those predicted from sequencing data. The major E1b T antigen detected in productively infected and transformed cells is a phosphoprotein with an apparent molecular weight in SDS-PAGE of 58–65K for Ad5 and 53–55K for Ad2 (values differ slightly among different reports, but 58K is the most commonly assigned value for Ad5). This protein is the product of the 22 S mRNA. A second major E1b product has been detected that has a molecular weight reported to be 15K (when measured against virion protein markers with incorrect molecular-weight assignments) (Harter and Lewis, 1978; Halbert *et al.*, 1979; Lassam *et al.*, 1979a,b; Ross *et al.*, 1980), but that migrates more like 19K on SDS-PAGE (Schrier *et al.*, 1979; Jochemsen *et al.*, 1981; Green *et al.*, 1982; Rowe *et al.*, 1984a). This protein is encoded by both the 22 and 13 S mRNAs (Bos *et al.*, 1981). The predicted molecular weights for these two E1b proteins from Ad2 and Ad5 sequence data are, respectively, 55 and 21K (Bos *et al.*, 1981; Gingeras *et al.*, 1982), reasonably close to the observed values. These two proteins are made from partially overlapping coding sequences, but in different reading frames, and consequently do not share common peptides. The 19K [21K] protein, which has been shown to be membrane-associated, has been purified to homogeneity, and a monospecific antiserum has been prepared (Persson *et al.*, 1982). Both the 58 and 19K proteins are clearly visible in Fig. 4A. In addition to these major species, a number of E1b-encoded polypeptides have been detected with molecular weights reported to range between 14 and 25K depending on markers and gel systems (Green *et al.*, 1979c). One of these, slightly larger than the 13 S mRNA product, is related to the larger E1b protein encoded by the 22 S mRNA. This protein has recently been purified to near homogeneity (Green *et al.*, 1982) and also translated *in vitro* from E1b-specific RNA, suggesting that it may be a primary translation product.

As mentioned in Section IV.A, the 58K [55K] polypeptide is not synthesized in cells infected with group II *hr* mutants (Lassam *et al.*, 1979a,b; Ross *et al.*, 1980; Rowe *et al.*, 1984a), and both 58K [55K] and 19K [21K] are absent in *dl*313 infections (Ross *et al.*, 1980; Rowe *et al.*, 1984a). Furthermore, the 58K protein is absent from cells transformed by DNA extracted from group II *hr* mutants (Rowe and Graham, 1983) and is not detected in cells transformed by the *Hind*III-G fragment (left 8%) of Ad5, but is found in most cells transformed by fragments that contain all of E1 (Schrier *et al.*, 1979; Rowe and Graham, 1983; Rowe *et al.*, 1983b). The 19K [21K] protein is detected in most cells transformed by *Hind*III-G or larger fragments, but is not detected in cells transformed by fragments covering only E1a (Schrier *et al.*, 1979). The N terminus of an E1b 15K [19K with correct molecular-weight markers] has been sequenced and shown to be identical to the predicted amino acid sequence for the 21K protein (Anderson and Lewis, 1980). All these observations are entirely consistent with the assignment of 58K to the E1b 22 S mRNA and 19K to the 13 S mRNA and the 5′ end of the 22 S mRNA. The fact that the 19K protein is detected in *Hind*III-G-transformed cells suggests that adjacent host-cell sequences may provide mRNA polyadenylation and processing signals that allow synthesis of a translatable mRNA species.

As indicated in Fig. 2 (Section III.B), there are two unassigned open reading frames on l-strand DNA that could encode polypeptides with molecular weights of 14 and 23K (encoded within sequences from 2290 to 2005 and 1712 to 1196, respectively, in the Ad2 genome) (Gingeras *et al.*, 1982) or polypeptides with molecular weights of 14K (2287–2008) and 29K (1715–1067) for Ad5. As mentioned previously, Katze *et al.* (1982) detected a low-molecular-weight protein that is made in Ad2-infected cells and can be translated off a 20 S mRNA species selected by hybridization to E1 DNA. This protein was reported to have a molecular weight of 11K but migrates just ahead of virion polypeptide IX on SDS-PAGE and thus, with corrected molecular-weight assignments, would probably be closer to 14K in size. Results from hybrid-arrest studies indicated that the protein is encoded off the l-strand of E1 (Katze *et al.*, 1982) and thus make it a likely candidate product of the smaller of the two l-strand open reading frames in E1. If these observations are confirmed, results from studies with *hr* and *dl* mutants may have to be reinterpreted in light of their possible effects on l-strand functions.

E. Tumor Antigens Encoded outside E1

Although E1 proteins are the only ones that are consistently expressed in adenovirus-transformed cells, some tumor sera contain antibody against other (usually early) virus-coded proteins. One of these is a low-molecular-weight protein [previously reported as a 10.5 or 11K protein and thought to be an E1a product (Lassam *et al.*, 1979a,b; Ross *et al.*, 1980)] that is now known to be an E4 product of molecular weight 14K (Downey *et al.*, 1983) (see Fig. 4A) and that is probably identical to an "11K" E4 protein detected in other studies *in vivo* or *in vitro* (J. B. Lewis *et al.*, 1976; Harter and Lewis, 1978; Chin and Maizel, 1977). This appears to be a major antigen in many transformed cells containing the right end of the viral genome, but is only one of several E4 proteins that have been detected in transformed and infected cells (Brackmann *et al.*, 1980; J. B. Lewis *et al.*, 1976; Harter and Lewis, 1978; Matsuo *et al.*, 1982). The E4 14K is a nuclear protein that is enriched in the nuclear matrix fraction (Chin and Maizel, 1977) and is found in association with chromatin (Rowe *et al.*, 1983c) and with a complex that is capable of synthesizing adenovirus DNA *in vitro* (Rho *et al.*, 1977). Other early adenovirus gene products detected in some transformed cell lines are an E3 glycoprotein of molecular weight 19–20K (Jeng *et al.*, 1978; Persson *et al.*, 1979) and the E2 72K DNA-binding protein (DBP) (Gilead *et al.*, 1975; Levinson *et al.*, 1976) (see Chapter 5 for a more detailed discussion).

It is clear that none of these proteins plays a major role in maintenance of transformation, since many, if not the majority, of fully transformed cells express no viral genes other than E1. This is not to say that products of other early regions would necessarily have no effect on the phenotype of transformed cells. The E3 glycoprotein in particular has been shown to form a ternary complex with the major histocompatibility antigen in the plasma membrane of infected and transformed cells (Kvist *et al.*, 1978; Signas *et al.*, 1982) and therefore might be expected to affect the host response to transplantation of cells expressing E3. Other proteins such as the 72K DBP or the 14K E4 protein, which interact with DNA or chromatin, could have profound effects on cells in which they are expressed even if these effects were not related to oncogenic transformation.

F. Tumor Antigens of Adenovirus Type 12

Since E1 of Ad12 has been sequenced and its mRNA species identified and mapped (Perricaudet *et al.*, 1980b; Kimura *et al.*, 1981; Bos *et al.*, 1981; Virtanen *et al.*, 1982), the structure of proteins encoded by the transforming genes can be predicted in considerable detail. Nevertheless, the Ad12 E1 proteins that are synthesized *in vivo* are not yet as well characterized as their Ad2 and Ad5 counterparts. The organization of Ad12 region E1 is quite similar to that of Ad2 and Ad5, except that in E1a of Ad12, there are two pairs of mRNAs, one pair analogous to the 13 S E1a mRNA of Ad2 but containing two mRNAs with different 5′ termini and a similar pair of mRNAs analogous to the 12 S mRNA. Heterogeneity at the 5′ end of E1a messages affects only the untranslated leader, so that E1a of Ad12 is predicted to encode only two related polypeptides with molecular weights of 26 and 30K. No mRNA analogous to the 9 S E1a mRNA of Ad2 has been reported for Ad12. The organization of E1b for Ad12 is slightly different from that of Ad2 or Ad5. Virtanen *et al.* (1982) detected a large mRNA corresponding to the 22 S mRNA seen in Ad2- and Ad5-infected cells and two smaller mRNA species that differ from the 13 S mRNA shown in Fig. 2 (Section III.B) in having two introns rather than a single intron. The large message could encode both a 54K and a 19K polypeptide read in different reading frames and from different start codons as for the subgroup C viruses (Bos *et al.*, 1981). The product predicted from both smaller messages would be a 12K polypeptide related to the 54K protein.

Besides organizational similarities and extensive homology between E1 genes of Ad12 and those of Ad2 and Ad5, the E1-coded proteins are of course functionally very similar as well. This is indicated by the fact that *hr* mutants of Ad5 can be complemented for growth in nonpermissive cells by coinfection with Ad12 (Rowe and Graham, 1981; Williams *et al.*, 1981; Brusca and Chinnadurai, 1981) and the Ad12 cytopathic (*cyt*) mutants can be complemented by coinfection with Ad5 (Lai Fatt and Mak, 1982). In addition, DNA segments containing E1a of Ad5 (or Ad12) can act together with E1b of Ad12 (or Ad5) to transform rat cells in DNA-mediated transformation assays (Bernards *et al.*, 1982). However, important differences must also exist, since Ad12-transformed rat cells (containing only E1) are tumorigenic in rats, whereas their Ad2- and Ad5-transformed counterparts generally are not.

Ad12 T antigens, made in infected or transformed cells or synthesized by *in vitro* translation, have been analyzed in a number of studies. In some cases, as many as six to eight proteins that may be encoded by E1 have been detected (Esche *et al.*, 1979; Wold *et al.*, 1979; Jochemsen *et al.*, 1980). The relationship of most of these proteins to polypeptides predicted from sequencing data is, however, difficult to decipher. Shiroki *et al.* (1980) could immunoprecipitate virus-specific products that formed a cluster of acidic proteins with molecular weights of 35–40K on 2D gels and that very likely are the Ad12 analogues of the acidic E1a proteins of Ad2 and Ad5. Similar observations were made with Ad12 E1a proteins synthesized by *in vitro* translation (Esche and Siegmann, 1982). Ad12-specific antigens have been shown to bind to double-stranded DNA (dsDNA) and have been partially purified on the basis of this property (Raska *et al.*, 1976; Biron and Raska, 1977; Biron *et al.*, 1978). Partially purified Ad12 T-antigen preparations contained immunoprecipitable protein with apparent molecular weights of 64, 50, and 11K on SDS-PAGE. Ledinko (1978b) detected a 60K protein in Ad12-transformed cells by immunoprecipitation with tumor sera and found that the transformation-defective Ad12 temperature-sensitive mutant *ts*401 was defective for expression of this antigen. It is not clear whether this protein or the DNA-binding T antigens are encoded by E1a or E1b, if indeed they are E1 products at all. Vasconcelos-Costa (1976) has characterized an Ad12-specific antigen that could be the Ad12 counterpart of the E1b 13 S product or perhaps the E3 glycoprotein. This antigen, called "S" antigen, was extensively purified (≈2000-fold) and shown to be a surface antigen in Ad12-infected and transformed cells.

G. Function of E1 Tumor Antigens

One might hope that most, if not all, of the Ad2 and Ad5 E1-encoded proteins have now been identified. The task of characterizing them, however, has hardly begun. Both E1a and E1b proteins are involved in viral replication in HeLa or KB cells and in transformation, but not all E1a- or E1b-encoded products are absolutely essential. The 12 S mRNA of E1a can be totally dispensed with for viral replication (Montell *et al.*, 1982) though the 12 S product seems to be needed in transformation (C. Montell, personal communication). The structural protein IX encoded in E1b seems not to be absolutely required for the assembly of infectious virions (Colby and Shenk, 1981), and neither protein IX mRNA nor E1a 9 S mRNA is expressed in transformed cells (Spector *et al.*, 1980; J. B. Lewis and Mathews, 1981; Green *et al.*, 1981). In addition, even *hr* and *dl* mutants, isolated on the basis of defective growth in HeLa cells, are extremely "leaky," especially at high multiplicities of infection. In the case of deletion mutants, *dl*312 or *dl*313, which lack essentially all of E1a or E1b, respectively, this leakiness clearly cannot be due to expression of partially functional virus-coded proteins, but must imply that the need for functional E1 proteins in viral replication in HeLa cells is not absolute. Either viral replication can bypass a deficiency in E1-coded products or perhaps host-cell functions can partially substitute for virus-coded functions. The fact that group II *hr* mutants can replicate efficiently in normal primary human embryo kidney cells (Harrison *et al.*, 1977) is consistent with the latter possibility.

One or more E1a products have been shown to play a key role in controlling expression of other early viral genes (Berk *et al.*, 1979a; N. Jones and Shenk, 1979b), possibly at the level of transcription (Nevins, 1981). It has been suggested that this control is manifested through inactivation of a cellular protein that acts to repress early viral transcription (Nevins, 1981) or to block accumulation of early mRNA (Katze *et al.*, 1981, 1983), since pretreatment of HeLa cells with inhibitors of protein synthesis permitted E1a *hr* or *dl* mutants to express other early regions. This model could help to explain why E1a mutants are leaky at high multiplicities of infection: leakiness might represent titration out of host-cell repressor proteins.

The importance of E1a in controlling other viral genes is reflected in what appears to be a central role in transformation as well. The fact that fragments of viral DNA that contain only E1a can immortalize primary rodent cells and induce a "partial" transformation was discussed at length in Section III, and the fact that *dl*313 (which lacks E1b) can, at high multiplicities of infection, induce a partial transformation has also been mentioned (Section IV). The ability of E1a to transform may be due to E1a-induced alterations in cellular control mechanisms. It has been known for several years that adenoviruses can induce cell DNA synthesis in quiescent cells (Shimojo and Yamashita, 1968; Takahashi *et al.*, 1969; Tsukamoto and Sugino, 1972), as well as the synthesis of a variety of cellular enzymes, many of them associated with DNA replication (Kit *et al.*, 1965; Takahashi *et al.*, 1966; Bresnick and Rapp, 1968; Ledinko, 1967, 1970; Strohl, 1973; Cheetham and Bellett, 1982). Infection with adenoviruses has also been found to induce a variety of chromosomal aberrations in mammalian cells as well as abnormal mitoses (Stich, 1973; zur Hausen, 1967, 1968; Bellett *et al.*, 1982a; Murray *et al.*, 1982a,b). In addition, induction of synthesis of mammalian heat-shock protein in adenovirus-infected HeLa cells has recently been reported (Nevins, 1982). Recent studies utilizing a variety of adenovirus mutants have shown that many of these effects—induction of thymidine kinase and cell DNA synthesis, heat-shock proteins, and cell-cycle abnormalities such as abnormal mitoses and chromosome aberrations—depend on expression of E1a (Cheetham and Bellett, 1982; Bellett *et al.*, 1982a,b; Nevins, 1982; Braithwaite *et al.*, 1983), although the E2a 72K DBP and one or more E1b functions have been implicated as well (Rossini *et al.*, 1981). All these observations tend to suggest that one of the major roles of E1a is not only to regulate expression of other early viral genes but also to regulate expression of many cellular genes. An attractive hypothesis is that continued control of cellular functions by E1a leads to immortalization of primary cells and results in acquisition by the host cell of at least a subset of the phenotypic properties associated with oncogenic transformation.

How E1a proteins act at the molecular level is still not clear, since their biochemical characterization is still in its infancy. Other than that they are highly acidic phosphorylated proteins, relatively little is known about them. Studies on the intracellular localization of E1a proteins have shown that they are found approximately equally in cytoplasmic and nucleoplasmic fractions of infected cells (Rowe *et al.*, 1983c). As mentioned in Section V.F, studies with Ad12 T antigens have shown that several of them bind efficiently to dsDNA, though it is not yet clear whether these DNA-binding proteins were encoded in E1a or E1b of Ad12 (Biron *et al.*, 1978; Biron and Raska, 1977; Raska *et al.*, 1976).

As discussed in the previous sections, in addition to E1a expression, one or more E1b functions are needed to confer on the transformed cells a full complement of transformed properties. The most important region of E1b in this respect seems to be the 5′ half, up to approximately 7–8 m.u. This segment contains the coding sequences for the 19K [21K] protein specified by the 13 S mRNA and an unassigned open reading frame on the l-strand (URF10) from approximately 5.5 to 6.5 m.u. (Gingeras *et al.*, 1982). It seems likely that either the 19K protein or the product of URF10 plays a key role in causing oncogenic transformation although a role for the N terminal region of the 58K [55K] protein cannot be ruled out.

The 19K protein that has been shown to be associated with the membrane of infected and transformed cells (Persson *et al.*, 1982; Rowe *et al.*, 1983c) is not a phosphoprotein (Levinson and Levine, 1977; Ross *et al.*, 1980; Gaynor *et al.*, 1982; Rowe *et al.*, 1983c) and is not glycosylated (Persson *et al.*, 1982). Indirect immunofluorescence of transformed cells using antibody against the 19K protein resulted in staining predominantly at the periphery of the cells, suggesting that the 19K protein may be localized at the plasma membrane (Rowe *et al.*, unpublished findings) and recent studies by Fohring *et al.* (1983) indicate that this protein may be exposed on the surface of Ad12 infected and transformed cells. Expression of the 19K T antigen at the cell surface may be a major factor in expression of the transformed phenotype, since many of the properties of transformed cells, such as lack of contact inhibition of growth, anchorage independence, and of course tumorigenicity may in large part be related to changes in the cell surface. Undoubtedly, the availability of purified 19K antigen and monospecific antibody against it (Persson *et al.*, 1982) will aid greatly in the elucidation of its role in transformation.

Only recently has a protein been detected that might be encoded within the unassigned open reading frame on the l-strand in E1b (Katze *et al.*, 1982). As discussed at length in the preceding sections, this open reading frame, like the coding sequences for 19K, is in a region of E1 that appears to be critical for obtaining complete transformation of primary rodent cells, and consequently more extensive characterization of this protein is certainly desirable.

Though the 58K antigen encoded by the 22 S mRNA of E1b may not play any major role in maintaining the transformed cell phenotype, it is not without interest, since it is definitely required for initiation of transformation by virions. Like the E1a proteins, the 58K protein is phosphorylated in both infected and transformed cells (Ross *et al.*, 1980; Gaynor *et al.*, 1982; Malette *et al.*, 1983) and is also found in both cytoplasmic and nucleoplasmic subcellular fractions, accumulating preferentially in the nucleus late in infection (Rowe *et al.*, 1983c). In agreement with these observations, fluorescent-antibody staining with a monoclonal antibody specific for the 58K antigen resulted in predominantly nuclear (with some cytoplasmic) staining (Sarnow *et al.*, 1982b).

Immunoprecipitates containing the 58K antigen have been shown to contain a protein kinase activity that can phosphorylate 58K as well as exogenous substrates (Lassam *et al.*, 1979c; Branton *et al.*, 1979, 1981). Kinase activity was produced in infected cells with the same kinetics as 58K, and activity was reduced in cells infected with group II *hr* mutants (Branton *et al.*, 1979). Protein kinase activity has also been found in immunoprecipitates of Ad12 T antigens (Raska *et al.*, 1979; Branton *et al.*, 1981). The significance of these observations is still not clear. It is not known whether the kinase activity is intrinsic to one or more virus-coded proteins or whether it represents binding of a cellular enzyme to adenovirus T antigens. In either event, in view of the key role that protein kinases may play in transformation by several retroviruses (Marx, 1981; Hunter, 1982), the association of protein kinase with adenovirus T antigens is certainly intriguing.

The Ad5 58K protein has been shown to be associated with a 54K cellular protein in a variety of different adenovirus-transformed rodent cells (Sarnow *et al.*, 1982a; Lane and Harlow, 1982). This cellular protein was shown to be identical or very closely related to the 54K cellular protein associated with SV40 T antigen, suggesting that the adenovirus 58K and SV40 T antigen may share similar functions in transformation. The 54K cellular protein is also complexed with Epstein–Barr nuclear antigen in Epstein–Barr-virus-transformed cells (Luka *et al.*, 1980) and is found in elevated amounts in cells transformed by chemical carcinogens as well as viruses (Linzer and Levine, 1979; Lane and Crawford, 1979; De Leo *et al.*, 1979; Lane and Harlow, 1982).

VI. FACTORS THAT AFFECT THE TUMORIGENICITY OF ADENOVIRUS-TRANSFORMED CELLS

A. *In Vitro* Properties of Transformed Cells

Generally, a number of growth properties are associated with malignant transformation of cells (reviewed in Tooze, 1981). Among these are the ability to grow to high saturation densities (loss of contact inhibition of growth), decreased dependence on serum growth factors, ability to grow in suspension in semisolid media, and, especially in the case of adenovirus-transformed cells, ability to grow in media containing low concentrations of calcium ions. In addition, a number of changes at the cell surface and in the cytoskeletal architecture may occur, including increased agglutinability by lectins, expression of virus-specific transplantation antigen, disruption of actin cables and reduced levels of fibronectin [formerly "large, external transformation-sensitive" (LETS) protein] at the cell surface. Of all these characteristics, probably the only one that correlates best with tumorigenicity for papovavirus transformation is ability to form colonies in semisolid media (Shin *et al.*, 1975; Steinberg *et al.*, 1979), though the correlation is by no means perfect and may not extend to other viruses. Absence of fibronectin on the cell surface of adenovirus-transformed cells has also been reported to correlate with tumorigenicity (Chen *et al.*, 1976; Gallimore and Paraskeva, 1979), though again the correlation is not perfect (Gallimore *et al.*, 1979).

The phenotypic properties of adenovirus-transformed cells can vary over a broad range and may encompass all or practically none of the aforementioned characteristics. As discussed previously, some of the properties of adenovirus-transformed cells can be correlated to a certain extent with expression of viral genes. For instance, cells transformed only by E1a of Ad5 or Ad12 do not grow well in medium containing low concentrations of calcium ions or serum, do not clone in soft agar or induce tumors, and have a fibroblastic morphology (Shiroki *et al.*, 1979b; Houweling *et al.*, 1980; McKinnon and Graham, unpublished observations), whereas transformation by larger fragments results in cells with growth properties more similar to those of oncogenically transformed cells. Thus, one or more E1b functions evidently can potentiate tumorigenicity. Expression of E1b alone, however, is not sufficient: several lines of rat cells transformed by group I host-range (*hr*) mutants have been shown to express E1b proteins, but the cells do not clone in soft agar and are nontumorigenic in nude mice (Ruben *et al.*, 1982).

Although E1 clearly plays a major role in controlling the properties of transformed cells, an important fact is that lines that contain and apparently express all of E1 do not necessarily display a uniform phenotype. In particular, tumorigenicity can vary greatly even among cell lines transformed by the same adenovirus serotype and containing the same viral genes. Indeed, even cell lines that apparently have all the *in vitro* properties associated with malignant transformation may nonetheless be nontumorigenic (Gallimore *et al.*, 1979). Probably the major factors in determining tumorigenicity are the interactions between transformed or tumor cells and the immune system of the host, and past failure to detect a clear correlation between *in vitro* properties of transformed cells and degree of tumorigenicity may be due to neglect of *in vivo* effects. Studies that are beginning to shed some light on the control of adenovirus-induced tumors by immune surveillance mechanisms are discussed in the following sections.

B. Immune Response to Adenovirus-Transformed Cells

It has long been apparent that the immune response of the host animal following injection of tumor viruses or transformed cells is of paramount importance in determining whether tumors will develop. In the case of adenoviruses, as for other tumor viruses, several observations indicated that the ability of virus or transformed cells to induce tumors requires that host immune surveillance mechanisms be circumvented: (1) Preimmunization of adult Syrian hamsters with Ad12 specifically increased their resistance to subsequent transplants of Ad12 tumor cells, implying the existence of tumor-specific transplantation antigens (TSTAs) in adenovirus-transformed or tumor cells. TSTAs of Ad12 cross-reacted with Ad7, but not with Ad5 (Trentin and Bryan, 1966; Sjogren *et al.*, 1967). (2) Thymectomized mice and mice treated with antilymphocyte serum were found to be more susceptible to tumor induction by Ad12 than were untreated mice (Allison *et al.*, 1967). (3) Rat cells transformed by "nononcogenic" Ad2, though generally unable to induce tumors in newborn or weanling rats, could cause tumors in animals treated with antithymocyte serum (Gallimore, 1972). Since Ad12-transformed cells can induce tumors in immunocompetent animals, a possible explanation for the differences in oncogenicity between subgroup C and subgroup A viruses is that their transformed cells differ in susceptibility to elimination by immune surveillance mechanisms.

Several studies with a variety of different virus-transformed cells derived from mouse, rat, or hamster suggest that the host immune response to challenge with transformed cells (or virus) may be largely comprised of two components: a slow but relatively specific induced immunity involving mainly the interaction of TSTAs and effector T lymphocytes and a more rapid but less specific response mediated by natural killer (NK) cells (Allison, 1980; Tevethia, 1980; Herberman and Holden, 1978). Although differences in tumorigenicity between different adenovirus-transformed cells are probably related to differences in sensitivity to both these immune surveillance mechanisms, the best correlation observed to date has been between degree of tumorigenicity and degree of resistance to NK cells (Raska and Gallimore, 1982) or resistance to NK cells as well as activated macrophages (Cook *et al.*, 1980, 1982). It was found that Ad2-transformed rat or hamster cell lines were nontumorigenic or weakly tumorigenic and highly susceptible to lysis by normal spleen cells, while more highly tumorigenic Ad12-transformed cells were relatively resistant to lysis. In parallel experiments (Raska and Gallimore, 1982), no significant differences were seen in the susceptibility of Ad2- and Ad12-transformed cells to lysis by sensitized T cells.

It is clear, however, that susceptibility to lysis by NK cells cannot be the only parameter that affects the degree of tumorigenicity. Many, if not most, Ad2- and Ad5-transformed rat cells, while nontumorigenic in rats, are able to induce tumors in nude mice that are not deficient in NK activity. This observation suggests that immune surveillance mechanisms other than NK cells may be responsible for preventing tumor induction by Ad2- and Ad5-transformed cells in rats, though it is possible that mouse NK cells are unable to lyse adenovirus-transformed rat cells.

C. Tumor-Specific Transplantation Antigens

One explanation for the inability of group-C-adenovirus-transformed rat cells to induce tumors in immunocompetent rats, first proposed by McAllister *et al.* (1969a), was that group C virus transformants might be more immunogenic than their group-A-transformed counterparts. TSTAs associated with adenovirus-transformed cells had already been demonstrated (Trentin and Bryan, 1966), and it has been shown that TSTA of Ad12 cross-reacted with TSTA of weakly oncogenic Ad7, but not with TSTA of nononcogenic Ad5 (Ankerst and Sjogren, 1970).

However, in the studies by Raska and Gallimore (1982) that showed a correlation between tumorigenicity of adenovirus-transformed cells and their susceptibility to lysis by NK cells, no correlation was seen with sensitivity to lysis by secondary cytotoxic T cells. Also, as pointed out by A. M. Lewis and Cook (1982), highly immunogenic SV40-transformed hamster cells can induce tumors in syngeneic and allogeneic hamsters, whereas Ad2-transformed cells are tumorigenic only in newborn syngeneic hamsters. Thus, differences in degree of immunogenicity alone may not suffice to explain differences in tumorigenicity of different adenovirus-transformed cells.

That TSTA of adenovirus-transformed cells was in all likelihood virus-coded was suggested from the fact that it was virus-specific; i.e., the common TSTA specified by Ad12 and Ad7 did not cross-react with that specified by Ad5 (Ankerst and Sjogren, 1970). On the other hand, Ad2 and Ad5 TSTAs do cross-react (Gallimore and Paraskeva, 1979). Although one component in adenovirus-specified TSTA may be the E3 19K glycoprotein shown to form a complex with the major histocompatibility antigen (Kvist *et al.*, 1978), cells that lack E3 (as do most adenovirus-transformed cells) nevertheless express TSTA. In particular, cell lines that contain only the left end of the viral genome have been shown to contain TSTA (Shiroki *et al.*, 1979a; Raska *et al.*, 1980, 1982), so that genes that encode it must be in E1. Shiroki *et al.* (1979a) looked at the ability of various Ad12-transformed cells to immunize rats against tumor induction by Ad12 and found that cells transformed by the left 7.2% of Ad12 could reduce tumor incidence, while cells transformed by the left 4.5% could not. If confirmed, these results would narrow the sequences that code for Ad12 TSTA to those that code for the 19K product of the E1b 13 S mRNA. The fact that the corresponding E1b protein of Ad2 and Ad5 has been shown to be a membrane-bound antigen (Persson *et al.*, 1982; Rowe *et al.*, 1983e) strengthens the case for this antigen being a component of TSTA.

Recent studies reported by Schrier *et al.* (1983) and Bernards *et al.* (1983) indicate that a major factor in determining the degree of tumorigenicity of adenovirus transformed cells may be related to the ability of viral gene functions to suppress the expression of the host cell major histocompatibility complex (MHC). Apparently Ad12 E1a functions, specifically those encoded by 13 S mRNA, inactivate MHC class I gene expression in Ad12 transformed cells whereas cells containing Ad5 E1a express normal levels of class I MHC antigens. Since foreign antigens are only recognized by cytotoxic T cells in the context of MHC class I antigens (Zinkernagel and Doherty, 1979), Ad12 transformed cells are able to avoid immune destruction in immunocompetent hosts under conditions in which Ad5 transformed cells are rejected. It still remains to be explained, however, why Ad2 and Ad5 transformed rat cells are nontumorigenic in rats while similarly transformed hamster cells are able to induce tumors in immunocompetent hamsters.

VII. ROLE OF ADENOVIRUSES IN NATURALLY OCCURRING CANCERS

A. Animal Models

There appears to be no known animal system in which adenoviruses play any clear role in causing malignancy under natural circumstances, even though the adenovirus family represents one of the most widely dispersed infectious agents, members having been isolated from over a dozen different species (Norrby *et al.*, 1976). Thus, outside the laboratory, there is no animal model to provide clues as to which, if any, human malignancies might have an adenovirus etiology. The only models available, therefore, are those that involve inoculation of experimental animals with adenoviruses or adenovirus-transformed cells. Tumors induced by human adenoviruses in rodents vary greatly in type, with undifferentiated sarcomas, undifferentiated tumors associated with adenocarcinomas, and very frequently tumors apparently derived from cells of the peripheral or central nervous system having been reported [see Mukai and Kobayashi (1972) and Ibelgaufts (1982) and references therein]. Mukai and collaborators have found that adenoviruses can induce retinal tumors in rats (Mukai and Murao, 1975) and have made the particularly intriguing observation that Ad12 could induce retinal tumors in newborn baboons, tumors that were undistinguishable from human retinoblastomas (Mukai *et al.*, 1980). Recently, human embryo retinoblasts transformed by Ad12 have been shown to induce, in nude mice, tumors that resemble retinoblastomas (Byrd *et al.*, 1982).

B. Human Tumors

Since, in essentially every case, tumors induced experimentally by adenovirus inoculation have been found to contain adenovirus-specific DNA sequences and since adenovirus sequences persist in transformed cells, it is not unreasonable to expect that tumors induced under natural conditions should also contain adenovirus DNA. However, despite very extensive and well-controlled experiments by Green and collaborators, using probes specific for several adenovirus serotypes, to detect adenovirus sequences in a large number of different human tumors, none has been found (Mackey *et al.*, 1976, 1979; Green *et al.*, 1979b, 1980). The level of sensitivity in many of these studies was such that if as little as one copy of only a small part of the viral genome per cell were present, it would have been detected. Thus, unless only a minority of cells in the tumors harbored viral genes, or unless adenovirus induces tumors by a hit-and-run mechanism, these studies suggest that tumors in man are rarely, if ever, caused by adenoviruses.

Following the reported induction of retinoblastomalike tumors in baboons by Mukai *et al.* (1980), an attempt was made to detect Ad12 sequences in DNA from six human retinoblastoma cell lines, without success (S. Mak *et al.*, 1982). These experiments used Southern blot hybridization with radioactive Ad12 DNA and should have detected as little as one copy per cell of a fraction of the viral genome, but these analyses have not yet been extended with other adenovirus serotypes.

Different techniques for detecting viral sequences in tumor cells were used by Ibelgaufts *et al.* (1982) in a study of human neurogenic tumors, with strikingly different results. With the use of adenovirus DNA labeled with ^{3}H by nick translation in *in situ* hybridizations of tumor biopsies and autoradiography, Ad2-, Ad12-, or bovine adenovirus type 3 (B Ad3)-specific nucleic acid sequences were detected in a majority of human neurogenic tumors. Not all cells in the tumor mass contained virus-specific sequences, with grains distributed only over tumor-cell islets and not over fibroblastic tumor stroma. The sensitivity and specificity of the procedure used to detect viral nucleic acids in the study are difficult to ascertain, and the interpretation is made still more difficult by the fact that K. W. Jones *et al.* (1979) reported hybridization of Ad2 and Ad5 DNA to RNA of normal human placenta.

Besides the mostly negative results obtained in molecular hybridization studies, there are other reasons that human adenoviruses may not seem likely candidates as etiological agents in human cancer. The failure to find any naturally occurring adenovirus-induced malignancy in animals is one reason. Another is that no epidemiological evidence has been reported that indicates a correlation between adenovirus infections and human malignancy. Finally, adenoviruses generally do not transform cells of their normal host, but rather transform nonpermissive cells from heterologous animals. Only two successful attempts at transforming human cells with human adenoviruses have been reported (Graham *et al.*, 1977; Byrd *et al.*, 1982), and both these attempts used adenovirus DNA fragments. Even with sheared, noninfectious Ad5 DNA, transformation of human embryo kidney cells was extraordinarily inefficient compared to transformation of rodent cells (discussed by Graham, 1977). It is of some interest that human embryo retinoblasts were at least 100-fold more readily transformed than were human embryo kidney cells, though still less easily transformed than were rat embryo cells (Byrd *et al.*, 1982).

In the absence of any epidemiological evidence linking adenoviruses with malignant disease, and in the absence of any clear demonstration of adenovirus DNA in any of the numerous human tumors analyzed to date, it may be questionable whether such efforts should continue. However, in view of the observations mentioned above, perhaps attention should be focussed on retinoblastomas, screening not only for human adenoviruses, which transform human cells only very inefficiently, but also for other animal viruses such as adenoviruses of canine or feline origin.

Whether adenoviruses are causative agents in human cancer is obviously of major clinical concern, but whatever the answer to that important question, adenoviruses are well established as very useful tools for probing the mysteries of oncogenic transformation. Essentially all the viral gene products that are involved in transformation have now been identified, at least for Ad2 and Ad5, and some have been purified. The task of characterizing these proteins is now under way, and that process, combined with continuing work on the development and characterization of mutants in the transforming functions, should lead to a clearer understanding of how adenoviruses and other agents transform mammalian cells.

ACKNOWLEDGMENTS. I am very grateful to the many colleagues who provided reprints and preprints for this review. I also want to thank Drs. S. Bacchetti, S. Mak, and W. Rawls for helpful discussions and comments on the manuscript, Janice Butera for preparing the manuscript, and D. Rowe and P. Branton for their comments and for preparing Fig. 4. Finally, I am indebted to my wife for displaying considerable patience during the preparation of this article. Work in my laboratory was supported by grants from the Medical Research Council and the National Cancer Institute of Canada, and I am a Research Associate of the latter.

REFERENCES

Aiello, L., Guilfoyle, R., Huebner, K., and Weinmann, R., 1979, Adenovirus 5 DNA sequences present and RNA sequences transcribed in transformed human embryo kidney cells (HEK-Ad5 or 293), *Virology* **94**:460–469.

Aleström, P., Akusjärvi, G., Perricaudet, M., Mathews, M.B., Klessig, D.F., and Pettersson, U., 1980, The gene for polypeptide IX of adenovirus type 2 and its unspliced messenger RNA, *Cell* **19**:671–682.

Allison, A.C., 1980, Immune responses to polyoma virus and polyoma virus-induced tumors, in: *Viral Oncology* (G. Klein, ed.), pp. 481–487, Raven Press, New York.

Allison, A.C., Berman, L.D., and Levey, R.H., 1967, Increased tumour production by adenovirus type 12 in thymectomized mice and mice treated with anti-lymphocyte serum, *Nature (London)* **215**:185–187.

Anderson, C.W., and Lewis, J.B., 1980, Amino-terminal sequence of the adenovirus type 2 proteins: Hexon, fibre, component IX, and early protein IB-15K, *Virology* **104**:27–41.

Ankerst, J., and Sjogren, H.O., 1969, Cross-reacting TSTAs in adeno 7 and 12 tumors demonstrated by ^{51}Cr-cytotoxicity and isograft rejection tests, *Int. J. Cancer* **4**:279–287.

Ankerst, J., and Sjogren, H.O., 1970, Demonstration of two group specific TSTAs in adenovirus induced tumours. *Int. J. Cancer* **6**:84–94.

Arrand, J.E., 1978, Mapping of adenovirus type 5 temperature-sensitive mutations by marker rescue in enhanced double DNA infection, *J. Gen. Virol.* **41**:573–586.

Babich, A., and Nevins, J.R., 1981, The stability of early adenovirus mRNA is controlled by the viral 72 Kd DNA binding protein, *Cell* **26**:371–379.

Babiss, L. E., Ginsberg, H.S., and Fisher, P.B., 1983, Cold-sensitive expression of transformation by a host range mutant of type 5 adenovirus, *Proc. Natl. Acad. Sci. U.S.A.* **80**:1352–1356.

Bellett, A.J.D., 1975, Covalent integration of viral DNA into cell DNA in hamster cells transformed by an avian adenovirus, *Virology* **65**:427–435.

Bellett, A.J.D., Waldron-Stevens, L.K., Braithwaite, A.W., and Cheetham, B.F., 1982a, Spermine and aminoguanidine protect cells from chromosome aberrations induced by adenovirus during the G2 phase of the cell cycle, *Chromosoma* **84**:571–583.

Bellett, A.J.D., Braithwaite, A.W., Peng, L., and Cheetham, B.F., 1982b, Adenovirus transforming gene(s) in early region E1A alter the program of gene expression during the cell growth cycle, in: *Manipulation and Expression of Genes in Eukaryotes* (P. Nagley, A.W., Liane, W.J. Peacock, and J.A. Puteman, eds.), Academic Press, Sydney.

Berk, A.J., Lee, F., Harrison, T., Williams, J., and Sharp, P.A., 1979a, Pre-early adenovirus 5 gene product regulates synthesis of early viral messenger RNAs, *Cell* **17**:935–944.

Berk, A.J., Lee, F., Harrison, T., Williams, J., and Sharp, P.A., 1979b, Phenotypes of adenovirus 5 host-range mutants for early mRNA synthesis, *Cold Spring Harbor Symp. Quant. Biol.* **44**:429–436.

Bernards, R., Houweling, A., Schrier, P.I., Bos, J.L., and van der Eb, A.J., 1982, Characterization of cells transformed by Ad5/Ad12 hybrid early region I plasmids, *Virology* **120**:422–432.

Bernards, R., Schrier, P.I., Houweling, A., Bos, J.L., van der Eb, A.J., Zijlstra, M., and Melief, C.J.M., 1983, Tumorigenicity of cells transformed by adenovirus 12 by evasion of T-cell immunity, *Nature* **305**:776–779.

Biron, K.K., and Raska, K., Jr., 1977, Purification of adenovirus type 12 tumor antigen from transformed hamster cells, *Virology* **76**:516–526.

Biron, K.K., Morrongiello, M.P., Raskova, J., and Raska, K., Jr., 1978, Adenovirus type 12 tumor antigen 1: Separation from DNA polymerase alpha and immunoprecipitation of tumor-antigen polypeptides, *Virology* **85**:464–474.

Blanton, R.A., and Carter, T.H., 1979, Autoregulation of adenovirus type 5 early gene expression. III. Transcription studies in isolated nuclei, *J. Virol.* **29**:458–465.

Bos, J.L., Polder, L.J., Bernards, R., Schrier, P.I., van den Elsen, P.J., van der Eb, A.J., and van Ormondt, H., 1981, The 2.2 kb E1b mRNA of human Ad12 and Ad5 codes for two tumor antigens starting at different AUG triplets, *Cell* **27**:121–131.

Brackmann, K.H., Green, M., Wold, W.S.M., Cartas, M., Matsuo, T., and Hashimoto, S., 1980, Identification and peptide mapping of human adenovirus type 2-induced early polypeptides isolated by two-dimensional gel electrophoresis and immunoprecipitation, *J. Biol. Chem.* **255**:6772–6779.

Braithwaite, A.W., Cheetham, B.F., Li, P., Parish, C.R., Waldron-Stevens, L.K., and Bellett, A.J.D., 1983, Adenovirus-induced alterations of the cell growth cycle: A requirement for expression of E1A but not of E1B, *J. Virol.* **45**:192–199.

Branton, P.E., Lassam, N.J., Graham, F.L., and Bayley, S.T., 1979, T-antigen related protein kinase activity in cells infected and transformed by human adenoviruses 5 and 12, *Cold Spring Harbor Symp. Quant. Biol.* **44**:487–491.

Branton, P.E., Lassam, N.J., Downey, J.F., Yee, S.-P., Graham, F.L., Mak, S., and Bayley, S.T., 1981, Protein kinase activity immunoprecipitated from adenovirus-infected cells by sera from tumor-bearing hamsters, *J. Virol.* **37**:601–608.

Bresnick, E., and Rapp, F., 1968, Thymidine kinase activity in cells abortively and productively infected with human adenovirus, *Virology* **34**:799–802.

Britten, R.J., and Kohne, D.E., 1968, Repeated sequences in DNA, *Science* **161**:529–540.

Brown, M., and Weber, J., 1982, Discrete subgenomic DNA fragments in incomplete particles of adenovirus type 2, *J. Gen. Virol.* **62**:81–89.

Brusca, J.S., and Chinnadurai, G., 1981, Transforming genes among three different oncogenic subgroups of human adenoviruses have similar replicative functions, *J. Virol.* **39**:300–305.

Burlingham, B.T., Brown, D.T., and Doerfler, W., 1974, Incomplete particles of adenovirus. I. Characteristics of the DNA associated with incomplete adenovirions of types 2 and 12, *Virology* **60**:419–430.

Burnett, J.P., and Harrington, J.A., 1968A, Simian adenovirus SA7 DNA: Chemical, physical and biological studies, *Proc. Natl. Acad. Sci. U.S.A.* **60**:1023–1029.

Burnett, J.P., and Harrington, J.A., 1968b, Infectivity associated with simian adenovirus type SA7 DNA, *Nature (London)* **220:**1245.

Byrd, P., Brown, K.W., and Gallimore, P.H., 1982, Malignant transformation of human embryo retinoblasts by cloned adenovirus 12 DNA, *Nature (London)* **298:**69–71.

Carlock, L.R., and Jones, N.C., 1981, Transformation-defective mutant of adenovirus type 5 containing a single altered E1a mRNA species, *J. Virol.* **40:**657–664.

Carter, T.H., and Blanton, R.A., 1978a, Possible role of the 72,000 dalton DNA-binding protein in regulation of adenovirus type 5 early gene expression, *J. Virol.* **25:**664–674.

Carter, T.H., and Blanton, R.A., 1978b, Autoregulation of adenovirus type 5 early gene expression, II. Effect of temperature-sensitive early mutations on virus RNA accumulation, *J. Virol.* **28:**450–456.

Carter, T. H., and Ginsberg, H.S., 1976, Viral transcription in KB cells infected by temperature-sensitive "early" mutants of adenovirus type 5, *J. Virol.* **18:**156–166.

Casto, B., 1968, Adenovirus transformation of hamster embryo cells, *J. Virol.* **2:**376–383.

Casto, B.C., 1969, Transformation of hamster embryo cells and tumor induction in newborn hamsters by simian adenovirus SV11, *J. Virol.* **3:**511–519.

Cheetham, B.F., and Bellett, A.J.D., 1982, A biochemical investigation of the adenovirus induced G1 to S phase progression: Thymidine kinase, ornithine decarboxylase and inhibitors of polyamine biosynthesis, *J. Cell. Physiol.* **110:**114–122.

Chen, L.B., Gallimore, P.H., and McDougall, J.K., 1976, Correlation between tumor induction and the large external transformation sensitive protein on the cell surface, *Proc. Natl. Acad. Sci. U.S.A.* **73:**3570–3574.

Chin, W.W., and Maizel, J.V., Jr., 1977, The polypeptides of adenovirus. VIII. The enrichment of E3 (11,000) in the nuclear matrix fraction, *Virology* **76:**79–89.

Chinnadurai, G., 1983, Adenovirus 2 lp^{+} locus codes for a 19kd tumor antigen that plays an essential role in cell transformation, *Cell* **33:**759–766.

Chinnadurai, G., Chinnadurai, S., and Brusca, J., 1979, Physical mapping of a large plaque mutation of adenovirus type 2, *J. Virol.* **32:**623–628.

Colby, W.W., and Shenk, T., 1981, Adenovirus type 5 virions can be assembled *in vivo* in the absence of detectable polypeptide IX, *J. Virol.* **39:**977–980.

Cook, J.L., Hibbs, J.B., Jr., and Lewis, A.M., Jr., 1980, Resistance of simian virus 40-transformed hamster cells to the cytolytic effect of activated macrophages: A possible factor in species-specific viral oncogenicity, *Proc. Natl. Acad. Sci. U.S.A.* **77:**6773–6777.

Cook, J.L., Hibbs, J.B., Jr., and Lewis, A.M., Jr., 1982, DNA virus-transformed hamster cell–host effector cell interactions: Level of resistance to cytolysis correlated with tumorigenicity, *Int. J. Cancer* **30:**795–803.

Daniell, E., 1976, Genome structure of incomplete particles of adenovirus, *J. Virol.* **19:**685–708.

Daniell, E., and Mullenbach, T., 1978, Synthesis of defective viral DNA in HeLa cells infected with adenovirus type 3, *J. Virol.* **26:**61–70.

De Leo, A.B., Jay, G., Apella, E., Dubois, G.C., Law, L.W., and Old, L.J., 1979, Detection of a transformation-related antigen in chemically induced sarcomas and other transformed cells of the mouse, *Proc. Natl. Acad. Sci. U.S.A.* **76:**2420–2424.

Deuring, R., Winterhoff, U., Tamanoi, F., Stabel, S., and Doerfler, W., 1981, Site of linkage between adenovirus type 12 and cell DNA in hamster tumour line CLAC3, *Nature (London)* **293:**81–84.

D'Halluin, J.C., Allart, C., Cousin, C., Boulanger, P.A., and Martin, G.R., 1979, Adenovirus early function required for protection of viral and cellular DNA, *J. Virol.* **32:**61–71.

D'Halluin, J.C., Cousin, C., and Boulanger, P., 1982, Physical mapping of adenovirus type 2 temperature-sensitive mutations by restriction endonuclease analysis of interserotypic recombinants, *J. Virol.* **41:**401–413.

Dijkema, R., Dekker, B.M.M., van der Feltz, M.J.M., and van der Eb, A.J., 1979, Transformation of primary rat kidney cells by DNA fragments of weakly oncogenic adenoviruses, *J. Virol.* **32:**943–950.

Dijkema, R., Dekker, B.M.M., and van Ormondt, H., 1980, The nucleotide sequence of the transforming BglIIIH fragment of adenovirus type 7 DNA, *Gene* **9**:141–156.

Doerfler, W., 1968, The fate of the DNA of adenovirus type 12 in baby hamster kidney cells, *Proc. Natl. Acad. Sci. U.S.A.* **60**:636–643.

Doerfler, W., 1969, Nonproductive infection of baby hamster kidney cells (BHK 21) with adenovirus type 12, *Virology* **38**:587–606.

Doerfler, W., Stabel, S., Ibelgaufts, H., Sutter, D., Neumann, R., Groneberg, J., Scheidtmann, K.H., Deuring, R., and Winterhoff, U., 1979, Selectivity in integration sites of adenovirus DNA, *Cold Spring Harbor Symp. Quant. Biol.* **44**:551–564.

Dorsch-Hasler, K., Fisher, P., Weinstein, B., and Ginsberg, H., 1980, Patterns of viral DNA integration in cells transformed by wild type or DNA-binding protein mutants of adenovirus type 5 and effect of chemical carcinogenesis on integration, *J. Virol.* **34**:305–314.

Downey, J.D., Rowe, D.T., Bacchetti, S., Graham, F.L., and Bayley, S.T., 1983, The mapping of a 14K antigen to early region 4 of the human adenovirus 5 genome, *J. Virol.* **45**:514–523.

Dunn, A.R., Gallimore, P.H., Jones, K.W., and McDouglall, J.K., 1973, *In situ* hybridization of adenovirus RNA and DNA. II. Detection of adenovirus-specific DNA in transformed and tumor cells, *Int. J. Cancer* **11**:628–636.

Dunsworth-Browne, M., Schell, R.E., and Berk, A.J., 1980, Adenovirus terminal protein protects single-stranded DNA from digestion by a cellular exonuclease, *Nucleic Acids Res.* **8**:543–554.

Eick, D., and Doerfler, W., 1982, Integrated adenovirus type 12 DNA in the transformed hamster cell line T637: Sequence arrangements at the termini of viral DNA and mode of amplification, *J. Virol.* **42**:317–321.

Eick, D., Stabel, S., and Doerfler, W., 1980, Revertants of adenovirus type 12-transformed hamster cell line T637 as tools in the analysis of integration patterns, *J. Virol.* **36**:41–49.

Enomoto, T., Lichy, J.H., Ikeda, J.E., and Hurwitz, J., 1981, Adenovirus DNA replication *in vitro*: Purification of the terminal protein in a functional form, *Proc. Natl. Acad. Sci. U.S.A.* **78**:6779–6783.

Ensinger, M.J., and Ginsberg, H.S., 1972, Selection and preliminary characterization of temperature-sensitive mutants of type 5 adenovirus, *J. Virol.* **10**:328–339.

Esche, H., and Siegmann, B., 1982, Expression of early viral gene products in adenovirus type 12-infected and -transformed cells, *J. Gen. Virol.* **60**:99–113.

Esche, H., Schilling, R., and Doerfler, W., 1979, *In vitro* translation of adenovirus type 12-specific mRNA isolated from infected and transformed cells, *J. Virol.* **30**:21–31.

Esche, H., Mathews, M.B., and Lewis, J.B., 1980, Proteins and messenger RNAs of the transforming region of wild-type and mutant adenoviruses, *J. Mol. Biol.* **142**:399–417.

Ezoe, H., Lai Fatt, R.B., and Mak, S., 1981, Degradation of intracellular DNA in KB cells infected with *cyt* mutants of human adenovirus type 12, *J. Virol.* **40**:20–27.

Fisher, P.B., Babiss, L.E., Weinstein, I.B., and Ginsberg, H.S., 1982, Analysis of type 5 adenovirus transformation with a cloned rat embryo cell line (CREF), *Proc. Natl. Acad. Sci. U.S.A.* **79**:3527–3531.

Flint, S.J., Sambrook, J., Williams, J., and Sharp, P.A., 1976, Viral nucleic acid sequences in transformed cells. IV. A study of the sequences of adenovirus 5 DNA and RNA in four lines of adenovirus 5 transformed rodent cells using specific fragments of the viral genome, *Virology* **72**:456–470.

Föhring, B. Gallimore, P.H., Mellow, G.H., and Raska, K, Jr., 1983, Adenovirus type 12 specific cell surface antigen in transformed cells is a product of the E1b early region. *Virology* **131**:463–472.

Freeman, A.E., Black, P.H., Vanderpool, E.A., Henry, P.H., Austin, J.B., and Huebner, R.J., 1967a, Transformation of primary rat embryo cells by adenovirus type 2, *Proc. Natl. Acad. Sci. U.S.A.* **58**:1205–1212.

Freeman, A.E., Black, P.H., Wolford, R., and Huebner, R.J., 1967b, Adenovirus type 12–rat embryo transformation system, *J. Virol. 1*, 362–367.

Freeman, A.E., Calisher, C., Price, P.J., Turner, H.C., and Huebner, R.J., 1966, Calcium sensitivity of cell cultures derived from adenovirus induced tumors. *Proc. Soc. Exp. Biol. Med.* **122**:835–840.Freeman, A.E., Vanderpool, E.A., Black, P.H., Turner, H.C., and Huebner, R.J., 1967c, Transformation of primary rat embryo cells by a weakly oncogenic adenovirus type 3, *Nature (London)* **216**:171–173.

Frost, E., and Williams, J., 1978, Mapping temperature-sensitive and host range mutations of adenovirus type 5 by marker rescue, *Virology* **91**:39–50.

Fujinaga, K., and Green, M., 1966, The mechanisms of viral carcinogenesis by DNA mammalian viruses. I. Viral specific RNA in polyribosomes of adenovirus tumor and transformed cells, *Proc. Natl. Acad. Sci. U.S.A.* **55**:1567–1574.

Fujinaga, K., and Green, M., 1967, Mechanism of viral carcinogenes is by deoxyribonucleic acid mammalian viruses. IV. Related virus-specific ribonucleic acids in tumor cells induced by "highly" oncogenic adenovirus types 12, 18 and 31, *J. Virol.* **1**:576–582.

Gahlmann, R., and Doerfler, W., 1983, Integration of viral DNA into the genome of the adenovirus type 2-transformed hamster cell line HE5 without loss or alteration of cellular nucleotides. *Nucl. Acids Res.* **11**:7347–7361.

Gahlmann, R., Leisten, R., Vardimon, L., and Doerfler, W., 1982, Patch homologies and the integration of adenovirus DNA in mammalian cells, *Eur. Mol. Biol. Org. J.* **1**:1101–1104.

Gallimore, P.H., 1972, Tumour production in immunosuppressed rats with cells transformed *in vitro* by adenovirus type 2, *J. Gen. Virol.* **16**:99–102.

Gallimore, P.H., 1974, Interactions of adenovirus type 2 with rat embryo cells: Permissiveness, transformation and *in vitro* characteristics of adenovirus transformed rat embryo cells, *J. Gen. Virol.* **25**:263–273.

Gallimore, P.H., and Paraskeva, C., 1979, A study to determine the reasons for differences in the tumorigenicity of rat cell lines transformed by adenovirus 2 and adenovirus 12, *Cold Spring Harbor Symp. Quant. Biol.* **44**:703–813.

Gallimore, P.H., Sharp, P.A., and Sambrook, J., 1974, Viral DNA in transformed cells. II. A study of the sequences of adenovirus 2 DNA in nine lines of transformed rat cells using specific fragments of the viral genome, *J. Mol. Biol.* **89**:49–72.

Gallimore, P.H., McDougall, J.K., and Chen, L.B., 1979, Malignant behaviour of three adenovirus 2 transformed brain cell lines and their methylcellulose selected subclones, *Int. J. Cancer* **24**:477.

Galos, R.S., Williams, J., Binger, M.H., and Flint, S.J., 1979, Location of additional early gene sequences in the adenoviral chromosome, *Cell* **17**:945–956.

Galos, R.S., Williams, J., Shenk, T., and Jones, N., 1980, Physical location of host range mutations of Ad5: Deletion and marker-rescue mapping, *Virology* **104**:510–513.

Gaynor, R.B., Tsukamoto, A., Montell, C., and Berk, A.J., 1982, Enhanced expression of adenovirus transforming proteins, *J. Virol.* **44**:276–285.

Gelb, L.D., Kohn, D.E., and Martin, M., 1971, Quantitation of simian virus 40 sequences in African green monkey, mouse and virus transformed cell lines, *J. Mol. Biol.* **57**:129–145.

Gilden, R.V., Kern, J., Freeman, A.E., Martin, C.E., McAllister, R.C., Turner, H.C., and Huebner, R.J., 1968, T and tumour antigens of adenovirus group C-infected and transformed cells, *Nature (London)* **219**:517–518.

Gilead, Z., Arens, M.Q., Bhadura, S., Shanmugan, G., and Green, M., 1975, Tumour antigen specificity of a DNA-binding protein from cells infected with adenovirus 2, *Nature (London)* **254**:533–536.

Gillam, S., and Smith, M., 1980, Site-specific mutagenesis using synthetic oligodeoxyribonucleotide primers. I. Optimum conditions and minimum oligodeoxyribonucleotide length, *Gene* **8**:81–97.

Gillam, S., Jahnke, P., Astell, C., Phillips, S., Hutchison, C.A., III, and Smith, M., 1979, Defined transversion mutations at a specific position in DNA using synthetic oligodeoxyribonucleotides as mutagens, *Nucleic Acids Res.* **6**:2973–2985.

Gingeras, T.R., Sciaky, D., Gelinas, R.E., Bing-Dong, J., Yen, C.E., Kelly, M.M., Bullock, P.A., Parsons, B.L., O'Neill, K.E., and Roberts, R.J., 1982, Nucleotide sequences from the adenovirus-2 genome, *J. Biol. Chem.* **257**:13,475–13,491.

Ginsberg, H.S., Ensinger, M.J., Kauffman, R.S., Mayer, A.J., and Lundholm, U., 1974, Cell transformation: A study of regulation with types 5 and 12 adenovirus temperature-sensitive mutants, *Cold Spring Harbor Symp. Quant. Biol.* **39**:419–426.

Ginsberg, H.S., Lundholm, U., and Linné, T., 1977, Adenovirus DNA-binding protein in cells infected with wild-type 5 adenovirus and two DNA-minus temperature-sensitive mutants, H5ts125 and H5ts149, *J. Virol.* **23**:142–151.

Girardi, A.J., Hilleman, M.R., and Zwickey, R.E., 1964, Tests in hamsters for oncogenic quality of ordinary viruses including adenovirus type 7, *Proc. Soc. Exp. Biol. Med.* **115**:1141–1155.

Graham, F.L., 1977, Biological activity of tumor virus DNA, *Adv. Cancer Res.* **25**:1–51.

Graham, F.L., and van der Eb, A.J., 1973a, A new technique for the assay of infectivity of human adenovirus 5 DNA. *Virol* **52**:456–467.

Graham, F.L., and van der Eb, A.J., 1973b, Transformation of rat cells by DNA of human adenovirus 5. *Virology* **54**:536–539.

Graham, F.L., van der Eb, A.J., and Heijneker, H.L., 1974a, Size and location of the transforming region in human adenovirus type 5 DNA, *Nature (London)* **251**:687–691.

Graham, F.L., Abrahams, P.J., Mulder, C., Heijneker, H.L., Warnaar, S.O., de Vries, F.A.J., Fiers, W., and van der Eb, A.J., 1974b, Studies on *in vitro* transformed by DNA and DNA fragments of human adenoviruses and simian virus 40, *Cold Spring Harbor Symp. Quant. Biol.* **39**:637–650.

Graham, F.L., Smiley, J., Russell, W.C., and Nairn, R., 1977, Characteristics of a human cell line transformed by DNA from human adenovirus 5, *J. Gen. Virol.* **36**:59–72.

Graham, F.L., Harrison, T., and Williams, J., 1978, Defective transforming capacity of adenovirus 5 host-range mutants, *Virology* **86**:10–21.

Graham, F.L., McKinnon, R., Ruben, M., Rowe, D.T., and Bacchetti, S., 1983, Studies on transformation of mammalian cells by human adenovirus type 5 (Ad5) and Ad5 DNA, *Proceedings of the 13th International Cancer Congress*, Alan R. Liss, New York **132**:313–326.

Green, M., 1970, Oncogenic viruses, *Annu. Rev. Biochem.* **39**:701–756.

Green, M., and Brackmann, K.H., 1982, The application of high-performance liquid chromatography for the resolution of proteins encoded by the human adenovirus type 2 cell transformation region, *Anal. Biochem.* **124**:209–216.

Green, M., Parsons, J.T., Piña, M., Fujinaga, K., Caffier, H., and Landgraf-Leurs, I., 1970, Transcription of adenovirus genes in productively infected and in transformed cells, *Cold Spring Harbor Symp. Quant. Biol.* **35**:803–818.

Green, M., Chinnadurai, G., Mackey, J., and Green, M., 1976, A unique pattern of integrated viral genes in hamster cells transformed by highly oncogenic human adenovirus 12, *Cell* **7**:419–428.

Green, M.R., Mackey, J.K., and Green, M., 1977, Multiple copies of human adenovirus 12 genomes are integrated in virus-induced hamster tumors, *J. Virol.* **22**:238–242.

Green, M., Mackey, J.K., Wold, W.S.M., and Rigden, P., 1979a, Thirty-one human adenovirus serotypes (Ad1–31) form five groups (A–E) based upon DNA genome homologies, *Virology* **93**:481–492.

Green, M., Wold, W.S.M., Mackey, J.K., and Rigden, P., 1979b, Analysis of human tonsil and cancer DNAs and RNAs for DNA sequences of group C (serotypes 1, 2, 5 and 6) human adenoviruses, *Proc. Natl. Acad. Sci. U.S.A.* **76**:6606–6610.

Green, M., Wold, W.S.M., Brackmann, K.H., and Cartas, M.A., 1979c, Identification of families of overlapping polypeptides coded by early "transforming" gene region 1 of human adenovirus type 2, *Virology* **97**:275–286.

Green, M., Wold, W.S.M., Brackman, K.H., Cartas, M.A., Sanders, P.R., Olson, K., Lee, T.K., Young, L., Matsuo, T., and Kapoor, Q., 1980, Human adenovirus transforming genes: Group relationship, integration, expression in transformed cells, and analysis of human cancers and tonsils, in: *Viruses in Naturally Occurring Cancers: Cold Spring Harbor Conferences on Cell Proliferation*, Vol. 7 (M. Essex, G. Todaro, H. zur Hausen, eds) pp 373–397, Cold Spring Harbor Press, Cold Spring Harbor, New York.

Green, M., Wold, W.S.M., and Buttner, W., 1981, Integration and transcription of group C human adenovirus sequences in the DNA of five lines of transformed rat cells, *J. Mol. Biol.* **151**:337–366.

Green, M., Brackmann, K.H., Cartas, M.A., and Matsuo, T., 1982, Identification and purification of a protein encoded by the human adenovirus type 2 transforming region, *J. Virol.* **42**:30–41.

Grodzicker, T., and Klessig, D.F., 1980, Expression of unselected adenovirus genes in human cells co-transformed with the HSV-1 tk gene and adenovirus 2 DNA, *Cell* **21**:453–463.

Groneberg, J., and Doerfler, W., 1979, Revertants of adenovirus type 12-transformed hamster cells have lost part of the viral genomes, *Int. J. Cancer* **24**:67–74.

Groneberg, J., Chardonnet, Y., and Doerfler, W., 1977, Integrated viral sequences in adenovirus type 12-transformed hamster cells. *Cell* **10**:101–111.

Groneberg, J., Sutter, D., Soboll, M., and Doerfler, W., 1978, Morphological revertants of adenovirus type 12-transformed hamster cells, *J. Gen. Virol.* **40**:635–645.

Haas, M., Vogt, M., and Dulbecco, R., 1972, Loss of simian virus 40 DNA–RNA hybrids from nitrocellulose membranes: Implication for the study of virus–host interaction, *Proc. Natl. Acad. Sci. U.S.A.* **69**:2160.

Halbert, D.N., and Raskas, H.J., 1982, Tryptic and chymotryptic methionine peptide analysis of the *in vitro* translation products specified by the transforming region of adenovirus type 2, *Virology* **116**:406–418.

Halbert, D.N., Spector, D.J., and Raskas, H.J., 1979, *In vitro* translation products specified by the transforming region of adenovirus type 2, *J. Virol.* **31**:621–629.

Halling, S.M., and Kleckner, N., 1982, A symmetrical six-base-pair target site sequence determines Tn10 insertion specificity, *Cell* **28**:155–163.

Hammarskjöld, M.L., and Winberg, G., 1980, Encapsidation of adenovirus 16 DNA is directed by a small DNA sequence at the left end of the genome, *Cell* **20**:787–795.

Hammarskjöld, M.L., Winberg, G., Norrby, E., and Wadell, G., 1977, Isolation of incomplete adenovirus 16 particles containing viral and host cell DNA, *Virology* **82**:449–446.

Harrison, T., Graham, F.L., and Williams, J., 1977, Host-range mutants of adenovirus type 5 defective for growth in HeLa cells, *Virology* **77**:319–329.

Harter, M.L., and Lewis, J.B., 1978, Adenovirus type 2 early proteins synthesized *in vitro* and *in vivo*: Identification in infected cells of the 38,000- to 50,000-molecular weight protein encoded by the left end of the adenovirus type 2 genome, *J. Virol.* **26**:736–749.

Herberman, R.B., and Holden, H.T., 1978, Natural cell-mediated immunity, *Adv. Cancer Res.* **27**:305–377.

Ho, Y.-S., Galos, R., and Williams, J., 1982, Isolation of type 5 adenovirus mutants with a cold-sensitive host range phenotype: Genetic evidence of an adenovirus transformation maintenance function, *Virology* **122**:109–124.

Houweling, A., van den Elsen, P.J., and van der Eb, A.J., 1980, Partial transformation of primary rat cells by the left-most 4.5% fragment of adenovirus 5 DNA, *Virology* **105**:537–550.

Huebner, R.J., 1967, Adenovirus-directed tumor and T-antigens, in: *Perspectives in Virology*, Vol. V (M. Pollard, ed.), pp. 147–166, Academic Press, New York.

Huebner, R.J., Rowe, W.P., and Lane, W.T., 1962, Oncogenic effects in hamsters of human adenovirus types 12 and 18, *Proc. Natl. Acad. Sci. U.S.A.* **48**:2051–2058.

Huebner, R.J., Rowe, W.P., Turner, H.C., and Lane, W.T., 1963, Specific adenovirus complement-fixing antigens in virus-free hamster and rat tumors, *Proc. Natl. Acad. Sci. U.S.A.* **50**:379–389.

Huebner, R.J., Pereira, H.G., Allison, A.C., Hollinshead, A.C., and Turner, H.C., 1964, Production of type-specific C antigen in virus-free hamster tumor cells induced by adenovirus type 12, *Proc. Natl. Acad. Sci. U.S.A.* **51**:432–439.

Huebner, R.J., Casey, M.J., Chanock, R.M., and Schell, K., 1965, Tumors induced in hamsters by a strain of adenovirus type 3: Sharing of tumor antigens and "neoantigens" with those produced by adenovirus type 7 tumors, *Proc. Natl. Acad. Sci. U.S.A.* **54**:381–388.

Hunter, J., 1982, Phosphotyrosine—a new protein modification, *Trends Biochem. Sci.* **7**:246–249.

Ibelgaufts, H., 1982, Are human DNA tumour viruses involved in the pathogenesis of human neurogenic tumours?, *Neurosurg. Rev.* **5**:3–24.

Ibelgaufts, H., Doerfler, W., Scheidtmann, K.H., and Wechsler, W., 1980, Adenovirus type 12-induced rat tumor cells of neuroepithelial origin: Persistence and expression of the viral genome, *J. Virol.* **33**:423–437.

Ibelgaufts, H., Jones, K.W., Maitland, N., and Shaw, J.F., 1982, Adenovirus-related RNA sequences in human neurogenic tumours, *Acta Neuropathol. (Berlin)* **56**:113–117.

Jeng, Y.-H., Wold, W.S.M., and Green, M., 1978, Evidence for an adenovirus type 2-coded early glycoprotein, *J. Virol.* **28**:314–323.

Jochemsen, H., Daniels, G.S.G., Lupker, J.H., and van der Eb, A.J., 1980, Identification and mapping of the early gene products of adenovirus type 12, *Virology* **105**:551–563.

Jochemsen, J., Hertoghs, J.L., Lupker, J.H., Davis, A., and van der Eb, A.J., 1981, *In vitro* synthesis of adenovirus type 5 T-antigens. II. Translation of virus-specific RNA from cells transformed by fragments of adenovirus type 5 DNA, *J. Virol.* **37**:530–534.

Jochemsen, H., Daniels, G.S.G., Hertoghs, J.J.L., Schrier, P.I., van den Elsen, P.J., and van der Eb, A.J., 1982, Identification of adenovirus type 12 gene products involved in transformation and oncogenesis, *Virology* **122**:15–28.

Johansson, K., Pettersson, U., Philipson, L., and Tibbetts, C., 1977, Reassociation of complementary strand-specific adenovirus type 2 DNA with viral DNA sequences of transformed cells, *J. Virol.* **23**:29–35.

Johansson, K., Persson, H., Lewis, A.M., Pettersson, U., Tibbetts, C., and Philipson, L., 1978, Viral DNA sequences and gene products in hamster cells transformed by adenovirus type 2, *J. Virol.* **27**:628–639.

Jones, K.W., Kinross, J., Maitland, N., and Norval, M., 1979, Normal human tissues contain RNA and antigens related to infectious adenovirus type 2, *Nature (London)* **277**:274–279.

Jones, N., and Shenk, T., 1979a, Isolation of adenovirus type 5 host range deletion mutants defective for transformation of rat embryo cells, *Cell* **17**:683–689.

Jones, N., and Shenk, T., 1979b, An adenovirus type 5 early gene function regulates expression of other early viral genes, *Proc. Natl. Acad. Sci. U.S.A.* **76**:3665–3669.

Katze, M.G., Persson, H., and Philipson, L., 1981, Control of adenovirus early gene expressions: A posttranscriptional control mediated by both viral and cellular gene products, *Mol. Cell. Biol.* **1**:807–813.

Katze, M.G., Persson, H., and Philipson, L., 1982, A novel mRNA and a low molecular weight polypeptide encoded in the transforming region of adenovirus DNA, *Eur. Mol. Biol. Org. J.* **1**:783–789.

Katze, M.G., Persson, H., Johansson, B.M., and Philipson, L., 1983, Control of adenovirus gene expression: Cellular gene products restrict expression of adenovirus host range mutants in nonpermissive cells, *J. Virol.* **46**:50–59.

Khittoo, G., and Weber, J., 1981, The nature of the DNA associated with incomplete particles of adenovirus type 2, *J. Gen. Virol.* **54**:343–355.

Kimura, T., Sawada, Y., Shinagawa, M., Shimuzu, Y., Shiroki, K., Shimojo, H., Sugiasaki, H., Takanami, M., Uemizu, Y., and Fujinaga, K., 1981, Nucleotide sequence of the transforming early region E1B of adenovirus type 12 DNA: Structural and gene organization, and comparison with those of adenovirus type 5 DNA, *Nucleic Acids Res.* **9**:6571–6589.

Kit, S., Dubbs, D.R., De Torres, R.A., and Melnick, J.L., 1965, Enhanced thymidine kinase activity following infection of green monkey kidney cells by simian adenoviruses, simian papovovirus SV-40, and an adenovirus SV40 "hybrid," *Virology* **27**:453–457.

Kitamura, I., van Hoosier, G., Jr., Samper, L., Taylor, G., and Trentin, J.J., 1964, Characteristics of human adenovirus type 12 induced hamster tumor cells in tissue culture, *Proc. Soc. Exp. Med. Biol.* **116**:563–568.

Kuhlmann, I., Achten, S., Rudolph, R., and Doerfler, W., 1982, Tumor induction by human adenovirus type 12 in hamsters: Loss of the viral genome from adenovirus type 12-induced tumor cells is compatible with tumor formation, *Eur. Mol. Biol. Org. J.* **1**:79–86.

Kuhlman, I., and Doerfler, W., 1982, Shift in the extent and patterns of DNA methylation upon explantation and subcultivation of adenovirus type 12 induced hamster tumor cells, *Virology* **118**:169–180.

Kvist, S., Ostberg, L., Persson, H., Philipson, L., and Peterson, P.A., 1978, Molecular association between transplantation antigens and cell surface antigen in adenovirus-transformed cell line, *Proc. Natl. Acad. Sci. U.S.A.* **75**:5674–5678.

Lai Fatt, R.B., and Mak, S., 1982, Mapping of an adenovirus function involved in the inhibition of DNA degradation. *J. Virol.* **42**:969–977.

Land, H., Parada, L.F., and Weinberg, R.A., 1983, Tumorigenic conversion of primary embryo fibroblasts requires at least two cooperating oncogenes, *Nature* **304**:596–602.

Landau, B.J., Larsson, V.M., Devers, G.A., and Hilleman, M.R., 1966, Studies on induction of virus from adenovirus and SV40 hamster tumors. I. Chemical and physical agents, *Proc. Soc. Exp. Biol. Med.* **122**:1174–1182.

Lane, D.P., and Crawford, L.V., 1979, T antigen is bound to a host protein in SV-40 transformed cells, *Nature (London)* **278**:261–263.

Lane, D., and Harlow, E., 1982, Two different viral transforming proteins bind the same host tumour antigen, *Nature (London)* **298**(5874):517.

Larsen, S.T., Margolskee, R.F., and Nathans, D., 1979, Alignment of the restriction map of mouse adenovirus FL with that of human adenovirus 2, *Virology* **97**:406–414.

Lassam, N.J., Bayley, S.T., and Graham, F.L., 1978, Synthesis of DNA, late polypeptides, and infectious virus by host-range mutants of Ad5 in non-permissive cells, *Virology* **87**:463–467.

Lassam, N.J., Bayley, S.T., and Graham, F.L., 1979a, Tumor antigens of human Ad5 in transformed cells and in cells infected with transformation-defective host-range mutants, *Cell* **18**:781–791.

Lassam, N.J., Bayley, S.T., and Graham, F.L., 1979b, Transforming proteins of human adenovirus type 5: Studies with infected and transformed cells, *Cold Spring Harbor Symp. Quant. Biol.* **44**:477–493.

Lassam, N.J., Bayley, S.T., Graham, F.L., and Branton, P.E., 1979c, Immunoprecipitation of protein kinase activity from adenovirus 5-infected cells using antiserum directed against tumor antigens, *Nature (London)* **277**:241–243.

Ledinko, N., 1967, Stimulation of DNA synthesis and thymidine kinase activity in human embryonic kidney cells infected by adenovirus 2 or 12, *Cancer Res.* **27**:1459–1469.

Ledinko, N., 1970, Transient stimulation of deoxyribonucleic acid-dependent ribonucleic acid polymerase and histone acetylation in human embryonic kidney cultures infected with adenovirus 2 or 12: Apparent induction of host ribonucleic acid synthesis, *J. Virol.* **6**:58–68.

Ledinko, N., 1974, Temperature-sensitive mutants of adenovirus type 12 defective in viral DNA synthesis, *J. Virol.* **14**:457–468.

Ledinko, N., 1978a, Adenovirus type 12 gene 401 function in transforming infection, *J. Virol.* **26**:828–831.

Ledinko, N., 1978b, Transformation-specific antigen induced by oncogenic human adenovirus, *Nature (London)* **274**:812–813.

Ledinko, N., Schaefele, J., and Soorma, O., 1979, Adenovirus type 12 gene 401 function and maintenance of transformation, *J. Virol.* **29**:250–260.

Lee, K., and Mak, S., 1977, Adenovirus type 12 DNA sequences in primary hamster tumors, *J. Virol.* **24**:408–411.

Levinson, A., and Levine, A.J., 1977, The isolation and identification of the adenovirus group C tumor antigens, *Virology* **76**:1–11.

Levinson, A., Levine, A.J., Anderson, S., Osborn, M., Rosenwirth, B., and Weber, K., 1976, The relationship between group C adenovirus tumor antigen and the adenovirus single-strand DNA-binding protein, *Cell* **7**:575–584.

Lewis, A.M., and Cook, J.L., 1982, Spectrum of tumorigenic phenotypes among adenovirus 2-, adenovirus 12- and simian virus 40-transformed Syrian hamster cells defined by host cellular immune–tumor cell interactions, *Cancer Res.* **42**:939–944.

Lewis, J.B., and Mathews, M.B., 1981, Viral messenger RNAs in six lines of adenovirus-transformed cells, *Virology* **115**:345–360.

Lewis, J.B., Atkins, J.F., Baum, P.R., Solem, R., Gesteland, R.F., and Anderson, C.W., 1976, Location and identification of the genes for adenovirus type 2 early polypeptides, *Cell* **7**:141–151.

Lewis, J.B., Esche, H., Smart, J.E., Stillman, B., Harter, M.L., and Mathews, M.B., 1979, Organization and expression of the left third of the genome of adenovirus, *Cold Spring Harbor Symp. Quant. Biol.* **44**:493–508.

Lichy, J.H., Field, J., Horwitz, M.S., and Hurwitz, J., 1982, Separation of the adenovirus terminal protein precursor from its associated DNA polymerase: Role of both proteins in the initiation of adenovirus DNA replication, *Proc. Natl. Acad. Sci. U.S.A.* **79**:5225–5229.

Linzer, D.I.H., and Levine, A.J., 1979, Characterization of a 54K dalton cellular SV-40 tumor antigen present in SV-40 transformed cells and uninfected embryonal carcinoma cells, *Cell* **17**:43–52.

Logan, J., Nicolas, J.C., Topp, W.C., Girard, M., Shenk, T., and Levine, A.J., 1981, Transformation by adenovirus early region 2A temperature-sensitive mutants and their revertants, *Virology* **115**:419–422.

Luka, J., Jornvall, H., and Klein, G., 1980, Purification and biochemical characterization of the Epstein–Barr virus determined nuclear antigen and as associated protein with a 53,000 dalton subunit, *J. Virol.* **35**:592–602.

Lupker, J.H., Davis, A., Jochemsen, H., and van der Eb, A.J., 1981, *In vitro* synthesis of adenovirus type 5 T-antigens. I. Translation of early region 1-specific RNA from lytically infected cells, *J. Virol.* **37**:524–529.

Maat, J., and van Ormondt, H., 1979, The nucleotide sequence of the transforming HindIII-G fragment of adenovirus type 5 DNA: The region between map positions 4.5 (HpaI site) and 8.0 (HindIII site), *Gene* **6**:75–90.

Maat, J., van Beveren, C.P., and van Ormondt, H., 1980, The nucleotide sequence of adenovirus type 5 early region E1: The region between map positions 8.0 (HindIII site) and 11.8 (SmaI site), *Gene* **10**:27–38.

Mackey, J.K., Rigden, P.M., and Green, M., 1976, Do highly oncogenic group A human adenoviruses cause human cancer? Analysis of human tumors for adenovirus 12 transforming DNA sequences, *Proc. Natl. Acad. Sci. U.S.A.* **73**:4657–4661.

Mackey, J.K., Green, M., Wold, W.S.M., and Rigden, P., 1979, Analysis of human cancer DNA for DNA sequences of human adenovirus type 4, *J. Natl. Cancer Inst.* **62**:23–26.

Mak, I., and Mak, S., 1983, Transformation of rat cells by cyt mutants of Ad12 and by mutants of Ad5, *J. Virol.* **45**:1107–1117.

Mak, S., 1969, Transcription and replication of viral deoxyribonucleic acid in cells coinfected with adenovirus types 2 and 12, *J. Virol.* **4**:651–656.

Mak, S., 1971, Defective virions in human adenovirus 12, *J. Virol.* **7**:426–433.

Mak, S., 1975, Virus specific RNA in hamster cells abortively infected with human adenovirus type 12, *Virology* **66**:474–480.

Mak, S., Mak, I., Smiley, J.R., and Graham, F.L., 1979, Tumorigenicity and viral gene expression in rat cells transformed by Ad12 virions or by the EcoRIC fragment of Ad12DNA, *Virology* **98**:456–460.

Mak, S., Mak, I., Gallie, B.L., Godbout, R., and Phillips, R.A., 1982, Adenovirus-12 genes undetectable in human retinoblastoma, *Int. J. Cancer* **30**:697–700.

Malette, P., Yee, S.-P., and Branton, P.E., 1983, Studies on the phosphorylation of the 58,000 dalton early region 1B protein of human adenovirus type 5, *J. Gen. Virol.* **64:**1069–1078.

Martin, G.R., Warocquier, R., Cousin, C., D'Halluin, J.C., and Boulanger, P., 1978, Isolation and phenotypic characterization of human adenovirus type 2 temperature-sensitive mutants, *J. Gen. Virol.* **41:**303–314.

Marx, J.L., 1981, Tumor viruses and the kinase connection, *Science* **211:**1336–1338.

Matsuo, T., Hashimoto, S., Wold, W.S.M., Symington, J., Rankin, A., and Green, M., 1982, Identification of adenovirus 2 early region 4 polypeptides by *in vitro* translation and tryptic peptide map analysis, *J. Virol.* **41:**334–339.

Mayne, N., Burnett, J.P., and Butler, L.K., 1971, Tumour induction by simian adenovirus SA7 DNA fragments, *Nature (London) New Biol.* **232:**182–183.

McAllister, R.M., and Macpherson, I., 1968, Transformation of a hamster cell line by adenovirus type 12, *J. Gen. Virol.* **2:**99–106.

McAllister, R.M., Nicolson, M.O., Lewis, A.M., Jr., Macpherson, I., and Huebner, R.J., 1969a, Transformation of rat embryo cells by adenovirus type 1, *J. Gen. Virol.* **4:**29–36.

McAllister, R.M., Nicholson, M.O., Reed, G., Kern, J., Gilden, R.V., and Huebner, R.J., 1969b, Transformation of rodent cells by adenovirus 19 and other group D adenoviruses, *J. Natl. Cancer Inst.* **43:**917–923.

McBride, W.D., and Weiner, A., 1964, *In vitro* transformation of hamster kidney cells by human adenovirus type 12, *Proc. Soc. Exp. Biol. Med.* **115:**870–874.

McKinnon, R.D., Bacchetti, S., and Graham, F.L., 1982, Tn5 mutagenesis of the transforming genes of human adenovirus type 5, *Gene* **19:**33–42.

Montell, C., Fisher, E.F., Caruthers, M.H., and Berk, A.J., 1982, Resolving the functions of overlapping viral genes by site-specific mutagenesis at a mRNA splice-site, *Nature (London)* **295:**380–384.

Mukai, N., and Kobayashi, S., 1972, Undifferentiated intraperitoneal tumors induced by human adenovirus type 12 in hamsters, *Am. J. Pathol.* **69:**331–348.

Mukai, N., and Murao, T., 1975, Retinal tumor induction by ocular inoculation of human adenovirus in 3 day old rats, *J. Neuropathol. Exp. Neurol.* **34:**28–35.

Mukai, N., Kalter, S.S., Cummins, L.B., Mathews, V.A., Nishida, T., and Nakajima, T., 1980, Retinal tumours induced in the baboon by human adenovirus 12, *Science* **210:**1023–1025.

Murray, J.D., Bellett, A.J.D., Braithwaite, A.W., Waldron, L.K., and Taylor, I.W., 1982a, Altered cell cycle progression and aberrant mitosis in adenovirus-infected rodent cells, *J. Cell. Physiol.* **111:**89–96.

Murray, J.D., Braithwaite, A.W., Taylor, I.W., and Bellett, A.J.D., 1982b, Adenovirus-induced alterations of the cell growth cycle: Effects of mutations in early regions E2A and E2B, *J. Virol.* **44:**1072–1075.

Nevins, J.R., 1981, Mechanism of activation of early viral transcription by the adenovirus E1A gene product, *Cell* **26:**213–220.

Nevins, J.R., 1982, Induction of the synthesis of a 70,000 dalton mammalian heat shock protein by the adenovirus E1A gene product, *Cell* **29:**913–919.

Norrby, E., Bartha, A., Boulanger, P., Dreizin, R.S., Ginsberg, H.S., Kalter, S.S., Kawamura, H., Rowe, W.P., Russell, W.C., Schlesinger, R.W., and Wigand, R., 1976, Adenoviridae, *Intervirology* **7:**117–125.

O'Farrell, P.H., 1975, High resolution two-dimensional electrophoresis of proteins, *J. Biol. Chem.* **250:**4007–4021.

Osborne, T.F., Gaynor, R.B., and Berk, A.J., 1982, The TATA homology and the mRNA 5' untranslated sequence are not required for expression of essential adenovirus E1A functions, *Cell* **29:**139–148.

Ostrove, J.M., and Berns, K.I., 1980, Adenovirus early region 1b gene function required for rescue of latent adeno-associated virus, *Virology* **104:**502–505.

Paraskeva, C., and Gallimore, P.H., 1980, Tumorigenicity and *in vitro* characteristics of rat liver epithelial cells and their adenovirus-transformed derivatives, *Int. J. Cancer* **25:**631–639.

Paraskeva, C., Brown, K.W., Dunn, A.R., and Gallimore, P.H., 1982, Adenovirus type 12-transformed rat embryo brain and rat liver epithelial cell lines: Adenovirus type 12 genome content and viral protein expression, *J. Virol.* **44**:759–764.

Pereira, M.S., Pereira, H.G., and Clarke, S.K., 1965, Human adenovirus type 31: A new serotype with oncogenic properties, *Lancet* **1**:21–23.

Perricaudet, M., Akusjärvi, G., Virtanen, A., and Pettersson, U., 1979, Structure of two spliced mRNAs from the transforming region of human subgroup C adenoviruses, *Nature (London)* **281**:694–696.

Perricaudet, M., Le Moullec, J.M., and Pettersson, U., 1980a, Predicted structure of two adenovirus tumor antigens, *Proc. Natl. Acad. Sci. U.S.A.* **77**:3778–3782.

Perricaudet, M., LeMoullec, J.M., Tiollais, P., and Pettersson, U., 1980b, Structure of two adenovirus type 12 transforming polypeptides and their evolutionary implications, *Nature (London)* **288**:174–176.

Persson, H., Kvist, S., Ostberg, L., Peterson, P.A., and Philipson, L., 1979, The early adenovirus glycoprotein E3-19K and its association with transplantation antigens, *Cold Spring Harbor Symp. Quant. Biol.* **44**:509–514.

Persson, H., Katze, M.G., and Philipson, L., 1982, Purification of a native membrane-associated adenovirus tumor antigen, *J. Virol.* **42**:905–917.

Pettersson, U., and Sambrook, J., 1973, Amount of viral DNA in the genome of cells transformed by adenovirus type 2, *J. Mol. Biol.* **73**:125–130.

Piña, M., and Green, M., 1965, Biochemical studies on adenovirus multiplication. IX. Chemical and base composition analysis of 28 human adenoviruses, *Proc. Natl. Acad. Sci. U.S.A.* **54**:547–551.

Piña, M., and Green, M., 1968, Base composition of the DNA of oncogenic simian adenovirus SA7 and homology with human adenovirus DNAs, *Virology* **36**:321–323.

Prage, L.S., Hoglund, S., and Philipson, L., 1972, Structural proteins of adenovirus. III. Characterization of incomplete particles of adenovirus 3, *Virology* **49**:745–757.

Rabson, A.S., Kirschstein, R.L., and Paul, F.J., 1964, Tumors produced by adenovirus 12 in *Mastomys* and mice, *J. Natl. Cancer Inst.* **32**:77–82.

Rajagopalan, S., and Chinnadurai, G., 1979, A unique subgenomic species of adenovirus 2 DNA generated under high multiplicities of infection, *Nucleic Acids Res.* **7**:1163–1174.

Raska, K., Jr., and Gallimore, P.H., 1982, An inverse relation of the oncogenic potential of adenovirus transformed cells and their sensitivity to killing by syngeneic natural killer cells, *Virology* **123**:8–18.

Raska, K., Jr., Rubio, V., and Biron, K.K., 1976, Double-stranded DNA binding of adenovirus type 12 tumor antigen, *FEBS Lett.* **62**:173–176.

Raska, K., Jr., Geis, A., and Fohring, B., 1979, Adenovirus type 12 tumor antigen. II. Immunoprecipitation of protein kinase from infected and transformed cells by antisera to T antigen and some normal rat sera, *Virology* **99**:174–178.

Raska, K., Jr., Morrongiello, M.P., and Fohring, B., 1980, Adenovirus type 12 tumor antigen. III. Tumorigenicity and immune response to syngeneic rat cells transformed with virions and isolated transforming fragments of adenovirus 12 DNA, *Int. J. Cancer* **26**:79–86.

Raska, K., Jr., Dougherty, J., and Gallimore, P.H., 1982, Product of adenovirus type 2 early gene block E1 in transformed cells elicits cytolytic response in syngeneic rats, *Virology* **117**:530–535.

Rho, H.M., Jeng, Y.-H., Wold, W.S.M., and Green, M., 1977, Association of adenovirus type 2 early proteins with a soluble complex that synthesizes adenovirus DNA *in vitro*, *Biochem. Biophys. Res. Commun.* **79**:422–428.

Ricciardi, R.L., Jones, R.L., Cepko, C.L., Sharp, P.A., and Roberts, B.E., 1981, Expression of early adenovirus genes requires a viral encoded acidic polypeptide, *Proc. Natl. Acad. Sci. U.S.A.* **78**:6121–6125.

Rosén, L., 1970, A hemagglutination-inhibition technique for typing adenoviruses, *Am. J. Hyg.* **71**:120–128.

Ross, S.R., Levine, A.J., Galos, R.S., Williams, J., and Shenk, T., 1980, Early viral proteins in HeLa cells infected with adenovirus type 5 host range mutants, *Virology* **103**:475–492.

Rossini, M., Jonak, G.J., and Baserga, R., 1981, Identification of adenovirus 2 early genes required for induction of cellular DNA synthesis in resting hamster cells, *J. Virol.* **38**:982–986.

Rowe, D.T., and Graham, F.L., 1981, Complementation of adenovirus type 5 host range mutants by adenovirus type 12 in coinfected HeLa and BHK-21 cells, *J. Virol.* **38**:191–197.

Rowe, D.T., and Graham, F.L., 1983, Transformation of rodent cells by DNA extracted from transformation-defective adenovirus mutants *J. Virol.* **46**:1039–1044.

Rowe, D.T., Ruben, M., Bacchetti, S., and Graham, F.L., 1983a, Identification of an adenovirus function required for initiation of cell transformation possibly at the level of DNA integration, in: *Gene Transfer and Cancer* (M. Pearson and N. Sternberg, eds.), pp. XX–XX, Raven Press, New York (in press).

Rowe, D.T., Branton, P.E., Bayley, S.T., and Graham, F.L., 1984a, The kinetics of synthesis of early viral proteins in KB cells infected with wild type and transformation-defective host range mutants of human adenovirus type 5 *J. Gen. Virol* (in press).

Rowe, D.T., Yee, S.-P., Otis, J., Graham, F.L., and Branton, P.E., 1983b, Characterization of human adenovirus type 5 early region 1A polypeptides using anti-tumor sera and an antiserum specific for the carboxy terminus *Virology* **127**:253–271.

Rowe, D.T., Graham, F.L., and Branton, P.E., 1983c, Intracellular localization of adenovirus type 5 tumor antigens in productively infected cells *Virology* **129**:456–468.

Rowe, D.T., Branton, P.E., Yee, S.-P., Bacchetti, S., and Graham, F.L., 1984b, Establishment and characterization of hamster cell lines transformed by restricton endonuclease fragments of adenovirus type 5 *J. Virol.* **49**:162–170.

Ruben, M., Bacchetti, S., and Graham, F.L., 1982, Integration and expression of viral DNA in cells transformed by host range mutants of adenovirus type 5, *J. Virol.* **41**:674–685.

Ruben, M., Bacchetti, S., and Graham, F.L., 1983, Covalently closed circles of adenovirus 5 DNA, *Nature (London)* **301**:172–174.

Ruley, H.E., Adenovirus early region 1A enables viral and cellular transforming genes to transform primary cells in culture, *Nature* **304**:602–606.

Sambrook, J., Botchan, M.B., Gallimore, P.H., Ozanne, B., Petterson, U., Williams, J.F., and Sharp, P.A., 1974, Viral DNA sequences in cells transformed by simian virus 40, adenovirus type 2 and adenovirus type 5, *Cold Spring Harbor Symp. Quant. Biol.* **39**:615–632.

Sambrook, J., Greene, R., Stringer, J., Mitchison, T., Hu, S.-L., and Botchan, M., 1979, Analysis of the sites of integration of viral DNA sequences in rat cells transformed by adenovirus 2 or SV40, *Cold Spring Harbor Symp. Quant. Biol.* **44**:569–584.

Sarnow, P., Ho, Y.S., Williams, J., and Levine, A.J., 1982a, Adenovirus E1B-58Kd tumor antigen and SV40 large tumor antigen are physically associated with the same 54kd cellular protein in transformed cells, *Cell* **28**:387–394.

Sarnow, P., Sullivan, C.A., and Levine, A.J., 1982b, A monoclonal antibody detecting the adenovirus type 5 E1B-58kd tumor antigen: Characterization of the E1B-58kd tumor antigen in adenovirus-infected and -transformed cells, *Virology* **120**:510–517.

Scangos, G.A., Huttner, K.M., Juricek, D.K., and Ruddle, F.H., 1981, Deoxyribonucleic acid-mediated gene transfer in mammalian cells: Molecular analysis of unstable transformants and their progression to stability, *Mol. Cell. Biol.* **1**:111–120.

Schirm, S., and Doerfler, W., 1981, Expression of viral DNA in adenovirus type 12 transformed cells, in tumor cells, and in revertants, *J. Virol.* **39**:694–702.

Schrier, P.I., van der Elsen, P.J., Hertoghs, J.J.L., and van der Eb, A.J., 1979, Characterization of tumor antigens in cells transformed by fragments of adenovirus type 5 DNA, *Virology* **99**:372–385.

Schrier, P.I., Bernards, R., Vaessen, R.T.M.J., Houweling, A., and van der Eb, A.J., 1983, Expression of class I major histocompatibility antigens switched off by highly oncogenic adenovirus 12 in transformed rat cells, *Nature* **305**:771–775.

Sekikawa, K., Shiroki, K., Shimojo, H., Ojima, S., and Fujinaga, K., 1978, Transformation of a rat cell line by an adenovirus type 7 DNA fragment, *Virology* **88**:1–7.

Sharp, P.A., Pettersson, U., and Sambrook, J., 1974, Viral DNA in transformed cells. 1. A study of the sequences of adenovirus 2 DNA in a line of transformed rat cells using specific fragments of the viral genome, *J. Mol. Biol.* **86**:709–726.

Shaw, A.R., and Ziff, E.B., 1982, Selective inhibition of adenovirus type 2 early region II and III transcription by an anisomycin block of protein synthesis, *Mol. Cell. Biol.* **2**:789–799.

Shenk, T., Jones, N., Colby, W., and Fowlkes, D., 1979, Functional analysis of adenovirus type 5 host range deletion mutants defective for transformation of rat embryo cells, *Cold Spring Habor Symp. Quant. Biol.* **44**:367–375.

Shimojo, H., and Yamashita, T., 1968, Induction of DNA synthesis by adenoviruses in contact inhibited hamster cells, *Virology* **36**:422–433.

Shin, S., Freedman, V.H., Risser, R., and Pollack, R., 1975, Tumorigenicity of virus-transformed cells in nude mice is correlated specifically with anchorage independent growth *in vitro, Proc. Natl. Acad. Sci. U.S.A.* **72**:4435–4439.

Shiroki, K., Handa, H., Shimojo, H., Yano, S., Ojima, S., and Fujinaga, K., 1977, Establishment and characterization of rat cell lines transformed by restriction endonuclease fragments of adenovirus 12 DNA, *Virology* **82**:462–471.

Shiroki, K., Shimojo, H., Maeta, Y., and Hamada, C., 1979a, Tumor-specific transplantation and surface antigen in cells transformed by the adenovirus 12 DNA fragments, *Virology* **99**:188–191.

Shiroki, K., Shimojo, H., Sawada, Y., Uemizu, Y., and Fujinaga, K., 1979b, Incomplete transformation of rat cells by a small fragment of adenovirus 12 DNA, *Virology* **95**:127–136.

Shiroki, K., Segawa, K., and Shimojo, H., 1980, Two tumor antigens and their polypeptides in adenovirus type 12-infected and transformed cells, *Proc. Natl. Acad. Sci. U.S.A.* **77**:2274–2278.

Shiroki, K., Maruyama, K., Saito, I., Fukui, Y., and Shimojo, H., 1981, Incomplete transformation of rat cells by a deletion mutant of adenovirus type 5, *J. Virol.* **38**:1048–1054.

Signas, C., Katze, M.G., Persson, H., and Philipson, L., 1982, An adenovirus glycoprotein is tightly bound to class I transplantation antigens, *Nature (London)* **299**:175–178.

Sjogren, H.O., Minowada, J., and Ankerst, J., 1967, Specific transplantation antigens of mouse sarcomas induced by adenovirus type 12, *J. Exp. Med.* **125**:689–701.

Smart, J.E., and Stillman, B.W., 1982, Adenovirus terminal protein precursor: Partial amino acid sequence and the site of covalent linkage to virus DNA, *J. Biol. Chem.* **257**:13,499–13,506.

Smart, J.E., Lewis, J.B., Mathews, M.B., Harter, M.L., and Anderson, C.W., 1981, Adenovirus type 2 early proteins: Assignment of the early region 1A proteins synthesized *in vivo* and *in vitro* to specific mRNAs, *Virology* **112**:703–713.

Solnick, D., 1981, An adenovirus mutant defective in splicing RNA from early region 1A, *Nature (London)* **291**:508–510.

Solnick, D., and Anderson, M.A., 1982, Transformation-deficient adenovirus mutant defective in expression of region 1A but not region 1B, *J. Virol.* **42**:106–113.

Southern, E.M., 1975, Detection of specific sequences among DNA fragments separated by gel electrophoresis, *J. Mol. Biol.* **98**:503–517.

Spector, D.J., Halbert, D.N., and Raskas, H.J., 1980, Regulation of integrated adenovirus sequences during adenovirus infection of transformed cells, *J. Virol.* **36**:860–871.

Stabel, S., Doerfler, W., and Friis, R.R., 1980, Integration sites of adenovirus type 12 DNA in transformed hamster cells and hamster tumor cells, *J. Virol.* **36**:22–40.

Steinberg, B., Rifkin, D., Shin, S., Boone, C., and Pollock, R., 1979, Tumorigenicity of revertants from an SV40 transformed cell line, *J. Supramol. Struct.* **11**:539–546.

Stich, H.F., 1973, Oncogenic and non-oncogenic mutants of adenovirus type 12: Induction of chromosome aberrations and cell division, *Prog. Exp. Tumor Res.* **18**:260–272.

Stillman, B.W., Lewis, J.B., Chow, L.T., Mathews, M.B., and Smart, J.E., 1981, Identification of the gene and mRNA for the adenovirus terminal protein procursor, *Cell* **23**:497–508.

Stillman, B.W., Tamanoi, F., and Mathews, M.B., 1982, Purification of an adenovirus coded DNA polymerase that is required for initiation of DNA replication, *Cell* **31**:613–623.

Stow, N.D., 1981, Cloning of a DNA fragment from the left-hand terminus of the adenovirus type 2 genome and its use in site-directed mutagenesis, *J. Virol.* **37**:171–180.

Stringer, J.R., 1982, DNA sequence homology and chromosomal deletion at a site of SV40 DNA integration, *Nature (London)* **296**:363–366.

Strohl, W.A., 1973, Alterations in hamster cell regulatory mechanisms resulting from abortive infection with an oncogenic adenovirus, *Prog. Exp. Tumor Res.* **18**:200–239.

Strohl, W.A., Rabson, A.S., and Rouse, H., 1967, Adenovirus tumorigenesis: Role of the viral genome in determining tumor morphology, *Science* **156**:1631–1633.

Sugisaki, H., Sugimoto, K., Takanami, M., Shiroki, K., Saito, I., Shimojo, H., Sawada, Y., Uemizu, Y., Uesugi, S., and Fujinaga, K., 1980, Structure and gene organization in the transforming HindIII G fragment of Ad12, *Cell* **20**:777–786.

Sutter, D., and Doerfler, W., 1979, Methylation of integrated viral DNA sequences in adenovirus type 12 transformed hamster cells, *Cold Spring Harbor Symp. Quant. Biol.* **44**:565–568.

Sutter, D., and Doerfler, W., 1980, Methylation of integrated adenovirus type 12 DNA sequences in transformed cells is inversely correlated with viral gene expression, *Proc. Natl. Acad. Sci. U.S.A.* **77**:253–256.

Sutter, D., Westphal, M., and Doerfler, W., 1978, Patterns of integration of viral DNA sequences in the genomes of adenovirus type 12-transformed hamster cells, *Cell* **14**:569–585.

Takahashi, M., Ueda, S., and Ogino, T., 1966, Enhancement of the thymidine activity of human embryonic kidney cells by infection with human adenovirus types 5 and 12, *Virology* **30**:742–743.

Takahashi, M., Ogino, T., Baba, K., and Onaka, M., 1969, Synthesis of deoxyribonucleic acid in human and hamster kidney cells infected with human adenovirus types 5 and 12, *Virology* **37**:513–520.

Takemori, N., Riggs, J.L., and Aldrich, C., 1968, Genetic studies with tumorigenic adenoviruses. 1. Isolation of cytocidal (cyt) mutants of adenovirus type 12, *Virology* **36**:575–586.

Takemori, N., Riggs, J.L., and Aldrich, C.D., 1969, Genetic studies with tumorigenic adenoviruses. II. Heterogeneity of cyt mutants of adenovirus type 12, *Virology* **38**:8–15.

Tevethia, S.S., 1980, Immunology of simian virus 40, in: *Viral Oncology* (G. Klein, ed.), pp. 581–602, Raven Press, New York.

Tibbetts, C., 1977, Viral DNA sequences from incomplete particles of human adenovirus type 7, *Cell* **12**:243–249.

Tooze, J. (ed.), 1981, *The Molecular Biology of Tumor Viruses*, 2nd rev. ed., Cold Spring Harbor Press, Cold Spring Harbor, New York.

Trentin, J.J., and Bryan, E., 1966, Virus induced transplantation immunity to human adenovirus type 12 tumours of the hamster and mouse, *Proc. Soc. Exp. Biol. Med.* **121**:1216–1219.

Trentin, J.J., Yabe, Y., and Taylor, G., 1962, The quest for human cancer viruses, **137**:835–841.

Trentin, J.J., Van Hoosier, G.L., and Samper, L., 1968, The oncogenicity of human adenoviruses in hamsters, *Proc. Soc. Exp. Biol. Med.* **127**:683–689.

Tsuei, D., Fujinaga, K., and Green, M., 1972, The mechanism of viral carcinogenesis by DNA mammalian viruses: RNA transcripts containing viral and highly reiterated cellular base sequences in adenovirus transformed cells, *Proc. Natl. Acad. Sci. U.S.A.* **69**:427–430.

Tsukamoto, K., and Sugino, Y., 1972, Nonproductive infection and induction of cellular deoxyribonucleic acid synthesis by bovine adenovirus type 3 in a contact inhibited mouse cell line, *J. Virol.* **9**:465–473.

van den Elsen, P., de Pater, S., Houweling, A., van der Veer, J., and van der Eb, A., 1982, The relationship between region E1A and E1B of human adenoviruses in cell transformation, *Gene* **18**:175–185.

van der Eb, A.J., Mulder, C., Graham, F.L., and Houweling, A., 1977, Transformation with specific fragments of adenovirus DNAs. I. Isolation of specific fragments with transforming activity of adenovirus 2 and 5 DNA, *Gene* **2:**115–132.

van der Eb, A.J., van Ormondt, H., Schrier, P.I., Lupker, H.J., Jochemsen, H., van der Elsen, P.J., De Leys, R.J., Maat, J., van Beveren, C.P., Dijkema, R., and De Waard, A., 1979, Structure and function of the transforming genes of human adenoviruses and SV40, *Cold Spring Harbor Symp. Quant. Biol.* **44:**383–399.

van der Eb, A.J., Bernards, R., van den Elsen, P.J., Bos, J.L., and Schrier, P.I., 1983, Studies on the role of adenovirus E1 genes in transformation and oncogenesis, in: *Human Carcinogenesis* (Harris and Autrup, eds.), Academic Press, New York (in press).

van der Vliet, P.C., Levine, A.J., Ensinger, M.J., and Ginsberg, H.S., 1975, Thermolabile DNA binding proteins from cells infected with a temperature-sensitive mutant of adenovirus defective in viral DNA synthesis, *J. Virol.* **15:**348–354.

van Ormondt, H., Maat, J., De Waard, A., and van der Eb, A.J., 1978, The nucleotide sequence of the transforming HpaI-E fragment of adenovirus type 5 DNA, *Gene* **4:**309–328.

van Ormondt, H., Maat, J., and van Beveren, C.P., 1980, The nucleotide sequence of the transforming early region E1 of adenovirus type 5 DNA, *Gene* **11:**299–309.

Vardimon, L., and Doerfler, W., 1981, Patterns of integration of viral DNA in adenovirus type 2-transformed hamster cells, *J. Mol. Biol.* **147:**227–246.

Vasconcelos-Costa, J., 1976, Solubilization and purification of surface antigen of cells transformed by adenovirus type 12, *Virology* **71:**122–133.

Virtanen, A., and Pettersson, U., 1983, The molecular structure of the 9S mRNA from early region 1A of adenovirus serotype 2, *J. Mol. Biol.* **165:**496–499.

Virtanen, A., Pettersson, U., Le Moullec, J.M., Tiollais, P., and Perricaudet, M., 1982, Different mRNAs from the transforming region (E1B) of highly- and non-oncogenic human adenoviruses, *Nature (London)* **295:**705–707.

Visser, L., Wassenaar, A.T.D.C., van Maarschalkerweerd, M.W., and Rozijn, T.H., 1981, Arrangement of integrated viral DNA sequences in cells transformed by adenovirus type 2 and 5, *J. Virol.* **39:**684–693.

Visser, L., Reemst, A.C.M.B., van Mansfeld, A.D.M., and Rozijn, T.H., 1982, Nucleotide sequence analysis of the linked left and right hand l regions of adenovirus type 5 DNA present in the transformed rat cell line 5RK20, *Nucleic Acids Res.* **10:**2189–2198.

Wall, R., Weber, J., and Darnell, J.E., 1973, Production of viral mRNA in adenovirus transformed cells by the post-transcriptional processing of heterogeneous nuclear RNA containing viral and cell sequences, *J. Virol.* **11:**953–960.

Westin, G., Visser, L., Zabielski, J., van Mansfeld, A.D.M., Pettersson, U., and Rozijn, T., 1982, Sequence organization of a viral DNA insertion present in the adenovirus type 5 transformed hamster line BHK268-C31, *Gene* **17:**263–270.

Wigler, M., Perucho, M., Levy, D., and Hanahan, D., 1980, DNA mediated gene transfer: Theory and application, in: *Ninth Annual ICN–UCLA Symposium: Animal Virus Genetics* (B. Fields and R. Jaenisch, eds.), p. 229, Academic Press New York.

Williams, J.F., Gharpure, M., Ustaçelebi, S., and McDonald, S., 1971, Isolation of temperature-sensitive mutants of adenovirus type 5, *J. Gen. Virol.* **11:**95–101.

Williams, J.F., Young, C.S.H., and Austin, P.E., 1974, Genetic analysis of human adenovirus type 5 in permissive and nonpermissive cells, *Cold Spring Harbor Symp. Quant. Biol.* **39:**427–437.

Williams, J.F., Galos, R.S., Binger, M.H., and Flint, S.J., 1979, Location of additional early regions within the left quarter of the adenovirus genome, *Cold Spring Harbor Symp. Quant. Biol.* **44:**353–366.

Williams, J., Ho, Y.S., and Galos, R., 1981, Evidence for functional relatedness of products encoded by the transforming sequences of human adenoviruses 5 and 12, *Virology* **110:**208–212.

Wold, W.S.M., Chinnadurai, G., Green, M., and Mak, S., 1979, Identification of adenovirus type 12 candidate transformation proteins by radioimmunoprecipitation with antisera to EcoRI-C-fragment transformed cells, *Virology* **94:**208–213.

Yabe, Y., Samper, L., Bryan, E., Taylor, G., and Trentin, J.J., 1964, Oncogenic effect of human adenovirus type 12 in mice, *Science* **143**:46.

Yano, S., Ojima, S., Fujinaga, K., Shiroki, K., and Shimojo, H., 1977, Transformation of a rat cell line by an adenovirus type 12 DNA fragment, *Virology* **82**:214–220.

Yee, S.-P., Rowe, D.T., Tremblay, M.L., McDermott, M., and Branton, P.E., 198?, Identification of human adenovirus early region 1 products using antisera against synthetic peptides corresponding to the predicted carboxy termini *J. Virol.* **46**:1003–1013.

Zinkernagel, R.M., and Doherty, P.C., 1979, MHC-restricted cytotoxic T cells: studies on the biological role of polymorphic major transplantation antigens determining T-cell restriction-specificity, function, and responsiveness, *Advances in Immunology* **27**:51–177.

zur Hausen, H., 1967, Induction of specific chromosomal abberations by adenovirus type 12 in human embryonic kidney cells, *J. Virol.* **1**:1174–1185.

zur Hausen, H., 1968, Chromosomal aberrations and cloning efficiency in adenovirus type 12-infected hamster cells, *J. Virol.* **2**:915–924.

CHAPTER 10

Adenovirus–Simian Virus 40 Interactions

DANIEL F. KLESSIG

I. INTRODUCTION

In this chapter, the numerous interactions between human adenovirus and simian virus 40 (SV40) are reviewed. These viruses can interact at several levels. They can recombine at the nucleotide (genome) level to form hybrid viruses. Many of these hybrid viruses induce the synthesis of chimeric messenger RNA (mRNA), while some even encode chimeric proteins that are functional.

Adenovirus and SV40 are also related functionally. In addition to their oncogenic properties, they interact to help or inhibit each other's growth. Thus, for example, SV40 relieves the block to adenovirus growth in monkey cells or complements a temperature-sensitive, DNA-replication-negative mutant of adenovirus at nonpermissive temperatures. Adenovirus, in turn, can help modulate the expression of SV40 early genes or, in contrast, inhibit SV40 DNA replication.

II. ADENOVIRUS–SIMIAN VIRUS 40 HYBRIDS

Adenovirus–SV40 hybrids are recombinants in which part of the adenovirus genome has been replaced with part or all of the SV40 genome. These recombinant genomes are enclosed in adenovirus capsids. Hybrids were originally isolated from stocks of human adenovirus types 1–5 (Ad1–5) and Ad7 that were adapted to grow in rhesus monkey kidney (RMK)

DANIEL F. KLESSIG • Department of Cellular, Viral and Molecular Biology, College of Medicine, University of Utah, Salt Lake City, Utah 84132.

cells for production of vaccines (Hartley *et al.*, 1956). Since then, more deliberate efforts to isolate or construct adenovirus–SV40 recombinants have yielded a variety of additional hybrids.

Because propagation of viruses in human cells for production of medical vaccines was prohibited, attempts were made in the mid-1950s to grow several serotypes of adenoviruses in rhesus monkey cells. Initially, these viruses grew very poorly, but after repeated passaging in cultured monkey cells, adapted viral strains able to multiply efficient in these cells appeared. This program of vaccine production had to be abandoned, however, when it was discovered that the viral stocks were contaminated with SV40, a virus capable of inducing tumors in newborn hamsters (Eddy *et al.*, 1961) and, at that time, of unknown consequence to humans. SV40 is ubiquitous in rhesus monkeys and in kidney cells derived from them (Sweet and Hilleman, 1960). It was subsequently discovered that the presence of SV40 greatly enhances the growth of human adenoviruses in cultured monkey fibroblasts (Rabson *et al.*, 1964; O'Connor *et al.*, 1963, 1965). Attempts to rid the viral stocks of contaminating SV40 by addition of SV40-specific neutralizing antiserum were not completely effective because the enhancing activity in adapted strains could be provided by adenovirus capsids containing both SV40 and adenovirus genetic information (Rapp *et al.*, 1964b; Huebner *et al.*, 1964; Rowe and Baum, 1964; Easton and Hiatt, 1965; Beardmore *et al.*, 1965; Lewis *et al.*, 1966). A variety of experimental approaches subsequently showed that (1) the chimeric genetic elements within these capsids were covalently joined and (2) the ability to enhance or help adenovirus growth in monkey cells could be provided either by the free SV40 genome present in coinfection or when part or all of the SV40 genetic information is covalently inserted into the adenovirus genome to form a hybrid virus.

An excellent and detailed review of adenovirus–SV40 hybrids has been prepared by Grodzicker (1980). For completeness, much of the same literature is briefly reviewed in this section; however, greater emphasis is placed on developments since 1979.

A. Adenovirus Type 2–Simian Virus 40 Hybrids

Although the Ad2–SV40 hybrid viruses were not the first described, they have been and continue to be the most extensively studied and are therefore discussed first. Most of the Ad2–SV40 hybrids have been isolated from a strain of Ad2, termed $Ad2^{++}$, that has been adapted for growth in RMK cells (Hartley *et al.*, 1956; Lewis *et al.*, 1966). $Ad2^{++}$ contains wild-type (Wt) Ad2, complete SV40 virions, and a mixture of Ad2–SV40 hybrids (Lewis *et al.*, 1969). For the purpose of discussion, the Ad2–SV40 hybrids have been divided into two groups–nondefective and defective hybrids.

1. Nondefective Hybrids

The nondefective hybrid viruses form plaques on monolayers of human cells [HeLa (cervical carcinoma) or human embryonic kidney (HEK)] with one-hit kinetics, indicating that only one infectious viral particle per cell is required for a productive infection. A family of nondefective hybrid viruses has been derived from the original $Ad2^{++}$ stock. A viral stock, termed pool B55, was grown from a single viral plaque that formed on a primary African green monkey kidney (AGMK) cell monolayer infected by the $Ad2^{++}$ adapted for growth in RMK cells (Grodzicker, 1980). Pool B55, which was propagated on AGMK cells, formed plaques with equal efficiency and one-hit kinetics on both HEK and AGMK cells (Lewis *et al.*, 1969, 1973). Five nondefective hybrids were isolated from pool B55 after a series of plaque isolations on either AGMK cells ($Ad2^{+}ND4$) or HEK cells ($Ad2^{+}ND1$, $Ad2^{+}ND2$, $Ad2^{+}ND3$, $Ad2^{+}ND5$) (for the pedigree of each, see Grodzicker, 1980). All five hybrids have a deletion of adenovirus sequences that are not essential for growth in HeLa or HEK cells and an insertion of SV40 sequences at the site of this deletion. Thus, these hybrids form plaques with one-hit kinetics on human cells. These viruses have been extensively characterized in terms of genomic structure and viral RNA and protein synthesis.

The genomic structures of the hybrids have been determined by restriction enzyme analysis, heteroduplex mapping, nuclease S1 and exonuclease VII mapping, and partial DNA sequence determination (Kelly and Lewis, 1973; Morrow *et al.*, 1973; Lebowitz *et al.*, 1974; Zain and Roberts, 1978; Westphal *et al.*, 1979; Khoury *et al.*, 1980; Westphal, 1981) (see Fig. 1). All five hybrids have varying amounts of Ad2 early region 3 (E3) replaced with different-size portions of the SV40 early region that encodes the large tumor (large-T) antigen. [Recently, it was discovered (Khoury *et al.*, 1980) that $Ad2^{+}ND2$ consist of two types of hybrid genomes (see Fig. 1).] One end point of the deletion in Ad2, at map position 85.5, is shared by all five (six) recombinants. Similarly, the inserted SV40 DNAs all begin at the same position on the SV40 genome [0.11 (positions are denoted on the SV40 maps as fractions of 1 and on the adenovirus map as fractions of 100)] and form an overlapping series that extends to various lengths into the SV40 early region. Thus, the hybrids contain a variable amount of SV40 early information and a constant amount of SV40 late DNA [~0.17–0.11 (see Fig. 1)]. Because the hybrids have a large portion of the E3 region of Ad2 deleted, yet are all viable, this region is not essential for viral growth, at least in tissue culture.

Although the mechanism or mechanisms by which the five (six) hybrids were formed are not completely clear, several observations provide adequate ground for reasonable speculation. First, all five (six) hybrids contain the same right-end Ad2–SV40 junction (map position 85.5 and 0.11 for Ad2 and SV40, respectively). Second, the B55 pool from which all the nondefective hybrids were derived has been shown to be identical to

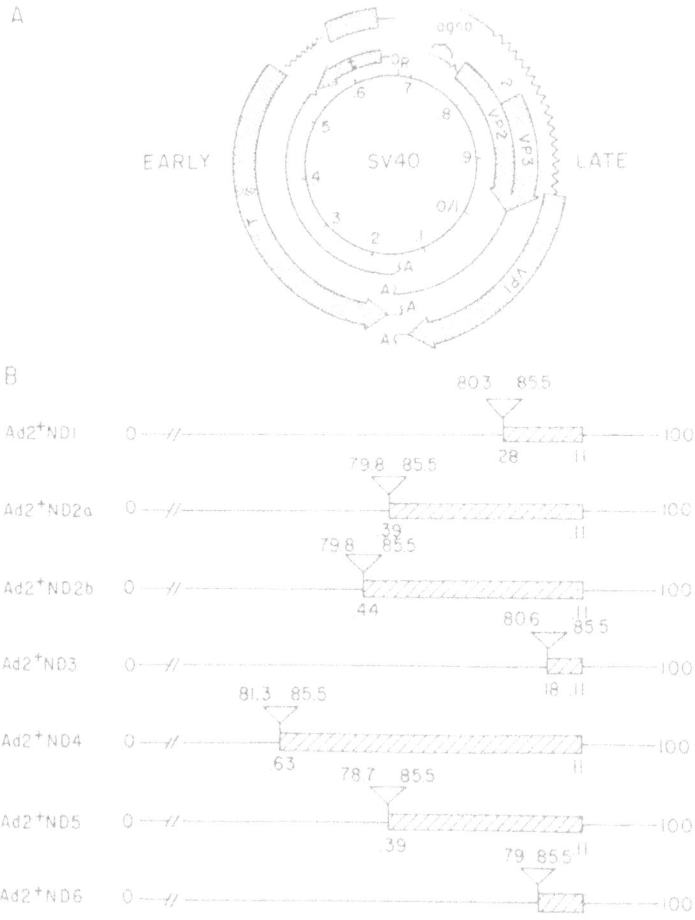

FIGURE 1. Map of SV40 and the nondefective Ad2–SV40 hybrids. (A) SV40 map of mRNAs and proteins. The 5′ termini of SV40 mRNAs map near the origin of DNA replication (O_R), which is located at position 0.67. Early mRNAs are transcribed counter-clockwise and late mRNAs clockwise. The two early species are coterminal (0.67–0.17), but differ in the positions of the internal splices. They encode the SV40 large-T (T) and small-t (t) antigens, which are amino-coterminal. The leaders spliced to the coding regions of the late mRNAs are heterogeneous, with multiple 5′ termini (···—). The 5′ terminus for the VP3 mRNA has not been well defined. The three late species are 3′-coterminal (~0.17) and encode the three capsid proteins, VP1, VP2, and VP3. VP2 and VP3 are carboxy-coterminal, and a small segment of their C terminus overlaps the N terminus of VP1, but different reading frames are used. (▭) protein coding regions; (〰) intervening sequences removed by RNA splicing. A small, basic, DNA-binding protein (called agno) encoded in the leader segment of the VP2 mRNA has recently been discovered (G. Jay *et al.*, 1981). (B) Genomic structures of nondefective Ad2–SV40 hybrids. All hybrids contain an insertion of SV40 DNA (▨) that replaces the Ad2 sequences (▽). All the insertions share a common end point at position 0.11 on the SV40 genomic map, while the deletions have a common end point at position 85.5 on the Ad2 physical map. The direction of transcription of the SV40 early sequences is from left to right, i.e., the same orientation of the Ad2 major late promoter. Map coordinates for SV40 are given as fractions of 1, those for adenovirus as fractions of 100. The SV40 and Ad2 segments of the hybrid genomes are not drawn to the same scale; i.e., the SV40 insertion is exaggerated for clarity.

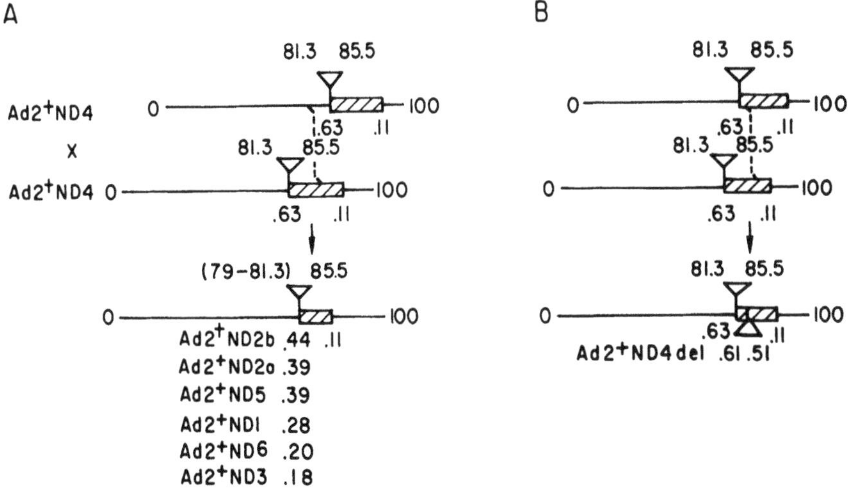

FIGURE 2. Model for the generation of the nondefective Ad2–SV40 hybrids. Unequal crossover between two molecules of Ad2$^+$ND4 generates new viruses that contain less genetic information. (A) Crossover between Ad2 sequences on one molecule and SV40 DNA on the second; (B) crossover within the SV40 insertion at different sites in the two molecules of Ad2$^+$ND4. Symbols and map coordinates are defined in the Fig. 1 caption. Modified from Grodzicker (1980).

Ad2$^+$ND4 (T. J. Kelly and A. M. Lewis, unpublished results). Third, all the nondefective hybrids contain less SV40 DNA than Ad2$^+$ND4 and are related to it as follows: they are all missing SV40 sequences from the left end of the SV40 insertion, as well as some Ad2 sequences located just to the left of the insertion. Fourth, there is no sequence homology between SV40 and Ad2 at or near the right junction site of Ad2$^+$ND1 (Zain and Roberts, 1978) or the left junction of Ad2$^+$ND4 (Westphal, 1981). These findings suggest a scheme in which SV40 genetic information equal to or greater than that present in Ad2$^+$ND4 was initially inserted into the adenovirus genome via nonhomologous or illegitimate recombination between complete Ad2 and SV40 genomes. Simultaneously or subsequently, Ad2 sequences from the E3 region, as well as perhaps extra SV40 DNA, were deleted again by illegitimate recombination to form Ad2$^+$ND4. The remaining nondefective hybrids could then have been generated by unequal crossing-over (again nonhomologous recombination) between two Ad2$^+$ND4 molecules (Grodzicker, 1980) (see Fig. 2). The process of unequal crossing-over between two Ad2$^+$ND4 genomes can also be invoked for the formation of at least two additional deletions of Ad2$^+$ND4 called Ad2$^+$ND4del (Morrow *et al.*, 1973; Westphal *et al.*, 1979; Khoury *et al.*, 1980; Westphal, 1981) and Ad2$^+$ND6 (Lewis and Westphal, 1983) (see Figs. 1 and 2). If a crossover event took place between Ad2 sequences to the left of the SV40 insertions on one DNA molecule and sequences in the middle of the SV40 insertion on a second DNA molecule, hybrids with genomic structures of Ad2$^+$ND1, Ad2$^+$ND2,

Ad2$^+$ND3, Ad2$^+$ND5, and Ad2$^+$ND6 could be formed. If, however, the crossover between two Ad2$^+$ND4 molecules occurred at different places within the SV40 insertions, a virus with the genomic structure of Ad2$^+$ND4del could result. Additional variants of Ad2$^+$ND4 have been uncovered but not characterized (Lewis and Westphal, 1983).

Why are so many variants of Ad2$^+$ND4 generated, particularly when propagated in human cells? The likely explanation concerns the synthesis of the multipartite adenovirus fiber mRNA, the leaders of which are encoded upstream from the SV40 insertion while its main body is located downstream. We will return to this question after discussing the structure and function of the chimeric Ad2–SV40 mRNAs.

Early studies using competition hybridization with RNA from hybrid-infected monkey cells showed that the amount of transcription of SV40 sequences is roughly proportional to the quantity of SV40 DNA contained in each hybrid (Levine *et al.*, 1973). Only the early strand of the SV40 genome is transcribed in the hybrid-infected cells, and cleavage and polyadenylation of these transcripts occur at the normal site (~0.17) for early SV40 mRNAs (Khoury *et al.*, 1973, 1980; Flint *et al.*, 1975; Westphal *et al.*, 1979). Thus, the SV40-specific transcripts encoded by the hybrids all contain varying lengths of the 3′ end of the SV40 early mRNA. Because the region that contains the SV40 early promoter is absent in all these hybrids, these SV40 sequences are probably under the control of one or more Ad2 promoters. A variety of experimental approaches have confirmed the chimeric nature of these RNAs; they include sequential hybridization of Ad2$^+$ND4 or Ad2$^+$ND1 RNA to Ad2 and SV40 DNA (Oxman *et al.*, 1974; Dunn and Hassell, 1977), saturation hybridization with Ad2$^+$ND1 and Ad2$^+$ND3 RNA (Flint *et al.*, 1975), cell-free translation of Ad2$^+$ND1 and Ad2$^+$ND4 mRNAs selected by hybridization to Ad2 or SV40 DNA (Anderson *et al.*, 1976), sequences determination of complementary DNA made from Ad2$^+$ND1 mRNA (Dunn *et al.*, 1978), Northern blot analysis of SV40-specific transcripts of Ad2$^+$ND4 selected by hybridization to Ad2 DNA (Khoury *et al.*, 1980), and electron-microscopic (EM) analysis of Ad2$^+$ND4 RNA (Westphal *et al.*, 1979).

The structure of the SV40 segments of the chimeric RNAs made in human cells late in the infectious cycle has been determined for Ad2$^+$ND1, Ad2$^+$ND2, and Ad2$^+$ND4 by nuclease-S1 and EM analyses. Ad2$^+$ND1 specifies chimeric transcripts with a unique SV40 segment that terminates at the normal SV40 early polyadenylation site at approximately 0.17 and is colinear (unspliced) with the entire early SV40 DNA present in this hybrid (Khoury *et al.*, 1980). Although the 3′ segment encoded by SV40 is homogeneous in this mRNA, the 5′ segment specified by Ad2 changes during the lytic cycle. First, the E3 promoter at map position 76, located just upstream from the SV40 insertion, is utilized (L. Chow, unpublished observations). But as the infection proceeds, transcription of the SV40 sequences is placed under control of the major late promoter (MLP) at 16.4. Thus, at late times after infection, these chimeric RNAs resemble late adenovirus mRNAs with the tripartite leader at their 5′ ends (Dunn *et al.*, 1978).

Ad2$^+$ND2 specifies RNAs with two types of SV40 sequences (Khoury *et al.*, 1980). Because the sizes of these SV40 RNA sequences are equivalent to the entire SV40 early region present in the two Ad2$^+$ND2 hybrids (see Fig. 1B), it is likely that one RNA is transcribed from each of the two hybrid genomes. The SV40 segment of these two RNAs is not spliced, though by analogy with Ad2$^+$ND1, we would predict that the Ad2 section would be altered by the splicing processes.

The situation is considerably more complex for Ad2$^+$ND4. This probably reflects (1) the extra SV40 information present in Ad2$^+$ND4, including the normal early region donor and acceptor slice sites between 0.59 and 0.54 map unit (m.u.), and (2) the heterogeneous nature of the hybrid genomes in human cells. Khoury *et al.* (1980) found that approximately 85% of the Ad2$^+$ND4 hybrid genomes in their stock contain a deletion between 0.61 and 0.51 m.u. (Ad2$^+$ND4del), while the remaining 15% have the normal Ad2$^+$ND4 structure. The chimeric RNAs are made up of 4–6 different types of SV40 segments coupled to 5 different types of Ad2 sections (Westphal *et al.*, 1979; Khoury *et al.*, 1980). Different permutations of these segments could generate up to 30 different chimeric RNAs, and in fact, a very complex pattern of SV40-specific RNAs has been observed by Northern blot analysis (Khoury *et al.*, 1980). Most of these RNAs are initiated at the MLP (16.4 m.u.) and contain, in addition to the tripartite leader, several additional leaders (exons) encoded in E3 (Westphal *et al.*, 1979). Two of these additional leaders correspond to the ancillary leaders x (~77 m.u.) and y (~79 m.u.) found on some of the fiber mRNAs (Chow and Broker, 1978). One of the predominant late chimeric RNAs is generated by splicing Ad2 sequences to an acceptor site in SV40 at 0.46. This was quite unexpected, since this splice-acceptor site is not utilized in either SV40-infected or -transformed cells.

Chimeric RNAs with similar SV40 segments have also been observed in AGMK cells late after infection with Ad2$^+$ND4 (Khoury *et al.*, 1980). The complex late pattern contrasts with a rather simple pattern of chimeric RNA made during the early phase of the lytic cycle in human cells. During this phase, the predominant RNA appears to have been transcribed off the Ad2$^+$ND4del template and is not spliced within the SV40 segment (Khoury *et al.*, 1980). On the basis of the observations with Ad2$^+$ND1, one might also expect that this early transcript would be initiated at the E3 promoter, though at present no data are available.

Several observations made with hybrid viruses suggest that viral infection may alter the pattern of RNA splicing. First, as just described, the pattern of splicing within the SV40 segment of the chimeric RNAs changes dramatically during infection with Ad2$^+$ND4. Second, a cryptic SV40 splice-acceptor site (0.46 m.u.) not detected in SV40-infected or -transformed cells is utilized late in the infectious cycle of Ad2$^+$ND4.

Third, chimeric mRNAs corresponding to the small-t antigen mRNA are not found in $Ad2^{+}ND4$-infected human cells even though all the sequences required for splicing of the RNA for small-t antigen are present (Westphal *et al.*, 1979). Reduced synthesis of small-t antigen mRNA has also been observed with defective Ad2–SV40 hybrids constructed for the purpose of overproduction of SV40 large-T and small-t antigens in which essentially all the SV40 early mRNA sequences are present (Thummel *et al.*, 1981, 1982) (see Section II.C). Because the latter observations were made in monkey cells (the normal host of SV40) as well as human cells, these differences in splicing pattern in SV40-infected vs. Ad2–SV40-hybrid-infected cells probably reflect the effect of viral infection on the splicing process, rather than possible cell-type differences in the splicing apparatus.

Returning now to the question of the instability of the $Ad2^{+}ND4$, let us begin by reviewing the pertinent facts: (1) $Ad2^{+}ND4$ propagated on human cells quickly gives rise to a variety of deletion mutants (e.g., $Ad2^{+}ND1$, $Ad2^{+}ND2$, $Ad2^{+}ND3$, $Ad2^{+}ND5$, $Ad2^{+}ND6$, $Ad2^{+}ND4del$) in which the normal SV40 early splice sites between 0.59 and 0.54 m.u. are deleted and the resulting mutants form plaques with one-hit kinetics in human cells. (2) $Ad2^{+}ND4$ will not grow on established lines of AGMK cells, such as CV-1 or BSC-1, without a Wt adenovirus helper, while in the presence of this helper, $Ad2^{+}ND4$ is readily propagated and stable (Lewis and Westphal, 1983). (3) In primary AGMK cells, $Ad2^{+}ND4$ is easily grown, forms plaques with one-hit kinetics, and is quite stable. (4) The $Ad2^{+}ND4$ chimeric mRNAs made late in infection and Ad2 fiber mRNAs have strikingly similar leader composition, suggesting a common pathway of maturation. Thus, it seems likely that the presence of the SV40 early splice sites immediately upstream from the coding region for fiber (an essential virion component) is detrimental to the expression of the fiber gene (Westphal *et al.*, 1979; Grodzicker, 1980; Westphal, 1981). Removal of these sites together with adjacent SV40 and Ad2 sequences results in the formation of two types of variants. One type (e.g., $Ad2^{+}ND1$ and $Ad2^{+}ND2$) forms plaques with one-hit kinetics on both human and monkey cells. Presumably, in these viruses, the genomic arrangement is such that functional chimeric mRNA (to provide the helper function) and fiber mRNA is made. Lawrence (1981) has recently shown that fiber mRNA and protein are made normally in $Ad2^{+}ND1$- and $Ad2^{+}ND2$-infected human cells. The other type of variant (e.g., $Ad2^{+}ND3$, $Ad2^{+}ND5$, $Ad2^{+}ND6$, $Ad2^{+}ND4del$), while still able to multiply in human cells, cannot do so in monkey cells either because too much of the SV40 information has been deleted or because the sequence arrangements do not allow efficient expression of the SV40 information. In the presence of Wt Ad2 or a deletion derivative of $Ad2^{+}ND4$, $Ad2^{+}ND4$ is now stable. Presumably, these viruses can supply the fiber gene product and thus relieve the strong selective pressure to remove the interfering SV40 splice sites from $Ad2^{+}ND4$.

A major anomaly is the ability of Ad2$^+$ND4 to multiply efficiently and form plaques with one-hit kinetics in primary AGMK cells. Lewis and Westphal (1983) have suggested that there may be factors in these cells that allow for the rapid formation of deletion mutants that can provide the fiber gene product. Alternatively, the splicing machinery of the primary AGMK cells might differ from that of established human cell lines or monkey cell lines in that it is capable of making both functional fiber and SV40-specific mRNAs from Ad2$^+$ND4 templates even in the presence of the SV40 early splicing signals.

The SV40-specific proteins from Ad2$^+$ND1- to Ad2$^+$ND5-infected human cells have been analyzed (Lopez-Revilla and Walter, 1973; Grodzicker *et al.*, 1974; Walter and Martin, 1975; Deppert and Walter, 1975; Deppert *et al.*, 1977; Mann *et al.*, 1977; Prives *et al.*, 1982). Their *in vitro* counterparts synthesized in cell-free translation extracts using mRNA selected by hybridization to SV40 DNA (Anderson *et al.*, 1976; Mann *et al.*, 1977; Lawrence *et al.*, 1979) or Ad2 DNA (Anderson *et al.*, 1976) have also been characterized. All these hybrids except Ad2$^+$ND3 encode at least one SV40-specific polypeptide (see Table I), and in general, the size of the polypeptide encoded by the hybrid is related to the size of the SV40 insertion. Thus, Ad2$^+$ND4, which carries the largest insertion, encodes the largest polypeptides, while Ad2$^+$ND1, which has the second smallest insertion (that of Ad2$^+$ND3 being smaller), specifies the smallest protein (see Fig. 3 and Table I).

All the SV40-specific proteins share common peptides that form a nested set; i.e., most of the peptides present in any one protein are also found in all larger-size hybrid proteins (Mann *et al.*, 1977). Comparison of the genomic structure of the hybrids together with the observation that most of the [^{35}S]methionine-labeled peptides from the hybrid-specific polypeptides are found in the SV40 tumor (T) antigen (Mann *et al.*, 1977) suggest that all the hybrid-specific proteins have a common SV40-encoded carboxylterminus. The shared sequences among the different proteins extend varying distances toward the amino terminus. In addition, the N termini as well as the C termini of the proteins of molecular weight 42,000 (42K) found in Ad2$^+$ND2- and Ad2$^+$ND4- infected human cells are identical, as are the N and C termini of the Ad2$^+$ND2- or Ad2$^+$ND4-specific 56K proteins (Lawrence *et al.*, 1979). Thus, not only are the proteins specified by the different hybrids related to each other, but also, in several cases, the same protein is specified by two different hybrid viruses.

Since each protein incorporates ^{35}S-labeled formyl methionine, each must result from a *de novo* initiation event rather than from cleavage of a common precursor polypeptide (Linke *et al.*, 1979; Lawrence *et al.*, 1979). Furthermore, because the different-size proteins, with the exception of the Ad2$^+$ND4-specific 95 and 74K polypeptides, have unique N termini, each protein has a unique initiation site (Lawrence *et al.*, 1979).

TABLE I. Simian-Virus-40-Specific Proteins, Antigens, and Functions Provided by the Nondefective Adenovirus Type 2–Simian Virus 40 Hybrids[a]

Hybrid virus	Size of SV40 insertion	Size of SV40-specific proteins	SV40 antigens induced[b]	Growth in AGMK cells	Ribosomal RNA gene activation[c]	DNA binding[d]	
						Non-sequence-specific	SV40 O_R
Ad2-+ND3	0.18–0.11	None	None	–	–	NT	NT
Ad2-+ND1	0.28–0.11	30K	U, TSTA[e]	+	–	NT	NT
Ad2-+ND2	0.39–0.11	42K	U, TSTA	+	+	–	–
	0.44–0.11	56K				+	–
Ad2-+ND5	0.39–0.11	42K[f]	None	–	+	NT	NT
Ad2-+ND4	0.63–0.11	95K				+	+
		74K				+	–
		72K				+	–
		64K	U, TSTA, T	+	+	+	–
		60K				+	–
		56K				g	g
		42K				g	g

[a] Compiled from Grodzicker (1980), Soprano *et al.* (1981), and Prives *et al.* (1982).
[b] (T) Tumor antigen; (TSTA) tumor-specific transplantation antigen.
[c] Reactivation in human–mouse hybrid cells.
[d] (O_R) SV40 origin of DNA replication; (NT) not tested.
[e] TSTA induced in BALB/c mice, but not in hamsters.
[f] Unstable.
[g] These proteins were not seen by Prives *et al.* (1982).

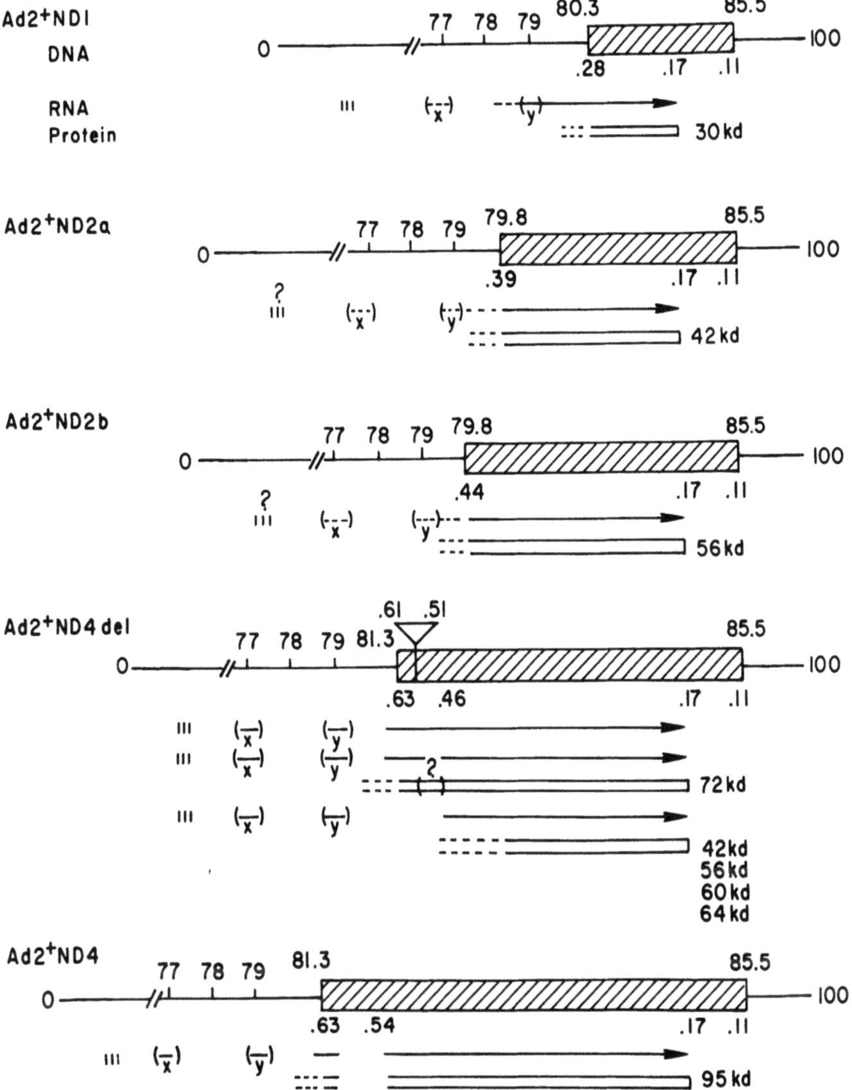

FIGURE 3. Putative map of the SV40-specific mRNAs and proteins in Ad2$^+$ND1, Ad2$^+$ND2, and Ad2$^+$ND4. (▨) SV40 insertion; (——) possible structures of the chimeric mRNAs synthesized during the late phase of infection in human cells; (||||) late Ad2 tripartite leader; (– – –) poorly defined 5′ end of the message or amino terminus of the protein; (→) direction of transcription; (— —) splices; (x, y) ancillary leaders that are sometimes also present on Ad2 fiber mRNA; (▭) proteins with molecular weights in kilodaltons (kd). The size of the SV40 insertion is exaggerated relative to that of the Ad2 sequence for clarity.

Cell-free translation of size-fractionated, SV40-specific RNAs from Ad2$^+$ND2- and Ad2$^+$ND4-infected human cells suggests that the various proteins are encoded by different-size mRNAs (Lawrence *et al.*, 1979). While in several cases the different-size mRNAs are probably transcribed from different DNA templates, differential use of splicing sites must be invoked to generate the large numbers of chimeric RNAs in $Ad2^+ND4$ infections.

At present, only a somewhat sketchy picture of the structure of the chimeric mRNAs and the proteins they encode has emerged (see Fig. 3). For example, we have no direct evidence that any of these proteins contain adenovirus peptides. Lawrence *et al.* (1979) suggested that the N-terminal sequence heterogeneity of the SV40-specific proteins makes it unlikely that the Ad2 sequences at the 5′ end of the chimeric RNAs could provide enough different translational start sites. However, EM analysis indicated that the 5′ Ad2 sequences in these hybrid RNAs are quite heterogeneous (Westphal *et al.*, 1979). In addition, studies on the defective hybrid virus have provided ample evidence that Ad2–SV40 chimeric proteins are functional (Tjian, 1978; Fey *et al.*, 1979) (see Sections II.A.2 and II.B). A final answer to this question awaits direct experimental assault.

Differences in the biological properties of the hybrids can be correlated with the sizes of (1) the deletion of Ad2 sequences, (2) the insertion of SV40 information, and (3) the SV40-specific proteins (see Table I). In SV40-infected and-transformed cells, three SV40-specific antigens are expressed: T antigen, U antigen, and tumor-specific transplantation antigen (TSTA) (Khera *et al.*, 1963; Rapp *et al.*, 1964a; Girardi and Defendi, 1970; Lewis and Rowe, 1971; Black *et al.*, 1963; Drapkin *et al.*, 1974). T and U antigens are both nuclear (Pope and Rowe, 1964; Lewis and Rowe, 1971; Deppert *et al.*, 1977; Deppert, 1978, 1979; Jay *et al.*, 1978), while TSTA is located in the plasma membrane (Tevethia and Rapp, 1966; Rogers *et al.*, 1977; Tevethia and Tevethia, 1976, 1977; Anderson *et al.*, 1977; Chang *et al.*, 1977b, 1979; Pretell *et al.*, 1979; Tevethia *et al.*, 1980; Ismail *et al.*, 1981). U and T antigens are distinguished on the basis of heat stability [U but not T antigen is heat-stable (Lewis and Rowe, 1971)] and location within subfractions of the nucleus (Deppert, 1978, 1979; Lewis and Rowe, 1971; Lewis *et al.*, 1969; Jay *et al.*, 1978). All three antigenic moieties appear to be on the same polypeptide (the 95K large-T antigen) and may reflect different subpopulations of this protein that assume different cellular locations (Anderson *et al.*, 1977; Chang *et al.*, 1977a; Prives and Beck, 1977; Robb, 1977; Rogers *et al.*, 1977; Deppert, 1979). The presence of each of the three SV40 antigenic determinants in hybrid-infected cells, in general, correlates with the size of the SV40 DNA insertion and SV40-specific protein(s) (see Table I). Thus, the hybrids can be used to map these determinants on the SV40 large-T protein.

In addition, three activities of T antigen have been mapped using this nondefective hybrid series: (1) enhancement of adenovirus growth in AGMK cells (see below), (2) reactivation of ribosomal RNA (rRNA) genes in human–mouse hybrid cells (Soprano *et al.*, 1981), and (3) DNA binding (Prives *et al.*, 1982). Here again, the sizes of the SV40-specific DNA and protein(s) can be correlated with the presence or absence of any one of these activities (see Table I). The order of the antigenic moieties from the C terminus of T antigen is: (U, TSTA); T. The order of activities is:

growth in AGMK cells; (rRNA gene activation, non-sequence-specific DNA binding); binding to the SV40 origin of replication.

The only anomaly in these correlations is Ad2$^+$ND5. This virus neither grows on AGMK cells nor induces SV40 antigens, yet the size of its SV40 insertion is larger than that of Ad2$^+$ND1 or Ad2$^+$ND2. Although it codes for a 42K protein that shares peptides with the other hybrid proteins, it is unstable in infected human cells and cannot be precipitated with some SV40 anti-T serums that precipitate the 42Kd protein encoded by Ad2$^+$ND5 (Deppert *et al.*, 1977; Jay *et al.*, 1978). Fortuitous acquisition of a mutation that renders the protein labile or nonfunctional may account for the properties of Ad2$^+$ND5. Equally likely, because this hybrid contains the largest deletion of Ad2 sequences, is the possiblity that Ad2$^+$ND5 may encode a chimeric mRNA that specifies a fusion protein that differs from the other hybrid proteins in stability or function or both. The ability of Ad2$^+$ND5 to reactivate rRNA genes suggests that either very small amounts of the 42K protein are required for this function or the functions of the 42K polypeptide are not equally labile.

2. Hybrid Derivatives of Ad2$^+$ND1

A large collection of Ad2$^+$ND1 host-range (*hr*) mutants and their revertants have been isolated and characterized by Grodzicker and colleagues (Grodzicker *et al.*, 1974, 1976; Gesteland *et al.*, 1977; Fey *et al.*, 1979; E. Lukanidin, J. Sambrook, and T. Grodzicker, unpublished results). This collection has been invaluable in showing that the very C terminus of SV40 large-T antigen carries the helper function required for growth of human adenoviruses in monkey cells.

The *hr* mutants of Ad2$^+$ND1 (Grodzicker *et al.*, 1974, 1976), which have lost their ability to grow on monkey cells yet still multiply efficiently in human cells, comprise a single complementation group. Coinfection with SV40 enhances their growth on monkey cells just as it does for Wt adenovirus.

Several of these mutants contain nonsense mutations. Instead of synthesizing the Ad2$^+$ND1 30K protein, each mutant encodes a new, shorter, unique polypeptide that can be elongated *in vitro* with the aid of yeast suppressor transfer RNAs to form the Ad2$^+$ND1 30K protein (Grodzicker *et al.*, 1976; Gesteland *et al.*, 1977). One class of revertants simultaneously regains the ability to grow in monkey cells and to synthesize the 30K protein and loses the ability to produce the shortened, mutant polypeptide. This result clearly indicates that the 30K protein provides the helper function.

A second class of revertants (1) express new SV40-specific proteins, (2) have easily detectable alterations of their genomes, and (3) often continue to synthesize their unique, shorter polypeptide (Fey *et al.*, 1979; Grodzicker, 1980; E. Lukanidin, J. Sambrook, and T. Grodzicker, unpublished results). This class of revertants falls into two categories: (1) those

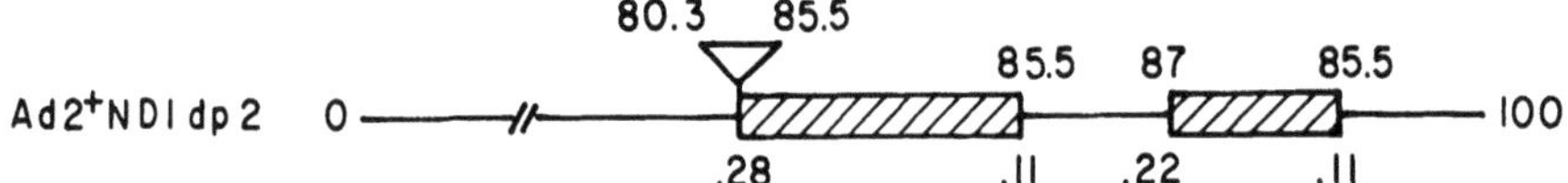

FIGURE 4. Genomic structures of Ad2$^+$ND1Δ7 and Ad2$^+$ND1dp2. Ad2$^+$ND1Δ7 contains a deletion of SV40 and Ad2 sequences that removes the nonsense mutation in the *hr* mutant, 71. Ad2$^+$ND1dp2 contains, in addition to the original SV40 insertion present in Ad2$^+$ND1 or 71 (0.28–0.11 m.u.), a new insertion comprised of Ad2 DNA (85.5–87 m.u.) and SV40 sequences located between 0.22 and 0.11 m.u. The arrangement in both these revertants of 71 allows expression of the 3′ end of the SV40 A gene, which provides the helper function required by adenovirus for growth on monkey cells. Symbols and map coordinates are defined in the Fig. 1 caption.

that carry small deletions of 50–200 base pairs (bp) within the SV40 insertion in Ad2$^+$ND1 and (2) those that harbor a partial duplication of the 5′ end of the Ad2 fiber gene and 3′ end of the SV40 A gene.

All the deletion revertants no longer synthesize the shorter polypeptide characteristics of the parent, but instead encode a new SV40-specific protein (Table II). Presumably, the nonsense codon in all these revertants is removed by an in-phase deletion that allows expression of the distal segment of the SV40 insertion. For example, the deletion in Ad2$^+$ND1Δ7 removes all but the very 3′ end of the SV40 A gene (Fig. 4). This results in the synthesis of a new 18K SV40-specific polypeptide in Ad2$^+$ND1Δ7-infected cells. Thus, the helper function must reside at the very C terminus (~90 amino acids) of large-T antigen.

In all the revertants that arose by duplication, the duplicated segment is located within the Ad2 fiber gene, which is just to the right of the original SV40 insertion in Ad2$^+$ND1. Thus, parts of both the SV40 insertion and the Ad2 fiber gene are duplicated. All the revertants continue to express the 10K fragment of the Ad2$^+$ND1 30K protein that is characteristic of the ocher mutant 71 from which the duplication revertants were derived (Table II). The prototype strain Ad2$^+$ND1dp2, for example, harbors a duplication comprised of Ad2 sequences from 85.5 to 87 m.u. joined to SV40 sequences from 0.22 to 0.11 m.u. (Fig. 4). New chimeric mRNAs are made in Ad2$^+$ND1dp2-infected cells that contain the Ad2 tripartite leader and duplicated fiber sequences attached to the duplicated 3′ end of the SV40 A gene (Dunn *et al.*, 1978). In agreement with this RNA structure is the appearance of a new SV40-specific 23K protein that contains fiber peptides at its N terminus end and a C terminus identical to that of large-T antigen (Fey *et al.*, 1979). This chimeric protein, like the D2 protein (see Section II.B), has been shown to carry the helper

TABLE II. Revertants of Ad2$^+$ND1 Host-Range Mutants 71 and 140[a]

	Revertant-specific proteins[b]		
	In vivo	SV40-selected mRNA translated *in vitro*	DNA[c]
Ad2$^+$ND1	30K	30K	
71	(10K)	10K	
140	19K	19K	
Deletions[d]: 71Δ			
Δ3b	29K	29K	Δ Very small (≈50 bp)
Δ4a	28K	28K	Δ <1–150 bp
Δ5a	—	Not done	Δ 1–150 bp
Δ6a	23K	23K	Δ >1–150 bp
Δ7a	—	18K	Δ 3–400 bp
Δ8a	115K	—	Δ Large: defective virus
Duplications[e]: 71dp			0.14 ~5% 0.14
dp1a	58K	58K + 10K	SV40 — SV40
dp2a	23K	23K + 10K	~2.5% SV40 — SV40
dp3a	21K	Not done	~2.2% SV40 — SV40
Deletions[d]: 140Δ			
Δ50a	25K	25K	Δ <1–150 bp

[a] Reprinted from Grodzicker (1980) with permission. The *hr* mutants contain chain-terminator triplets in the SV40 region of Ad2$^+$ND1 and fail to grow in monkey cells. Revertants all grow efficiently in monkey cells.

[b] The proteins are coded for by the SV40 insertion in the hybrid virus. Many, if not all, also contain adenovirus-coded peptides.

[c] All changes in the DNA, relative to Ad2$^+$ND1, refer to alterations within (in the case of deletions) or adjacent to (in the case of duplications) the original SV40 insertion in Ad2$^+$ND1.

[d] All deletions have the same general structure as Ad2$^+$ND1Δ7 shown in Fig. 4.

[e] All duplications have the same general structure as Ad2$^+$ND1dp2 shown in Fig. 4.

function by microinjection into Ad2-infected monkey cells, in which it increases the yield of virus (Tjian *et al.*, 1978). Thus, the helper function resides at the very C terminus of SV40 large-T antigen and appears to be independent of N-terminal peptides to which it is attached. Microinjection of DNA fragments covering different segments of the SV40 early region supports the conclusion that the C terminus of SV40 large-T antigen provides helper function (Mueller *et al.*, 1978). Isolation and characterization of the deletion mutants that are located in the 3′ end of the SV40 A gene and that destroy the ability of SV40 to enhance adenovirus multiplication in monkey cells have mapped this function even more precisely to the C-terminal 35 amino acids (Cole *et al.*, 1979; van Heuverswyn *et al.*, 1979; Feunteun *et al.*, 1981; van Roy *et al.*, 1981; Tornow and Cole, 1983). The nature of the block to adenovirus growth in monkey cells and possible roles played by SV40 T antigen in overcoming this block are discussed in Section III.B.

The deletion and duplication revertants may have been generated from the Ad2$^+$ND1 *hr* mutants by the same type of recombination invoked to explain the derivation of the nondefective hybrids from Ad2$^+$ND4, namely, unequal crossing-over between two Ad2$^+$ND1 mutant molecules. A crossover event that occurred to the left of the nonsense mutation in one Ad2$^+$ND1 mutant molecule and to the right of the nonsense mutation in the SV40 insertion of the second molecule would have generated a virus that carries a deletion covering the original *hr* mutation. Similarly, a partial duplication could have been produced by a crossover that occurred to the right of the *hr* mutation in the SV40 insertion of one molecule and within the Ad2 fiber gene of the second mutant genome.

B. Defective Adenovirus Type 2–Simian Virus 40 Hybrids

Two populations of stable hybrids have been isolated from the original Ad2^{++} strain after plaquing and passaging on AGMK cells (for their derivation, see Grodzicker, 1980). Each is a mixture of hybrid viruses and Wt Ad2. In AGMK cells, in addition to Wt Ad2 and hybrid viruses, infectious SV40 virions are produced. Their respective names are derived from the efficiency with which SV40 is produced: Ad2^{++}HEY (high-efficiency yielder) and Ad2^{++}LEY (low-efficiency yielder) (Lewis and Rowe, 1970).

The structure of each mixture of hybrid viruses has been determined by EM heteroduplex (Kelly *et al.*, 1974, 1975; van Roy and Fiers, 1978a) and restriction enzyme analyses (van Roy and Fiers, 1978a) and by partial DNA sequence determination (Ling *et al.*, 1982). The Ad2^{++}HEY population contains three hybrids (I, II, III,) that differ by integral numbers of SV40 genomes contained in the hybrid genome. SV40 genome lengths of 0.4 (HEYI), 1.4 (HEYII), and 2.4 (HEYIII) are inserted into the Ad2 genome, replacing the Ad2 sequences between 30 and 75 m.u. (see Fig. 5). The left (positions 30 and 0.26) and right (positions 75 and 0.66) Ad2–SV40 junctions are at identical positions in the three hybrids. The SV40 DNA is arranged in tandem repeats in the viruses with more than one genome length of SV40.

The Ad2^{++}LEY population contains only two hybrids. The Ad2 sequence between 81.3 and 93 is replaced with 0.02 (LEYI) and 1.02 (LEYII) genome lengths of SV40, the end points of which are at 0.61 and 0.63 m.u. (see Fig. 5). As in HEY, the Ad2–SV40 junctions are at the same position (81.3 and 0.63 for left, 93 and 0.61 for right) for both LEYI and LEYII, and the SV40 sequences are arranged in tandem.

Within each population, there appears to be interconversion among the various hybrid molecules as well as free SV40 (van Roy and Fiers,

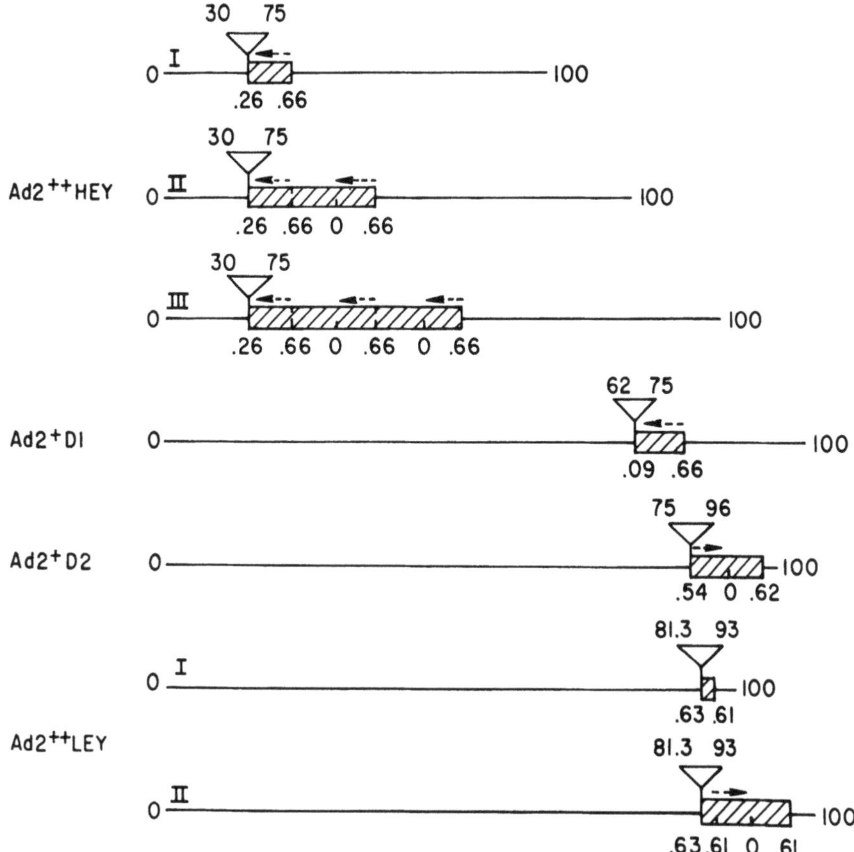

FIGURE 5. Genomic structures of the defective Ad2–SV40 hybrids. Ad2^{++}HEY and Ad2^{++}LEY each consist of more than one hybrid species and are thus designated I, II, or III. (←– – , – –→) Direction of SV40 early transcription. Symbols and map coordinates are defined in the Fig. 1 caption.

1978a). An equilibrium among members of the HEY or LEY population is probably maintained by intra- and intermolecular reciprocal recombination within the SV40 sequences of the different hybrids or free SV40 genomes or both (Kelly *et al.*, 1974). This intercoversion of hybrid genomes is almost certainly facilitated by homologous recombination. In contrast, the initial events leading to the formation of these two hybrid strains must have involved illegitimate recombination between Ad2 and SV40 sequences, since there is no homology between the Ad2 and SV40 sequences at the junction sites (Ling *et al.*, 1982).

Ad2^{+}ND4 (pool B55), Ad2^{++}HEY, and Ad2^{++}LEY originated from the same adapted Ad2^{++} strain. The identical left Ad2–SV40 junction found for Ad2^{+}ND4 and Ad2^{++}LEY by DNA sequence analysis suggests that either Ad2^{+}ND4 was generated by an unequal crossover event between Wt Ad2 and Ad2^{++}LEYII (Ling *et al.*, 1982) or Ad2^{+}ND4 and Ad2^{++}LEY were derived from a common parental hybrid virus via recombination with Wt Ad2.

Because both $Ad2^{++}HEY$ and $Ad2^{++}LEY$ contain defective adenovirus genomes, replication of the hybrid viruses depends on the presence of a Wt Ad2 virus. Thus, while each population forms plaques with one-hit kinetics on human cells (due to the Wt Ad2), two-hit kinetics are observed on AGMK cells, in which the hybrids or free SV40 provide the helper function for Ad2 growth in these cells (Lewis and Rowe, 1970).

Both $Ad2^{++}HEY$- and $Ad2^{++}LEY$-infected cells express SV40 U and T antigens, but only $Ad2^{++}HEY$ infections produce V antigens, which requires SV40 late gene expression (Lewis and Rowe, 1970; Siegel *et al.*, 1975). The failure of $Ad2^{++}LEY$ to induce V antigen is surprising, since the late SV40 genes are present and intact in LEYII. This difference between $Ad2^{++}HEY$ and $Ad2^{++}LEY$ may reflect either the different arrangements of SV40 sequences within the hybrids or the amount of free SV40 in the population that may then act as the template for late gene expression, or both.

Two new hybrids ($Ad2^{+}D1$ and $Ad2^{+}D2$) were isolated from $Ad2^{++}HEY$- infected AGMK cells by Hassell *et al.* (1978). Since both hybrid viruses are defective, they must be propagated with a helper adenovirus. As a result, each stock of $Ad2^{+}D1$ or $Ad2^{+}D2$, like $Ad2^{++}HEY$, contains both Wt Ad2 and hybrid virions and plaques with one-hit kinetics on human cells (due to Wt Ad2) and two-hit kinetics on monkey cells.

In contrast to $Ad2^{++}HEY$, no free SV40 is present, and each strain has a unique genome structure. EM heteroduplex analysis, restriction enzyme mapping, Southern blot analysis, and partial DNA sequence determination (Hassell *et al.*, 1978; Ling *et al.*, 1982) have been employed to determine their structures. Both $Ad2^{+}D1$ and $Ad2^{+}D2$ contain less than one copy of SV40 DNA and are therefore unable to generate complete, free SV40 virions. In $Ad2^{+}D1$, Ad2 sequences between positions 62 and 75 have been replaced with SV40 information encoded between 0.09 and 0.66 m.u. on the SV40 genome (see Fig. 5). Nearly the complete SV40 early region is present in a contiguous segment that has the same orientation as in $Ad2^{++}HEY$. The observation that $Ad2^{+}D1$ and $Ad2^{++}HEY$ contain an identical right Ad2–SV40 junction (75 and 0.66 m.u.) (Ling *et al.*, 1982) strongly suggests that $Ad2^{+}D1$ arose by unequal crossing over between Wt Ad2 and either HEYII or HEYIII. The lack of sequence homology between Ad2 and SV40 at this junction again indicates that illegitimate recombination was involved.

In $Ad2^{+}D2$, Ad2 sequences between 75 and 96 have been replaced with almost the entire SV40 genome (0.54–0.62 m.u.) (see Fig. 5). The SV40 information is in the opposite orientation to that found in $Ad2^{++}HEY$, and the early region is no longer contiguous. The structure of $Ad2^{+}D2$ suggests that it was formed by either recombination between free SV40 and Wt Ad2 in the $Ad2^{+}HEY$ population or multiple recombination events between Wt Ad2 and $Ad2^{++}HEY$ hybrids. $Ad2^{+}D1$ and $Ad2^{+}D2$ are both defective, presumably because they lack some of the essential information encoded between 62 and 95 (E2, L4, and L5 Ad2 genes).

The early segment of SV40 (0.66–0.17 m.u.) in Ad2$^+$D1 is expressed at both early and late times after infection, whereas the late region (0.17–0.09 m.u.) is unexpressed (Hassell *et al.*, 1978). This is somewhat surprising, since Ad2$^+$D1 lacks the bona fide SV40 early promoter including the TATA box and upstream sequences necessary for transcription (Ling *et al.*, 1982). Thus, an Ad2 promoter must be controlling expression of the SV40 information. The only Ad2 promoter in this area with the correct orientation is the E2 early promoter located several hundred base pairs upstream. Several observations suggest that this promoter is in control. First, the SV40 information is expressed early in the productive cycle. Second, expression of T antigen in rat cells transformed by a plasmid that contains the SV40 genetic material and adjacent Ad2 sequence from Ad2$^+$D1 is enhanced 5-fold after infection with Ad2 (R. Frisque and Y. Gluzman, unpublished results). Presumably, the infecting Ad2 provides the E1a gene product that is required for efficient transcription from E2. A protein similar to SV40 large-T antigen in size, antigenic determinants, and phosphorylation is produced in Ad2$^+$D1-infected human or monkey cells (Hassell *et al.*, 1978).

In contrast to Ad2$^+$D1, the SV40-specific mRNA and protein made in Ad2$^+$D2-infected cells are synthesized only during the late phase of the lytic cycle. While SV40 early sequences are expressed, neither strand of the SV40 late region is transcribed. As in Ad2^{++}LEY, the normal termination signal at 0.17 m.u. is operative, whereas the late SV40 promoter is nonfunctional.

The structure of the Ad2$^+$D2 genome together with the mode of expression of SV40 information suggests that the SV40 early region transcript is fused to the 5′ end of the mRNA for the L4 gene family and is thus under the control of the MLP of Ad2 at 16.4 m.u. (Hassell *et al.*, 1978). Ad2$^+$D2 specifies a protein (D2) that is related to SV40 T antigen structurally, antigenically, and functionally. Characteristics of the D2 protein are that it: (1) is immunoprecipitable with SV40 anti-T serum (Hassell *et al.*, 1978), (2) carries the TSTA antigenic determinants (Tevethia *et al.*, 1980; Ismail *et al.*, 1981; Baumann *et al.*, 1981), (3) shares peptides with authentic T antigen (Tjian, 1978), (4) specifically binds to the SV40 origin of replication (Tjian, 1978; McKay, 1981; Tenen *et al.*, 1982), (5) contains an ATPase activity (Tjian and Robbins, 1979), (6) activates ribosomal genes (Soprano *et al.*, 1980), (7) regulates SV40 early transcription (Rio *et al.*, 1980; Hansen *et al.*, 1981), and after microinjection into appropriate cells (8) stimulates cellular DNA synthesis and (9) provides the helper function that allows Ad2 to multiply in monkey cells (Tjian *et al.*, 1978). All these SV40 T-antigen functions thus must be encoded in the contiguous segment of SV40 early sequences located between 0.54 and 0.17 m.u. The D2 protein (115K) is larger than authentic T antigen (95K). The larger size of this protein and the location of the

Ad2–SV40 left junction suggest that D2 is a chimeric protein in which the N-terminal segment of T antigen (0.67–0.54 m.u.) is replaced with the N-terminal region of the Ad2-specific pVIII protein (Hassell *et al.*, 1978). Because (1) the D2 protein is made in large quantities (it is controlled by the very strong MLP of Ad2), (2) adenovirus shuts down host protein synthesis late in infection, and (3) the D2 protein is functionally similar to authentic T antigen, this protein has been purified and used extensively for the study of SV40 T antigen (Tjian, 1978; Tjian *et al.*, 1978; Graessmann *et al.*, 1980; McCormick *et al.*, 1981; Soprano *et al.*, 1980; S. S. Tevethia *et al.*, 1980; Rio *et al.*, 1980; Tjian and Robbins, 1979; McKay, 1981; Ismail *et al.*, 1981; Hansen *et al.*, 1981; Tenen *et al.*, 1982).

C. Adenovirus Expression Vectors

Recently, several adenovirus expression vectors have been constructed (Solnick, 1981; Thummel *et al.*, 1981, 1982; 1983). The basic approach has been to replace *in vitro* a segment of the adenovirus genome with SV40 early region devoid of its normal promoter. Because the resulting hybrid is missing adenovirus information, it is defective and must be grown in the presence of nondefective, helper adenovirus. To select for and maintain the defective hybrid virus in the mixed population, advantage is taken of the ability of SV40 T antigen to help adenovirus growth in monkey cells. To provide the helper function, the SV40 sequences must be properly positioned so that an adenovirus promoter can substitute for the missing SV40 promoter. A second requirement of this approach is that the size of the genome not be altered more than about 5–10% by either the deletion or the substitution or both, so that the hybrid can still be efficiently packaged into the adenovirus virion.

In the first approach taken by Thummel *et al.* (1981), a segment of the SV40 genome (0.67–0.14 m.u.) that contains the entire early region except for the promoter was ligated *in vitro* to either of two Ad2–Ad5 interserotypic hybrid viruses (1 × 5li and 4 × 225b). Because each of these two viruses contains only two of the three *Bam*HI sites present in the Ad2 genome at positions 29, 42, and 60, the SV40 segment with flanking *Bam*HI linkers can be readily inserted into the *Bam*HI-cleaved recipient genomes. After transfection into receptive human cells together with parental, helper virus DNA, the mixed population was passaged in monkey cells to enhance the amount of defective adenovirus–SV40 hybrids. Four hybrids have been isolated and partially characterized (Ad-SVR1, 5, 6, and 15).

The genomic structures of Ad-SVR5 and 15 have been determined (see Fig. 6). Ad-SVR5 consists of two hybrid species; the genome of the predominant species has the *Bam*HI fragment between 42 and 59.5 replaced by a head-to-tail tandem insertion of two SV40 early segments (0.65–0.14 m.u.). The SV40 early region and the Ad2 MLP are in the same

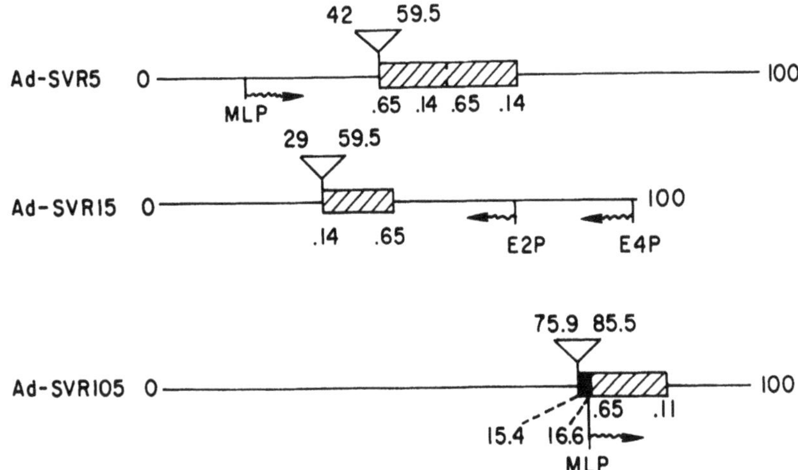

FIGURE 6. Genomic structures of Ad2–SV40 expression vectors. Symbols and map coordinates are defined in the Fig. 1 caption. (⤳) Directions of transcription from the major late (MLP), early region 2 (E2P) and 4 (E4P) promoters; (■) 15.4–16.6 fragment that contains the ectopic MLP.

orientation. In contrast, in Ad-SVR15, the 29–59.5 Ad2 fragment is replaced with a single SV40 early segment with orientation opposite to that of MLP. Ad-SVR1 and Ad-SVR6 are related or identical adenovirus–SV40 hybrid populations consisting of a complex mixture of approximately ten different hybrid molecules. These hybrids have SV40 and calf thymus DNA (used as carrier) inserted behind the E1b promoter. The large chimeric insertions replace adenovirus sequences between coordinates of approximately 5 and 60 and probably resulted from extensive rearrangement of viral sequences in the transfected cell (C. Thummel, R. Tjian, T. Grodzicker, and L. Chow, unpublished data).

Ad-SVR15-infected cells produce only very small amounts of large-T antigen. This is perhaps not surprising when its genomic structure is considered. The only adenovirus promoters with the proper orientation for expression of SV40 T antigen are those that control E2 and E4. The proximity of the SV40 genetic material to the E2 promoter suggests that it is in control. Ad-SVR1, 5, and 6 express small-t antigen as well as the large-T antigen, though the amount of small-t antigen is less than that found in SV40-infected cells. T antigen is expressed late in the lytic cycle of Ad-SVR5 and 6, consistent with its being under MLP control.

Contrary to the complex mixture of hybrid genomes for Ad-SVR6, only one prominent SV40-specific mRNA is seen in Ad-SVR6-infected cells. Its size is intermediate between those of the normal SV40 early 16 S (2200n; encoding T) and 19 S (2500n; encoding t) mRNAs. This RNA may consist of more than one species, since nuclease-S1 analysis indicates that the splice junction for both T and t mRNA are utilized. Two major SV40-specific RNAs (3000n or 4000n) are seen in Ad-SVR5-infected cells.

Analysis of the Ad-SVR5 RNA using the sandwich hybridization technique (Dunn and Hassell, 1977) indicates that these mRNAs contain, in addition to the normally spliced SV40 segment, the adenovirus tripartite leader and adenovirus sequences located just upstream from the SV40 insertion. While both the SV40 large-T and small-t mRNA splice sites appear to be used in Ad-SVR5- and Ad-SVR6-infected monkey cells, the ratio of small-t to large-T mRNA is reduced relative to the ratio of the authentic SV40 messages in SV40 infections. This may reflect the influence on splicing of the adenovirus sequences in the chimeric transcript. Alternatively, the host splicing apparatus may be altered during viral infection.

In a second variation of this basic approach, the SV40 insertion carries with it the adenovirus MLP. Solnick (1981) has employed this modification to construct a defective hybrid virus (Ad-SVR105) in which the Ad2 sequences between 75.9 and 85.5 are substituted with the SV40 early region (0.65–0.11 m.u.) together with a small segment of Ad2 sequence (15.4–16.6 m.u.) that contains the MLP (see Fig. 6). Ad-SVR105 thus contains two copies of MLP in the same orientation. The two SV40-specific mRNAs made in Ad-SVR105-infected cells are analogous to the authentic 16 and 19 S early SV40 mRNAs. They are slightly larger (~35n) than the authentic large-T and small-t mRNAs, consistent with their initiation from the ectopic MLP. Both large-T and small-t are made in Ad-SVR105-infected monkey or human cells. Again, as seen for the other adenovirus–SVR hybrids, the ratio of small-t to large-T is much less than in SV40 infections. With Ad-SVR105, however, this difference is not reflected at the RNA level, implying that translation or posttranslational regulation may be involved.

A third strategy for construction of adenovirus expression vectors depends on both *in vitro* and *in vivo* recombination (Thummel *et al.*, 1982). The procedure involves (1) constructing a bacterial plasmid that contains the foreign gene of interest (SV40 A gene) linked to a large fragment of adenovirus DNA and (2) inserting the recombinant insert into the appropriate region of the recipient genome. Once the adenovirus–SV40 segment is inserted into the recipient, a duplication is created in the recombinant genome. Because adenovirus can undergo homologous recombination extremely efficiently, the adenovirus DNA located between the duplicated sequences is lost, while at the same time the foreign DNA is moved into the desired position (see Fig. 7A).

By this strategy, a series of adenovirus–SV40 hybrids have been created in which the SV40 early A gene is located at different positions within the adenovirus late leader region (Thummel *et al.*, 1982; 1983) (see Fig. 7B). One well characterized hybrid, Ad-SVR26, was constructed by *in vitro* insertion into the recipient genome (1 × 5li) at position 29 of a recombinant segment of DNA that contained the SV40 early region attached to the adenovirus sequences located between 11.6 and 17 m.u. (see Fig. 7A). The sequences between 17 and 29 were then deleted *in vivo*

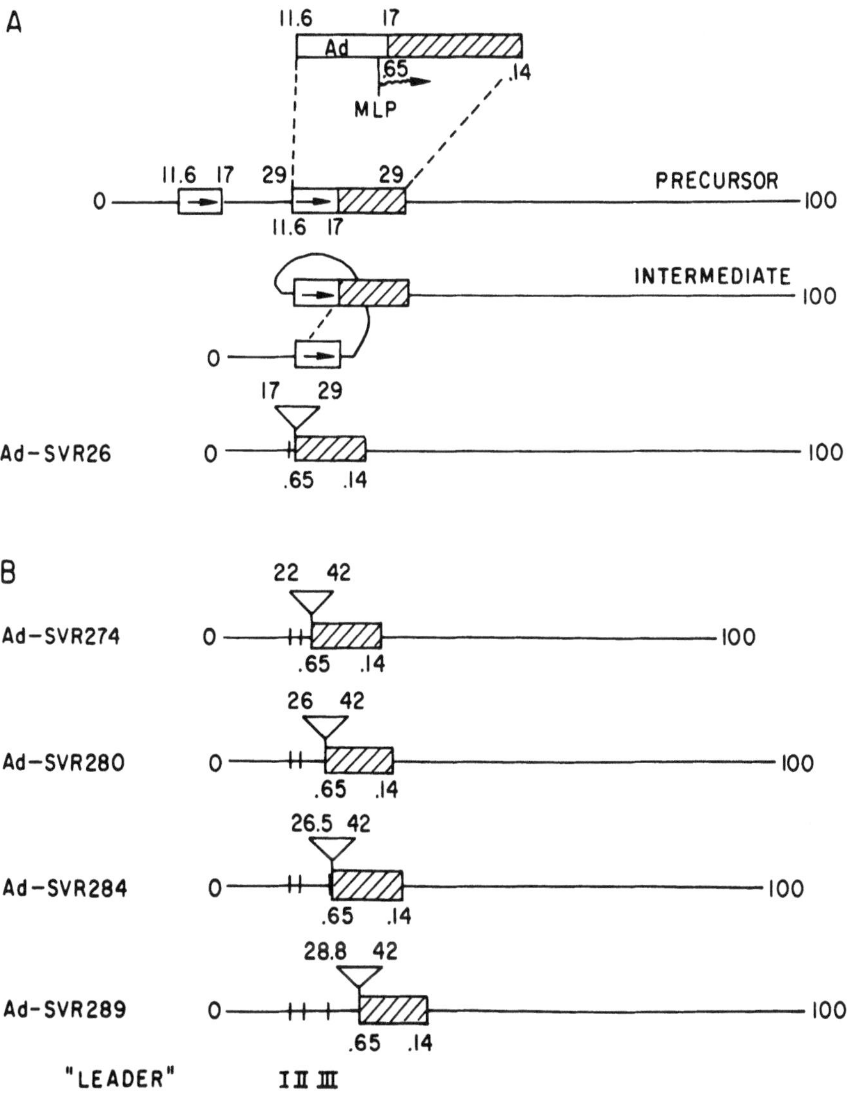

FIGURE 7. Construction and genomic structures of Ad2–SV40 expression vectors. (A) The Ad2–SV40 insertion fragment shown at the top contains Ad2 DNA that extends from map positions 11.6 to 17 (▭) and includes the major late promoter (MLP) and the first of the tripartite leader segments. This is attached to the SV40 early region (0.65–0.14 m.u.). Insertion of this fragment into adenovirus at the *Bam*HI site at map coordinate 29 results in the proposed precursor to Ad-SVR26. Intramolecular recombination between the duplicated regions of adenovirus DNA (emphasized as ⇥ would result in the formation of Ad-SVR26. The recombination could proceed via the alternative intermolecular intermediate (not shown). Symbols and map coordinates are defined in the Fig. 1 caption. (B) Genomic structure of addition adenovirus–SV40 hybrids constructed in a similar fashion using both *in vitro* and *in vivo* recombination. Each segment (I, II, and III) of the tripartite leader is indicated as a heavy vertical dash (—|—) on the adenovirus genome.

by recombination between the duplicated regions (11.6–17 m.u.) in the hybrid genome. This recombination could have occurred either between separate viral genomes or intramolecularly. The final result is that the SV40 early region is positioned just downstream from MLP in the intervening sequence between the first (16.4 m.u.) and second (19.6 m.u.) leaders.

Although transcription of the Ad-SVR26 T-antigen mRNA initiates at MLP, there are two alternative 5′ segments on these messages that are present in equal abundance. In one case, mRNA sequences proceed uninterrupted through the first segment of the tripartite leader into adenovirus intervening sequence DNA and the SV40 insertion. In the second type of chimeric mRNA, the first segment of the tripartite leader is, most likely, spliced to a cryptic acceptor site within the beginning of the coding region for large-T antigen. In both RNAs, the early SV40 intervening sequences between 0.59 and 0.54 m.u. are removed. Minor mRNA species are also seen where other cryptic splice-acceptor sites within the 3′ segment of the SV40 early RNA are used.

Translation of the two abundant mRNAs is apparently initiated from novel AUG codons to synthesize two species of proteins, both of which differ from Wt SV40 T antigen at their amino terminus. The T-antigen message that is unspliced within the adenovirus sequences codes for a protein 5K larger than authentic SV40 T antigen. Translation of this RNA is initiated at an adenovirus AUG that is located distal to the first leader segment and is in phase with the SV40 A gene. Translation of the second abundant chimeric message probably begins at an AUG codon 39 bases downstream from the normal SV40 initiating triplet, which has been removed from the transcript by splicing.

Although the T mRNAs are highly overproduced in Ad-SVR26-infected cells, translation of the T-antigen-related proteins is not correspondingly enhanced. One possibility is that these novel initiators do not serve as efficient starts for translation. Alternatively, all three segments of the tripartite leader may be required for efficient translation during the late phase of adenovirus infection. That both may be important factors for determining the efficiency of translation is suggested by the observations that (1) the SV40 T antigens are overproduced at high levels when their mRNAs contain the tripartite leader at the 5′ end, while (2) their overproduction is modified by the nature of the sequence surrounding the initiating AUG codon (Thummel, *et al.*, 1983).

Neither mRNA with structure analogous to SV40 small-t mRNA nor small-t antigen is seen in Ad-SVR26-infected cells. The fact that the normal small-t splice-donor sites are not utilized, even though essentially the entire SV40 early information is present in these transcripts while cryptic SV40 acceptor sites are utilized, again suggests that the specificity of the splicing machinery may be altered by viral infection.

Another type of adenovirus expression vector not reliant on the helper function provided by SV40 T antigen has recently been constructed

FIGURE 8. Genomic structure of the Ad7–SV40 hybrid PARA. (← – –) Direction of transcription of the SV40 early region. The SV40 sequences in the triangle below the SV40 insertion, from positions 0.71 to 0.11, are not present in the hybrid. Symbols and map coordinates are defined in the Fig. 1 caption.

(Y. Gluzman, unpublished data). In this vector, foreign genes can be placed under the control of an ectopic MLP that replaces E1b. This vector, though defective, can be propagated without helper virus in human 293 cells, which contain and express the E1a and E1b genes.

D. Adenovirus Type 7–Simian Virus 40 Hybrids

The first adenovirus–SV40 hybrid discovered was isolated after 28 passages of an Ad7 vaccine strain, termed L.L., in monkey kidney cells. This strain of virus, which has been designated E46$^+$ or PARA (an acronym for "particles aiding replication of adenovirus") contains both Wt Ad7 and an Ad7–SV40 hybrid genome enclosed in an Ad7 capsid (Huebner *et al.*, 1964; Rapp *et al.*, 1964b; Rowe and Baum, 1964, 1965; Baum *et al.*, 1966).

The structure of the PARA hybrid genome has been determined by EM heteroduplex and restriction enzyme analyses (Kelly and Rose, 1971; Kelly, 1975). The Ad7 sequences between 79 and 95 have been replaced by essentially the entire SV40 early region (see Fig. 8). The SV40 insertion has a complex arrangement that suggests that PARA arose by multiple recombination events between SV40 and Ad7 sequences. The SV40 sequences at the right Ad7–SV40 junction start at position 0.66 on the SV40 genome and continue uninterrupted through the early region to 0.11 m.u. Sequences at 0.11 are joined to SV40 sequences at 0.71 and continue to 0.50 on the SV40 genome, where they are attached to Ad7 DNA (79) to form the left junction.

Because most of the SV40 early region is represented at least once while few SV40 late sequences are present, it is not surprising that SV40 U, TSTA, and T but not V antigens are expressed in PARA-infected cells (Rowe and Baum, 1964; Rapp *et al.*, 1966b; Lausch *et al.*, 1968). SV40 early RNA and anti-late RNA, though only a small amount of the latter, are detected in these cells (Lebowitz and Khoury, 1975). The genomic structure of PARA suggests that the SV40 sequences are controlled by either the normal SV40 early promoter or the adenovirus E4 promoter. SV40 T antigen normally autoregulates the expression of its gene (Tegtmeyer *et al.*, 1975; Reed *et al.*, 1976; Alwine *et al.*, 1977; Khoury and

May, 1977). The failure to do so in PARA-infected cells (Lanford and Butel, 1980a) suggests that either the E4 promoter is in control or the insertion event has altered the T-antigen-binding site(s) of the SV40 early promoter required for effective feedback regulation by T antigen (Rio *et al.*, 1980; Hansen *et al.*, 1981).

As expected from its genomic structure, the hybrid is defective and requires a helper virus to supply the adenovirus functions deleted in the hybrid. The hypothesis that a primary function provided by the helper adenovirus is to supply a capsid for the PARA genome is strengthened by the discovery of the phenomenon of "transcapsidation" (Rapp *et al.*, 1965). During transcapsidation, which occurs in cells coinfected with the PARA–Ad7 mixed population and a heterologous adenovirus serotype (e.g., Ad1, Ad2, Ad5, Ad6, or Ad12), the hybrid PARA genome is enclosed in the heterologous adenovirus capsid (Rapp *et al.*, 1965, 1968; Rowe, 1965). The properties of the transcapsidated PARA populations do not appear to be altered by changing the associated helper virus (Rapp, 1973). The defective hybrid is maintained in the mixed population presumably because it provides the helper function required by adenovirus to grow in monkey cells (Boeye *et al.*, 1966; Rowe and Baum, 1965).

Mutants of the PARA population have been isolated by plaque purification on AGMK cells (Rapp *et al.*, 1969). Several of these variants were characterized by differences in their ability to transform cells *in vitro* or to form tumors when injected into newborn hamsters (Rapp *et al.*, 1969; Duff *et al.*, 1970). A second class of mutants, PARA(cT), induces the synthesis of SV40 large-T-antigen-related proteins that accumulate in the cytoplasm rather than in the nucleus of infected or transformed cells (Butel *et al.*, 1969; Duff *et al.*, 1970; Richardson and Butel, 1971). Like the first class of variants, the PARA(cT) mutants also transform and induce tumors in newborn hamsters less efficiently than the parental PARA population (Rapp *et al.*, 1969; Duff *et al.*, 1970; Richardson and Butel, 1971). The PARA(cT)-transformed cells appear to be minimally transformed: they grow to lower saturation density, have decreased capacity for anchorage-independent growth, and are less oncogenic *in vivo* than cells transformed by the parental PARA population (Lanford and Butel, 1981).

All PARA-transformed cells contain SV40 T antigen, and most exhibit the morphology typical of SV40 transformants, whereas only a minority express adenovirus T antigen (Diamond, 1967; Duff and Rapp, 1970). These observations, together with the correlation between altered SV40 T-antigen location and transformation efficiency, strongly implicate the involvement of SV40 T antigen in cellular transformation by PARA. While the PARA(cT) mutation does not affect the lytic growth properties of PARA (Butel *et al.*, 1969), the absence of SV40 T antigen in the nucleus of PARA(cT)-transformed cells impairs their growth in culture and sets up a strong selection for variants of PARA(cT) that at least transiently accumulate SV40 T antigen in their nucleus during the S phase of the cell cycle (Lanford and Butel, 1981).

Both PARA- and PARA(cT)-infected cells synthesize normal-size Wt SV40 large-T and small-t antigen. A slightly larger T, which is again restricted to the cytoplasm, is also found in PARA(cT) infections. Comparisons of T antigen from PARA(cT)- vs. PARA- or SV40-infected cells indicate similar DNA-binding activities and isoelectric focusing points, but different levels of phosphorylation (Lanford and Butel, 1980a). The PARA(cT) mutant does not alter the association of SV40 T antigen with the cell surface (Lanford and Butel, 1982), but quite surprisingly acts in *trans* to prevent migration of SV40- or PARA-encoded T antigen to the nucleus of coinfected cells (Lanford and Butel, 1980b). This inhibition of transport to the nucleus is specific for SV40 T antigen, since no other viral proteins are affected. Because SV40 replication occurs in these coinfected cells, very little nuclear T antigen appears to be required for lytic SV40 growth.

Several alternative hypotheses have been suggested to account for the *trans*-acting nature of the PARA(cT) mutant (Lanford and Butel, 1980b). These include (1) an altered pattern of phosphorylation, perhaps via the T antigen-associated kinase or ATPase (Tjian and Robbins, 1979; Griffin *et al.*, 1979; Baumann and Hand, 1979); (2) changes in the type of T antigen aggregation (Carroll *et al.*, 1974; Osborn and Weber, 1974); or (3) irreversible binding of PARA(cT) antigen to a cellular factor required for normal transport of T antigen to the nucleus.

As in the case of the Ad2–SV40 hybrids, the Ad7–SV40 hybrids and their mutants are valuable tools in dissecting the multiple roles played by SV40 T antigen. Continued study of the PARA(cT) mutants should help segregate the cytoplasmic and cell-surface functions of T antigen from its nuclear activities.

E. Other Adenovirus–Simian Virus 40 Hybrids

A recombinant between Ad3 and SV40 has been described (Lewis *et al.*, 1966). It resembles PARA in that SV40 T antigen is induced but no free SV40 is produced. A second group of adenovirus–SV40 hybrids yield SV40 progeny and are thus similar to $Ad2^{++}HEY$ and $Ad2^{++}LEY$ populations. These include Ad1–SV40 (Lewis *et al.*, 1966), Ad4–SV40 (Easton and Hiatt, 1965; Beardmore *et al.*, 1965), Ad12–SV40 (Schell *et al.*, 1966; Rhim *et al.*, 1981), and Ad5–SV40 (Hartley *et al.*, 1956; Lewis *et al.*, 1966).

Recently, an Ad5–SV40 hybrid population ($Ad5^{++}D1$) has been isolated that gives rise to infectious SV40 virions and closely resembles the defective $Ad2^{++}HEY$ and $Ad2^{++}LEY$ populations (Gluzman and Van Doren, 1983). $Ad5^{++}D1$ forms plaques with two-hit kinetics on monkey cells. It consists of a mixed population of Wt Ad5, free SV40, and several hybrids that differ incrementally from one another by a full-length copy of SV40 DNA. The genome of the most abundant hybrid is comprised of a giant inverted repeat about 35,000 bp in length. Beginning from one end, the genome consists of sequences derived from the left 3534 bp of the Ad5 genome joined to 2.7 copies of SV40 DNA arranged as head-to-tail tandems. The entire structure is repeated in reverse orientation and results in a large inverted repeat of the following structure: $_{10\%}$Ad5-SV40-SV40-04VS-04VS-5dA$_{10\%}$. Other hybrids contain 4.4, 3.4, 2.4, and 1.4 copies of SV40 DNA and appear to be in equilibrium with each other as well as with the abundant recombinant.

III. HUMAN ADENOVIRUS GROWTH IN MONKEY CELLS

A. Methods of Enhancement

Human adenoviruses grow very poorly in primary African green monkey kidney (AGMK) cells and cell lines derived from them (e.g., BSC-1, CV-1). Several alternative means for overcoming the block to productive infection have been uncovered during the past 15 years. One of the best characterized is the helper function provided by SV40. Extensive analysis of Ad2–SV40 hybrid virus as well as mapping of SV40 deletion mutants that affect this function indicate that this activity resides in the C-terminal 35 amino acids of the SV40 large tumor (T) antigen (see Section II.A.2). This is consistent with the observations that temperature-sensitive (*ts*) mutations in the late genes of SV40 continue to enhance adenovirus growth at the nonpermissive temperature while some but not all *ts*A mutants lose the helper function at elevated temperatures (Jerkofsky and Rapp, 1973; Jerkofsky, 1975; Kimura, 1974). Since large-T is expressed in SV40-transformed cells, it is not surprising that adenoviruses grow readily in SV40-transformed monkey cells (Shiroki and Shimojo, 1971; Hashimoto *et al.*, 1973).

Butel and co-workers (Butel, 1967; Butel and Rapp, 1967) detected a monkey-adapting component (MAC) in strains of Ad7 that were capable of replicating efficiently in monkey cells. MAC replicated in both human and monkey cells only in the presence of Wt Ad7 helper and was lost on plaque purification in human cells. In this respect, it is similar to PARA (see Section II.D) and to Ad2^{++}HEY and Ad2^{++}LEY (see Section II.B). However, no SV40 T or V antigens were detected in MAC + Ad7-infected cells, nor did DNA extracted from these cells hybridize *in vitro* to SV40 complementary RNA. Although these data suggest that SV40 sequences are not required to overcome the block to adenovirus multiplication, in light of the finding that as little as a few hundred base pairs of SV40 can encode helper activity, a final conclusion must await more refined analysis of the genome of MAC using more sensitive hybridization probes or DNA sequence determination or both.

In 1971, Gardner *et al.* (1971) discovered a human papovavirus called BKV. Genetic and biochemical studies have established its close relationship to SV40. There is immunological cross-reactivity between the V antigens as well as the T antigens of both viruses (Takemoto and Mullarkey, 1973). This similarity, particularly for the late genes (V antigen), is also seen at the DNA sequence level (Khoury *et al.*, 1975; Howley *et al.*, 1975; Osborn *et al.*, 1976). Furthermore, a comparison by tryptic peptide analysis of the T antigens of these two viruses indicated that there is extensive amino acid homology in both the large-T and small-t antigens (Crawford *et al.*, 1978; Simons and Martin, 1978). BKV can provide both the SV40 DNA replication and early gene autoregulation functions that are disrupted at the nonpermissive temperature in SV40*ts*A-infected cells (Mason and Takemoto, 1976; Lai *et al.*, 1979). Thus, it is not too surprising that BKV, like SV40, enhances human adenovirus growth in monkey cells (Miyamura and Takemoto, 1979; Lai *et al.*, 1979).

Initial attempts to show that other viruses beside SV40 and the related BKV (e.g., herpes simplex virus, measles virus, rabbit papilloma virus, and human wart papovavirus) could enhance the growth of adenovirus in monkey cells were unsuccessful (Feldman *et al.*, 1966). In 1967 and 1968, however, Naegele and Rapp (1967) and Alstein and co-workers (Alstein and Dodonova, 1968; Alstein *et al.*, 1968) found that several simian adenoviruses (SA7, SV15, SV32, and SV38) are capable of enhancing human adenovirus replication in monkey cells. The level of enhancement is similar to that seen with SV40. In contrast to SV40 coinfections, the replication of the simian adenovirus is not inhibited by the presence of the human virus (Naegele and Rapp, 1967) (see below). Little is known concerning the mechanism of or the simian adenovirus gene(s) required for enhancement except that passage of SA7, together with human Ad2, often results in a mixed population of virus that forms plaques with two-hit kinetics on monkey cells (Alstein and Dodonova, 1968; Alstein *et al.*, 1968; Kaplan *et al.*, 1971). These populations contain Wt Ad2 and defective SA7 genomes encapsidated in the human virion. The structures of the defective SA7 genomes have not been elucidated.

Jerkofsky and Rapp (1975) and Staal and Rowe (1975) reported that multiplication in monkey cells by Ad7 could be increased approximately 10-fold by pretreatment of the cells with 5-iodo-2′-deoxyuridine (IUdR). Whether the mechanism of enhancement is similar to that seen with SV40, in which a 1000-fold increase is observed, remains unclear. Staal and Rowe (1975) suggested that different mechanisms were involved, since IUdR treatment increased the number of cells showing Ad7 T antigen by 10-fold, while coinfection with SV40 had no effect on the number of T-antigen-positive cells, but rather increased the yield of virus per cell. In addition, the enhancements with IUdR and SV40 were additive. In contrast, Jerkofsky and Rapp (1975) argued that since both IUdR treatment and SV40 infection stimulated host-cell DNA synthesis but did not affect the number of adenovirus infectious centers, the enhancement mechanism may be similar. They postulated that the stimulation of monkey-cell DNA synthesis may be very important for the enhancement of adenovirus multiplication. However, since stimulation of DNA synthesis after nutrient deprivation does not enhance the multiplication of human adenovirus in monkey cells (Jerkofsky and Rapp, 1975), other factors must be involved. A similar low level of enhancement was obtained by treatment of monkey cells with dimethylsulfoxide (G. A. Zorn and C. W. Anderson, unpublished results).

Several groups have shown that multiplication of human adenovirus in monkey cells can be enhanced by increasing the multiplicity of infection (m.o.i.) (Friedman *et al.*, 1970; Eron *et al.*, 1975; Baum, 1977; Cole *et al.*, 1979; van Roy and Fiers, 1978b; Goldman and Khoury, 1980). The level of enhancement is rarely above 10-fold and appears to depend on both the virus serotype and the monkey cell line employed. In primary AGMK cells, Baum (1977) found no effect of m.o.i. [range of 2–100 plaque-forming units (PFU)/cell] with Ad5 and Ad12, a small level of enhancement for Ad2 and Ad3, and a modest increase in virus yield for Ad7. Little or no enhancement of Ad2 or Ad5 growth was observed in CV-1 cells when the m.o.i. was raised from 5 to 2000 PFU/cell (Klessig and Hassell, 1978; Rice and Klessig, 1984).

A cell line derived from primary AGMK cells called Vero is permissive for Ad2 growth (Kelly and Lewis, 1973; Eron *et al.*, 1975), but not for Ad12 growth. Serial passage of Ad12 in Vero cells leads to an adapted strain that grows well in both Vero and primary AGMK cells (Werner and zur Hausen, 1978). This increased host range is accompanied by two alterations in the viral genome; deletions of 69 bp at the left end of the Ad12 genome and insertions at the right end of 190–290 bp of virus sequence originating from different regions of the virus genome (Werner and zur Hausen, 1978; E. Salewski, E. Schwarz, G. Werner, B. Hafemann, and H. Schulte-Holthausen, unpublished results). Terminal rearrangements of the genome appear to be a general mechanism employed by Ad12 to adapt itself to new environments, since Ad12 mutants adapted to growth in two different human tumor-cell lines also exhibit this type of alteration (Schwarz *et al.*, 1982).

Finally, the block can be very efficiently overcome by point mutations in the 72K DNA-binding protein (DBP) encoded by the E2a region of adenovirus. Several host-range (*hr*) mutants (Ad2*hr*400-3) have been isolated after several passages of nitrous-acid-mutagenized Wt Ad2 in CV-1 cells (Klessig, 1977; Klessig and Hassell, 1978). Spontaneous *hr* mutants of Ad5 [Ad5*hr*404 (Klessig and Grodzicker, 1979)] and Ad2$^+$ND3 [Ad2$^+$ND3*hr*600–3 (Anderson, 1981)] have also been obtained by selection in monkey cells. These *hr* mutants grow well in both human and monkey cells. The mutations responsible for the increased host range have been physically mapped by marker-rescue experiments to the promoter proximal half of the gene that encodes DBP (61.3–66) (Klessig and Grodzicker, 1979; Klessig and Quinlan, 1982; Anderson *et al.*, 1983). The *hr* alteration has been determined by DNA sequence analysis for the nine mutants (Kruijer *et al.*, 1981; Anderson *et al.*, 1983; D. F. Klessig and D. Brough, unpublished results). Quite surprisingly, the same substitution of tyrosine for histidine at amino acid residue 130 is seen in all nine mutants. This suggests that the structure of DBP can be modified in only an extremely limited number of ways for it to correctly perform its functions in monkey cells as well as human cells. The different mechanisms for altering host range used by group A (e.g., Ad12) and group C (e.g., Ad2 and Ad5) adenoviruses emphasize the distinct biologies of the different adenovirus groups.

B. Nature of the Block

1. Early Events

Although much progress has been made in identifying the portion of the SV40 genetic information required for helper function and in identifying alternative methods of enhancement, the precise nature of the block to adenovirus multiplication in monkey cells remains somewhat obscure. Adsorption and entry of adenovirus into monkey cells appear normal (Hashimoto *et al.*, 1973). The synthesis of the few early viral proteins that have been studied (T antigens and DBP) is normal in abortively infected monkey cells (Feldman *et al.*, 1966; Friedman *et al.*, 1970; Malmgren *et al.*, 1966; Ensinger and Ginsberg, 1972; van der Vliet *et al.*, 1975). Northern blot analyses of the early RNAs made in abortively (Ad2-) or productively (Ad2 plus SV40- or *hr*400-)infected monkey cells indicate that similar quantities of normal-size viral RNA are made in both types of infections (Anderson and Klessig, 1982). Furthermore, these RNAs can be translated *in vitro* (E. A. Oosterom-Dragon and C. W. Anderson, unpublished results). Virus-associated RNA (VA-RNA) is also produced in normal amounts in abortive infections (Fox and Baum, 1972; Anderson and Klessig, 1982).

Both the onset and rate of viral DNA synthesis are comparable in abortively and productively infected monkey cells (Reich *et al.*, 1966; Rapp *et al.*, 1966a; Friedman *et al.*, 1970; Hashimoto *et al.*, 1973; K. P. Anderson and D. F. Klessig, unpublished results). Although as much as 40% of the virus-specific DNA in the nuclei of monkey cells consists of subgenomic pieces in which internal regions of the virus genome are missing, the production of subgenomic DNAs is not correlated with abortive infections. The same amounts of defective virus genomes are seen in productive infections of monkey cells in which the block has been relieved by an *hr* mutation (Ad5*hr*404) or by coinfection with SV40 (Daniell, 1981).

2. Late Events

In contrast to the normal synthesis of early virus mRNA and proteins as well as virus DNA, the production of late adenovirus proteins is adversely affected in the abortively infected monkey cell. All investigators agree that the synthesis of these proteins is significantly reduced; however, the extent of reduction of any particular protein remains in question (Feldman *et al.*, 1966; Malmgren *et al.*, 1966; Friedman *et al.*, 1970; Henry *et al.*, 1971; Baum *et al.*, 1972; Lucas and Ginsberg, 1972; Hashimoto *et al.*, 1973; Grodzicker *et al.*, 1974, 1976; Eron *et al.*, 1975; Klessig and Anderson *et al.*, 1975). No doubt some of the variation is due to the different methods employed for detection. Even use of the same standard sodium dodecyl sulfate–polyacrylamide gel electrophoresis techniques in recent years, however, has not eliminated some of these differences. The amount of reduction in the synthesis of any particular late protein probably reflects several factors including (1) virus serotype, (2) host cell (primary AGMK, BSC-1, CV-1, CVC), (3) growth conditions, and (4) time after infection during which protein synthesis is analyzed. Nonetheless, it can be safely concluded that the synthesis of at least several late proteins is markedly depressed in abortive infections. The marked reduction in the level of synthesis of at least several late virus proteins, particularly fiber, is believed to be responsible for the 1000-fold-lower level of virus production in abortive infections.

A molecular description of the block(s) to the synthesis of virus late proteins in these cells is beginning to emerge. Several groups have compared the RNA made in human or in abortively vs. productively infected monkey cells to determine whether the complexity and quantity of adenovirus specified RNA are similar. While the rate of total virus RNA synthesis was similar in abortively and productively infected cells (Lucas and Ginsberg, 1972; Hashimoto *et al.*, 1973), the percentage of RNA that is virus-specified was reduced severalfold in abortively infected cells (Baum *et al.*, 1968; Lucas and Ginsberg, 1972; Hashimoto *et al.*, 1973; Fox and Baum, 1974). Hashimoto *et al.* (1973) have suggested that this apparent discrepancy, similar rate of synthesis but different percentage of virus-specific RNA, may be due to less efficient turnoff of host RNA synthesis during abortive infections. Alternatively, it may reflect some of the shortcomings of these early approaches, since recent analysis with cloned probes covering different segments of the late transcriptional unit indicates that both initiation and termination of transcription from the major late promoter (MLP) is affected (K. P. Anderson and D. F. Klessig, unpublished results).

Competition hybridization studies indicated that the complexity of total virus RNA in abortively infected monkey cells was comparable to that found in productive infections of monkey or human cells (Lucas and Ginsberg, 1972; Fox and Baum, 1972, 1974; Hashimoto *et al.*, 1973). More elaborate analyses using labeled, separated strands of restriction-enzyme-generated fragments of the Ad2 genome as hybridization probes (C_rt analysis) indicated that all or nearly all of the virus late RNA species made in productive infections are also found in abortive infections. However, the concentrations of many of the late mRNAs are reduced in Ad2-infected monkey cells, and this appears to partially account for the depressed synthesis of the corresponding proteins (Klessig and Anderson, 1975; Farber and Baum, 1978). Northern blot analysis has confirmed these findings and in addition has shown that most of the late mRNAs found in abortive infections are of normal size (D. F. Klessig, unpublished results).

The half-lives of several late RNAs (encoded by L3, L4, and L5) have been determined in productive and abortive infection. Because the turnover of these mRNAs was the same in both types of infections, the reduced concentration of late RNAs is not the result of their more rapid degradation in abortive infections (K. P. Anderson and D. F. Klessig, unpublished results). Rather, the lower level of virus RNA is due to two effects on transcription. First, transcripts are initiated 2 to 4-fold less efficiently at the MLP in abortively vs. productively infected monkey cells. Second, in abortive vs. productive infections, more of the RNA polymerases prematurely terminate, before reaching the end of this long, late transcriptional unit (K. P. Anderson and D. F. Klessig, unpublished results). The combined effect is to reduce the rate of fiber gene transcription (this gene is the last member of this transcriptional unit, L5) 5- to 10-fold. This is in good agreement with the relative steady-state levels for this mRNA determined by C_rt and Northern blot analyses.

The reasons for the poor synthesis of at least some late proteins are not entirely clear. Certainly the depressed transcription rate that results in lower concentrations of the mRNAs can account for the reduced synthesis of many of the late proteins. This, however, appears to be only part of the block. The reduced level of synthesis of some proteins, in particular of fiber, cannot be due solely to reduced RNA concentration. While fiber mRNA is depressed 5- to 10-fold, the production of fiber protein is reduced several-hundred-fold (Klessig and Anderson, 1975; Klessig and Chow, 1980; Anderson and Klessig, 1983).

The nature of this second block is a matter of some controversy. Because different laboratories have often used different virus serotypes and host cells as well as different techniques, the results have sometimes been inconsistent, and comparison of them may not always be justifiable. The interaction between these viruses and host cells appears to be very sensitive to a number of external factors. Changes in any one of them may alter the interaction in seemingly subtle ways and in so doing alter the block.

The relative quantity and complexity of the virus RNA on polysomes in the two types of infections are in dispute. Early studies using competition hybridization indicated a lower complexity as well as concentration of virus late RNA on polysomes in abortive infections (Hashimoto *et al.*, 1973; Fox and Baum, 1974). In contrast, Eron *et al.* (1975), studying the kinetics of reassociation of polysomal RNA with total Ad2 DNA, failed to detect either difference. Recent studies using (1) cloned probes for Northern blot analysis and (2) *in vitro* runoff translation of polysomes showed the same distribution in abortive vs. productive infections of fiber RNA among free ribonucleoproteins and large or small polysomelike structures. The amount of fiber RNA in the different cytoplasmic fractions, however was depressed in abortive infections. Similar results were obtained for the hexon mRNA, which does not appear to be subjected to this second block (Anderson and Klessig, 1983).

The early studies of Nakajima and co-workers (Hashimoto *et al.*, 1973; Nakajima *et al.*, 1974; Nakajima and Oda, 1975) suggested that a defect in the translational machinery of the abortively infected monkey cells was responsible for poor synthesis of several late proteins. These workers found that whereas cell-free lysates prepared from monkey cells infected or transformed with SV40 were able to translate exogenous, functional, Ad2 mRNA isolated from infected human cells, lysates prepared from mock- or Ad2-infected monkey cells were inactive. Unfortunately, since the activities of these lysates were extremely low, and they were not tested with other mRNAs as a control to show that the reduced activity of the systems prepared from mock- or Ad2-infected monkey cells was specific for Ad2 mRNAs, interpretation of these results is difficult. For example, since SV40 infection is known to stimulate the metabolic activity of host cells (Acheson, 1980), the greater translational activity of cell-free lysates prepared from SV40-infected cells may simply reflect the metabolic state of the cell, irrespective of the block to human adenovirus multiplication.

Nakajima and colleagues suggested that translational initiation was blocked in abortive infections, since the high-salt washes (initiation factors) of ribosomes from SV40-infected or -transformed cells, but not from mock- or Ad2-infected cells, were capable of forming 80 S initiation complexes with exogenous Ad2 mRNA and ribosomal subunits from monkey cells. The initiation factors from abortive infections were able to form 80 S complexes when SV40 mRNA was substituted for Ad2 mRNA.

In contrast, several more recent *in vitro* studies suggest that the translational machinery of the mock- or abortively infected monkey cells can utilize functional Ad2 mRNA. (1) Quinlan and Klessig (1982) observed that a variety of mRNAs, including Ad2 mRNAs isolated from infected human cells, were translated with similar efficiencies in nuclease-treated, cell-free lysates prepared from productively and abortively infected monkey cells. (2) Heterokaryons formed by fusing enucleated mock- or abortively infected monkey cells with nuclei from Ad2-infected human cells are capable of synthesizing late viral proteins, including fiber (Zorn and Anderson, 1981). (3) The synthesis of late proteins, including fiber, in mock- or abortively infected monkey cells after microinjection of Ad2 mRNA isolated from infected human cells also suggests that the translational apparatus of the monkey cell is competent to translate functional Ad2 late mRNA (C. W. Anderson and W. Richardson, unpublished results; K. P. Anderson, E. Wong, and D. F. Klessig, unpublished results).

3. Hypotheses

If the monkey cell is capable of translating functional Ad2 mRNA and yet the dramatically reduced synthesis of at least one late protein, fiber, cannot be entirely accounted for by a lower concentration of its message, then it is probable that the residue fiber mRNA made in abortive infections is defective. However, since the fiber mRNA made in abortively infected monkey cells can be translated efficiently *in vitro* (Eron *et al.*, 1975; Anderson and Klessig, 1983), such a defect must alter the functionality of this mRNA only *in vivo*. Although fiber mRNAs are polyadenylated and capped and contain the tripartite leader, a portion (10–25%) of these messages harbor, between the tripartite leader and the coding sequences, additional long segments of RNA that are normally removed by splicing (Klessig and Chow, 1980). While these splicing abnormalities cannot account for the differences in translatability *in vivo* of apparently mature-size fiber mRNA, they do show that processing of fiber mRNA in abortively infected CV-1 cells is somewhat anomalous. It is possible that other less easily detected aberrations in processing that could affect the translatability of fiber message *in vivo* might be occurring in abortively infected cells.

One possibility for such aberrant processing might be found at the level of the ancilary x, y, and z leader sequences of the fiber message (Chow and Broker, 1978). Using primer extension analysis it has recently been discovered that fiber mRNA isolated from abortively infected monkey cells is processed differently than that made in productively infected cells (Anderson and Klessig, 1984). Analysis of the 5′ ends of fiber messages from several different productive and abortive infections shows a direct correlation between synthesis of the fiber polypeptide *in vivo*, and the presence of the "x" and/or "y" ancillary leaders on messages encoding the fiber polypeptide. Interestingly, of all the mRNAs encoded by the major late transcriptional unit of Ad2 only the fiber message can contain the "x" and "y" leaders, and the fiber protein is the only late Ad2 protein reported to be glycosylated. Perhaps these leader sequences play a role in the synthesis of this glycoprotein, as well as that of the Ad2 19K glycoprotein encoded by early region 3 whose mRNA also contains the "x" and "y" leaders.

It is possible that in abortive infections fiber mRNA and its attached ribosomes are prevented from entering a compartment of the cell where translation of proteins to be glycosylated normally occurs. The ancillary leader "x" or "y", present only on fiber mRNAs made in productive infections, may direct the message and its associated ribosomes to this hypothesized compartment. Disruption of the cell would destroy this compartmentalization and could explain how ribosomes bound to but not translating fiber mRNA *in vivo* could synthesize fiber protein from these attached messages once the cells were opened (Anderson and Klessig, 1983).

It is also possible that processing of fiber mRNA in both productively and abortively infected CV-1 cells results in a slightly altered message that translates efficiently *in vitro*, but for translation *in vivo* requires interaction with an altered adenovirus DBP (as in cells infected with the adenovirus *hr* mutants) or with SV40 T-antigen helper function (as in Ad2–SV40 hybrid infections or coinfections with SV40). The observation that the two SV40-specific proteins (56 and 42K) induced by Ad2^{+}ND2 are found tightly associated with the 40 S ribosomal subunits is consistent with this possibility (Jay *et al.*, 1977).

The possible involvement (directly or indirectly) of the adenovirus DBP and SV40 T antigen in translation of virus mRNA has also been suggested by work with adeno-associated virus (AAV). AAV is a defective virus that requires one or more functions provided by adenovirus for growth in human cells. While cells coinfected with Wt AAV and a *ts* mutant of Ad5 located in the DBP gene (Ad5*ts*125) produce mRNA that encodes the AAV capsid proteins at the nonpermissive temperature, no capsid proteins are made *in vivo*, even though they can be synthesized *in vitro* from RNA extracted from these cells (F. T. Jay *et al.*, 1981). Furthermore, monkey cells infected with AAV and Wt adenovirus synthesize AAV RNA and DNA, but no AAV-specific polypeptides (Buller *et al.*, 1979). AAV multiples efficiently in these cells only in the presence of an *hr* mutant of adenovirus (Ad2*hr*400, Ad5*hr*404) or Wt adenovirus plus SV40 (F. T. Jay *et al.*, 1981; Buller *et al.*, 1979; McPherson *et al.*, 1982).

A situation similar to that for AAV occurs when human cells are infected with deletion mutants of Ad5 that are missing the VA1 RNA gene. Even though the late mRNAs are produced and are translatable *in vitro*, very little late viral proteins are made in the cell. This suggests that adenovirus changes the translational machinery of the infected cell (Thimmappaya *et al.*, 1982). Thus, it is not unreasonable to speculate that there may be alterations of a message that affect its translation in infected cells but not *in vitro*.

An alternative explanation for the discrepancy between level of reduction of fiber protein vs. its mRNA is that this protein is made but rapidly degraded in abortive infections. Pulse-chase experiments, however, indicate that the small amount of fiber protein synthesized in abortive infections, like that made in productive infections, is very stable (K. P. Anderson and D. F. Klessig, unpublished results). Furthermore, the observation that large amounts of fiber are produced in mock- or abortively infected monkey cells after microinjection of cytoplasmic, polyadenylated RNA from infected human cells suggest that protein degradation is not the explanation (C. W. Anderson and W. Richardson, unpublished results; K. P. Anderson, E. Wong, and D. F. Klessig, unpublished results).

Much more experimental work is required before a clear choice can be made among these alternatives. Whatever the answer, however, it will provide insight into how the expression of genes is regulated posttranscriptionally.

Alterations in the distribution of viral late proteins, particularly the noncapsid, late 100K protein, have been correlated with changes in the permissivity of infection (Cepko and Sharp, 1983). During Ad2 infection of monkey cells, the 100K protein is located exclusively in the nucleus, whereas in productively infected human cells or monkey cells, it is distributed over the cytoplasm. While preliminary studies have been unable to detect changes in the pattern of phosphorylation of this phosphoprotein in the two types of infections, speculation continues that aberrant modification of virus or cellular proteins, probably by phosphorylation, may be responsible for the block(s) seen in monkey cells (Cepko and Sharp, 1983).

The aberrant phosphorylation model relates the 100K with the two proteins that alter the host range, namely, the adenovirus DBP (a highly phosphorylated protein) and the carboxy terminus of SV40 T antigen (site of phosphorylation). The correlation between loss of helper function and loss of the last nine amino acids of large-T antigen including a phosphorylation site in the deletion mutant SV40*dl*1265 is consistent with the model (van Roy *et al.*, 1981). Also in accordance with this model is the observation that all the mutations in the DBP that extend the host range of the virus are replacements of the same histidine residue by the phosphorylatable amino acid tyrosine. Unfortunately, little, if any of the DBP is phosphorylated at this tyrosine in monkey cells (S. Sell and D. F. Klessig, unpublished results).

IV. COMPLEMENTATION BETWEEN SIMIAN VIRUS 40 AND ADENOVIRUS MUTANTS

Similarity in function and possible interaction of the SV40 large T antigen and the adenovirus DNA-binding protein were first suggested by studies on adenovirus growth in monkey cells in which the C terminus of T antigen provides the helper function(s) or an altered DBP allows adenovirus to grow in these cells without the aid of SV40. Although the evidence accumulated to date suggests that large-T and the host-range DBP act by the same mechanism to overcome the block in monkey cells, the possibility that they may act by different mechanisms has not been excluded.

Why should SV40 T antigen carry a function required by human adenoviruses — a virus it rarely encounters in nature? A possible clue to this mystery comes from studies on a deletion mutation (d1A2459) in the C terminus of SV40 T antigen which destroys the adenovirus helper function and also makes the virus nonviable. Viral DNA is replicated and late mRNAs are made in d1A2459-infected cells but late protein synthesis is dramatically reduced (Tornow and Cole, 1983; Cole, unpublished results). Perhaps T antigen is required for proper processing and/or translation of SV40 late mRNAs just as it is needed for synthesis and utilization of adenovirus fiber mRNA in infected monkey cells.

T antigen and DBP share several additional features. Both are early gene products that are modified by phosphorylation (Tegtmeyer *et al.*, 1977; Mann and Hunter, 1980; McCormick and Harlow, 1980; Walter and Flory, 1980; van Roy *et al.*, 1981; Jeng *et al.*, 1977; Levinson *et al.*, 1977; Linné *et al.*, 1977; Russell and Blair, 1977; Levinson and Levine, 1977; Axelrod, 1978; Klein *et al.*, 1979; Linné and Philipson, 1980). Each controls the expression of its own gene as well as that of other early virus genes (Tegtmeyer, *et al.*, 1975; Reed *et al.*, 1976; Khoury and May, 1977; Alwine *et al.*, 1977; Rio *et al.*, 1980; Hansen *et al.*, 1981; Carter and Ginsberg, 1976; Carter and Blanton, 1978a,b; Blanton and Carter, 1979; Nevins and Winkler, 1980; Babich and Nevins, 1981). Both are DNA-binding proteins, though their specificities differ; T antigen preferentially binds sites around the SV40 origin of DNA replication (Carroll *et al.*, 1974; Reed *et al.*, 1975; Jessel *et al.*, 1976; Tjian, 1978; Tenen *et al.*, 1982), while DBP binds single-stranded DNA (ssDNA) in a non-sequence-specific manner (van der Vliet and Levine, 1973; van der Vliet *et al.*, 1975; Fowkles *et al.*, 1979; Schechter *et al.*, 1980). Each is required for replication of its virus genome (Tegtmeyer, 1972; Ensinger and Ginsberg, 1972; Rubenstein and Ginsberg, 1974).

Possible interchange of functions between these two proteins has been suggested (Levine *et al.*, 1974; Rabek *et al.*, 1981; Goldman *et al.*, 1981). Ad5*ts*125 is a temperature-sensitive mutant that fails to replicate its DNA at the nonpermissive temperature due to a defect in the DBP gene. This defect can be complemented in primary African green monkey kidney cells, but not in established lines of AGMK cells, by coinfection with Wt SV40 but not SV40*ts*A mutants. Thus, the SV40 large-T antigen or a gene product induced by its presence can, at least partially, play the role of DBP in adenovirus DNA replication. Since another early adenovirus DNA replication-minus mutant (Ad5*ts*36) that alters the Ad5 DNA polymerase, is not complemented by SV40, the T antigen–DBP interrelationship appears to be quite specific.

In return for SV40's help, adenovirus coinfection can remedy the inability of SV40*ts*A mutants to turn off their early genes at the nonpermissive temperature. While it is implied that DBP is substituting for the thermolabile T antigen, this has not been shown directly using mutants in the DBP of adenovirus (Goldman *et al.*, 1981).

V. ADENOVIRUS INHIBITION OF SIMIAN VIRUS 40 GROWTH

A less intensively studied corollary to the enhancement of adenovirus multiplication is the inhibition of SV40 growth that occurs during coinfection of monkey cells. These interactions were first observed by O'Conor and co-workers (O'Conor *et al.*, 1963, 1965; Rabson *et al.*, 1964) in EM studies of monkey cells infected with adenovirus, SV40, or both. These investigators noted that while large numbers of adenovirus virion particles were present only in dually infected cells, very few SV40 virion particles were observed in these same cells. They concluded that SV40 provided a helper function essential for adenovirus growth in monkey cells and, concomitantly, some adenovirus function(s) acted to suppress SV40 replication in the same cells.

The early events is in the lytic cycle of SV40, such as adsorption, penetration, and uncoating, occur normally in coinfections (van Roy and Fiers, 1978b; Goldman and Khoury, 1980). The expression of large-T and small-t antigen is unaltered (Malmgren *et al.*, 1966; Friedman *et al.*, 1970; Goldman and Khoury, 1980; Patch *et al.*, 1981). This might have been anticipated, since SV40 T antigen is required to help adenovirus growth.

SV40 DNA replication is inhibited in coinfections. The severity of the block is proportional to the multiplicity of infection (m.o.i.) of adenovirus (Friedman *et al.*, 1970; van Roy and Fiers, 1978b; Goldman and Khoury, 1980). The time interval between SV40 infection and adenovirus superinfection also affects the amount of repression of SV40 DNA replication—repression increases as the interval is shortened (van Roy and Fiers, 1978b; Patch *et al.*, 1981).

The mechanism of inhibition is only partially defined. The infecting SV40 genome is not replicated even though this template remains intact during adenovirus coinfection (van Roy and Fiers, 1978b). Because SV40 induces host-cell DNA synthesis (for a review, see Acheson, 1980) while adenovirus suppressed it under certain conditions (Ginsberg *et al.*, 1967; Hodge and Scharff, 1969), adenovirus might interfere with SV40 induction of factor(s) required for both host and SV40 DNA replication. This possibility is less likely, however, in view of the observation that host-cell DNA replication is induced in coinfections (van Roy and Fiers, 1978b).

Adenovirus appears to actively supply a function required for inhibition of SV40 DNA replication. Neither adenovirus capsid proteins nor hydroxylamine-inactivated adenovirus can inhibit SV40 propagation (Friedman *et al.*, 1970). At the nonpermissive temperature, each of the two DNA-replication-negative mutants of adenovirus, Ad5*ts*125 (E2a) and Ad5*ts*36 (E2b), still prevents SV40 DNA replication (van Roy and Fiers, 1978b). Thus, adenovirus DNA replication is not required, and inhibition is probably induced directly or indirectly by an adenovirus early gene because most late genes are expressed only after DNA replication.

In addition to the E2a and E2b gene products debilitated by the *ts* mutants, the 19 and 14K proteins encoded in E3 are unlikely to be involved in this inhibition, since coinfection with Ad2^{+}ND3 or *dl*–Ad2 also prevents SV40 replication even though most of E3 is deleted in these viruses (van Roy *et al.*, 1979; Patch *et al.*, 1981).

Goldman and Khoury (1980) failed to detect any alteration in early or late SV40 RNA or protein synthesis when monkey cells were superinfected at low m.o.i. with adenovirus 24 hr after SV40 infection. Under these conditions, SV40 DNA replication is partially blocked. At higher m.o.i. with adenovirus, where inhibition is more complete, early and late SV40 mRNA synthesis is reduced 10-fold. The residual SV40-specific RNA is not, however, sequestered in the nucleus as occurs for host RNA late in adenovirus infection (Beltz and Flint, 1979).

Several hypotheses have been presented to explain the inhibition phenomenon (van Roy and Fiers, 1978b; Goldman and Khoury, 1980; Patch *et al.*, 1981). Perhaps the most intriguing possibility was proposed by Patch *et al.* (1981) based on the aforecited findings plus the observation that preinfection with SV40 delays adenovirus DNA replication. They hypothesized that the adenovirus and SV40 genomes compete for one or more early adenovirus proteins that block SV40 replication but are essential for adenovirus DNA replication. A probable early adenovirus protein is the DBP because of its involvement in DNA replication. Although the DBP encoded by Ad5*ts*125 has reduced capacity to bind ssDNA and cannot perform its role in adenovirus DNA replication at the nonpermissive temperature (van der Vliet *et al.*, 1975), it still may retain its inhibitory functions. Precedence for partial retention of function is provided by studies on the mutant Ad2*ts*400. It contains a *ts* mutation in the C-terminal half of DBP that destroys its DNA-replication function and an *hr* mutation in the N-terminal half that extends the host range of the virus to monkey cells. At the nonpermissive temperature, Ad2*ts*400 can grow in monkey cells coinfected with Wt Ad2 (Rice and Klessig, 1984). Presumably, the Ad2 DBP provides the DNA replication function and the Ad2*ts*400 DBP supplies the activities required to fully express the virus late genes in these cells. The observation that superinfection with Ad2 partially restores modulation of SV40 early gene expression in SV40*ts*A-infected cells at the nonpermissive temperature (Goldman *et al.*, 1981) suggests interaction of an adenovirus protein, perhaps the DBP, with the SV40 genome. However the culprit adenovirus protein interacts with SV40, the specificity of this interaction must be sufficiently broad to cover human papovavirus (assuming the same mechanism of inhibition is used), since Ad2 blocks BKV DNA replication in coinfected human cells (Patch *et al.*, 1981).

VI. SUMMARY

The amount of interaction between adenovirus and SV40 and the different forms these interactions take is quite extraordinary. Why these two very different viruses should interact so extensively even to the point of complementing each other's defects is puzzling. Continued research in this area should supply additional insight into this question as well as provide greater understanding of the biology of these two DNA tumor viruses.

ACKNOWLEDGMENTS. I would like to thank those who graciously communicated their results prior to publication. Special thanks are given to Carl Anderson and Terri Grodzicker for critical review of this chapter.

REFERENCES

Acheson, N.H., 1980, Lytic cycle of SV40 and polyomavirus, in: *The Molecular Biology of Tumor Viruses*, Part 2, *DNA Tumor Viruses*, 2nd rev. ed. (J. Tooze, ed.), pp. 125–204, Cold Spring Harbor Press, Cold Spring Harbor, New York.

Alstein, A.D., and Dodonova, N.N., 1968, Interaction between human and simian adenoviruses in simian cells: Complementation, phenotypic mixing and formation of monkey cell "adapted" virions, *Virology* **35**:248.

Alstein, A.D., Dodonova, N.N., Vassilyeva, N.N., and Tsetlin, E.M., 1968, Hybrid of monkey and human adenoviruses, *J. Virol.* **2**:488.

Alwine, J.C., Reed, S.I., and Stark, G.M., 1977, Characterization of the autoregulation of simian virus 40 gene A, *J. Virol.* **24**:2.

Anderson, C.W., 1981, Spontaneous mutants of the adenovirus–SV40 hybrid Ad2$^+$ND3 that grow efficiently in monkey cells, *Virology* **111**:263.

Anderson, C.W., Hardy, M.M., Dunn, J.J., and Klessig, D.F., 1983, Independent spontaneous mutants of adenovirus type 2-Simian virus 40 hybrid Ad2$^+$ND3 that grow efficiently in monkey cells possess identical mutations in the adenovirus type 2 DNA-binding protein gene, *J. Virol.* **48**:31.

Anderson, C.W., Lewis, J.B., Baum, P.R., and Gesteland, R.F., 1976, Simian virus 40-specific polypeptides in Ad2$^+$ND1- and Ad2$^+$ND4-infected cells, *J. Virol.* **18**:685.

Anderson, J.L., Martin, R.G., Chang, C., and Mora, P.T., 1977, Tumor-specific transplantation antigen is expressed during SV40 lytic infection with wild-type and tsA mutant viruses, *Virology* **76**:254.

Anderson, K.P., and Klessig, D.F., 1982, Synthesis of human adenovirus early RNA species is similar in productive and abortive infections of monkey and human cells, *J. Virol.* **42**:748.

Anderson, K.P., and Klessig, D.F., 1983, Post-transcriptional block to synthesis of a human adenovirus capsid protein in abortively infected monkey cells, *J. Mol. Appl. Genet.* **2**:

Anderson, K.P. and Klessig, D.F., 1984, Altered mRNA splicing in monkey cells abortively infected with human adenovirus may be responsible for inefficient synthesis of the virion fiber polypeptide, *Proc. Natl. Acad. Sci. U.S.A.* (in press).

Axelrod, N., 1978, Phosphoproteins of adenovirus 2, *Virology* **87**:366.

Babich, A., and Nevins, J.R., 1981, The stability of early adenovirus mRNA is controlled by the viral 72kd DNA-binding protein, *Cell* **26**:371.

Baum, S.G., 1977, Persistent adenovirus infections of nonpermissive monkey cells, *J. Virol.* **23**:412.

Baum, S.G., Reich, P.R., Huebner, R.J., Rowe, W.P., and Weissman, S.M., 1966, Biophysical evidence for linkage of adenovirus and SV40 DNAs in adenovirus 7–SV40 hybrid particles, *Proc. Natl. Acad. Sci. U.S.A.* **56**:1509.

Baum, S.G., Wiese, W.H., and Reich, P.R., 1968, Studies on the mechanism of enhancement of adenovirus 7 infection in African green monkey cells by simian virus 40: Formation of adenovirus-specific RNA, *Virology* **34**:373.

Baum, S.G., Horwitz, M.S., and Maizel, J.V., Jr., 1972, Studies of the mechanism of enhancement of human adenovirus infection in monkey cells by simian virus 40, *J. Virol.* **10**:211.

Baumann, E.A., and Hand, R., 1979, Protein kinase activity associated with D2 hybrid proteins related to simian virus 40 T antigen: Some characteristics of the related products, *Proc. Natl. Acad. Sci. U.S.A.* **76**:3688.

Baumann, E.A., Stedman, D., Raptis, L., Fuks, A., and Hand, R., 1981, Immunological cross reactivity between SV40 large T antigen and D-2 hybrid T antigen, *J. Virol.* **38**:1090.

Beardmore, W.B., Havlick, M.J., Serafini, A., and McLean, I.W., Jr., 1965, Interrelationship of adenovirus (type 4) and papovavirus (SV40) in monkey kidney cell cultures, *J. Immunol.* **95**:422.

Beltz, G.A., and Flint, S.J., 1979, Inhibition of HeLa cell protein synthesis during adenovirus infection: Restriction of cellular messenger RNA sequences to the nucleus, *J. Mol. Biol.* **131**:353.

Black, P.H., Rowe, W.P., Turner, H.C., and Huebner, R.J., 1963, A specific complement-fixing antigen present in SV40 tumors and transformed cells, *Proc. Natl. Acad. Sci. U.S.A.* **50**:1148.

Blanton, R.A., and Carter, T.H., 1979, Autoregulation of adenovirus type 5 early gene expression. III. Transcription studies on isolated nuclei, *J. Virol.* **29**:458.

Boeye, A., Melnick, J.L., and Rapp, F., 1966, SV40-adenovirus "hybrids": Presence of two genotypes and the requirement of their complementations for viral replication, *Virology* **28**:56.

Buller, R.M.L., Straus, S.E., and Rose, J.A., 1979, Mechanism of host restrictoin of adenovirus associated virus replication in African green monkey kidney cells, *J. Gen. Virol.* **43**:663.

Butel, J.S., 1967, Characterization of the strain of adenovirus type 7 carrying the defective monkey cell-adapting component, *J. Virol.* **1**:876.

Butel, J.S., and Rapp, F., 1967, Complementation between a defective monkey cell-adapting component and human adenoviruses in simian cells, *Virology* **31**:573.

Butel, J.S., Guntzel, M., and Rapp, F., 1969, Variants of defective simian papovavirus 40 (PARA) characterized by cytoplasmic localization of the simian papovavirus 40 tumor antigen, *J. Virol.* **4**:632.

Carroll, R.B., Hager, L., and Dulbecco, R., 1974, Simian virus 40 T antigen binds to DNA, *Proc. Natl. Acad. Sci. U.S.A.* **71**:3754.

Carter, T.H., and Blanton, R.A., 1978a, Possible role of the 72,000 dalton DNA binding protein on regulation of adenovirus type 5 early gene expression, *J. Virol.* **25**:664.

Carter, T.H., and Blanton, R.A., 1978b, Autoregulation of adenovirus type 5 early gene expression. II. Effect of temperature-sensitive early mutations on virus RNA accumulation, *J. Virol.* **28**:450.

Carter, T.H., and Ginsberg, H.S., 1976, Viral transcription of KB cells infected by temperature-sensitive "early" mutants of adenovirus type 5, *J. Virol.* **18**:156.

Cepko, C.L., and Sharp, P.A., 1983, Aberrant distribution of human adenovirus type 2 late proteins in monkey kidney cells, *J. Virol.* **46**:302.

Chang, C., Luborsky, S.W., and Mora, P.T., 1977a, Tumour-specific transplantation antigen from SV40-transformed cells binds to DNA, *Nature (London)* **269**:438.

Chang, C., Pancake, S.J., Lubonsky, S.W., and Mora, P.T., 1977b, Detergent solubilization and partial purification of tumor specific surface and transplantation antigens from SV40-transformed mouse cells, *Int. J. Cancer* **19**:258.

Chang, C., Martin, R.G., Livingston, D.M., Luborsky, S.W., Cheng-Po, H., and Mora, P.T., 1979, Relationship between T-antigen and tumor specific transplantation antigen in simian virus 40-transformed cells, *J. Virol.* **29**:69.

Chow, L.T., and Broker, T.R., 1978, The spliced structures of adenovirus 2 fiber message and other late mRNAs, *Cell* **15**:497.

Cole, C.N., Crawford, L.V., and Berg, P., 1979, Simian virus 40 mutants with deletions at the 3′ end of the early region are defective in adenovirus helper function, *J. Virol.* **30**:683.

Crawford, L.V., Cole, C.N., Smith, A.E., Paucha, E., Tegtmeyer, P., Rundell, K., and Berg, P., 1978, Organization and expression of early genes of simian virus 40, *Proc. Natl. Acad. Sci. U.S.A.* **75**:117.

Daniell, E., 1981, Subgenomic viral DNA species synthesized in simian cells by human and simian adenoviruses, *J. Virol.* **37**:620.

Deppert, W., 1978, Simian virus 40 (SV40)-specific proteins associated with the nuclear matrix isolated from adenovirus type 2–SV40 hybrid virus-infected HeLa cells carry SV40 U-antigen determinants, *J. Virol.* **26**:165.

Deppert, W., 1979, SV40 tumor antigen and U antigen: Immunological characterization and localization in different nuclear subfractions of SV40-transformed cells, *J. Virol.* **29**:576.

Deppert, W., and Walter, G., 1975, Simian virus 40 (SV40) tumor-specific proteins in nucleus and plasma membrane of HeLa cells infected by the adenovirus 2–SV40 hybrid virus, $Ad2^+ND2$, *Proc. Natl. Acad. Sci. U.S.A.* **73**:2505.

Deppert, W., Walter, G., and Linke, H., 1977, Simian virus 40 tumor-specific proteins: Subcellular distribution and metabolic stability in HeLa cells infected with nondefective adenovirus type 2–simian virus 40 hybrid viruses, *J. Virol.* **21**:1170.

Diamond, L., 1967, Transformation of SV40-resistant hamster cells with an adenovirus 7–SV40 hybrid, *J. Virol.* **1**:1109.

Drapkin, M.S., Appella, E., and Law, L.W., 1974, Immunogenic properties of a soluble tumor specific transplantation antigen induced by simian virus 40, *J. Natl. Cancer Inst.* **52**:259.

Duff, R., and Rapp, F., 1970, Quantitative characteristics of the transformation of hamster cells with PARA (defective SV40)–adenovirus 7, *J. Virol.* **5**:568.

Duff, R., Rapp, F., and Butel, J.S., 1970, Tranformation of hamster cells by variants of PARA–adenovirus 7 able to induce SV40 tumor antigen in the cytoplasm, *Virology* **42**:273.

Dunn, A.R., and Hassell, J.A., 1977, A novel method to map transcripts: Evidence for homology between an adenovirus mRNA and discrete multiple regions of the viral genome, *Cell* **12**:23.

Dunn, A.R., Mathews, M.B., Chow, L.T., and Sambrook, J., 1978, A supplementary adenoviral leader sequence and its role in messenger translation, *Cell* **15**:511.

Easton, J.M., and Hiatt, C.W., 1965, Possible incorporation of SV40 genome within capsid proteins of adenovirus 4, *Proc. Natl. Acad. Sci. U.S.A.* **54**:1100.

Eddy, B.E., Berman, G.S., Berkeley, W., and Young, A.D., 1961, Tumors induced in hamsters by injection of rhesus monkey kidney extracts, *Proc. Soc. Exp. Biol. Med.* **107**:191.

Ensinger, M.J., and Ginsberg, H.S., 1972, Selection and preliminary characterization of temperature-sensitive mutants of type 5 adenovirus, *J. Virol.* **10**:328.

Eron, L., Westphal, H., and Khoury, G., 1975, Postranscriptional restriction of human adenovirus expression in monkey cells, *J. Virol.* **15**:1256.

Farber, M.S., and Baum, S.G., 1978, Transcription of adenovirus RNA in permissive and nonpermissive infections, *J. Virol.* **27**:136.

Feldman, L.A., Butel, J.S., and Rapp, F., 1966, Interaction of a simian papovavirus and adenovirus. I. Induction of adenovirus tumor antigen during abortive infection of simian cells, *J. Bacteriol.* **91**:813.

Feunteun, J., Carmichael, G., Nicolas, J.C., and Kress, M., 1981, Mutant carrying deletions in the 2 SV40 early genes, *J. Virol.* **40**:625.

Fey, G., Lewis, J.B., Grodzicker, T., and Bothwell, A., 1979, Characterization of a fused protein specified by the adenovirus 2–simian virus 40 hybrid $Ads^+ND1dp2$, *J. Virol.* **30**:201.

Flint, S.J., Wewerka-Lutz, Y., Levine, A.S., Sambrook, J., and Sharp, P.A., 1975, Adenovirus transcription, II. RNA sequences complementary to simian virus 40 and adenovirus 2 DNA in $Ad2^+ND1$ and $Ad2^+ND3$ infected cells, *J. Virol.* **16**:662.

Fowlkes, D.M., Lord, S.T., Linné, T., Pettersson, U., and Philipson, L., 1979, Interaction between the adenovirus DNA-binding protein and double stranded DNA, *J. Mol. Biol.* **132**:163.

Fox, R.I., and Baum, S.G., 1972, Synthesis of viral ribonucleic acid during restricted adenovirus infection, *J. Virol.* **10**:220.

Fox, R.I., and Baum, S.G., 1974, Posttranscriptional block to adenovirus replication in nonpermissive monkey cells, *Virology* **60**:45.

Friedman, M.P., Lyons, M.J., and Ginsberg, H.S., 1970, Biochemical consequences of type 2 adenovirus and simian virus 40 double infections of African green monkey cells, *J. Virol.* **5**:586.

Gardner, S.D., Field, A.M., Coleman, D.V., and Hulme, B., 1971, New human papovavirus (B.K.) isolated from urine after renal transplantation, *Lancet* **1**:1253.

Gesteland, R.F., Wills, N., Lewis, J.B., and Grodzicker, T., 1977, Identification of amber and ochre mutants of the human virus Ad2$^+$ND1, *Proc. Natl. Acad. Sci. U.S.A.* **74**:4567.

Ginsberg, H.S., Bello, L.J., and Levine, A.J., 1967, Control of biosynthesis of host macromolecules in cells infected with adenovirus, in: *The Molecular Biology of Viruses* (J. S. Colter and W. Paranchych eds.), pp. 547–572, Academic Press, New York.

Girardi, A.J., and Defendi, V., 1970, Induction of SV40 transplantation antigen (TrAg) during the lytic cycle, *Virology* **42**:688.

Gluzman, Y. and Van Doren, Y., 1983, Palindromic adenovirus type 5-simian virus 40 hybird, *J. Virol.* **45**:91.

Goldman, N.D., and Khoury, G., 1980, Adenovirus regulation of SV40 macromolecular synthesis, *J. Virol.* **34**:658.

Goldman, N.D., Howley, P., and Khoury, G., 1981, Functional interaction between the early viral proteins of simian virus 40 and adenovirus, *Virology* **109**:303.

Graessmann, A., Graessmann, M., Tjian, R., and Topp, W.C., 1980, SV40 small t protein is required for loss of actin cable networks in rat cells, *J. Virol.* **33**:1182.

Griffin, J.D., Spangler, G., and Livingston, D.M., 1979, Protein kinase activity associated with simian virus 40 T antigen, *Proc. Natl. Acad. Sci. U.S.A.* **76**:2610.

Grodzicker, T., 1980, Adenovirus–SV40 hybrids, in: *The Molecular Biology of Tumor Viruses*, Part 2, *DNA Tumor Viruses*, 2nd rev. ed. (J. Tooze, ed.), pp. 577–614, Cold Spring Harbor Press, Cold Spring Harbor, New York.

Grodzicker, T., Anderson, C., Sharp, P.A., and Sambrook, J., 1974, Conditional lethal mutants of adenovirus 2–simian virus 40 hybrids. I. Host range mutants of Ad2$^+$ND1, *J. Virol.* **13**:1237.

Grodzicker, T., Lewis, J.B., and Anderson, C.W., 1976, Conditional lethal mutants of adenovirus type 2–simian virus 40 hybrids. II. Ad2$^+$ND1 host-range mutants that synthesize fragments of the Ad2$^+$ND1 30K protein, *J. Virol.* **19**:559.

Hansen, U., Tenen, D.G., Livingston, D.M., and Sharp, P.A., 1981, Repression of SV40 early transcription from 2 promoters, *Cell* **27**:603.

Hartley, J.W., Heubner, R.J., and Rowe, W.P., 1956, Serial propagation of adenoviruses (APC) in monkey kidney tissue cultures, *Proc. Soc. Exp. Biol. Med.* **92**:667.

Hashimoto, K., Nakajima, K., Oda, K., and Shimojo, H., 1973, Complementation of translational defect for growth of human adenovirus type 2 in simian cells by a simian virus 40-induced factor, *J. Mol. Biol.* **81**:207.

Hassell, J.A., Lukanidin, E., Fey, G., and Sambrook, J., 1978, The structure and expression of two defective adenovirus 2–simian virus 40 hybrids, *J. Mol. Biol.* **120**:209.

Henry, C.J., Slifkin, M., and Merkow, L., 1971, Mechanism of host cell restriction in African green monkey kidney cells abortively infected with human adenovirus type 2, *Nature (London) New Biol.* **233**:39.

Hodge, L.D., and Scharff, M.D., 1969, Effect of adenovirus on host cell DNA synthesis in synchronized cells, *Virology* **37**:554.

Howley, P.M., Khoury, G., Byrne, J.C., Takemoto, K.K., and Martin, M.A., 1975, Physical map of the BK virus genome, *J. Virol.* **16**:959.

Huebner, R.J., Chanock, R.M., Rubin, B.A., and Casey, M.J., 1964, Induction by adenovirus type 7 of tumors in hamsters having the antigenic characteristics of SV40 virus, *Proc. Natl. Acad. Sci. U.S.A.* **52**:1333.

Ismail, A., Bauman, E.A., and Hand, R., 1981, Cell surface T antigen in cells infected with SV40 or an adenovirus–SV40 hybrid, Ad2$^+$D2, *J. Virol.* **40**:615.

Jay, F.T., Laughlin, C.A., and Carter, B.J., 1981, Eukaryotic translational control: Adeno-associated virus protein synthesis is affected by a mutation in the adenovirus DNA-binding protein, *Proc. Natl. Acad. Sci. U.S.A.* **78**:2927.

Jay, G., Jay, F.T., Friedman, R.M., and Levine, A.S., 1977, Simian virus 40-specific ribosome binding proteins induced by a nondefective adenovirus 2–simian virus 40 hybrid, *J. Virol.* **23:**692.

Jay, G., Jay, F.T., Friedman, R.M., and Levine, A.S., 1978, Biosynthesis, immunological specificity and intracellular distribution of the simian virus 40-specific protein induced by the nondefective hybrid, $Ad2^{+}ND1$, *J. Virol.* **26:**411.

Jay, G., Nomura, S., Anderson, C.W., and Khoury, G., 1981, Identification of the SV40 agnogene product: A DNA binding protein, *Nature* (London) **291:**346.

Jeng, Y.-H., Wold, W.S., Sugawara, K., Gilead, Z., and Green, M., 1977, Adenovirus type 2 coded single-stranded DNA binding protein: *In vivo* phosphorylation and modification, *J. Virol.* **22:**402.

Jerkofsky, M., 1975, Enhancement of the replication of human adenovirus in simian cells by a series of temperature sensitive mutants of simian virus 40, *Virology* **65:**579.

Jerkofsky, M., and Rapp, F., 1973, Host cell DNA synthesis as a possible factor in the enhancement of replication of human adenoviruses in simian cells by SV40, *Virology* **51:**466.

Jerkofsky, M., and Rapp, F., 1975, Stimulation of adenovirus replication in simian cells in the absence of a helper virus by pretreatment of the cells with iododeoxyuridine, *J. Virol.* **15:**253.

Jessel, D., Landau, T., Hudson, J., Lalor, T., Tenen, D., and Livingston, D.M., 1976, Identification of regions of the SV40 genome which contain preferred SV40 T-antigen binding sites, *Cell* **8:**535.

Kaplan, P.M., Melnick, J.L., and Tevethia, S.S., 1971, Development of nononcogenic SA7–adenovirus 2 populations that immunize against SA7-transformed cells, *J. Natl. Cancer Inst.* **46:**565.

Kelly, T.J., Jr., 1975, Structure of the DNA of the adenovirus 7–simian virus 40 hybrid, *J. Virol.* **15:**1267.

Kelly, T.J., Jr., and Lewis, A.M., 1973, Use of nondefective adenovirus–simian virus 40 hybrids for mapping the simian virus 40 genome, *J. Virol.* **12:**643.

Kelly, T.J., Jr., and Rose, J.A., 1971, Simian virus 40 integration site in an adenovirus 7–SV40 hybrid DNA molecule, *Proc. Natl. Acad. Sci. U.S.A.* **68:**1037.

Kelly, T.J., Jr., Lewis, A.M., Jr., Levine, A.S., and Siegel, S., 1974, Structure of two adenovirus–simian virus 40 hybrids which contain the entire SV40 genome, *J. Mol. Biol.* **89:**113.

Kelly, T.J., Jr., Lewis, A.M., Jr., Levine, A.S., and Siegel, S., 1975, Structural studies on two adenovirus 2–SV40 hybrids containing the entire SV40 genome, *Cold Spring Harbor Symp. Quant. Biol.* **39:**409.

Khera, K.S., Ashkenazi, A., Rapp, F., and Melnick, J.L., 1963, Immunity in hamsters to cells transformed *in vitro* and *in vivo* by SV40, *J. Immunol.* **91:**604.

Khoury, G., and May, E., 1977, Regulation of early and late SV40 transcription: Overproduction of early viral RNA in the absence of a functional T-antigen, *J. Virol.* **23:**167.

Khoury, G., Lewis, A.M., Jr., Oxman, M.N., and Levine, A.S., 1973, Strand orientation of SV40 transcription in cells infected by non-defective adenovirus 2–SV40 hybrid viruses, *Nature (London) New Biol.* **246:**202.

Khoury, G., Howley, P.M., Garon, C., Mullarkey, M.F., Takemoto, K.K., and Martin, M.A., 1975, Homology and relationship between the genomes of papovaviruses, BK virus and simian virus 40, *Proc. Natl. Acad. Sci. U.S.A.* **72:**2563.

Khoury, G., Alwine, J., Goldman, N., Gruss, P., and Jay, G., 1980, Hybrid viral messenger RNA, *J. Virol.* **36:**148.

Kimura, G., 1974, Genetic evidence for SV40 gene function in enhancement of replication of human adenovirus in simian cells, *Nature (London)* **248:**590.

Klein, H., Maltzman, W., and Levine, A.J., 1979, Structure–function relationships of the adenovirus DNA binding protein, *J. Biol. Chem.* **254:**11051.

Klessig, D.F., 1977, Isolation of a variant of human adenovirus serotype 2 that multiplies efficiently on monkey cells, *J. Virol.* **21:**1243.

Klessig, D.F., and Anderson, C.W., 1975, Block to multiplication of adenovirus serotype 2 in monkey cells, *J. Virol.* **16**:1650.

Klessig, D.F., and Chow, L.T., 1980, Incomplete splicing and deficient accumulation of fiber messenger RNA in monkey cells infected by human adenovirus type 2, *J. Mol. Biol.* **139**:221.

Klessig, D.F., and Grodzicker, T., 1979, Mutations that allow Ad2 and Ad5 to express late genes in monkey cells map in the viral gene encoding the 72kd DNA binding protein, *Cell* **17**:957.

Klessig, D.F., and Hassell, J.A., 1978, Characterization of a variant of human adenovirus type 2 which multiplies efficiently in simian cells, *J. Virol.* **28**:945.

Klessig, D.F., and Quinlan, M.P., 1982, Genetic evidence for separate functional domains on the human adenovirus specified, 72kd, DNA binding protein, *J. Mol. Appl. Genet.* **1**:263.

Kruijer, W., van Schaik, F.M.A., and Sussenbach, J.S., 1981, Structure and organization of the gene coding for the DNA binding protein of adenovirus type 5, *Nucleic Acids Res.* **9**:4439.

Lai, C.-J., Goldman, N.D., and Khoury, G., 1979, Functional similarity between the early antigens of SV40 and human papovavirus BK, *J. Virol.* **30**:141.

Lanford, R.E., and Butel, J.S., 1980a, Biochemical characterization of nuclear and cytoplasmic forms of SV40 tumor antigens encoded by parental and transport defective mutant of SV40–adenovirus 7 hybrid viruses, *Virology* **105**:314.

Lanford, R.E., and Butel, J.S., 1980b, Inhibition of nuclear migration of wild-type SV40 tumor antigen: By a transport defective mutant of SV40 adenovirus 7 hybrid virus, 1980, *Virology* **105:**303.

Lanford, R.E., and Butel, J.S., 1981, Effect of nuclear localization of large tumor antigen on growth potential of SV40 transformed cells, *Virology* **110**:147.

Lanford, R.E., and Butel, J.S., 1982, Intracellular transport of SV40 large tumor antigen: A mutation which abolishes migration to the nucleus does not prevent association with the cell surface, *Virology* **119**:169.

Lausch, R.N., Tevethia, S.S., and Rapp, F.E., 1968, Evidence of SV40-specific transplantation and surface antigens in cells transformed by PARA–adenovirus 12, *J. Immunol.* **101**:845.

Lawrence, C.B., 1981, Expression of adenovirus type 2 late genes in HeLa cells infected with adenvirus type 2 or adenovirus type 2–SV40hybrid viruses, *J. Virol.* **40**:610.

Lawrence, C., Hunter, T., and Walter, G., 1979, Characterization of the SV40 specific messenger RNA isolated from HeLa cells infected with the nondefective adenovirus 2–SV40 hybrid viruses, Ad2$^+$ND2 and Ad2$^+$ND4, *J. Mol. Biol.* **134**:95.

Lebowitz, P., and Khoury, G., 1975, Simian virus 40 DNA segment of the adenovirus 7–simian virus 40 hybrid, E46$^+$, and its transcription during permissive infection of monkey kidney cells, *J. Virol.* **15**:1214–1221.

Lebowitz, P., Kelly, T.J., Jr., Nathans, D., Lee, T.H.N., and Lewis, A.M., Jr., 1974, A colinear map relating the simian virus 40 (SV40) DNA segments of six adenovirus–SV40 hybrids to the DNA fragments produced by restriction endonuclease cleavage of SV40 DNA, *Proc. Natl. Acad. Sci. U.S.A.* **71**:441.

Levine, A.J., van der Vliet, P.C., Rosenwirth, B., Rabek, J., Frenkel, G., and Ensinger M., 1974, Adenovirus-infected, cell-specific, DNA binding proteins, *Cold Spring Harbor Symp. Quant. Biol.* **39**:559.

Levine, A.S., Levin, M.J., Oxman, M.N., and Lewis, A.M., Jr., 1973, Studies of nondefective adenovirus 2–simian virus 40 hybrid viruses. VII. Characterization of the simian virus 40 RNA species induced by five nondefective hybrid viruses, *J. Virol.* **11**:672–681.

Levinson, A., and Levine, A.J., 1977, The isolation and identification of the adenovirus group C tumor antigens, *Virology* **76**:1–11.

Levinson, A.D., Postel, E.H., and Levine, A.J., 1977, *In vivo* and *in vitro* phosphorylation of the adenovirus type 5 single strand-specific DNA binding protein, *Virology* **79**:144–159.

Lewis, A.M., Jr., and Rowe, W.P., 1970, Isolation of two plaque variants from the adenovirus type 2–simian virus 40 hybrid populations which differ in their efficiency in yielding simian virus 40, *J. Virol.* **5**:413.

Lewis, A.M., Jr., and Rowe, W.P., 1971, Studies on nondefective adenovirus–simian virus 40 hybrid viruses. I. A newly characterized simian virus 40 antigen induced by the Ad2$^+$ND1 virus, *J. Virol.* **7**:189.

Lewis, A.M., Jr., and Westphal, H., 1983, The adenovirus type 2-simian virus 40 hybrid virus Ad2$^+$ ND4 requires deletion variants to grow in monkey cells, *J. Virol.* **47**:344.

Lewis, A.M., Jr., Baum, S.G., Prigge, K.O., and Rowe, W.P., 1966, Occurrence of adenovirus–SV40 hybrids among monkey kidney cell adapted strains of adenovirus, *Proc. Soc. Exp. Biol. Med.* **122**:214–218.

Lewis, A.M., Jr., Levin, M.J., Wiese, W.H., Crumpacker, C.S., and Henry, P.H., 1969, A nondefective (competent) adenovirus–SV40 hybrid isolated from the Ad2–SV40 hybrid population, *Proc. Natl. Acad. Sci. U.S.A.* **63**:1128–1135.

Lewis, A.M., Jr., Levine, A.S., Crumpacker, C.S., Levin, M.J., Samaha, R.J., and Henry, P.H., 1973, Studies of nondefective adenovirus 2–simian virus 40 hybrid viruses. IV. Isolation of additional hybrids which differ in their simian virus 40-specific biological properties, *J. Virol.* **11**:655.

Ling, L.E., Manos, M.M., and Gluzman, Y., 1982, Sequence of the junction in adenovirus 2–SV40 hybrids: examples of illegitimate recombination, *Nucleic Acids Res.* 10:

Linke, H.L., Hunter, T., and Walker, G., 1979, Structural relationship between the 100,000 and 17,000 molecular weight T antigens of simian virus 40 (SV40) as deduced by comparison with the SV40-specific proteins coded by the nondefective adenovirus type 2–SV40 hybrid viruses, *J. Virol.* **29**:390–394.

Linné, T., and Philipson, L., 1980, Further characterization of the phosphate moiety of the adenovirus type 2 DNA binding protein, *Eur. J. Biochem.* **103**:259–270.

Linné, T., Jornvall, H., and Philipson, L., 1977, Purification and characterization of the phosphorylated DNA binding protein from adenovirus type 2-infected cells, *Eur. J. Biochem.* **76**:481–490.

Lopez-Revilla, R., and Walter, G., 1973, Polypeptide specific for cells with adenovirus 2–SV40 hybrid, Ad2$^+$ND1, *Nature (London) New Biol.* **244**:165–167.

Lucas, J.J., and Ginsberg, H.S., 1972, Transcription and transport of virus-specific ribonucleic acids in African green monkey kidney cells abortively infected with type 2 adenovirus, *J. Virol.* **10**:1109–1117.

Malmgren, R.A., Rabson, A.S., Carney, P.G., and Paul, F.J., 1966, Immunofluorescence of green monkey kidney cells infected with adenovirus 12 and with adenovirus 12 plus simian virus 40, *J. Bacteriol.* **91**:262.

Mann, K., and Hunter, T., 1980, Phosphorylation of SV40 large T antigen in SV40 nucleoprotein complexes, *Virology* **107**:526.

Mann, K., Hunter, T., Walter, G., and Linke, H., 1977, Evidence for simian virus 40 (SV40) coding of SV40 T antigen and the SV40-specific proteins in HeLa cells infected with nondefective adenovirus type 2–SV40 hybrid viruses, *J. Virol.* **24**:151.

Mason, D.H., Jr., and Takemoto, K.K., 1976, Complementation between BK human papovavirus and a simian virus 40 tsA mutant, *J. Virol.* **17**:1060.

McCormick, F., and Harlow, E., 1980, Association of a murine 53,000 dalton phosphoprotein with simian virus 40 large T antigen in transformed cells, *J. Virol.* **34**:213.

McCormick, F., Clark, R., Harlow, E., and Tjian, R., 1981, SV40 T antigen binds specifically to a cellular 53 kilodalton protein *in vitro*, *Nature (London)* **292**:63.

McKay, R.D.G., 1981, Binding of a SV40 T antigen related protein to DNA, *J. Mol. Biol.* **145**:471.

McPherson, R.A., Ginsberg, H.S., and Rose, J.A., 1982, Adenovirus associated virus helper activity of adenovirus DNA binding protein, *J. Virol.* **44**:666.

Miyamura, T., and Takemoto, K.K., IV, 1979, Helper function for adenovirus replication in monkey cells by BK human papovavirus, *Virology* **98**:279.

Morrow, J.F., Berg, P., Kelly, T.J., Jr., and Lewis, A.M., Jr., 1973, Mapping of simian virus 40 early functions on the viral chromosome, *J. Virol.* **12**:653.

Mueller, C., Graessmann, A., and Graessmann, M., 1978, Mapping of early SV40 specific functions by microinjection of different early viral DNA fragments, *Cell* **15**:579.

Myers, M.W., Laughlin, C.A., Jay, F.T., and Carter, B.J., 1980, Adenovirus helper function for growth of adeno-associated virus: Effect of temperature sensitive mutations in adenovirus early gene region 2, *J. Virol.* **35**:65.

Naegele, R.F., and Rapp, F., 1967, Enhancement of the replication of human adenoviruses in simian cells by simian adenovirus, SV15, *J. Virol.* **1**:838.

Nakajima, K., and Oda, K., 1975, The alteration of ribosomes for mRNA selection concerned with adenovirus growth in SV40-infected simian cells, *Virology* **67**:85.

Nakajima, K., Ishitsuka, H., and Oda, K., 1974, An SV40 induced initiation factor for protein synthesis concerned with the regulation of permissiveness, *Nature (London)* **252**:649.

Nevins, J.R., and Winkler, J.J., 1980, Regulation of early adenovirus transcription: A protein product of early region 2 specifically represses region 4 transcription, *Proc. Natl. Acad. Sci. U.S.A.* **77**:1893.

O'Conor, G.T., Rabson, A.S., Berezesky, I.K., and Paul, F.J., 1963, Mixed infections with simian virus 40 and adenovirus 12, *J. Natl. Cancer Inst.* **31**:903.

O'Connor, G.T., Rabson, A.S., Malmgren, R.A., Berezesky, I.K., and Paul, F.J., 1965, Morphologic observations of green monkey kidney cells after single and double infection with adenovirus 12 and simian virus 40, *J. Natl. Cancer Inst.* **34**:679.

Osborn, J.E., Robertson, S.M., Padgett, B.L., Walker, D.L., and Weisblum, B., 1976, Comparison of JC and BK human papovaviruses with simian virus 40: DNA homology studies, *J. Virol.* **19**:675.

Osborn, M., and Weber, K., 1974, SV40: T antigen, the A function and transformation, *Cold Spring Harbor Symp. Quant. Biol.* **39**:267.

Oxman, M.N., Levin, M.J., and Lewis, A.M., Jr., 1974, Control of simian virus 40 gene expression in adenovirus–simian virus 40 hybrid viruses: Synthesis of hybrid adenovirus 2–simian virus 40 RNA molecules in cells infected with a nondefective adenovirus 2–simian virus 40 hybrid virus, *J. Virol.* **13**:322.

Patch, C.T., Howley, P.M., Hauser, J., and Levine, A.S., 1981, Kinetics of inhibition of papovavirus DNA synthesis by super-infection with adenovirus 2 and nondefective adenovirus 2–SV40 hybrid viruses, *J. Gen. Virol.* **55**:355.

Pope, J.H., and Rowe, W.P., 1964, Detection of specific antigens in SV40 transformed cell by immunofluorescence, *J. Exp. Med.* **120**:121.

Pretell, J., Greenfield, R.S., and Tevethia, S.S., 1979, Biology of simian virus 40 (SV40) transplantation rejection antigen (TrAg). V. *In vitro* demonstration of SV40 TrAg in SV40 infected nonpermissive mouse cells by the lymphocyte mediated cytotoxicity assay, *Virology* **97**:32.

Prives, C., and Beck, Y., 1977, Characterization of SV40 T-antigen polypeptides synthesized *in vivo* and *in vitro*, *Collog. INSERN* **69**:175.

Prives, C., Barnet, B., Scheller, A., Khoury, G., and Jay, G., 1982, Discrete regions for nonspecific and viral origin-specific DNA binding, *J. Virol.* **43**:73.

Quinlan, M.P., and Klessig, D.F., 1982, Normal translation of human adenovirus mRNA in cell-free lysates prepared from abortively as well as productively infected monkey cells, *J. Virol.* **44**:426.

Rabek, J.P., Zakian, V.A., and Levine, A.J., 1981, The SV40 A gene product suppresses the adenovirus H5ts125 defect in DNA replication, *Virology* **109**:290.

Rabson, A.S., O'Conor, G.T., Berezeksy, I.K., and Paul, F.J., 1964, Enhancement of adenovirus growth in African green monkey kidney cell cultures by SV40, *Proc. Soc. Exp. Biol. Med.* **116**:187.

Rapp, F., 1973, The PARA–adenoviruses, *Prog. Exp. Tumor Res.* **18**:104.

Rapp, F., Kitakaro, T., Butel, J.S., and Melnick, J.L., 1964a, Synthesis of SV40 tumor antigen during replication of simian papovavirus (SV40), *Proc. Natl. Acad. Sci. U.S.A.* **52**:1138.

Rapp, F., Melnick, J.L., Butel, J.S., and Kitahara, T., 1964b, The incorporation of SV40 genetic material into adenovirus 7 as measured by intranuclear synthesis of SV40 tumor antigen, *Proc. Natl. Acad. Sci. U.S.A.* **52:**1348.

Rapp, F., Butel, J.S., and Melnick, J.L., 1965, SV40–adenovirus "hybrid" populations: Transfer to SV40 determinants from one type of adenovirus to another, *Proc. Natl. Acad. Sci. U.S.A.* **54:**717.

Rapp, R., Feldman, L.A., and Mandel, M., 1966a, Synthesis of virus deoxynucleic acid during abortive infection of simian cells by human adenoviruses, *J. Bacteriol.* **92:**931.

Rapp, F., Tevethia, S.S. and Melnick, J.L., 1966b, Papovavirus SV40 transplantation immunity conferred by an adenovirus–SV40 hybrid, *J. Natl. Cancer Inst.* **7:**75.

Rapp, F., Jerkofsky, M., Melnick, J.L., and Levy, B., 1968, Variation in the oncogenic potential of human adenoviruses carrying a defective SV40 genome (PARA), *J. Exp. Med.* **127:**77.

Rapp, F., Pauluzzi, S., and Butel, J.S., 1969, Variation in properties of plaque progeny of PARA (defective simian papovavirus 40)–adenovirus 7, *J. Virol.* **4:**626.

Reed, S.I., Ferguson, J., Davis, R.W., and Stark, G.R., 1975, T-antigen binds to simian virus 40 DNA at the origin of replication, *Proc. Natl. Acad. Sci. U.S.A.* **72:**1605.

Reed, S.I., Stark, G., and Alwine, J.C., 1976, Autoregulation of simian virus 40 gene A by T-antigen, *Proc. Natl. Acad. Sci. U.S.A.* **73:**3083.

Reich, P.R., Baum, S.G., Rose, J.A., Rowe, W.P., and Weissman, S.M., 1966, Nucleic acid homology studies of adenovirus type 7–SV40 interactions, *Proc. Natl. Acad. Sci. U.S.A.* **55:**336.

Rhim, J.S., Trimmer, R., Arnstein, P., and Huebner, R.J., 1981, Neoplastic transformation of chimpanzee cells induced by adenovirus type 12–SV40 hybrid virus, *Proc. Natl. Acad. Sci. U.S.A.* **78:**313.

Rice, S.A., and Klessig, D.F., 1984, The function(s) provided by the adenovirus-specified, DNA-binding-protein required for viral late gene expression is independent of the role of the protein in viral DNA replication, *J. Virol.* **49:**35.

Richardson, L.S., and Butel, J.S., 1971, Properties of the transformed hamster cells containing SV40 tumor antigen in the cytoplasm, *Int. J. Cancer* **7:**75.

Rio, D., Robbins, A., Myers, R., and Tijian, R., 1980, Regulation of SV40 early transcription *in vitro* by a purified tumor antigen, *Proc. Natl. Acad. Sci. U.S.A.* **77:**5706.

Robb, J.A., 1977, Identification of simian virus 40 tumor and U antigens, *Proc. Natl. Acad. Sci. U.S.A.* **74:**447.

Rogers, M.J., Law, L.W., and Apella, E., 1977, Subcellular distribution of the tumor-specific transplantation antigen of SV40 transformed cells, *J. Natl. Cancer Inst.* **59:**1291.

Rowe, W.P., 1965, Studies of adenovirus–SV40 hybrid viruses. III. Transfer of SV40 genes between adenovirus types, *Proc. Natl. Acad. Sci. U.S.A.* **54:**711.

Rowe, W.P., and Baum, S.G., 1964, Evidence for a possible genetic hybrid between adenovirus type 7 and SV40 viruses, *Proc. Natl. Acad. Sci. U.S.A.* **52:**1340.

Rowe, W.P., and Baum, S.G., 1965, Studies of adenovirus–SV40 hybrid viruses. II. Defectiveness of the hybrid particles, *J. Exp. Med.* **122:**955.

Rubenstein, F.E., and Ginsberg, H.S., 1974, *Intervirology* **3:**170.

Russell, W.C., and Blair, G.E., 1977, Polypeptide phosphorylation in adenovirus-infected cells, *J. Gen. Virol.* **34:**19.

Schechter, N.M., Davies, W., and Anderson, C.W., 1980, Adenovirus coded deoxyribonucleic acid binding protein: Isolation, physical properties, and effects of proteolytic digestion, *Biochemistry* **19:**2802.

Schell, K., Lane, N.T., Casey, M.J., and Huebner, R.J., 1966, Potentiation of oncogenicity of adenovirus type 12 grown in African green monkey kidney cell cultures preinfected with SV40 virus: Persistence of both T antigens in the tumors and evidence for possible hybridization, *Proc. Natl. Acad. Sci. U.S.A.* **55:**81.

Schwarz, E., Reinke, C., Yamamoto, N., and zur Hausen, H., 1982, Terminal rearrangements in the genome of adenovirus type 12 mutants adapted to growth in two human tumor cell lines, *Virology* **116:**284.

Shiroki, K., and Shimojo, H., 1971, Transformation of green monkey kidney cells by SV40 genome: The establishment of transformed cell lines and the replication of human adenoviruses and SV40 in transformed cells, *Virology* **45:**163.

Siegel, S.E., Patch, C.T., Lewis, A.M., Jr., and Levine, A.S., 1975, Simian virus 40 DNA replication, transcription and antigen induction during infection with two adenovirus 2–SV40 hybrids that contain the entire SV40 genome, *J. Virol.* **16:**43.

Simmons, D.T., and Martin, M.A., 1978, Common methionine-tryptic peptides near the amino-terminal end of primate papovavirus tumor antigens, *Proc. Natl. Acad. Sci. U.S.A.* **75:**1131.

Solnick, D., 1981, Construction of an adenovirus–SV40 recombinant producing SV40 T antigen from an adenovirus late promoter, *Cell* **24:**135.

Soprano, K.J., Rossini, M., Croce, C., and Baserga, R., 1980, The role of large T antigen in SV40 induced reactivation of silent ribosomal RNA genes in human–mouse hybrid cells, *Virology* **102:**317.

Soprano, K.J., Jonak, G.J., Galanti, N., Floros, J., and Baserga, R., 1981, Identification of an SV40 DNA sequence related to the reactivation of silent ribosomal RNA genes in human and mouse hybrid cells, *Virology* **109:**127.

Staal, S.P., and Rowe, W.P., 1975, Enhancement of adenovirus infection in WI38 and AGMK cells by pretreatment of cells with 5-iododeoxyuridine, *Virology* **64:**513.

Sweet, B.H., and Hilleman, M.R., 1960, The vacuolating virus, SV40, *Proc. Soc. Exp. Biol. Med.* **105:**420.

Takemoto, K.K., and Mullarkey, M.F., 1973, Human papovavirus, BK strain: Biological studies including antigenic relationship to simian virus 40, *J. Virol.* **12:**625.

Tegtmeyer, P., 1972, Simian virus 40 deoxyribonucleic acid synthesis: the viral replican, *J. Virol.* **10:**591.

Tegtmeyer, P., Rundell, K., and Collins, J.K., 1977, Modification of simian virus 40 protein A, *J. Virol.* **21:**647.

Tegtmeyer, P., Schwartz, M., Collins, J.K., and Rundell, K., 1975, Regulation of tumor antigen synthesis by virus 40 gene A, *J. Virol.* **16:**168.

Tenen, D.G., Haines, L.L., and Livingston, D.M., 1982, Binding of an analog of the simian virus 40 T antigen to wild-type and mutant viral replication origins, *J. Mol. Biol.* **157:**473.

Tevethia, M.J., and Tevethia, S.S., 1976, Biology of SV40 transplantation antigen (TrAg). I. Demonstration of SV40 TrAg on glutaraldehyde fixed, SV40-infected African green monkey kidney cells, *Virology* **69:**474.

Tevethia, M.J., and Tevethia, S.S., 1977, Biology of simian virus 40 (SV40) transplantation antigen (TrAg). III. Involvement of SV40 gene A in the expression of TrAg in permissive cells, *Virology* **81:**212.

Tevethia, S.S., and Rapp, F., 1966, Prevention and interruption of SV40 induced transplantation immunity with tumor cell extracts, *Proc. Soc. Exp. Biol. Med.* **123:**612.

Tevethia, S.S., Flyer, D.C., and Tjian, R., 1980, Biology of SV40 transplantation antigen. VI. Mechanism of induction of SV40 transplantation immunity in mice by purified SV40 T antigen D2 protein, *Virology* **107:**13.

Thimmappaya, B., Weinberger, C., Schneider, R.J., and Shenk, T., 1982, Adenovirus VA1 RNA is required for efficient translation of viral mRNAs at late times after infection, *Cell* **31:**543.

Thummel, C., Tjian, R., and Grodzicker, T., 1981, Expression of SV40 large T antigen under control of adenovirus promoters, *Cell* **23:**825.

Thummel, C., Tjian, R., and Grodzicker, T., 1982, Construction of adenovirus expression vectors by site-directed *in vivo* recombination, *J. Mol. Appl. Genet.* **1:**435.

Thummel, C., Tijian, R., Hu, S.L., and Grodzicker, T., 1983, Translational control of SV40 T-antigen expressed from the adenovirus late promoter, *Cell* **33:**455.

Tjian, R., 1978, The binding site on SV40 DNA for a T-antigen related protein, *Cell* **13:**165.

Tjian, R., and Robbins, A., 1979, Enzymatic activities associated with a purified SV40 tumor antigen related protein, *Proc. Natl. Acad. Sci. U.S.A.* **76:**610.

Tjian, R., Fey, G., and Graessemann, A., 1978, Biological activity of purified simian virus 40 T antigen proteins, *Proc. Natl. Acad. Sci. U.S.A* **75**:1279.

Tornow, J., and Cole, C.N., 1983, Nonviable mutants of simian virus 40 with deletions near the 3′ end of gene A define a function for large T-antigen required after onset of viral DNA replication, *J. Virol.* **47**:487.

Van der Vliet, P.C., and Levine, A.J., 1973, DNA binding proteins specific for cells infected by adenovirus, *Nature (London) New Biol.* **246**:170.

Van der Vliet, P.C., Levine, A.J., Ensinger, M.J., and Ginsberg, H.S., 1975, Thermolabile DNA binding proteins from cells infected with a temperature-sensitive mutant of adenovirus defective in viral DNA synthesis, *J. Virol.* **15**:348.

Van Heuverswyn, H., Cole, C., Berg, P., and Fiers, A., 1979, Nucleotide sequence analysis of two simian virus 40 mutants with deletions in the region coding for the carboxyl terminus of T-antigen, *J. Virol.* **30**:936.

Van Roy, F., and Fiers, W., 1978a, Structure of the components of the adenovirus–SV40 hybrid population $Ad2^{++}$Hey and reciprocal molecular interactions, *J. Mol. Biol.* **126**:691.

Van Roy, F., and Fiers, W., 1978b, Interference with SV40 DNA replication by adenovirus type 2 during mixed infection of monkey cells, *J. Virol.* **27**:275.

Van Roy, R., Engler, F., and Fiers, W., 1979, Isolation and characterization of a specific deletion mutant of human adenovirus type 2, *Virology* **96**:486.

Van Roy, F., Fransen, L., and Fiers, W., 1981, Phosphorylation patterns of tumor antigens in cells lytically infected or transformed by SV40, *J. Virol.* **40**:28.

Walter, G., and Flory, P.J., Jr., 1980, Phosphorylation of SV40 large T antigen, *Cold Spring Harbor Symp. Quant. Biol.* **44**:165.

Walter, G., and Martin, H., 1975, Simian virus 40-specific proteins in HeLa cells infected with nondefective adenovirus 2–simian virus 40 hybrid viruses, *J. Virol.* **16**:1236.

Werner, G., and zur Hausen, H., 1978, Deletions and insertions in adenovirus type 12 DNA after viral replication in Vero cells, *Virology* **86**:66.

Westphal, H., 1981, Genomic arrangement of an adenovirus–SV40 hybrid virus, $Ad2^{+}$ND4del, *J. Virol.* **40**:526.

Westphal, H., Lai, S.-P., Lawrence, C., Hunter, T., and Walter, G., 1979, Mosaic adenovirus–SV40 RNA specified by the nondefective hybrid virus, $Ad2^{+}$ND4, *J. Mol. Biol.* **130**:337.

Zain, B.S., and Roberts, R.J., 1978, Characterization and sequence analysis of a recombination site in the hybrid virus $Ad2^{+}$ND1, *J. Mol. Biol.* **120**:13.

Zorn, G.A., and Anderson, C.W., 1981, Adenovirus type 2 expresses fiber in monkey–human hybrids and reconstructed cells, *J. Virol.* **37**:759.

CHAPTER 11

Adenovirus Infections in Humans

STEPHEN E. STRAUS

I. INTRODUCTION

Adenoviruses are ubiquitous agents that infect humans of all ages. The discovery of the first adenovirus types three decades ago paved the way for countless studies that continue to uncover additional viral strains and ever expand our comprehension of their significance to man. Whereas some human adenoviruses are being examined at the increasingly sophisticated molecular and cellular levels, the means by which they infect man, provoke illness, or are handled by the body's immune system remain very poorly understood.

This chapter represents an attempt to catalogue the known human adenovirus agents, to summarize the data pertaining to their acquisition and transmission, to review the range of illness with which they are associated, and to highlight those aspects of adenovirus infection that are in need of further investigation.

II. ADENOVIRUSES RECOVERED FROM HUMANS

The 41 distinct adenovirus types that have been recovered from humans thus far are listed in Table I. Most of these agents were isolated during an extremely fruitful decade of studies that followed the initial discovery of adenoviruses. A wide range of illnesses has been associated with the better-defined, lower-numbered virus types, but the predilection

STEPHEN E. STRAUS • Medical Virology Section, Laboratory of Clinical Investigation, National Institutes of Health, Bethesda, Maryland 20205.

TABLE I. Adenovirus Immunotypes[a]

Type	Prototype strain	Source	Patient diagnosis	Major associated diseases[b]
1	Ad71	Adenoid	Hypertrophied tonsils and adenoids	Respiratory
2	Ad6	Adenoid	Hypertrophied tonsils and adenoids	Respiratory
3	G.B.	Nasal washing	Common cold	Respiratory
4	RI67	Throat washing	Influenzalike or viral (atypical) pneumonia	Respiratory
5	Ad75	Adenoid	Hypertrophied tonsils and adenoids	Respiratory
6	Ton99	Tonsil	Hypertrophied tonsils and adenoids	Respiratory
7	Gomen	Throat washing	Pharyngitis	Respiratory
8	Trim.	Eye swab	Epidemic keratoconjunctivitis	Ocular
9	Hicks	Stool	Rheumatoid arthritis	—[c]
10	J.J.	Eye swab	Conjunctivitis	—
11	Slobitski	Stool	Polio	Cystitis
12	Huie	Stool	Polio?	—
13	A.A.	Stool	Healthy	—
14	DeWit	Throat swab	Acute respiratory illness	Respiratory
15	Ch38	Eye swab	Conjunctivitis	—
16	Ch79	Eye swab	Conjunctivitis	—
17	Ch22	Eye swab	Conjunctivitis	—
18	D.C.	Rectal swab	Niemann–Pick disease	—
19	587	Conjunctival scraping	Trachoma	Ocular
20	931	Conjunctival scraping	Trachoma	—
21	1645	Conjunctival scraping	Trachoma	Respiratory
22	2711	Conjunctival scraping	Trachoma	—
23	2732	Conjunctival scraping	Trachoma	—
24	3153	Conjunctival scraping	Trachoma	—
25	BP-1	Rectal swab	Healthy	—
26	BP-2	Rectal swab	Healthy	—
27	BP-4	Rectal swab	Healthy	—
28	BP-5	Rectal swab	Healthy	—
29	BP-6	Rectal swab	Healthy	—
30	BP-7	Rectal swab	Healthy	—
31	1315163	Stool	Healthy	—
32	H.H.	Rectal swab	Healthy	—
33	D.J.	Rectal swab	Healthy	—
34	Compton	Urine	Renal transplant	—
35	Holden	Lung, kidney	Renal transplant	—
36	275	Stool	Gastroenteritis	—
37	G.W.	Conjunctival scraping	Keratoconjunctivitis	Ocular, genital?

(Continued)

TABLE I. (*Continued*)

Type	Prototype strain	Source	Patient diagnosis	Major associated diseases[b]
38	No assignment	—	—	—
39	D335	Stool	Acute respiratory illness	—
40	Dugan	Stool	Gastroenteritis	Enteric
41	Tak	Stool	Gastroenteritis	Enteric

[a] Modified from Kasel (1979) with permission. The type status of recently identified adenoviruses is according to the current (October 1982) American Type Culture Collection nomenclature.
[b] Major infectious syndrome associated with each immunotype.
[c] Human disease not proved to be associated with the immunotype.

of these human strains for ocular and respiratory tissues is most prominent. The nature of the few more recently recognized adenoviruses remains uncertain, and it is possible that these strains represent variants or hybrids of previously known virus types, though some of the newer agents have been linked to illnesses with which other adenoviruses were never associated. Adenovirus type 37 (Ad37), for example, has been associated with urethritis and cervicitis and represents the first adenovirus for which the suggestion of sexual transmission has been seriously entertained (de Jong *et al.*, 1981). In addition, a number of unusually fastidious adenoviruses have been associated with diarrheal disease and perhaps respiratory disease as well in infants and immune-compromised patients (Retter *et al.*, 1979; Gary *et al.*, 1979; Yolken *et al.*, 1982). These recent observations challenge our complacency about adenoviruses and suggest that we have yet to complete the list of all medically important types and the diseases with which they are associated.

III. CLASSIFICATION OF HUMAN ADENOVIRUSES

A. General

All human adenoviruses are members of the genus *Mastadenovirus* of the family Adenoviridae to indicate their similarity to other vertebrate (but nonavian) adenoviruses (Norrby *et al.*, 1976). Human adenoviruses have been classified into types and subgroups in many different ways according to biological, chemical, immunological, or structural properties (Table II) (Rosen, 1960; Piña and Green, 1965; Norrby, 1969; Green, 1970; Green *et al.*, 1979a; Kasel, 1979; Wadell *et al.*, 1980a). The most widely employed subgrouping scheme is based on the ability of specific immunotypes to partially or completely hemagglutinate rat or rhesus erythrocytes (Rosen, 1960). Other subgroupings have been defined by percentage of guanine plus cytosine (GC) content (Piña and Green, 1965),

TABLE II. Some Subgroup Classifications for Human Adenoviruses[a]

Type	HA	GC	ONCO	PROT	*Sma*I
1	III	H	C	C	M
2	III	H	C	C	M
3	I	M	B	B	M
4	III	H	C	E	H
5	III	H	C	C	M
6	III	H	C	C	M
7	I	M	B	B	M
8	II	H	C	D	H
9	II	H	C	D	H
10	II	H	C	D	H
11	I	M	B	B	M
12	III	L	A	A	L
13	II	H	C	D	H
14	I	M	B	B	M
15	II	H	C	D	H
16	I	M	B	B	M
17	II	H	C	D	H
18	IV	L	A	A	L
19	II	H	C	D	H
20	II	H	C	D	H
21	I	M	B	B	M
22	II	H	C	D	H
23	II	H	C	D	H
24	II	H	C	D	H
25	II	H	C	D	H
26	II	H	C	D	H
27	II	H	C	D	H
28	II	H	C	D	H
29	II	H	C	D	H
30	II	H	C	D	H
31	IV	L	A	A	L
32	II	—	C	D	H
33	II	—	C	D	H
34	II[b]	—	B	B	M
35	I[c]	—	B	B	M
36	II[d]	—	—	—	H[d]
37	II[e]	—	—	D	—
38	—	—	—	—	—
39	II[f]	—	—	—	H[f]
40	—	—	C	F	H[g]
41	—	—	C[g]	G	H[g]

[a] (HA) Hemagglutination grouping (data from Kasel, 1979): (I) absent rat erythrocyte (RBC) HA, complete rhesus RBC HA; (II) complete rat RBC HA, complete or absent rhesus RBC HA; (III) partial rat RBC HA, absent rhesus RBC HA; (IV) absent rat RBC HA, absent rhesus RBC HA. (GC) Percentage guanine plus cytosine content (data from Piña and Green, 1965): (L) low (48–49%); (M) medium (50–52%); (H) high (57–60%). (ONCO) Oncogenicity (grouping data adapted from Green, 1970; Wadell *et al.*, 1980a). (PROT) Protein analyses (grouping adapted from Wadell, 1979; Wadell *et al.*, personal communication). (*Sma*I) Number of DNA fragments induced by cleavage with restriction endonuclease *Sma* I (data adapted from Wadell *et al.*, 1980a): (H) high number (≧14); (M) medium number (8–12); (L) low number (4–5).

[b] Hierholzer *et al.* (1975). [c] Myerowitz *et al.* (1975). [d] Wigand *et al.* (1980).

[e] De Jong *et al.* (1981). [f] Hierholzer *et al.* (1982).

[g] H. E. Takiff and S. E. Straus (unpublished observations).

morphology (Norrby, 1969), oncogenicity (Green, 1970), electrophoretic mobility of virion proteins (Wadell *et al.*, 1980a), genome homology (Green *et al.*, 1979a), and restriction endonuclease cleavage profiles (Wadell *et al.*, 1980a). A complete review of all classification schemes is beyond the scope of this chapter. In general, however, the various classification schemes are highly concordant and merely reflect different facets of relatively few biochemical properties shared by adenoviruses. There is also a limited tendency for virus types within the same subgroups to be associated with similar host-tissue tropisms and pathogenicity. Sections III. B–E summarize briefly the data regarding four of the more interesting or recent approaches to adenovirus subgrouping (Table II).

B. Oncogenicity

A number of human adenoviruses have been subgrouped according to their ability to induce tumors in newborn hamsters or to transform rodent cells *in vitro* (Huebner, 1967) (see Table II and Chapter 9). Subgroup A includes highly tumorigenic strains (serotypes 12, 18, and 31) and subgroup B includes weakly tumorigenic strains (serotypes 3, 7, 11, 14, 16, 21, 34, and 35), while the members of subgroup C (serotypes 1, 2, 5, 6, and others) are nontumorigenic strains capable of transformation only *in vitro*. Some have proposed dividing subgroup C into two distinct groups that differ in the antigenicity of some early proteins [tumor (T) antigens] that are expressed in the transformed cells (McAllister *et al.*, 1969). Since adenoviruses have not been implicated in the development of human neoplasms (see Sections VI, VII. I, and XI), and there are no general differences in the clinical expression of members of each subgroup, this scheme of classification has no current relevance to human disease.

C. Electrophoretic Mobility of Virion Polypeptides

The hemagglutinating, serological, and oncogenic properties of adenoviruses probably reflect features of relatively few of the viral proteins. Analyses of virion polypeptides by sodium dodecyl sulfate–polyacrylamide gel electrophoresis (SDS-PAGE) allow one to assess simultaneously the similarities or differences among many of the proteins of various adenovirus strains. Wadell has classified all known adenovirus immunotypes into seven subgroups [A–G (see Table II)] based on the apparent molecular weights of virion polypeptides as assessed by SDS-PAGE (Wadell *et al.*, 1980a; Wadell, 1979 and personal communication). Interestingly, the composition of these groups closely parallels that defined by oncogenicity. For example, Ad12, 18, and 31 are all highly oncogenic, and their polypeptides display similar electrophoretic patterns. Moreover, there appears to be some clinical relevance to this classification scheme.

Three members of subgroup D (Ad8, 19 and 37) are most associated with severe ocular infection and the members of subgroup C (Ad1, 2, 5, and 6) with commonly occurring mild respiratory illnesses of children, while strains assigned to (the tentatively defined) subgroups F (Ad40) and G (Ad41) are fastidious in cell culture and linked to diarrheal disease.

D. Genome Homologies

Most of the human adenovirus immunotypes have been grouped by methods that can detect significant differences or similarities in base sequences throughout the genome. By heteroduplex mapping and liquid-phase DNA–DNA hybridizations, 31 adenovirus types were placed in five subgroups (A–E) that correlate highly with their classifications by polypeptide analysis and oncogenicity (Garon *et al.*, 1973; Green *et al.*, 1979a). Similarly, the homology of enteric adenoviruses to other human adenoviruses has recently been examined (H. Takiff and S. Straus, unpublished observations). Strains of Ad40 and 41, which possess restriction patterns similar to those of (Wadell's) F and G subgroups, exhibit partial to high levels of DNA homology throughout the genome, but little homology (and that largely limited to the left half of the genomes) to other lower-numbered adenovirus types.

E. Restriction Endonuclease Analysis

A highly sensitive probe of adenovirus genome sequences involves agarose-gel electrophoresis of viral DNA digested by restriction endonucleases. Mutations in one DNA base will alter the cleavage profile of the genome if the change occurs within the recognition site for the enzyme. This technique has allowed the rather artificial grouping of adenoviruses according to the number of restriction endonuclease *Sma*I-generated cleavage segments, a reflection of GC content (Wadell *et al.*, 1980a). More important, restriction endonuclease analysis has become the definitive method for typing strains and for studying the molecular epidemiology of adenovirus outbreaks (Wadell and Varsanyi, 1978; Wadell and de Jong, 1980; Wadell *et al.*, 1980a, 1981). Endonuclease analysis has been most fruitfully applied to resolving problems pertaining to the latency, transmission, and reactivation of herpes simplex viruses (Buchanan *et al.*, 1978). Restriction endonuclease analysis has yet to be utilized to determine whether adenoviruses can undergo cycles of latency and reactivation.

IV. EPIDEMIOLOGY AND TRANSMISSION

Our current comprehension of the epidemiology and modes of transmission of adenoviruses derives largely from two major groups of studies.

The first group involved careful observations of naturally occurring or experimentally induced outbreaks of specific adenovirus-induced clinical syndromes such as pharyngoconjunctival fever (PCF), epidemic keratoconjunctivitis (EKC), and acute respiratory disease (ARD) of military recruits. The epidemiological features of these syndromes are discussed in Section VII together with the clinical description of the illnesses themselves. The second group of studies entailed a series of longitudinal community and institutional studies (Jordan, 1957; J. A. Bell *et al.*, 1961; Sterner, 1962; Hamre *et al.*, 1966; Fox *et al.*, 1969; Hall *et al.*, 1971; Cooney *et al.*, 1972; Brandt *et al.*, 1972; Foy and Grayston, 1976). These two groups of seroepidemiological and viroepidemiological studies established that adenoviruses are ubiquitous. These viruses can infect humans of all ages, races, and nationalities, and, other than for ARD of military recruits (who are primarily males), there is no apparent difference in attack rate according to sex. In addition, the combined data from these studies suggests that adenovirus types can be divided for clinical purposes into three overlapping groups: those types that are endemic particularly in early childhood, those that are epidemic, and those that are predominantly sporadic. All the known human adenoviruses have been recovered on one or more occasions from patients with a wide range of sporadically appearing illnesses.

The lower-numbered adenoviruses, Ad1, 2, 5, and 6, appear to be endemic in many areas of the world. In western Europe and the United States, between one third and one half of children have acquired antibodies to one or more of these adenovirus types by the age of 1 year (Fig. 1). (Huebner *et al.*, 1954; Sterner, 1962; Hall *et al.*, 1971; Foy and Grayston, 1976). The prevalence of adenovirus-specific neutralizing antibodies by age 3 has been estimated to be about 37% in Sweden, and about 80% in the United States, and close to 100% in Taiwan (Sterner, 1962; Hall *et al.*, 1971; Cooney *et al.*, 1972; Foy and Grayston, 1976). Institutionalized children tend to acquire adenoviruses at an even earlier age than children living at home (J. A. Bell *et al.*, 1961; Sterner, 1962). In general, conditions of crowding and poor hygiene lead to earlier acquisition of adenovirus-specific antibodies. Few individuals enter adult life lacking immunological experience with these endemic adenovirus types.

Ad3, 4, 7, 8, and occasionally other types, such as Ad11, 14, 19, and 21, tend to be associated with epidemic infections. Ad3, 4, and 7 are commonly associated with epidemics of PCF (J. A. Bell *et al.*, 1955, 1956; D'Angelo *et al.*, 1979). Ad4 and 7, and less often Ad11, 14, and 21 are associated with epidemic ARD of recruits (Hilleman *et al.*, 1954; Van der Veen and Kok, 1957; Tai *et al.*, 1960; Van der Veen *et al.*, 1969), while Ad8 and 19 are associated with EKC (Sprague *et al.*, 1973; O'Day *et al.*, 1976; D'Angelo *et al.*, 1981). More recently, outbreaks of gastroenteritis have been attributed to fastidious enteric adenoviruses (Gary *et al.*, 1979; Retter *et al.*, 1979; Yolken *et al.*, 1982).

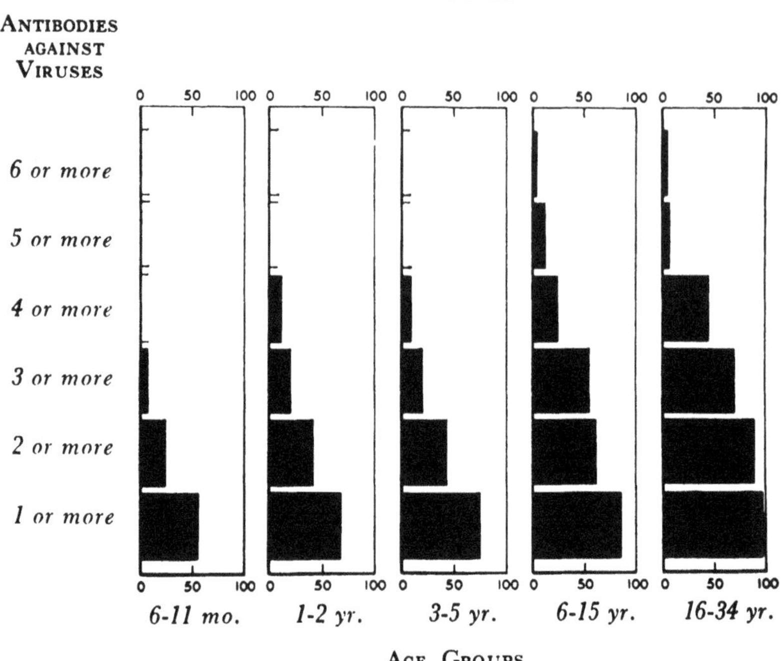

FIGURE 1. Percentage of Washington, D.C., area residents possessing antibodies to one or more of Ad1–6, by age. These data were acquired using a rather insensitive neutralization assay and are now thought to considerably underestimate the prevalance of seropositivity (R. Chanock, personal communication). Reprinted from Huebner *et al.* (1954) with permission.

Adenovirus infections have been documented in tropical (S. D. Bell, Jr., *et al.*, 1960; Kurian *et al.*, 1966) and temperate climates and occur throughout the year. Seasonal differences in infection rates are not as marked in the tropical settings as they are in temperate zones. Brandt *et al.* (1972) studied inpatient and outpatient children in Washington, D.C., over a 10-year period. Their data (Fig. 2) revealed two annual peaks in adenovirus-associated respiratory illness: midwinter and midsummer. Whereas there were more children from whom adenoviruses were recovered in the winter, the percentage of isolates that were believed to reflect respiratory illness was greater in the summer. Hilleman (1957) showed that adenovirus-induced ARD in military recruits is largely confined to one broad peak in the cooler months of the year [see Fig. 7 (Section VII. C.4)].

The fecal–oral route is thought to be the major means by which adenoviruses are transmitted, at least during childhood (Fox *et al.*, 1969). This belief is based on the observation that adenoviruses continue to be excreted for months or years following the initial infection (Table III). Chronic respiratory shedding occurs, but fecal shedding appears to be more intense and prolonged. Longitudinal studies of families in the New

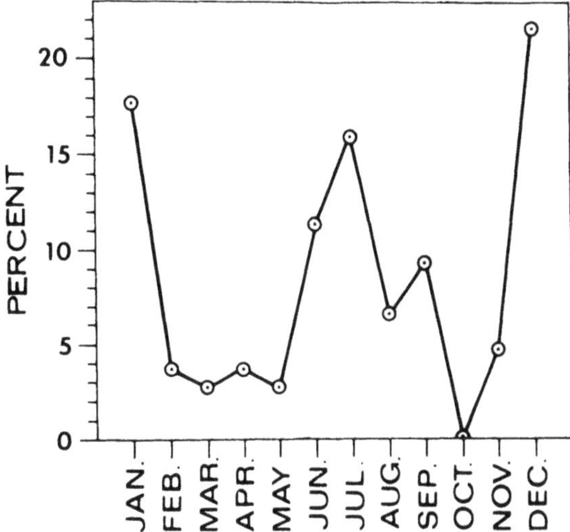

FIGURE 2. Percentage distribution of adenovirus-associated respiratory illness in inpatients, by month (12 months = 100%). Data reprinted from Brandt *et al.* (1972) with permission.

York Virus Watch documented a number of infants who became infected shortly after birth into families in which there were older siblings who chronically excreted the same virus type in their stools (Fig. 3).

ARD can be transmitted by oral instillation of infected respiratory secretions (Commission on Acute Respiratory Diseases, 1947a,b). PCF is thought to be transmissible by instillation of contaminated swimming pool water or other infected materials into the nose or eyes (J. A. Bell *et al.*, 1956; D'Angelo *et al.*, 1979). Direct conjunctival inoculation is im-

TABLE III. Percentage of Persons Who Excreted Adenoviruses for at Least the Numbers of Days Indicated following an Initial Infection[a]

Days	%
1	100
2	52
16	44
31	36
61	32
91	26
181	20
361	9
721	2

[a] Modified from Fox *et al.* (1969) with permission. Percentages are those of 133 patients from whom stool specimens remained positive.

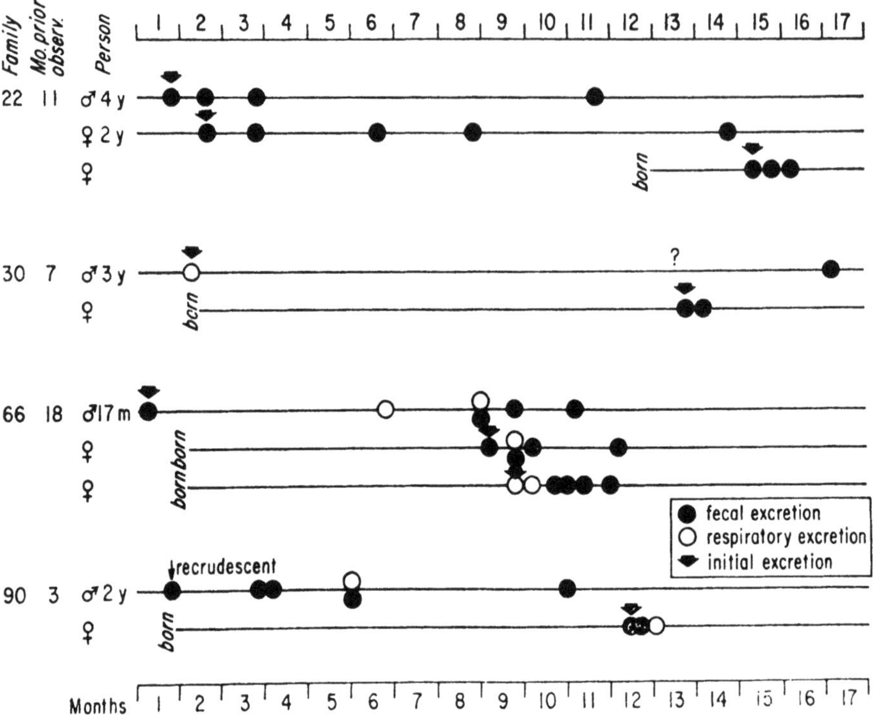

FIGURE 3. Spread of Ad2 infection to infants born into families in which older siblings were previously infected. Reprinted from Fox *et al.* (1969) with permission.

portant for transmission of Ad8 in cases of EKC (Dawson *et al.*, 1970; D'Angelo *et al.*, 1981).

The incubation period of adenovirus infections has been defined best in volunteer trials. In studies of PCF, J. A. Bell *et al.* (1955) estimated an incubation period of 5–10 days with a mean of about 7 days. Military studies of ARD suggest an incubation period of about 4–10 days with a mean of 5–6 days for that syndrome (Commission on Acute Respiratory Diseases, 1947a,b). Careful observations of natural outbreaks of ARD and of EKC indicated mean incubation periods of 7–10 days (Hilleman, 1957; Sprague *et al.*, 1973).

V. PATHOLOGY

The pathological alterations that attend typical adenovirus infections are not very well defined. Only the more severe infections have generally justified acquisition of tissue specimens. Most of the reported pathological studies have described the results of postmortem examinations of patients who succumbed to pulmonary, hepatic, or disseminated adenovirus infections. Becroft (1967), in a study of fatal adenovirus respiratory tract infection of infants, observed extensive destruction of bronchial epithelium and bronchial glands. Alveoli were distended with an eosinophilic exudate. All involved tissues displayed a cellular, predominantly mononuclear infiltration. Two types of nuclear changes were described: The first involved a diffuse accumulation of strongly Feulgen-positive basophilic or amphophilic material. The second entailed more discrete Feulgen-negative, amphophilic, or eosinophilic intranuclear inclusions. Similar patterns of tissue necrosis, mononuclear infiltration, and irregular amphophilic or basophilic inclusions were observed in an acutely infected liver (Aterman *et al.*, 1973). Myerowitz *et al.* (1975) reported similar pathological alterations in several organs in a patient with a fatal disseminated infection (Fig. 4). Electron-microscopic (EM) studies of involved tissues from fatal infections demonstrate typical paracrystalline arrays of adenovirus particles (Fig. 5).

Thus, much of the pathological alteration observed in adenovirus infections is mediated by direct tissue damage from productive virus replication in susceptible tissues (Becroft, 1971; Henson and Mufson, 1971; Aterman *et al.*, 1973). The results of studies in which animals were inoculated with large numbers of human adenoviruses have suggested that abortive infectious cycles may also participate in tissue injury, but there are no data to support this in the proper (human) host (Duncan *et al.*, 1978). There is also some evidence for cellular injury via immunopathological mechanisms. Chronic inflammatory sequelae of adenovirus-induced keratitis appear to evolve in the absence of cultivable virus and respond to corticosteroid therapy (Laibson *et al.*, 1970). It has been postulated that persistent viral antigen–antibody complexes incite an inflammatory response that is capable of injury to the cornea (Boniuk *et al.*, 1965).

VI. PATHOPHYSIOLOGY AND IMMUNE RESPONSES

Since there are no satisfactory animal models of human adenovirus infection, our understanding of what occurs following initial infection is sketchy and derives largely from speculation or extrapolation from better-defined virus–host systems. It is presumed that adenoviruses establish productive infections in gastrointestinal, respiratory, or ocular epithelial cells (Fishaut *et al.*, 1980). The lower-numbered adenovirus types are commonly recovered from ocular and respiratory tissues. The higher-numbered types (Ad25–41) have almost always been recovered from stools (see Table I). For these latter types, the major and primary site of infection may be in the gastrointestinal tract. From the primary sites of infection, the newly replicated virus may then enter the bloodstream and spread to other body tissues, where the ultimate events will depend on a complex interplay of host and viral factors (Andiman *et al.*, 1977). It is clear that different adenoviruses possess unique tissue tropisms. Most cases of severe lower respiratory or disseminated infections have involved

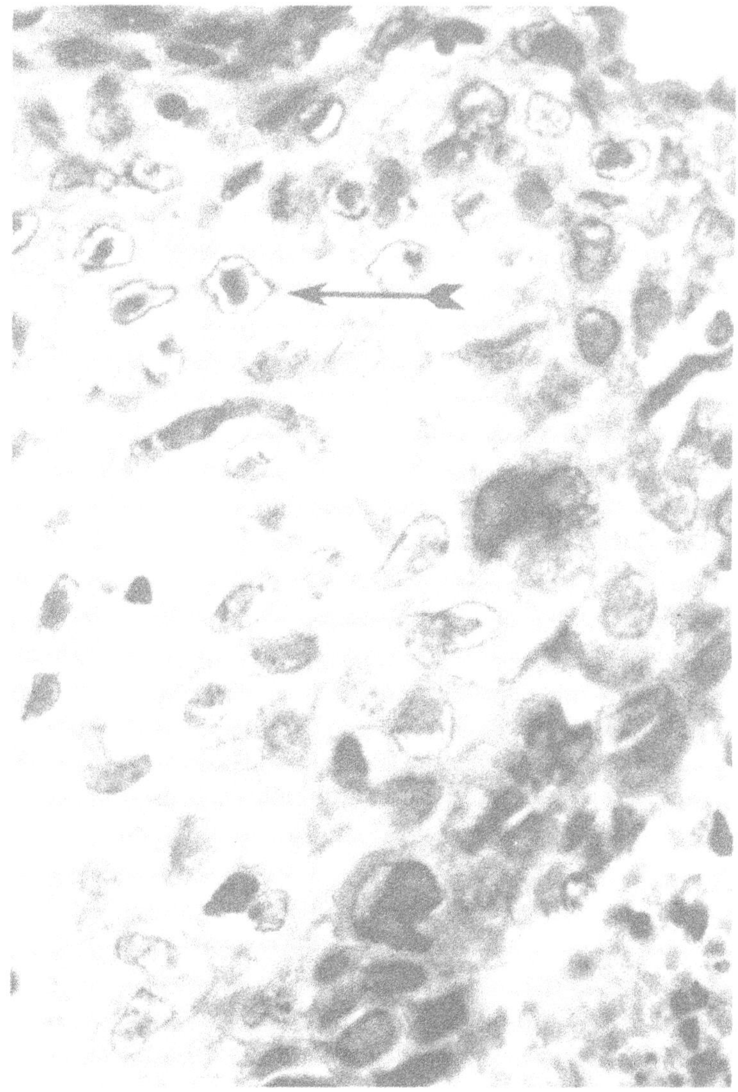

FIGURE 4. Photomicrograph of esophageal epithelium of a patient with disseminated Ad34 infection demonstrating cytoplasmic swelling and haloed intranuclear inclusions (←). Original magnification: ×400. Reprinted from Myerowitz *et al.* (1975) with permission.

Ad1–5, 7, and 21 (Lang *et al.*, 1969; Henson and Mufson, 1971; Dudding *et al.*, 1972; Aterman *et al.*, 1973; Field *et al.*, 1978; Carmichael *et al.*, 1979; Zahradnik *et al.*, 1980). Ocular infections are common with Ad3, 4, 7, 11, and 37 and are likely to be severe with Ad8 and 19 (Dawson *et al.*, 1970; O'Day *et al.*, 1976). Some Ad7 strains may be particularly neurotropic (Gabrielson *et al.*, 1966; Huttunen, 1970; Simulä *et al.*, 1970).

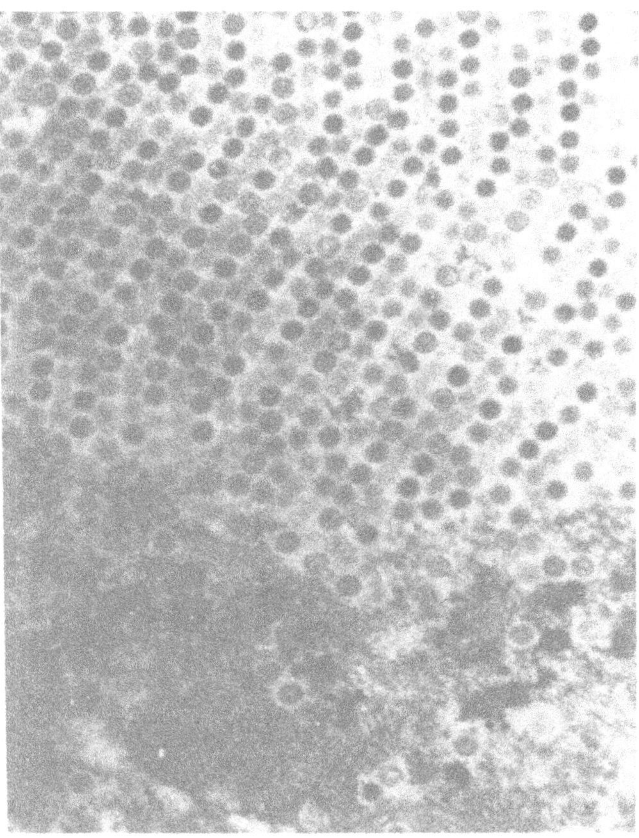

FIGURE 5. Electron micrograph of a pulmonary alveolar cell from the patient in Fig. 4 demonstrating 80-nm particles in a paracrystalline array. Original magnification: ×46,500. Reprinted from Myerowitz *et al.* (1975) with permission.

Immune clearance of infecting virus presumably requires successful interactions among macrophages, lymphocytes, natural killer cells, and humoral factors including antibody, complement, interferon, and other lymphokines (Fishaut *et al.*, 1980). Although there is some evidence that humoral responses modify the course of adenovirus infections (McCormick *et al.*, 1972; Siegal *et al.*, 1981), it is obvious that cell-mediated immune mechanisms are of paramount importance for the containment and resolution of these infections. Most cases of devastating adenovirus infections have been documented in individuals devoid of adequate cellular immunity, including neonates, immune-depressed cancer or transplant patients, or children with congenital thymic aplasia or other T-cell disorders (Henson and Mufson, 1971; Aterman *et al.*, 1973; Hierholzer *et al.*, 1975a; Myerowitz *et al.*, 1975; Andiman *et al.*, 1977; Carmichael *et al.*, 1979; Zahradnik *et al.*, 1980).

Infections with adenoviruses are analogous to infections with herpesviruses, which are better understood in terms of host–virus interactions. The ultimate outcome of infection with either virus is dependent on cellular immunity. Importantly, members of both virus families induce persistent or latent infections of human tissues. Herpesviruses and adenoviruses can be shed chronically and asymptomatically, indicating a failure of immune clearance. Unlike the case with the herpesvirus, however, it is not known whether adenoviruses can reactivate from a latent infectious state and induce recurrent disease. In addition, the results of *in vitro* and animal studies of viral transformation have led to speculation regarding the potential of both kinds of viruses for human oncogenesis, but the evidence has been far less supportive for such an association with adenoviruses than for an association with herpesviruses (Huebner, 1967; McAllister *et al.*, 1969; Green, 1970).

VII. INFECTIOUS SYNDROMES ASSOCIATED WITH ADENOVIRUSES

A. General

Among the many adenovirus serotypes, only about a dozen (including Ad1–8, 11, 14, 19, 21, and 37) have been proved to produce defined clinical syndromes (Chanock, 1974). This is not to say that the remaining serotypes are devoid of pathogenic potential. Rather, proving causality for those serotypes has been difficult for two reasons: First, many adenovirus-associated infections are subclinical (Hilleman, 1957; J. A. Bell *et al.*, 1961; Forsyth *et al.*, 1964; Fox *et al.*, 1969). Second, it is not uncommon for adenoviruses, especially the lower-numbered serotypes, to be persistently shed for months or years following an initial infection (see Table III) (Fox *et al.*, 1969). Thus, recovery of an adenovirus from an individual with a particular clinical syndrome does not imply a causal relationship between the syndrome and that virus. It must be demonstrated that individuals with a specific illness are statistically more likely to be shedding a specific virus type than are well controls. This has been feasible only in challenge studies, in very large population studies, and with epidemics of certain infectious syndromes.

B. Asymptomatic Infections

The proportion of adenovirus infections that result in clinical illness has been estimated in a number of studies involving continuing surveillance for infection and illness in defined populations (J. A. Bell *et al.*, 1961; Fox *et al.*, 1969). The onset of illness in close temporal relationship to the first detection of excretion of virus has been considered to be one measure of viral pathogenicity. In the Junior Village study, 53% of children who showed a rise in homologous neutralizing-antibody titer in association with shedding of virus had a febrile illness (J. A. Bell *et al.*, 1961). In a study of Marine Corps recruits at Camp Lejeune, North Carolina, who were commonly infected with Ad4, first recovery of virus within the week of onset of illness was assumed to be evidence for pathogenicity. Forsyth *et al.* (1964) estimated that about 48% of infected recruits experienced a compatible clinical illness. In another study, Hilleman (1957) showed that as many as 80% of military recruits acquired adenovirus infections (particularly Ad4 and 7) during basic training. About half these infections were asymptomatic. In the Virus Watch Program in New York City, in which surveillance cultures were taken every other week, first recovery of virus between 14 days before and 7 days after onset of an illness was considered evidence of pathogenicity. By comparison with the incidence of first recovery of virus in the absence of clinical illness, it was estimated that a maximum of 45% of adenovirus infections were associated with symptoms (Fox *et al.*, 1969). This estimate may be further reduced by additional data generated in the Virus Watch Study. Other agents, predominantly rhinoviruses, were also recovered from many of the subjects in close temporal relationship to illnesses. Assuming that these other agents were primarily responsible for the symptoms observed, it was estimated that as few as 15% of all adenovirus infections were symptomatic.

The differences in the aforecited estimates of the proportion of asymptomatic infections may be attributed, in part, to the system employed in each study to define illness. In the Junior Village Study, temperatures were recorded for all subjects twice daily and brief physical examinations were performed daily. Minor illness could be readily detected. In the New York Virus Watch Program, only those children whose parents recognized a problem and documented a fever were followed. For this reason, the rate of symptomatic illness may be underestimated in this latter study (15–45%).

The proportion of asymptomatic adenovirus infections is also dependent on serotype. In a controlled study in Washington, D.C., involving over 900 pediatric respiratory patients with adenovirus infections, it was estimated that among those with throat or anal swab isolates or both, adenoviruses were causally related to respiratory illness in 32% overall, including 47% with Ad1, 20% with Ad2, 61% with Ad3, 30% with Ad5, 93% with Ad6, and 43% with Ad7 infection (Brandt *et al.*, 1969).

C. Respiratory Infections

1. Spectrum

Data on the relationship between adenoviruses and symptomatic illness give one measure of the pathogenic potential of these agents. An-

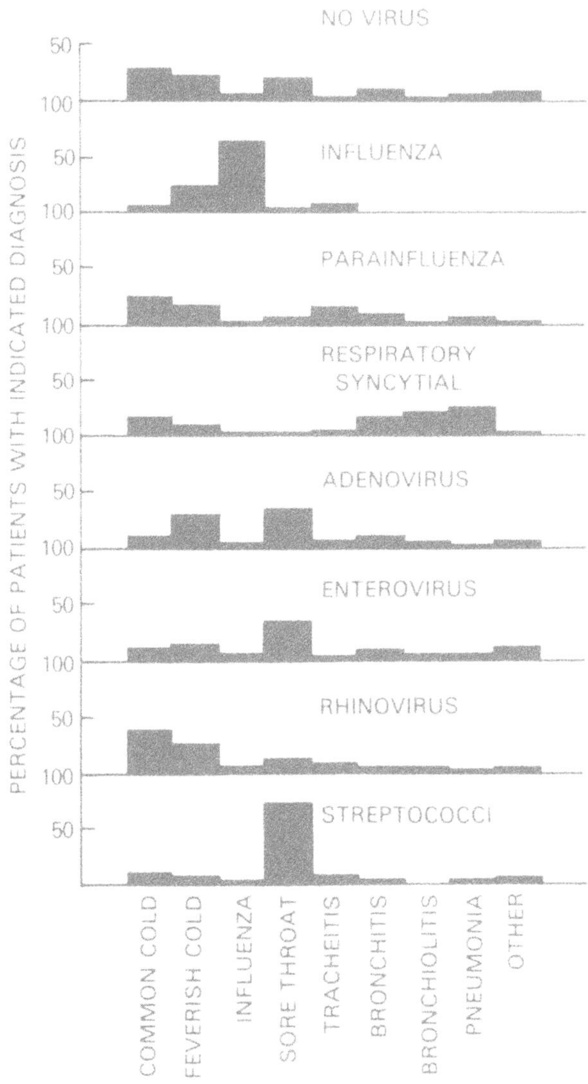

FIGURE 6. Relative frequency of various clinical syndromes among patients infected with respiratory pathogens. Redrawn from the Medical Research Council Working Party on Acute Respiratory Virus Infections (1965) report with permission.

other way of examining this problem is to assess the proportions of various clinical syndromes that are associated with adenovirus infection.

Adenoviruses have been clearly shown to contribute significantly to the total community morbidity, especially from respiratory infections. A Medical Research Council (1965) study in Britain demonstrated that in the general community, adenoviruses are associated with the entire spectrum of respiratory illness, but are important causes of relatively mild illnesses such as febrile colds and sore throats (Fig. 6). In Tecumseh, Michigan, adenoviruses accounted for about 4.5% of pathogens recovered from patients of all ages with respiratory infections (Monto and Ullman, 1974). During a decade of study of infants and children at D.C. Children's Hospital, adenovirus was recovered from throat or anal swab specimens or both from 12.4% of 6700 respiratory outpatients and 10.2% of 4700 respiratory inpatients (Brandt *et al.*, 1969). On the whole, the adenovirus group ranked third behind respiratory syncytial virus and parainfluenza viruses as a cause of pediatric respiratory disease (Kim *et al.*, 1979).

2. Coryzal Syndromes

Monto and Ullman (1974) reported that the average number of upper respiratory illnesses experienced each year decreases with age. Averages of 6, 4, 2, and 1 infection/year were recorded for individuals aged 0–2, 3–9, 10–39, and 40 or more, respectively. Coryzal syndromes consisting of rhinorrhea, nasal congestion, and sneezing accounted for most of the illnesses at all ages. Hamre *et al.* (1966) showed that medical students experience almost 2 upper respiratory infections per year, of which 83% are coryzal. Only a few (2–4%) of all coryzal illnesses are associated with adenoviruses; the rhinoviruses, coronaviruses, parainfluenza viruses, and respiratory syncytial virus being more prevalent.

3. Pharyngoconjunctival Fever

Parrott *et al.* (1954) reported that a syndrome characterized by fever, sore throat, and conjunctivitis, was commonly associated with adenoviruses. The illness is a relatively mild one that follows a 5- to 7-day incubation period. Fever, lasting 4–5 days, seldom exceeds 102–103°F. Headache, malaise, anorexia, and sore throat dominate the early phase of the illness. Hoarseness, nonproductive cough, and dull chest pain become evident later in the illness. Unilateral or bilateral follicular conjunctivitis may be present. Examination of the pharynx reveals mild erythema, tonsillar enlargement, and a modest exudate. The syndrome is distinguished from streptococcal pharyngitis in several regards. Adenovirus infection is more likely to result in constitutional symptoms, coryza, a mild tracheobronchitis, conjunctivitis, and preauricular adenopathy. The β-hemolytic streptococci tend to produce a more severe pharyngitis, more intense erythema, and more extensive exudates and do not cause conjunctivitis. An elevated leukocyte count with a shift toward immature forms is more characteristic of streptococcal disease. One unique and characteristic epidemiological feature of pharyngoconjunctival fever (PCF) has been its appearance in common-source outbreaks of individuals exposed by bathing in inadequately chlorinated swimming pools (D'Angelo *et al.*, 1979). The illness observed in the swimming-pool outbreaks is very similar to that observed in non-water-related PCF cases (Table IV). In one recent epidemic involving 72 individuals with PCF,

TABLE IV. Pharyngoconjunctival Fever: Swimming-Pool Outbreak[a]

Signs and symptoms	N (%)
Fever	50 (69)
Conjunctivitis	49 (68)
Sore throat	31 (43)
Headache	31 (43)
Runny nose	13 (18)
Chills	12 (17)
Myalgias	8 (11)
Vomiting	7 (10)
Earache	7 (10)
Nausea	6 (8)
Arthralgia	6 (8)
Cough	5 (7)
Swollen lymph nodes	5 (7)
Diarrhea	5 (7)
Rash	0 (0)
Sputum	0 (0)

[a] Modified from D'Angelo *et al.* (1979) with permission.

D'Angelo *et al.* (1979) reported that acquisition of adenovirus infection correlated with the amount of time spent in the water and that the virus could be recovered from the inadequately chlorinated pool water. The viral infection apparently can be acquired by nasal or oral aspiration of contaminated pool water or by direct conjunctival inoculation (J. A. Bell *et al.*, 1956).

4. Acute Respiratory Disease of Recruits

Among the first clinical syndromes attributed to adenoviruses was that of an epidemic respiratory infection occurring in closed populations, most notably among new military recruits. The history of this entity is a fascinating one that has been reviewed extensively (Dingle and Langmuir, 1968).

Briefly, it was recognized for many years that epidemics of respiratory illness befall companies of new recruits, but not of seasoned troops, and that similar problems occur in schools. Stuart-Harris *et al.* (1938) termed the illness "febrile catarrh," postulating it to be distinct from influenza. In the winter of 1941–1942, an epidemic of an illness labeled "primary atypical pneumonia" was described in Camp Claiborne, Louisiana. To further define this entity and to better understand the nature of the illness that produced such extensive morbidity in new recruits, the Surgeon General of the U.S. Army created the Commission on Acute Respiratory Diseases (ARDs). The studies of ARD by the commission during and following World War II formed the basis of our current comprehension of this entity. The commission's studies defined an epidemic illness of recruits that was most prevalent in the winter months, resulted in sustained immunity, and could be transmitted to volunteers by bacteria-free filtered respiratory secretions (Commission on Acute Respiratory Diseases, 1947a,b).

Studies in the early 1960s showed that *Mycoplasma pneumoniae* was one cause of the primary atypical pneumonia syndrome recognized during World War II (Chanock *et al.*, 1962). Hilleman and Werner (1954) reported that Ad4 was a major cause of the other respiratory problem of recruits (ARD). Analysis of paired sera saved from World War II studies documented that most of the ARD was associated with agents serologically related to Ad4. It is now known that AD3, 4, 7, 14, and 21 are the major causes of ARD in military recruits around the world (Van der Veen and Kok, 1957; Tai *et al.*, 1960; Van der Veen *et al.*, 1969). The predictability and impact of ARD in military troops led to the development of an effective live enteric vaccine that is now widely utilized (see Section X.C).

ARD is usually a mild illness. Hilleman (1957) showed that up to 80% of susceptible new recruits became infected, with most cases occurring during the cooler months (Fig. 7). About half the infected recruits were asymptomatic. Most of the remainder experienced a mild illness that interfered with training. About 20% of all recruits were admitted for ARD because on American military bases a fever higher than 100°F necessitated hospitalization.

The incubation period of ARD is 5–7 days, with a gradual onset of fever and chills over 2–3 days (Fig. 8). The symptoms are commonly accompanied by headache, malaise, and anorexia. Many individuals manifest upper respiratory signs of nasal discharge, mild sore throat, and a nonproductive cough. Some patients possess pharyngeal injection and tonsillar and cervical adenopathy. Fever is generally of a low grade and lasts for 2–4 days. Constitutional symptoms are similarly of a mild nature, but a postviral "neurasthenia" and troublesome cough may persist for 1–2 weeks or longer. Most laboratory tests are within normal limits except in rare, severely ill patients.

5. Pneumonia

As indicated, adenovirus infections most commonly involve the upper respiratory tract. However, lower respiratory tract infection with frank pneumonia is observed with some frequency in special settings. Infants, military recruits, and immune-deficient patients account for most documented cases of adenovirus pneumonia. Ad3, 4, 7, and 21 have been most prevalent in adults. There are no pathognomonic features of adenovirus pneumonias that permit them to be distinguished from other viral pneumonitides on purely clinical grounds.

The natural history of adenovirus pneumonia has been best defined in military personnel (see Section VII.C.4). Because mild fever and ina-

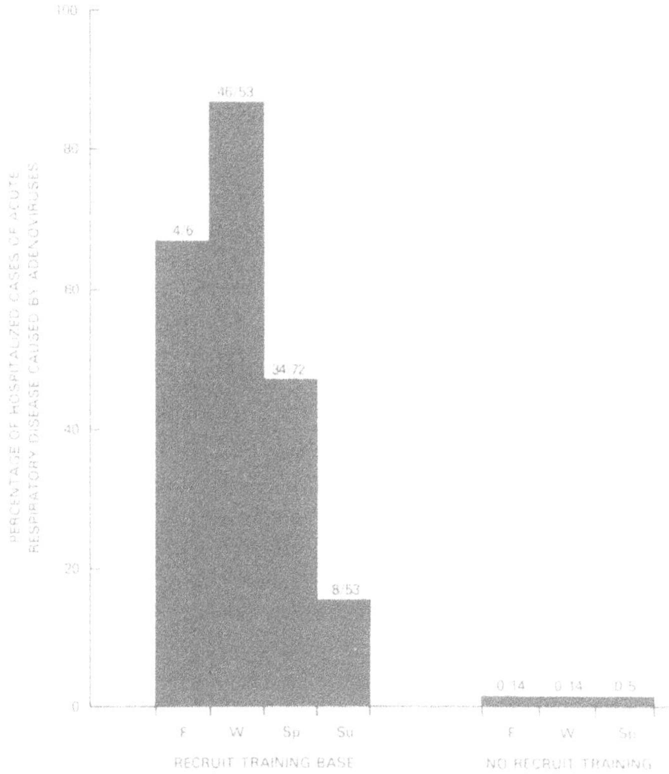

FIGURE 7. Percentage of hospitalized cases with ARD caused by adenoviruses at recruit training bases or other bases, by season. Redrawn from Hilleman (1957) with permission.

bility to perform assigned duties necessitate hospitalization of American military personnel, the entire spectrum of adenovirus pneumonias has been observed. Dascomb and Hilleman (1956) showed that as many as 20% of all hospitalized adenovirus-infected recruits develop objective clinical and radiological features of adenovirus pneumonia (Fig. 8). The signs and symptoms associated with adenovirus pneumonia in 12 recruits are shown graphically in Figs. 9 and 10 (Bryant and Rhoades, 1967). Pneumonic infiltrates are usually limited to one or a few lower pulmonary segments. Excellent recovery is expected without significant chronic pulmonary sequelae (Klocke *et al.*, 1966). In some individuals, the pneumonias have been more aggressive, and rare fatalities have been observed (Dudding *et al.*, 1972).

The appearance of adenovirus pneumonia in the general medical community is in striking contrast to the military experience. The cases tend to be sporadic rather than epidemic and involve infants more often than adults (Lang *et al.*, 1969; Henson and Mufson, 1971; Field *et al.*, 1978). Because hospitalization of patients in the general community is usually dictated by significant illness, the clinical features of adenovirus

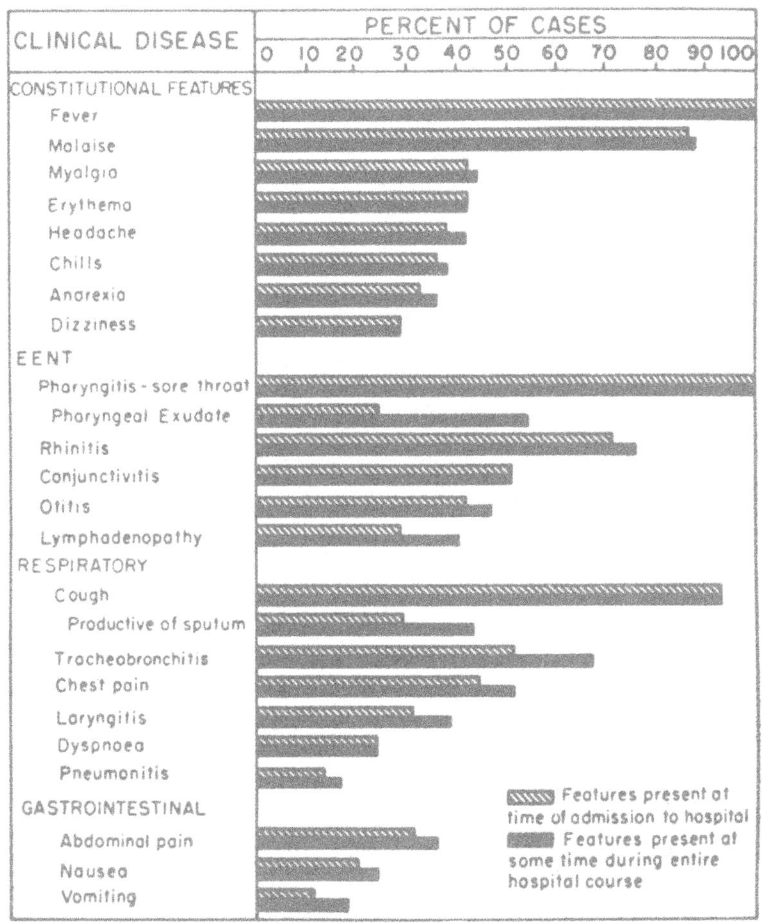

FIGURE 8. Clinical features of 45 military recruits hospitalized with acute respiratory disease. Reprinted from Dascomb and Hilleman (1956) with permission.

pneumonias in that population are rather more dramatic. Lang *et al.* (1969) reviewed the histories of 25 infants with Ad21 pneumonia. Most patients demonstrated moderate to severe dyspnea at the time of admission. Some of the more severely ill patients experienced a protracted illness lasting several weeks with exacerbations attributable to secondary bacterial infections. The acute radiological abnormalities were quite variable, including diffuse or patchy bronchial, peribronchial, or interstitial infiltrates. Significant pleural effusions were not observed. Ultimately, 2 (8%) of these 25 children died, while 8 (32%) showed full recovery. Importantly, 15 (60%) developed bronchiectasis or other chronic pulmonary sequelae.

Becroft (1971) reported histological findings in five children who died of acute Ad7 pneumonia. Their lungs displayed varying degrees of necrotizing bronchitis, bronchiolitis, and pneumonia. As in the adult and

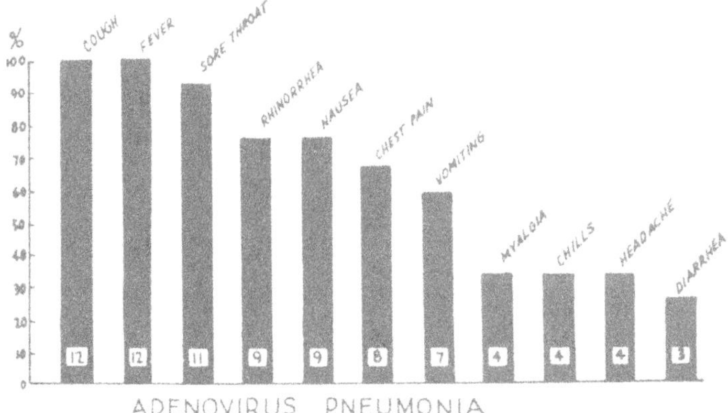

FIGURE 9. Frequency of symptoms in 12 recruits with adenovirus pneumonia. Reprinted from Bryant and Rhoades (1967) with permission.

military patients who succumbed to adenovirus pneumonia, the lungs of these infants also showed characteristic basophilic or amphophilic intranuclear inclusions [see Fig. 4 (Section V)]. The prominent feature of adenovirus pneumonia is ulceration and scarring of bronchial epithelium with destruction of bronchial glands. Becroft (1971) postulated that these changes contribute directly to the high incidence of chronic pulmonary sequelae observed in the children who survived infantile adenovirus pneumonia. Bronchiolar obliteration and bronchiectasis were demonstrated histologically in 5 of 25 survivors of Ad21 pneumonia (Lang *et al.*, 1969).

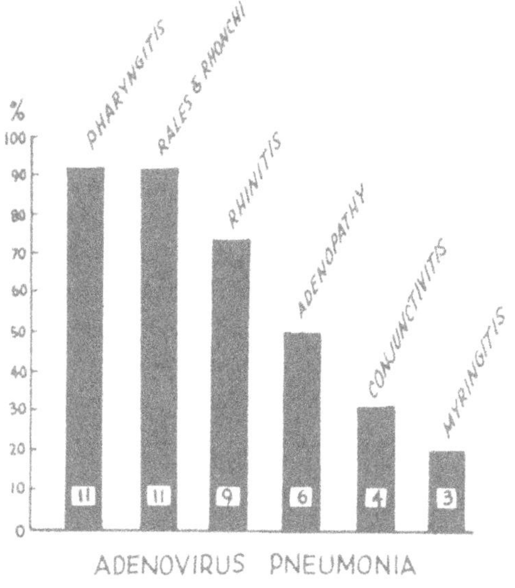

FIGURE 10. Frequency of physical findings in 12 recruits with adenovirus pneumonia. Reprinted from Bryant and Rhoades (1967) with permission.

Thus, adenovirus pneumonia may be a major cause of postinfectious bronchiectasis of childhood. Presumably, the unique vulnerability of developing airways of infants permits expression of chronic pulmonary sequelae that are not readily apparent in adults who have recovered from adenovirus pneumonia.

6. Pertussis Syndrome

Pertussis represents a classic syndrome that begins rather inauspiciously as a minor upper respiratory illness (Wilkins and Wehrle, 1979) After about 1–2 weeks, bronchial involvement becomes apparent with the onset of paroxysmal bouts of coughing that are terminated by vigorous inspirations resulting in a characteristic whooping sound. This syndrome is primarily caused by *Bordetella pertussis*, with occasional cases caused by the related bacteria *B. parapertussis* and *B. bronchiseptica*. A common laboratory observation in this syndrome is a profound lymphocytosis. Despite good immunization practices, pertussis has not been totally eradicated, suggesting that the immunization is of inadequate efficacy or that other organisms also cause pertussis. In the context of this latter hypothesis, adenoviruses have been implicated in some patients with a pertussislike syndrome. Connor (1970) reported 13 infants and children with pertussis, most of whom lacked microbiological or serological evidence for *Bordetella* infection. Of these 13 children, 11 shed Ad1, 2, 3, or 5. Sera from 7 of the children also showed rising titers of adenovirus antibodies. Klenk *et al.* (1972) reported 19 children with the pertussis syndrome. *Bordetella* species were recovered from 14 patients, 7 of whom had virological or serological evidence of concomitant adenovirus infections. Nelson *et al.* (1975) studied 134 patients with the pertussis syndrome. Adenoviruses were recovered from 22 patients, while 8 additional patients exhibited at least 4-fold rise in antibody titers to adenoviruses (Table V). Adenovirus infections were more common in patients from whom *Bordetella* species were also recovered.

The role of adenoviruses in the pertussis syndrome remains uncertain. The syndrome is not reproduced in volunteer studies. As indicated earlier, and as was correctly concluded by Nelson *et al.* (1975) and Baraff *et al.* (1978), the adenoviruses that are isolated may have been reactivated endogenously, or chronically shed, and may have played no part in the etiopathogenesis of pertussis. Alternatively, adenoviruses may be capable of provoking a pertussislike syndrome when acting alone or in combination with *Bordetella* species or other agents.

D. Adenovirus Infections in the Immune-Suppressed Patient

Life-threatening localized or disseminated adenovirus infections are uncommon and involve primarily the lower-numbered adenovirus ser-

TABLE V. Numbers of Patients with the Pertussis Syndrome and of Healthy Controls from Whom Various Viruses Were Isolated[a]

Virus	Pertussis (N = 134)	Control (N = 101)
Adenovirus	22	5
Coxsackie A	2	1
Coxsackie B	2	0
Poliovirus	8	1
Parainfluenza	5	0
Herpes simplex	3	4
ECHO	6	14
No viruses	90	76
TOTAL:	138[b]	101

[a] Modified from Nelson *et al.* (1975) with permission.
[b] More than one virus was isolated from 4 patients.

otypes. The progression of an adenovirus infection to the point at which life is endangered is most likely to take place in the setting of immune deficiency. Many of the fatal adenovirus infections have occurred in infants and young children with an immature or congenitally deficient immune system or with coexisting malnutrition (Aterman *et al.*, 1973; Andiman *et al.*, 1977; Zahradnik *et al.*, 1980). Additional fatal adenovirus infections have developed in adults or older children who are immune-compromised by virtue of either underlying diseases or immunosuppressive treatments involving steroids, cytotoxic drugs, or radiation (Zahradnik *et al.*, 1980; Carmichael *et al.*, 1979).

The lung is the most frequent target of severe adenovirus infections in the immune-compromised host, but disseminated infection involving multiple organ systems, particularly the liver, occurs as well. It is presumed that involvement of other organs follows a viremic event (Andiman *et al.*, 1977). In the largest series of immune-deficient patients with adenovirus infection, Zahradnik *et al.* (1980) described 15 individuals ranging in age from 2 days (premature infant) to 65 years. In this series, 7 patients had underlying malignancies, 3 were recipients of bone marrow or kidney transplants, and 4 were treated with steroids or cytotoxic drugs for other reasons. Eighty percent of the patients had severe systemic symptoms and pneumonia, while moderate to severe hepatitis occurred in nearly half. Nine patients died. At autopsy, adenovirus involvement was most commonly demonstrated in the lung and liver. One feature shared by 5 of these patients and many other immune-deficient individuals with viral infections is the presence of simultaneous infection with multiple organisms. Coexisting infections with fungi, bacteria, pneumocystis, and even cytomegalovirus make it difficult to determine how much the adenoviruses contributed to the total morbidity and mortality.

An understanding of the role of adenovirus in the immune-deficient host is confounded because recovery of the virus from respiratory secretions and even from lung tissue is not sufficient evidence for active infection. Recovery of virus from other viscera or demonstration of viral inclusions in the tissues, or both, are more indicative of active infection.

E. Keratoconjunctivitis

1. General

Ocular involvement during respiratory infection with adenovirus is common. Thus, conjunctivitis is a defining feature of classic PCF (see Section VII.C.3). Other patterns of adenovirus-induced eye infection are also recognized. Inflammation of the cornea (keratitis) accompanies conjunctivitis in sporadic cases of infection with certain adenoviruses including Ad3, 11, and 37 (Tai *et al.*, 1974; Yin-Murphy *et al.*, 1974; Schaap *et al.*, 1979).

2. Epidemic Keratoconjunctivitis

The keratitis that occasionally accompanies the aforementioned sporadic infections must be distinguished from that which occurs in a more virulent form of ocular adenovirus infection, termed "epidemic keratoconjunctivitis" (EKC). Typical epidemics of severe keratoconjunctivitis were first described nearly a century ago (Fuchs, 1889) and in 1955 were shown to be associated with Ad8 (Javetz *et al.*, 1955). This syndrome is distinguished by its greater ocular morbidity, its potential for chronicity and permanent visual impairment, its unique epidemiological setting, and its association with Ad8, 19, and rarely others (Dawson *et al.*, 1970; Sprague *et al.*, 1973; O'Day *et al.*, 1976).

Direct inoculation of virus into the eye and ocular trauma appear to predispose to EKC. The prevalence of antibodies to Ad8 or 19 in the general community is low, so that most individuals are considered susceptible (Sprague *et al.*, 1973; O'Day *et al.*, 1976). Proven outbreaks, each involving as many as 200 individuals, have been associated with industrial settings in which an eye injury is common and with eye-care facilities. A typical case would involve an industrial worker who presents to the plant nurse with the complaint that a foreign particle has lodged in his eye. The nurse removes the particle, but 2 weeks later the typical features of EKC appear. Careful investigation would reveal that the eyewash solution or the nurse's hands or medical instruments were contaminated with adenoviruses.

The clinical features of the illness that this worker may expect to endure have been well defined (Dawson *et al.*, 1970; Sprague *et al.*, 1973; O'Day *et al.*, 1976; D'Angelo *et al.*, 1982) (Table VI). Following an in-

TABLE VI. Signs and Symptoms of Epidemic Keratoconjunctivitis[a]

Signs and Symptoms	Patients (%)
Redness	98.4
Discharge	90.6
Eye pain	70.3
Visual change	70.3
Light sensitivity	68.2
Swollen lymph nodes	38.5
Headache	26.0
Fever	14.1
Sore throat	12.0
Cough	7.3
Nausea or vomiting	5.7
Diarrhea	3.1

[a] Modified from D'Angelo *et al.* (1981) with permission.

cubation period of 3–21 days, unilateral or bilateral conjunctival infection develops, and the patient senses ocular discomfort similar to that caused by an irritating foreign body. The pattern of conjunctival and corneal inflammation is often sufficiently characteristic to suggest a diagnosis. On an average, the illness lasts about 2 weeks, but permanent visual impairment may result. Moreover, prolonged viral shedding and even chronic or relapsing ocular pathology have been described (Boniuk *et al.*, 1965). Pettit and Holland (1979) reported recovery of virus from the conjunctiva for up to 25 months after the onset of EKC. Most individuals with chronic infection do not shed virus, but display a papillary conjunctivitis that is quite distinct from the original keratoconjunctivitis. These later manifestations may be the result of immunopathological rather than infectious mechanisms.

F. Meningoencephalitis

Adenoviruses have occasionally been associated with aseptic meningitis, meningoencephalitis, and encephalitis. The distinction in each case rests on the presence or absence of stiff neck and headache, of other signs of meningeal irritation, or of altered mental status and other signs of brain dysfunction. Most patients have some encephalitic component in their illness, rather than a pure aseptic meningitis (LeLong *et al.*, 1956; Faulkner and van Rooyen, 1962; Gabrielson *et al.*, 1966; Huttunen, 1970; Simulä *et al.*, 1970; Kelsey, 1978). In addition, most have been children or immune-compromised individuals in whom central nervous system involvement was but part of a generalized infection. The mortality and prospects for neurological sequelae are substantial (Similä *et al.*, 1970).

Ad3 and 7 are most commonly implicated, and virus has been recovered from the brain or spinal fluid (LeLong *et al.*, 1956; Faulkner and van Rooyen, 1962; Similä *et al.*, 1970). Whereas nearly all cases have involved an acute illness, Roos *et al.* (1972) reported a single, unconvincing case of a subacute encephalitis associated with Ad32.

The medical literature records the histories of but a few sporadic cases or small epidemics in which neurological symptoms accompanied adenovirus infection. Unfortunately, our understanding of this syndrome is clouded by a lack of rigorous documentation in many of these cases. Altered mental status is not necessarily a sign of encephalitis in febrile children with severe systemic illness. Brain tissue obtained at the time of postmortem examination of patients with putative adenovirus encephalitis has in some instances been surprisingly normal (Similä *et al.*, 1970). Moreover, several patients reported as having adenovirus encephalitis had recent bouts of varicella or zoster, each of which is well recognized to cause a postinfectious encephalitis. Thus, it would appear that adenovirus meningoencephalitis can occur, but it is rare, difficult to document, and of inadequately defined pathophysiology and prognosis.

G. Gastrointestinal Infections

1. Gastroenteritis

Adenoviruses are commonly recovered from stools of patients with recent respiratory illness even when symptoms of gastroenteritis are absent, indicating that the detection of fecal virus does not necessarily imply intestinal illness (Brandt *et al.*, 1969; Fox *et al.*, 1969). This observation has confounded our understanding of the significance of adenoviruses that are recovered from stools of patients with acute nonbacterial gastroenteritis (Joncas and Pavilanis, 1960; Dunkin and Hutchinson, 1961). In one pediatric study, about as many individuals with gastroenteritis shed cultivable adenoviruses in their stool as did control patients with respiratory disease (Brandt *et al.*, 1980). Thus, an etiological relationship between adenoviruses and gastroenteritis had not been as apparent as with the rotaviruses and Norwalk-like agents (Flewett *et al.*, 1974: Kapikian *et al.*, 1974. Recently, however, infantile gastroenteritis has been associated with fastidious adenoviruses that could be detected by EM, but failed to grow in conventional cell-culture lines (Flewett *et al.*, 1975; G. B. B. White and Stancliffe, 1975; Madeley *et al.*, 1977; Richmond *et al.*, 1979; Brandt *et al.*, 1980). These agents, termed *enteric adenoviruses*, have become a subject of several recent attempts to determine their significance and relationship to cultivatable adenoviruses (Gary *et al.*, 1979; Retter *et al.*, 1979). Waddell *et al.* (1980, personal communication) showed that enteric adenoviruses represent unique serogroups (F and G) of the human adenoviruses that have structural and some antigenic features in common with the other adenoviruses. Takiff and coworkers (Takiff *et al.*, 1981; Takiff and Straus, 1982) reported that in many primary or continuous cell lines, these viruses are blocked at an early stage in their replicative cycle, but that they can grow efficiently in 293 cells, a human embryonic kidney cell line transformed by Ad5. Using an enzyme immunoassay to detect viral antigen, and by cultivation in 293 cells, Yolken *et al.* (1982) showed that enteric adenoviruses are significantly associated with outbreaks of diarrheal illness in infants. In this study, most of the infected infants also demonstrated significant respiratory symptoms including frank pneumonia. The relationship of these unique adenoviruses to other infectious syndromes in normal or immune-deficient patients has yet to be assessed.

2. Other Gastrointestinal Disorders

Adenoviruses are reported to be associated with three gastrointestinal syndromes other than gastroenteritis: mesenteric adenitis, appendicitis, and intussusception. Mesenteric adenitis refers to a well-recognized pediatric condition in which inflamed mesenteric lymph nodes provoke abdominal pain that may be sufficiently severe to mimic that of acute appendicitis. Adenoviruses have been isolated from mesenteric nodes in up to 28% of cases of severe mesenteric adenitis (Kjellén *et al.*, 1957; P. M. Bell and Steyn, 1962). In a study of children with concomitant appendicitis, pharyngitis, and mesenteric adenitis, adenoviruses were cultured from 6 of 20 appendices (Kulcsar *et al.*, 1970). In addition, adenoviruslike particles have been observed in electron micrographs of inflamed appendices (Yunis and Hashida, 1973).

Much effort has been expended in studies of the relationship between adenoviruses and intussusception. Intussusception refers to a fixed telescoping of one bowel segment into that of an adjacent segment. As the segments become increasingly inflamed, the vascular supply of the segments is compromised, resulting in infarction of the bowel. The condition most commonly involves the distal ileum of children under age 2. In three studies, adenoviruses were isolated from 26–68% of stools, throat swabs, or inflamed mesenteric nodes of children with intussusception, but from only 3–8% of specimens from control patients (P. M. Bell and Steyn, 1962; Potter, 1964; Clarke *et al.*, 1969). D. O. White and Solomon (1966) observed a seasonal distribution of intussusception and postulated an association with adenovirus infection. Adenoviruses may contribute to the pathogenesis of intussusception by stimulating intestinal motility or by provoking mesenteric adenitis or in both ways. Enhanced motility may increase the possibility that bowel segments will telescope during active peristalsis, while the adenitis may prevent spontaneous release of the telescoped segments.

Mesenteric adenitis, appendicitis, and intussusception have multiple etiologies, of which adenovirus infection may be but one. Because of the high incidence of prolonged shedding of virus following ubiquitous adenovirus infections of childhood (Fox *et al.*, 1969), the actual significance of recovering virus from patients with these disorders remains uncertain.

H. Hemorrhagic Cystitis

Numazaki *et al.* (1968) reported that acute hemorrhagic cystitis in children is significantly associated with Ad11. Since that report, they and other workers have confirmed and extended this observation (Numazaki *et al.*, 1973; Mufson and Belshe, 1976). The syndrome occurs more frequently in male than in female children, is associated with several days of urinary frequency, bladder pain, gross or microscopic hematuria, minimal pyuria, and viruria. Replication of the virus appears to be in the bladder itself, because exfoliated bladder cells have been shown by immunofluorescence microscopy to contain adenovirus antigens (Belshe and Mufson, 1974). Moreover, cytoscopy demonstrates small hemorrhagic foci in the bladder mucosa. In general, there are no concomitant respiratory symptoms to suggest involvement of other organs. This syndrome is unusual among adenovirus infections in being almost totally limited to a single serotype. Only a few cases have been associated with Ad21 (Mufson and Belshe, 1976).

I. Cancer

Since adenoviruses produce latent infection in man and are known to transform cells both *in vitro* and *in vivo*, they have been considered as a possible cause of human cancers (Pereira *et al.*, 1965). When the oncogenic potential of Ad7 was first recognized, development of a live vaccine containing this virus was halted (see Section X.C). A special collaborative effort was mounted under the auspices of the National Center Institute to assess the role of adenoviruses in human tumors. The investigators ultimately concluded that development of the Ad7 vaccine could safely be resumed because they found no antibody directed against tumor (T) antigens in sera of cancer patients (Gilden *et al.*, 1970) and no RNA homologous to Ad7 DNA in a large number of human neoplasms (McAllister *et al.*, 1972). Recent studies employing even more sensitive hybridization techniques capable of detecting latent adenovirus DNA sequences in human tonsils have failed to detect similar sequences in human tumors (Mackey *et al.*, 1976, 1979; Green *et al.*, 1979b; Wold *et al.*, 1979). The issue is not completely closed, however, because not all human cancers have been scrutinized and not all adenovirus serogroups have been used as probes.

J. Congenital Anomalies

While some viruses, such as cytomegalovirus and rubella virus, have been clearly associated with congenital anomalies, the association has been less compelling for most other viruses. In one large study, Evans and Brown (1963) could find no significant association between maternal adenovirus infection and the ultimate appearance of congenital anomalies in the offspring.

K. Other Syndromes

Adenoviruses have been suggested as etiological agents in numerous other clinical syndromes on one or more occasions. Hatch and Siem (1966) suggested that adenoviruses may have been responsible for an outbreak of infectious hepatitis. Rodriguez-Torres *et al.* (1969) studied two patients with isolated myocarditis and adenovirus infection. In a seroepidemiological study, Stechenberg *et al.* (1975) suggested that recent adenovirus infection may be a factor in the etiopathogenesis of Reye's syndrome. As a whole, these isolated reports are not compelling; they suggest future areas of investigation, but add little to the current understanding of adenovirus disease.

VIII. DIAGNOSIS

Other than epidemic keratoconjunctivitis (EKC) and possibly acute respiratory disease (ARD) of recruits or pharyngoconjunctival fever (PCF), no adenovirus-associated syndromes display clinical features sufficiently distinctive to immediately suggest their etiology. Therefore, more specific diagnostic methods are required (Taylor, 1977; Kasel, 1979). Traditionally, adenoviruses have been detected by growth in cell-culture systems, especially from specimens taken during the acute phase of illness. For viral isolation, throat, anal, or conjunctival swabs, nasal washes, conjunctival scrapings, or urine, spinal fluid, or pathological specimens should be inoculated promptly into susceptible cell lines. Primary or continuous epithelial lines such as human embryonic kidney, HeLa, KB, or Hep-2 are appropriate. Characteristic cytopathic alterations develop a few days to weeks later, depending on the viral type and multiplicity, and consist of rounding, ballooning, increasing refractility, and aggregation of cells in grapelike clusters. Newly developed antigen-detection systems, including enzyme immunoassays, are far faster, simpler, and less expensive and may eliminate the necessity for viral cultures in many cases (Yolken *et al.*, 1982). A means of establishing a diagnosis rapidly will certainly be useful when effective antiviral treatments become available.

Light microscopy of hematoxylin- and eosin-stained tissues can help determine whether adenoviruses are implicated in patients with diffuse pneumonitides or unexplained multisystem infections. Adenoviruses induce the appearance of basophilic or amphophilic intranuclear inclusions (Becroft, 1971). Indirect fluorescence microscopy of urine sediments, conjunctival scrapings, or tissue sections demonstrates both nuclear and cytoplasmic staining of viral antigens and serves, where available, as a rapid diagnostic method (Knight *et al.*, 1975).

A number of successful EM techniques have been applied to the diagnosis of adenovirus infections. One method involves examination of fixed and sectioned tissues for intranuclear paracrystalline arrays of viral capsids (Myerowitz *et al.*, 1975). Another procedure entails examination of particles concentrated by precipitation with specific antisera (Edwards *et al.*, 1975). Recently, Boerner *et al.* (1981) have reported rapid identification of adenoviruses by EM of tear samples from patients with suspected viral conjunctivitis. Enteric adenoviruses can also be detected rapidly by direct EM (Brandt *et al.*, 1981).

It is rarely necessary to type adenoviruses except for epidemiological investigations or research purposes. The schemata for grouping, subgrouping, and typing adenoviruses are relatively cumbersome and, in general, are not well suited to processing large numbers of specimens at one time. Confirmation of a cultured isolate as an adenovirus is performed by fluorescence microscopy or complement fixation (Dowdle *et al.*, 1971). Hemagglutination reactions using panels of erythrocytes from different species provide subgrouping data (Hierholzer, 1973). Assignment of specific types is performed by either hemagglutination-inhibition or neutralization methods (Rosen, 1970; Hierholzer *et al.*, 1975b). Some adenoviruses cross-react at low to modest titer with more than one typing serum (Wigand *et al.*, 1965; Stevens *et al.*, 1967; Hierholzer *et al.*, 1975). Adenoviruses that show an unusually high serological cross-reactivity have been described, but have not proved to be a serious practical problem (Wigand *et al.*, 1965; Parks *et al.*, 1967; Crandell *et al.*, 1968; Hierholzer and Pumsrola, 1976; Hierholzer and Rodriguez, 1981). The marker of cross-reactivity has even proved to be a useful epidemiological tool for investigating the long-term shedding of adenovirus of a single serotype (Brandt *et al.*, 1966).

Because adenoviruses can usually be shed in an intermittent pattern from months to years after an initial illness, recovery of virus from respiratory secretions or stool does not imply their association with a given intercurrent illness. Therefore, serological methods are often used to confirm the diagnosis of adenovirus infections. A 4-fold or greater rise in titers of complement fixing, hemagglutination-inhibiting, or neutralizing antibodies provides presumptive evidence of recent adenovirus infection (Hierholzer *et al.*, 1969; Dowdle *et al.*, 1971; Kasel, 1979). A firm diagnosis is ideally based on the combination of compatible clinical, virological, and serological findings.

IX. TREATMENT

A. General

Unfortunately, the treatment of adenovirus infections is at present limited to symptomatic and supportive care. Furthermore, physicians caring for patients with common respiratory infections generally cannot discriminate adenoviral from other infectious etiologies on clinical grounds, and rapid diagnostic techniques are not widely available. The therapeutic armamentarium consists of one or more of the following: antipyretics, antihistamines, decongestants, antitussives, hydration, and rest. Pharyngitis often warrants culturing for β-hemolytic streptococci so that antibiotics can be prescribed appropriately. Serious and life-threatening adenovirus infections may require hospitalization for laboratory diagnosis, closer observation, intravenous hydration, pulmonary toilet, assisted ventilation, or other modalities as the case demands.

B. Antiviral Therapy

Specific effective antiviral therapy is now available for herpes and influenza virus infections, but not for adenovirus infections. Drugs such as phosphonoacetic acid (Overby *et al.*, 1977), 5-iodo-2′-deoxyuridine (IUdR) (Dudgeon *et al.*, 1969), arabinofuranosyl adenine (ARA-A) (Sidwell *et al.*, 1973), acyclovir (Collins and Bauer, 1979), trifluorothymidine (Lennette and Eiferman, 1978), ribavirin (Sidwell, 1980), and amantidine (Neumayer *et al.*, 1965) exhibit little or no *in vitro* activity as inhibitors of adenovirus replication. When used in desperate clinical situations, the clinical responses to these agents have been equivocal at best. Interferons, with their broader antiviral spectrum, may hold more promise, but have not been available for administration in suitable amounts until recently.

Serious efforts to exploit the known biochemical machinery of adenovirus replication for therapeutic purposes have not been made. Inhibitors of adenovirus DNA-binding proteins or polymerase, for example, should be sought. The clinical activities of future candidate antiviral compounds that exhibit *in vitro* activity against adenoviruses are going to be difficult to assess because serious and life-threatening infections are uncommon and sporadic. EKC, ARD of recruits, and gastroenteritis would seem to be appropriate areas for controlled investigations. To date, only the treatment of adenovirus keratoconjuctivitis has been approached in an organized and deliberate fashion.

C. Keratoconjunctivitis

Ocular viral infections have been the subject of numerous therapeutic trials. The relative ease of diagnosis and sequential assessment by noninvasive examination has facilitated these studies. Definitive evidence for the efficacy of antiviral compounds in treatment of herpes simplex keratoconjunctivitis led naturally to similar investigations with adenovirus-induced keratoconjunctivitis. Unfortunately, but not surprisingly, each agent in turn showed little if any positive impact on the courses of these infections. IUdR and ARA-A have little or no activity in inhibiting the replication of adenoviruses *in vitro* and in carefully controlled trials have not shown an alteration in the course of adenovirus-induced keratoconjunctivitis (Dudgeon *et al.*, 1969; Pavan-Langston and Dohlman, 1972; Waring *et al.*, 1976). On the basis of the results of rather preliminary uncontrolled studies, Maichuk (1980) suggests that human leukocyte interferon or interferon inducers are beneficial.

Topical corticosteroids are commonly employed in inflammatory eye disease, but are best avoided in ocular infections except when immunopathological processes are most important. For example, subepithelial corneal infiltrates are common sequelae of EKC. These are postulated to represent leukocytic infiltrates that evolve in response to viral antigens. At the time of their appearance, virus generally can no longer be recovered from the eye. In a carefully controlled study, Laibson *et al.* (1970) showed that topical corticosteroids halt the progression and speed the resolution of these infiltrates (Laibson *et al.*, 1970).

X. PREVENTION

A. General

Preventing exposure to adenoviruses can be reasonably accomplished in only a few settings. Proper sterilization of ophthalmological instruments and solutions and good hand-washing will help eliminate spread of keratoconjunctivitis. Establishing either respiratory-secretion or stool precautions or both for hospitalized patients with suspected respiratory or enteric infections may prevent nosocomial transmission of adenoviruses. In some other settings, active prophylactic measures in the form of vaccines have been required and found beneficial. The first study to demonstrate transfer of active immunity to adenoviruses was performed under the auspices of the Commission on Acute Respiratory Diseases of the United States Army during World War II, at a time when the etiology of ARD was not yet defined (Commission on Acute Respiratory Diseases, 1947a,b). Volunteers who were given filtrates of respiratory secretions developed typical symptoms of ARD, but were protected on rechallenge. Obviously, this filtrate contained unattenuated live material and could not be construed as a vaccine candidate.

B. Killed Vaccines

Shortly after the discovery of adenoviruses, Huebner *et al.* (1955) and Hilleman *et al.* (1955) reported successful preparation of formalin- or heat-killed vaccines from monkey cell cultures infected with Ad3, 4, or 7. Parenteral injection of these reagents stimulated the appearance of type-specific neutralizing antibody and conferred immunity to later homotypic challenge. In one trial, 624 new military recruits were followed through basic training (Hilleman, 1957). Of these recruits, 311 were vaccinated with a killed Ad4 preparation. From the 2nd through the 8th week of training, 23.8% of controls and 4.8% of vaccinees required hospitalization for ARD, suggesting an 80% reduction of significant illness due to vaccination.

These early successes stimulated much wider field testing of killed vaccine, initially with bivalent vaccine (Ad4 and 7) and, after 1962, with trivalent vaccine (Ad3, 4, and 7). Several problems emerged, however (Buescher, 1967). First, the need to develop multivalent vaccines resulted from unpredictable shifts in the virus types causing epidemics among recruits. Recruits who were protected against Ad4 were still susceptible to Ad7. Thus, the incidence of ARD did not diminish sufficiently. Second, there were enormous variations in antigenic potency among vaccine lots. In all, the vaccine appeared to be only 40–70% effective. Third, the adenovirus seed stocks prepared in monkey cells were found to be contaminated with simian virus 40 (SV40). Following removal of SV40 contamination, the viral pools were found to be contaminated with adenovirus–SV40 hybrid that could not be removed (O'Conor *et al.*, 1963; Rowe and Baum, 1964; Lewis *et al.*, 1966). Thus, in 1963, the killed vaccines were withdrawn from study and licensure.

C. Live Vaccines

At about the same time, Couch and Chanock at the National Institutes of Health had begun to test live adenovirus vaccines prepared in primary human cell cultures. Volunteer trials showed that ingestion of enteric-coated capsules containing Ad4 and 7 caused inapparent enteric infection that did not spread to or cause symptoms in the respiratory tract, rarely spread to susceptible contacts, and elicited serum but not nasal secretory neutralizing antibodies (Couch *et al.*, 1963; Chanock *et al.*, 1966; Top *et al.*, 1971a–c). Over the next several years, field trials of monovalent (Ad4) and later bivalent (Ad4 and 7) live enteric vaccines demonstrated their safety and efficacy (Edmondson *et al.*, 1966; Smith *et al.*, 1970; Top *et al.*, 1971a–c). On an average, the bivalent vaccines have reduced adenovirus infection in recruits by over 80% and ARD by over 50%. Since the early 1970s, bivalent vaccine has been regularly used by the U.S. Army as well as other military groups around the world for basic combat training posts during the cooler months in which adenovirus disease is anticipated.

Three problems have beset the live adenovirus vaccine program: First, as in the killed vaccine trials, other viral types have emerged. Outbreaks of Ad21 have occurred in American, Dutch, Russian, and Indian trainees (Zhdanov and Dreizin, 1961; Kurian *et al.*, 1966; Van der Veen *et al.*, 1969; Rose *et al.*, 1970). Initial studies of an Ad21 live enteric vaccine were successful, suggesting that the enteric immunization approach will prove effective against this agent as well (Takafuji *et al.*, 1979). Ad14 has also been detected, but its impact remains undefined. Second, the vaccines used by the military may not have practical utility for the general populace. It is conceivable that vaccines for the lower-numbered adenovirus types that commonly infect children can also be protective, but the diseases associated with most such infections may be too mild to warrant such effort. Third is the problem of the potential oncogenicity of the vaccine strains. To address this question, extensive efforts were made prior to the initiation of the live adenovirus vaccine trials to ensure that these viruses were not associated with human cancer (see Section VII.I) (Gilden *et al.*, 1970; McAllister *et al.*, 1972).

XI. ADENOVIRUS–SIMIAN VIRUS 40 HYBRIDS

Human adenoviruses do not propagate well in most simian cell lines. It is now recognized that the ability of adenoviruses to grow in monkey cells during preparation of an activated vaccine in the late 1950s resulted from helper activity provided by contaminating SV40 (O'Conor *et al.*, 1963). Suppression of the replication of SV40 by repeated passage in the presence of antibody to SV40 was accompanied by the emergence of adenovirus–SV40 hybrids (see Chapter 10). These viruses contain SV40 sequences that are inserted into the adenovirus genome (Rowe and Baum, 1964). Hybrids between SV40 and Ad1–7 and 12 have been prepared (Lewis *et al.*, 1966). Most hybrids lack or interrupt critical adenovirus sequences, so they can replicate only in the presence of coinfecting nonhybrid adenoviruses. These unusual genetic chimeras have posed difficult and as yet unresolved clinical problems. For example, does the introduction of highly oncogenic SV40 sequences into human cells in a vehicle that permits their replication in human cells predispose to human malignancy? The answer is not known, nor is there any current evidence for disease resulting from early trials with these vaccines. Nonetheless, the problem will remain more than an academic one as long as the recipients of SV40-contaminated vaccines live, because SV40-like viruses are associated with progressive multifocal leukoencephalopathy (JC virus) and other problems in the immune-compromised host (BK virus) (Gardner *et al.*, 1971; Padgett *et al.*, 1971; Weiner *et al.*, 1972).

XII. ADENO-ASSOCIATED VIRUSES

Adeno-associated viruses (AAVs) are defective parvoviruses that replicate efficiently in the presence of helper adenoviruses (Ward and Tattersall, 1978). As the name implies, AAV strains were first found as contaminants in adenovirus-infected cell cultures (Atchison *et al.*, 1965). There are four AAV serotypes. AAV 2 and 3 have been recovered from anal and throat swab specimens in close association with adenoviruses (Blacklow *et al.*, 1968a). Serological responses to AAV 1, 2, and 3 commonly develop during early childhood (Blacklow *et al.*, 1968b; Boucher *et al.*, 1970). AAV type 4 appears to be a simian virus (Boucher *et al.*, 1970). In an 8-year survey of 551 hospitalized children, Blacklow *et al.* (1971) reported that about 30% of uninfected or adenovirus-infected well or ill children possess antibodies to AAV 2 or 3 or both. All but 1 of 32 children who seroconverted to AAV 2 or 3 or both showed evidence of a recent adenovirus infection.

Although AAV inhibits adenovirus growth in cell culture and adenovirus oncogenicity in hamsters (Kastow *et al.*, 1967; Parks *et al.*, 1968), it is not known whether AAV has any impact on adenovirus-induced human diseases. In the study of Blacklow (1971), children coinfected with AAV and adenoviruses were not more likely to have experienced a severe respiratory illness. Finally, even though related parvoviruses produce disease in a variety of animal species, it is now known whether AAV is associated with any human disease. Studies suggesting such associations have been unconvincing (Kasczak *et al.*, 1977).

XIII. LATENCY

The ability to establish and sustain a latent infection is a well-recognized property of adenoviruses. Adenoviruses were initially discovered because of the spontaneous expression of latent virus in human tonsillar tissues (Rowe *et al.*, 1953). Although the observation of adenovirus latency was made about 30 years ago, we know less about the process than we do about herpesvirus latency, which was proved about a dozen years ago (Bastian *et al.*, 1972). In part, the failure is due to a lack of established animal models of adenovirus infection. As discussed in Sections IV and VII.A, it is known that an individual can shed virus intermittently for months or even years after an initial adenovirus infection (Fox *et al.*, 1969). We do not know whether the intermittency of shedding results from repeated virus reactivations or from transient reductions in virus production to undetectable levels. Moreover, there are no data to determine whether reactivation can induce disease. These questions are difficult but not impossible to tackle in the absence of animal models. Restriction endonuclease analysis provides the definitive tool to distinguish reactivation from reinfection. The implications of adenovirus latency, the factors that regulate and sustain the latent state, and the extent of expression of the latent sequences are largely unexplored, but are worthy of investigation.

XIV. FURTHER RESEARCH

Three aspects of adenovirus infection are in greatest need of further research efforts. First, better animal models need to be developed so that the pathophysiology, biology, and immunology of acute and latent adenovirus infections can be examined. Second, an effort must be made to identify agents that interfere with adenovirus-specific synthetic function. Suitable animal models can then serve as the initial subject for exploring the *in vivo* activity and tolerance of such agents. Third, the fastidious enteric adenoviruses need to be further defined, in terms of their unique biology and their epidemiology, pathophysiology, and the range of illness that they provoke. The first three decades of research have obviously not closed the book on adenovirus infections of man.

ACKNOWLEDGEMENTS. The author gratefully thanks Dr. Robert M. Chanock and Dr. Carl Brandt for helpful suggestions and review of this manuscript. He also thanks Karen Leighty for manuscript preparation and editing.

REFERENCES

Andiman, W.A., Jacobson, R.I., and Tucker, G., 1977, Leukocyte-associated viremia with adenovirus type 2 in an infant with lower respiratory tract disease, *N. Engl. J. Med.* **297**:100.

Atchison, R.W., Casto, B.C., and Hammond, W. McD., 1965, Adenovirus associated defective virus particles, *Science* **149**:754.

Aterman, K., Embil, J., Easterbrook, K.B., Haldane, E.V., and Crosby, J., 1973, Liver necrosis, adenovirus type 2 and thymic dysplasia, *Virchows Arch. A* **360**:155.

Baraff, L.J., Wikins, J., Wehrle, P.F., 1978, The role of antibiotics, immunizations, and adenoviruses in *Pertussis Pediatrics* **6**:224–230.

Bastian, F.O., Rabson, A.S., Yee, C.L., and Tralka, T.S., 1972, *Herpesvirus hominis*: Isolation from human trigeminal ganglia, *Science* **178**:306.

Becroft, D.M.O., 1967, Histopathology of fatal adenovirus infections of the respiratory tract of young children, *J. Clin. Pathol.* **20**:561.

Becroft, D.M.O., 1971, Bronchiolitis obliterans, bronchiectasis, and other sequelae of adenovirus type 21 infection in young children, *J. Clin. Pathol.* **24**:72.

Bell, J.A., Rowe, W.P., Engler, J.I., Parrott, R.H., and Huebner, R.J., 1955, Pharyngoconjunctival fever: Epidemiological studies of a recently recognized disease entity, *J. Am. Med. Assoc.* **175**:1083.

Bell, J.A., Ward, D.G., Huebner, R.J., Rowe, W.P., Suskind, R.G., and Paffenbarger, R.S., 1956, Studies of adenoviruses (APC) in volunteers, *Am. J. Public Health* **46**:1130.

Bell, J.A., Huebner, R.J., Rosén, L., Rowe, W.P., Cole, R.M., Mastrota, F.M., Floyd, T.M., Chanock, R.M., and Shvedoff, R.A., 1961, Illness and microbial experiences of nursery children at Junior Village, *Am. J. Hyg.* **74**:267.

Bell, P.M., and Steyn, J.H., 1962, Viruses in lymph nodes of children with mesenteric adenitis and intussusception, *Br. Med. J.* **2**:700.

Bell, S.D., Jr., Rota, T.R., and McComb, D.E., 1960, Adenoviruses isolated from Saudi Arabia. III. Six new serotypes, *Am. J. Trop. Med. Hyg.* **9**:523.

Belshe, R.B., and Mufson, M.A., 1974, Identification by immunofluorescence of adenoviral antigen in exfoliated bladder epithelial cells from patients with acute hemorrhagic cystitis, *Proc. Soc. Exp. Biol. Med.* **146**:754.

Blacklow, N.R., Hoggan, M.D., Kapikian, A.Z., Austin, J.B., and Rowe, W.P., 1968a, Epidemiology of adenovirus-associated virus infection in a nursery population, *Am. J. Epidemiol.* **88**:368.

Blacklow, N.R., Hoggan, M.D., and Rowe, W.P., 1968b, Serologic evidence for human infection with adenovirus-associated viruses, *J. Natl. Cancer Inst.* **40**:319.

Blacklow, N.R., Hoggan, M.D., Sereno, M.S., Brandt, C.D., Kim, H.W., Parrott, R.H., and Chanock, R.M., 1971, A seroepidemiologic study of adenovirus-associated virus infection in infants and children, *Am. J. Epidemiol.* **94**:359.

Boerner, C.F., Lee, F.K., Wickliffe, C.L., Nehmias, A.J., Cavanagh, H.D., and Straus, S.E., 1981, Electron microscopy for the diagnosis of ocular viral infections, *Ophthalmology* **88**:1377.

Boniuk, M., Phillips, C.A., and Friedman, J.B., 1965, Chronic adenovirus type 2 keratitis in man, *N. Engl. J. Med.* **273**:924.

Boucher, D.W., Parks, W.P., and Melnick, J.L., 1970, A sensitive neutralization test for the adeno-associated satellite viruses, *J. Immunol.* **104**:555.

Brandt, C.D., Wassermann, F.E., and Fox, J.P., 1966, The Virus Watch Program. IV. Recovery and comparison of two serological varieties of adenovirus type 5, *Proc. Soc. Exp. Biol. Med.* **123**:513.

Brandt, C.D., Kim, H.W., Vargosko, H.A., Jeffries, B.C., Arrobio, J.O., Rindge, B., Parrott, R.H., and Chanock, R.M., 1969, Infections in 18,000 infants and children in a controlled study of respiratory tract disease. I. Adenovirus pathogenicity in relation to serologic type and illness syndrome, *Am. J. Epidemiol.* **90**:484.

Brandt, C.D., Kim, H.W., Jeffries, B.C., Pyles, G., Christmas, E.E., Reid, J.L., Chanock, R.M., and Parrott, R.H., 1972, Infections in 18,000 infants and children in a controlled study of respiratory tract disease. II. Variation in adenovirus infections by year and season, *Am. J. Epidemiol.* **95**:218.

Brandt, C.D., Kim, H.W., Yolken, R.H., Kapikian, A.Z., Arrobio, J.O., Rodriguez, W.J., Wyatt, R.G., Chanock, R.M., and Parrott, R.H., 1980, Comparative epidemiology of two rotavirus serotypes and other viral agents associated with pediatric gastroenteritis, *Am. J. Epidemiol.* **110**:243.

Brandt, C.D., Kim, H.W., Rodriguez, W.J., Thomas, L., Yolken, R.H., Arrobio, J.D., Kapikian, A.Z., Parrott, R.H., and Chanock, R.M., 1981, Comparison of direct electron microscopy, immune electron microscopy, and rotavirus enzyme-linked immunosorbent assay for detection of gastroenteritis viruses in children, *J. Clin. Microbiol.* **13**:976.

Bryant, R.E., and Rhoades, E.R., 1967, Clinical features of adenoviral pneumonia in Air Force recruits, *Am. Rev. Respir. Dis.* **96**:717.

Buchanan, T.G., Roizman, B., Adams, G., and Stover, B.H., 1978, Restriction endonuclease fingerprinting of herpes simplex virus DNA: A novel epidemiologic tool applied to a nosocomial outbreak, *J. Infect. Dis.* **138**:488.

Buescher, E.L., 1967, Respiratory disease and the adenoviruses, *Med. Clin. N. Am.* **51**:769.

Carmichael, G.P., Zahvadnik, J.M., Moyer, G.H., 1979, Adenovirus hepatitis in an immunosuppressed adult patient, *Am. J. Clin. Pathol.* **71**:352.

Chanock, R.M., 1974, Impact of adenoviruses in human disease, *Prev. Med.* **3**:466.

Chanock, R.M., Hayflick, L., and Barile, M.F., 1962, Growth on artificial medium of an agent associated with atypical pneumonia and its identification as a PPLO, *Proc. Natl. Acad. Sci. U.S.A.* **48**:41.

Chanock, R.M., Ludwig, M., Huebner, R.J., Cate, T.R., and Chu, L.W., 1966, Immunization by selective infection with type 4 adenovirus grown in human diploid tissue culture. I. Safety and lack of oncogenicity and tests for potency in volunteers, *J. Am. Med. Assoc.* **195**:445.

Clarke, E.J., Jr., Phillips, I.A., and Alexander, E.R., 1969, Adenovirus infection in intussusception in children in Taiwan, *J. Am. Med. Assoc.* **208**:1671.

Collins, P., and Bauer, D.J., 1979, The activity *in vitro* against herpesvirus of 9-(2-hydroxyethoxymethyl)guanine (acycloguanosine), a new antiviral agent, *J. Microb. Chemother.* **5**:431.

Commission on Acute Respiratory Diseases, 1947a, Experimental transmission of minor respiratory illness to human volunteers by filter passing agents. I. Demonstration of two long types of illness characterized by long and short incubation periods and different clinical features, *J. Clin. Invest.* **26**:957.

Commission on Acute Respiratory Diseases, 1947b, Experimental transmission of minor respiratory illness to human volunteers by filter passing agents. II. Immunity on reinoculation with agents from the two types of minor respiratory illness and from primary atypical pneumonia, *J. Clin. Invest.* **26**:974.

Connor, J.D., 1970, Evidence for an etiologic role for adenoviral infections in pertussis syndrome, *N. Engl. J. Med.* **283**:390.

Cooney, M.K., Hall, C.E., and Fox, J.P., 1972, Seattle virus watch. III. Evaluation of isolation methods and summary of infections detected by virus isolations, *Am. J. Epidemiol.* **96**:286.

Couch, R.B., Chanock, R.M., Cate, T.R., Lang, D.J., Knight, V., and Huebner, R.J., 1963, Immunization with types 4 and 7 adenovirus by selective infection of the intestinal tract, *Am. Rev. Respir. Dis.* **88**:394.

Crandell, R.A., Dowdle, W.R., Holcomb, T.M., and Dahl, E.V., 1968, A fatal illness associated with two viruses: An intermediate adenovirus type (21–16) and influenza A2, *J. Pediatr.* **72**:467.

D'Angelo, L.J., Hierholzer, J.C., Keenlyside, R.A., Anderson, L.J., and Martone, W.J., 1979, Pharyngoconjunctival fever caused by adenovirus type 4: Report of a swimming pool-related outbreak with recovery of virus from pool water, *J. Infect. Dis.* **140**:42.

D'Angelo, L.J., Hierholzer, J.C., Holman, R.C., and Smith, J.D., 1981, Epidemic keratoconjunctivitis caused by adenovirus type 8: Epidemiologic and laboratory aspects of a large outbreak, *Am. J. Epidemiol.* **113**:44.

Dascomb, H.E., and Hilleman, M.R., 1956, Clinical and laboratory studies in patients with respiratory diseases caused by adenoviruses (RI-APC-ARD agents), *Am. J. Med.* **21**:161.

Dawson, C.R., Hanna, L., Wood, T.R., and Despain, R., 1970, Adenovirus type 8 keratoconjunctivitis in the United States. III. Epidemiologic, clinical, and microbiologic features, *Am. J. Ophthalmol.* **69**:473.

De Jong, J.C., Wigand, R., Wadell, G., Keller, D., Muzerie, C.J., Wermenbol, A.G., and Schaap, G.J.P., 1981, Adenovirus 37: Identification and characterization of a medically significant new adenovirus type of subgroup D, *J. Med. Virol.* **7**:105.

Dingle, J., and Langmuir, A.D., 1968, Epidemiology of acute respiratory diseases in military recruits, *Am. Rev. Respir. Dis.* **97**:1.

Dowdle, W.R., Labrieux, M., and Hierholzer, J.C., 1971, Production and evaluation of a purified adenovirus group specific (hexon) antigen for use in the diagnostic complement fixation test, *Appl. Microbiol.* **21**:718.

Dudding, B.A., Wagner, S.C., Zeller, J.A., Gmelich, J.T., French, G.R., and Top, F.H., Jr., 1972, Fatal pneumonia associated with adenovirus type 7 in three military trainees, *N. Engl. J. Med.* **286**:1289.

Dudgeon, J., Bhargava, S.K., and Ross, C.A.C., 1969, Treatment of adenovirus infection of the eye with 5-iodo-2′-deoxyuridine: A double-blind trial, *Br. J. Ophthalmol.* **53**:530.

Duncan, S.J., Gordon, F.C.A., Gregory, D.W., McPhie, J.L., Postlethwaite, R., White, R., and Wilcox, H.N.A., 1978, Infection of mouse liver by human adenovirus type 5, *J. Gen. Virol.* **40**:45.

Dunkin, I.B.R., and Hutchison, J.G.P., 1961, Type 3 adenovirus with gastrointestinal symptoms, *Lancet* **1**:530.

Edmondson, W.P., Purcell, R.H., Gundelfinger, B.F., Love, J.W.P., Ludwig, W., and Chanock, R.M., 1966, Immunization by selective infection with type 4 adenovirus grown in diploid tissue culture II. Specific protective effect against epidemic disease, *J. Am. Med. Assoc.* **195**:453.

Edwards, E.A., Valters, W.A., Boehm, L.G., and Rosenbaum, M.J., 1975, Visualization by immunoelectron microscopy of viruses associated with acute respiratory disease, *J. Immunol. Methods* **8**:159.

Evans, T.N., and Brown, G.C., 1963, Congenital anomalies in virus infections, *Am. J. Obstet. Gynecol.* **87**:749.

Faulkner, R., and van Rooyen, C.E., 1962, Adenovirus types 3 and 5 isolated from the cerebrospinal fluid of children, *Can. Med. Assoc. J.* **87**:1123.

Field, P.R., Patwardhan, J., McKenzie, J.A., and Murphy, A.M., 1978, Fatal adenovirus type 7 pneumonia in an adult, *Med. J. Aust.* **2**:445–447.

Fishaut, M., Tubergen, D., and McIntosh, K., 1980, Medical progress: Cellular responses to respiratory viruses with particular reference to children with disorders of cell-mediated immunity, *J. Pediatr.* **96**:179.

Flewett, T.H., Davies, H., Bryden, A.S., and Robertson, N.J., 1974, Diagnostic electron microscopy of feces. II. Acute gastroenteritis associated with reovirus-like particles, *J. Clin. Pathol.* **27**:608.

Flewett, T.H., Bryden, A.S., Davies, H., and Morris, C.A., 1975, Epidemic viral enteritis in a long-stay children's ward, *Lancet* **1**:4.

Forsyth, B.R., Bloom, H.H., Johnson, K.M., and Chanock, R.M., 1964, Patterns of adenovirus infections in Marine Corps personnel. II. Longitudinal study of successive advanced recruit training companies, *Am. J. Hyg.* **80**:343.

Fox, J.P., Brandt, C.D., Wasserman, F.E., Hall, C.E., Spigland, I., Kogon, A., and Elveback, L.R., 1969, The Virus Watch Program: A continuing surveillance of viral infections in metropolitan New York families. VI. Observations of adenovirus infections: Virus excretion patterns, antibody response, efficiency of surveillance, patterns of infection, and relation to illness. *Am. J. Epidemiol.* **89**:25.

Foy, H.M., and Grayston, J.T., 1976, Adenovirus, in: *Viral Infections of Man* (A.S. Evans, ed.), p. 53, Raven Press, New York.

Fuchs, E., 1889, Keratitis punctata superficialis, *Wein. Klin. Wochenschr.* **2**:837.

Gabrielson, M.O., Joseph, C., and Hsiung, G.D., 1966, Encephalitis associated with adenovirus type 7 occurring in a family outbreak, *J. Pediatr.* **68**:142.

Gardner, S.D., Field, A.M., Coleman, D.V., and Hulme, B., 1971, New Human papovavirus (BK) isolated from urine after renal transplantation, *Lancet* **1**:1253.

Garon, C.F., Berry, K.W., Hierholzer, J.C., and Rose, J.A., 1973, Mapping of base sequence heterologies between genomes from different adenovirus serotypes, *Virology* **54**:414.

Gary, G.W., Hierholzer, J.C., and Black, R.E., 1979, Characteristics of noncultivable adenoviruses associated with diarrhea in infants: A new subgroup of human adenoviruses, *J. Clin. Microbiol.* **10**:96.

Gilden, R.Z., Kern, J., Lee, Y.K., Rapp, F., Melnick, J.L., Riggs, J.L., Lennette, E.H., Zbar, B., Rapp, H.J., Turner, H.C., and Huebner, R.J., 1970, Serologic surveys of human cancer patients for antibody to adenovirus T antigens. *Am. J. Epidemiol.* **91**:500.

Green, M., 1970, Oncogenic viruses, *Annu. Rev. Biochem.* **39**:701.

Green, M., Mackey, J.K., Wold, W.S.M., and Rigden, P., 1979a, Thirty-one human adenovirus serotypes (Ad1–Ad31) form five groups based upon DNA genome homologies, *Virology* **93**:481.

Green, M., Wold, W.S.M., Mackey, J.K., and Rigden, P.M., 1979b, Analysis of the human tonsil and cancer DNAs and RNAs for DNA sequences of group C (serotypes 1, 2, 5, and 6) human adenoviruses, *Proc. Natl. Acad. Sci. U.S.A.* **76**:6606.

Hall, C.E., Brandt, C.D., Frothingham, T.E., Spigland, I., Cooney, M.K., and Fox, J.P., 1971, The Virus Watch Program: A continuing surveillance of viral infections in metropolitan New York families. IX. A comparison of infections with several respiratory pathogens in New York and New Orleans families, *Am. J. Epidemiol.* **94**:367.

Hamre, D., Connelly, A.P.V., and Procknow, J.J., 1966, Virologic studies of acute respiratory diseases in young adults. IV. Virus isolates during four years of surveillance, *Am. J. Epidemiol.* **83**:238.

Hatch, M.H., and Siem, R.A., 1966, Viruses isolated from children with infectious hepatitis, *Am. J. Epidemiol.* **84:**495.

Henson, D., and Mufson, M.A., 1971, Myocarditis and pneumonitis with type 21 adenovirus infection: Association with fatal myocarditis and pneumonitis, *Am. J. Dis. Child.* **121:**334.

Hierholzer, J.C., 1973, Further subgrouping of the human adenoviruses by differential hemagglutination, *J. Infect. Dis.* **128:**541.

Hierholzer, J.C., and Pumsrola, A., 1976, Antigenic characterization of intermediate adenovirus 14-11 strains associated with upper respiratory illness in a military camp, *Infect. Immun.* **13:**354.

Hierholzer, J.C., and Rodriguez, F.H., 1981, Antigenically intermediate human adenovirus strain associated with conjunctivitis, *J. Clin. Microbiol.* **13:**395.

Hierholzer, J.C., Suggs, M.T., and Hall, E.C., 1969, Standardized viral hemagglutination and hemagglutination-inhibition tests. II. Description and evaluation, *Appl. Microbiol.* **18:**824.

Hierholzer, J.C., Atuk, N.O., and Gwaltney, J.M., 1975a, New human adenovirus isolated from a renal transplant recipient: Description and characterization of candidate adenovirus type 34, *J. Clin. Microbiol.* **1:**366.

Hierholzer, J.C., Gamble, W.C., and Dowdle, W.R., 1975b, Reference equine antisera to 33 human adenovirus types: Homologous and heterologous titers, *J. Clin. Microbiol.* **1:**65.

Hierholzer, J.C., Kemp, M.C., Gary, G.W., Jr., and Spencer, H.C., 1982, New human adenovirus associated with respiratory illness: Candidate adenovirus type 39, *J. Clin. Microbiol.* **16:**15.

Hilleman, M.R., 1957, Epidemiology of adenovirus respiratory infection in military recruit populations, *Ann. N.Y. Acad. Sci.* **62:**262.

Hilleman, M.R., and Werner, J.H., 1954, Recovery of new agent from patients with acute respiratory illness. *Proc. Soc. Exp. Biol. Med.* **85:**183.

Hilleman, M.R., Werner, J.H., Dascomb, H.E., and Butler, R.J., 1955, Epidemiologic investigations with respiratory disease virus RI-67, *Am. J. Public Health* **45:**203.

Huebner, R.J., 1967, Adenovirus-directed tumor and T antigens, in: *Perspectives in Virology*, Vol. 5 (M. Pollard, ed.), p. 147, Academic Press, New York.

Huebner, R.J., Rowe, W.R., Ward, T.G., Parrott, R.H., and Bell, J.A., 1954, Adenoidal–pharyngeal–conjunctival agents: A newly recognized group of common viruses of the respiratory system, *N. Engl. J. Med.* **251:**1077.

Huebner, R.J., Bell, J.A., Rowe, W.P., Ward, T.G., Suskind, R.F., Hartley, J.W., and Paffenbarger, R.S., 1955, Studies of adenoidal–pharyngeal–conjunctival vaccines in volunteers, *J. Am. Med. Assoc.* **159:**986.

Huttunen, L., 1970, Adenovirus type 7 associated encephalitis, *Scand. J. Infect. Dis.* **2:**151.

Javetz, E., Kimura, S.J., Nicholas, A.N., Thygeson, P., and Hanna, L., 1955, New type of APC virus from epidemic keratoconjunctivitis, *Science* **122:**1190.

Joncas, J., and Pavilanis, V., 1960, Diarrhea and vomiting in infancy and childhood: Viral studies, *Can. Med. Assoc. J.* **83:**1108.

Jordan, W.S., Jr., 1957, The frequency of infection with adenoviruses in a family study population, *Ann. N.Y. Acad. Sci.* **67:**273.

Kapikian, A.Z., Kim, H.W., Wyatt, R.G., Rodriguez, W.J., Ross, S., Cline, W.L., and Parrott, R.H., 1974, Reovirus-like agents in stool: Association with infantile diarrhea and development of serological tests, *Science* **185:**1049.

Kasczak, R.J., Carp, R.I., Donnenfeld, H., and Bartfeld, H., 1977, Association between amyotrophic lateral sclerosis (ALS) and adeno-associated viruses (AAV), *Abstr. Annu. Meet. Am. Soc. Microbiol.*, No. 5144.

Kasel, J.A., 1979, Adenovirus, in: *Diagnostic Procedures for Viral, Rickettsial, and Chlamydial Infections*, 5th ed. (E. H. Lennette and N.J. Schmidt, eds.), p. 229, American Public Health Association, Washington, D.C.

Kastow, B.C., Atchison, R.W., and Hammond, W. McD., 1967, Studies on the relationship between adeno-associated virus type 1 (AAV-1) and adenovirus. I. Replication of AAV-1 in certain cell cultures and its effects on helper adenovirus, *Virology* **32:**52.

Kelsey, D.S., 1978, Adenovirus meningoencephalitis, *Pediatrics* **61**:291.

Kibrick, S., Melendez, L., and Enders, J.F., 1957, Clinical association of enteric viruses with particular reference to agents exhibiting properties of the ECHO group, *Ann. N.Y. Acad. Sci.* **67**:311.

Kim, H.W., Brandt, C.D., Arrobio, O., Murphy, B., Chanock, R.M., and Parrott, R.H., 1979, Influenza A and B virus infection in infants and young children during the years 1957–1976, *Am. J. Epidemiol.* **109**:464.

Kjellén, L., Sterner, G., and Svedmyr, A., 1957, On the occurrence of adenovirus in Sweden, *Acta Paediatr. Scand.* **46**:164.

Klenk, E.L., Gaultney, J.V., and Bass, J.W., 1972, Bacteriologically proved pertussis and adenovirus infection: Possible association, *Am. J. Dis. Child.* **124**:203.

Klocke, R.A., Artenstein, M.S., Green, R.W., Dennehy, J.J., and Richert, J.H., 1966, The effect of acute respiratory infection on pulmonary function in military recruits, *Am. Rev. Respir. Dis.* **93**:549.

Knight, E.V., Brasier, F., Greenberg, S.B., and Jones, D.B., 1975, Immunofluorescent diagnosis of acute viral infection, *South. Med. J.* **68**:764.

Kulcsar, G., Vutskits, Z., Nasz, I., Dan, P., and Leb, J., 1970, Viruses isolated from appendicitis cases in childhood, *Zentralbl. Bakteriol.* **215**:506.

Kurian, P.V., Lal, R., and Pandit, V., 1966, Adenovirus infection in Indian army personnel, *Indian J. Med. Res.* **54**:812.

Laibson, P.R., Dhiri, S., Oconer, J., and Ortolan, G., 1970, Corneal infiltrates in epidemic keratoconjunctivitis: Response to double-blind corticosteroid therapy, *Arch. Ophthalmol.* **84**:36.

Lang, W.R., Howden, C.W., Laws, J., and Burton, J.F., 1969, Bronchopneumonia with serious sequelae in children with evidence of adenovirus type 21 infection, *Br. Med. J.* **1**:73.

Lelong, m., Lépine, P., Alison, F., Vinh, L.T., Satgé, P., and Chany, C., 1956, La pneumonia à virus du groupe A.P.C. chez de Nourrison isolement du virus: Les lesions anatomohistologiques, *Arch. Fr. Pediatr.* **13**:1092.

Lennette, D.A., and Eiferman, R.A., 1978, Inhibition of adenovirus replication *in vitro* by trifluridine, *Arch. Ophthalmol.* **96**:1662.

Lewis, A.M., Jr., Baum, S.G., Prigge, K.O., and Rowe, W.P., 1966, Occurrence of adenovirus–SV40 hybrids among monkey kidney cell adapted strains of adenoviruses, *Proc. Soc. Exp. Biol. Med.* **122**:214.

Mackey, J.K., Rigden, P.M., and Green, M., 1976, Do highly oncogenic group A human adenoviruses cause human cancer? Analysis of human tumors for adenovirus 12 transforming DNA sequences, *Proc. Natl. Acad. Sci. U.S.A.* **73**:4651.

Mackey, J.K., Green, M., Wold, W.S.M., and Rigden, P.M., 1979, Analysis of human cancer DNA for DNA sequences of human adenovirus type 8, *J. Natl. Cancer Inst.* **62**:23.

Madeley, C.R., Cosgrove, B.P., Bell, E.J., Fallon, R.J., 1977, Stool viruses in babies in Glasgow. I. Hospital admissions with diarrhea, *J. Hyg. Camb.* **78**:261.

Madeley, C.R., Middleton, P.J., Szymanski, M.T., Abbott, G.D., Bortolussi, R., and Hamilton, J.R., 1974, Orbivirus acute gastroenteritis of infancy, *Lancet* **1**:1241.

Maichuk, Y.F., 1980, Interferon and interferon inducers in the treatment of herpesvirus and adenovirus eye diseases, *J. Rev. Int. Trach. Path. Ocul.* **57**:61.

McAllister, R.M., Nicholson, M.O., Reed, G., Kern, J., Gilden, R.V., and Huebner, R.J., 1969, Transformation of rodent cells by adenovirus 19 and other group D adenoviruses, *J. Natl. Cancer Inst.* **43**:917.

McAllister, R.M., Gilden, R.V., and Green, M., 1972, Adenoviruses in human cancer, *Lancet* **1**:831.

McCormick, D.P., Wenzel, R.P., Davies, J.A., and Beam, W.E., 1972, Nasal secretion protein responses in patients with wild-type adenovirus disease, *Infect. Immun.* **6**:282.

Medical Research Council Working Party on Acute Respiratory Virus Infections, 1965, a collaborative study of the etiology of acute respiratory infections in Britain 1961–4: A report of the Medical Research Council Working Party on Acute Respiratory Virus Infections, *Br. Med. J.* **2**:319.

Monto, A.S., and Ullman, B.M., 1974, Acute respiratory illness in an American community: The Tecumseh study, *J. Am. Med. Assoc.* **227**:164.

Mufson, M.A., and Belshe, R.B., 1976, A review of adenoviruses in the etiology of acute hemorrhagic cystitis, *J. Virol.* **115**:191.

Myerowitz, R.L., Stalder, H., Oxman, M.N., Levin, M.J., Moore, M., Leith, J.D., Gantz, N.M., Pellegrini, J., and Hierholzer, J.C., 1975, Fatal disseminated adenovirus infection in a renal transplant recipient, *Am. J. Med.* **59**:591.

Nelson, K.E., Gavitt, F., Batt, M.D., Kallick, C.A., Reddi, K.T., and Levin, S., 1975, The role of adenoviruses in the pertussis syndrome, *J. Pediatr.* **86**:335.

Neumayer, E.M., Haff, R.F., and Hoffman, C.E., 1965, Antiviral activity of amantidine hydrochloride in tissue culture and *in vivo*, *Proc. Soc. Exp. Biol. Med.* **119**:393.

Norrby, E., 1969, The structural and functional diversity of adenovirus capsid components, *J. Gen. Virol.* **5**:221.

Norrby, E., Bartha, A., Boulanger, T., Dreizin, R., Ginsberg, H.S., Kalter, S.S., Kawamura, H., Rowe, W.P., Russell, W.C., Schlesinger, W., and Wigand, R., 1976, Adenoviridae, *Intervirology* **7**:117.

Numazaki, Y., Shigeta, S., Kumasaka, T., Miyazawa, T., Yamanaka, M., Yano, N., Takai, S., and Ishida, N., 1968, Acute hemorrhagic cystitis in children: Isolation of adenovirus type II, *N. Engl. J. Med.* **278**:700.

Numazaki, Y., Kumasaka, T., Yano, N., Yamanaka, M., Miyazawa, T., Takai, S., and Ishida, N., 1973, Further study on acute hemorrhagic cystitis due to adenovirus type 11, *N. Engl. J. Med.* **289**:344.

O'Conor, G.T., Rabson, A.S., Berezesky, I.K., and Paul, F.J., 1963, Mixed infection with simian virus 40 and adenovirus 12, *J. Natl. Cancer Inst.* **31**:903.

O'Day, D.M., Guyer, B., Hierholzer, J.C., Rosing, K.J., and Schaffner, W., 1976, Clinical and laboratory evaluation of epidemic keratoconjunctivitis due to adenovirus types 8 and 19, *Am. J. Ophthalmol.* **81**:207.

Olsen, L.C., Miller, G., and Hanshaw, J.B., 1964, Acute infectious lymphocytosis presenting as a pertussis-like illness: Its association with adenovirus type 12, *Lancet* **1**:200.

Overby, L.R., Duff, R.G., and Mao, J.C.-H., 1977, Antiviral potential of phosphonoacetic acid, *Ann. N.Y. Acad. Sci.* **284**:310.

Padgett, B.L., Zu Rhein, G.M., Walker, D.L., Eckroade, R.J., and Dessel, B.H., 1971, Cultivation of a papova-like virus from human brain with progressive multifocal leukoencephalopathy, *Lancet* **1**:1257.

Parks, W.P., Queiroga, L.M., Melnick, J.L., and Pereira, H.G., 1967, Recent adenovirus isolates exhibiting broad intratypic and intertypic antigenicity, *Proc. Soc. Exp. Biol. Med.* **125**:498.

Parks, W.P., Casazza, A.M., Alcott, J., and Melnick, J., 1968, Adeno-associated satellite virus interferes with the replication of its helper adenovirus, *J. Exp. Med.* **127**:91.

Parrott, R.H., Rowe, W.P., Huebner, R.M., Bernton, H.W., and McCullough, N.B., 1954, Outbreak of febrile pharyngitis and conjunctivitis associated with type 3 adenoidal–pharyngeal–conjunctival virus infection, *N. Engl. J. Med.* **251**:1087.

Pavan-Langston, D., and Dohlman, C.H., 1972, A double-blind clinical study of adenine arabinoside therapy of keratoconjunctivitis, *Am. J. Ophthalmol.* **74**:81.

Pereira, M.S., Pereira, H.G., and Clarke, S.K.R., 1965, Human adenovirus type 31: A new serotype with oncogenic properties, *Lancet* **1**:21.

Pettit, T.H., and Holland, G.N., 1979, Chronic keratoconjunctivitis associated with ocular adenovirus infection, *Am. J. Ophthalmol.* **88**:748.

Piña, M., and Green, M., 1965, Biochemical studies on adenovirus multiplication. IX. Chemical and base composition analyses of 28 human adenoviruses, *Proc. Natl. Acad. Sci. U.S.A.* **54**:547.

Potter, C.W., 1964, Adenovirus infection as an etiological factor in intussusception of infants and young children, *J. Pathol. Bacteriol.* **88**:263.

Retter, M., Middleton, P.J., Tam, J.S., and Petric, M., 1979, Enteric adenoviruses: Detection, replication, and significance, *J. Clin. Microbiol.* **10**:574.

Richmond, S.J., Caul, E.O., Dunn, S.M., Ashley, C.R., Clarke, S.K.R., Seymour, N.R., 1979, An outbreak of gastroenteritis in young children caused by adenoviruses, *Lancet* **1:**1178.

Rodriguez-Torres, R., Lin, J.-S., and Berkovich, S., 1969, A sensitive electrocardiographic sign in myocarditis associated with viral infection, *J. Pediatr.* **43:**846.

Roos, R., Chou, S.N., Rodgers, N.G., Basnight, M., and Gojdusek, D.C., 1972, Isolation of adenovirus 32 strain from human brain in a case of subacute encephalitis, *Proc. Soc. Exp. Biol. Med.* **139:**636.

Rose, H.M., Lamson, T.H., and Buescher, E.L., 1970, Adenoviral infection in military recruits: Emergence of type 7 and 21 infections in recruits immunized with type 4 oral vaccine, *Arch. Environ. Health* **21:**356.

Rosen, L., 1960, A hemagglutination-inhibition technique for typing adenoviruses, *Am. J. Hyg.* **71:**121.

Rowe, W.P., and Baum, S.G., 1964, Evidence of a possible genetic hybrid between adenovirus type 7 and SV40 viruses. *Proc. Natl. Acad. Sci. U.S.A.* **52:**1340.

Rowe, W.P., Huebner, R.J., Gilmore, L.K., Parrott, R.H., and Ward, T.G., 1953, Isolation of a cytopathogenic agent from human adenoids undergoing spontaneous degeneration in tissue culture, *Proc. Soc. Exp. Biol. Med.* **84:**570.

Schaap, G.J.P., de Jong, j.C., van Büsterveld, O.P., and Beekhuis, W.H., 1979, A new intermediate adenovirus causing conjunctivitis, *Arch. Ophthalmol.* **97:**2336.

Sidwell, R.W., 1980, Ribavirin: *In vitro* antiviral activity, in: *Ribavirin: A Broad Spectrum Antiviral Agent* (R.A. Smith and W. Kirkpatrick, eds.), p. 23, Academic Press, New York.

Sidwell, R.W., Allen, L.B., Huffman, J.H., Khwaja, T.A., Tolman, R.L., and Robins, R.K., 1973, Anti DNA virus activity of the 5′-nucleotide and 3′,5′-cyclic nucleotide of 9-beta-D-arabinofuranosyladenine, *Chemotherapy* **19:**325.

Siegal, F.P., Dikman, S.H., Arayata, R.B., and Botione, E.J., 1981, Fatal disseminated adenovirus 11 pneumonia in an agammaglobulinemic patient, *Am. J. Med.* **71:**1062.

Similä, S., Jouppila, R., Salmi, A., and Pohjonen, R., 1970, Encephalomeningitis in children associated with an adenovirus type 7 epidemic, *Acta Paediatr. Scand.* **59:**310.

Smith, J.J., Buescher, E.L., and Top, F.H., Jr., 1970, Experimental respiratory infection with type 4 adenovirus vaccine in volunteers: Clinical and immunological responses, *J. Infect. Dis.* **122:**239.

Sprague, J.B., Hierholzer, J.C., Currier, R.W., II, Hattwick, M.A.W., and Smith, M.D., 1973, Epidemic keratoconjunctivitis: A severe industrial outbreak due to adenovirus type 8, *N. Engl. J. Med.* **289:**1341.

Stechenberg, B.W., Keating, J.P., Koslov, S., Schechter, M., Chang, M., Haymond, M.W., and Feigin, R.D., 1975, Epidemiologic investigations of Reye's syndrome, *J. Pediatr.* **87:**234.

Stevens, D.A., Schaeffer, M., Fox, J.P., Brandt, C.D., and Romano, M., 1967, Standardization and certification of reference antigens and antisera to 30 human adenovirus serotypes, *Am. J. Epidemiol.* **86:**617.

Sterner, G., 1962, Adenovirus infections in childhood: An epidemiological and clinical survey among Swedish children, *Acta Paediatr. Scand. Suppl.* **142:**1.

Stuart-Harris, C.H., Andrewes, C.H., Smith, W., Chalmers, D.K.M., Cowen, E.G.H., and Hughes, D.L., 1938, A study of epidemic influenza: With special reference to 1936–37 epidemic, Special Reports Series 228, Medical Research Council, London.

Tai, F.H., Grayston, J.T., Johnson, P.B., and Woolridge, R.L., 1960, Adenovirus infections in Chinese army recruits on Taiwan, *J. Infect. Dis.* **107:**160.

Tai, F.H., Chu, S., Chi, W.H., Wei, H.Y., and Hierholzer, J.C., 1974, Epidemic hemorrhagic conjunctivitis associated with adenovirus type 11 in Taiwan, *Southeast Asian J. Trop. Med. Public Health* **5:**342.

Takafuji, E.T., Gaydos, J.C., Allen, R.G., and Top, F.H., Jr., 1979, Simultaneous administration of live, enteric-coated adenoviruses types 4, 7, and 21 vaccines: Safety and immunogenicity, *J. Infect. Dis.* **140:**48.

Takiff, H.E., and Straus, S.E., 1982, An early replicative block prevents the efficient growth of fastidious diarrhea-associated adenoviruses in cell culture, *J. Med. Virol.* **9:**93.

Takiff, H.E., Straus, S.E., and Garon, C.F., 1981, Propagation and *in vitro* studies of previously noncultivable enteral adenoviruses in 293 cells, *Lancet* **2**:832.

Taylor, P.E., 1977, Adenoviruses: Diagnosis of infections, in: *Comparative Diagnosis of Viral Diseases*, Vol. 1 (E. Kurstak and C. Kurstak, eds), p. 85, Academic Press, New York.

Top, F.H., Jr., Buescher, E.L., Bancroft, W.H., and Russell, P.K., 1971a, Immunozation with live types 7 and 4 adenovirus vaccines. II. Antibody response and protective effect against acute respiratory disease due to adenovirus type 7, *J. Infect. Dis.* **124**:155.

Top, F.H., Jr., Dudding, B.A., Russell, P.K., and Buescher, E.L., 1971b, Control of respiratory disease in recruits with types 4 and 7 adenovirus vaccines, *Am. J. Epidemiol.* **94**:142.

Top, F.H., Jr., Grossman, R.A., Bartelloni, P.J., Segal, H.E., Dudding, B.A., Russell, P.K., and Buescher, E.L., 1971c, Immunization with live types 7 and 4 adenovirus vaccines. I. Safety, infectivity, antigenicity, and potency of adenovirus type 7 vaccine in humans, *J. Infect. Dis.* **124**:148.

Van der Veen, J., and Kok, G., 1957, Isolation and typing of adenovirus recovered from military recruits with acute respiratory disease in the Netherlands, *Am. J. Hyg.* **65**:119.

Van der Veen, J., Oei, K.G., and Abarbanel, M.F.W., 1969, Patterns of infections with adenovirus types 4, 7, and 21 in military recruits during a 9-year survey, *J. Hyg.* **67**:255.

Wadell, G., 1979, Classification of human adenoviruses by SDS–polyacrylamide gel electrophoresis of structural proteins, *Intervirology* **11**:47.

Wadell, G., and de Jong, J.C., 1980, Restriction endonucleases in identification of a genome type of adenovirus 19 associated with keratoconjunctivitis, Infect. Immun. **27**:292.

Wadell, G., and Varsanyi, T.M., 1978, Demonstration of three different subtypes of adenovirus type 7 by DNA restriction site mapping, *Infect. Immun.* **21**:238.

Wadell, G., Hammarskjöld, M.-L., Winberg, G., Varsanyi, T.M., and Sundell, G., 1980a, Genetic variability of adenoviruses, *Ann. N.Y. Acad. Sci.* **354**:16.

Wadell, G., de Jong, J.C., and Wolondis, S., 1981, Molecular epidemiology of adenoviruses: Alternating appearance of two different genome types of adenovirus 7 during epidemic outbreaks in Europe from 1958 to 1980, *Infect. Immunol.* **34**:368.

Ward, D.C., and Tattersall, P. (eds.), 1978, *Replication of Mammalian Parvoviruses*, Cold Spring Harbor Laboratory Press, Cold Spring Harbor, New York.

Waring, G.O., III, Liabson, P.R., Satz, J.E., and Joseph, N.H., 1976, Use of vidarabine in epidemic keratoconjunctivitis due to adenovirus types 3, 7, 8, and 19, *Am. J. Ophthalmol.* **82**:781.

Weiner, L.P., Herndon, R.M., Narayan, O., Johnson, R.T., Shah, K., Ruberstein, L.J., Preziosi, T.J., and Conley, F.K., 1972, Isolation of virus related to SV40 from patients with progressive multifocal leukoencephalopathy, *N. Engl. J. Med.* **286**:385.

White, D.O., and Solomon, J.R., 1966, Adenovirus and intussusception, *Med. J. Aust.* **1**:447.

White, G.B.B., and Stancliffe, D., 1975, Viruses and gastroenteritis, *Lancet* **2**:703.

Wigand, R., Bauer, H., Lang, E., and Adam, W., 1965, Neutralization of the adenovirus types 1 to 28: Specificity and antigenic relationships, *Arch. Gesamte. Virusforsch.* **15**:188.

Wigand, R., Geldenbloom, H., and Wadell, G., 1980, New human adenoviruses (candidate adenovirus 36): A novel member of subgroup D, *Arch. Virol.* **64**:225.

Wilkins, J., and Wehrle, P.F., 1979, *Bordetella* species, in: *Principles and Practices of Infectious Diseases* (G.L. Mandell, R.G. Douglas, Jr., and J.E. Bennett, eds.), p. 1800, John Wiley, New York.

Wold, W.S.M., Mackey, J.K., Rigden, P.M., and Green, M., 1979, Analysis of human cancer DNA's for DNA sequences of human serotypes 3, 7, 11, 14, 16, and 21 in group B, *Cancer Res.* **39**:3479.

Yin-Murphy, M., Lim, K.H., and Chua, P.H., 1974, Adenovirus type II epidemic conjunctivitis in Singapore, *Southeast Asian J. Trop. Med. Public Health* **5**:333.

Yolken, R.H., Lawrence, F., Leister, F., Takiff, H.E., and Straus, S.E., 1982, Enteric type adenovirus: An important cause of gastrointestinal and respiratory disease in hospitalized infants, *J. Pediatr.* **101**:21.

Yunis, E.J., and Hashida, Y., 1973, Electron microscopic demonstration of adenovirus in appendix vermiformis in a case of ileocecal intussusception, *Pediatrics* **51**:566.

Zahradnik, J.M., Spencer, G.J., and Porter, D.D., 1980, Adenovirus infection in the immune-compromised patient, *Am. J. Med.* **68**:725.

Zhdanov, V.M., and Dreizin, R.S., 1961, A group of strains of a new serologic type of adenovirus, *Probl. Virol.* **6**:98.

CHAPTER 12

Adenoviruses of Animals

MASAHIDE ISHIBASHI AND HIROSHI YASUE

> During the fall of that year we observed a violent epizootic in the fur ranges of Fromm Brothers in Wisconsin, where 4,000 foxes were running free in four wooded enclosures approximately fifty acres in size. Under fur-range conditions, where direct contact was possible, deaths mounted rapidly to a maximum and then quickly subsided to an occasional death. As foxes are placed in fur ranges only during the fall, prior to pelting, an opportunity for well organized study of such a large epizootic was not again offered until the fall of 1926. The Fromm fur range epizootic of 1926 included a population of about 6,500 foxes, of which 747 died.
>
> —R.G. Green *et al.* (1930)

I. INTRODUCTION

The opening quotation is a part of the description of the first epizootic of adenovirus, in which the etiological agent then named fox encephalitis virus, and now known as canine adenovirus type 1, was isolated. It was more than 20 years later that a group of human viruses were isolated and reported as adenoid-degenerating agents (Rowe *et al.*, 1953) or respiratory-illness agents (Hilleman and Werner, 1954), and the name *adenovirus* was devised for the group of viruses (Enders *et al.*, 1956). In retrospect, therefore, the canine adenovirus is the first member of the adenovirus family.

MASAHIDE ISHIBASHI AND HIROSHI YASUE • Laboratory of Viral Oncology, Research Institute, Aichi Cancer Center, Chikusa-ku, Nagoya 464, Japan.

At the time when Sohier *et al.* (1965) wrote their review titled "Adenoviruses: Status of current knowledge," the number of adenovirus types totaled 49. According to the third report of the International Committee on Taxonomy of Viruses (Matthews, 1979), 89 serotypes were definitely classified as adenoviruses, 5 strains were probable members, and 14 strains were possible members. (The data available at present have led the authors of this chapter to include still another 8 strains.) This large increase in the number of adenoviruses, especially the increase in nonhuman adenoviruses, and the finding that mammalian adenoviruses more or less share a common antigen allowed this group of viruses to obtain a higher taxonomic rank, that is, in the family called Adenoviridae. This family is divided into two genera, *Mastadenovirus* and *Aviadenovirus*. In addition, there are some candidate viruses for this family among poikilothermic animals.

When a filterable infective agent is isolated in a cell-culture system, the following presumptive pieces of evidence must be present to classify the agent as an adenovirus:

1. The agent in the infected cell lysate tolerates the prescribed heat treatment of 56°C for 10 min or more, as well as lipid solvent such as chloroform or ether.
2. The agent grows well in cultures of cells such as primary kidney or testicle cells originating from a presumptive natural host animal, accompanied by a characteristic microscopically visible cytopathic effect (CPE). However, the agent does not grow well or at all even in cells of animals pylogenetically close to the presumptive natural host.
3. Characteristic intranuclear inclusion bodies are observed when infected cells that show CPE are fixed and stained with hematoxylin–eosin.
4. When tested under the electron microscope (EM) after negative staining, the infected cell lysate or the purified sample shows a large number of icosahedral particles, 70–80 nm in diameter, consisting of 252 capsomers.
5. The agent cannot grow in a medium containing thymidine analogues such as 5-bromo-2'-deoxyuridine because of the DNA nature of its genome.

When these presumptive pieces of evidence are provided, more precise information is required, such as presence of the antigen common to mammalian adenoviruses when the presumptive natural host is mammalian, serotypic distinctiveness by the neutralization test, pathogenicity in the presumptive natural host, polypeptide composition of the capsid proteins, size of the viral DNA, and the presence of an inverted terminal repetition in its DNA sequence.

Today, the adenovirus is one of the best-studied viruses. Much precise information, especially in the field of molecular biology, has been accumulated, mostly on certain human adenoviruses. Therefore, studies on other than human members of the family must be conducted not only for advancement of veterinary medicine but also for thorough understanding of the family of Adenoviridae, including their evolutionary aspect. Actually, some information that contributes to a general understanding of this family of adenoviruses as a whole has been obtained by taking advantage of particular conditions of some nonhuman adenoviruses, for instance: (1) Many rodent tumors induced with fowl adenovirus type 1 carry the viral DNA as repetitive virus–cellular sequences of more than a hundred copies per cell (Yasue *et al.*, 1980). With the use of these cell lines, the tumor cells in the mitotic phase were analyzed by the *in situ* DNA hybridization technique. The analysis has provided "seeing-is-believing"-type evidence that the viral DNA in an adenovirus-induced tumor cell is an integral part of the cellular chromosomes (Yasue and Ishibashi, 1982), as is discussed in Section VIII. (2) Some temperature-sensitive mutants of fowl adenovirus type 1 have provided very clear genetic evidence (Ishibashi, 1970; 1971) supporting the inference, obtained mainly by an autoradiographic analysis of human adenovirus type 5, that capsid proteins are synthesized in the cytoplasm of the infected cells and rapidly transported into the nucleus (Velicer and Ginsberg, 1968; 1970). (3) More important, egg drop syndrome 1976 virus, an avian adenovirus, provided evidence that in some special conditions an adenovirus was initially transmitted almost exclusively by the vertical route and caused serious pathogenic effects in the successive generations of host animals (McFerran *et al.*, 1977), whereas adenoviruses were usually transmitted through the horizontal route without vectors. This quite new epizootiological situation seems to be due to the unexpected contamination of the adenovirus in the preparation of live vaccine, as is explained more precisely in Section VII. This prompts the taking of more precautionary measures before application of live vaccines for disease prevention in humans and animals. (4) Infection of laboratory mice with murine adenoviruses seems to be the best model system for studies of viral pathogenesis, since many well-known genetic, more particularly immunogenetic, phenotypes and several indigenous and invading microorganisms of the mouse are familiar and controllable. A few lines of experimental studies on pathogenesis are developing with murine adenoviruses, as is noted in Section VII. In addition, cell-mediated immunity against adenoviruses has been studied best in the mouse system (Inada and Uetake, 1978), though the body of information on it is still very limited in comparison with that on humoral immunity.

This chapter on animal adenoviruses complements the information and discussions in several of the preceding chapters. The various sections deal only with those subjects concerning which there is an existing body of information sufficient for discussion. Some overlapping with other chapters is unavoidable. Section VII provides a treatment of the infection of natural host animals with nonhuman adenoviruses, because it follows, and complements, the information in Chapter 11 on adenovirus infection in humans.

II. NOMENCLATURE AND ABBREVIATIONS

The conventional nomenclature for most adenoviruses [e.g., bovine adenovirus type 3 (BAV-3)] consists of an abbreviation of the name of the adenovirus group and an arabic number denoting the individual serotype in the group. For the sake of conciseness, individual adenoviruses are so abbreviated in this chapter, with three exceptions:

1. A new and slightly different nomenclature has been proposed for simian adenoviruses (e.g., adenovirus S-2) to unify several conventional nomenclatures (Kalter *et al.*, 1980), and simian adenovirus serotypes are designated both by the newer nomenclature and in the aforedescribed manner, e.g., simian adenovirus type 2 (SAV-2). Additionally, since older nomenclatural designations such as simian virus (SV) or simian agent (SA), followed by an isolation number (e.g., SV11), are still popular, the newer designations are followed by the older ones in brackets, e.g., SAV-2 [SV11].
2. Human adenoviruses are designated as they are throughout this volume, e.g., adenovirus type 2 (Ad2).
3. Adenoviruses that have not yet been assigned serotype numbers are designated by abbreviations only, e.g., hemorrhagic enteritis virus of turkey (HEV).

III. STRUCTURE AND COMPONENTS OF THE VIRAL CAPSID

Since it is a prerequisite for classification as an adenovirus that isolated viral particles show icosahedra comprising 252 capsomers and that the diameter of the icosahedra be about 70–80 nm, electron micrographs showing these characteristics have been provided for most nonhuman adenoviruses. However, only a few have demonstrated the presence of fiber projections. Equine adenovirus (EAV) was shown to have the longest fiber projection, 50 nm (Fatemie-Nainie and Marusyk, 1979). Canine adenovirus type 1 (CAV-1) and CAV-2 were demonstrated to have fibers 25–27 and 35–37 nm long, respectively (Marusyk *et al.*, 1970). The distinctive difference in fibers between these canine adenoviruses was one of the pieces of evidence that prompted researchers to consider that the strain now called CAV-2 was not just an attenuated variant of CAV-1. Mouse adenovirus type 1 (MAV-1) was shown to have fiber projections 29 nm long (Wigand *et al.*, 1977); bovine adenovirus type 3 (BAV-3) was reported to have fibers 12–18 nm long (Niiyama *et al.*, 1975). In contrast to each penton having a single fiber, as shown for other adenoviruses, fowl adenovirus type 1 (FAV-1) was found to have unique pentons, each with a penton base and two attached fibers—one short (8.5 nm) and the other long (42.5 nm). The long fiber was in many cases bent sharply, with an "elbow" at about 10 nm from the base (Laver *et al.*, 1971). A more recent EM study (Gelderblom and Maichle-Lauppe, 1982) not only confirmed the aforecited finding of FAV-1 but also demonstrated that all the other types of fowl adenoviruses also had pentons with two fibers. However, the paired fibers in these viruses did not display such a distinctive difference in length as those of FAV-1. The length of fibers varied from one strain to another within the range from 16 to 30 nm. Since egg drop syndrome 1976 virus (EDS76V), a duck adenovirus, was shown to have pentons with a single fiber 25–42 nm long (Todd and McNulty, 1978; Kraft *et al.*, 1979; Gelderblom and Maichle-Lauppe, 1982), like mammalian adenoviruses, the possession of pentons with double fibers does not seem to be a general feature of the avian adenoviruses; it may be a unique feature of the fowl adenovirus group.

The adoption of sodium dodecyl sulfate–polyacrylamide gel electrophoresis (SDS-PAGE) for analysis of the capsid proteins of the virus has not only permitted determination of the approximate size and number of the component polypeptides, but also facilitated identification of the location of individual polypeptides in the capsid. With human adenoviruses such as Ad2, most polypeptides that form the viral capsid have been located, as reviewed in detail in the previous chapters. Catalogues of polypeptides that make up the viral capsid are also available for a few animal adenoviruses, namely, simian adenovirus type 16 (SAV-16) [SA7] (Wadell *et al.*, 1980), BAV-3 (Niiyama *et al.*, 1975), EAV (Fatemie-Nainie and Marusyk, 1979), *Tupaia* adenovirus (TupAV) (Faissner *et al.*, 1980), FAV-1 (Yasue and Ishibashi, 1977), and EDS76V (Todd and McNulty, 1978). In the report on capsid polypeptides of SAV-16 [SA7], the electrophoreis was done in parallel with that of the polypeptides of Ad2 and Ad18. The electrophoretic patterns of these viruses were so similar that one may assume that simian polypeptides showing mobilities similar to those of their human counterparts are located at the same positions in the architecture of the capsid. The more remote the phylogenetic relationship between two host animals becomes, the more the electrophoretic patterns of polypeptides of the corresponding adenoviruses seem to differ. The safest inference to be made only on the basis of the electropherograms is that the thickest band showing the slowest mobility is that of hexon polypeptides.

It may be of interest to compare the electrophoretic patterns of the capsid polypeptides of all the adenoviruses under defined conditions. Catalogues of such patterns will be of considerable assistance in the identification of unknown adenovirus isolates, possibly contributing to epizootiological studies of possible intraserotypic as well as interserotypic variation. Actually, Wadell *et al.* (1980) attempted to classify human adenoviruses into subgroups on the basis of their electrophoretic similarity and concluded that polypeptides of the externally located hexon and vertex capsomers displayed a larger variation in apparent molecular weights than the internal polypeptides, when the ranges of the apparent molecular weights of several capsid polypeptides of the 36 human adenovirus serotypes were compared.

The fiber protein of Ad2 was demonstrated to be a glycoprotein, with approximately two *N*-acetylglucosamines being linked to a fiber polypeptide (Ishibashi and Maizel, 1974). This was the first evidence that glycoprotein was a part of the composition of a nonenveloped virus. It is therefore intriguing for comparative virology to ask whether similar sugar residues are present in fiber proteins of all the adenoviruses.

IV. VIRAL DNA

A. General Properties

For all the nonhuman adenoviruses, simple investigations to determine whether viral growth is prevented by a DNA replication inhibitor have been conducted to show that nucleic acid in the viral particle is DNA rather than RNA. However, there are few nonhuman adenovirus serotypes for which the DNA has been precisely analyzed. The linearity and double-strandedness of viral DNA have been observed by EM, by cleavage mapping with sequence-specific (restriction) endonucleases, and by sucrose density-gradient centrifugation for a number of nonhuman adenoviruses belonging to both genera (Table I).

The presence of terminal proteins covalently linked to the 5′ end of the two strands has also been indirectly indicated for several nonhuman adenoviruses belonging to the two genera, such as simian adenovirus type 15 (SAV-15) [SV38] (Dimitrov *et al.*, 1979), SAV-16 [SA7] (Naroditsky *et al.*, 1978), mouse adenovirus type 1 (MAV-1) (Larsen and Nathans, 1977), *Tupaia* adenovirus type 1 (TupAV-1) (Darai *et al.*, 1980), fowl adenovirus type 1 (FAV-1) (Robinson *et al.*, 1973; Denisova *et al.*, 1979), and egg drop syndrome 1976 virus (EDS76V) (Zsák and Kisary, 1981).

The presence of inverted terminal repetition (ITR) sequences has been demonstrated for only three serotypes of nonhuman adenoviruses belonging to the two genera, SAV-16 [SA7] (Tolun *et al.*, 1979), MAV-1 (Temple *et al.*, 1981), and FAV-1 (Robinson and Bellett, 1975; Aleström *et al.*, 1982a). The lengths of the ITR sequences of SAV-16 [SA7] and MAV-1 are reported to be 120 base pairs (bp) and 93 bp, respectively. The ITR sequence of FAV-1 is the shortest ITR sequence thus far reported for an adenovirus, and it is no longer than 63 bp, even if one disregards the mismatches between the left and right terminal sequences at the 55th and 56th bases from each of the two termini.

TABLE I. Physicochemical Properties of DNA of Nonhuman Adenoviruses

Group and subgroup	Serotypes	Molecular weight $(\times 10^{-6})$	GC content (%)	Density in CsCl	References
Simian					Burnett *et al.* (1972)
I	SAV-1 [SV1]	22.5	55	1.714	
	SAV-2 [SV11]	22.5	55	1.714	
	SAV-7 [SV25]	20.7	56	1.715	
	SAV-11 [SV33]	19.9	55	1.714	
	SAV-12 [SV34]	17.9	55	1.714	
	SAV-15 [SV38][a]	17.6	55	1.714	
II	SAV-3 [SV15]	17.7	55	1.714	
	SAV-4 [SV17]	17.0	55	1.714	
	SAV-6 [SV23]	16.9	56	1.715	
	SAV-9 [SV31]	17.6	55	1.714	
	SAV-10 [SV32]	20.2	55	1.714	
	SAV-14 [SV37]	21.2	55	1.714	
III	SAV-16 [SA7][a]	21.7	58	1.717	
	SAV-5 [SV20][a]	20.4	62	1.721	
	SAV-8 [SV30]	18.7	60	1.719	
Bovine	BAV-3[a]	24	58	1.717	Niiyama *et al.* (1975)
Equine	EAV[a]	21–22	—	—	Fatemie-Nainie and Marusyk (1979)
Mouse	MAV-1[a]	20.7	44	1.702	Larsen and Nathans (1977)
Tupaia	TupAV	21.5	47	1.706	Darai *et al.* (1980)
Fowl	FAV-1[a]	28.3	54	1.713	Laver *et al.* (1971), Robinson *et al.* (1973)
Duck	EDS76V[a]	22.6	—	—	Zsák and Kisary (1981)

[a] A cleavage map of the virus DNA has been constructed.

Thus the basic features of adenovirus DNA are considered to be (1) the linearity of the double strands, (2) the presence of terminal proteins covalently linked to the 5′ ends of both strands, and (3) the presence of the ITR sequence.

The sizes (or molecular weights) of viral DNAs have been estimated for 15 serotypes of simian adenoviruses and for bovine adenovirus type 3 (BAV-3), equine adenovirus (EAV), MAV-1, TupAV, FAV-1, and EDS76V (Table I). These show a wide variation from 17 to 28 $\times$ 10^6 daltons. FAV-1 is the only virus among the 11 types of the fowl adenovirus group for which the DNA size has been measured. It is estimated to be 28 $\times$ 10^6 daltons, distinctively larger than those of mastadenoviruses and that of a duck adenovirus, EDS76V, and suggests that the average-size DNA of this group is clearly larger than those of other groups of adenoviruses in the two genera. One may assume that several repetition sequences other than an ITR sequence are present in FAV-1 DNA. However, the Southern blot hybridization experiment using the nick-translated DNA probe, which can detect even the ITR sequence of FAV-1 under particular conditions, has never indicated such a repetition sequence (Yasue *et al.*, 1980; Yasue and Ishibashi, 1982).

The guanine–cytosine (GC) contents of nonhuman adenoviruses thus far reported range from 44 to 62%. A clear correlation between GC content and oncogenicity has been observed in human adenoviruses (Piña and Green, 1965), but not in simian adenoviruses (Goodheart, 1971), indicating that the correlation observed in human adenoviruses may be only apparent.

B. DNA Homology

From the outset, the relatedness of adenoviruses has been deduced from several of their phenotypes that have been considered to be controlled by the viral genome, such as hemagglutination (HA) activity, hemagglutination-inhibition antigen, and serum neutralization (SN) antigen, together with the range of host-cell species in which individual viruses can grow. The conventional classification of adenoviruses is based on these phenotypic properties of the viruses. However, this classification may be reviewed from the viewpoint of genotype homology among viruses. Such an approach may provide important information that has not been obtained by analyses of several phenotypic properties.

A knowledge of the complete base sequences of individual adenovirus DNAs is ideal for comparative analysis of genotype homology. However, determination of a 30- to 40-kilobase (kb) DNA sequence of adenoviruses requires a huge amount of labor and time with the techniques available at present. In lieu of this ideal type of analysis, genotype homology has been examined by a few simplified methods, though the information obtainable with these simplified methods is severely limited. These methods are discussed in the following sections.

1. Homology Observed in ITR Sequences

Although the functional significance of the ITR sequence has not yet been elucidated, this sequence seems to be a basic feature of the adenovirus family, as mentioned in Section IV.A. Therefore, the base sequence homology has been compared among the ITR sequences of several adenoviruses thus far determined, and it can be summarized as follows: A comparison of two primate adenoviruses, Ad2 and SAV-16 [SA7], revealed large portions of the ITRs with an identical base sequence. A comparison of three mastadenoviruses (Ad2, SAV-16 [SA7], and MAV-1) showed a common sequence from the 9th to the 17th base from the termini, 5′-

ATAATATAC-3′. When the ITR sequence of FAV-1, an aviadenovirus, was included in this comparison, only the sequence from the 9th to the 14th base, 5′-ATAATA-3′, was found to be a common one. So far, the more remote the taxonomic relationship of host animal species becomes, the shorter is the common base sequence in ITRs of the corresponding adenoviruses. This correlation would be well substantiated if a few additional ITR sequences were determined in aviadenoviruses in particular.

2. Homology Observed in Sequence-Specific Endonuclease Cleavage Patterns

Wadell *et al.* (1980) attempted to classify all the human adenoviruses on the basis of the cleavage patterns of the viral DNA by a six-base sequence-specific restriction endonuclease, *Sma*I. Their classification was found to be compatible with classifications based on other principles such as viral HA activity. Their report also indicated that this method was very useful to evaluate genotypic relationships between a field isolate expressing intermediate serotypes and its hypothetical parents of authentic serotypes. Moreover, certain chimpanzee adenoviruses known to be very similar to human adenoviruses in several phenotypic properties were found to have a cleavage pattern very similar to that of Ad4.

Cleavage maps or patterns of DNA with sequence-specific restriction endonucleases such as *Eco*RI have been developed for several nonhuman adenoviruses, SAV-5 [SV20] (Denisova *et al.*, 1980), SAV-15 [SV38] (Dimitrov *et al.*, 1979), SAV-16 [SA7] (Naroditsky *et al.*, 1978), BAV-3 (Kurokawa *et al.*, 1978), MAV-1 (Larsen and Nathans, 1977; Larsen *et al.*, 1979; Temple *et al.*, 1981), TupAV (Darai *et al.*, 1980), FAV-1 (Denisova *et al.*, 1979; Shimada *et al.*, 1983), and EDS76V (Zsák and Kisary, 1981). However, it is impossible to find any similarity among these cleavage patterns, indicating that this type of analysis may be useful only when a number of adenovirus strains to be compared are so close as to belong to the same group (i.e., have the same host range of growth) or subgroup. Actually, the Phelps and Ote strains belonging to the same serotype, FAV-1, although indistinguishable from each other by the SN test, have been demonstrated to show minor differences in patterns when cleaved by a four-base sequence-specific restriction endonuclease (Aleström *et al.*, 1982a; Yasue *et al.*, unpublished data). Similar observations were made for Ad7 (Wadell *et al.*, 1980).

3. Homology Observed in DNA–DNA Hybridization

Simplified DNA–DNA hybridization procedures have often been utilized as a means to measure the degree of genotypic homology of two viruses (Denhardt, 1966). In these procedures, one of the two viral DNAs is radioactively labeled, sheared into pieces approximately 800 bp long,

TABLE II. Simian Adenovirus Subgrouping Based on DNA Homology Index[a]

Subgroup	Serotypes	DNA homology index
I	SAV-1 [SV1], SAV-2 [SV11], SAV-7 [SV25], SAV-11 [SV33], SAV-12 [SV34], SAV-15 [SV38]	78–95% within subgroup 28–55% with other subgroups
II	SAV-3 [SV15], SAV-4 [SV17], SAV-6 [SV23], SAV-9 [SV31], SAV-10 [SV32], SAV-14 [SV37]	80–100% within subgroup 28–48% with other subgroups
III	SAV-5 [SV20], SAV-8 [SV30], SAV-16 [SA7]	64–87% within subgroup 38–55% with other subgroups

[a] Data reported by Burnett *et al.* (1972) are rearranged.

and denatured. Under particular conditions, the pieces are hybridized with a large excess of the other DNA, which has been previously denatured and immobilized on a filter. The homology index between the two DNA is the relative efficiency (expressed in percent), which is calculated by dividing the efficiency of the aforedescribed heterologous DNA hybridization (e.g., FAV-1 vs. Ad2) by the efficiency of the control homologous DNA hybridization (e.g., FAV-1 vs. FAV-1 or Ad2 vs. Ad2).

Using this index as the measure of DNA homology, M. Green *et al.* (1979) have divided a number of serotypes of human adenoviruses into five subgroups, A–E. This classification shows good accordance with the classification based on several phenotypic properties of the viruses. The homology indices of any pair of serotypes within the subgroups except for subgroup A are more than 90%. The homology indices of pairs of serotypes in subgroup A range from 48 to 69%. Homology indices of pairs of serotypes belonging to different subgroups range from 4 to 23%. In nonhuman adenoviruses, studies based on the DNA homology indices have been limited to the cases discussed below. In simian adenoviruses, the homology index between every pair of the 15 serotypes has been measured (Burnett *et al.*, 1972). On the basis of homology indices, they were divided into three subgroups (Table II). Homology indices of any pair of serotypes within subgroups I, II, and III ranged from 78 to 95%, 80 to 100%, and 64 to 87% respectively. The homology indices of any pair of serotypes belonging to different subgroups ranged from 28 to 55%. As in the case of human adenoviruses, this classification of the serotypes was in good agreement with the classification based on phenotypic properties of the viruses (Burnett *et al.*, 1972). The homology index between canine adenovirus type 1 (CAV-1) and CAV-2 was reported to be 70% (Marusyk and Hammerskjöld, 1972), this value seems to be relatively lower than might be expected in light of the close relationship of their phenotypic properties.

The homology index between two serotypes belonging to different groups has been measured in the following pairs: The homology index of any pair of simian adenovirus (SAV-16 [SA7] or SAV-3 [SV15]) and human adenovirus (Ad2, 5, or 12) was found to range between 9 and 23% (Piña and Green, 1968). The homology index between BAV-3 and Ad2 was reported to be 25% (Niiyama *et al.*, 1975). A homology index of less than 10% has been demonstrated in any pair between MAV-1 and human adenovirus (Ad2, 7, or 12) (Larsen and Nathans, 1977). The homology index of any pair between canine adenovirus (CAV-1 or CAV-2) and Ad16 was reported to be 1% (Marusyk and Hammerskjöld, 1972).

From the results cited above, it is clear that the range of homology indices observed among serotypes belonging to the same subgroup is distinctively higher than that observed among serotypes belonging to different subgroups. However, the range of indices observed among serotypes belonging to the same group, but to different subgroups, does not differ from the range of indices observed among serotypes belonging to different groups. Therefore, this index seems to be useful only to examine whether several adenoviruses are closely related in the same group. It is not an accurate means to express the degree of genotypic homology among several adenoviruses with not very close phylogenetic relations. The inaccuracy is attributed first of all to the fact that the difference in DNA size is considerable in many cases in which the two adenoviruses compared belong to different subgroups or groups. This difference has a considerable effect on the apparent efficiency of heterologous hybridization, which in turn affects the homology index. Second, most of the hybridization experiments discussed above were conducted under conditions such that even if an 800-bp-long radioactive single-stranded DNA piece were mismatched with the immobilized DNA up to approximately 25% of its base sequence, the piece could form a heteroduplex with the immobilized DNA. Therefore, this range of mismatch in individual pieces is not reflected in the homology index. Precise measurement of the melting temperature of the heteroduplex will clarify the degree of mismatch.

The homology index in the pair Ad2 and MAV-1 was less than 10% (Larsen and Nathans, 1977). Hence, the segment on the viral DNAs forming the heteroduplex was identified for this pair. The principle of the experiment was similar to that for the homology index described above, but the analysis was done for individual fragments produced by six-base sequence-specific endonuclease treatment of viral DNAs. MAV-1 DNA segments from 6 to 16 map units (m.u.) and from 54 to 66 m.u.) were found to form heteroduplexes with Ad2 DNA segments from 12 to 18 m.u. and from 51 to 62 m.u., respectively (Temple *et al.*, 1981). As noted in Section VI, the common antigen of mastadenovirus is considered to be ascribed to a part of the antigenic determinants of the hexon protein. The afore described heteroduplex formation is compatible with this notion, since the hexon gene of Ad2 is reported to be located in the segment from 51.2 to 59.7 m.u. (Akusjärvi and Pettersson, 1978a,b).

Similarly, FAV-1 DNA segments from 12 to 20 m.u. and from 41 to 49 m.u. were found to form heteroduplexes to Ad2 DNA segments from 18.1 to 19.3 m.u. and from 57 to 58 m.u., respectively (Aleström *et al.*,

1982b). No protein or proteins encoded in the segment of Ad2 from 18.1 to 19.3 m.u. have so far been detected, though an open reading frame for the translation has been observed in this region. The DNA segment of Ad2 from 57 to 58 m.u. is considered to code for the amino acid sequence of the carboxy-terminal region of the hexon protein. No antigen, however, has yet been reported to be shared between mastadenoviruses and aviadenoviruses. Nevertheless, DNA sequence analysis may serve to clarify the extent to which the amino acid sequence in hexon protein is common to the two genera of this family.

V. VIRAL ANTIGENS AND HEMAGGLUTININS

It might be ideal for taxonomy, or for the study of the evolution of the adenoviruses, to compare the origins of viruses from several animals on the bases of the viral DNA sequence homology and the structural similarities of the viral proteins encoded in comparable viral genes. However, it will take considerable time for molecular biology to accumulate analytical data sufficient to review adenoviruses of animals on these bases. Until that time, one must rely, by and large, on the antigenic specificities and hemagglutination (HA) ability determined by the limited part of amino acid sequences of the external capsid proteins such as the hexon and penton. Serum neutralization (SN) is the reaction that inactivates viral infectivity with antibodies in the serum. The antigenic determinants for it are indicated to be located on the surface of the hexon in the case of human adenoviruses (Wilcox and Ginsberg, 1963; Willcox and Mautner, 1976), so presumably the same would hold for adenoviruses of animals. It is empirically known in the case of adenoviruses that no other conventional serological test distinguishes a number of the viral isolates as clearly as the SN test. Indeed, Adair and McFerran (1976) carried out reciprocal SN tests among 9 serotypes of bovine adenoviruses, 5 serotypes of ovine adenoviruses, and 4 serotypes of porcine adenoviruses, and they found no cross-reaction. Furthermore, they attempted to neutralize the infectivity of these serotypic viruses with each of the antisera for 28 different serotypes of human adenoviruses, but failed to demonstrate a cross-reaction. More recently, Willimzik *et al.* (1981), in a larger-scale experiment, attempted to neutralize 35 serotypes of human adenoviruses with each of the antisera prepared against 25 simian adenoviruses and 31 types of adenoviruses from various nonprimate mammals and birds, seeking to neutralize that 25 serotypes of simian adenoviruses with each of the antisera prepared against the 35 serotypes of human adenoviruses. The results showed that only 2 chimpanzee adenoviruses had a cross-reaction of any degree with some human adenoviruses, but there was no significant cross-reaction between other pairs.

In the area of nonhuman adenoviruses, the HA reaction, the hemagglutination-inhibition reaction, and the hemagglution-enhancement reaction are also often utilized to classify or identify serotypes, as discussed in Section VI. However, only the reports cited herein have attempted to analyze structures of hemagglutinins in detail. Soluble antigens of canine adenoviruses were fractionated by sucrose-gradient ultracentrifugation (Marusyk *et al.*, 1970), and the rapidly sedimenting complete hemagglutinins were found to be penton oligomers. Two incomplete hemagglutinins were also found. the more rapidly sedimenting one was pentons, which agglutinated red cells in the presence of the antiserum directed against pentons of subgroup I, II, or III of human adenoviruses; the more slowly sedimenting one was fibers, which agglutinated red cells in the presence of the antiserum against fiber of subgroup II and III of human adenoviruses. Similar complete and incomplete HA antigens were recognized in simian adenovirus type 3 (SAV-3) [SV15] and SAV-16 [SA7]. In the case of SAV-5 [SV20], the trypsin-sensitive and -resistant complete HA antigens sedimenting slower than hexons were recognized (Norrby *et al.*, 1971). These findings indicate that pentons or fibers of canine and monkey adenoviruses carry the antigenic determinants common to those of human adenoviruses. A complete hemagglutinin was found in the soluble antigen fraction of egg drop syndrome 1976 virus (EDS76V) -infected cell lysate. The hemagglutinin was found by EM to consist of up to 11 rodlike elements and to form two bands of polypeptides corresponding to No. 2 and No. 3 virion polypeptides on SDS-PAGE (Todd and McNulty, 1978).

The complement-fixation (CF) test, the agar–gel precipitin test, and immunofluorescence microscopy are also very often used in identification and classification of nonhuman adenoviruses. Depending on the specificity of the antiserum used, these tests can detect the antigenic determinants of any range of specificities from serotype to genus on the surface of viral capsomers.

Norrby *et al.* (1971) purified proteins of four serotypes of human adenoviruses belonging to different subgroups, two serotypes of monkey adenoviruses, and two serotypes of canine adenoviruses and prepared antisera against them. By doing cross-CF tests and cross-absorption tests with these hexon preparations and antisera, they demonstrated clearly that the antigen common to mastadenoviruses was located on hexon protein.

VI. CLASSIFICATION OF ANIMAL ADENOVIRUSES (FAMILY ADENOVIRIDAE)

A variety of antigenic determinants have been found to be located on the surface of the external capsomers of adenoviruses, and they show different ranges of specificity. Like the serum neutralization (SN) antigen, some show a very narrow range of specificities, whereas others such as the determinants common to mastadenoviruses show a very wide range;

the remainder reveal an intermediate range. The hemagglutination (HA), hemagglutination-inhibition (HI), and hemagglutination-enhancement (HE) reactions have also been recognized for many adenoviruses. Individual serotypes can grow in only a single or a few related species of animals. A narrow host range is in fact one of the characteristic features of the adenoviruses, this being true not only for the possible transmission of an adenovirus from animals of one species to those of another, but also when appropriate cell-culture systems for viral propagation are under consideration. The conventional classification of adenoviruses is based on these specificities, as reviewed in the following sections. Since epizootiological aspects of viral transmission are discussed in Section VII, host ranges in cell-culture systems are covered mainly in this section.

All the viral isolates that provide the five presumptive pieces of evidence or their equivalents mentioned in Section I are included in this family, which now comprises 89 or more serotypes of viruses that can be distinguished by the SN test. The family is largely divided into two authentic genera (*Mastadenovirus* and *Aviadenovirus*) and a minor group of what may well be Adenoviridae growing in poikilothermic animals. Neither an antigen common to both genera nor the family-specific antigen has been found.

A. Genus *Mastadenovirus*

This genus consists of 78 or more serotypes of the adenoviruses. These viruses are included in this genus not only because their natural host animals are mammalian but also because they share an antigen that is usually detected by the complement-fixation (CF) or agar–gel precipitin (AGP) test. Though the range of species from which mastadenoviruses have been isolated (i.e., from the marsupial oppossum to man) is very wide, the mastadenoviruses contain the genus-specific antigen.

1. Bovine Adenoviruses

Klein *et al.* (1959) reported the first isolation of bovine adenovirus, now known as bovine adenovirus type 1 (BAV-1). Mohanty (1971), who attempted to classify more than a dozen bovine adenovirus isolates from several locations in the world by cross-SN tests, reported that they were divided into 10 serotypes. However, it has been more widely accepted that the bovine adenoviruses so far isolated comprise 9 serotypes (Matumoto *et al.*, 1969; Guenov *et al.*, 1970; Adair and McFerran, 1976; Matthews, 1979).

Bartha (1969) proposed that bovine adenoviruses could be divided into two subgroups, subgroup I (serotypes 1–3) and subgroup II (serotypes 4–9), on the basis of the following characteristics: Subgroup I carried the antigen common to mastadenoviruses, as clearly detectable in the CF or

AGP test. Subgroup I grew in bovine kidney cells as efficiently as in bovine testicle cells. In contrast, subgroup II carried so little of the common antigen that it was hardly detectable. Subgroup II viruses failed to grow in the bovine kidney cell culture and could be propagated only in bovine testicle cell cultures. The infectivity yield was relatively low even in the testicle cell cultures. In addition, the viral particles of subgroup I were more vulnerable to heat treatment than those of subgroup II. This proposal was in essence supported by subsequent reports (Adair and Curran, 1978–1979; Mohanty, 1971), though the small amount of the common antigen was found to be detectable in subgroup II. The latter report showed, in addition, that the subgroup-specific antigens of these two subgroups can be observed by the CF test and by immunofluoroscence microscopy of infected cells.

The subgroup I viruses can grow as efficiently in embryo lamb kidney cell cultures fed a medium containing lamb serum as in bovine kidney cell cultures, but they cannot replicate in goat kidney cell cultures. They failed to grow in cells of human, monkey, dog, rabbit, and chicken origin and also did not produce cytopathic effect (CPE) in these cells (Mohanty, 1971).

The HA or HI test does not seem to have been used extensively in classification or identification of the bovine adenoviruses. Klein *et al.* (1960) reported, however, that BAV-1 agglutinated rat erythrocytes and that BAV-2 agglutinated rat and mouse erythrocytes. Ceccarelli *et al.* (1979) reported that the erythrocytes of guinea pig and cattle were best for the HA test on the subgroup 1 viruses and that rat erythrocytes were the most suitable for the HA test on BAV-4, BAV-6, and BAV-7. The Fukuroi strain of BAV-7 was reported to agglutinate erythrocytes of cattle, sheep, goat, guinea pig, rat, and hamster (Inaba *et al.*, 1968).

2. Ovine Adenoviruses

Since the first virus from sheep was reported in 1969, five serotypes have been recognized as ovine adenoviruses to date (McFerran *et al.*, 1969; Sharp *et al.*, 1974; Bauer *et al.*, 1975). These serotypes grow well in lamb kidney cell cultures. Findings are, at least apparently, contradictory in two reports regarding the growth of ovine adenovirus type 5 (OAV-5) in the bovine kidney or testicle cell culture. Bauer *et al.* (1975) reported that the virus lost infectivity during three successive passages in these cells, whereas Belák *et al.* (1978) reported that it did not.

Belák (1980) found that all the ovine adenoviruses agglutinated rat erythrocytes, but not monkey, human O, ovine, or chicken erythrocytes, and that ovine adenovirus serotypes could be identified by the HI test more quickly than by the SN test.

In an outbreak of mild respiratory disease in sheep in southern Hungary, adenovirus strains indistinguishable from BAV-2 by the SN test were isolated (Bélak and Pálfi, 1974), but they were considered to be distinguished by the following findings: (1) like other ovine adenoviruses, the BAV-2-like ovine adenoviruses failed to agglutinate calf erythrocytes, while BAV-2 did so; (2) the BAV-2-like ovine adenovirus DNA was split into more fragments by the restriction enzymes than was BAV-2 and was 5–6 megadaltons greater in molecular weight than the latter (Bélak, 1980). The evolutionary origin of BAV-2-like ovine adenovirus remains to be elucidated. Molecular analyses of this virus and the related viruses would be highly useful in this regard.

In addition, two ovine adenovirus strains have been isolated in New Zealand using cultured ovine testicle cells. At least one may be a new serotype, since this strain has been reported to agglutinate chicken erythrocytes (Davies and Humphreys, 1977).

3. Caprine Adenoviruses

Two viral strains were isolated from primary lamb kidney cell cultures inoculated with mucosal scrapings from the large intestine of two goats (Gibbs *et al.*, 1977). In addition to the morphological and physicochemical findings, the presence of the common antigen between these viruses and Ad5 in the AGP test verifies these viruses as adenoviruses. They did not produce CPE in primary bovine kidney cultures or in the established baby hamster kidney 21 (BHK-21) and Vero cell lines. The viruses were distinguished from each other by the cross-neutralization test. They could not be typed with antisera raised against the nine bovine and five ovine adenoviruses, but they showed a degree of unidirectional positive results in cross-neutralization with some bovine or ovine adenoviruses. These results indicate that such viruses are new serotypes of mastadenoviruses. They may be termed caprine adenoviruses.

4. Porcine Adenoviruses

Since the first isolation was reported (Haig *et al.*, 1964), four serotypes have been widely recognized (Hafez and Liess, 1980; Matthews, 1979). As might be expected, they grow best in porcine kidney cell cultures. Porcine adenovirus type 4 (PAV-4) was reported to grow relatively well even in canine melanoma cell cultures, but very poorly in bovine and human kidney cell cultures (Kasza, 1966). PAV-1 was found to agglutinate guinea pig, human group O, and rat erythrocytes at both 37 and 4°C, but agglutinated monkey and mouse erythrocytes only at 4°C. PAV-4 showed strong agglutination with rats and chickens erythrocytes. No HA could be detected for PAV-2 or PAV-3 (Clarke *et al.*, 1967; Kasza, 1966).

More recently, however, eight additional strains have been isolated in the United Kingdom (Derbyshire *et al.*, 1975). In SN tests, there was no relationship between any of these additional strains and PAV-4, but unidirectional cross-neutralization reactions did occur between seven of the new strains and one or the other of PAV-1–3. In addition, unidirectional cross-reactions occurred among all the new strains. A similar report from France was published in the same year (Chappuis and Tektoff, 1975). Twelve strains were newly isolated and displayed no significant cross-neutralization with PAV-1–4. Of course, they showed no cross-neutralization with canine adenovirus type 1 (CAV-1), CAV-2, or BAV-3. More studies on these new isolates would increase the number of authentic serotypes and elucidate intratypic variations or recombination among this group of viruses.

5. Equine Adenoviruses

Since the first equine adenovirus (EAV) was reported in Pennsylvania in 1969 (Todd, 1969), a large number of strains have been isolated throughout the world. However, examination of these strains with SN tests as well as HI tests has indicated unanimously that EAV consists of no more than one serotype (Studdert *et al.*, 1974; Moorthy and Spradbrow, 1978). Human O group and rat erythrocytes are reported to be the best for the HA test, but horse and cattle erythrocytes also serve well. This virus grows very well in equine kidney cell culture. It can also grow in bovine kidney cell culture and a line of BSC-1 cells, but fails to grow in cells of many other mammals, such as African green monkey kidney (AGMK) cells, in which they produce only CPE (England *et al.*, 1973; Shahrabadi *et al.*, 1977; Fatemi-Nainie and Marusyk, 1979).

6. Canine Adenoviruses

It was more than 50 years ago that CAV-1 was recognized as the causative filterable agent of fox encephalitis (R.G. Green *et al.*, 1930), as noted in Section I. Independently, the same agent was later recognized as infectious canine hepatitis virus (Rubarth, 1947). It was one of the viruses that Horne (1962) utilized to elucidate the principal structure of the adenovirus capsid. CAV-2 was isolated as the virus that was partially cross-neutralized with anti-CAV-1 serum (Ditchfield *et al.*, 1962). Guinea pig and human erythrocytes are agglutinated with these viruses (Shimizu *et al.*, 1960). Complete and incomplete hemagglutinins of these viruses were well described and already referred to in Section V. The HI test can also distinguish the two viruses. The positive results in the HE test with antisera raised against human adenoviruses indicate that the antigen related to human adenoviruses is located on the penton structure of these viruses.

The canine viruses grow well in canine kidney cell cultures, but fail to grow in cells of human, monkey, or bovine origin (Yamamoto, 1966; Gehle and Smith, 1969).

7. Chimpanzee Adenoviruses

Because chimpanzee adenoviruses show some properties distinguishable from those of other simian adenoviruses, they are discussed here separately from other simian adenoviruses. The latter are called monkey adenoviruses (see Section VI.A.8).

Since the first reported isolation of an adenovirus from chimpanzees in 1958 (Rowe *et al.*, 1958), at least 7 serotypes of chimpanzee adenoviruses have been recognized by cross-SN tests between one another and between them and 33 serotypes of human adenoviruses or 22 serotypes of monkey adenoviruses (Basnight *et al.*, 1971; Asher *et al.*, 1978). These 7 serotypes comprise strains C1, Pan 5–7, and Pan 9–11. The first 4 serotypes have been termed adenoviruses S-21–S-24 (Kalter *et al.*, 1980). The last three, not included in the proposal, may be called adenoviruses S-25–S-27. They grow efficiently in human kidney cells as well as in chimpanzee kidney cells, but they do not propagate in cultures of AGMK cells or in cells of several nonprimate origins. Their growth inefficiency in cells of monkey origin would appear to distinguish these viruses from monkey adenoviruses. Morphological and physicochemical evidence and the antigen common to human adenoviruses observed in the CF test provided the bases for these viruses to be counted as adenoviruses. SAV-21 (C1) agglutinates rhesus and vervet money erythrocytes (Hillis and Goodman, 1969), but no appropriate erythrocyte has been found for the HA test with other serotypes.

In addition to the aforementioned strains, more than 20 adenovirus strains were isolated from chimpanzees (Hillis *et al.*, 1969; Hillis and Goodman, 1969). They were classified into three subgroups on the basis of HA patterns and further classified by the HI test. Some of them showed partial relatedness with human adenoviruses according to the HI test. However, a large number of human and monkey adenoviruses should be investigated by the cross-SN test and their possible identity with the 7 serotypes of chimpanzee adenoviruses mentioned above checked before these isolates are recognized as separate entities.

8. Monkey Adenoviruses

Since the first reported isolation in 1956 (Hull *et al.*, 1956), a large number of adenoviruses have been isolated from monkeys of Asian, African, and New World origin. Not only physicochemical and morphological evidence but also immunological data showing the presence of an antigen related to one from human adenoviruses verified these viruses as adenoviruses. Many of these adenovirus isolates have been designated by consecutive SV, SA, M, or P numbers. They were compared with one another and with human and chimpanzee adenoviruses by cross-SN tests for identification purposes, as reviewed by Hull (1968) and Heberling (1972). The most recent proposal (Kalter *et al.*, 1980) recognized 20 monkey serotypes and recommended that they be called adenoviruses S-1 [SV1] to S-20 [V340]. The prototype strains of adenoviruses S-1 [SV1] to S-15 [SV38] were isolated from Asian macaques; those of adenoviruses S-16 [SA7] to S-18 [SA18] and of S-20 [V340], from African green monkeys;

and that of adenovirus S-19 [AA153], from an African baboon. These serotypes were recently confirmed by cross-SN tests to be separate from 35 human serotypes. (Willimzik *et al.*, 1981). Unlike many human adenoviruses, such monkey adenoviruses grow well in rhesus monkey or AGMK cells. However, some monkey adenoviruses, such as SAV-1 [SV1] and SAV-2 [SV11[grow in cells of human origin, whereas many others do not (Hull *et al.*, 1956; 1958; Hull, 1968).

Sixteen serotypes of monkey adenoviruses were classified into four subgroups by the HA reaction with rhesus monkey, rat, or guinea pig erythrocytes at different temperatures (Rapoza, 1967).

Strain sq M-1 (isolated from squirrel monkey) is not included in the aforementioned proposal for simian adenoviruses (Kalter *et al.*, 1980). However, it must be mentioned that this virus was reported to be the most prevalent in New World monkeys (Heberling, 1972; Shroyer *et al.*, 1979).

9. *Tupaia* Adenovirus

The tree shrew (*Tupaia belangeri*) is a prosimian, located in a phylogenetic position intermediate between primates and rodents, which has encouraged scientists to utilize this animal as an experimental model. An adenovirus was isolated from a kidney culture of an apparently healthy *Tupaia* (Darai *et al.*, 1980). *Tupaia* adenovirus (TupAV) grows very efficiently in *Tupaia* kidney cell culture, but less so in owl monkey kidney, mink lung, and bovine kidney cell cultures. It is almost impossible to grow in kidney cells originating from man, rabbit, monkey, pig, or lamb. In addition to morphological and physicochemical characters, the presence of the common antigen between TupAV and Ad1–7 was detected with the AGP test, and some characteristics of viral DNA such as the presence of terminal proteins were shown. Thus, TupAV is classified as a new member of the mastadenovirus group. This virus agglutinates rat and human type O erythrocytes.

10. Murine Adenoviruses

Two adenovirus strains now known as mouse adenovirus type 1 (MAV-1) and MAV-2 were isolated from laboratory mice in the United States (Hartley and Rowe, 1960) and Japan (Hashimoto *et al.*, 1966), respectively. The two strains were distinguished by the SN test, and even by the CF test, probably because the sera employed were very type-specific (Van der Veen and Mes, 1974). They did not agglutinate the erythrocytes of several animals tested. They grow well in mouse kidney cell culture, but reportedly show no signs of growth in a variety of other tissue-culture preparations including human embryo, rat embryo, and several lines of malignant mouse tissues.

In addition to the aforenamed two serotypes, several strains were isolated from laboratory mice in Germany (Missal, 1969). They grew well in mouse kidney cell cultures, but did not produce CPE in cultures of FL or BHK-21 cells, primary cultures of calf kidney or calf testicle cells, or mouse embryo fibroblast cultures. CPE was observed only in cultures of piglet kidney cells. The viruses showed no HA ability. These properties are similar to those of MAV-1 and MAV-2, but the serological identity of these German isolates with the authentic serotypes must be investigated.

11. An Adenovirus Isolate from Opossum: A Probable New Mastadenovirus

A virus appeared spontaneously in a kidney cell culture of an opossum, a primitive mammal, the fetus of which has no placenta (Morales-Ayala *et al.*, 1964). Morphological and physicochemical findings indicated that this virus should be regarded as an adenovirus. The virus cannot grow in rabbit, hamster, rhesus monkey, or human kidney cell cultures. It was CF-positive in 8 of 10 human sera and 1 of 5 oppossum sera, suggesting the presence of the common antigen between this virus and human adenoviruses. However, more studies on the host range and more extensive research with defined sera on the antigenic specificities are necessary to distinguish this virus from other mammalian adenoviruses.

12. An Adenovirus Isolate from Rabbit

Rabbit kidney cell cultures inoculated with visceral material of rabbits suffering from diarrhea produced detectable virus after several passages (Bodon *et al.*, 1979). Some morphological and physicochemical findings and the demonstration of the antigen common to human adenoviruses proved that this virus was a mastadenovirus. The virus did not replicate in primary pig kidney, calf kidney, or calf testicle cells or in HEL (human embryo lung) cell lines. Antisera against PAV-1–4 and BAV-1, -2, and -6, failed to neutralize this virus. The same virus agglutinates mixed erythrocytes from several rabbits at 4°C. More cross-SN and host-range studies should be undertaken to confirm beyond doubt that this virus is a new serotype of mastadenovirus that has the rabbit as a natural host.

13. An Adenovirus that Possibly Originates from the Feral Rodent *Peromyscus maniscultatus*

A viral agent was isolated from primary hamster kidney cell cultures that had shown CPE after inoculation with blood from a white-footed deer mouse (*Peromyscus maniscultatus*) captured in the Calfornia desert

(Rĕeves *et al.*, 1967). The agent was resistant to deoxycholate and chloroform and passed through a Seitz filter (type ST, L3 pad). Serologically, the agent showed a positive CF reaction with the sera of patients in the convalescent stage of Ad4 or Ad7 infection, but not with the sera of the same patients in the acute phase of infection. The virus was not neutralized by rabbit antisera to Ad1–30 or MAV-1. These findings indicate that this agent is a mastadenovirus. However, 90 sera of the white-footed deer mice captured in the same area showed a negative CF reaction with this virus. The results suggest that the feral mice are not the natural host. The virus induced characteristic adenovirus CPE in rhesus and AGMK cell cultures and in an HEp-2 cell culture (human origin), but it could not be carried in serial passage in these cells. It produced a slight CPE in human embryonic kidney cell cultures, but no CPE in mouse embryo or WI38 cell cultures. The possibility that the virus was derived from the hamster kidney cells used for the initial isolation was considered in this report to be very remote, since both inoculated cultures developed CPE, whereas none of the more than 100 additional tubes prepared from the same kidneys on the same date and used as controls or inoculated with other specimens evidenced similar changes. The question remains as to whether the kidney cell culture of the white-footed deer mouse can support the viral growth that cells of the hamster, which belongs to a different genus than the white-footed deer mouse, do in fact promote. More precise serological and host-range analyses have to be performed if this orphan adenovirus is to be considered a new mastadenovirus.

14. Adenoviruslike Particles Observed in Guinea Pigs, Rats, and Sea Lions

This section discusses three reports that provide only the morphological and some other circumstantial evidence that suggests the presence of adenoviruses in three species of mammals that have never been described as the natural host of any adenovirus.

At necropsy of guinea pigs that died during an epizootic of lethal pneumonia, the nuclei of desquamated epithelial cells of bronchi were often found to contain basophilic inclusion bodies. These inclusion bodies were Feulgen-positive. EM demonstrated many 58 to 72-nm viruslike particles in the areas in which basophilic inclusion bodies had been found (Naumann *et al.*, 1981).

At necropsy of two rats experimentally injected with immunosuppressive cancer drugs, basophilic inclusion bodies were found in the nuclei of epithelial cells of small intestinal villi. The inclusion bodies were Feulgen-positive and were found to contain many viruslike particles averaging 65 nm in diameter (Ward and Young, 1976).

Five California sea lions were found stranded along the Los Angeles coast within a period of only 10 days. They were pathologically diagnosed as suffering from acute viral hepatitis. Nuclei in some hepatic cells contained basophilic inclusion bodies. Ultrastructural analysis revealed that the inclusion bodies carried many nonenveloped, viruslike structures 70–75 nm in diameter (Britt *et al.*, 1979).

It is hoped that these three case reports will prompt more extensive studies, especially virological ones, on similar materials to determine whether the three aminal species involved are natural hosts of previously unreported mastadenoviruses.

B. Genus *Aviadenovirus*

This genus consists of 15 or more serotypes. The number of species of host animals from which the viruses have been isolated is still very much limited. A genus-specific antigen, or an antigen common to all members of the genus, has not been found, though intergroup-specific antigens such as the common antigens between fowl adenoviruses and turkey adenoviruses have been reported.

1. Fowl Adenoviruses

Large numbers of adenoviruses were isolated from fowls (or chickens) in Japan (Kawamura *et al.*, 1964), Hungary (Khanna, 1966), the United Kingdom (McFerran *et al.*, 1972), and the United States (Calnek and Cowen, 1975). Individual laboratories classified their own isolates into 5–9 serotypes by cross-SN tests. Prototype strains of individual laboratories were transferred to other laboratories and compared. The most recent study (McFerran and Connor, 1978) reported that this group of viruses consisted of 11 serotypes with some intratypic variants in terms of SN antigen. Though these viruses were earlier described by the general term avian adenoviruses, it was proposed that this group be called fowl adenoviruses, as opposed to, for example, turkey adenoviruses (McFerran and Adair, 1977).

These fowl adenoviruses grow very well in chicken kidney cell cultures. At least fowl adenovirus type 1 (FAV-1) [or chick embryo lethal orphan (CELO) virus] can also grow very well in turkey kidney cells (Blalock *et al.*, 1975) and less efficiently, though still well, in turkey, duck, or chicken embryo fibroblast cultures (Ishibashi, 1971). Such fowl adenoviruses carry antigens common to all the members of this group (Kawamura *et al.*, 1964; McFerran and Adair, 1977). FAV-1 was reported to agglutinate rat erythrocytes (El Mishad *et al.*, 1975), whereas other types of fowl adenoviruses have been reported not to agglutinate erythrocytes of several animal species.

The appellation fowl adenoviruses may be appropriate at least until more data are available on aviadenoviruses, at which time the present classification may be reconsidered. Some circumstantial evidence indicates, however, that natural hosts of some of these viruses are not limited to the fowls and closely related birds. For instance, the duck is relatively remote from the fowl in phylogeny, but FAV-1 can grow in duck embryo cells as well as in chicken embryo cells. FAV-1 and some other fowl adenoviruses are isolated not only from domestic birds, but also from several wild birds. All these viruses are discussed in Section VII.

2. Egg Drop Syndrome 1976 Virus

This virus was first isolated in Northern Ireland (prototype, strain 127) and identified as the etiological agent of egg drop syndrome 1976 (McFerran *et al.*, 1977) (see Section VII.I). EDS76V did not show any antigenic identity with FAV-1–11 by the cross-SN test. The AGP test also failed to demonstrate the common antigen between this virus and FAV-1. However, it was demonstrated that antibody production in FAV-1 infected chickens that had already passed the peak of infection could be recalled by subsequent infection with EDS76V. The specificity of this so-called "recall phenomenon" was shown by the failure of avian infectious bronchitis virus to recall the production of FAV-1 antibody. These experiments indicate that EDS76V and FAV-1 share a cross-reactive antigen that is not directly detected by the AGP test (McFerran *et al.*, 1978a). EDS76V strongly agglutinates duck, turkey, and chicken erythrocytes, and a specific HI test is useful for identification and titration (Todd and McNulty, 1978). Nothing is more characteristic of this virus than that it grows most efficiently in duck kidney or hepatic cell cultures, though the virus was isolated from a hen. It also grows well in chicken kidney or hepatic cells, but hardly at all in cells of turkey origin. Of course, it does not grow in mammalian cells (Adair *et al.*, 1979). The characteristic host range of this virus and several lines of other evidence discussed in the Section VII.I suggest strongly that this virus is in fact a duck adenovirus.

3. Turkey Adenoviruses

Two serotypes of adenoviruses were isolated from turkeys (Scott and McFerran, 1972; McFerran *et al.*, 1975). They were distinguished from each other and from the first eight fowl adenoviruses by SN tests and shown by the AGP test to share an antigen common to fowl adenoviruses. The host range of these turkey adenoviruses makes it possible to distinguish them from fowl adenoviruses. They grow well in turkey kidney cell cultures, but not in chicken kidney cell cultures or chicken embryo fibroblasts. They also fail to grow in pig kidney cell cultures and lamb kidney cell cultures. No HA activity has been found for these viruses to date.

Independently of the aforecited isolations, five adenoviruses were isolated from turkeys and classified into four serotypes by a reciprocal neutralization kinetic experiment (Easton and Simmons, 1977). One of them was identical with FAV-1 (Blalock *et al.*, 1975); the other three must be examined for identity with fowl adenoviruses already reported and the turkey adenoviruses mentioned above. Another adenovirus strain has been isolated from a turkey with a severe respiratory disease (Cheville and Sato, 1977). Differentiation of this virus from other avian adenoviruses must also be undertaken.

4. Hemorrhagic Enteritis Virus of Turkey or Marble Spleen Disease Virus of Pheasant

This virus and some human enteric adenoviruses (Flewett *et al.*, 1975) may be the only adenoviruses that have not been propagated in a cell-culture system. Therefore, all propagation and the passage of the virus have been done by using virus-sensitive birds. The identity of hemorrhagic enteritis virus of turkey (HEV) with marble spleen disease virus of pheasant (MSDV) was demonstrated by a series of experiments such as induction of hemorrhagic enteritis of turkey with MSDV and its prevention by the antisera against HEV (Iltis *et al.*, 1975; Domermuth *et al.*, 1975). In addition to morphological and physicochemical evidence (Itakura and Carlson, 1975; Iltis *et al.*, 1977), partial identity of the antigen in the AGP test of this virus with turkey adenoviruses discussed in Section VII.J allows this virus to be classified in the adenovirus family (Iltis and Daniels, 1977). However, AGP tests failed to demonstrate a common antigen between this virus and FAV-1 or FAV-5 (Silim *et al.*, 1978; Darbyshire, 1980).

C. Adenoviruslike Agent and Particles Observed in Poikilothermal Animals

1. A Probable Adenovirus among Poikilothermal Animals

An *adenoviruslike* agent designated frog adenovirus 1 was isolated from a turtle cell line (TH-1) culture inoculated with cells of a granuloma-bearing kidney of the leopard frog (Clark *et al.*, 1973). The ultrastructure of the infected cells and the viral particles, the resistance of the infectivity to the chemicals, and the inhibition of the viral growth by 5-bromo-2'-deoxyuridine indicated that the virus was an adenovirus. It had no antigen common with mastadenovirus and was not neutralized with antiserum to aviadenovirus FAV-1 or FAV-2. Oddly enough, the turtle cell line TH-1 was the only one capable of supporting viral growth. Not only the cells originating from other tested reptiles, fish, avians, and mammals but also two cell lines originating from the same genus as the leopard frog failed to support growth.

The possibility that this virus was a virus latent in TH-1 cells was considered in the afore-cited report, but this possibility contradicts the demonstrated failure of such a virus to appear in TH-1 cells under any other circumstances during more than 10 years of experimental manipulation. It is conceivable, however, that a virus latent in TH-1 cells might have been induced to multiply by having had contact with granuloma cells of the frog in the tissue culture. If the latter speculation is correct, the virus may well be called a turtle adenovirus. Whichever is the case, this virus is the first probable adenovirus among poikilothermic animals. More precise characterization is desirable.

2. Adenoviruslike Particles Observed in Poikilothermal Animals

An ultrastructure very similar to that of adenovirus was observed in a pronephric cell line of the leopard frog that had been inoculated with a mitochondrial fraction of Lücké tumor cells (Wong and Tweedell, 1974). However, the number of viruslike structures in the cells was reported to decrease gradually during several passages. Hence, more work is required to ascertain the biological conditions under which the viruslike structures can multiply more constantly to provide materials sufficient for several kinds of analyses.

A type of epidermal hyperplasia was observed in a few cod captured in the Baltic Sea, and EM examination revealed that the nuclei of the epithelial cells in the outermost cell layers of the thickened epidermis contained particles averaging 77 nm in diameter and considered to be icosahedral. Furthermore, fibers approximately 20–25 nm long were often connected with corners of the particles. These and other ultrastructural features suggested that these particles were adenoviruses (Jensen and Bloch, 1980). More research, especially virological, will result in accurate identification of these adenoviruslike particles.

VII. NATURAL AND EXPERIMENTAL INFECTION OF ANIMALS

This section deals with infection of natural host animals with individual nonhuman adenoviruses, with reference to the association of these viruses with diseases observed in host animals. Many of the following subsections first refer to clinicopathological reports that provide circumstantial evidence to indicate the association of adenovirus infection with the disease. They then devote considerable space to the experimental attempts to reproduce the disease for the purpose of evaluating the etiological contribution of the adenovirus infection to the disease. Problems of pathogenesis and epizootiology are also noted in passing. Some adenoviruses for which the association with diseases has not been studied to any degree are discussed collectively, rather than in separate subsections.

A. Bovine Adenoviruses

Several serotypes of bovine adenoviruses were isolated from calves suffering from pneumoenteritis, an acute febrile disease with rhinorrhea, or keratoconjunctivitis (Bartha and Áldásy, 1966; Inaba *et al.*, 1968; Bartha *et al.*, 1970; Saxegaard and Bratberg, 1971; Pavlov, 1979; Descoteaux *et al.*, 1980). However, the isolation of adenoviruses from sick animals provided only circumstantial evidence that these serotypes are etiological agents of these diseases, since these and other serotypes were also isolated from apparently healthy calves on different occasions. Actually, most prototype strains of nine serotypes of bovine adenoviruses were first isolated from apparently healthy calves (Klein *et al.*, 1959, 1960; Darbyshire *et al.*, 1965; Rondhuis, 1968; Guenov *et al.*, 1970; Burki *et al.*, 1978).

To confirm this circumstantial etiological evidence, experimental infections were induced. Darbyshire *et al.* (1969) inoculated colostrum-deprived newborn calves with bovine adenovirus type 1 (BAV-1) or BAV-2 by the respiratory route, and examined several parameters extensively. All but one calf showed mild respiratory symptoms and diarrhea, but overt clinical pneumonia was not observed. One calf died following an acute clinical pneumonia and enteritis 2 days after inoculation with BAV-2. At necropsy 4 and 7 days postinfection (p.i.), gross lesions of pneumonia were observed in all calves, and histopathologically significant lesions were found in the respiratory system and the regional lymphatic tissues. Intranuclear inclusion bodies were found in cells of bronchioles, alveolar cells, and reticulum cells of the lymphatic tissues. The virus was recovered from swabs of nasal and conjunctival mucosa and of feces of all calves during the course of clinical observation, and at necropsy the virus was recovered from virtually all organs, including brain, and from cardiac muscles. Thus, the experiments indicate strongly that these serotypes can induce pneumonia when they infect calves. It is remarkable that although viremia presumably distributed the virus to all organs, gross lesions were observed only in the lungs. This "organ tropism" deserves further study.

A similar experiment was conducted with BAV-3 (Darbyshire *et al.*, 1966), and similar pathogenetic findings with more extensive lesions in the lung were observed. In some experiments with older calves, inoculation with BAV-1, BAV-2, or BAV-3 produced no symptoms or only mild ones without any significant pathological lesions (Ide *et al.*, 1969; Lehmkuhl *et al.*, 1975). However, in another experiment using similar older calves, intranasal or intratracheal inoculation with BAV-1 produced respiratory illness, whereas inoculation into the conjunctiva produced no illness (Mohanty and Lillie, 1965). Intranasal and intratracheal inoculation of 8-week-old calves with BAV-4 or BAV-5 produced bronchopneumonia, but inoculation with BAV-8 did not induce changes (Mohanty, 1971).

All such experiments indicate that the pathogenicity of bovine adenoviruses is weak and that the viral infection produces clinical illness only when some superimposing factors coincide.

The first isolation of BAV-7 was from the blood sample of a cow suffering from an acute febrile disease (Inaba *et al.*, 1968). This result may be good evidence that at least transient viremia is a common pathogenic process of bovine adenoviruses. BAV-2 was reported to be isolated from a commercially purchased primary bovine kidney cell culture prepared from 4- to 16-week-old bovine embryos (Mohanty and Lillie, 1970). This finding is also noteworthy, since it may be evidence that there is vertical transmission of mastadenoviruses similar to the egg transmission of aviadenoviruses.

While BAV-1–9 have been isolated from cattle in several parts of the world, epizootiological studies showed that Idaho mule deer and pronghorns carried antibodies against BAV-7 and BAV-3, respectively (Stauber *et al.*, 1977, 1980). Free-living African buffaloes were also found to be infected with BAV-3, BAV-4, and BAV-8 (Baber and Condy, 1981).

B. Ovine Adenoviruses

Since ovine adenoviruses can be isolated from apparently healthy lambs (Bauer *et al.*, 1975) as well as those suffering from respiratory disease (Sharp *et al.*, 1974), experimental infection of lambs with ovine adenoviruses was carried out to determine whether, under more defined conditions, they can induce respiratory disease similar to the natural infection. Sharp *et al.* (1976) intranasally inoculated specific-pathogen-free, 2-week-old lambs with ovine adenovirus type 4 (OAV-4). Although infection did not produce clinical disease, it appeared to be associated with the development of pulmonary edema observable at necropsy 7–11 days p.i. The virus was recovered from nasal and rectal swabs, lung, upper respiratory tract, and intestine from 1 to 11 days p.i. Antibodies increased significantly after 8 days p.i. Belák *et al.* (1980) inoculated colostrum-deprived 10-day-old lambs with each of two strains belonging to OAV-5. The route of the inoculation was through the nose and trachea. Both strains caused mild respiratory and enteric disease. At necropys 9 days p.i., interstitial pneumonia, rhinitis, enteritis, and general proliferation of the cells belonging to the reticuloendothelial system were observed. The lesions were similar, but more severe in one of the two strains. Histopathological changes of nephrosis and interstitial nephritis were observed with the more virulent strain. A similar experiment with OAV-1 was reported by Palya *et al.* (1977). As with OAV-5, respiratory systems were remarkably affected. Clinical symptoms of pneumonia with temperature increase, diarrhea, and conjunctivitis were observed on the 4th day p.i. The pathological changes observed 6–14 days p.i. were interstitial pneumonia, bronchopneumonia, and catarrhal enteritis. Histopathological observations demonstrated renal and hepatic lesions in addition to the pulmonary and intestinal lesions. Intranuclear inclusion bodies were observed primarily in the epithelial and reticular cells of the respiratory system. In this experiment, infection spread even to a contact control animal not inoculated with the virus.

Symptoms and lesions of experimentally infected lambs may not be as severe as those of a natural infection, but these experimental studies strongly support the epizootiological conclusion that ovine adenoviruses are one of the causative agents in ovine respiratory illness.

Experimental infection of lambs with BAV-2-like ovine adenovirus also produced respiratory illness, indicating that this virus was the causative agent in the outbreak of a respiratory disease in Hungary (Belák *et al.*, 1975).

C. Porcine Adenoviruses

To verify the pathogenicity of the porcine adenoviruses, conventional and germ-controlled piglets were experimentally infected with these viruses. Sharpe and Jessett (1967) intranasally inoculated conventional and gnotobiotic piglets with porcine adenovirus type 2 (PAV-2) or PAV-3. When conventional piglets 32–36 days of age were inoculated with PAV-3, neither clinical symptoms nor pathological lesions at necropsy were observed. No virus was recovered from the tissues, but a serological response was observed 2–3 weeks p.i. When gnotobiotic piglets 1, 22, and 32 days of age were inoculated intratracheally with PAV-2 or PAV-3, infections of the tonsil and lower intestinal tract were established, so that the virus could be isolated from these tissues at necropsy 5–18 days p.i. A serological response was also observed 2 weeks p.i., but neither clinical symptoms nor gross lesions developed in the animals.

Shadduck *et al.* (1967) conducted a similar experiment with PAV-4. When 10 germ-free and 17 pathogen-free piglets less than 48 hr of age were inoculated intranasally with PAV-4, they developed interstitial pneumonia without symptoms. At necropsy 10 and 13 days, p.i., lesions with intranuclear inclusion bodies were seen in kidneys, thyroid glands, and lymph nodes in addition to lungs. The virus was frequently isolated from cerebral cortex, nasal mucosa, tonsil, lung, and ileum from 5 to 48 days p.i. (Shadduck *et al.*, 1968). Later, Edington *et al.* (1972) inoculated gnotobiotic piglets orally and intranasally with PAV-4 and confirmed the observation of Shadduck *et al.* (1967). They further found histological lesions of the central nervous system due to PAV-4 infection. However, the inoculated piglets showed no clinical symptoms.

These experimental results indicate that porcine adenoviruses alone do not produce clinically severe diseases accompanying gross pathological lesions and that, if superimposed with some agents, the viruses may cause clinically noticeable pneumoenteritis or encephalitis.

D. Equine Adenovirus

When McChesney *et al.* (1973) reviewed case reports in the United States, 31 foals (29 Arabian and 2 non-Arabian) less than 3 months of age were recognized as having had adenovirus infection. These foals developed a disease characterized by pneumonia, lymphopenia, and intermittent fever. Of the 31, 28 Arabian foals died; 1 Arabian and the 2 non-Arabian foals recovered. Diseased foals originated in 22 herds in 11 states. In most herds, infected foals in these case reports were in contact with other foals of the susceptible age group and with adult horses; only in two herds, however, did more than one foal develop the disease in a 1-year period. Studdert *et al.* (1974) carried out an extensive seroepizootiological analysis and made it clear that equine adenovirus (EAV) was distributed worldwide with a high frequency of antibody in all breeds and all adult horses.

From these two studies, one can speculate that EAV infection must have occurred very often among foals of any breed and that most foals, when infected, must have tolerated the infection by developing few remarkable symptoms and lesions. A small fraction of Arabian foals, however, must have had a special susceptibility to the infection and suffered from a fatal disease like severe pneumonia. Actually, in these case reports, the severe absolute lymphopenia was recorded in the dead Arabian foals, whereas only moderate lymphopenia was observed in convalescent Arabian and non-Arabian foals.

Comparative experimental infection of foals with EAV was undertaken to examine how parameters that define host animals affect their response to the infection (McChesney *et al.*, 1974); foals of different breeds (Arabian or not), age (neonatal or 2–4 months of age), and colostrum status (deprived or not) were used as the experimental animals. All but control foals were similarly inoculated with the virus into the tracheal, nasal, and oral cavities. Of 7 Arabian foals, 6 responded to the infection, as did all 18 non-Arabian foals. All the colostrum-deprived and the neonatal colostrum-fed foals developed similar clinical symptoms and postmortem lesions, but the colostrum-deprived ones usually had more apparent clinical symptoms and more extensive postmortem lesions. Symptoms such as intermittent high fever, nasal and ocular discharges, polypnea, and cough appeared 5–7 days p.i. Gross pathological lesions at 7 days p.i. were pulmonary atelectases and interstitial pneumonia and serous inflammation in many somatic and visceral lymph nodes. Swelling, hyperplasia, and intranuclear inclusions were consistently found in epithelial cells of the respiratory system and were occasionally observed in epithelial cells of conjuctiva, pharynx, salivary gland, urinary tract, or pancreas. Both colostrum-deprived and colostrum-fed foals showed similar antibody responses 3 weeks after the viral inoculation, but colostrum-fed foals, especially Arabian foals, had higher adenoviral antibody titers than colostrum-deprived couterparts before the viral inoculation. At 11 days p.i., the virus was recovered from lungs of many of experimental foals, but not from the colostrum-fed Arabian foals, the antibody titers of which were very high even before the viral inoculation.

The foals that were 2–4 months old at the time of viral inoculation developed few or no clinical signs and were free from gross lesions at necropsy. Those that were not sacrificed for necropsy developed normally.

In an exceptional case, an Arabian foal (neonatal colostrum-fed), a sibling of a foal that died with naturally occurring adenovirus infection, developed a cough, nasal discharge, and diarrhea, but had no fever after viral inoculation. The foal had absolute lymphopenia even before the viral inoculation, and it persisted throughout the experimental period, but total neutrophil count remained within the normal limit. Adenovirus antibody titer in this foal did not show any increase in 7 days p.i. When killed 9 days p.i., the foals showed gross lesions of lungs at necropsy with scattered purple-brown lobules of atelectasis and bronchopneumonia. In addition to pulmonary lesions, the lymph nodes, spleen, and thymus were very small and thin and found to be histologically deficient in lymphocytes and plasma cells. These findings from the exceptional Arabian foal were quite similar to those from the 28 dead Arabian foals in the case reports cited above.

These experimental results verify the aforestated speculation and indicate that almost all foals, regardless of their breed, can tolerate EAV infection, developing only mild respiratory symptoms. The hypersensitivity to adenovirus infection observed in a small portion of Arabian foals was later elucidated to be due to congenital deficiency of the T- and B-lymphocyte system genetically controlled by an autosomal recessive immune-response (*Ir*) gene (McGuire and Poppie, 1973; Thompson *et al.*, 1975).

E. Canine Adenoviruses

As noted in Section I, R. G. Green *et al.* (1930, 1931, 1933, 1934, 1936) studied the epizootic of fox encephalitis that took place in a fur range during the fall for several consecutive years beginning in 1925. They found a virus that they named fox encephalitis virus, which is now known as canine adenovirus type 1 (CAV-1). Their observations of the epizootic and experimental infection can be summarized as follows: Fox pups 4–5 months of age that had been raised with parent foxes in small pens were taken into a large range where thousands of young foxes were reared. Foxes suddenly began dying after about 4 weeks, and during this period, 15–20% of the population perished. The major symptoms of the foxes were neurological: convulsion, paralysis, and lethargy. The most striking finding at necropsy was the paucity of marked gross lesions. The typical gross lesions of the disease were hemorrhage or extravasation of blood.

Such hemorrhages were often observed in the adrenal gland, pleural and abdominal cavities, pancreas, thymus, salivary glands, lungs, and central nervous system. Major symptoms of the disease seemed to be due to brain hemorrhage. The general histopathological findings in these organs were large or multiple small hemorrhages, or perivascular infiltration of the round cells, and endothelial proliferation. The characteristic intranuclear inclusions were consistently found in endothelial cells. Morbidity in the epizootic was about half the population of foxes, though mortality was 15–20%. The virus was recovered from blood, spleen, spinal cord, and brain; it seemed to be distributed throughout the body.

Experimental infection of apparently healthy foxes was attempted by using the homogenized brains, spinal cords, or spleens of the foxes that died in the epizootic as the presumptive viral preparation. Both intramuscular and intracranial injection were equally effective. The first attempt to infect adult foxes was inconclusive. On the other hand, the use of red fox pups dug from their dens in the wild and raised in quarantine gave uniform and consistent reactions to the virus. The incubation period was usually 2 or 3 days. Most of the pups died 3–5 days p.i. A mortality of 80% was observed in experimentally infected foxes less than 6 months old. The clinical symptoms and postmortem pathological findings were similar to those of natural infection. The homogenate of the brain was passed through a Berkefeld filter, and the filtrate was also found to be infectious. A very high degree of immunity was exhibited by the foxes that survived the experimental infection. Later, experimental reproduction of the disease by inoculation of virus through the respiratory route became possible. The disease could be transmitted to dogs and coyotes, but not to rabbits, white rats, or ferrets.

Rubarth (1947) studied for about 20 years a canine disease that he named "hepatitis contagiosa canis." There was no characteristic symptom for this disease, but many cases showed vomiting and diarrhea followed by fever and apathy; sometimes convulsions were observed. Postmortem, the disease was characterized by edema of most organs (liver, lymph nodes, spleen, thymus, heart, brain, stomach, and adrenal cortex). Microscopically, hyperemia, edema, and hemorrhage and increased numbers of leukocytes, monocytes, and lymphocytes were often observed. The characteristic inclusion bodies were often seen in nuclei of hepatic and endothelial cells. As a consequence of natural disease, more than 50% of the mortality took place within 10 days. Death owing to natural disease usually occurred sporadically, but severe attacks of an almost enzootic character were observed rather frequently in dog kennels.

Experimental transmission of the disease to apparently healthy dogs was attempted with an inoculum of liver extracts from dogs dying from the natural disease. Intraperitoneal injection and oral administration were used. The experiment did not yield conclusive evidence, however, until puppies bred in the laboratory were used as the experimental animals. With this protocol, the puppies died 3–5 days p.i., and the clinical symptoms and pathological findings in successfully induced cases were essentially similar to those seen in the natural disease. Even when the liver extract had been filtered through a Seitz E.K. filter, the disease was transmitted. The disease-specific antigen was found in liver by the complement-fixation test. No fewer than 70 of 100 apparently healthy dogs were found to be seropositive to this antigen, indicating that the virus was widely disseminated and explaining why commercially available adult dogs were resistant to the experimental infection.

If one is overly concerned with the names of diseases in foxes and dogs, one might consider that one and the same virus (CAV-1) exhibits a different "organ tropism" in the two animals; however, from the precise pathological findings, one realizes that both diseases are essentially the same.

Later, development of experimental techniques (especially in virology) provided occasions to reexamine the pathogenesis of experimental infection (Salenstedt, 1963; Hamilton *et al.*, 1966). It was found that oral administration of CAV-1, 40–400 median tissue-culture infectious doses ($TCID_{50}$) in titer, was enough to develop disease of various severity in dogs. The titration of the viral infectivity recoverable from several organs, blood, and lymph allowed the following conclusions: The initial multiplication of the virus took place in the tonsils and the regional lymph nodes, and the virus was transferred to the blood through the lymph vessels. The virus was then disseminated to all tissues of the body. At 5–6 days p.i., the virus could be isolated from adrenals, kidneys, spleen, liver, and lungs. In contrast to the relatively high rate of viral isolation from pharyngeal swabs in the early stage of infection, recovery of the virus from the feces was quite rare throughout the course of infection.

CAV-2 was first isolated from a group of dogs in an outbreak of laryngotracheitis (kennel cough) and was presumed to be one of the etiological agents of this disease (Ditchfield *et al.*, 1962). Serological relatedness and distinctiveness between CAV-1 and CAV-2 were clarified as discussed in Section VI.A.6. However, to elucidate the presumed difference in the pathogenesis of CAV-2 from that of CAV-1, comparative experimental inoculations of dogs with CAV-1 and CAV-2 were carried out (Swango *et al.*, 1970). Results of the experiment showed that irrespective of the route of infection, all the CAV-1-inoculated dogs showed typical signs of infectious canine hepatitis (or hepatitis contagiosa canis) and died, but CAV-2-inoculated dogs showed milder signs of illness: fever, depression, and nasal discharge appeared in all dogs 2 or 3 days p.i. and continued for a few days. A cough with paroxyms appeared 3 days p.i. in a dog inoculated with the virus by the ororespiratory route and persisted through day 10. Nevertheless, no dogs died. Since CAV-2 inoculation by the ororespiratory route led to symptoms of disease more conspicious than those induced by inoculation by other routes, intranasal and intratracheal inoculation of 13- to 14-week-old beagles with CAV-2 (40,000 $TCID_{50}$) was attempted. All 10 inoculated dogs developed respiratory disease with manifestations varying from a cough of a week's duration to one case of fatal pneumonia. In addition, fever was frequently recorded. Virus was recovered from the throats and rectums of all dogs through day 7 p.i. Viremia was detected in 9 dogs some time between days 2 and 5 p.i. In necropsy at several stages of infection, the virus was recovered from most organs, but the respiratory tract was most frequently affected. Gross pathological changes were pulmonary atelectasis and lung congestion and hemorrhage of bronchia lymph nodes. There were no gross lesions in the liver. Microscopically, proliferative interstitial pneumonia with intranuclear inclusion bodies in alveolar and bronchial cells was observed.

These experiments clearly indicate that CAV-2 is not only serologically distinct from CAV-1, but also different in pathogenesis, corroborating the presumption based on clinicoepizootiological studies that CAV-2 is one of the etiological agents of kennel cough.

Several animals related to dog and fox, such as timber wolf, coyote, bear, and polar bear, were reported to be naturally infected with CAV-1 (Chaddock and Carlson, 1950).

F. Simian Adenoviruses

The classification of chimpanzee adenoviruses (Section VI.A.7) and that of other monkey adenoviruses (Section VI.A.8) were discussed separately, but all these viruses are discussed as simian adenoviruses in this section.

Since many serotypes of simian adenoviruses were isolated from apparently healthy monkeys and apes, as with the other animal adenoviruses, isolation of adenoviruses from a sick monkey or ape does not warrant its designation as the etiologic agent of the illness. Only when case reports of isolation of adenoviruses are accompanied by some supporting evidence (i.e., increase of antibody titer during the course of infection, specific cytopathological changes of the postmortem gross lesions, absence of any other pathogenic microorganism) can they be regarded as the causative viruses. Case reports containing these supporting data are presented below.

Simian adenovirus type 20 SAV-20 [V340] was consistently isolated from newborn baboons suffering from pneumoenteritis (Eugster *et al.*, 1969). SAV-3 [SV15] or SAV-10 [SV30] was isolated from several macaques suffering from an acute disease accompanied by maculopapular skin rash, conjunctivitis, facial edema, and erythema (Espana, 1974). SAV-9 [SV31] was isolated from a rhesus monkey that died as the result of acute necrotizing pancreatitis (McClure *et al.*, 1978). SAV-2 [SV11] was isolated from three of four infant macaques housed together in an isolator and suffering from severe to mild pneumonia in an outbreak (Boyce *et al.*, 1978).

Although there has been a relatively large number of epizootiological and clinical reports, there have been few studies of experimental infections of monkeys with simian adenoviruses. Heath *et al.* (1966) inoculated African velvet monkeys intranasally with SAV-4 [SV17] after keeping the monkeys in quarantine for at least 1 month. All six monkeys developed a disease of the upper respiratory tract with accompanying symptoms of reddening and swelling of pharyngeal mucous membranes and tonsils. The virus was recovered from nasal and pharyngeal swabs of all monkeys from days 3 to 5 p.i. Neutralizing antibodies increased very significantly. Landon and Bennett (1969) observed a naturally occurring mild form of nonfebrile conjunctivitis in a group of six infant macaques and isolated SAV-3 [SV15] from conjunctival swabs of sick monkeys. Thereafter, they inoculated eight monkeys with a prototype strain of SAV-3 [SV15] using several routes of inoculation; only monkeys inoculated directly into the conjunctiva developed conjunctival involvement similar to the naturally occurring outbreak. All monkeys showed increased antibody titers after inoculation, and the virus was isolated from all monkeys. These two experimental reports indicate that these viruses are causative agents for experimentally produced diseases. However, the lack of postmortem pathological and virological surveys on several organs of sick animals limits somewhat the value of these experiments.

Extensive seroepizootiological studies on several simian adenoviruses were carried out by Kalter and Heberling (1971). It is noteworthy that such apes as gorillas, orangutans, and gibbons were usually free of adenovirus antibody, whereas chimpanzees were often seropositive to several simian adenoviruses. It must be kept in mind, however, that these data were obtained with apes in captivity and do not necessarily reflect the real situation of wild apes. Naturally occurring intermittent viruria with chimpanzee adenovirus Pan 11 was observed in many chimpanzees for more than 1 year (Asher *et al.*, 1978). A small amount of neutralizing antibodies of the immunoglobulin G (IgG) and IgA classes were also found in the urine, and they were considered to be locally produced in the infected urinary system.

G. Murine Adenoviruses

Until Hartley and Rowe (1960) frist isolated mouse adenovirus type 1 (MAV-1) as a passenger virus in the cell-culture system of Friend mouse leukemia virus, the contamination of laboratory mouse colonies with adenoviruses had not been realized. This finding suggests that laboratory mice usually tolerate viral infection without developing conspicious diseases and that mouse adenoviruses may be indigenous to laboratory mice.

The pathogenesis of MAV-1 infection was experimentally analyzed by a few groups. Hartley and Rowe (1960) inoculated suckling, weanling and adult mice with MAV-1; the site of viral injections was the peritoneal cavity, brain, or nasal cavity. A fatal disease characterized by general weakness followed by death 5–7 days p.i. was produced in suckling mice. Disseminated pathological changes, with intranuclear inclusion bodies, were observed in brown fat, myocardium, adrenal cortex, and, less extensively, in salivary glands and kidneys. Suckling mice born of mothers with antibodies were not susceptible. Weanling and adult mice did not develop overt disease, but readily responded with antibodies following the infection.

Heck *et al.* (1972) undertook a similar experiment with suckling mice. Mice were inoculated intraperitoneally with MAV-1. New findings in this experiment were as follows: (1) inoculation with median lethal doses 100 (LD_{50}) of virus produced viremia as early as 1 day p.i., and viremia persisted for 6 additional days; (2) virus was recovered from brain tissue in addition to the heart, kidney, and spleen on days 3 and 5 p.i. (numerous intranuclear inclusions were observed in endothelial cells of blood vessels in the brain and also in a few microglial cells and small neurons in the cerebral cortex from days 6 through 13 p.i.).

Van der Veen and Mes (1973) carried out a similar experiment on adult (4-week-old) mice. Mice infected with MAV-1 were reared and kept under observation for 2 years. Viruria was first detected around 2 weeks p.i. and found to persist for 24 months in mice inoculated intraperitoneally with 2000 $TCID_{50}$ or intranasally with 500 $TCID_{50}$ of the virus, but virus was never recovered from feces. A small number of intraperitoneally inoculated mice became lethargic and died by the 16th day p.i.

Winters and Brown (1980) inoculated adult nude (*nu/nu* mice intraperitoneally with MAV-1. A viral inoculum of as little as 100 plague-forming units (PFU) resulted in "wasting disease" and death after a varied period of time. All the mice hemorrhaged into the lumen of the duodenum shortly before death. EM observations revealed the presence of the virus in endothelial cells of duodenal villi. Heterozygous (+/*nu*) mice showed neither "wasting disease" nor duodenal hemorrhage. These results indicate that T lymphocytes play an important role in resistance to adenovirus infection.

MAV-1 showed an "adrenal and cardiac tropism" (Hartley and Rowe, 1960; Margolis *et al.*, 1974; Blailock *et al.*, 1967), but a mutant of MAV-1 showing apparent "pulmonary tropism" was isolated as a clear plaque mutant in L-cell monolayer cultures (Winters *et al.*, 1981). Intraperitoneal or intranasal inoculation of adult mice with 10^8 PFU of this mutant virus resulted in marked physical inactivity and death. Death occurred 3–5 days p.i. Collapsed hemorrhagic lungs were observed at necropsy; extensive infiltration of erythrocytes and lymphocytes into the interstices was observed histologically in hemorrhagic lungs. Lower doses of virus resulted in decreased mortality, and the affected lungs showed interstitial pneumonia. The LD_{50} of the mutant and that of the wild-type (Wt) virus were not significantly different. Mice inoculated with a comparable dose of Wt virus also showed clinical disease and died, but hemorrhagic lungs were not observed. As mentioned in Section I, the laboratory mouse system has advantages over other animals for several types of analytical studies; for example, more precise studies on this "lung tropic" mutant or similar mutants, if isolated, may pave the way for a better understanding of the underlying mechanism of "organ tropism" of adenoviruses in general.

Pathogenesis of MAV-2 infection was also experimentally analyzed (Sugiyama *et al.*, 1967). Adult and suckling mice were inoculated with MAV-2; in many experiments, the virus was injected orally through a metal tube inserted into the stomach. No clinical manifestations were observed in infected mice; however, virus was recovered from the intestinal tract for 7 days p.i. and from the feces for at least 2 weeks, but not from oropharyngeal washings, nasal tissue, lung, spleen, or urine. Neutralizing antibody was detected in the serum 3 weeks after primary inoculation. These results were not greatly influenced by either the titer of the input virus or the route of virus injection (intraperitoneal, subcutaneous, intracerebral, or intranasal); however, intramuscularly injected virus was almost impossible to recover from feces. Hashimoto *et al.* (1970, 1971) studied this experimental system more precisely and found that the virus-producing cells were scattered in the epithelial layer of the mucous membrane of the small intestine. They found that the intestinal resistance to the virus was controlled mainly by locally produced IgA. In striking contrast, MAV-2 was excreted exclusively into the feces, whereas MAV-1 was not. The "intestinal tropism" of MAV-2 also appears to be in marked contrast to the "adrenal or cardiac tropism" of MAV-1.

H. Fowl Adenoviruses

Fowl adenoviruses were isolated from tracheal and rectal swabs of apparently healthy chickens (Kawamura *et al.*, 1964) or tissue-culture cells prepared from apparently healthy chickens (Burke *et al.*, 1965). They were also isolated from chickens ill from diseases such as inclusion body hepatitis (Fadly and Winterfield, 1973; McFerran *et al.*, 1976b; Otsuki *et al.*, 1976; Grimes *et al.*, 1977). In earlier days, fowl adenovirus type 1 (FAV-1) was often isolated as an indigenous contaminant virus in eggs used for virological studies and was found to kill chicken embryos when inoculated into normal embryonated eggs. The chick embryo lethal orphan (CELO) virus was partly responsible for this finding (Yates and Fry, 1957). Kawamura *et al.* (1964) reported that of their eight serotypes of fowl adenoviruses, FAV-1 was the most virulent for embryonating chicken eggs. More recently, the egg transmissibility of FAV-1 was studied by exposing parent chickens to natural or experimental infection (Dawson *et al.*, 1979, 1981). Thus, experimentally infected hens were found to transmit virus to eggs for about a week after inoculation and to stop transmitting it when hen antibodies to the virus appeared to increase. The virus also appeared in nonfertilized eggs, but not in the semen of experimentally infected roosters. Furthermore, the experiment seemed to all but eliminate the possibility that egg transmission of the virus was due to contamination from fecal materials from the intestinal tract of infected hens, since virus stopped entering eggs 8 days after hens were infected, but continued to appear in feces even 15 days p.i.

A syndrome now called "inclusion body hepatitis of chickens" was first described by Helmboldt and Frazier (1963). The syndrome appeared in two large flocks of 7-week-old chickens. Though the mortality rate was only a few percent, the sick chickens succumbed rather suddenly—within 2 days. Consistent pathological findings reported were gross lesions in the liver: the organ was slightly swollen, had a bright orange hue, and, in some advanced cases, showed prominent stellate hemorrhages. Microscopically, the parenchyma of the liver was almost unrecognizable and showed marked fatty metamorphosis. Focal hemorrhage and infiltration by round cell types were common. However, the most characteristic feature was a number of mauve-colored (or eosinophilic) intranuclear inclusion bodies in hepatic cells. outbreaks of similar disease, sometimes with variations (e.g., higher mortality rate, lesions of the bursa of Fabricius and of bone marrow) were reported from several laboratories (Howell *et al.*, 1970; Pettit and Carlson, 1972; Fadly and Winterfield, 1973; Bickford *et al.*, 1973; Itakura *et al.*, 1974b; Macpherson *et al.*, 1974; Rosenberger *et al.*, 1974; McFerran *et al.*, 1976b; Otsuki *et al.*, 1976; Grimes *et al.*, 1977). Fowl adenoviruses were reported to be the predominant viruses isolated from sick birds in the virological surveys reported in the aforecited literature. The surveys made by McFerran *et al.* (1976b) are the most extensive. In 16 outbreaks of the syndrome studied in Northern Ireland, they isolated 46 adenovirus strains, identified as FAV-2, FAV-3, FAV-4, FAV-5, and FAV-8, as well as 6 reovirus strains (in one outbreak, only reoviruses were isolated). Some of the authors cited above determined whether the isolated adenoviruses could reproduce the syndrome when inoculated into young chickens. As shown by comparative experimental studies (McCracken *et al.*, 1976; Grimes *et al.*, 1978; Lee *et al.*, 1978), several serotypes of fowl adenoviruses variously affected the liver or sometimes other organs of inoculated chickens. However, only a limited number of viruses (Fadly and Winterfield, 1973; Winterfield *et al.*, 1973; Gallina *et al.*, 1973) could successfully produce hepatic lesions by viral inoculation through a natural route such as conjunctival sac and tracheal cavity. It is noteworthy that in most cases, intranuclear inclusion bodies observed in experimentally produced hepatic lesions were basophilic, not the eosinophilic types described in the original study (Helmboldt and Frazier, 1963).

At present, it seems clear that fowl adenoviruses, when inoculated parenterally, produce lesions in the liver and, occasionally, in some other organs of experimental chickens and that the viruses are recoverable from several organs (Kawamura and Horiuchi, 1964). However, these adenoviruses are still not established as the principal etiological agents of this hepatic disease.

A type of quail bronchitis not due to avian pneumoencephalitis virus or infectious bronchitis virus was reported from West Virginia by Olson (1951). Symptoms were observed in quails about 3 weeks of age. Feed consumption was reduced markedly at the start. Coughing, sneezing, and stertorous breathing were the main symptoms. A mortality of 80% was common. A virus was isolated by inoculating the tracheal mucus of sick birds into embryonated chicken eggs. Yates and Fry (1957) demonstrated, through a neutralization test with the hyperimmune rabbit serum against FAV-1, that the virus from quail bronchitis (or quail bronchitis virus) belonged to the same serotype as CELO virus (or FAV-1). A similar virus was also isolated from birds with quail bronchitis in Texas by DuBose *et al.* (1958). Both Olson's and DuBose's viruses were shown to be neutralized by the pooled serum collected from quails recovered from the bronchitis. DuBose's virus was also shown to be FAV-1 by the cross-serum neutralization (SN) test (Dubose and Grumbles, 1959). Experimental inoculation of quails with these viruses was reported to produce symptoms similar to those of the natural disease; in one of the experiments, contact control birds also appeared to be diseased, and the virus was recovered even from the lungs and abdomonial air sacs of the contact control birds. However, the mortality rate due to the experimental disease was low, and this somewhat compromised the presumption that quail bronchitis virus or FAV-1 was the principal agent of the quail bronchitis that Olson first described. Since the stock of the quail bronchitis virus was later shown to contain avian adeno-associated virus (AAV), Yates *et al.* (1973) and Bagshaw *et al.* (1980) attempted experimental infection of quails in which FAV-1 was inoculated simultaneously with various doses of the AAV. The mortality of the experimental birds inoculated with FAV-1 alone was 2 of 15, as compared with 7 of 15 deaths in the birds inoculated with FAV-1 plus 10^4 $TCID_{50}$ of AAV. The results suggest that an appropriate dose of AAV accompanied by FAV-1 causes severe respiratory disease in quails observed in nature. They are qualified, however, by the facts that the major postmortem lesions appeared in the liver and heart, not in the respiratory system, and that the viruses were inoculated intraperitoneally, not through the natural route.

Though isolation of quail bronchitis virus was reported from two laboratories on different occasions, both laboratories used embryonated chicken eggs for propagation of the presumed etiological agent in the days when embryonated eggs, even for experimental purposes, were not infrequently contaminated with FAV-1. The reported quail bronchitis virus may therefore have originated from FAV-1 in the eggs.

Fowl adenoviruses seem to distribute widely in several species of birds other than fowls (or chickens). FAV-1 was found not only in chickens and quails, as discussed above, but also in turkeys and guinea fowls. Blalock *et al.* (1975) isolated a strain of FAV-1 from turkey poults with respiratory signs by using specific-pathogen-free embryonated chicken eggs for viral propagation. Pascucci *et al.* (1970) isolated FAV-1 from a guinea fowl suffering from pancreatitis and reportedly produced experimental pancreatitis in guinea fowls by inoculating the virus through various routes. McFerran *et al.* (1976a) isolated FAV-2 from a budgerigar, FAV-5 from a bantam and a mallard, and FAV-8 from a budgerigar and pigeons. Furthermore, viruses carrying an antigen common to FAV-1 in the agar–gel precipitin (AGP) test were found in pheasants (Cakala, 1966), ducks (Ahmed *et al.*, 1969), geese (Kaleta, 1969), and migratory birds such as shovelers and gadwalls (Mathur *et al.*, 1972).

I. Egg Drop Syndrome 1976 Virus

Van Eck *et al.* (1976) reported decreased egg production and production of soft-shelled and shell-less eggs, which were not considered to be due to factors previously described to affect egg production. The outbreaks took place in flocks of parent and laying chickens in parts of the Netherlands. McFerran *et al.* (1978b) also reported a similar syndrome in Northern Ireland in 1976. Clinically, birds appeared healthy, and the syndrome began to appear when hens of flocks became 29–31 weeks of age on average. Among several microbial agents isolated from the affected flocks, there was a previously undescribed adenovirus that strongly agglutinated chicken red cells. A relationship was also established between the development of an antibody to the hemagglutinating adenovirus and depressed egg production. Similar syndromes were later reported in several areas of Europe and Japan (Baxendale, 1978; Meulemans *et al.*, 1979; Yamaguchi *et al.*, 1981a), and quite similar viruses were isolated.

To ascertain whether these viruses were the etiological agent in the aforedescribed egg drop syndrome, adult female birds were inoculated orally with the prototype strain of the hemagglutinating virus (McCracken and McFerran, 1978). First, loss of egg-shell pigmentation appeared 7 days p.i., then soft-shelled or shell-less eggs started to appear. From 13 to 16 days p.i., 25–40% of eggs laid were abnormal. Then the majority of hens returned to laying normal eggs by 24 days p.i. Despite the marked egg-shell changes, overall egg production was not much affected, and experimental hens were apparently healthy during the 17 weeks of the experimental period.

Similar experimental inoculation of hens with a Japanese isolate of the virus not only confirmed the experimental results described above but also provided additional data (Yamaguchi *et al.*, 1981b; Taniguchi *et al.*, 1981): Experimental hens were sacrificed on various days postinfection and examined by pathological, virological, and serological methods.

At necropsy, significant changes were observed in the uterus, which showed remarkable edema and swelling of the mucosal folds 14 days p.i. Microscopically, intranuclear inclusion bodies were observed in epithelial cells of the uterus, isthmus, and vaginal gland region 10 and 14 days p.i. Desquamation of the epithelial cells, atrophy of uterine glands, and cellular infiltration were also observed in the uterus 10 and 14 days p.i. Immunofluorescence microscopy demonstrated viral antigens in the epithelial cells of uterus and isthmus at the same stage of infection. The virus was recovered from several organs (oviduct, liver, spleen, and pancreas) 3–7 days p.i. The hemagglutination-inhibition antibody of the experimental hens started to be detected 10 days p.i.

Baxendale (1978) also produced this experimental disease by ocular inoculation of 23- and 29-week-old hens with an English strain of the virus. He also succeeded in preventing the experimental disease by subcutaneously injecting the inactivated viral vaccine at 19 weeks of age. These experimental results strongly indicate that the newly isolated hemagglutinating adenovirus named egg drop syndrome 1976 virus (EDS76V) is the etiological agent of the egg drop syndrome.

Seroepizootiological surveys conducted in Northern Ireland (McFerran *et al.*, 1977) yielded the following remarkable findings: None of the sera sampled from chickens before importation of the virus-infected chickens contained any antibody to the virus; horizontal transmission of the virus from chicken to chicken, though possible, almost never took place; and wire-mesh fence was enough to prevent contamination from flock to flock. These and other epizootiological findings led these workers to postulate that the virus was transmitted vertically from parent chickens to embryos and that sudden and widespread occurrences of this syndrome in Europe could best be explained by contamination of a vaccine with this virus and vertical transmission (or egg transmission) of the virus.

The absence of the antibody to this virus in the sera of chickens collected before the outbreaks, and the observation that this virus grew more efficiently in cells of duck origin than in those of chicken origin, prompted further epizootiological studies on ducks. Baxendale (1978) carried out a survey of 24 duck flocks and in England and found 20 of them to have the antibodies to the virus. He also found that 10 of 10 white Peking ducklings were seropositive. Calnek (1978) examined 7 flocks of white Peking ducks in New York and 2 flocks in Michigan and found all the duck flocks carrying antibodies to the virus. These findings strongly indicate that EDS76V is indigenous in the duck population and may well be called a duck virus. Kaleta *et al.* (1980) further reported that antibodies to EDS76V were found in serum samples of 2 of 381 owls, in 1 of 4 storks, in 1 of 2 swans, and in 1 of 18 wild geese. They also reported that the 2 antibody-positive owls and 1 antibody-positive stork laid eggs with malformed shells.

J. Hemorrhagic Enteritis Virus of Turkey (or Marble Spleen Disease Virus of Pheasant)

Since hemorrhagic enteritis of turkey poults was first described by Pomeroy and Fenstermacher (1937), outbreaks of what has been considered to be the same disease have occurred in several places. Turkeys 6 to 12 weeks old were most often affected by the disease. Clinical signs were sudden depression and bloody droppings; after appearance of these clinical signs, about 10% of the sick birds died within 1 day. At necropsy, the entire intestinal tract was distended, extremely cyanotic, and filled with sanguineous material. The mucosae were markedly congested. The inflammatory changes were most prominent in the duodenal portion. Small hemorrhages were often observed in breast and thigh muscles, apex of the heart, and surface of liver and kidney. The spleen showed a dark discoloration with varying degrees of enlargement. Lungs were congested. Histologically, significant pathological changes were observed in the mucosa of the intestinal tract. Some part of the mucosa showed an earlier stage of pathological changes; the villi became distended due to infiltration of lymphatic cells, and several intranuclear inclusion bodies were observed in reticuloendothelial cells. Some other parts showed a more advanced stage: massive desquamation of the epithelium from the surface of the villi. Formation of the intranuclear inclusion bodies in reticuloendothelial cells with lymphatic cell hyperplasia was also observed in other organs such as spleen, liver, and lung.

Since the disease seemed to be transmitted quickly in large or small flocks of turkey poults, experimental transmission of the disease was attempted. Gross and Moore (1967) were the first to show that the etiological agent was obtained from the extract of affected intestines and passed through 220-nm pores. Later, the disease was shown to be transmittable to 6- to 8-week-old poults not only with sera of sick poults that had been filtered through 220-nm filters but also with extracts of several other organs of the sick poults. Intravenous, oral, and cloacal inoculation were all effective in producing the disease. Clinical and pathological changes were most prominent 5 and 6 days p.i. In an experiment, contact control birds were also shown to be affected. The specific AGP reaction was observed between the presumptive antigen prepared from the spleen of poults 4–6 days p.i. and convalescent sera of the sick poults. Other organs of the sick poults did not usually give a detectable amount of the antigen. However, few birds died from the experimental infection (Domermuth and Gross, 1971; Domermuth *et al.*, 1972; Itakura *et al.*, 1974a).

All these findings indicate that the etiological agent is a virus and that the virus, called hemorrhagic enteritis virus of turkeys (HEV), is most abundant in the spleens of sick poults. Furthermore, EM of ultrathin sections of the spleen, intestine, and bone marrow and negative staining of the preparation purified from the spleen demonstrated a large number of adenoviruslike particles (Carlson *et al.*, 1974; Tolin and Domermuth, 1975; Itakura and Carlson, 1975).

Early reports of marble spleen disease of pheasants called it "a disease involving the spleen and lungs in pheasant" (Mandelli *et al.*, 1966) or "marble spleen and pulmonary edema syndrome of pheasants" (Rothenbacher *et al.*, 1970), significant pathological changes of this disease being observed in spleen and lungs. According to Wyand *et al.* (1972), the age incidence of the fatal cases ranged from 4 to 8 months. Clinically, the disease was very acute; death was apparently due to anorexia resulting from severe pulmonary edema. At necropsy, all birds had spleens enlarged two to three times, with gray confluent foci resulting in a mottled appearance; the lungs were firm, wet, and liver-colored. No other gross lesions were noted. Histologically, the spleen showed an inflammatory reaction, necrosis, and extended amyloidosis. Numerous intranuclear inclusion bodies were seen in reticuloendothelial cells of the spleen. Small scattered foci of coagulation necrosis of epithelial and endothelial cells were observed in the lung parenchyma. Occasionally, cells containing intranuclear inclusions were observed. The tertiary bronchi were often filled with red blood cells. Proteinaceous materials, fibrin, and red cells were seen in the respiratory tubules. EM demonstrated that the intranuclear inclusions contained various numbers of nonenveloped nucleocapsids with an average diameter of 70 nm. Jakowski and Wyand (1972) recognized the specific AGP reaction between the splenic extract of a pheasant dying of marble spleen disease and 31 of 283 nonpooled serum samples obtained from a flock of 9-month-old pheasants that had experienced an outbreak of marble spleen disease 3–4 months previously. The AGP test indicated that the splenic antigen preparation of turkeys suffering from hemorrhagic enteritis was very similar to that of pheasants suffering from marble spleen disease (Iltis *et al.*, 1975).

Inasmuch as these findings of natural disease were enough to consider that the etiological agent of disease was transmissible, experiments to transmit the disease to turkey poults and young pheasants were conducted (Domermuth *et al.*, 1975; Iltis *et al.*, 1975, 1977). When young pheasants or turkey poults were inoculated orally or through other routes with the splenic extracts of pheasants that had died from disease, increase of the splenic antigen, splenomegaly, and appearance of intranuclear inclusion bodies in many organs were observed in many experimental animals. The treatment of the inocula with chloroform or filtration of the inocula through 220-nm pores did not significantly affect the experimental results. Simultaneous inoculation of convalescent pheasant marble spleen disease antiserum or convalescent turkey hemorrhagic enteritis antiserum completely inhibited the experimental changes described above. However, the transmission experiments did not cause mortality due to the disease except in one pheasant inoculated with the crude extract of the spleen, whereas naturally occurring marble spleen disease and hemorrhagic enteritis usually led to significant mortality. These experimental results suggest that both naturally occurring diseases require some superimposing factors in addition to marble spleen disease virus of pheasant (MSDV) or HEV.

A virus indistinguishable from HEV (or MSDV) was isolated from chickens showing splenomegaly and considered to have Marek's disease (Domermuth *et al.*, 1979). Extensive seroepizootiological studies indicated that the viral infection is also very common among broiler breeder chicken flocks (Domermuth *et al.*, 1980). The experimental infection of chickens with the virus obtained from naturally occurring hemorrhagic enteritis of turkey produced splenic and intestinal lesions similar to, but milder than, those observed in the naturally occurring hemorrhagic enteritis of turkey (Silim *et al.*, 1978; Darbyshire, 1980).

K. General Considerations

As already noted in the individual cases described above and in Section VI, adenoviruses are not infrequently isolated from cultured cells of kidney, testicles, and pharyngeal or rectal swabs of apparently healthy animals. Therefore, isolation of a particular type of adenovirus from clinically sick animals does not necessarily mean that the isolated virus plays an etiological role in the observed disease. To consider that a given type of adenovirus is the etiological agent of a particular disease, at least the following circumstantial evidence must be presented:

1. The titer of the virus-specific SN antibody must be low before the onset of the disease or at the time when early clinical signs appear. Then, when sick animals start to recover from the disease, the titer of the SN antibody in many of them should also start to increase.

2. At necropsy of euthanatized sick animals, virus must be isolatable from several organs, especially from one showing a gross lesion, at the time when clinical symptoms appear most prominently. Since most adenovirus infections seem to show at least a transient viremia, a blood sample taken from an animal at an early stage of the disease not infrequently contain the virus. However, viral isolations from organs gradually become difficult after the titer of the SN antibody starts to increase.

3. Characteristic intranuclear inclusion bodies must be observable in cells, predominantly in the reticuloendothelial cells, of organs that show gross lesions at the time clinical symptoms are most prominent. The cells carrying intranuclear inclusion bodies must also stain with fluorescein-conjugated antibodies to the viral antigens.

4. Other pathogenic microorganisms may not be isolated from sick animals, especially from the organs that show the gross lesions. However, if such a microorganism is isolated, its possible etiological role must be clearly eliminated—e.g., by inducing experimental infection with it.

5. In enzootic or epizootic cases, these findings must be observed in many animals, and even if two outbreaks are geographically or chronologically separated, the findings must be similar.

Not many clinical cases are examined so extensively as to present all this circumstantial evidence. When a considerable portion of it can be accumulated, however, experimental reproduction of the disease must be attempted to determine the etiological role of the isolated virus from the sick animal. Use of a clone-purified viral preparation is preferable for injection into the experimental animal. Viral inoculation by a nonparenteral route, such as oral or intranasal inoculation, is preferred, since natural invasion seems to be through the nonparenteral route. Several parameters that influence serological, microbiological, physiological, and genetic conditions of experimental animals must be controlled to reproduce the disease:

1. Younger animals generally show more severe clinicopathological changes due to the adenovirus infection.
2. The genetic background of animals, especially the immunogenetic background, greatly influences viral pathogenesis, as noted earlier in regard to immunoinsufficient Arab foals (Section VII.D) and nude mice (Section VII.G).
3. The presence or absence of the virus-neutralizing antibodies (IgG and blood and local IgA) also greatly affects pathogenesis. Previous infection with the same or related serotypes of adenovirus is also a factor in pathogenesis. In young animals, transfer of maternal antibodies through the placenta and colostrum has a significant influence on pathogenesis.
4. Contamination by indigenous or invading microorganisms must be controlled to obtain reproducible observations. To monitor the pathogenesis of experimental infection, pathological, serological, and microbiological examinations must be carried out successively before and after the experimental inoculation of virus into animals.

Transmission of adenovirus is usually horizontal, from animal to animal, without vectors. Coughing and licking mate animals are considered to be means of transmitting the virus. As referred to in the preceding sections, contact control animals kept in the same cages with experimentally infected animals often provided evidence that horizontal transmission took place. Especially when the population density of sensitive animals is high, transmissiblity becomes very high, as observed in the epizootic of fox encephalitis quoted at the beginning of this chapter. When excreted in urine or feces, the viral particles do not seem to lose infectivity immediately, so food and drinking water are probably contaminated with urine or feces and become a source of viral transmission to the animals in the same enclosure or cage. Airborne transmission of FAV-1 was experimentally demonstrated in a group of uninoculated sensitive chickens located 153 cm away from a group of virus-inoculated chickens in a room so that the direction of air flow was from the inoculated to the uninoculated (Cook, 1974). Although rather exceptional, some adenoviruses are also transmitted vertically. FAV-1 is known to be transmitted through the egg, although not as frequently as by horizontal transmission (see Section VII.H). Egg transmission is postulated to be a major route of EDS76V spreading in chickens, though horizontal transmission by oral inoculation with the virus is experimentally possible (see Section VII.I). Even mastadenoviruses may be transmitted vertically by intrauterine infection, as in the case of BAV-2 (see Section VII.A).

When adenoviruses are transmitted horizontally, the major points of entry are the mouth and nose. The mucous membranes of the oral and nasopharyngeal cavity, especially pharyngeal lymphatic follicles, are considered to be the most frequent site of initial multiplication. Conjunctival and, in the case of birds, the cloacal mucous membranes also serve as the initial site of multiplication. Although not yet precisely analyzed, it is considered that the presence or absence of the local IgA-type antibodies, and of other nonspecific surveillance mechanisms, will determine the degree of initial multiplication of the virus in the mucous membrane and regional lymphatic organs. Whether or not the replicated virus enters the bloodstream is largely determined by the concentration of humoral neutralizing antibodies. Since viremia has been observed in infection with several adenoviruses [BAV-7 (Section VII.A), CAV-1 (Section VII.E), HEV (Section VII.J), and FAV-1 (Section VII.H)], most adenoviruses are considered to produce at least a transient viremic state so that they can enter into several organs of animals and multiply in reticuloendothelial cells in individual organs to produce intranuclear inclusion bodies. As pointed out in the cases of BAV-1, CAV-1, MAV-1, MSDV, and EDS76V, the virus can be recovered in several organs at the peak of disease, but the gross lesions with inflammatory reactions are observed in particular organs in consequence of infection with a particular type of adenovirus. This tropism is called "organ tropism." It is well known that interstitial nephritis, proliferative glomerulonephritis, and corneal uveitis are often complications in dogs recovering from CAV-1 infection. The first one is considered to be due to Type IV hypersensitivity, and the remaining are attributed to Type III hypersensitivity (Wright, 1976). However, this kind of humoral antibody reaction to the viral particles or cell-mediated hypersensitive reaction to the surface of the virus-infected cells is considered to take place in any organ. Hence, it is not considered to be the primary cause of the organ tropism in an adenovirus infection. More regional "specificities" of the surveillance mechanism and the "specificities" of viral virulence seem to determine the apparent organ tropism. The age and genetic composition of the animal and genetic changes of the virus are considered to provide a direct or indirect influence on the variations of organ tropism. The mechanisms of viral pathogenesis re-

TABLE III. Nonhuman Adenoviruses Oncogenic in Hamsters

Adenoviruses	Serotype	Reference
Simian		
Subgroup I	SAV-1 [SV1]	Huebner *et al.* (1962)
	SAV-2 [SV11]	Gilden *et al.* (1968)
	SAV-7 [SV25]	Gilden *et al.* (1968)
	SAV-11 [SV33]	Hull *et al.* (1965)
	SAV-12 [SV34]	Hull *et al.* (1965)
	SAV-15 [SV38]	Hull *et al.* (1965)
Subgroup II	SAV-6 [SV23][a]	Gilden *et al.* (1968)
	SAV-14 [SV37][a]	Hull *et al.* (1965)
Subgroup III	SAV-5 [SV20][b]	Hull *et al.* (1965)
	SAV-8 [SV30]	Slifkin *et al.* (1968)
	SAV-16 [SA7][b]	Hull *et al.* (1965)
Canine	CAV-1	Sarma *et al.* (1967)
	CAV-2	Dulac *et al.* (1970)
Bovine	BAV-3[b]	Darbyshire (1966)
	BAV-8	Rondhuis (1973)
Fowl	FAV-1[b]	Sarma *et al.* (1965)
	FAV-2	Ishibashi *et al.* (1980)
	FAV-3	Ishibashi *et al.* (1980)
	FAV-4	Ishibashi *et al.* (1980)
	FAV-6	Ishibashi *et al.* (1980)

[a] The oncogenicity of the virus is still problematic.
[b] The virus reportedly has the ability to transform cultured cells.

main an intriguing problem to be elucidated in the future at the cellular and, it is to be hoped, the molecular level.

VIII. ONCOGENICITY

A. Induction of Tumors

Of 55 or more serotypes of nonhuman adenoviruses so far isolated, those reported to induce tumors in newborn hamsters are 11 serotypes of monkey adenoviruses, 2 serotypes of bovine adenoviruses, 2 serotypes of canine adenoviruses, and 5 serotypes of fowl adenoviruses (see Table III). Simian adenovirus type 16 (SAV-16) [SA7] was also reported to induce tumors in suckling rats (Hull *et al.*, 1965), while SAV-5 [SV20] (Hull *et al.*, 1965) and fowl adenovirus type 1 (FAV-1) (Yasue and Ishibashi, 1982) also induced tumors in mice and rats, respectively. However, the oncogenicity of SAV-6 [SV23] and SAV-14 [SV37] is a matter of question, since Burnett *et al.* (1972) presented several lines of evidence against the oncogenicity of these viruses.

It is difficult to compare the level of oncogenicity in the individual serotypes listed in Table III because of the different number of plaque-forming units injected into hamsters. Even if the number of plaque-forming units injected were adjusted to a certain value, viral particles without complete infectivity, but with oncogenicity, could be present in different numbers in the viral preparation (Igarashi *et al.*, 1975). However, since there seems to be no great difference in the volume of infected-cell lysate (or crude viral preparation) injected into hamsters among these viral serotypes, one may make a rough comparison of their oncogenicity on the basis of an equal volume of crude viral preparation. This comparison leads to the conclusion that SAV-16 [SA7] and bovine adenovirus type 3 (BAV-3) are highly oncogenic among the nonhuman adenoviruses.

When the fragments obtained by cleavage of SAV-16 [SA7] DNA with some of the 6-base sequence-specific restriction endonucleases were injected into newborn hamsters subcutaneously, some hamsters were reported to develop tumors (Burnett *et al.*, 1975; Ponomareva *et al.*, 1979); furthermore, Ponomareva *et al.* (1979) succeeded in inducing tumors in hamsters by injecting the preparation of the left-end fragment of SAV-16 [SA7] DNA (0–20 m.u.). They also found that the deproteinized viral DNA had a higher oncogenicity in hamsters than the viral DNA–terminal protein (TP) complex, whereas the latter had a higher plaque-forming activity in cultured cells prepared from the natural host. One possible explanation for the latter finding is that covalent blocking of viral DNA 5′ ends with TP precludes integration of viral DNA into cellular DNA as discussed by the authors cited.

B. Transformation of Cultured Cells

Since the viral property of transforming cultured cells is closely related to the property of inducing tumors in experimental animals, all the serotypes of nonhuman adenoviruses tumorigenic in animals (Table III) are also considered to be capable of transforming cultured cells. Only a few of them, however, have so far been reported to transform cultured cells by inoculation of the viral particles into the cell culture: SAV-16 [SA7] transformed cells of mice, rats, and hamsters (Altstein *et al.*, 1967; Casto, 1968; Riggs and Lennette, 1967; McAllister *et al.*, 1969) and SAV-5 [SV20] transformed cells of rats (Riggs and Lennette, 1967; Parkhomenko *et al.*, 1981). BAV-3 transformed cells of hamsters and mice (Motoi *et al.*, 1972; Igarashi *et al.*, 1978). FAV-1 transformed cells of hamsters, rats, mice, and humans (Anderson *et al.*, 1969a–c; Ishibashi *et al.*, 1980). Though the dissociation of the two related properties—namely, the ability to transform cultured cells but the inability to induce tumors in animals—has been observed in some human adenoviruses such as Ad2 and Ad5, a similar dissociation has not so far been documented in nonhuman adenoviruses. Only the transforming ability has so far been reported for *Tupaia* adenovirus type 1 (TupAV-1), one of the most recently discovered adenoviruses (Darai *et al.*, 1980); it remains to be investigated, however, whether or not this virus is tumorigenic in animals.

By using techniques such as the calcium method to introduce the viral DNA or its fragments directly into cells, transformations of culture cells were performed with SAV-16 [SA7], BAV-3, and FAV-1. With SAV-16 [SA7], the viral DNA fragments produced by sequence-specific restriction endonucleases (mixture of all the DNA fragments) induced transformation of hamster cells, indicating that intact viral DNA is not necesary for the transformation of cells (Ponomareva *et al.*, 1979). Since these endonucleases are known not to cleave SAV-16 [SA7] DNA in the segment between 0 and 20 m.u., this observation is compatible with the finding that the viral DNA segment (0–20 m.u.) induces tumors when injected into hamsters. Attempts to transform cells of mice with various DNA fragments from BAV-3 revealed that the viral DNA segment ranging from 3.6 to 11.9 m.u. was necessary for transformation of cells (Igarashi *et al.*, 1978; Niiyama *et al.*, 1981). In FAV-1, intact viral DNA was found to transform cells of rats. However, some attempts to transform cells with fragments of FAV-1 DNA have failed (Yasue and Ishibashi, unpublished data).

C. Tumor Antigens

As with human adenovirus tumors, the virus-specific tumor (T) antigens and anti-T sera were described in tumors induced by nonhuman adenoviruses such as SAV-8 [SV30], canine adenovirus type 1 (CAV-1), and FAV-1. Sera collected from animals bearing tumors induced by these adenoviruses often showed specific complement-fixation-positive reactions to the crude extract of the tumor induced by the same serotypes as the presumed antigen preparation (Slifkin *et al.*, 1968; Sarma *et al.*, 1967; Potter and Oxford, 1969; McCormick *et al.*, 1971). These sera, called anti-T sera, were also found to react with the productively infected cells, namely, the cultured cells of natural host animals infected with the corresponding serotypes of the virus (Slifkin *et al.*, 1968; Sarma *et al.*, 1967; Dulac *et al.*, 1970; Oxford and Potter, 1970). This result strongly indicated that tumor antigens recognized by these sera were identical with some of the virus-coded proteins synthesized in the productively infected cells. With FAV-1, the following evidence indicated that the tumor antigens were similar to the viral proteins synthesized before the onset of viral DNA replication in the productive infection: immunofluorescence tests with the anti-T serum showed that the T antigen appeared in the nuclei of the infected chicken kidney cells cultured even in the presence of an inhibitor of DNA synthesis, cytosine arabinoside (Oxford and Potter, 1970; Ishibashi *et al.*, 1980); immunoprecipitation tests (Ishibashi *et al.*, 1980) revealed that at least four viral polypeptides are synthesized in these cells.

It was demonstrated by immunofluorescence studies with the anti-T sera of the corresponding viruses that T antigens are also present in cultured cells transformed by nonhuman adenoviruses such as SAV-16 [SA7] (McAllister *et al.*, 1969), BAV-3 (Motoi *et al.*, 1972), and FAV-1 (Anderson *et al.*, 1969c; Ishibashi *et al.*, 1980). In addition, hamsters in which tumors had been induced by inoculation of BAV-3-transformed cells yielded sera that precipitated proteins possibly related to cell transformation by BAV-3 (Niiyama, 1981). Two polypeptides with molecular weights of 15 and 54K were consistently precipitated when cultured bovine kidney cells of early-phase infection were used, but one polypeptide with a molecular weight of 80K was detected when the transformed cells were used as the antigen. The relationship between the 80K polypeptide and the 15 and 54K polypeptides has yet to be clarified.

It can be inferred from the preceding discussion that T antigen is one of several early polypeptide(s) of individual nonhuman adenoviruses that are synthesized not only in natural-host cells of early-phase infection but also in tumor cells and transformed cells, similar to human adenoviruses. Yet it is still unknown which of these polypeptides is the T antigen(s) in the strict sense of the term, i.e., a product(s) of a viral gene(s) indispensable for oncogenesis or transformation. Some possibility also remains that the polypeptide(s) not detected with the anti-T sera used in the studies cited above is a T-antigen(s) in the strict sense.

D. Viral DNA Persisting in Tumor and Transformed Cells

The demonstration of virus-specific T antigens in some lines of nonhuman-adenovirus-induced tumor cells and transformed cells, and the analogy with the cases of human adenoviruses, suggest that viral DNA persists in all lines of tumor cells and transformed cells induced by nonhuman adenoviruses. However, the persistence of viral DNA in both tumor cells and transformed cells induced by nonhuman adenoviruses has been directly confirmed in very limited cases such as SAV-16 [SA7], BAV-3, and FAV-1. From the analyses using DNA–DNA reassociation kinetics (Cot analysis), hamster cells transformed by SAV-16 [SA7] (Casto *et al.*, 1979) and mouse cells transformed by BAV-3 (Niiyama *et al.*, 1981) were estimated to contain up to 10 and 2 copies of viral DNA per cell (or per diploid amount of cellular DNA), respectively. With the use of the same technique, one tumor line and two transformed cell lines induced by the Phelps strain of FAV-1 were found to contain up to 4 copies of viral DNA per cell (May *et al.*, 1975, 1978). The existence of viral DNA in 24 lines of tumor and transformed cells induced by the Ote strain of FAV-1 was also investigated with both Cot analysis and the Southern blot method (Ishibashi *et al.*, 1980; Yasue *et al.*, 1980). Many of the cell lines were thus found to contain hundreds of copies of viral DNA per cell, though the cell lines displayed a wide variation in the number of copies of viral DNA per cell, ranging from 1 to 920, as shown in Fig. 1 (Ishibashi *et al.*, 1980; Yasue *et al.*, 1980; Ishibashi and Yasue, unpub-

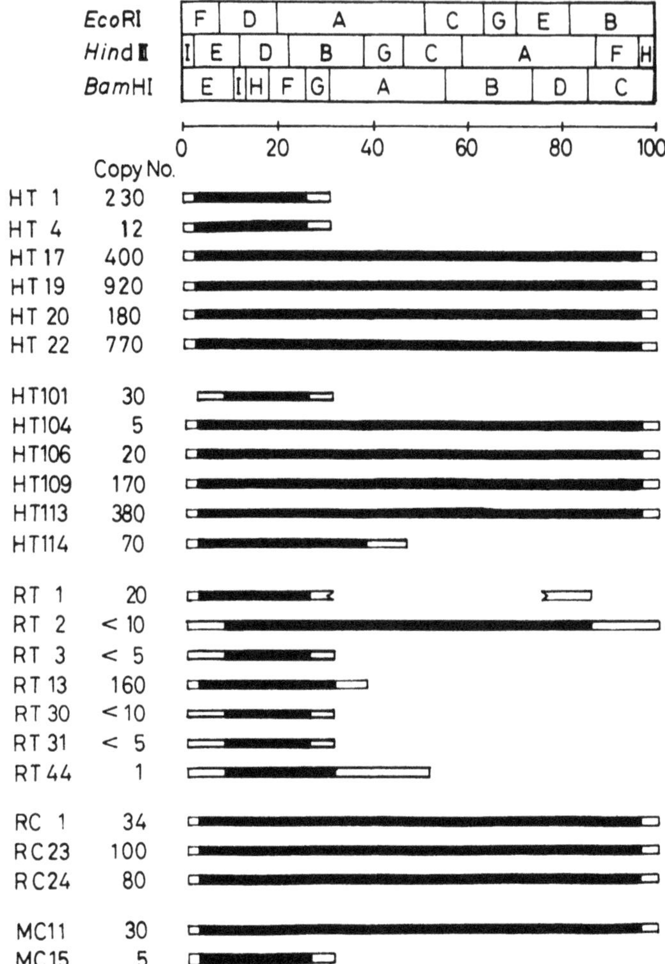

FIGURE 1. Viral DNA segments persisting in FAV-1-induced tumor and transformed cells. A total of 24 tumor and transformed cell lines induced by FAV-1 (Ote strain) were subjected to Southern blot analysis using ^{32}P-labeled viral DNA fragments as probes. DNAs prepared from cells were digested with any one of the restriction endonucleases *Eco*RI, *Hind*III, and *Bam*HI. Viral DNA fragments used as probes were fragments produced by cleavage of viral DNA with *Eco*RI (*Eco*RI-A–G) and fragments produced by cleavage of viral DNA with *Hind*III (*Hind*III-D, -F, -H, and -I). (■) Viral DNA segments colinear with viron DNA; (□) viral DNA undergoing recombination with cellular DNA somewhere in the segments; (⊏ ⊐) two separate segments of viral DNA that were indicated to fuse. (HT) Tumor line of hamster (Syrian); (RT) tumor line of rat (Fischer); (RC) transformed cell line of cultured cell line (3Y1) of rat (Fischer); (MC) transformed cell line of cultured cell line [A31 (a clone of 3T3)] of mouse (BALB/c). The detailed experimental procedure is to be described by Ishibashi and Yasue (unpublished data). The cleavage map of viral DNA and the number of copies of viral DNA per diploid amount of cellular DNA in individual lines are shown for reference.

lished data). Since the number of copies of viral DNA present in almost all the tumor or transformed cell lines induced by other adenoviruses does not exceed 20 per cell, the Ote-strain-induced tumor and transformed cell line findings were unusual and may be unique to FAV-1 or the Ote strain of FAV-1.

E. Arrangement of Viral DNA in Tumor or Transformed Cells

The arrangement at the molecular level of viral DNA in tumor cells or transformed cells has been investigated only in the aforementioned 24 lines of tumor and transformed cells induced by the Ote strain of FAV-1 (Yasue *et al.*, 1980; Yasue and Ishibashi, 1982; Ishibashi and Yasue, unpublished data). As summarized in Fig. 1, the investigation with Southern blot analysis revealed the following: (1) Viral DNA in tumor and transformed cells underwent recombination with cellular DNA. (2) Though many lines of tumor cells and transformed cells carried a large number of copies of the viral DNA in individual cells, most of the recombination with cellular DNA took place at two sites along the unit length of the viral DNA. At one site, the viral DNA was linked to the cellular DNA of the left flank; at the other site, to the cellular DNA of the right flank. (3) The viral DNA between the right and left recombination sites was colinear with the virion DNA. (4) Distribution of these recombination sites along the viral DNA was very limited, when plotted for the 24 cell lines. The left recombination sites were distributed mainly within 0–7.8 viral m.u. and the right recombination sites mainly within 25.8–31 or 96.5–100 m.u. (5) The viral DNA sequence (e.g., 14.1kb in RT13 cells) with a certain length of cellular DNA sequence at the both flanks (e.g., more than 18.2 kb in RT13 cells) formed a virus–cellular sequence unit, and a large number of copies of this unit were present in individual tumor cells or transformed cells carrying a large number of the viral DNA. The multiple repetition of the virus–cellular sequences might be explained in one of two ways: (a) When the virus was inoculated into the animal (or culture cells), a few molecules of viral DNA succeeded in forming recombinations at "random" positions on the cellular DNA of the cell from which the tumor (or transformed cell line) would originate. Subsequently, some of the integrated viral DNA molecules and their flanking cellular DNA sequences formed amplification units, which were amplified and sometimes mobilized like prokaryotic transposons, during the course of events by which the original cell progressed to the visible tumor (or the established transformed cell line). (b) When the virus was inoculated into the animal (or cultured cells), as many viral DNA molecules as found in a tumor cell (or a cell of the transformed cell line) underwent recombination at particular sites along the viral DNA and the repeated DNA sequences of the cell from which the tumor (or the transformed cell line) would originate. These two alternatives have yet to be

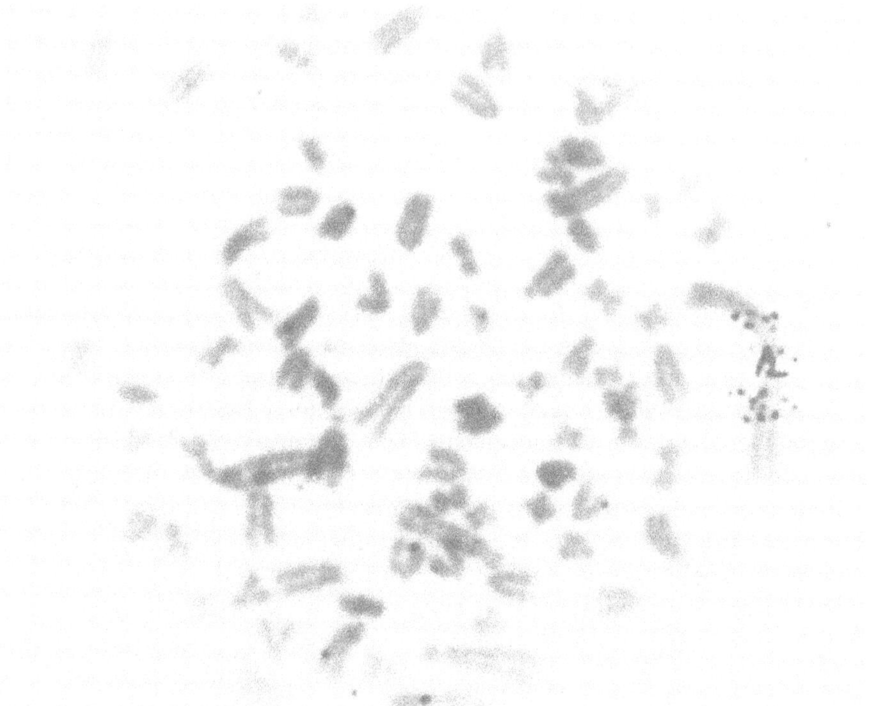

FIGURE 2. *In situ* hybridization of FAV-1-induced tumor cell (RT13) DNA with ^{3}H-labeled-FAV-1 DNA. The photomicrograph shows *in vitro*-cultured RT13 cells in mitotic phase. Note that almost all the silver grains cluster over a part of an abnormally long chromosome. The cluster of silver grains is roughly estimated to represent 160–320 copies of the viral DNA segment, though the diversity of karyotypes among individual cells of this line does not allow more precise estimation. Several control experiments and technical details were described previously (Yasue and Ishibashi, 1982).

clarified by utilizing recently obtained molecularly cloned virus–cellular sequences (Yasue and Ishibashi, unpublished data).

To study, at the microscopic level, the location of a large number of viral DNAs in FAV-1-induced tumor cells (e.g., on authentic chromosomes or double minute chromosomes that do not appear in normal cells), Yasue and Ishibashi (1982) applied *in situ* hybridization, using ^{3}H-labeled FAV-1 DNA as probe, to the cells in mitotic phase of an FAV-1-induced rat tumor line (RT13). The experiment showed that silver grains were clustered over a limited part of one to three chromosomes per cell, as shown in Fig. 2. Many chromosomes clustered with silver grains were abnormally longer than chromosome No. 1 of the rat. This observation, together with the results of the Southern blot analysis, indicated strongly that viral DNAs recombined with cellular DNA and did not form anything like double minute chromosomes or extrachromosomal plasmids, but remained in constituents of chromosomal DNA. In addition, this *in situ* DNA hybridization experiment provided the first clearly visible form of microscopic evidence that the genome of the tumor virus is incorporated into the chromosomal DNA. Yasue *et al.* (unpublished data) have also analyzed a transformed cell line (RC24) more precisely, applying both G-banding and *in situ* hybridization to one and the same cell in metaphase. The analysis has revealed that viral DNA resides in a particular part of particular chromosomes.

F. Indispensable Segment of Viral DNA for Tumor Induction or Cell Transformation

As discussed in Section VIII.B, experiments with inoculation of restriction-endonuclease-generated DNA fragments indicated that the DNA segments of SAV-16 [SA7] and BAV-3 indispensable for tumor induction and cell transformation were localized within the range from 0 to 20 and from 3.6 to 11.9 viral DNA m.u., respectively (Ponomareva *et al.*, 1979; Igarashi *et al.*, 1978; Niiyama *et al.*, 1981). However, since these inoculation experiments were carried out with viral DNA fragments that were prepared without the DNA cloning technique and hence were possibly contaminated with other fragments of the viral DNA, the conclusion remains a tenuous one, unless it is first determined whether or not the transformed cells induced in the aforecited experiments contain viral DNA segments other than those of DNA fragments actually intended for inoculation.

Though there have been no FAV-1 studies that succeeded in inducing tumors or transformation with the segmental DNA of the virus, analysis of viral DNA persisting in the tumor and transformed cells allows one to surmise what part of viral DNA is necessary for at least the maintenance of cells in the tumorous or the transformed state. Studies on a total of 24 lines of tumor cells and transformed cells induced with the Ote strain of FAV-1 (Yasue *et al.*, 1980; Yasue and Ishibashi, 1982; Ishibashi and Yasue, (unpublished data) have shown that all lines of tumor cells and transformed cells carry at least the viral DNA segment from 2.5 to 31 m.u. (see Fig. 1), though most of them carry additional segments. Accordingly, viral genes necessary for cell maintenance in the tumorous or transformed state are considered to be located in such a segment. Also, all three cell lines (RT13, RC1, and RC24) so far tested produce the viral mRNA complementary to a part of the *Eco*RI-A and -F fragments (Yasue and Ishibashi, unpublished data), supporting the conclusions of the DNA study cited above. However, studies by Cot analysis on three tumor and transformed cell lines induced with the Phelps strain of FAV-1 (May *et al.*, 1975, 1978) have provided results that may not necessarily agree with this view. In any case, further exploration is required to conclude which segment of FAV-1 DNA is indispensable for tumor induction and cell transformation.

In the aforenamed three nonhuman adenoviruses, the viral DNA maps are oriented so that viral DNA segments involved in tumor induction or cell transformation (oncogenic genes) are located in the left-hand side. This may lead one to think that oncogenic genes of nonhuman adenoviruses resemble one another and those of the human adenoviruses. However, the orientation of the DNA map for these viruses must be considered arbitrary until the basic arrangement of transcriptional units along the viral DNAs is elucidated as much as those of human adenoviruses. Actually, the orientation of the DNA map of FAV-1 shown in Fig. 1 is opposite that of Denisova *et al.* (1979) and Aleström *et al.* (1982b) discussed in Section IV.

IX. CONCLUDING REMARKS

The number of host animal species or genera from which adenoviruses have been isolated is still very small. Many host animal species are, as a matter of course, domestic animals, the existence of which has special significance for human life. However, for a more thorough understanding of the adenovirus family, more especially the evolutionary aspect of this virus, it is desirable to isolate more viruses from animals of different species that have phylogenetic relationships further removed from the currently known species of host animals. When such viruses are isolated, their phenotypic and genotypic features should be analyzed at the molecular level.

It is worth noting that two adenoviruses that show intestinal "organ tropism," namely, human enteric adenovirus and HEV of turkey, do not propagate in cultured cells (see Section VI.B.4). It is also of interest that by using cultured testicle cells rather than the more widely used kidney cells, several bovine and ovine adenoviruses, which may otherwise not have been discovered, were isolated (see Sections VI.A.1 and VI.A.2). These findings suggest that undescribed adenoviruses may exist even in species of animals from which several serotypes have already been isolated.

ACKNOWLEDGMENTS. The authors are thankful to Drs. D. M. Asher, J. O. Britt, R. L. Heberling, W. D. Hillis, J. B. McFerran, and D. G. Simmons for their kind responses to authors' inquiries about information relevant to this chapter and for their donation of reprints. The authors are also grateful to Ms. T. Shibata for her typewriting.

Note added in proof

The authors would like to call reader's attention to the fact that the whole nucleotide sequence of ITRs was reported for CAV-1, EAV, and

TupAV (Shinagawa *et al.*, 1983, *Virology* **25**:491–495; Brinckmann *et al.*, 1983, *Gene* **24**:131–135), after the manuscript of this chapter was submitted for publication.

REFERENCES

Adair, B. M., and Curran, W., 1978–1979, Studies on the cytopathology of bovine adenoviruses in calf testis and calf kidney cell cultures, *Vet. Microbiol.* **3**:241–254.

Adair, B. M., and McFerran, J. B., 1976, Comparative serological studies with mammalian adenoviruses, *Arch. Virol.* **51**:319–325.

Adair, B. M., McFerran, J. B., Connor, T. J., McNulty, M. S., and McKillop, E. R., 1979, Biological and physical properties of a virus (strain 127) associated with the egg drop syndrome, 1976, *Avian Pathol.* **8**:249–264.

Ahmed, A. A. S., El-Sisi, M. A., and Saber, M. S., 1969, Ein Beitrag zur Epizootiologie einer adeno-ähnlichen Virus-(CELO)-Infection beim Geflügel in Ägypten, *Tieraerztl. Umsch.* **24**:111–115.

Akusjärvi, G., and Pettersson, U., 1978a, Nucleotide sequence analysis of adenovirus DNA. I. Nucleotide sequence at the carboxy terminal end of the gene for adenovirus type 2 hexon, *Virology* **91**:235–238.

Akusjärvi, G., and Pettersson, U., 1978b, Nucleotide sequence analysis of adenovirus DNA. II. Nucleotide sequence at the junction between the coding part of the hexon m-RNA and its leader sequence, *Proc. Natl. Acad. Sci. U.S.A.* **75**:5822–5827.

Aleström, P., Stenlund, A., Li, P., and Pettersson, U., 1982a, A common sequence in the inverted terminal repetitions of human and avian adenoviruses, *Gene* **18**:193–197.

Aleström, P., Stenlund, A., Li, P., Bellett, A., and Pettersson, U., 1982b, Sequence homology between avian and human adenoviruses, *J. Viol.* **42**:306–310.

Alstein, A. D., Sarycheva, O. F., and Dodonova, N. N., 1967, Transforming activity of green monkey SA7(C8) adenovirus in tissue culture, *Science* **158**:1455–1457.

Anderson, J., Yates, V. J., Jasty, V., and Mancini, L. O., 1969a, *In vitro* transformation by an avian adenovirus (CELO). I. Hamster-embryo fibroblastic cultures, *J. Natl. Cancer Inst.* **42**:1–7.

Anderson, J., Yates, V. J., Jasty, V., and Mancini, L. O., 1969b, *In vitro* transformation by an avian adenovirus (CELO). II. Hamster kidney cell cultures, *J. Natl. Cancer Inst.* **43**:65–68.

Anderson, J., Yates, V. J., Jasty, V., and Mancini, L. O., 1969c, The *in vitro* transformation by an avian adenovirus (CELO). III. Human amnion cell cultures, *J. Natl. Cancer Inst.* **43**:575–580.

Asher, D. M., Hooks, J. J., Amyx, H. L., Luber, N. P., Asher, L. V. S., Gibbs, C. J., and Gajdusek, D. C., 1978, Persistent shedding of adenovirus in urine of chimpanzees, *Infect. Immun.* **21**:129–134.

Baber, D. J., and Condy, J. B., 1981, Isolation and characterisation of bovine adenoviruses types 3, 4 and 8 from free-living African buffaloes (*Syncerus caffer*), *Res. Vet. Sci.* **31**:69–75.

Bagshaw, C., Yates, V. J., Pronovost, A. D., and Chang, P. W., 1980, Enhancement and inhibition of CELO virus pathogenicity in quail by avian adenovirus-associated virus, *J. Wildl. Dis.* **16**:287–291.

Bartha, A., 1969, Proposal for subgrouping of bovine adenoviruscs, *Acta Vet. Acad. Sci. Hung.* **19**:319–321.

Bartha, A., and Áldásy, P., 1966, Further two serotypes of bovine adenoviruses (serotype 4 and 5), *Acta Vet. Acad. Sci. Hung.* **16**:107–108.

Bartha, A., Máthé, S., and Áldásy, P., 1970, New serotype 8 of bovine adenoviruses, *Acta Vet. Acad. Sci. Hung.* **20**:399–400.

Basnight, M., Rogers, N. G., Gibbs, C. J., and Gajdusek, D. C., 1971, Characterization of four new adenovirus serotypes isolated from chimpanzee tissue explants, *Am. J. Epidemiol.* **94**:166–171.

Bauer, K., Müller, H., and Gürtürk, S., 1975, Isolierung eines Virus von Schafen und seine Einordnung als neuer Serotyp oviner Adenoviren, *Zentralbl. Veterinaermed. Reihe B* **22**:656–665.

Baxendale, W., 1978, Egg drop syndrome 76, *Vet. Rec.* **102**:285–286.

Belák, S., 1980, Properties of ovine adenoviruses, *Acta Vet. Acad. Sci. Hung.* **28**:47–55.

Belák, S., and Pálfi, V., 1974, An adenovirus isolated from sheep and its relationship to type 2 bovine adenovirus, *Arch. Gesamte Virusforsch.* **46**:366–369.

Belák, S., Pálfi, V., and Tury, E., 1975, Experimental infection of lambs with an adenovirus isolated from sheep and related to bovine adenovirus type 2, *Acta Vet. Acad. Sci. Hung.* **25**:91–95.

Belák, S., Pálfi, V., and Tuboly, S., 1978, Replication of ovine adenoviruses, *Acta Microbiol. Acad. Sci. Hung.* **25**:253–264.

Belák, S., Vetési, F., Pálfi, V., and Papp, L., 1980, Isolation of a pathogenic strain of ovine adenovirus type 5 and a comparison of its pathogenicity with that of another strain of the same serotype, *J. Comp. Pathol.* **90**:169–176.

Bickford, A. A., Krasovich, M. A., and Fradly, A. M., 1973, Demonstration of virus particles in hepatic cells of chickens with inclusion body hepatitis, *Avian Dis.* **17**:629–638.

Blailock, Z. R., Rabin, E. R., and Melnick, J. L., 1967, Adenovirus endocarditis in mice, *Science* **157**:69–70.

Blalock, H. G., Simmons, D. G., Muse, K. E., Gray, J. G., and Derieux, W. T., 1975, Adenovirus respiratory infection in turkey poults, *Avian Dis.* **19**:707–716.

Bodon, L., Prohászka, L., Ádám, É., and Nász, I., 1979, Isolation of an adenovirus from rabbits, *Acta Vet. Acad. Sci. Hung.* **27**:73–75.

Boyce, J. T., Giddens, W. E., and Valerio, M., 1978, Simian adenoviral pneumonia, *Am. J. Pathol.* **91**:259–271.

Britt, j. O., Nagy, A. Z., and Howard, E. B., 1979, Acute viral hepatitis in California sea lions, *J. Am. Vet. Med. Assoc.* **175**:921–923.

Burke, C. N., Luginbuhl, R. E., and Helmboldt, C. F., 1965, The isolation of a latent adenolike virus from chicken kidney cell cultures, *Avian Dis.* **9**:31–43.

Burki, F., Wege, H., Reich-Rohrwig, C., and Hinaidy, B., 1978, Bovine adenovirus. I. Characterization and serological classification as types 4 of two isolates from latently infected calf testicles, *Zentralbl. Veterinaermed. Reihe B* **25**:555–565.

Burnett, J. P., Mayne, N., Butler, L. K., and Harrington, J. A., 1972, Chemical and physical relationships of oncogenic and non-oncogenic simian adenovirus DNA's, *J. Gen. Virol.* **17**:245–253.

Burnett, J. P., Mayne, N., and Helton, L., 1975, Retention of tumour-inducing capacity by adenovirus DNA after cleavage by restriction endonucleases, *Nature (London)* **254**:158–159.

Cakala, A., 1966, Szczep wirusa CELO wyosobniony z bażantów, *Med. Weter.* **22**:261–264.

Calnek, B.W., 1978, Hemagglutination-inhibition antibodies against an adenovirus (virus-127) in white Peking ducks in the United States, *Avian Dis.* **22**:798–801.

Calnek, B.W., and Cowen, B.S., 1975, Adenoviruses of chickens: Serologic groups, *Avian Dis.* **19**:91–103.

Carlson, H.C., Al-Sheikhly, F., Pettit, J.R., and Seawright, G.L., 1974, Virus particles in spleens and intestines of turkeys with hemorrhagic enteritis, *Avian Dis.* **18**:67–73.

Casto, B.C., 1968, Adenovirus transformation of hamster embryo cells, *J. Virol.* **2**:376–383.

Casto, B.C., Miyagi, M., Meyers, J., and DiPaolo, J.A., 1979, Increased integration of viral genome following chemical and viral treatment of hamster embryo cells, *Chem.–Biol. Interact.* **25**:255–269.

Ceccarelli. A., Ferina, R., and Mani, P., 1979, First seroepidemiological and virological observations on cattle adenoviruses in intensive breeding in Italy, *Arch. Vet. Ital.* **30**:28–32.

Chaddock, T.T., and Carlson, W.E., 1950, Fox encephalitis (infectious canine hepatitis) in the dog, *North Am. Vet.* **31**:35–41.

Chappuis, G., and Tektoff, J., 1975, Isolation et identification de plusieurs souches d'adénovirus chez le porc. *Rec. Med. Vet.* **151**:223–230.

Cheville, N., and Sato, S., 1977, Pathology of adenoviral infection in turkeys (*Meleagris gallopavo*) with respiratory disease and colisepticemia, *Vet. Pathol.* **14**:567–581.

Clark, H.F., Michalski, F., Tweedell, K.S., Yohn, D., and Zeigel, R.F., 1973, An adenovirus, FAV-1, isolated from the kidney of a frog (*Rana pipiens*), *Virology* **51**:392–400.

Clarke, M.C., Sharpe, H.B.A., and Derbyshire, J.B., 1967, Some characteristics of three porcine adenoviruses, *Arch. Gesamte Virusforsch.* **21**:91–97.

Cook, J.K.A., 1974, Spread of an avian adenovirus (CELO virus) to uninoculated fowls, *Res. Vet. Sci.* **16**:156–161.

Darai, G., Matz, B., Flügel, R.M., Grafe, A., Gelderblom, H., and Delius, H., 1980, An adenovirus from *Tupaia* (tree shrew): Growth of the virus, characterization of viral DNA, and transforming ability, *Virology* **104**:122–138.

Darbyshire, J.H., 1966, Oncogenicity of bovine adenovirus type 3 in hamsters, *Nature* **211**:102.

Darbyshire, J.H., 1980, Infection of chickens with hemorrhagic entertis virus of turkeys, *Vet. Rec.* **106**:270–271.

Darbyshire, J.H., Dawson, P.S., and Lamont, P.H., 1965, A new adenovirus serotype of bovine origin, *J. Comp. Pathol.* **75**:327–330.

Darbyshire, J.H., Jennings, A.R., Dawson, P.S., Lamont, P.H., and Omar, A.R., 1966, The pathogenesis and pathology of infection in calves with a strain of bovine adenovirus type 3, *Res. Vet. Sci.* **7**:81–93.

Darbyshire, J.H. Kinch, D.A., and Jennings, A.R., 1969, Experimental infection of calves with bovine adenovirus types 1 and 2, *Res. Vet. Sci.* **10**:39–45.

Davies, D.H., and Humphreys, S., 1977, Characterization of two strains of adenovirus isolated from New Zealand sheep, *Vet. Microbiol.* **2**:97–107.

Dawson, G.J., Yates, V.J., Chang, P.W., Orsi, L.N., and Pronovost, A.D., 1979, Egg transmission of avian adenovirus-associated virus and CELO virus during a naturally occurring infection, *Am. J. Vet. Res.* **40**:1624–1627.

Dawson, G.J., Yates, V.J., Chang, P.W., and Wattanavijarin, W., 1981, Egg transmission of avian adenovirus-associated virus and CELO virus during experimental infections, *Am. J. Vet. Res.* **42**:1833–1837.

Denhardt, D., 1966, A membrane-filter technique for the detection of complementary DNA, *Biochem. Biophys. Res. Commun.* **23**:641–646.

Denisova, T.S., Sitnikov, B.S., and Gibadulin, R.A., 1979, Study of the fragmentation of chicken adenovirus CELO DNA with specific endonucleases R. HpaI, B. EcoRI, and HindIII, *Mol. Biol.* **13**:1021–1034 (in Russian).

Denisova, T.S., Gibadulin, R.A., Sitnikov, B.S., Senkevich, T.G., and Medvedkina, O.A., 1980, Study of the genome of simian SV20 adenovirus. I. Construction of physical maps with the aid of restriction enzymes BamHI, EcoRI, XbaI, SalI, and HindIII, *Mol. Biol.* **14**:708–720 (in Russian).

Derbyshire, J.B., Clarke, M.C., and Collins, A.P., 1975, Serological and pathogenicity studies with some unclassified porcine adenoviruses, *J. Comp. Pathol.* **85**:437–442.

Descoteaux, J.P., Whissel, K., and Assaf, R., 1980, Detection of antibody titers to bovine viruses in rabbit sera, *Lab. Anim. Sci.* **30**:581–582.

Dimitrov, D.H., Dubitchev, A.G., Naroditsky, B.S., Dreizin, R.S., and Tikchonenko, T.I., 1979, Physicochemical properties and restriction maps of simian adenovirus type 38 DNA, *J. Gen. Virol.* **44**:69–80.

Ditchfield, J., Macpherson, L.W., and Zbitnew, Audrey, 1962, Association of a canine adenovirus (Toronto A 26/61) with an outbreak of laryngotracheitis ("kennel cough"), *Can. Vet. J.* **3**:238–247.

Domermuth, C.H., and Gross, W.B., 1971, Effect of disinfectants and drying on the virus of hemorrhagic enteritis of turkeys, *Avian Dis.* **15**:94–97.

Domermuth, C.H., Gross, W.G., DuBose, R.T., Douglass, C.S., and Reubush, C.B., 1972, Agar gel diffusion precipitin test for hemorrhagic enteritis of turkeys, *Avian Dis.* **16**:882–857.

Domermuth, C.H., Gross, W.B., DuBose, R.T., and Mallinson, E.T., 1975, Experimental reproduction and antibody inhibition of marble spleen disease of pheasants, *J. Wildl. Dis.* **11**:338–342.

Domermuth, C.H., Harris, J.R., Gross, W.B., and DuBose, R.T., 1979, A naturally occurring infection of chickens with a hemorrhagic enteritis/marble spleen disease type of virus, *Avian Dis.* **23**:479–484.

Domermuth, C.H., Weston, C.R., Cowen, B.S., Colwell, W.M., Gross, W.B., and DuBose, R.T., 1980, Incidence and distribution of "avian adenovirus group II splenomegaly of chickens," *Avian Dis.* **24**:591–594.

DuBose, R.T., and Grumbles, L.C., 1959, The relationship between quail bronchitis virus and chicken embryo lethal orphan virus, *Avian Dis.* **3**:321–344.

DuBose, R.T., Grumbles, L.C., and Flowers, A.I., 1958, The isolation of a nonbacterial agent from quail with a respiratory disease, *Poult. Sci.* **37**:654–658.

Dulac, G.C., Swango, L.J., and Burnstein, T., 1970, Tumors induced in hamsters by a respiratory-associated canine adenovirus, *Can. J. Microbiol.* **16**:391–394.

Easton, G.D., and Simmons, D.G., 1977, Antigenic analysis of several turkey respiratory adenoviruses by reciprocal-neutralization kinetics, *Avian Dis.* **21**:605–611.

Edington, N., and Kasza, L., and Christofinis, G.J., 1972, Meningo-encephalitis in gnotobiotic pigs inoculated intranasally and orally with porcine adenovirus 4, *Res. Vet. Sci.* **13**:289–291.

El Mishad, A.M., McCormick, K.J., Stenback, W.A., Yates, V.J., and Trentin, J.J., 1975, Hemagglutinating properties of CELO, an oncogenic avian adenovirus, *Avian Dis.* **19**:761–772.

Enders, J.F., Bell, J.A., Dingle, J.H., Francis, T., Hilleman, M.R., Huebner, R.J., and Payne, A.M.M., 1956, "Adenoviruses," group name proposed for new respiratory-tract viruses, *Science* **124**:119–120.

England, J.J., McChesney, A.E., and Chow, T.L., 1973, Characterization of an equine adenovirus, *Am. J. Vet. Res.* **34**:1587–1590.

Espana, C., 1974, Viral epizootics in captive nonhuman primates, *Lab. Anim. Sci.* **24**:167–176.

Eugster, A.K., Kalter, S.S., Kim, C.S., and Pinkerton, M.E., 1969, Isolation of adenoviruses from baboons (*Papio* sp.) with respiratory and enteric infections, *Arch. Gesamte Virusforsch.* **26**:260–270.

Fadly, A.M., and Winterfield, R.W., 1973, Isolation and some characteristics of an agent associated with inclusion body hepatitis, haemorrhages, and aplastic anaemia in chickens, *Avian Dis.* **17**:182–193.

Faissner, A., Darai, G., and Flugel, R.M., 1980, Identification of polypeptides of the tree shrew (*Tupaia*) adenovirus, *Intervirology* **14**:272–276.

Fatemie-Nainie, S., and Marusyk, R., 1979, Biophysical and serologic comparison of four equine adenovirus isolates, *Am. J. Vet. Res.* **40**:521–528.

Flewett, T.H., Bryden, A.S., Davies, H., and Morris, C.A., 1975, Epidemic viral enteritis in a long-stay children's ward, *Lancet* **1**:4–5.

Gallina, A.M., Winterfield, R.W., and Fadly, A.M., 1973, Adenovirus infection and disease. II. Histopathology of natural and experimental disease, *Avian Dis.* **17**:343–353.

Gehle, W.D., and Smith, K.O., 1969, Abortive infection of human cell cultures by a canine adenovirus, *Proc. Soc. Exp. Biol. Med.* **131**:87–93.

Gelderblom, H., and Maichle-Lauppe, I., 1982, The fibers of fowl adenoviruses, *Arch. Virol.* **72**:280–298.

Gibbs, E.P.J., Taylor, W.P., and Lawman, M.J.P., 1977, The isolation of adenoviruses from goats affected with peste des petits ruminants in Nigeria, *Res. Vet. Sci.* **23**:331–335.

Gilden, R.V., Kern, J., Heberling, R.L., and Huebner, R.J., 1968, Serological studies of the T and tumor antigens of the oncogenic simian adenoviruses, *Appl. Microbiol.* **16**:1015–1018.

Goodheart, C.R., 1971, DNA density of oncogenic and non-oncogenic simian adenoviruses, *Virology* **44**:645–648.

Green, M., Mackey, J.K., Wold, W.S.M., and Rigden, P., 1979, Thirty-one human adenovirus serotypes (Ad–Ad31) from five groups (A–E) based upon DNA genome homologies, *Virology* **93**:481–492.

Green, R.G., Ziegler, N.R., Green, B.B., and Dewey, E.T., 1930, Epizootic fox encephalitis. I. General description, *Am. J. Hyg.* **12**:109–129.

Green, R.G., Ziegler, N.R., Dewey, E.T., and Shillinger, J.E., 1931, Epizootic fox encephalitis. III. Experimental transmission, *Am. J. Hyg.* **14**:353–373.

Green, R.G., Katter, M.S., Shillinger, J.E., and Hanson, K.B., 1933, Epizootic fox encephalitis. IV. The intranuclear inclusions, *Am. J. Hyg.* **18**:462–481.

Green, R.G., Ziegler, N.R., Carlson, W.E., Shillinger, J.E., Tyler, S.H., and Dewey, E.T., 1934, Epizootic fox encephalitis. V. General and pathogenic properties of the virus, *Am. J. Hyg.* **19**:343–361.

Green, R.G., Green, B.B., Carlson, W.E., and Shillinger, J.E., 1936, Epizootic fox encephalitis. VIII. The occurrence of the virus in the upper respiratory tract in natural and experimental infections, *Am. J. Hyg.* **24**:57–70.

Grimes, T.M., King, D.J., Kleven, S.H., and Fletcher, O.J., 1977, Involvement of a type-8 avian adenovirus in the etiology of inclusion body hepatitis, *Avian Dis.* **21**:26–38.

Grimes, T.M., Fletcher, O.J., and Munnell, J.F., 1978, Comparative study of experimental inclusion body hepatitis of chickens caused by two serotypes of avian adenovirus, *Vet. Pathol.* **15**:249–263.

Gross, W.B., and Moore, W.E.C., 1967, Hemorrhagic enteritis of turkeys, *Avian Dis.* **11**:296–307.

Guenov, I., Sartmadshiev, K., Schopov, I., Shljabinkov, Z., and Fjodorov, W., 1970, Neuer Serotyp 9 der Adenoviren des Rindes, *Zentralbl. Veterinaermed. Reihe B* **17**:1064–1066.

Hafez, S.M., and Liess, B., 1980, Immunodiffusion and neutralization studies on porcine adenoviruses, *Vet. Microbiol.* **5**:101–111.

Haig, D.A., Clarke, M.C., and Pereira, M.S., 1964, Isolation of an adenovirus from a pig, *J. Comp. Pathol.* **74**:81–84.

Hamilton, J.M., Cornwell, H.J.C., McCusker, H.B., and Campbell, R.S.F., 1966, Studies on the pathogenesis of canine virus hepatitis, *Br. Vet. J.* **122**:225–238.

Hartley, J.W., and Rowe, W.P., 1960, A new mouse virus apparently related to the adenovirus group, *Virology* **11**:645–647.

Hashimoto, K., Sugiyama, T., and Sasaki, S., 1966, An adenovirus isolated from the feces of mice. I. Isolation and identification, *Jpn. J. Microbiol.* **10**:115–125.

Hashimoto, K., Sugiyama, T., Yoshikawa, M., and Sasaki, S., 1970, Intestinal resistance in the experimental enteric infection of mice with a mouse adenovirus. I. Growth of the virus and appearance of a neutralizing substance in the intestinal tract, *Jpn. J. Microbiol.* **14**:381–395.

Hashimoto, K., Yoshikawa, M., Sugihara, Y., and Sasaki, S., 1971, Intestinal resistance in the experimental enteric infection of mice with a mouse adenovirus. II. Determination of the neutralizing substance in the intestinal tract as an IgA antibody, *Jpn. J. Microbiol.* **15**:499–508.

Heath, R.B., El Falaky, I., Stark, J.E., Herbst-Laiter, R.H., and Larin, N.M., 1966, Induced upper respiratory infection in the vervet monkey, *Br. J. Exp. Pathol.* **47**:93–98.

Heberling, R.L., 1972, The simian adenoviruses, in: *Pathology of Simian Primates*, Part II (R. N. Fiennes, ed.), pp. 572–591, S. Karger, Basel.

Heck, F.C., Sheldon, W.G., and Gleiser, C.A., 1972, Pathogenesis of experimentally produced mouse adenovirus infection in mice, *Am. J. Vet. Res.* **33**:841–846.

Helmboldt, C.F., and Frazier, M.N., 1963, Avian hepatic inclusion bodies of unknown significance, *Avian Dis.* **7**:446–450.

Hilleman, M.R., and Werner, J.H., 1954, Recovery of new agent from patients with acute respiratory illness, *Proc. Soc. Exp. Biol. Med.* **85**:183–188.

Hillis, W.D., and Goodman, R., 1969, Serologic classification of chimpanzee adenoviruses by hemagglutination and hemagglutination inhibition, *J. Immunol.* **103:**1089–1095.

Hillis, W.D., Garner, A.C., and Hillis, A.I., 1969, A new simian adenovirus serologically related to human adenovirus type 2 from a chimpanzee with "viral hepatitis," *Am. J. Epidemiol.* **90:**344–353.

Horne, R.W., 1962, The comparative structure of adenoviruses, *Ann. N. Y. Acad. Sci.* **101:**475–484.

Howell, J., MacDonald, D.W., and Christian, R.G., 1970, Inclusion body hepatitis in chickens, *Can. Vet. J.* **11:**99–101.

Huebner, R.J., Rowe, W.P., and Lane, W.T., 1962, Oncogenic effects in hamsters of human adenovirus types 12 and 18, *Proc. Natl. Acad. Sci. U.S.A.* **48:**2051–2058.

Hull, R.N., 1968, The simian viruses, in: *Virology Monographs 2* (S. Gard and C. Hallauer, eds.), pp. 1–66, Springer-Verlag, Vienna and New York.

Hull, R.N., Minner, J.R., and Smith, J.W., 1956, New viral agents recovered from tissue cultures of monkey kidney cells, *Am. J. Hyg.* **63:**204–215.

Hull, R.N., Minner, J.R., and Mascoli, C.C., 1958, New viral agents recovered from tissue cultures of monkey kidney cells, *Am. J. Hyg.* **68:**31–44.

Hull, R.N., Johnson, I.S., Culbertson, C.G., Reimer, C.B., and Wright, H.F., 1965, Oncogenicity of the simian adenoviruses, *Science* **150:**1044–1046.

Ide, P.R., Thomson, R.G., and Ditchfield, W.J.B., 1969, Experimental adenovirus infection in calves, *Can. J. Comp. Med.* **33:**1–9.

Igarashi, K., Niiyama, Y., Tsukamoto, K., Kurokawa, T., and Sugino, Y., 1975, Biochemical studies on bovine adenovirus type 3. II. Incomplete virus, *J. Virol.* **16:**634–641.

Igarashi, K., Sasada, R., Niiyama, K., Tsumamoto, K., and Sugino, Y., 1978, Biochemical studies on bovine adenovirus type 3. IV. Transformation by viral DNA and DNA fragments, *J. Virol.* **28:**219–226.

Iltis, J.P., and Daniels, S.B., 1977, Adenovirus of ring-necked pheasants: Purification and partial characterization of marble spleen disease virus, *Infect. Immun.* **16:**701–705.

Iltis, J.P., Jacowski, R.M., and Wyand, D.S., 1975, Transmission of marble spleen disease in turkeys and pheasants, *Am. J. Vet Res.* **36:**97–101.

Iltis, J.P., Daniels, S.B., and Wyand, D.S., 1977, Demonstration of an avian adenovirus as the causative agent of marble spleen disease, *Am. J. Vet. Res.* **38:**95–100.

Inaba, Y., Tanaka, Y., Sato, K., Ito, H., Ito, Y., Omori, T., and Matumoto, M., 1968, Bovine adenovirus. II. A serotype, Fukuroi, recovered from Japanese cattle, *Jpn. J. Microbiol.* **12:**219–229.

Inada, T., and Uetake, H., 1978, Cell-mediated immunity assayed by ^{51}Cr release test in mice infected with mouse adenovirus, *Infect. Immun.* **20:**1–5.

Ishibashi, M., 1970, Retention of viral antigen in the cytoplasm of cells infected with temperature-sensitive mutants of an avian adenovirus, *Proc. Natl. Acad. Sci. U.S.A.* **65:**304–309.

Ishibashi, M., 1971, Temperature-sensitive conditional-lethal mutants of an avian adenovirus (CELO), *Virology* **45:**42–52.

Ishibashi, M., and Maizel, J.V., 1974, The polypeptides of adenovirus. VI. Early and late glycopolypeptides, *Virology* **58:**345–361.

Ishibashi, M., Yasue, H., Fujinaga, K., and Kawamata, J., 1980, The oncogenicity of avian adenoviruses. I. An unusually large number of viral DNA molecules in some tumors, and virus-specific T-antigenic proteins, *Virology* **106:**349–360.

Itakura, C., and Carlson, H.C., 1975, Electron microscopic findings of cells with inclusion bodies in experimental hemorrhagic enteritis of turkeys, *Can. J. Comp. Med.* **39:**299–304.

Itakura, C., Carlson, H.C., and Lang, G.N., 1974a, Experimental transmission of hemorrhagic enteritis of turkeys, *Avian Pathol.* **3:**279–292.

Itakura, C., Yasuba, M., and Goto, M., 1974b, Histopathological studies on inclusion body hepatitis in broiler chickens, *Jpn. J. Vet. Sci.* **36:**329–340.

Jakowski, R.M., and Wyand, D.S., 1972, Marble spleen disease in ring-necked pheasants: Demonstration of agar gel precipitin antibody in pheasants from an infected flock, *J. Wildl. Dis.* **8**:261–263.

Jensen, N.J., and Block, B., 1980, Adenovirus-like particles associated with epidermal hyperplasia in cod (*Gadus morhua*), *Nord. Veterinaermed.* **32**:173–175.

Kaleta, E.F., 1969, CELO-virus bei Gänseküken, *Dtsch. Tierarztl. Wochenschr.* **76**:421–444.

Kaleta, E.F., Khalaf, S.E.D., and Siegmann, O., 1980, Antibodies to egg-drop syndrome 76 virus in wild birds in possible conjunction with egg-shell problems, *Avian Pathol.* **9**:587–590.

Kalter, S.S., and Heberling, R.L., 1971, Comparative virology of primates, *Bacteriol. Rev.* **35**:310–364.

Kalter, S.S., Ablashi, D., Espana, C., Heberling, R.L., Hull, R.N., Lennette, E.H., Malherbe, H.H., McConnel, S., and Yohn, D., 1980, Simian virus nomenclature 1980, *Intervirology* **13**:317–330.

Kasza, L., 1966, Isolation of an adenovirus from the brain of a pig, *Am. J. Vet. Res.* **27**:751–758.

Kawamura, H., and Horiuchi, T., 1964, Pathological changes in chickens inoculated with CELO virus, *Natl. Inst. Anim. Health Q.* **4**:31–39.

Kawamura, H., Shimizu, F., and Tsubahara, H., 1964, Avian adenovirus: Its properties and serological classification, *Natl. Inst. Anim. Health Q.* **4**:183–193.

Khanna, P.N., 1966, Occurrence of avian adenoviruses in Hungary, *Acta Vet. Acad. Sci. Hung.* **16**:351–356.

Klein, M., Earley, E., and Zellat, J., 1959, Isolation from cattle of a virus related to human adenovirus, *Proc. Soc. Exp. Biol. Med.* **102**:1–4.

Klein, M., Zellat, J., and Michaelson, T.C., 1960, A new bovine adenovirus related to human adenovirus, *Proc. Soc. Exp. Biol. Med.* **105**:340–342.

Kraft, V., Grund, S., and Monreal, G., 1979, Ultrastructural characterisation of isolate 127 of egg drop syndrome 1976 virus as an adenovirus, *Avian Pathol.* **8**:353–361.

Kurokawa, T., Igarashi, K., and Sugino, Y., 1978, Biochemical studies on bovine adenovirus type 3. III. Cleavage maps of viral DNA by restriction endonucleases EcoRI, BamHI, and HindIII, *J. Virol.* **28**:212–218.

Landon, J.C., and Bennett, D.G., 1969, Viral induced simian conjunctivitis, *Nature (London)* **222**:683–684.

Larsen, S.H., and Nathans, D., 1977, Mouse adenovirus: Growth of plaque-purified FL virus in cell lines and characterization of viral DNA, *Virology* **82**:182–195.

Larsen, S.H., Margolskee, R.F., and Nathans, D., 1979, Alignment of the restriction map of mouse adenovirus FL with that of human adenovirus 2, *Virology* **97**:406–414.

Laver, W.G., Younghusband, H.B., and Wrigley, N.G., 1971, Purification and properties of chick embryo lethal orphan virus (an avian adenovirus), *Virology* **45**:598–614.

Lee, K.P., Henry, N.W., and Rosenberger, J.K., 1978, Comparative pathogenicity of six avian adenovirus isolants in the liver, *Avian Dis.* **22**:610–619.

Lehmkuhl, H.D., Smith, M.H., and Dierks, R.E., 1975, A bovine adenovirus type 3: Isolation, characterization, and experimental infection in calves, *Arch. Virol.* **48**:39–46.

Macpherson, I., McDougall, J.S., and Laursen-Jones, A.P., 1974, Inclusion body hepatitis in a broiler integration, *Vet. Rec.* **95**:286–289.

Mandelli, G., Rinaldi, A., and Cervio, G., 1966, A disease involving the spleen and lungs in pheasants: Epidemiology, symptoms, and lesions, *Clin. Vet. (Milan)* **89**:129–138 (in Italian).

Margolis, G., Kilham, L., and Hoenig, E.M., 1974, Experimental adenovirus infection of the mouse adrenal gland, *Am. J. Pathol.* **75**:363–371.

Marusyk, R.G., and Hammarskjöld, M.-L., 1972, The genetic relationship of two canine adenoviruses as determined by nucleic acid hybridization, *Microbiology* **5**:259–264.

Marusyk, R.G., Norrby, E., and Lundqvist, U., 1970, Biophysical comparison of two canine adenoviruses, *J. Virol.* **5**:507–512.

Mathur, B.B.L., Verma, K.C., Agarwal, K.K., and Kumar, S., 1972, Serological survey for the detection of certain common respiratory infections in migratory birds: A note, *Indian J. Anim. Sci.* **42:**144–145.

Matthews, R.E.F., 1979, Classification and nomenclature of viruses, *Intervirology* **12:**177–179.

Matumoto, M., Inaba, Y., Tanaka, Y., Sato, K., Ito, H., and Omori, T., 1969, Serological typing of bovine adenovirus, Nagano and Fukuroi, as type 4 and new type 6, *Jpn. J. Microbiol.* **13:**131–132.

May, J.T., Robinson, A.J., and Eaton, B.T., 1975, A study of the sequences of chicken embryo lethal orphan (CELO) virus DNA present in a transformed hamster cell line with use of specific fragments of the virus genome, *Virology* **68:**483–489.

May, J.T., Welsh, J.K., Asch, B.B., and McCormick, K.J., 1978, Chicken embryo lethal orphan (CELO) virus DNA present in hamster cell lines derived from CELO-induced hepatomas, *Intervirology* **9:**321–324.

McAllister, R.M., Riggs, J.L., Reed, G., and Macpherson, I., 1969, Transformation of rodent cells by simian adenovirus SA-7, *Proc. Soc. Exp. Biol. Med.* **131:**1442–1445.

McChesney, A.E., England, J.J., and Rich, L.J., 1973, Adenovirus infection in foals, *J. Am. Vet. Med. Assoc.* **162:**545–549.

McChesney, A.E., England, J.J., Whiteman, C.E., Adcock, J.L., Rich, L.J., and Chow, T.L., 1974, Experimental transmission of equine adenovirus in Arabian and non-Arabian foals, *Am. J. Vet. Res.* **35:**1015–1023.

McClure, H.M., Chandler, F.W., and Hierholzer, J.C., 1978, Necrotizing pancreatitis due to simian adenovirus type 31 in a rhesus monkey, *Arch. Pathol. Lab. Med.* **102:**150–153.

McCormick, K.J., Anderson, J.P., Stenback, W.A., and Trentin, J.J., 1971, Occurrence of antibodies to the T antigen of chicken embryo lethal orphan virus, *Cancer Res.* **31:**981–984.

McCracken, R.M., and McFerran, J.B., 1978, Experimental reproduction of the egg drop syndrome 1976 with a haemagglutinating adenovirus, *Avian Pathol.* **7:**483–490.

McCracken, R.M., McFerran, J.B., Evans, R.T., and Connor, T.J., 1976, Experimental studies on the aetiology of inclusion body hepatitis, *Avian Pathol.* **5:**325–339.

McFerran, J.B., and Adair, B.M.C., 1977, Avian adenoviruses—a review, *Avian Pathol.* **6:**189–217.

McFerran, J.B., and Connor, T.J., 1978, Further studies on the classification of fowl adenoviruses, *Avian Dis.* **21:**585–595.

McFerran, J.B., Neison, R., McCracken, J.M., and Ross, J.G., 1969, Virus isolated from sheep, *Nature (London)* **221:**194–195.

McFerran, J.B., Clarke, J.K., and Connor, T.J., 1972, Serological classification of avian adenoviruses, *Arch. Gesamte Virusforsch.* **39:**132–139.

McFerran, J.B., Adair, B., and Connor, T.J., 1975, Adenoviral antigens (CELO, QBV, GAL), *Am. J. Vet. Res.* **36:**527–529.

McFerran, J.B., Connor, T.J., and McCracken, R.M., 1976a, Isolation of adenoviruses and reoviruses from avian species other than domestic fowl, *Avian Dis.* **20:**519–524.

McFerran, J.B., McCracken, R.M., Connor, T.J., and Evans, R.T., 1976b, Isolation of viruses from clinical outbreaks of inclusion body hepatitis, *Avian Pathol.* **5:**315–324.

McFerran, J.B., Rowley, H.M., McNulty, M.S., and Montgomery, L.J., 1977, Serological studies on flocks showing depressed egg production, *Avian Pathol.* **6:**405–413.

McFerran, J.B., Connor, T.J., and Adair, B.M., 1978a, Studies on the antigenic relationship between an isolate (127) from the egg drop syndrome 1976 and a fowl adenovirus, *Avian Pathol.* **7:**629–636.

McFerran, J.B., McCracken, R.M., McKillop, E.R., McNulty, M.S., and Collins, D.S., 1978b, Studies on a depressed egg production syndrome in Northern Ireland, *Avian Pathol.* **7:**35–47.

McGuire, T.C., and Poppie, M.J., 1973, Hypogammaglobulinemia and thymic hypoplasia in horses: A primary combined immunodeficiency disorder, *Infect. Immun.* **8:**272–277.

Meulemans, G., Dekegel, D., Peeters, J., Van Meirhaeghe, E., and Halen, P., 1979, Isolation of an adeno-like virus from laying chickens affected by egg drop syndrome 1976, *Viaams Diergeneeskd. Tijdschr.* **48:**151–157.

Missal, O., 1969, Isolierung von Adenovirus aus einer Mäusekolonie und dessen Charakterisierung, *Arch. Exp. Vet. Med.* **23:**783–791.

Mohanty, S.B., 1971, Comparative study of bovine adenoviruses, *Am. J. Vet. Res.* **32:**1899–1905.

Mohanty, S.B., and Lillie, M.G., 1965, Experimental infection of calves with a bovine adenovirus, *Proc. Soc. Exp. Biol. Med.* **120:**679–682.

Mohanty, S.B., and Lillie, M.G., 1970, Type 2 bovine adenovirus as an adventitious contaminant in primary bovine embryonic kidney cell cultures, *Appl. Microbiol.* **19:**381–382.

Moorthy, A.R.S., and Spradbrow, P.B., 1978, Adenoviral infection of Arab foals with respiratory tract disease, *Zentralbl. Veterinaermed. Reihe B* **25:**469–477.

Morales-Ayala, F., Nuñez-Montiel, O., Merino, F., Morcano, E., and Vitelli-Flores, J., 1964, A spontaneous isolation and characteristics of a new oppossum adenovirus, *Bacteriol. Proc.* **64:**117.

Motoi, M., Fukui, H., and Ogawa, K., 1972, Neoplastic transformation of hamster cells *in vitro* by bovine adenovirus type-3, *Gann* **63:**415–418.

Naroditsky, B.S., Zavizion, B.A., Karamov, E.V., and Tikhonenko, T.I., 1978, Analysis of the genome of type 7 simian adenovirus using restrictases, *Nucleic Acids Res.* **5:**999–1011.

Naumann, S., Kunstýř, I., Langer, I., Maess, J., and Hörning, R., 1981, Lethal pneumonia in guineapigs associated with a virus, *Lab. Anim.* **15:**235–242.

Niiyama, Y., 1981, Biochemical studies on bovine adenovirus type 3. VI. Identification of virus-specific early proteins and tumor antigens, *Cell Struct. Funct.* **6:**133–146.

Niiyama, Y., Igarashi, K., Tsukamoto, K., Kurokawa, T., and Sugino, Y., 1975, Biochemical studies on bovine adenovirus type 3. I. Purification and properties, *J. Virol.* **16:**621–633.

Niiyama, Y., Sasada, R., Igarashi, K., Kurokawa, T., and Sugino, Y., 1981, Biochemical studies on bovine adenovirus type 3. V. Some properties of mouse cells transformed with viral DNA fragments, *Cell Struct. Funct.* **6:**121–131.

Norrby, E., Marusyk, R.G., and Wadell, G., 1971, Immunological relationships between capsid components of adenoviruses of different host cell species origin, *Can. J. Microbiol.* **17:**1227–1237.

Olson, N.O., 1951, A respiratory disease (bronchitis) of quail caused by a virus, *Vet. Med.* **46:**22.

Otsuki, K., Tsubokura, M., Yamamoto, H., Imamura, M., Sakagami, Y., Saito, H., and Hosokawa, D., 1976, Some properties of avian adenoviruses isolated from chickens with inclusion body hepatitis in Japan, *Avian Dis.* **20:**693–705.

Oxford, J.S., and Potter, C.W., 1970, Immunofluorescence studies of T and virus capsid antigens induced by chick embryo lethal orphan virus, *J. Gen. Virol.* **8:**33–41.

Palya, V., Belák, S., and Pálfi, V., 1977, Adenovirus infection in lambs. II. Experimental infection of lambs, *Zentralbl. Veterinaermed. Reihe B* **24:**529–541.

Parkhomenko, I.I., Eliseeva, L.B., and Bogdanov, G.H., 1981, Transformation and kinetic regularities of the cell growth *in vitro* after adenovirus SV-20 DNA application, *Vopr. Onkol.* **27:**68–71.

Pascucci, S., Quaglio, G.L., Maestrini, N., and Prati, A., 1970, Pancreatitis of guinea fowl by CELO virus, *Atti Soc. Ital. Sci. Vet.* **24:**667–668 (in Italian).

Pavlov, N., 1979, Pathohistologisches Bild der Adenovirusinfektion bei Kalbern, *Arch. Exp. Veterinaermed.* **33:**253–261.

Pettit, J.R., and Carlson, H.C., 1972, Inclusion body hepatitis in broiler chickens, *Avian Dis.* **16:**858–863.

Piña, M., and Green, M., 1965, Biochemical studies on adenovirus multiplication. IX. Chemical and base composition analysis of 28 human adenoviruses, *Proc. Natl. Acad. Sci. U.S.A.* **54:**547–551.

Piña, M., and Green, M., 1968, Base composition the DNA of oncogenic simian adenovirus SA7 and homology with human adenovirus DNA's, *Virology* **36**:321–323.
Pomeroy, B.S., and Fenstermacher, R., 1937, *Hemorrhagic enteritis in turkeys, Poult. Sci.* **16**:378–382.
Ponomareva, T.I., Grodnitskaya, N.A., Goldberg, E.E., Chaplygina, N.M., Naroditsky, B.S., and Tikchonenko, T.I., 1979, Biological activity of the intact and cleaved DNA of the simian adenovirus 7, *Nucleic Acid Res.* **6**:3119–3131.
Potter, C.W., and Oxford, J.S., 1969, Specific tumour antigen induced by chick embryo lethal orphan (CELO) virus, *J. Gen. Virol.* **4**:287–289.
Rapoza, N.P., 1967, A classification of simian adenoviruses based on hemagglutination, *Am. J. Epidemiol.* **86**:736–745.
Rěeves, W.C., Scrivani, R.P., Pugh, W.E., and Rowe, W.P., 1967, Recovery of an adenovirus from a feral rodent *Peromyscus maniculatus, Proc. Soc. Exp. Biol. Med.* **124**:1173–1175.
Riggs, J.L., and Lennette, E.H., 1967, *In vitro* transformation of newborn-hamster kidney cells by simian adenoviruses, *Proc. Soc. Exp. Biol. Med.* **126**:802–806.
Robinson, A.J., and Bellett, A.J.D., 1975, Complementary strands of CELO virus DNA, *J. Virol.* **15**:458–465.
Robinson, A.J., Younghusband, H.B., and Bellett, A.J.D., 1973, A circular DNA–protein complex from adenoviruses, *Virology* **56**:54–69.
Rondhuis, P.R., 1968, A new bovine adenovirus, *Arch. Gesamte Virusforsch.* **25**:235–236.
Rondhuis, P.R., 1973, Induction of tumors in hamsters with a bovine adenovirus strain (serotype 8), *Arch. Gesamte Virusforsch.* **41**:147–149.
Rosenberger, J.K., Eckroade, R.J., Klopp, S., and Krauss, W.C., 1974, Characterization of several viruses isolated from chickens with inclusion body hepatitis and aplastic anemia, *Avian Dis.* **18**:399–409.
Rothenbacher, H., Mallinson, E.T., Gentry, R.F., and Schwartz, L.D., 1970, Marble spleen and pulmonary edema syndrome of pheasants, *Game Breeders Bull.* **3**:99–101.
Rowe, W.P., Huebner, R.J., Gilmore, L.K., Parrott, R.H., and Ward, T.G., 1953, Isolation of a cytopathogenic agent from human adenoids underoing spontaneous degeneration in tissue culture, *Proc. Soc. Exp. Biol. Med.* **84**:570–573.
Rowe, W.P., Hartley, J.W., and Huebner, R.J., 1958, Serotype composition of the adenovirus group, *Proc. Soc. Exp. Biol. Med.* **97**:465–470.
Rubarth, S., 1947, An acute virus disease with liver lesion in dogs (hepatitis contagiosa canis), *Acta Pathol. Microbiol. Scand. Suppl.* **69**:1–222.
Salenstedt, C.R., 1963, Studies on the virus of hepatitis contagiosa canis (HCC). IV. The pathogenesis of hepatitis contagiosa canis in dogs, *Acta Vet. Scand.* **4**:371–383.
Sarma, P.S., Huebner, R.J., and Lane, W.T., 1965, Induction of tumors in hamsters with an avian adenovirus (CELO), *Science* **149**:1108.
Sarma, P.S., Vass, W., Huebner, R.J., Igel, H., Lane, W.T., and Turner, H.C., 1967, Induction of tumours in hamsters with infectious canine hepatitis virus, *Nature (London)* **215**:293–294.
Saxegaard, F., and Bratberg, B., 1971, Isolation of bovine adenovirus type 1 from a calf with pneumo-enteritis, *Acta Vet. Scand.* **12**:464–466.
Scott, M., and McFerran, J.B., 1972, Isolation of adenoviruses from turkeys, *Avian Dis.* **16**:413–420.
Shadduck, J.A., Koestner, A., and Kasza, L., 1967, The lesions of porcine adenoviral infection in germfree and pathogen-free pigs, *Pathol. Vet.* **4**:537–552.
Shadduck, J.A., Kasza, L., and Koestner, A., 1968, Distribution of a porcine adenovirus after inoculation of experimental animals, *Zentralbl. Bakteriol.* **207**:152–157.
Shahrabadi, M.S., Marusyk, R.G., and Crawford, T.B., 1977, Electron-microscopic study of the development of an equine adenovirus in cultured fetal equine kidney cells, *Can. J. Microbiol.* **23**:497–509.
Sharp, J.M., McFerran, J.B., and Rae, A., 1974, A new adenovirus from sheep, *Res. Vet. Sci.* **17**:268–269.

Sharp, J.M., Rushton, B., and Rimer, R.D., 1976, Experimental infection of specific pathogen-free lambs with ovine adenovirus type 4, *J. Comp. Pathol.* **86:**621–628.

Sharpe, H.B.A., and Jessett, D.M., 1967, Experimental infection of pigs with 2 strains of porcine adenovirus, *J. Comp. Pathol.* **77:**45–50.

Shimada, K., Yoshida, T., Kuroishi, T., and Ishibashi, M., 1983, Restriction map of CELO virus DNA, *Biochim. Biophys. Acta* **740:**169–178.

Shimizu, Y., Kunishige, T., and Hirato, K., 1960, Immunological studies on the infectious canine hepatitis virus. II. Hemagglutinin, *Jpn. J. Vet. Res.* **8:**271–278.

Shroyer, E.L., Kelley, S.T., Taylor, P.C., Vanderloo, P., and Lester, T.L., 1979, Three serologic types of adenovirus infections of owl monkeys, *Am. J. Vet. Res.* **40:**532–536.

Silim, A., Thorsen, J., and Carson, H.C., 1978, Experimental infection of chickens with hemorrhagic enteritis virus, *Avian Dis.* **22:**106–114.

Slifkin, M., Merkow, L., and Rapoza, N., 1968, Tumor induction by simian adenovirus 30 and establishment of tumor cell lines, *Cancer Res.* **28:**1173–1179.

Sohier, R., Chardonnet, Y., and Prunieras, M., 1965, Adenoviruses: Status of current knowledge, *Prog. Med. Virol.* **7:**253–325.

Stauber, E.H., Nellis, C.H., Magonigle, R.A., and Vaughn, H.W., 1977, Prevalence of reactors to selected livestock pathogens in Idaho mule deer, *Wildl. Management* **41:**515–519.

Stauber, E.H., Autenrieth, R., Markham, O.D., and Whitbeck, V., 1980, A seroepidemiologic survey of three pronghorn (*Antilocapra Americana*) populations in southeastern Idaho, 1975–1977, *J. Wildl. Dis.* **16:**109–115.

Studdert, M.J., Wilks, C.R., and Coggins, L., 1974, Antigenic comparisons and serologic survey of equine adenoviruses, *Am. J. Vet. Res.* **35:**693–699.

Sugiyama, T., Hashimoto, K., and Sasaki, S., 1967, An adenovirus isolated from the feces of mice. II. Experimental infection, *Jpn. J. Microbiol.* **11:**33–42.

Swango, L.J., Wooding, W.L., and Binn, L.N., 1970, A comparison of the pathogenesis and antigenicity of infectious canine hepatitis virus and the A26/61 virus strain (Toronto), *J. Am. Vet. Med. Assoc.* **156:**1687–1696.

Taniguichi, T., Yamaguchi, S., Maeda, M., Kawamura, H., and Horiuchi, T., 1981, Pathological change in laying hens inoculated with the JPA-1 strain of egg drop syndrome-1976 virus, *Jpn. Natl. Inst. Anim. Health Q.* **21:**83–93.

Temple, M., Antoine, G., Delius, H., Stahl, S., and Winnacker, E.-L., 1981, Replication of mouse adenovirus strain FL DNA, *Virology* **109:**1–12.

Thompson, D.B., Studdert, M.J., Beilharz, R.G., and Littlejohns, I.R., 1975, Inheritance of a lethal immunodeficiency disease of Arabian foals, *Aust. Vet. J.* **51:**109–113.

Todd, J.D., 1969, Comments on rhinoviruses and parainfluenza viruses of horses, *J. Am. Vet. Med. Assoc.* **155:**387–390.

Todd, D., and McNulty, M.S., 1978, Biochemical studies on a virus associated with egg drop syndrome 1976, *J. Gen. Virol.* **40:**63–75.

Tolin, S-A., and Domermuth, C.H., 1975, Hemorrhagic enteritis of turkeys: Electron microscopy of the causal virus, *Avian Dis.* **19:**118–125.

Tolun, A., Aleström, P., and Pettersson, U., 1979, Sequence of inverted terminal repetitions from different adenoviruses: Demonstration of conserved sequences and homology between SA7 termini and SV40 DNA, *Cell* **17:**705–713.

Van der Veen, J., and Mes, A., 1973, Experimental infection with mouse adenovirus in adult mice, *Arch. Gesamte Virusforsch.* **42:**235–241.

Van der Veen, J., and Mes, A., 1974, Serological classification of two mouse adenoviruses, *Arch. Gesamte Virusforsch.* **45:**386–387.

Van Eck, J.H.H., Davelaar, F.G., Van den Heuvel-Plesman, T.A.M., Van Kol, N., Kouwenhoven, B., and Guldie, F.H.M., 1976, Dropped egg production, soft shelled and shell-less eggs associated with appearance of precipitins to adenovirus in flocks of laying fowls, *Avian Pathol.* **5:**261–272.

Velicer, L.F., and Ginsberg, H.S., 1968, Ctyoplasmic synthesis of type 5 adenovirus capsid proteins, *Proc. Natl. Acad. Sci. U.S.A.* **61:**1264–1271.

Velicer, L.F., and Ginsberg, H.S., 1970, Synthesis, transport, and morphogenesis of type 5 adenovirus capsid proteins, *J. Virol.* **5**:338–352.

Wadell, G., Hammarskjöld, M., Winberg, G., Varsanyi, T.M., and Sundell, G., 1980, Genetic variability of adenoviruses, *Ann. N.Y. Acad. Sci.* **354**:16–42.

Ward, J.M., and Young, D.M., 1976, Latent adenoviral infection of rats: Intranuclear inclusions induced by treatment with a cancer chemotherapeutic agent, *J. Am. Vet. Med. Assoc.* **169**:952–953.

Wigand, R., Gelderblom, H., and Ozel, M., 1977, Biological and biophysical characteristics of mouse adenovirus, strain FL, *Arch. Virol.* **54**:131–142.

Wilcox, W.C., and Ginsberg, H.S., 1963, Production of specific neutralizing antibody with soluble antigens of type 5 adenovirus, *Proc. Soc. Exp. Biol. Med.* **114**:37–42.

Willcox, N., and Mautner, V., 1976, Antigenic determinants of adenovirus capsids. I. Measurement of antibody cross-reactivity, *J. Immunol.* **116**:19–24.

Willimzik, H.F., Kalter, S.S., Lester, T.L., and Wigand, R., 1981, Immunological relationship among adenoviruses of humans, simians, and nonprimates as determined by the neutralization test, *Intervirology* **15**:28–36.

Winterfield, R.W., Fadly, A.M., and Gallina, A.M., 1973, Adenovirus infection and disease. I. Some characteristics of an isolate from chickens in Indiana, *Avian Dis.* **17**:334–342.

Winters, A.L., and Brown, H.K., 1980, Duodenal lesions associated with adenovirus infection in athymic "nude" mice, *Proc. Soc. Exp. Biol. Med.* **164**:280–286.

Winters, A.L., Brown, H.K., and Carlson, J.K., 1981, Interstitial pneumonia induced by a plaque-type variant of mouse adenovirus, *Proc. Soc. Exp. Biol. Med.* **167**:359–364.

Wong, W.Y., and Tweedell, K.S., 1974, Two viruses from the Lücké tumor isolated in a frog pronephric cell line, *Proc. Soc. Exp. Biol. Med.* **145**:1201–1206.

Wright, N.G., 1976, Canine adenovirus: Its role in renal and ocular disease: A review, *J. Small Anim. Pract.* **17**:25–33.

Wyand, D.S., Jakowski, R.M., and Burke, C.N., 1972, Marble spleen disease in ring-necked pheasants: Histology and ultrastructure, *Avian Dis.* **16**:319–329.

Yamaguchi, S., Imada, T., Kawamura, H., Taniguchi, S., Saio, H., and Shimamatsu, K., 1981a, Outbreaks of egg-drop syndrome-1976 in Japan and its etiological agent, *Avian Dis.* **25**:628–641.

Yamaguchi, S., Imada, T., Kawamura, H., Taniguchi, T., and Kawakami, M., 1981b, Pathogenicity and distribution of egg-drop syndrome-1976 virus (JPA-1) in inoculated laying hens, *Avian Dis.* **25**:642–649.

Yamamoto, T., 1966, Some physical and growth characteristics of a canine adenovirus isolated from dogs with laryngotracheitis, *Can. J. Microbiol.* **12**:303–311.

Yasue, H., and Ishibashi, M., 1977, Chick embryo lethal orphan (CELO) virus-induced early and late polypeptides, *Virology* **78**:216–233.

Yasue, H., and Ishibashi, M., 1982, The oncogenicity of avian adenoviruses. III. *In situ* DNA hybridization of tumor line cells localized a large number of a virocellular sequence in few chromosomes, *Virology* **116**:99–115.

Yasue, H., Ishibashi, M., and Fujinaga, K., 1980, The oncogenicity of avian adenoviruses. II. The arrangement of the viral DNA in tumors and transformed cells, *Virology* **106**:361–373.

Yates, V.J., and Fry, D.E., 1957, Observation on a chicken embryo lethal orphan (CELO) virus, *Am. J. Vet. Res.* **18**:657–660.

Yates, V.J., El Mishad, A.M., McCormick, K.J., and Trentin, J.J., 1973, Isolation and characterization of an avian adenovirus-associated virus, *Infect. Immun.* **7**:973–980.

Zsák, L., and Kisary, J., 1981, Studies on egg drop syndrome (EDS) and chick embryo lethal orphan (CELO) avian adenovirus DNAs by restriction endonucleases, *J. Gen. Virol.* **56**:87–95.

CHAPTER 13

Adeno-Associated Virus

KENNETH I. BERNS

I. INTRODUCTION

Adeno-associated virus (AAV) was discovered as a contaminant of purified adenovirus preparations. Although the small icosahedral particles (diameter 18–28 nm) were initially considered to be either precursors to the mature adenovirus particle or breakdown products of the virion, they were rapidly demonstrated to represent a distinct virus that was physically, chemically, and immunologically unrelated to adenovirus (Atchison *et al.*, 1965; Hoggan *et al.*, 1966; Parks *et al.*, 1967a,b). AAV was found to be absolutely defective and dependent on coinfection with the adenovirus for a productive infection. With additional studies, the reciprocity of the relationship between the two viruses has become clearer. Adenovirus not only helps AAV replication in a coinfection, but also serves to efficiently rescue the AAV genome from an integrated state in the DNA of cells latently infected by AAV (Hoggan *et al.*, 1972; Berns *et al.*, 1975; Handa *et al.*, 1977). On the other hand, AAV inhibits the replication of adenovirus in a lytic coinfection (Hoggan *et al.*, 1966; Smith *et al.*, 1966; Casto *et al.*, 1967; Parks *et al.*, 1968) and can also inhibit the oncogenicity of both adenovirus (Kirchstein, *et al.*, 1968; Casto and Goodheart, 1972; Mayor *et al.*, 1973) and adenovirus-transformed cells (Ostrove *et al.*, 1981; de la Maza and Carter, 1981).

Because of the closely intertwined biology of the two viruses, knowledge derived from a study of one is frequently applicable to the other. Thus, a study of the inhibition of adenovirus oncogenicity by AAV may serve to pinpoint adenovirus functions required for tumor formation. Similarly, some adenovirus functions required for transcription and DNA

KENNETH I. BERNS • Department of Immunology and Medical Microbiology, University of Florida College of Medicine, Gainesville, Florida 32610.

replication also subserve these functions for AAV. Because the two viruses are physically quite distinct and the details of replication differ markedly in some aspects, those functions of adenovirus that also help AAV replication take on the nature of being more general in a biological sense. The concept of potential generality is enhanced by the fact that AAV replication and rescue from the latent state may also be helped by herpes-viruses of several types. Although it was originally reported that herpesviruses were only partial helpers and defective in the process of assembly of AAV virions (Atchison, 1970; Blacklow *et al.*, 1970; Dolin and Rabson, 1973; Blacklow, 1975), more recent data indicate that at least herpes simplex virus type 1 (HSV-1) and HSV-2 can serve as complete helpers for AAV replication (Buller *et al.*, 1981). Thus, the study of AAV has served to link the Parvoviridae, Adenoviridae, and Herpesviridae, the three virus families with linear DNA genomes that replicate in the nuclei of animal cells.

II. VIRION

In 1981, the International Committee on the Taxonomy of Viruses changed the name of the AAV genus of the Parvoviridae family to *Dependovirus* (G. Siegl, personal communication). Because of its close association with adenovirus, AAV has also been referred to in the past as "adeno-satellite virus" and "adenovirus-associated virus." The virion is icosahedral with diameters ranging from 18 to 28 nm reported for various serotypes (Atchison *et al.*, 1965; Mayor *et al.*, 1965; Hoggan *et al.*, 1966; Parks *et al.*, 1967a,b). The genome is a linear single-stranded DNA (ssDNA) molecule with a molecular weight of 1.5×10^6 (Gerry *et al.*, 1973). There are three coat proteins with molecular weights of 90,000 (90K) (VP1), 75K (VP2), and 62K (VP3) (Johnson *et al.*, 1971; Rose *et al.*, 1971). VP3 is the most abundant species, representing approximately 80% of the total coat protein by mass. No cellular histones or corresponding viral proteins have been found in the virion.

There are four AAV serotypes that have been isolated from primates. AAV 2 and 3 are clearly of human origin and AAV 4 is of monkey origin. Although the origin of AAV 1 is quite uncertain, it seems more likely to be of human origin (Hoggan, 1970). Antibodies raised against virions of these four serotypes do not significantly cross-react; however, after sodium dodecyl sulfate denaturation, there is cross-reactivity among the proteins of all four serotypes (Johnson *et al.*, 1972). These data are in accord with the extent of DNA homology observed in cross-hybridization experiments (Rose *et al.*, 1968; Koczot, 1970). Approximately 35–50% homology was observed in such studies, and in the electron microscopy (EM), the homology in heteroduplex molecules was in the terminal regions of the genome.

III. BIOLOGY OF ADENO-ASSOCIATED VIRUS

AAV has been isolated from many vertebrates, including man, monkey, cow, dog, and birds (Atchison *et al.*, 1965; Mayor *et al.*, 1965; Hoggan *et al.*, 1966; Parks *et al.*, 1967a; Luchsinger and Wellemans, 1971; Domoto and Yanagawa, 1969; Dutta and Pomeroy, 1967). The serotype or serotypes from each species are distinctive, but the host range *per se* does not seem to be species-limited. Productive infections of canine cells in culture by human AAV have been reported (Casto *et al.*, 1967). In each case, the helper was a species-specific adenovirus. From these data, it seems likely that an AAV serotype may replicate in the cells of any species as long as there is coinfection with a helper virus able to replicate in the target cells. Although not completely documented, it would also seem likely that any species for which there is an adenovirus will also have AAV.

AAV is unusual in that it not only is a defective virus, but also has never been associated with disease in the infected host. In this sense, AAV may approach being a perfect parasite. In fact, the possibility exists that there may be a symbiotic relationship between AAV and the host. Because of the ability of AAV to inhibit lytic infection by adenovirus and the oncogenicity of both adenovirus and HSV-2 (Blacklow *et al.*, 1978), AAV may have a protective function for its host. Only preliminary studies to assess these possibilities have been done, and although they are far from conclusive (see Section VI.B.1), the data are in accord with the idea.

The most extensive studies of the biology of AAV have involved human infections (Blacklow *et al.*, 1976, 1968a,b, 1971b). AAV has been isolated from humans only occasionally, and then only in the presence of a concomitant adenovirus infection (Blacklow *et al.*, 1967). However, by middle age, approximately 85% of people have antibodies to AAV 2 or AAV 3 or both. In one study of children raised in an orphanage, the mean age of seroconversion was 7–8 years (Blacklow *et al.*, 1971b). Because AAV latent infection was first discovered in primary cell cultures of human origin (1% of embryonic kidney cell lots positive) and of monkey origin (20% of African green monkey kidney cell lots positive) (Hoggan *et al.*, 1972), it is assumed that latent infection with AAV may also be a common occurrence, but this point has not yet been documented.

Additional studies have also been done with AAV infections of chickens. Such infections are very common in some flocks, and there have been suggestions not only that AAV latent infection occurs in chickens, but also that there may be germ-line transmission of the AAV genome (Dawson *et al.*, 1981).

IV. MACROMOLECULAR STRUCTURE AND SYNTHESIS

AAV is absolutely defective in terms of its replication. In the absence of a helper-virus coinfection, the virion can attach to the cell and penetrate to the nucleus, where the DNA is uncoated, but no virus-specific macromolecular synthesis is detectable (Rose and Koczot, 1972). There are specific adenovirus helper functions for both AAV DNA and RNA synthesis. AAV RNA synthesis can occur in the presence of adenovirus early region 1a (E1a) and E1b (Ostrove and Berns, 1980), but AAV DNA replication requires adenovirus E1a, E2, and E4 (Richardson *et al.*, 1980; Janik *et al.*, 1981) (see Section IV.A.5). There is some evidence that an additional helper function is required for protein synthesis as well (Myers and Carter, 1981). AAV macromolecular synthesis is reviewed in detail below, with special attention to the possible adenovirus functions required.

A. DNA

The AAV genome is a linear, ssDNA molecule with an inverted terminal repetition (ITR) (Koczot *et al.*, 1973). The ITR is 145 nucleotides long, and the terminal 125 nucleotides constitute a self-complementary or palindromic sequence that can form a hydrogen-bonded hairpin structure (Lusby *et al.*, 1981). The current model of AAV DNA replication is derived from one initially proposed by Cavalier-Smith (1974) to explain the replication of the ends of eukaryotic chromosomal DNA. Linear DNA molecules that do not cyclize during replication must have special mechanisms to maintain the integrity of 5′-terminal sequences during replication (Watson, 1972), the reason being that all known DNA polymerases synthesize in the 5′ → 3′ direction and require a primer with a free 3′-OH group as well as a template. The Cavalier–Smith model is adaptable to AAV DNA replication and consistent with all known aspects of the process. As modified for the specific case of AAV, the model is illustrated in Fig. 1. There are five steps in the process: (1) The parental DNA forms a single-stranded circle stabilized by hydrogen bonding of the ITRs. This is not an essential step, but allows the first round of replication to initiate in the same manner as subsequent rounds in which the complementary strands at the ends of duplex replicative intermediates have to be separated. (2) The hydrogen-bonded termini are separated by a protein–nucleic acid interaction to allow the 3′ terminus to hairpin and serve as a primer for DNA replication. (3) The hairpin structure is cleaved by a site-specific nuclease at a position on the parental strand directly opposite the original 3′ terminus of the DNA. In this process, the hairpin structure of 125 nucleotides, which is of parental origin, is transferred to the progeny strand, and the transfer results in the inversion of the hairpin sequence. (4) The resultant gap at the 3′ end of the parental strand is filled in by repair synthesis. (5) The replicative intermediate is formally equivalent to the structure found in step 1, and another round of synthesis can occur.

In summary, the model predicts a single-strand displacement type of synthesis in which the origin of synthesis is at the 3′ terminus of the

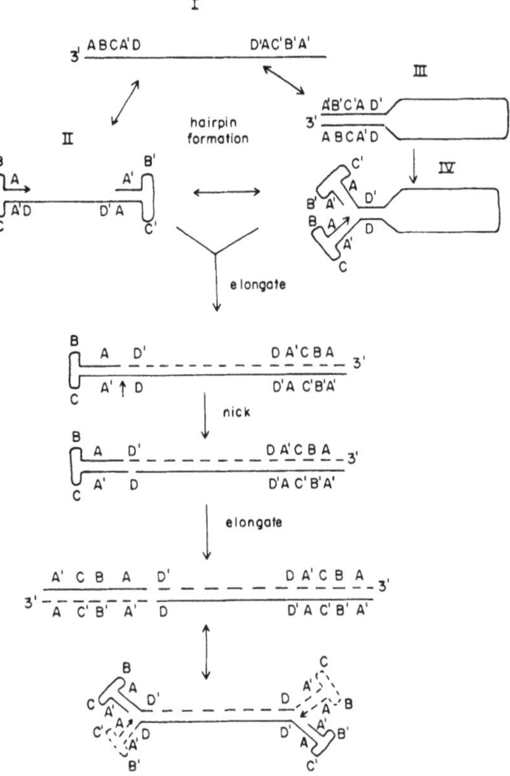

FIGURE 1. Model of AAV DNA replication.

parental strand. The model makes a great many predictions about both the types of replicative intermediates to be anticipated and the fine structure of mature viral DNA. In the following sections, the available data are critically evaluated with regard to these predictions.

1. Structure of Virion DNA

The study of the structure of the mature viral genome has played a critical role in our understanding of AAV DNA replication. As the final product of the replicative process, the details of its sequence arrangements must be able to be explained by any model of DNA replication. Although the genome is a linear single strand, AAV DNA purified from virions is usually double-stranded. The seeming paradox is explained by the fact that both plus and minus strands are encapsidated with equal frequency into separate particles (Mayor *et al.*, 1969; Rose *et al.*, 1969; Berns and Rose, 1970; Berns and Adler, 1972). Unless special precautions are taken during the DNA purification, the complementary strands anneal. This rather unusual property has greatly facilitated the physicochemical characterization of the DNA. Although considered to be unique when first discovered, the phenomenon of encapsidation of both plus and

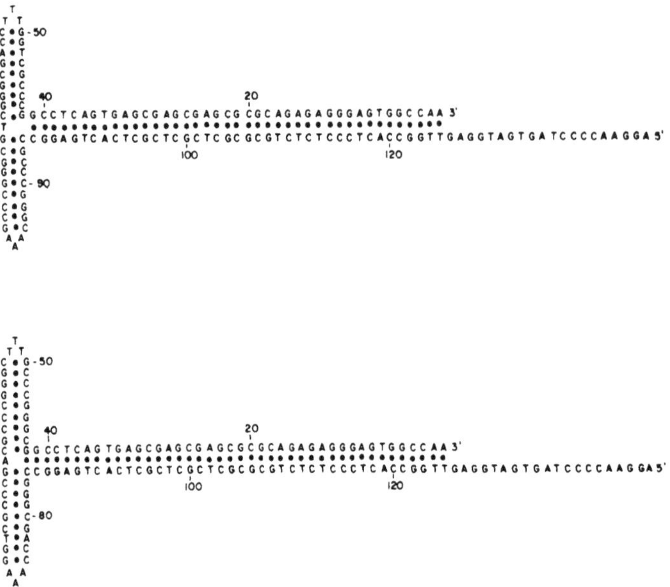

FIGURE 2. Sequences of the inverted terminal repetition in the 3′ → 5′ direction. Both orientations of the inversion are shown.

minus strands at relatively equal frequencies has been found to occur in the cases of several paramyxoviruses (East and Kingsbury, 1971; Portner and Kingsbury, 1971; Kelly *et al.*, 1977), of at least one mammalian autonomous parvovirus (Siegl, personal communication), and of the unusual densonucleosis virus of butterflies (Kurstak *et al.*, 1971).

Operationally, AAV DNA has properties consistent with both a natural terminal repetition (AB———AB, where A and B represent nucleotide sequences) and ITR (AB———B′A′, where A′ and B′ are sequences complementary to A and B, respectively). Thus, duplex molecules of AAV DNA can be annealed to form hydrogen-bond-stabilized circles after limited digestion with exonuclease III (Gerry *et al.*, 1973), and AAV linear single strands can be self-annealed to form hydrogen-bond-stabilized single-stranded circles (Koczot *et al.*, 1973). Because early physical measurements had indicated that the natural and inverted terminal repetitions both represented about 3% of the genome, it was suggested that the terminal repeat was palindromic (ABB′A′———ABB′A′) (Gerry *et al.*, 1973). This notion was confirmed by studies that showed that the termini could fold over to form a duplex structure that would protect either the 5′ or the 3′-terminal nucleotides against digestion with a single-strand-specific deoxyribonuclease (DNase) (Fife *et al.*, 1977).

The detailed structure of the terminal repetition was determined by nucleotide sequencing (Fig. 2) and proved to be more complex than a simple palindrome (Lusby *et al.*, 1980). The ITR was 145 nucleotides long. The first 125 nucleotides contained three palindromic sequences. Nucleotides 1–41 were complementary to nucleotides 125–85. The sequence from 42 to 84 contained two shorter palindromes, nucleotides 42–50 were complementary to 62–54, and nucleotides 64–72 were complementary to 84–76. When the first 125 bases were folded over to permit maximum base-pairing, a T-shaped structure resulted (Fig. 2) in which only 7 bases were unpaired. Interestingly, all the unpaired bases were A or T and 6 of the unpaired bases were required to allow the short internal palindromes to fold back to base pair. The 7th unpaired base served to separate the short internal palindromes. With one exception, all the base pairs formed by the short internal palindromes were GC pairs.

Our current hypothesis is that the palindromic sequence folds to form the T-shaped structure as a specific substrate for interaction with one or more proteins involved in the initiation of DNA synthesis, or for the specific cleavage of the hairpin predicted by the model for DNA replication or for both. The hypercycle theory of Eigen and Schuster (1978) may explain the origin of the T-shaped structure. This theory suggests that secondary structure in polynucleotides tends to asymptote to a common structure evolutionarily. The basis of the theory is that substitutions that enhance base-pairing are favored and that the formation of GC base pairs is favored over that of AT base pairs. When applied to a polyribonucleotide of chain length 80, the theory correctly predicts the structure of transfer RNA.

As presented in Fig. 1, the model for DNA replication makes a further prediction about the terminal structure of AAV DNA. During the transfer of the hairpin from the parental to the progeny strand, the transferred sequence is inverted (step 3). If the hairpin were a perfect palindrome, no overall change in the sequence of the terminal 125 nucleotides would result; however, the actual terminal sequence would predict that the inversion should lead to the existence of a second sequence in the region of nucleotides 42–84, which are not part of the overall palindrome. This prediction has been borne out by several types of experiments: (1) When the complementary strands of AAV DNA are annealed, two types of terminal structures are formed. Half of the termini have a normal Watson–Crick double-helical structure; i.e., both strands are in the same orientation (either flip-flip or flop-flop). The other half of the termini have an aberrant secondary structure (mispairing between a flip and a flop orientation) that causes terminal restriction fragments that contain the entire sequence to migrate anomalously on polyacrylamide gels (Denhardt *et al.*, 1976; Spear *et al.*, 1977). Denaturation and reannealing of either species of terminal fragment (normal or aberrant double-helical structure) leads to an equal number of both species of fragment, as would be predicted (Berns *et al.*, 1979). (2) Restriction fragments that make a terminal cut in the affected region (42–84) produce two species of fully duplex terminal fragments from either end of the molecule when termini with normal duplex structure are cleaved (Spear *et al.*, 1977; Lusby *et al.*, 1980). (3) Finally, sequencing of the terminal fragments described above has directly demonstrated both the existence of two sequences in the region predicted and the fact that the two sequences represent an inversion (Lusby *et al.*, 1980). The model further predicts that the orientation at one end of the molecule should be independent of the orientation at the other end, and this prediction has also been borne out experimentally (Lusby *et al.*, 1981).

The sequence of the entire AAV 2 genome has been determined. The DNA contains 4675 nucleotides (Srivastava, *et al.*, 1983). There are two long open reading frames, the first extending from 323 to 2184 and the second from 2810 to 4322. The fine structure of these regions is described in Sections IV.B and IV.C. Additionally, the initial sequencing of the termini of the DNA revealed a very limited heterogeneity involving the terminal 2 nucleotides (Fife *et al.*, 1977). The terminal sequences were 5′-TTGGCC (35%), 5′-TGGCC (50%), and 5′-GGCC (15%). Whether the heterogenity was physiological or not was uncertain for a long time, but recent experiments using AAV DNA cloned in pBR322 indicate that the heterogeneity is physiological and represents a slight variability (± 1 nucleotide) in the site at which the hairpin is cleaved (Samulski *et al.*, 1983).

In summary, detailed studies on the termini of mature AAV DNA have confirmed all the predictions implicit in the model for DNA replication.

2. DNA Replication *in Vivo*

In a simultaneous coinfection with adenovirus, AAV DNA synthesis can first be detected approximately 7 hr postinfection (Rose and Koczot, 1972) and is the first AAV-specific macromolecular synthesis detected by filter hybridization. If adenovirus infection precedes AAV infection by several hours, AAV DNA synthesis can be detected 3 hr earlier (Carter *et al.*, 1973), demonstrating the requirement for adenovirus early function(s) for AAV DNA replication.

Sedimentation of DNA isolated from infected cells through neutral sucrose gradients revealed three major peaks of AAV-specific sequences [the assay is made somewhat easier because AAV inhibits adenovirus DNA replication (see Section VI.A)]. There was a fast peak containing a large fraction of replicative intermediates, a peak corresponding to mature DNA in duplex form (and some apparent replicative intermediate), and a slower-sedimenting species representing defective variant molecules. It has been possible by means of pulse-label experiments to chase some of the putative replicative intermediate into unit-length mature DNA (Straus *et al.*, 1976b; Hauswirth and Berns, 1979).

The outstanding feature of the replicative intermediate DNA isolated from neutral sucrose gradients was that it was shown to contain complementary strands covalently cross-linked at the termini in a head-to-head tandem. Such DNA would spontaneously reanneal after denaturation (snap back) unless the original duplex structure was first digested with a single-strand-specific DNase. The finding of such intermediates prompted Straus *et al.* (1976b) to first suggest that AAV DNA replication proceeded by essentially the model illustrated in Fig. 1. Subsequently, Hauswirth and Berns (1979) were able to isolate the terminal hairpin structure from replicative intermediate molecules and to show that it had the physical properties predicted.

A salient prediction made by the DNA replication model is that the 3′ terminus of the DNA folds over to serve as a primer for DNA synthesis. To directly test this hypothesis, AAV-infected cells were pulse-labeled with [^{3}H]thymidine, and mature unit-length AAV DNA that had been labeled during the pulse was isolated and digested with *Hpa*II, which cleaves the DNA at 23 sites. The radioactivity incorporated into each strand of most fragments was measured and a gradient of labeling thus determined for both plus and minus strands. For strands of both polarities, the gradient indicated that synthesis of each progeny strand started at or near the 5′ end and terminated at the 3′ end, exactly as would have been expected if the 3′ terminus of the parental strand acted as a primer (Hauswirth and Berns, 1977). These data indicated that AAV DNA replicates by a continuous single-strand displacement mechanism that is very similar in a general sense to the mechanism of DNA replication proposed for adenovirus (Lechner and Kelly, 1977).

As noted above, a considerable portion of label added to infected cells was incorporated in AAV DNA of less than unit length. These variant molecules were both single- and double-stranded. Characterization of the double-stranded species isolated from infected cells showed that the DNA contained primarily terminal sequences and was cross-linked by a terminal hairpin (Hauswirth and Berns, 1979). The majority of the hairpins were at internal positions on the genome as though there had been a template-strand switch during replication, although there was also some evidence for hairpinning of the normal termini. Variant molecules can be encapsidated, and those that are have also been characterized (Laughlin *et al.*, 1979a; de la Maza and Carter, 1980). The single-stranded encapsidated variant species contained long deletions of variable lengths of the internal sequences. The duplex species were cross-linked, but the majority of cross-linking reported for the encapsidated double-stranded species was at the position of the normal termini. Virions with variant genomes required intact AAV as helper in addition to adenovirus for DNA replication to occur. In turn, they inhibited normal AAV DNA replication, but did not directly inhibit adenovirus DNA replication in the absence of intact AAV. Other interesting biological properties are reviewed below.

3. Cloned AAV DNA

Recently, it has been possible to clone the intact AAV genome in duplex form into pBR322 (Samulski *et al.*, 1982). When the recombinant plasmid was transfected into human cells in the presence of adenovirus coinfection, the AAV DNA was rescued from the plasmid and replicated, and infectious virus was produced. This system has allowed initiation of a systematic study of the details of AAV DNA replication and of the consequences of specific lesions in the genome on DNA replication. To date, the following observations have been made: (1) A clone, by definition, has a specific orientation of the invertible sequence in each terminal repetition. When a clone in which both terminal repetitions were in the flop orientation was transfected into human cells, the progeny DNA molecules had terminal repetitions of both orientations at both ends. This was direct confirmation that the inversion is a consequence of DNA replication. (2) Clones in which both termini were 5′ GGCC produced progeny that had 5′-TTGGCC or 5′-TGGCC terminal sequences. (3) Clones that contained deletions of up to 115 terminal nucleotides at either end (but not at both ends) were rescued, and viable virus was produced. The DNA isolated from progeny virions was intact; during the replication process, the deletion had been repaired. How this repair took place is unknown, but it is undoubtedly a reflection of the ITR. Because some of the ITR remained in the clone tested, repair may have taken place via the circular intermediate proposed in Fig. 1 or via a recombinational mechanism. (4) From the previous data, it is clear that it is necessary for a given deletion to be present in both terminal repeats to assess potential effects of a deletion in the terminal repetition. One small deletion from nucleotides 47–57 (Fig. 2) was created in a clone with a 115-nucleotide deletion in the other terminal repeat. In this case, rescue took place along with DNA synthesis, but very little, if any, monomer-length mature DNA was produced, and no infectious virions could be detected (Samulski *et al.*, 1983). Apparently, this deletion selectively affected processing of the replicative intermediate. There would seem to be less, if any, effect on initiation of DNA synthesis. (5) Larger deletions, sufficiently long to prevent hairpin formation, on both ends of the genome prevented rescue and replication, as did deletions on either end that extended beyond the terminal repetition. This system holds great promise for elucidating much of the molecular biology of AAV.

4. AAV Gene Products Required for DNA Replication

Because of the small size of the AAV genome and the requirement for adenovirus helper function, it is not clear that any AAV gene product is required for DNA replication. However, some experiments have been performed that do indicate the existence of an AAV gene product that can function in *trans*. The most direct approach has been to create a deletion in the AAV genome from 0.38 to 0.42 by digestion of the DNA with *Eco*RI followed by ligation of the terminal fragments. The deleted DNA did not replicate after transfection in a coinfection with adenovirus. However, addition of intact AAV DNA to the transfection mixture did lead to the replication of the deleted molecules (Cheung, Tan, and Berns, unpublished data).

Other suggestive evidence for a transacting AAV gene function comes from studies of AAV defective interfering (DI) particles (Laughlin *et al.*, 1979a). The DI particles apparently replicated, but only if coinfected with both intact AAV and adenovirus. This ability was inhibited by UV irradiation of the DI particles prior to infection.

The nature of the required AAV helper function is unknown. One possibility would be that the coat proteins play an active role in DNA replication, either by the encapsidation process or by also serving to cleave the hairpin in replicative intermediates. A second possibility is the existence of an as yet unidentified nonstructural protein that might also be a candidate to cleave the hairpin.

5. Adenovirus Helper Functions for AAV DNA Replication

Elucidation of adenovirus helper functions required for the synthesis of specific AAV macromolecules has proved to be a difficult task. At this point, three adenovirus early regions have been implicated as being specifically required for AAV DNA replication: E1, E2a, and E4. The pertinent experiments have involved the use of temperature-sensitive (*ts*) and host-range (*hr*) mutants as helpers under nonpermissive conditions, transfection of specific regions of the adenovirus genome, and microinjection of both specific adenovirus DNA fragments and messenger RNAs (mRNAs). From these studies, a complex picture is beginning to emerge in which the clearest conclusion is that no late adenovirus functions seem to be required. The manner in which the implicated early functions interact is still very much a matter of debate.

E1 of adenovirus has been subdivided into E1a and E1b (Philipson, 1979). Host-range mutants in E1a (Harrison *et al.*, 1977) do not help AAV replication in nonpermissive cells (Ostrove and Berns, 1980; Janik *et al.*, 1981); thus, one or more E1a functions are clearly required. The story for E1b is very much less clear. Both Ostrove and Berns (1980) and Janik *et al.* (1981) found that *hr*6 and *hr*7, respectively, served as good helpers for AAV DNA replication in HeLa cells (nonpermissive for *hr*6 and *hr*7). Ostrove and Berns (1980) did report that *hr*6 failed to rescue AAV DNA from its integrated state in latently infected Detroit 6 cells. The mutant also failed to rescue AAV DNA from a recombinant AAV–pBR322 plasmid after transfection in HeLa cells (Samulski, Berns, Srivastava, and Muzyczka, unpublished data). Laughlin *et al.* (1982) have studied the ability of a series of deletion mutants in E1b to help AAV replication. These mutants all contain a 58–base-pair deletion at the right end of E1a (Jones and Shenk 1979b; Berk *et al.*, 1979; Shenk, *et al.*, 1980) and extend for various distances into E1b. None could help AAV replication, and when all of E1b was deleted, AAV DNA replication was not detectable (partial deletions were apprently not characterized). Similarly, a deletion from the right into E1a alone failed to help AAV DNA replication. The reason or reasons for the different results obtained with the deletion mutants as opposed to the *hr* mutants in E1b are not known. E1b codes for at least two polypeptides, a 15K protein and the 58K major tumor (T) antigen. Both *hr*6 and *hr*7 are point mutants that have been mapped to the 58K T antigen. Thus, it is possible that only the 15K protein is required for AAV DNA replication in a lytic coinfection or that 58 K T antigen is required, but that neither the *hr*6 nor the *hr*7 mutation alters this function of the protein. In microinjection experiments, E1 DNA or RNA was required. In these experiments, however, there was no separation of E1a RNA from E1b RNA or of the DNAs (Richardson and Westphal, 1981). The actual function supplied by E1b is totally unknown in this context, as well as that which may be necessary for rescue of AAV from the latent state.

The original DNA-negative mutants used to study possible adenovirus helper function were adenovirus type 5 (Ad5) mutants that have now been mapped in E2a and E2b (Straus *et al.*, 1976a). It seems certain that the E2b mutants [H5*ts*149 and H5*ts*36 (Ensinger and Ginsberg, 1972)] define a region that is not required for AAV replication. By the same token, E2a is involved, but there has been a controversy concerning the functional basis of involvement and specifically whether H5*ts*125 can serve as a helper. Straus *et al.* (1976a) originally reported that H5*ts*125 (E2a) was as good a helper as wild-type (Wt) adenovirus for production of infectious AAV. But in a more recent paper, Myers *et al.* (1980) presented data that contradicted the original conclusion. These authors found that synthesis of duplex forms of AAV DNA was at nearly normal levels with H5*ts*125 coinfection, but that production of virion ssDNA and capsid proteins was inhibited 100-fold. They hypothesized that adenovirus E2a was required for both AAV and adenovirus late protein synthesis and that the former was required for production of normal amounts of AAV virion ssDNA. The requirement for E2a for late-protein synthesis as well as AAV DNA replication does seem likely from other work (see Section IV.A), but the question is still unanswered as to whether the lesion in H5*ts*125 defines a domain in the 72K adenovirus ssDNA-binding protein (DBP) that has the effect noted above. In a very recent paper, McPherson and Rose (1982) have reinvestigated the helper activity of H5*ts*125 and found again that at the nonpermissive temperature, the mutant was an effective helper for AAV virion production. Thus, at the level of the mutation represented by H5*ts*125, the data are inconclusive with regard to the necessity for E2a function.

The question has been approached much more directly in transfection and microinjection experiments. Janik *et al.* (1981) found that E2a sequences were required for AAV replication in experiments in which specific fragments were transfected into 293 cells [Ad5-transformed human cells that contain functional E1a and E1b (Graham *et al.*, 1977)]. Richardson and Westphal (1981) found the same in microinjection of specific adenovirus DNA fragments into Vero cells (monkey cells permissive for human adenovirus replication). Thus, E2a gene product seems to be required for AAV replication. Whether it is directly needed for DNA replication or simply serves to turn on another adenovirus early region is not completely clear because the experiments of Janik *et al.* (1981) did not directly measure AAV DNA synthesis.

Janik *et al.* (1981) also found evidence that adenovirus VA1 is required for AAV replication. They suggested that this small RNA might be significant in mRNA splicing or play some other role related to protein synthesis. This is an important point that bears on the results of Richardson and Westphal (1981). These investigators found that microinjection of DNA representing E1, E2a, and E4 into Vero cells supported AAV replication. Most significantly, they also reported that injection of E4-specific mRNA into the Vero cells was sufficient to help AAV DNA synthesis and protein synthesis. From these data, they concluded that E1 turned on E2a, which in turn permitted E4 expression, and that the latter was all that was required for AAV replication.

These data provide several anomalies that are difficult to reconcile with other work. Data presented in Section IV.A. suggest that only E1 function may be needed for at least some AAV transcription, whereas the data of Richardson and Westphal (1981) suggested that no specific adenovirus function or functions are required for AAV transcription, except possibly E4. Data described above suggested that E2a provides a function important at least for late-protein synthesis, and in the experiments of Richardson and Westphal (1981), there was no E2a function present. The same question can be raised concerning the data of Janik *et al.* (1981) suggesting that VA1 is required. One possible explanation would be that the Vero cells are an anomalous system in which not all normal adenovirus helper functions are required.

In summary, it is possible that E4 is responsible for the only direct helper function required for AAV DNA replication. Unresolved questions, however, cast some doubt on this seemingly straightforward explanation.

B. RNA

The AAV genome has been of interest, in part, because its small size makes it amenable to detailed analysis. It has now become apparent that the small size of the genome has forced the virus to adopt some special strategies to be able to synthesize all its proteins. To date, three coat proteins have been described (see Section IV.C). Were the sequences that code for these proteins not to overlap, the information required for capsid proteins alone would exceed the potential coding capacity by a factor of 2.

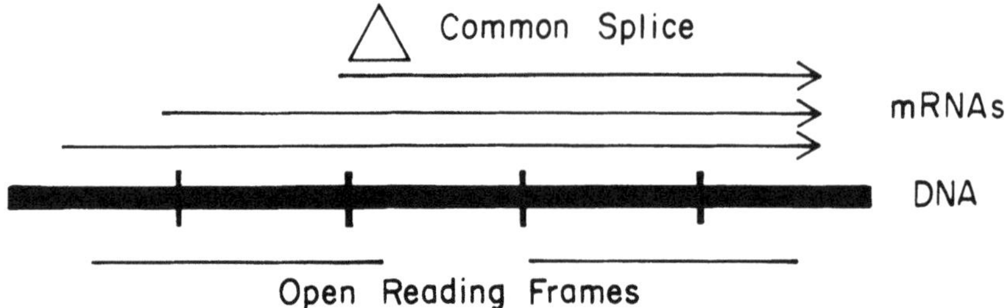

FIGURE 3. Map of the AAV genome illustrating the three transcripts and the major open reading frames.

1. Transcripts

Even though all three AAV coat proteins appear to be coded for by one species of RNA, three AAV-specific RNAs have been described and characterized through use of Berk and Sharp (1977) analysis of RNA–DNA hybrids and by heteroduplex analysis of such hybrids in the EM (Laughlin *et al.*, 1979b). The three species had different start points on the genome (0.06, 0.19, 0.38) (Green and Roeder, 1980b; Lusby and Berns, 1982), but all seemed to terminate at 0.96 (see Fig. 3). A fourth species starting at 0.13 was also found, but not reproducibly, and is now considered not to represent a distinct transcript. Each species was found in both spliced and unspliced forms in total cytoplasmic RNA, and all were polyadenylated. The 5′ ends of each species have now been mapped at the nucleotide level on the genome, and the start points are at nucleotides 286, 972, and 1852, respectively (Lusby and Berns, 1982; Green and Roeder, 1980b). The splice appears to involve the same intron for each species. The intron has been determined at the nucleotide level for the shortest transcript and extends from nucleotides 1907 to 2207 (Green and Roeder, 1980a).

AAV RNA has been reported to be capped (Carter, 1976). The majority of the RNA had type I cap structures and the rest type II. There was also a limited amount of internal methylation (approximately 5 bases/molecule). These data represent total polyadenylated RNA, and the 5′ terminal structures of the individual species of transcripts have not been reported.

It appears likely that the three transcripts result from independent initiation events, rather than represent a precursor–product relationship. Each is preceded by a TATA box 31 ± 2 bases upstream from the 5′ end of the transcript. Although definitive experiments have not been done to rule out processing of a common precursor, it would indeed be fortuitous if such processing involved several sites in specific proximity to TATA boxes.

2. Transcript Function

The transcripts have been localized in cells only late in the infection process. Under these conditions, Laughlin *et al.* (1979b) reported finding the spliced and unspliced forms of all three transcripts in a polyadenylated form in the cytoplasm. However, Green and colleagues (Green *et al.*, 1980; Green and Roeder, 1980b) could not find evidence for splicing of the larger transcripts. Additionally, they found only the spliced form of the shortest transcript [2.3 kilobases (kb)] and possibly the unspliced form of the same transcript (2.6 kb) associated with polysomes. In a sense, considering that RNA was isolated at a time late in infection and that all the structural proteins are apparently translated from the shortest transcript, this result is not too surprising and does not rule out the possibility that the larger transcripts may be translated earlier in replication. It is not clear why these authors could not see spliced forms of the larger transcripts in the cytoplasmic fraction. Recent data indicate that all three coat proteins are translated from the shortest transcript. Jay *et al.* (1981) separated the different RNA species on sucrose gradients and assayed part of each fraction by gel electrophoresis to determine which RNA species were present. Not until the shortest species (2.3–2.6 kb) appeared were the AAV coat proteins able to be translated in an *in vitro* assay. All three coat proteins could be detected. It was not clear whether this implied different subspecies in the shortest transcripts, different start points for translation, or a proteolytic activity in the reticulocyte assay system used. It would seem, however, that if the larger transcripts are translated, the proteins synthesized are in addition to the coat proteins.

3. Adenovirus Helper Functions for AAV Transcription

In early experiments, involving simultaneous coinfection by adenovirus and AAV, AAV DNA synthesis was first detected at 7–9 hr in conjunction with adenovirus DNA synthesis, and RNA synthesis first appeared 2–3 hr later at the time late adenovirus mRNA appeared (Rose and Koczot, 1972). However, if adenovirus infection preceded AAV infection by 6 hr, then both AAV DNA and RNA could be detected by 4 hr (Carter *et al.*, 1973). Thus, it was suggested that there are separate adenovirus functions that are required for AAV DNA and RNA synthesis. More recent data support this notion in a more defined way. In the absence of adenovirus coinfection, there is no detectable AAV RNA synthesis (Rose and Koczot, 1972). Recently, this has also been confirmed using the much more sensitive Northern blot technique (Laughlin *et al.*, 1982). Although many different adenovirus-transformed cell lines have been tested, none has ever been reported to support AAV multiplication in the absence of helper coinfection. At a molecular level, using a variety of sensitive assays, none of the transformed cells will support AAV DNA replication. However, AAV infection of two lines of adenovirus-transformed cells led to the synthesis of AAV RNA. H14b cells are an Ad5-transformed hamster embryo cell line that contains 5.5 copies of the left-hand 40% of the adenovirus genome (Williams, 1973). The transcripts represent 0.02–0.11 on the genetic map (i.e., E1a and E1b). Cells of the

293 line, an Ad5-transformed human cell line, contain the left 12% and the right 9% terminal sequences of the adenovirus genome (Graham *et al.*, 1977). Only transcripts that map to the left 12% have been detected in these cells. Infection of both the aforenamed cell lines by AAV alone led to AAV RNA synthesis as determined by northern blotting, but not to DNA synthesis (Ostrove *et al.*, 1981; Labow and Berns, unpublished data). Similar experiments using cloned AAV DNA missing only the terminal 125 nucleotides on each end have been carried out by Janik *et al.* (1982). Transfection of the cloned DNA into 293 cells again led to AAV transcripts detectable by Northern blotting.

Because both H14b and 293 cells contain functional E1a and E1b, it has not yet been determined whether one or both of these regions are required for AAV transcription. It is known that *hr*1 does not help AAV replication and is negative for the E1a 13 S messenger but positive for the 12 S messenger (Montell *et al.*, 1982). Ad 215*pm*975, in contrast, is missing the 12 S mRNA and functions as wild-type (Carter, personal communication). Therefore, it would appear that E1a is required and it is the 13 S messenger that is critical. Because no experiments to assess transcription have been done in the absence of E1b function, it is not known whether this is also required for AAV transcription.

The observation of transcription in the presence of only E1a and E1b raises several interesting questions. First, because AAV DNA synthesis was not normally detectable under these conditions, either the transcription occurred from a single-stranded template in the experiments involving AAV infection of the transformed cells or there was a limited amount of replication such as would be expected if the single-stranded parental DNA were converted to a duplex form but could not continue replication without further helper factors. Second, in several of the cases described above, the relative amounts of the various AAV transcripts differed from those seen late in a productive infection. The least prominent species were the 2.3 to 2.6-kb forms thought to code for the coat proteins. In a normal infection, these are the most prominent species (Laughlin *et al.*, 1979b). If these preliminary observations are confirmed, the data suggest a temporal regulation of differential RNA synthesis that would be at least indirectly dependent on additional adenovirus helper functions.

C. Proteins

To date, only the AAV structural proteins have been identified. There are three coat proteins, VP1–3, that vary slightly in molecular weight in different serotypes (Johnson *et al.*, 1971). For AAV 2, the reported molecular weights are 87, 73, and 62K, respectively (Rose *et al.*, 1971). No amino acid sequence data are available for any of the proteins, but all three have similar isoelectric points and all cross-react with specific antisera after denaturation (Johnson *et al.*, 1977). Finally, all three have very similar tryptic peptide maps, suggesting that all three are encoded from the same region of the genome. This would fit well with the *in vitro* translation data (Jay *et al.*, 1981) cited in Section IV.B.2. Whether the three proteins result from posttranslational cleavage of a common precursor or from independent initiation or termination events, or both, is unknown. One potential problem is that the open reading frame in AAV 2 DNA, which presumably codes for the coat proteins, is not large enough to accommodate an 87K protein (Srivastava *et al.*, unpublished data). Therefore, either the molecular-weight estimates of the coat proteins are somewhat too high, the DNA sequence is incorrect, or there is a novel type of translation. Until an amino acid sequence is determined, it will be difficult to resolve this problem.

There is no evidence for glycosylation or phosphorylation of any of the coat proteins, but evidence has been presented that suggests that the coat proteins are acetylated (Johnson and Hoggan, 1979).

The question of the coat protein molecular weight and the generation of multiple coat proteins from essentially the same sequence is further complicated by the fact that use of high-resolution stacking gels or two-dimensional gel systems has revealed that each class of coat proteins can be further subdivided into several discrete species of varying mobilities that may represent either additional differences in molecular weight or modification (McPherson and Rose, 1982). Similar results have been obtained for the autonomous parvovirus minute virus of mice (Ward, personal communication).

1. Proteins from Infected Cells

Additional proteins have been observed in labeled extracts of AAV-infected cells. Buller and Rose (1978a) observed smaller proteins with molecular weights of 24.9K and 15.8K. When AAV was used to infect African green monkey kidney cells with human adenovirus as a helper, there was an abortive infection in which synthesis of the coat proteins was greatly depressed (Buller and Rose, 1978b). Under these conditions, an additional protein was noted that was slightly smaller than VP2. Johnson *et al.* (1977) reported that additional proteins with molecular weights of 56, 100, and 110K could be immunoprecipitated.

Buller and Rose (1978b) also translated AAV RNA in an *in vitro* system and observed synthesis of proteins with molecular weights of 13.2, 22.4, 30.2, 44.5, 61.4, and 75K in addition to VP1–3. Whether any of these represent physiologically significant proteins is unknown.

The origin of the additional proteins observed in infected cells was also uncertain. The smaller 25 and 16K proteins might have been the products of posttranslational cleavage of a large precursor to the coat proteins or might have represented independent translation products. In the same vein, the larger immunoprecipitable polypeptides observed by

Johnson *et al.* (1977) might have corresponded to precursors of the coat proteins.

2. Possible Additional Proteins

Nonstructural proteins might be predicted for two reasons. The first suggestion for additional proteins comes from the DNA sequence. On the left half of the genome, there is a large open reading frame that could correspond to a protein with a molecular weight greater than 70K (Lusby and Berns, 1982). This open reading frame is contained entirely within the largest AAV transcript. Additionally, a significant part of the reading frame is also contained within the intermediate RNA species. In addition to the large open reading frame just described, tbere are several additional open reading frames on the left side of the genome, none of which corresponds to a protein with a molecular weight greater than 7K. On the right side of the genome, there is an additional open reading frame that could code for a protein with a molecular weight greater than 21K (Srivastava, *et al.*, 1983).

The second reason for suspecting the existence of nonstructural proteins comes from our current knowledge of the molecular biology of AAV. As suggested in Section IV.A, it would appear that an AAV function that acts in *trans* is required. A possible function for such a protein could be to specifically cleave the hairpin. AAV can also establish a latent infection by integrating its DNA into host-cell DNA (Cheung *et al.*, 1980). It is conceivable that the virus might code for the equivalent of a transposase to facilitate DNA integration.

Not a single nonstructural protein has yet been identified for any parvovirus, although a protein covalently attached to the 5′ termini of replicating molecules might represent such a protein (Ward, personal communication). This area of study thus remains of great interest and significance in fully understanding the replication of these viruses.

3. Possible Adenovirus Helper Functions

Whether there is one or more specific adenovirus helper functions for AAV protein synthesis is unclear. Myers *et al.* (1980) reported that although Ad5*ts*125, a *ts* mutant in the adenovirus DBP that is essential for adenovirus DNA replication, supported AAV DNA replication at 20–50% of the normal level at the nonpermissive temperature, AAV protein synthesis and virion assembly were reduced 100-fold. Unfortunately, these data were in direct conflict with the data of Straus *et al.* (1976a), who reported *ts*125 to be a good helper for AAV infectious virion production. There is also evidence that the adenovirus virus-associated RNAs that are crucial to adenovirus late-protein synthesis but map near position 30 on the adenovirus genome (Mathews, 1975; Petterson and

Philipson, 1975; Soderlund *et al.*, 1976) may be necessary for AAV capsid-protein synthesis (J. A. Rose, personal communication).

Additional data that suggest a requirement for helper virus function for AAV protein synthesis was seen when AAV infected monkey cells with human adenovirus as a helper. Again, a significant amount of AAV DNA synthesis was seen, albeit processing was not normal, but AAV coat-protein synthesis and virion assembly were greatly inhibited (Buller and Rose, 1978b). A mutation in E2a, *hr*400 (Klessig, 1977) permits human adenovirus growth in monkey cells. This mutant also supports AAV replication in monkey cells (B. J. Carter and C. Laughlin, personal communication). However, since this mutant corrects a defect in late RNA synthesis rather than protein synthesis, the significance for AAV protein synthesis is not clear.

V. LATENT INFECTIONS

Because it is absolutely defective, AAV has a special problem in teleological terms of ensuring the biological continuity of its genome. An appropriate helper virus is not always going to be present to support AAV multiplication. The virus has neatly sidestepped this problem by being able to establish a latent infection in the absence of a helper (Hoggan *et al.*, 1972). By integration of the AAV genome into cell DNA, the genetic information of the virus is conserved until the cell is infected by a suitable helper virus, at which time the AAV genome is rescued with high efficiency and viral multiplication ensues.

A. Latent Infection *in Vivo*

Latent infection *in vivo* was discovered by Hoggan *et al.* (1972) as a consequence of a United States government program to screen for the presence of cryptic viruses in primary cell cultures intended for vaccine production. As would have been anticipated from the defective nature of the virus, none of the cultures were positive for the presence of AAV antigens or infectivity. However, when the cells were infected with adenovirus, 1–2% of the human embryonic kidney cell lots and 20% of the African green monkey kidney cell lots produced infectious AAV. Thus, it appeared that AAV latent infection might be a rather common phenomenon in primates. Unfortunately, these original studies have not been followed by significant epidemiological studies in humans. There have been numerous studies in chickens that indicate that AAV latent infection is a rather common event (Dawson *et al.*, 1981). There has also been the suggestion of germ-line transmission of AAV in chickens.

B. Latent Infection in Cell Culture

To study the underlying mechanisms at the molecular level, it was extremely helpful to be able to establish the latent infection in a cell-culture model. This was first done in Detroit 6 cells, a continuous line of human cells (Berns *et al.*, 1975). Infection at high multiplicity (250 tissue-culture infectious doses/cell) led to a cell culture that would produce AAV for over 100 passages on infection with adenovirus. In the absence of adenovirus infection no AAV antigen or infectivity was detectable, and the latent infection was not inhibited by addition of anti-AAV antiserum to the culture medium. At the 39th passage, the culture was cloned and 30% of the clones were inducible for AAV on challenge with adenovirus. Subclones of the initially positive clones remained positive for AAV for more than 150 passages. In a similar manner, KB cells have also been latently infected with AAV (Handa *et al.*, 1977).

All DNA animal viruses that replicate in cell nuclei have been reported to establish latent infections *in vivo*. Detailed studies of the mechanisms of such infections have proved nearly impossible because autonomous DNA viruses in cell culture undergo a lytic infection. The unique advantage with AAV is its absolute defectivenss, which allows it to latently infect cells from the normal host in culture in the absence of helper coinfection. The cell-culture system is rendered even more attractive as a model by virtue of the high efficiency with which the latent virus may be rescued. On challenge with adenovirus, up to 10–30% of the cells become positive for AAV antigen (Handa *et al.*, 1977; Hoggan, personal communication), a number that approximates the percentage of positive cells in a lytic coinfection and even compares favorably with the induction of bacterial lysogens.

Although no AAV antigens are detectable in latently infected cells, it is readily possible to detect AAV DNA. By measuring reassociation kinetics in solution, it was demonstrated that there were 3–5 AAV genome equivalents per diploid amount of cell DNA in two clones of latently infected Detroit 6 cells (Berns *et al.*, 1975). The viral DNA in latently infected cells was covalently associated with the cellular DNA. Handa *et al.* (1977) analyzed latently infected KB cells using the technique of network association. Cellular DNA was denatured and briefly annealed, so that only DNA that contained highly reiterated sequences could reanneal to form large networks that were sedimented at low *g* forces under high-salt conditions. A large majority of the AAV sequences (80%) were associated with the cellular DNA in this assay. In other studies using Southern blot analysis in early passages of latently infected Detroit 6 cells, all the AAV sequences migrated as high-molecular-weight species in agarose gels (Southern, 1975; Cheung *et al.*, 1980). Digestion of the DNA with restriction enzymes that do not cleave AAV DNA reduced the molecular weight of the DNA that contained AAV sequences, thus demonstrating linkage of the AAV DNA to cellular DNA. From these data, it was concluded that AAV DNA had integrated into chromosomal DNA in latent infection, although other explanations were theoretically possible.

In a more detailed analysis of the organization of the integrated AAV sequences (Cheung *et al.*, 1980), it was found that the joints between viral and cellular DNAs always involved the termini of the viral DNA, but did not involve specific sites in cellular DNA (Berns *et al.*, 1982). The integrated viral sequences were arranged as a head-to-tail tandem repeat, and in some instances, short stretches of additional DNA of uncertain origin were inserted between the termini of the different copies of the viral genome. In the clones studied in detail, there was evidence for a limited amount of deletion of viral sequences in some copies of the genome.

In early passages (8–9) of cloned cells, free copies of the viral DNA were not found. However, in a late passage (118), free AAV DNA was detectable and was identical in mobility to the duplex form of virion DNA. The arrangement of sequences within the free DNA was identical to that of virion DNA as measured by the number and size of fragments produced by several restriction enzymes. In more recent studies of intermediate-passage cells, additional species that appeared to be free AAV DNA were also observed (Bohenzky and Berns, unpublished data). It is not known whether the free DNA can replicate or is simply diluted out with continued passage of the cells. It would seem, however, that the excision process necessary for free DNA to occur must also involve the termini of the viral DNA because of the correct orientation of the sequences in the unit-length free species.

The aforestated notion was supported by studies of the arrangement of the integrated viral sequences in late- as compared to early-passage cells. Studies with six enzymes revealed no differences in the integrated sequences. However, *Sma*I, which cuts AAV DNA only in the middle of the palindrome in the inverted terminal repetition, produced different patterns for early- and late-passage cells. This indicated active rearrangement in this region of the genome, which would have been compatible with recombinational activity to produce free copies of AAV DNA seen in late-passage cells (Cheung *et al.*, 1980).

C. Viral Functions in Rescue

Although both adenovirus and many herpesviruses can rescue AAV from the latent state, studies on the details of rescue have concentrated on adenovirus because of its better-characterized genetic map. The functions required for AAV DNA replication seem to be encoded by adenovirus E1a and possibly E1b, E2, and E4 (Richardson *et al.*, 1980; Janik *et al.*, 1981). The adenovirus host-range (*hr*) mutant *hr*6 in E1b is DNA-positive (Lassam *et al.*, 1978) and can also support AAV DNA replication in a simultaneous coinfection. However, *hr*6 cannot rescue AAV DNA from the latent state at a multiplicity of infection of 5–10 plaque-forming units/cell (Ostrove and Berns, 1980). Thus, it would appear that there is a separate function involved in rescue from the latent state in addition to those required to support AAV DNA replication. It is interesting that *hr*6 is also absolutely defective in cell transformation and produces no detectable 58K tumor antigen (Schrier *et al.*, 1979; Lassam *et al.*, 1979).

D. Cloned Adeno-Associated Virus DNA as a Model System

In Section IV.A.3, a recombinant clone of the intact AAV genome in pBR322 was described (Samulski *et al.*, 1982). The rescue of AAV DNA from this clone by adenovirus on transfection into human cells may well approximate the latent state. If so, it would constitute a completely defined model that would allow very detailed studies on the mechanism of rescue of the integrated DNA. One experiment tends to support the notion that this is, in fact, a good model. The *hr*6 mutant described in Section V.C, which does not rescue AAV DNA from latently infected Detroit 6 cells, also does not rescue AAV DNA from the recombinant pBR322 plasmid in HeLa cells, but does do so in adenovirus-transformed 293 cells (Samulski *et al.* unpublished data).

Two models for the rescue process have been suggested. The first would be a simple excision prior to the initiation of DNA replication. The second would require that replication *per se* be needed to get rescue (Sambrook *et al.*, 1974). To test for simple excision, the recombinant plasmid was transfected into 293 cells, which contain a functional adenovirus E1b. No excision of the AAV sequences was detectable by Southern blotting under conditions in which the input DNA was easily seen. These experiments are continuing to see whether DNA replication *per se* is required for rescue or whether only some of the additional adenovirus early functions are required.

E. Biological Consequences

The consequences of AAV latent infection may be considered with regard to both the virus and the host. As stated initially, a latent infection permits maintenance of the viral genome until advent of the helper and thus assures biological continuity. No known harmful effects accrue to the host as a consequence of the latent infection. A more interesting possibility is that the latent infection may be protective to the host by leading to inhibition of the pathogenic effects of the helper virus. A particularly intriguing example involves carcinomas of the cervix or prostate, both of which are thought to be associated with HSV-2 (Rawls *et al.*,

1973, 1977). In one study (Mayor *et al.*, 1976), only 14% of patients with these diseases had antibodies to AAV and thus were possibly latently infected, whereas 85% of matched, control patients were seropositive for AAV. Both groups were seropositive for HSV-2 at the 80–90% level.

VI. INHIBITION OF ADENOVIRUS

Throughout this review, specific functions supplied by adenovirus that are required for AAV replication have been emphasized. However, not only does AAV depend on adenovirus coinfection, but also AAV replication is inhibitory to adenovirus.

A. Inhibition of Lytic Infection

AAV replication inhibits adenovirus replication (Hoggan *et al.*, 1966; Smith *et al.*, 1966; Casto *et al.*, 1967; Parks *et al.*, 1967a, 1968). At high enough multiplicities of infection (>20 tissue-culture infectious doses/cell), AAV not only can completely inhibit adenovirus replication, but also inhibits its own replication. The step in adenovirus replication that has been implicated in inhibition by AAV is DNA synthesis (Carter *et al.*, 1979). The inhibition apparently results from competition for an adenovirus early gene product that is limiting for both AAV and adenovirus early in adenovirus infection. This idea is supported by the observation that AAV infection of a cell in which adenovirus DNA replication has already been initiated did not affect adenovirus replication, no matter how high the AAV multiplicity of infection (Parks *et al.*, 1968; Carter *et al.*, 1973).

It is not known which adenovirus gene product is limiting. If the observations by Richardson and Westphal (1981) that in Vero cells only adenovirus E4 gene product(s) is required to help AAV, then presumably this is the adenovirus protein for which there is competition. On the other hand, if an E1a product is required for AAV transcription, then this protein(s) might also be a candidate.

In any event, the ability of AAV to inhibit adenovirus replication in cell culture raises the question of whether this occurs *in vivo* and has significant effects in an epidemiological sense on adenovirus infection. This possibility has been suggested with regard to adenovirus infections in both people (Berns *et al.*, 1982) and chickens (Dawson *et al.*, 1981).

B. Inhibition of Oncogenicity

One of the most interesting aspects of the interaction between AAV and adenovirus is the observed ability of AAV to inhibit the oncogenicity of adenovirus. While the significance *in vivo* of both adenovirus oncogenicity and its inhibition might be questionable, the system does offer the opportunity to learn more of the mechanisms underlying adenovirus oncogenicity in model systems.

1. Inhibition *In Vivo*

AAV can inhibit adenovirus oncogenicity in the hamster model system. If AAV and adenovirus are coinjected into newborn Syrian hamsters, the incidence of tumors is significantly decreased and the time of onset is significantly lengthened (Kirchstein *et al.*, 1967; Mayor *et al.*, 1973). Although this observation has been reproduced several times by different laboratories, the timing of AAV injection seems to be critical (Mayor *et al.*, 1973). If the AAV inoculation was delayed by more than a day, then it was not possible to demonstrate the inhibition even though the time to tumor onset was several weeks.

It has also been reported that the inhibition can be reproduced by purified AAV DNA (de la Maza and Carter, 1981). Surprisingly, the efficiency of inhibition per AAV DNA molecule inoculated was nearly the same as that observed for intact virions. Inhibition was equally efficient with fragmented DNA and DNA isolated from AAV DI particles which consisted primarily of the normal terminal repeat sequences. Because DI particles were also effective at inhibition but did not inhibit Ad replication (Laughlin *et al.*, 1979a), the conclusion was that inhibition of adenovirus oncogenicity was by a different mechanism then the one involved in inhibition of adenovirus DNA replication.

2. Inhibition in Cell Culture

Development of a model system in which AAV can inhibit the oncogenicity of adenovirus-transformed cells has allowed studies of the inhibition at the molecular level (Ostrove *et al.*, 1981). H14b cells are a line of adenovirus 5 transformed hamster embryo cells which can induce sarcomas in 100% of newborn Syrian hamsters. The cells contain 5.5 copies of the left 40% of the adenovirus genome but only early regions Ia and Ib are expressed. Infection of these cells from 3 h to 3 days prior to inoculation into hamsters significantly decreased the number of animals which developed tumors (by a factor of 2–3) and increased the time of onset (again by a factor of 2–3). Those tumors which did develop from AAV-infected cells had a volume 1000 fold smaller than control tumors. Chemically transformed cells and SV40-transformed cells were not affected; however, HSV II-transformed cells have been reported to have reduced oncogenicity after AAV infection (Blacklow *et al.*, 1978). The decrease in H14b cell oncogenicity after AAV infection was overcome by increasing the inoculum size fivefold. Thus, the inhibition was not absolute.

The properties of the H14b cells in culture were also affected after AAV infection (Ostrove *et al.*, 1981). The ability of the cells to form colonies in methyl cellulose decreased by 50–90% and the saturation density in monolayers also decreased about twofold.

At the molecular level the integrated copies of the Ad DNA were unaffected. The major Ib transcript thought to represent the 58K t-antigen also appeared to be unaffected but two minor high m.w. transcripts were replaced by three species with altered mobilities. The major alteration noted was an 80% decrease in the amount of the 58K t-antigen after AAV infection, in direct correlation to the decreased tumorigenicity observed in the H14b cells.

The mechanism whereby AAV infection of the H14b cells decreased the amount of the 58K adenovirus t-antigen is unknown. AAV DNA replication did not occur but there was transcription. Because nonstructural AAV proteins have not been identified it is impossible to say whether any were synthesized. Also it is unknown whether the decrease in the 58K t-antigen represented decreased synthesis, an increased rate of degradation, or a sequestration.

Because the inhibition in oncogenicity could be overcome by increasing the inoculum fivefold a likely explanation for the inhibition would be that the decrease in t-antigen led to a slower growth of the inoculated cells. Thus, the host would have had an increased period of time to mount an effective immune response before the inoculated cells increased to a critical number which would have ensured tumor formation. Although not directly assessed, it seemed likely that the inhibition was transient. The decrease in saturation density disappeared after a week in culture. This observation could have reflected the inability of the AAV to replicate in the cells; thus, continued cell growth would have diluted the input AAV. At no time was the growth rate of the cells in culture affected.

VII. CONCLUSION

It may be appropriate to consider AAV and adenovirus as a single unified animal virus system which represents a unique relationship between two viruses which are unrelated in the physicochemical sense. Although a good start has been made on the understanding of the molecular biology of both viruses, almost nothing is known of the consequences *in vivo* of the relationship. It is to be hoped that there will be more studies of these effects in the future. The CELO-AAAV (avian AAV) system could be of particular utility in this respect (Dawson *et al.*, 1981).

The details of the molecular biology of AAV are reviewed in greater detail in *The Parvoviruses*, a volume in this series.

ACKNOWLEDGMENTS. The helpful comments of Drs. N. Muzyczka and W. Hauswirth are appreciated. The author's research is supported by U.S. Public Health Service Research Grant AI-16326.

REFERENCES

Atchison, R.W., 1970, The role of herpesviruses in adenovirus-associated virus replication *in vitro, Virology* **42:**155.

Atchison, R.W., Casto, B.C., and Hammon, W.McD., 1965, Adenovirus-associated defective virus particles, *Science* **149:**754.

Berk, A.J., and Sharp, P.A., 1977, Sizing and mapping of early adenovirus mRNAs by gel electrophoresis of S1 endonuclease-digested hybrids, *Cell* **12:**721.

Berk, A.J., Lee, F., Harrison, T., Williams, J., and Sharp, P.A., 1979, Pre-early Ad5 gene product regulates synthesis of early viral mRNAs, *Cell* **17:**935.

Berns, K.I., and Adler, S., 1972, Separation of two types of adeno-associated virus particles containing complementary polynucleotide chains, *J. Virol.* **9:**394.

Berns, K.I., and Rose, J.A., 1970, Evidence for a single-stranded adenovirus-associated virus genome: Isolation and separation of complementary single strands, *J. Virol.* **5:**693.

Berns, K.I., Pinkerton, T.C., Thomas, G.F., and Hoggan, M.D., 1975, Detection of adeno-associated virus (AAV)-specific nucleotide sequences in DNA isolated from latently infected Detroit 6 cells, *Virology* **68:**556.

Berns, K.I., Hauswirth, W.W., Fife, K.H., and Lusby, E., 1979, Adeno-associated virus DNA replication, *Cold Spring Harbor Symp. Quant. Biol.* **43:**781.

Berns, K.I., Cheung, A.K.-M., Ostrove, J.M., and Lewis, M., 1982, Adeno-associated virus latent infection, in: *Virus Persistence* (B.W.J. Mahy, A.C. Mirson, and G.K. Darby, eds.), p. 249, Cambridge University Press.

Blacklow, N.R., 1975, Potentiation of an adenovirus-associated virus by herpes simplex virus type 2 transformed cells, *J. Natl. Cancer Inst.* **54:**241.

Blacklow, N.R., Hoggan, M.D., and Rose, W.P., 1967, Isolation of adenovirus-associated viruses from man, *Proc. Natl. Acad. Sci. U.S.A.* **58:**1410.

Blacklow, N.R., Hoggan, M.D., and Rowe, W.P., 1968a, Serologic evidence for human infection with adenovirus-associated viruses, *J. Natl. Cancer Inst.* **40:**319.

Blacklow, N.R., Hoggan, M.D., Kapikian, A.Z., Austin, J.B., and Rowe, W.P., 1968b, Epidemiology of adenovirus-associated virus infection in a nursery population, *Am. J. Epidemiol.* **88:**368.

Blacklow, N.R., Hoggan, M.D., and McClanahan, M.S., 1970, Adenovirus-associated viruses: Enhancement by human herpesvirus, *Proc. Soc. Exp. Biol. Med.* **134:**952.

Blacklow, N.R., Dolin, R., and Hoggan, M.D., 1971a, Studies of enhancement of an adenovirus-associated virus by herpes simplex virus, *J. Gen. Virol.* **10:**29.

Blacklow, N.R., Hoggan, M.D., Sereno, M.S., Brandt, C.D., Kim, H.W., Parrott, R.H., and Chanock, R.M., 1971b, A seroepidemiologic study of adenovirus-associated virus infection in infants and children, *Am. J. Epidemiol.* **94:**359.

Blacklow, N.R., Cukor, G., Kibrick, S., and Quinman, G., 1978, Interactions of adeno-associated viruses with cells transformed by herpes simplex virus, in: *Replication of Mammalian Parvoviruses* (D.C. Ward and P. Tattersall, eds.), p. 87, Cold Spring Harbor Laboratory Press, Cold Spring Harbor, New York.

Buller, R.M.L., and Rose, J.A., 1978a, Characterization of adenovirus-associated virus-induced polypeptides in KB cells, *J. Virol.* **25:**331.

Buller, R.M.L., and Rose, J.A., 1978b, Characterization of adeno-associated virus polypeptides synthesized *in vivo* and *in vitro*, in: *Replication of Mammalian Parvoviruses* (D.C. Ward and P. Tattersall, eds.), p. 399, Cold Spring Harbor Laboratory Press, Cold Spring Harbor, New York.

Buller, R.M.L., Janik, J.E., Sebring, E.D., and Rose, J.A., 1981, Herpes simplex virus types 1 and 2 help adeno-associated virus replication, *J. Virol.* **40:**241.

Carter, B.J., 1976, Intracellular distribution and polyadenylate content of adeno-associated virus RNA sequences, *Virology* **73:**273.

Carter, B.J., Koczot, F.J., Garrison, J., Rose, J., and Dolin, J.A., 1973, Separate function provided by adenovirus for adeno-associated virus multiplication, *Nature (London) New Biol.* **244:**71.

Carter, B.J., Laughlin, C.A., de la Maza, L.M., and Myers, M., 1979, Adeno-associated virus autointerference, *Virology* **92:**449.

Casto, B.C., and Goodheart, C.R., 1972, Inhibition of adenovirus transformation *in vitro* by AAV-1, *Proc. Soc. Exp. Biol. Med.* **140:**72.

Casto, B.C., Armstrong, J.A., Atchison, R.W., and Hammon, W.McD., 1967, Studies on the relationship between adeno-associated virus type 1 (AAV-1) and adenoviruses. II. Inhibition of adenovirus plaques by AAV: Its nature and specificity, *Virology* **33:**452.

Cavalier-Smith, T., 1974, Palindromic base sequences and replication of eukaryotic chromosome ends, *Nature (London)* **250:**467.

Cheung, A.K.-M., Hoggan, M.D., Hauswirth, W.W., and Berns, K.I., 1980, Integration of the adeno-associated virus genome into cellular DNA in latently infected human Detroit 6 cells, *J. Virol.* **33:**739.

Dawson, G.J., Yates, V.J., Chang, P.W., and Wattanavijarin, W., 1981, Egg transmission of avian adeno-associated virus and CELO virus during experimental infections, *Am. J. Vet. Res.* **42:**1833.

de la Maza, L.M., and Carter, B.J., 1980, Molecular structure of adeno-associated virus variant DNA, *J. Biol. Chem.* **255:**3194.

de la Maza, L.M., and Carter, B.J., 1981, Inhibition of adenovirus oncogenicity by adeno-associated virus DNA, *J. Natl. Cancer Inst.* **67:**1323.

Denhardt, D.T., Eisenberg, G., Bartok, K., and Carter, B.J., 1976, Multiple structure of adeno-associated virus DNA: Analysis of terminally-labelled molecules with endonuclease R.·*HaeIII*, *J. Virol.* **18:**672.

Dolin, R., and Rabson, A.S., 1973, Herpesvirus saimiri: Enhancement of adenovirus-associated virus, *J. Natl. Cancer Inst.* **50:**205.

Domoto, K., and Yanagawa, R., 1969, Properties of a small virus associated with infectious canine hepatitis virus, *Jpn. J. Vet. Res.* **17:**32:

Dutta, S.K., and Pomeroy, B.S., 1967, Electron microscopic studies of quail bronchitis virus, *Am. J. Vet. Res.* **28:**296.

East, D.L., and Kingsbury, D.W., 1971, Mumps virus replication in chick embryo lung cells: Properties of RNA species in virions and infected cells, *J. Virol.* **8:**161.

Eigen, M., and Schuster, P., 1978, The hypercycle, a principle of natural self organization. Part C. The realistic hypercycle, *Naturwissenschaften* **65:**341.

Ensinger, M.J., and Ginsberg, H.S., 1972, Selection and preliminary characterization of temperature-sensitive mutants of type 5 adenovirus, *J. Virol.* **10:**328.

Fife, K.H., Berns, K.I., and Murray, K., 1977, Structure and nucleotide sequence of the terminal regions of adeno-associated virus DNA, *Virology* **78:**475.

Gerry, H.W., Kelly, T.J., Jr., and Berns, K.I., 1973, Arrangement of nucleotide sequences in adeno-associated virus DNA, *J. Mol. Biol.* **79:**207.

Graham, F.L., Smiley, J., Russell, W.C., and Nairn, R., 1977, Characteristics of a human cell line transformed by DNA from human adenovirus type 5, *J. Gen. Virol.* **36:**59.

Green, M.R., and Roeder, R.G., 1980a, Transcripts of the adenovirus-associated virus genome: Mapping of the major RNAs, *J. Virol.* **36:**79.

Green, M.R., and Roeder, R.G., 1980b, Definition of a novel promoter for the major adeno-associated virus mRNA, *Cell* **22:**231.

Green, M.R., Straus, S.E., and Roeder, R.G., 1980, Transcripts of the adenovirus-associated virus genome: Multiple polyadenylated RNAs including a potential primary transcript, *J. Virol.* **35:**560.

Handa, H., Shiroki, K., and Shimojo, H., 1977, Establishment and characterization of KB cell lines latently infected with adeno-associated virus type 1, *Virology* **82:**84.

Harrison, T., Graham, F., and Williams, J., 1977, Host range mutants of adenovirus type 5 defective for growth in HeLa cells, *Virology* **77:**319.

Hauswirth, W.W., and Berns, K.I., 1977, Origin and termination of adeno-associated virus DNA replication, *Virology* **78:**488.

Hauswirth, W.W., and Berns, K.I., 1979, Adeno-associated virus DNA replication: Non unit-length molecules, *Virology* **93:**57.

Hoggan, M.D., 1970, Adenovirus-associated viruses, *Prog. Med. Virol.* **12:**112.

Hoggan, M.D., Blacklow, N.R., and Rowe, W.P., 1966, Studies of small DNA viruses found in various adenovirus preparations: Physical, biological and immunological characteristics, *Proc. Natl. Acad. Sci. U.S.A.* **55:**1467.

Hoggan, M.D., Thomas, G.F., Thomas, F.B., and Johnson, F.B., 1972, Continuous "carriage" of adenovirus associated virus genome in cell cultures in the absence of helper adenoviruses, in: Proceedings of the fourth Lepetit colloquium, Cocoyac, Mexico, p. 243, North-Holland, Amsterdam.

Janik, J.E., Houston, M.M., and Rose, J.A., 1981, Locations of adenovirus genes required for the replication of adenovirus-associated virus, *Proc. Natl. Acad. Sci. U.S.A.* **78:**1925.

Janik, J.E., Houston, M.M., Cho, K., and Rose, J.A., 1982, Abstract, 11th Annual UCLA Symposia, No. 0580, p. 209.

Jay, F.T., Laughlin, C.A., and Carter, B.J., 1981, Eukaryotic translational control: Adeno-associated virus protein synthesis is affected by a mutation in the adenovirus DNA binding protein, *Proc. Natl. Acad. Sci. U.S.A.* **78:**2927.

Johnson, F.B., and Hoggan, M.D., 1979, Labeling of adeno-associated virus proteins with radioactive acetate, *Abstr. Annu. Meet. Am. Soc. Microbiol.*, p. 243.

Johnson, F.B., Ozer, H.L., and Hoggan, M.D., 1971, Structural proteins of adenovirus-associated virus type 3, *J. Virol.* **8:**860.

Johnson, F.B., Blacklow, N.R., and Hoggan, M.D., 1972, Immunological reactivity of antisera prepared against the sodium dodecyl sulfate-treated structural polypeptides of adenovirus-associated virus, *J. Virol.* **9:**1017.

Johnson, F.B., Thomson, T.A., Taylor, P.A., Vlazny, D.A., 1977, Molecular similarities among the adenovirus-associated virus polypeptides and evidence for a precursor protein, *Virology* **82:**1.

Jones, N., and Shenk, T., 1979, Isolation of Ad5 host range deletion mutants defective for transformation of rat embryo cells, *Cell* **17:**683.

Kelly, D.C., Barwise, A.H., and Walker, I.O., 1977, DNA contained by two densonucleosis viruses, *J. Virol.* **21:**396.

Kirschstein, R.L., Smith, K.O., and Peters, E.A., 1968, Inhibition of adenovirus 12 oncogenicity by adeno-associated virus, *Proc. Soc. Exp. Biol. Med.* **128:**670.

Klessig, D.F., 1977, Isolation of a variant human adenovirus serotype 2 that multiplies efficiently on monkey cells, *J. Virol.* **21:**1243.

Koczot, F., 1970, Homology among different adeno-associated virus serotypes, Masters thesis, University of Maryland, College Park.

Koczot, F.J., Carter, B.J., Garon, C.F., and Rose, J.A., 1973, Self-complementarity of terminal sequences within plus or minus strands of adenovirus-associated virus DNA, *Proc. Natl. Acad. Sci. U.S.A.* **79:**215.

Kurstak, E., Vernoux, J.P., Niveleau, A., and Onju, P.A., 1971, Visualisation de la densonucleose (VDN) a chaines monocatenaires complementaries de polarities inverses plus et moins, *C. R. Acad. Sci. Paris* **272:**762.

Lassam, N.J., Bayley, S.T., and Graham, F.L., 1978, Synthesis of DNA, late polypeptides, and infectious virus by host-range mutants of adenovirus 5 in nonpermissive cells, *Virology* **87:**463.

Lassam, N.J., Bayley, S.T., and Graham, F.L., 1979, Tumor antigens of human Ad5 in transformed cells and in cells infected with transformation-defective host-range mutants, *Cell* **18:**781.

Laughlin, C.A., Myers, M.W., Risin, D.L., and Carter, B.J., 1979a, Defective interfering particles of the human parvovirus adeno-associated virus, *Virology* **94:**162.

Laughlin, C.A., Westphal, H., and Carter, B.J., 1979b, Spliced adenovirus-associated virus RNA, *Proc. Natl. Acad. Sci. U.S.A.* **76:**5567.

Laughlin, C.A., Jones, N., and Carter, B.J., 1982, Effect of deletions in adenovirus early region II genes upon replication of adeno-associated virus, *J. Virol.* **41:**868.

Lechner, R.L., and Kelly, T.J., Jr., 1977, The structure of replicating adenovirus 2 DNA molecules, *Cell* **12:**1007.

Luchsinger, E., and Wellemans, G., 1971, A study on the presence of antibodies against adenovirus and associated virus X_7 in the sera of cattle from Eynatten and of calves with respiratory disease, *Arch. Gesamte Virusforsch.* **35**:203.

Lusby, E.W., and Berns, K.I., 1982, Mapping of the 5′ termini of two adeno-associated virus 2 RNAs in the left half of the genome, *J. Virol.* **41**:518.

Lusby, E., Fife, K.H., and Berns, K.I., 1980, Nucleotide sequence of the inverted terminal repetition in adeno-associated virus DNA, *J. Virol.* **34**:402.

Lusby, E., Bohenzky, R., and Berns, K.I., 1981, The inverted terminal repetition in adeno-associated virus DNA: Independence of orientation at either end of the genome, *J. Virol.* **37**:1083.

Mathews, M.B., 1975, Genes for VA RNA in adenovirus 2, *Cell* **6**:223.

Mayor, H.D., Jamison, R.M., Jordan, L.E., and Melnick, J.L., 1965, Structure and composition of a small particle prepared from a simian adenovirus, *J. Bacteriol.* **90**:235.

Mayor, H.D., Torikai, K., Melnick, J.L., and Mandel, M., 1969, Plus and minus single-stranded DNA separately encapsidated in adeno-associated satellite virions, *Science* **166**:1280.

Mayor, H.D., Houlditch, G.S., and Mumford, D.M., 1973, Influence of adeno-associated satellite virus on adenovirus-induced tumors in hamsters, *Nature (London) New Biol.* **241**:44.

Mayor, H.D., Drake, S., Stahmann, J., and Mumford, D.M., 1976, Antibodies to adeno-associated satellite virus and herpes simplex in sera from cancer patients and normal adults, *Am. J. Obstet. Gynecol.* **126**:100.

McPherson, R.A., and Rose, J.A., 1982, Adeno-associated virus helper activity of adenovirus DNA binding protein, *Abstr. Annu. Meet. Am. Soc. Microbiol.*, p. 247.

Montell, C., Fisher, E.F., Caruthers, M., and Berk, A.J., 1982, Resolving the functions of overlapping viral genes by site specific mutagenesis at a mRNA splice site, *Nature (London)* **295**:380.

Myers, M.W., and Carter, B.J., 1981, Adeno-associated virus replication: The effect of L-canavanine or a helper virus mutation on accumulation of viral capsids and progeny single-stranded DNA, *J. Biol. Chem.* **256**:567.

Myers, M.W., Laughlin, C.A., Jay, F.T., and Carter, B.J., 1980, Adenovirus helper function for growth of adeno-associated virus: Effect of temperature sensitive mutations in adenovirus early gene region 2, *J. Virol.* **35**:65.

Ostrove, J.M., and Berns, K.I., 1980, Adenovirus early region Ib gene function required for rescue of latent adeno-associated virus, *Virology* **104**:502.

Ostrove, J.M., Duckworth, D.H., and Berns, K.I., 1981, Inhibition of adenovirus-transformed cell oncogenicity by adeno-associated virus, *Virology* **113**:521.

Parks, W.P., Melnick, J.L., Rongey, R., and Mayor, H.D., 1967a, Physical assay and growth cycle studies of a defective adeno-satellite virus, *J. Virol.* **1**:171.

Parks, W.P., Green, M., Piña, M., and Melnick, J.L., 1967b, Physiochemical characterization of adeno-associated satellite virus type 5 and its nucleic acid, *J. Virol.* **1**:980.

Parks, W.P., Casazza, A.M., Alcott, J., and Melnick, J.L., 1968, Adeno-associated satellite virus interference with the replication of its helper adenovirus, *J. Exp. Med.* **127**:91.

Pettersson, U., and Philipson, L., 1975, Location of sequences on the adenovirus genome coding for the 5.5S RNA, *Cell* **6**:1.

Philipson, L., 1979, Adenovirus proteins and their messenger RNAs, *Adv. Virus Res.* **25**:357.

Portner, A., and Kingsbury, D.W., 1970, Complementary RNAs in paramyxovirions and paramyxovirus-infected cells, *Nature (London)* **228**:1196.

Rawls, W.E., Adam, E., and Melnick, J.L., 1973, An analysis of seroepidemiological studies of herpesvirus type 2 and carcinoma of the cervix, *Cancer Res.* **33**:1477.

Rawls, W.E., Bacchetti, S., and Graham, F.L., 1977, Relation of herpes simplex viruses to human malignancies, *Curr. Top. Microbiol. Immunol.* **77**:71.

Richardson, W.D., and Westphal, H., 1981, A cascade of adenovirus early function is required for expression of adeno-associated virus, *Cell* **27**:133.

Richardson, W.D., Carter, B.J., and Westphal, H., 1980, Vero cells injected with adenovirus type 2 mRNA produce authentic viral polypeptide patterns: Early mRNA promotes growth of adenovirus-associated virus, *Proc. Natl. Acad. Sci. U.S.A.* **77**:931.

Rose, J.A., and Koczot, F., 1972, Adenovirus-associated virus multiplication. VII. Helper requirement for viral deoxyribonucleic acid and ribonucleic acid synthesis, *J. Virol.* **10**:1.

Rose, J.A., Hoggan, M.D., Koczot, F.J., and Shatkin, A.J., 1968, Genetic relatedness studies with adenovirus-associated viruses, *J. Virol.* **2**:999.

Rose, J.A., Berns, K.I., Hoggan, M.D., and Koczot, F.J., 1969, Evidence for a single-stranded adenovirus-associated virus genome: Formation of a DNA density hybrid on release of viral DNA, *Proc. Natl. Acad. Sci. U.S.A.* **64**:863–869.

Rose, J.A., Maizel, J.V., Jr., Inman, J.K., and Shatkin, A.J., 1971, Structural proteins of adenovirus-associated viruses, *J. Virol.* **8**:766.

Sambrook, J., Botchan, M., Gallimore, P., Ozanne, B., Petterson, U., Williams, J., and Sharp, P.A., 1974, Viral DNA sequences in cells transformed by simian virus 40, adenovirus type 2, and adenovirus type 5, *Cold Spring Harbor Symp. Quant. Biol.* **39**:615.

Samulski, R.J., Berns, K.I., Tan, M., and Muzyczka, N., 1982, Cloning of AAV into pBR322: Rescue of intact virus from the recombinant plasmid in human cells, *Proc. Natl. Acad. Sci. U.S.A.* **79**:2077.

Samulski, R.J., Srivastava, A., Berns, K.I. and Muzyczka, N., 1983, Rescue of adeno-associated virus from recombinant plasmids: gene correction within the terminal repeats of AAV. *Cell* **33**:155.

Schrier, P.I., van den Elsen, P.J., Hertoghs, J.J.L., and Van der Eb, A.J., 1979, Characterization of tumor antigens in cells transformed by fragments of adenovirus type 5 DNA, *Virology* **99**:372.

Shenk, T., Jones, N., Colby, W., and Fowlkes, D., 1980, Functional analysis of adenovirus-5 host-range deletion mutants defective for transformation of rat embryo cells, *Cold Spring Harbor Symp. Quant. Biol.* **44**:367.

Smith, K.O., Gehle, W.D., and Thiel, J.F., 1966, Properties of a small virus associated with adenovirus type 4, *J. Immunol.* **97**:754.

Soderlund, J., Pettersson, U., Venustrom, B., Philipson, L., and Mathews, M.B., 1976, A new species of virus-coded low molecular weight RNA from cells infected with adenovirus 2, *Cell* **7**:585.

Southern, E.M., 1975, Detection of specific sequences among DNA fragments separated by gel electrophoresis, *J. Mol. Biol.* **98**:503.

Spear, I.S., Fife, K.H., Hauswirth, W.W., Jones, C.J., and Berns, K.I., 1977, Evidence for two nucleotide sequence orientations within the terminal repetition of adeno-associated virus DNA, *J. Virol.* **24**:627.

Srivastava, A., Lusby, E.W. and Berns, K.I., 1983. Nucleotide sequence and organization of the adeno-associated virus 2 gerome. *J. Virol.* **45**:555.

Straus, S., Ginsberg, H., and Rose, J.A., 1967a, DNA-minus temperature sensitive mutants of adeno-associated virus replication, *J. Virol.* **17**:149.

Straus, S.E., Sebring, E.D., and Rose, J.A., 1976b, Concatemers of alternating plus and minus strands are intermediates in adenovirus-associated virus DNA synthesis, *Proc. Natl. Acad. Sci. U.S.A.* **73**:742.

Watson, J.D., 1972, Origin of concatemeric T7 DNA, *Nature (London) New Biol.* **239**:197.

Williams, J.F., 1973, Oncogenic transformation of hamster embryo cells *in vitro* by adenovirus type 5, *Nature (London)* **243**:162.

Index

GPSR Compliance
The European Union's (EU) General Product Safety Regulation (GPSR) is a set of rules that requires consumer products to be safe and our obligations to ensure this.

If you have any concerns about our products, you can contact us on

ProductSafety@springernature.com

In case Publisher is established outside the EU, the EU authorized representative is:

Springer Nature Customer Service Center GmbH
Europaplatz 3
69115 Heidelberg, Germany

www.ingramcontent.com/pod-product-compliance
Ingram Content Group UK Ltd.
Pitfield, Milton Keynes, MK11 3LW, UK
UKHW051131260726
13967UKWH00010B/2991

* 9 7 8 1 4 6 8 4 7 9 3 6 2 *